Geophysics for the Mineral Exploration Geoscientist

High global demand for mineral commodities has led to increasing application of geophysical technologies to a wide variety of mineral deposits. Co-authored by a university professor and an industry geophysicist, this state-of-the-art overview of geophysical methods provides a careful balance between principles and practice. It takes readers from the basic physical phenomena, through the acquisition and processing of geophysical data, to the creation of subsurface models and their geological interpretation.

- Presents detailed descriptions of all the main geophysical methods, including gravity, magnetic, radiometric, electrical, electromagnetic and seismic methods.
- Explains the cutting-edge current practice in exploration and mining geophysics for the discovery of 'blind' mineral deposits.
- Describes techniques in a consistent way and without the use of complex mathematics, enabling easy comparison between the various methods.
- Gives a practical guide to data acquisition, processing and accurate interpretation of geophysical datasets.
- Includes presentation and analysis of new petrophysical data, giving geologists and geophysicists key information on the physical properties of rocks.
- Emphasises extraction of maximum geological information from geophysical data, providing explanations of data modelling, and common interpretation pitfalls.
- Provides examples from all the main types of mineral deposit around the world, giving students exposure to real geophysical data.
- Richly illustrated with over 300 full-colour figures, with access to electronic versions for instructors.

Designed for advanced undergraduate and graduate courses in minerals geoscience and geology, this book is also a valuable reference for geologists and professionals in the mining industry wishing to make greater use of geophysical methods.

Michael Dentith is Professor of Geophysics at The University of Western Australia and a research theme leader in the Centre for Exploration Targeting. He has been an active researcher and teacher of university-level applied geophysics and geology for more than 25 years, and he also consults to the minerals industry. Professor Dentith's research interests include geophysical signatures of mineral deposits (about which he has edited two books), petrophysics and terrain scale analysis of geophysical data for exploration targeting. He is a member of the American Geophysical Union, Australian Society of Exploration Geophysicists, Society of Exploration Geophysicists and Geological Society of Australia.

Stephen Mudge has worked as an exploration geophysicist in Australia for more than 35 years, and currently works as a consultant in his own company, Vector Research. He has worked in many parts of the world and has participated in a number of new mineral discoveries. Mr Mudge has a keen interest in data processing techniques for mineral discovery and has produced several publications reporting new developments. He is a member of the Australasian Institute of Mining and Metallurgy, Australian Institute of Geoscientists, Australian Society of Exploration Geophysicists, Society of Exploration Geophysicists and European Association of Engineers and Geoscientists.

'More and more, great ore deposits are being found under cover and knowledge of exploration geophysics provides a distinct advantage in their discovery. Dentith and Mudge provide a clear, comprehensive, up-to-date, and (very significantly) applied approach for the general geologist, demonstrating how to locate concealed orebodies by employing modern-day geophysical techniques.'
Richard J. Goldfarb, *Fellow, Society of Economic Geologists*

'Readers will really appreciate the up-to-date system descriptions, examples and case histories presented in this new book. In particular, the diagrams in this textbook are superb; the explanatory diagrams have been drawn professionally and the geophysical data and images are shown in full colour.'
Professor Richard Smith, *Laurentian University, Ontario, Canada*

Geophysics for the Mineral Exploration Geoscientist

Michael Dentith
The University of Western Australia, Perth

Stephen T. Mudge
Vector Research Pty Ltd, Perth

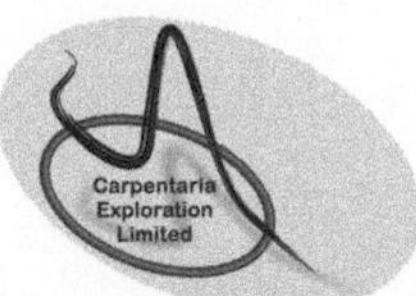

Shaftesbury Road, Cambridge CB2 8EA, United Kingdom

One Liberty Plaza, 20th Floor, New York, NY 10006, USA

477 Williamstown Road, Port Melbourne, VIC 3207, Australia

314–321, 3rd Floor, Plot 3, Splendor Forum, Jasola District Centre, New Delhi – 110025, India

103 Penang Road, #05–06/07, Visioncrest Commercial, Singapore 238467

Cambridge University Press is part of Cambridge University Press & Assessment, a department of the University of Cambridge.

We share the University's mission to contribute to society through the pursuit of education, learning and research at the highest international levels of excellence.

www.cambridge.org
Information on this title: www.cambridge.org/9780521809511

First published 2014 (version 7, June 2024)

Printed in Great Britain by CPI Group (UK) Ltd, Croydon CR0 4YY, June 2024

A catalogue record for this publication is available from the British Library

Library of Congress Cataloging-in-Publication data
Dentith, M. C. (Michael C.)
Geophysics for the mineral exploration geoscientist / Michael Dentith, Stephen T. Mudge.
pages cm
ISBN 978-0-521-80951-1 (Hardback)
1. Geophysics. 2. Earth scientists. I. Mudge, Stephen T., 1952– II. Title.
QC807.D46 2014
550–dc23 2013034215

ISBN 978-0-521-80951-1 Hardback

Additional resources for this publication at www.cambridge.org/dentith

CONTENTS

ONLINE APPENDICES

Available at www.cambridge.org/dentith

FIGURE CREDITS

The following institutions, publishers and authors are gratefully acknowledged for their kind permission to use redrawn figures based on illustrations in journals, books and other publications for which they hold copyright. We have cited the original sources in our figure captions. We have made every effort to obtain permissions to make use of copyrighted materials and apologise for any errors or omissions. The publishers welcome errors and omissions being brought to their attention.

Institutions and publishers	**Figure number(s)**
Allen & Unwin (Taylor and Francis)	
Image Interpretation in Geology (Fig. 5.42, page 147)	2.31b
American Association of Petroleum Geologists	
AAPG Bulletin	5.62
Australasian Institute of Mining and Metallurgy	
Geology of the Mineral Deposits of Australia and Papua New Guinea	3.76c, 5.55c
Australian Society of Exploration Geophysicists	
Exploration Geophysics (CSIRO Publishing)	3.17, 5.57b, 5.81a,b,c, 5.89a,b,c, 5.93, A3.3, A5.6, A5.8, A6.9, A6.10d
Geophysical Signatures of South Australian Mineral Deposits	5.55a, 5.61
Geophysical Signatures of Western Australian Mineral Deposits	4.24d, 4.30, 5.59
Preview (CSIRO Publishing)	3.54
Burval Working Group, Leibniz Institute for Applied Geosciences	
Groundwater Resources in Buried Valleys. A Challenge for the Geosciences	2.19
Cambridge University Press	
Fundamentals of Geophysics	4.2
Canadian Institute of Mining, Metallurgy and Petroleum	
CIM Bulletin	3.74, 5.49, A5.3
Methods and Case Histories in Mining Geophysics, Proceedings of the Sixth Commonwealth Mining and Metallurgical Congress	3.77a
Canadian Society of Exploration Geophysicists	
CSEG Recorder	5.89d
Centre for Exploration Targeting, The University of Western Australia	
Geophysical Signatures of South Australian Mineral Deposits	5.55a 5.61
Geophysical Signatures of Western Australian Mineral Deposits	4.24d, 4.30, 5.59
SEG 2004: Predictive Mineral Discovery Under Cover, (Extended Abstracts)	6.41a,c
Department of Manufacturing, Innovation, Trade, Resources and Energy – South Australia	
Geophysical Signatures of South Australian Mineral Deposits	5.55a, 5.61

Institutions and publishers	Figure number(s)
Elsevier	
Earth and Planetary Science Letters	3.49
Geochimica et Cosmochimica Acta	3.34
Geoexploration	2.43b, 3.77d, 5.29g
Journal of Applied Geophysics	6.19, A5.5a,b
Journal of Geodynamics	2.8, 2.13
Tectonophysics	3.63a
European Association of Geoscientists and Engineers	
First Break	2.37c
Geological Association of Canada	
Geophysics in Mineral Exploration: Fundamentals and Case Histories	2.37a
Geological Society of America	
Geological Society of America Bulletin	3.51
Geological Society of London	
Journal of the Geological Society	3.47
Geological Survey of India	
Indian Minerals	5.31
Geometrics	
Applications Manual for Portable Magnetometers	3.22
Geonics	
Technical Note TN-7	5.72
Geoscience Australia	
© Commonwealth of Australia (Geoscience Australia) 2013. These products are released under the Creative Commons Attribution 3.0 Australia Licence.	
AGSO Journal of Australian Geology and Geophysics	3.39, 3.41, 3.42, 3.43, 4.3a, 4.6, 4.16, 4.18, 4.19
Airborne Gravity 2010 – ASEG-PESA Airborne Gravity 2010 Workshop (Abstracts)	3.11
International Research Centre for Telecommunications, Transmission and Radar; Delft University of Technology	
Proceedings of the Second International Workshop on Advanced Ground Penetrating Radar	A5.5c
Mineralogical Society of America	
Elements	3.7
Natural Resources Canada	
© Department of Natural Resources Canada. All rights reserved.	
Geophysics and Geochemistry in the Search for Metallic Ores	2.37b, 5.56
Mining and Groundwater Geophysics 1967	3.77c
Uranium Prospecting Handbook (Maney Publishing on behalf of the Institute of Materials, Minerals and Mining)	4.23
Northwest Mining Association	
PG III Northwest Mining Association's 1998 Practical Geophysics Short Course: Selected Papers	1.2, 1.3
Practical Geophysics for the Exploration Geologist II	A4.4, A4.5

Institutions and publishers	Figure number(s)
NRC Research Press	
© Canadian Science Publishing or its licensors.	
Canadian Journal of Earth Sciences	3.44, 3.56, 3.69
Pergamon (Elsevier)	
Applied Geophysics for Geologists and Engineers	2.49a
Physical Properties of Rocks: Fundamentals and Principles of Petrophysics	3.33, 5.18
Plenum Press (Springer Science + Business Media)	
Electrical Properties of Rocks (Fig. 24, page 90)	5.13
Prospectors and Developers Association of Canada (Society of Exploration Geophysicists)	
Proceedings of Exploration '97: Fourth Decennial International Conference on Mineral Exploration	5.90, 5.96, 5.100, A5.1, A5.4
Society of Economic Geologists	
Economic Geology	3.40, 3.45, 3.76c, 4.20, 4.21, 4.25, A4.9
Society of Exploration Geophysicists	
An Overview of Exploration Geophysics in China	3.64
Electromagnetic Methods in Applied Geophysics	5.14, 5.80, A4.2b,c,d, A4.6, A4.10a,b,c
Seventy-second Annual Conference (Expanded Abstracts)	6.51
Geophysics	3.77b, 4.9b,d, 5.17, 5.21, 5.26a, b, 5.83, 5.88, 6.38, 6.40, 6.47, 6.48, 6.49, A3.2, A6.2a
Geotechnical and Environmental Geophysics, Volume 1	5.24
Hardrock Seismic Exploration	2.26, 3.37a, 6.13c, 6.14c, 6.41b
Springer Science + Business Media	
Pure and Applied Geophysics	2.43c
Studia Geophysica et Geodaetica	6.4d
Handbook of Geochemistry (Fig. 26-G-1, page 26-G-3)	3.53a
Taylor & Francis (www.tandfonline.com on behalf of the Geological Society of Australia)	
Australian Journal of Earth Sciences	3.74
© Geological Society of Australia	
Wiley-Blackwell	
A Petroleum Geologist's Guide to Seismic Reflection	2.21
Geophysical Prospecting	A5.2
Authors	
Bolt, B.	6.2, 6.3
Francke, J.C. and Yelf, R.	A5.5c
Garrels, R.M. and Christ, C.L.	3.53b,c,d
Heithersay, P.S.	3.76c
Reeve, J.S. (Cross, K.C.)	5.55c
Stanley, J.M.	2.9
Titley, S.R.	5.32
Whiteley, R.J.	5.66

PREFACE

This book is about how geophysics is used in the search for mineral deposits. It has been written with the needs of the mineral exploration geologist in mind and for the geophysicist requiring further information about data interpretation, but also for the mining engineer and other professionals, including managers, who have a need to understand geophysical techniques applied to mineral exploration. Equally we have written for students of geology, geophysics and engineering who plan to enter the minerals industry.

Present and future demands for mineral explorers include deeper exploration, more near-mine exploration and greater use of geophysics in geological mapping. This has resulted in geophysics now lying at the heart of most mineral exploration and mineral mapping programmes. We describe here modern practice in mineral geophysics, but with an emphasis on the geological application of geophysical techniques. Our aim is to provide an understanding of the physical phenomena, the acquisition and manipulation of geophysical data, and their integration and interpretation with other types of data to produce an acceptable geological model of the subsurface. We have deliberately avoided presenting older techniques and practices not used widely today, leaving descriptions of these to earlier texts. It has been our determined intention to provide descriptions in plain language without resorting to mathematical descriptions of complex physics. Only the essential formulae are used to clarify the basis of a geophysical technique or a particular point. Full use has been made of modern software in the descriptions of geophysical data processing, modelling and display techniques. The references cited emphasise those we believe suit the requirements of the exploration geologist.

We have endeavoured to present the key aspects of each geophysical method and its application in the context of modern exploration practice. In so doing, we have summarised the important and relevant results of many people's work and also included some of our own original work. Key features of the text are the detailed descriptions of petrophysical properties and how these influence the geophysical response, and the descriptions of techniques for obtaining geological information from geophysical data. Real data and numerous real-world examples, from a variety of mineral deposit types and geological environments, are used to demonstrate the principles and concepts described. In some instances we have taken the liberty of reprocessing or interpreting the published data to demonstrate aspects we wish to emphasise.

M.D. has been an active researcher and teacher of university-level geology and applied geophysics for more than 25 years. S.M. has been an active minerals exploration geophysicist and researcher for more than 35 years. We hope this book will be a source of understanding for, in particular, the younger generation of mineral explorers who are required to embrace and assimilate more technologies more rapidly than previous generations, and in times of ever-increasing demand for mineral discoveries.

ACKNOWLEDGEMENTS

This project would not have been possible without the great many individuals who generously offered assistance or advice or provided materials. Not all of this made it directly into the final manuscript, but their contributions helped to develop the final content and for this we are most grateful. They are listed below and we sincerely apologise for any omissions:

Ray Addenbrooke, Craig Annison, Theo Aravanis, Gary Arnold, William Atkinson, Leon Bagas, Simon Bate, Kirsty Beckett, Jenny Bevan, John Bishop, Tim Bodger, Miro Bosner, Barry Bourne, Justin Brown, Amanda Buckingham, Andrew Calvert, Malcolm Cattach, Tim Chalke, Gordon Chunnett, David Clark, John Coggon, Jeremy Cook, Kim Cook, Gordon Cooper, Jun Cowan, Terry Crabb, Pat Cuneen, Giancarlo Dal Moro, Heike Delius, Mike Doyle, Mark Dransfield, Joseph Duncan, Braam Du Ploy, David Eaton, Donald Emerson, Nicoleta Enescu, Brian Evans, Paul Evans, Shane Evans, Derek Fairhead, Ian Ferguson, Keith Fisk, Andrew Fitzpatrick, Marcus Flis, Catherine Foley, Mary Fowler, Jan Francke, Kim Frankcombe, Peter Fullagar, Stefan Gawlinski, Don Gendzwill, Mark Gibson, Howard Golden, Neil Goulty, Bob Grasty, Ronald Green, David Groves, Steffen Hagemann, Richard Haines, Greg Hall, Michael Hallett, Craig Hart, John Hart, Mike Hatch, Phil Hawke, Nick Hayward, Graham Heinson, Bob Henderson, Larissa Hewitt, Eun-Jung Holden, Terry Hoschke, David Howard, Neil Hughes, Ross Johnson, Steven Johnson, Gregory Johnston, Aurore Joly, Leonie Jones, John Joseph, Christopher Juhlin, Maija Kurimo, Richard Lane, Terry Lee, Michael Lees, Peter Leggatt, James Leven, Ted Lilley, Mark Lindsay, Andrew Lockwood, Andrew Long, Jim Macnae, Alireza Malehmir, Simon Mann, Jelena Markov, Christopher Martin, Keith Martin, Charter Mathison, Cam McCuaig, Steve McCutcheon, Ed McGovern, Stephen McIntosh, Katherine McKenna, Glen Measday, Jayson Meyers, John Miller, Brian Minty, Shane Mule, Mallika Mullick, Jonathan Mwenifumbo, Helen Nash, Adrian Noetzli, Jacob Paggi, Derecke Palmer, Glen Pears, Allan Perry, Mark Pilkington, Sergei Pisarevski, Louis Polome, Rod Pullin, Des Rainsford, Bret Rankin, Emmett Reed, James Reid, Robert L. Richardson, Mike Roach, Brian Roberts, Chris Royles, Greg Ruedavey, Michael Rybakov, Lee Sampson, Gilberto Sanchez, Ian Scrimgeour, Gavin Selfe, Kerim Sener, Nick Sheard, Rob Shives, Jeff Shragge, Richard Smith, John Stanley, Edgar Stettler, Barney Stevens, Ian Stewart, Larry Stolarczyk, Ned Stolz, Rob Stuart, Nicolas Thebaud, Ludger Timmen, Allan Trench, Jarrad Trunfell, Greg Turner, Ted Tyne, Phil Uttley, Simon van der Wielen, Frank van Kann, Lisa Vella, Chris Walton, Herb Wang, Tony Watts, Daniel Wedge, Bob Whiteley, Chris Wijns, Ken Witherley, Peter Wolfgram, Faye Worrall and Binzhong Zhou. Particular thanks are due to Duncan Cowan of Cowan Geodata Services for creating almost every image in the book and to Andrew Duncan of EMIT for creating the EM model curves.

We also thank Simon Tegg for his work 'colourising' the figures. From Cambridge University Press, we thank Laura Clark, Susan Francis, Matthew Lloyd, Lindsay Nightingale and Sarah Payne.

We are also very grateful to the following organisations for providing, or allowing the use of, their data or access to geophysical software:

Barrick (Australia Pacific) Limited
CGG
Department of Manufacturing, Innovation, Trade, Resources and Energy, South Australia
EMIT Electromagnetic Imaging Technology
Evolution Mining
Geological Survey of Botswana
Geological Survey of NSW, NSW Trade & Investment
Geological Survey of Western Australia, Department of Mines and Petroleum
Geometrics

Geonics
Geoscience Australia
Geotech Geophysical Surveys
GPX Surveys
Ground Probe (SkyTEM)
Haines Surveys
Mines Geophysical Services
Montezuma Mining Company
Natural Resources Canada, Geological Survey of Canada
Northern Territory Geological Survey
Ontario Geological Survey
University of British Columbia, Geophysical Inversion Facility (UBC-GIF)

Finally we are most grateful to the six industry sponsors: Carpentaria Exploration, First Quantum Minerals, MMG, Rio Tinto Exploration, AngloGold Ashanti and St Barbara, plus the Centre for Exploration Targeting at The University of Western Australia, whose financial support has allowed us to produce a textbook with colour throughout, greatly improving the presentation of the data.

Mike Dentith and Stephen Mudge

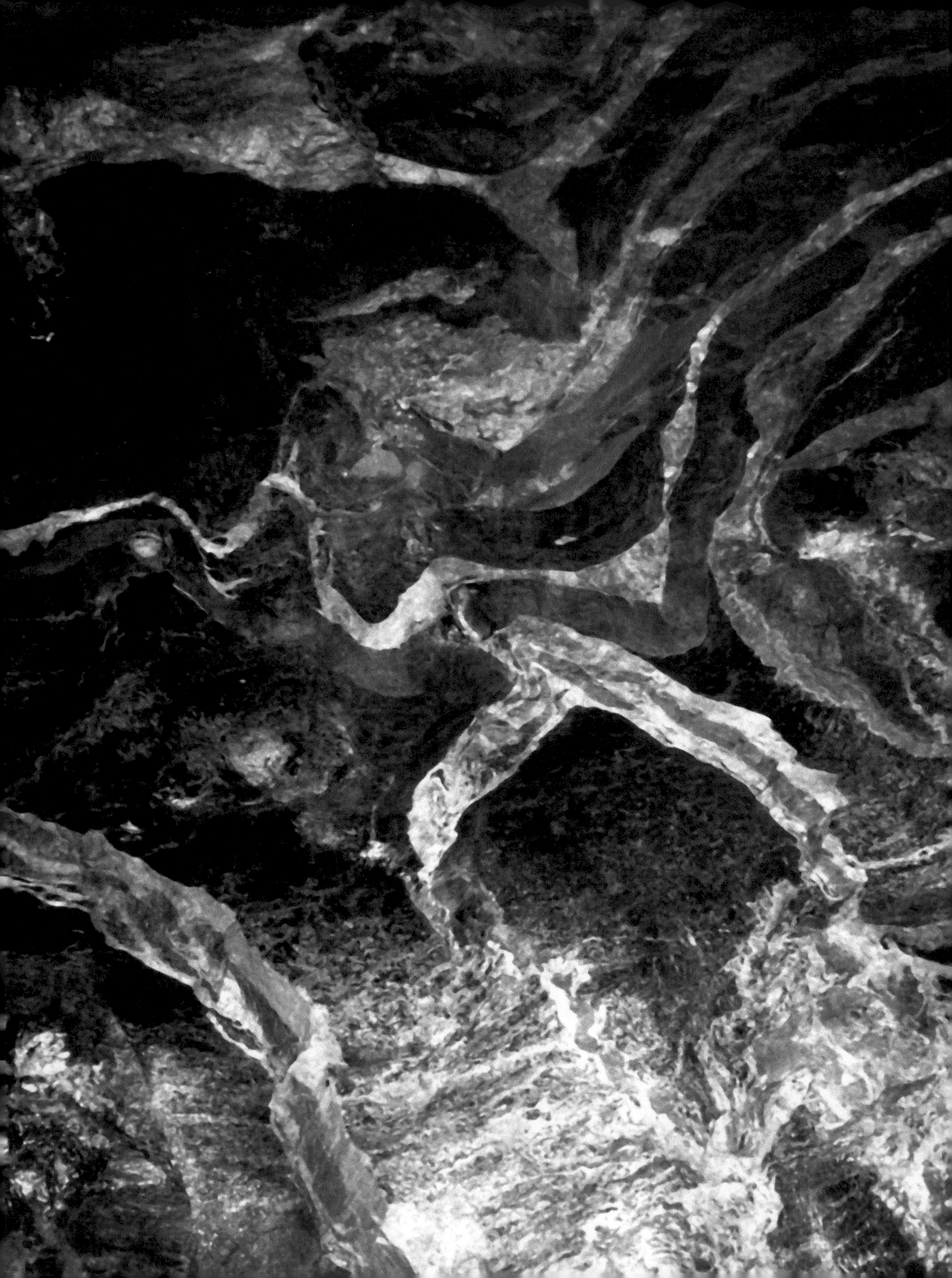

CHAPTER

1 Introduction

Geophysical methods respond to differences in the physical properties of rocks. Figure 1.1 is a schematic illustration of a geophysical survey. Over the area of interest, instruments are deployed in the field to measure variations in a physical parameter associated with variations in a physical property of the subsurface. The measurements are used to infer the geology of the survey area. Of particular significance is the ability of geophysical methods to make these inferences from a distance, and, for some methods, without contact with the ground, meaning that geophysics is a form of remote sensing (*sensu lato*). Surveys may be conducted on the ground, in the air or in-ground (downhole). Information about the geology can be obtained at scales ranging from the size of a geological province down to that of an individual drillhole.

Geophysics is an integral part of most mineral exploration programmes, both greenfields and brownfields, and is increasingly used during the mining of orebodies. It is widely used because it can map large areas quickly and cost effectively, delineate subtle physical variations in the geology that might otherwise not be observed by field geological investigations and detect occurrences of a wide variety of mineral deposits.

It is generally accepted that there are few large orebodies remaining to be found at the surface, so mineral exploration is increasingly being directed toward searching for covered and deep targets. Unlike geochemistry and other remote sensing techniques, geophysics can see into the subsurface to provide information about the concealed geology. Despite this advantage, the interpretation of geophysical data is critically dependent on their calibration against geological and geochemical data.

◀ Folded massive nickel sulphide mineralisation in the Maggie Hays mine, Western Australia. The field of view is 1.2 m wide. Photograph: John Miller.

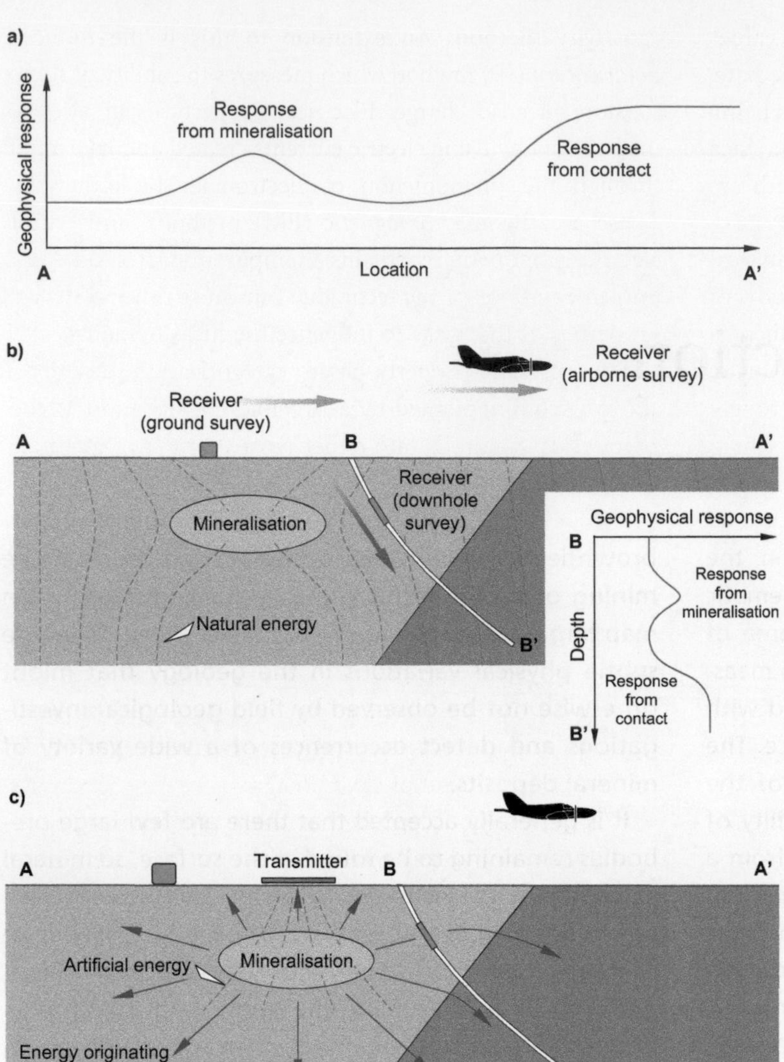

Figure 1.1 Geophysical surveying schematically illustrated detecting mineralisation and mapping a contact between different rock types. Instruments (receivers) make measurements of a physical parameter at a series of locations on or above the surface (A–A′) or downhole (B–B′). The data are plotted as a function of location or depth down the drillhole (a). (b) Passive geophysical surveying where a natural source of energy is used and only a receiver is required. (c) Active geophysical surveying where an artificial source of energy (transmitter) and a receiver are both required.

1.1 Physical versus chemical characterisation of the geological environment

The geophysical view of the geological environment focuses on variations in the physical properties within some volume of rock. This is in direct contrast with the geological view, which is primarily of variations in the bulk chemistry of the geology. The bulk chemistry is inferred from visual and chemical assessment of the proportions of different silicate and carbonate minerals at locations where the geology happens to be exposed, or has been drilled. These two fundamentally different approaches to assessing the geological environment mean that a particular area of geology may appear homogeneous to a geologist but may be geophysically heterogeneous, and vice versa. The two perspectives are complementary, but they may also appear to be contradictory. Any contradiction is resolved by the 'chemical' versus 'physical' basis of investigating the geology. For example, porosity and pore contents are commonly important influences on physical properties, but are not a factor in the various schemes used by geologists to assign a lithological name, these schemes being based on mineralogical content and to a lesser extent the distribution of the minerals.

Some geophysical methods can measure the actual physical property of the subsurface, but all methods are sensitive to physical property contrasts or relative changes in properties, i.e. the juxtaposition of rocks with different physical properties. It is the changes in physical properties that are detected and mapped. This relativist geophysics approach is another fundamental aspect that differs from the absolutist geological approach. For example, one way of geologically classifying igneous rocks is according to their silica content, with absolute values used to define categories such as felsic, intermediate, mafic etc. The geophysical approach is equivalent to being able to tell that one rock contains, say, 20% more silica than another, without knowing whether one or both are mafic, felsic etc.

The link between the geological and geophysical perspectives of the Earth is *petrophysics* – the study of the physical properties of rocks and minerals, which is the foundation of the interpretation of geophysical data. Petrophysics is a subject that we emphasise strongly throughout this book, although it is a subject in which some important aspects are not fully understood and more research is urgently required.

1.2 Geophysical methods in exploration and mining

Geophysical methods are used in mineral exploration for geological mapping and to identify geological environments favourable for mineralisation, i.e. to directly detect, or target, the mineralised environment. During exploitation of mineral resources, geophysics is used both in delineating and evaluating the ore itself, and in the engineering-led process of accessing and extracting the ore.

There are five main classes of geophysical methods, distinguished according to the physical properties of the geology to which they respond. The *gravity* and *magnetic* methods detect differences in density and magnetism, respectively, by measuring variations in the Earth's gravity and magnetic fields. The *radiometric* method detects variations in natural radioactivity, from which the radioelement content of the rocks can be estimated. The *seismic* method detects variations in the elastic properties of the rocks, manifest as variations in the behaviour of seismic waves passing through them. Seismic surveys are highly effective for investigating layered stratigraphy, so they are the mainstay of the petroleum industry but are comparatively rarely used by the minerals industry.

The electrical methods, based on the electrical properties of rocks and minerals, are the most diverse of the five classes. Electrical conductivity, or its reciprocal resistivity, can be obtained by measuring differences in electrical potentials in the rocks. When the potentials arise from natural processes the technique is known as the *spontaneous potential* or *self-potential* (SP) method. When they are associated with artificially generated electric currents passing through the rocks, the technique is known as the *resistivity* method. An extension to this is the *induced polarisation* (IP) method which measures the ability of rocks to store electric charge. Electrical properties can also be investigated by using electric currents created and measured through the phenomenon of electromagnetic induction. These are the *electromagnetic* (EM) methods, and whilst electrical conductivity remains an important factor, different implementations of the technique can cause other electrical properties of the rocks to influence the measurements.

The physical-property-based categorisation described above is complemented by a two-fold classification of the geophysical methods into either *passive* or *active* methods (Fig. 1.1b and c).

Passive methods use natural sources of energy, of which the Earth's gravity and magnetic fields are two examples, to investigate the ground. The geophysical measurement is made with some form of instrument, known as a *detector*, *sensor* or *receiver*. The receiver measures the *response* of the local geology to the natural energy. The passive geophysical methods are the gravity, magnetic, radiometric and SP methods, plus a form of electromagnetic surveying known as *magnetotellurics* (described in online Appendix 4).

Active geophysical methods involve the deliberate introduction of some form of energy into the ground, for example seismic waves, electric currents, electromagnetic waves etc. Again, the ground's response to the introduced energy is measured with some form of detector. The need to supplement the detector with a *source* of this energy, often called the *transmitter*, means that the active methods are more complicated and expensive to work with. However, they do have the advantage that the transmission of the energy into the ground can be controlled to produce responses that provide particular information about the subsurface, and to focus on the response from some region (usually depth) of particular interest. Note that, confusingly, the cause of a geophysical response in the subsurface is also commonly called a *source* – a term and context we use extensively throughout the text.

1.2.1 Airborne, ground and in-ground surveys

Geophysical surveying involves making a series of measurements over an area of interest with survey parameters appropriate to the scale of the geological features being investigated. Usually, a single survey instrument is used to traverse the area, either on the ground, in the air or within a drillhole (Fig. 1.1). Surveys from space or on water are also possible but are uncommon in the mining industry. In general, airborne measurements made from a low-flying aircraft are more cost-effective than ground measurements for surveys covering a large area or comprising a large number of readings. The chief advantages of airborne surveying relative to ground surveying are the greater speed of data acquisition and the completeness of the survey coverage.

As exploration progresses and focuses on smaller areas, there is a general reduction in both the extent of geophysical surveys and the distances between the individual readings in a survey. Airborne surveys are usually part of the reconnaissance phase, which is often the initial phase of exploration, although some modern airborne systems offer higher resolution by surveying very close to the ground and may find application in the later stages of exploration. Ground and drillhole surveys, on the other hand, offer the highest resolution of the subsurface. They are mostly used for further investigation of areas targeted from the reconnaissance work for their higher prospectivity, i.e. they are used at the smaller prospect scale.

Methods that can be implemented from the air include magnetics, known as *aeromagnetics*; gravity, sometimes referred to as *aerogravity* or as currently implemented for mineral exploration as *airborne gravity gradiometry*; radiometrics; and electromagnetics, usually referred to as *airborne electromagnetics* (AEM). All the geophysical methods can be implemented downhole, i.e. in a drillhole. Downhole surveys are a compact implementation of conventional surface surveying techniques. There are two quite distinct modes of making downhole measurements: *downhole logging* and *downhole surveying*.

Downhole logging is where the *in situ* physical properties of the rocks penetrated by a drillhole are measured to produce a continuous record of the measured parameter. Downhole logs are commonly used for making stratigraphic correlations between drillholes in the sedimentary sequences that host coal seams and iron formations. Measurements of several physical parameters, producing a suite of logs, allow the physical characterisation of the local geology, which is useful for the analysis of other geophysical data and also to help plan future surveys, e.g. Mwenifumbo *et al.* (2004). Despite the valuable information obtainable, *multiparameter logging* is not ubiquitous in mineral exploration. However, its use is increasing along with integrated interpretation of multiple geophysical datasets.

Downhole surveying is designed to investigate the larger region surrounding the drillhole, with physical property variations obtained indirectly, and to indicate the direction and even the shape of targets. That is, downhole electrical conductivity logging measures the conductivity of the rocks that form the drillhole walls, whereas a downhole electromagnetic survey detects conductivity variations, perhaps owing to mineralisation, in the volume surrounding the drillhole. Downhole geophysical surveys increase the radius of investigation of the drillhole, increase the depth of investigation and provide greater resolution of buried targets.

Geophysical surveys are sometimes conducted in open-pit and underground mines; measurements are made in vertical shafts and/or along (inclined) drives, usually to detect and delineate ore horizons. There exists a rather small literature describing underground applications of geophysics, e.g. Fallon *et al.* (1997), Fullagar and Fallon (1997) and McDowell *et al.* (2007), despite many successful surveys having been completed. Application and implementation of geophysics underground tend to be unique to a particular situation, and survey design requires a fair degree of ingenuity to adapt the arrangement of transmitter and receiver to the confines of the underground environment. They are usually highly focused towards determining a specific characteristic of a small volume of ground in the immediate surrounds. Electrical and mechanical interference from mine infrastructure limits the sensitivity of surveys, which require a high level of planning and coordination with mining activities. Also, data from in-mine surveys require particular skills to interpret the more complex three-dimensional (3D) nature of the responses obtained: for example, the response may emanate from overhead, or the survey could pass through the target. The generally unique nature of underground geophysical surveys and our desire to emphasise the principles and common practices of geophysics in mineral exploration restrict us from describing this most interesting application of geophysics, other than to mention, where appropriate, the possibilities of using a particular geophysical method underground.

1.2.2 Geophysical methods and mineral deposits

The physical properties of the geological environment most commonly measured in mining geophysics are density, magnetism, radioactivity and electrical properties. Elastic (seismic) properties are not commonly exploited. In general, density, magnetism and radioactivity are used to map the geology, the latter when the nature of the surface materials is important. The limited use of electrical properties is due to their non-availability from an airborne

Table 1.1 Geophysical methods commonly used in the exploration and exploitation of some important types of mineral deposits. Brackets denote lesser use. Also shown, for comparison, are methods used for petroleum exploration and groundwater studies. L – downhole logging, M – geological mapping of prospective terrains, D – detection/delineation of the mineralised environment. The entries in the density column reflect both the use of ground gravity surveys and anticipated future use of aerogravity. Developed from a table in Harman (2004).

Deposit type	Density	Magnetism	Electrical properties	Radioactivity	Elastic properties
Iron formation associated Fe ores	M D L	M D	D	M (L)	
Coal	(M) L	M D	L	L	M D L
Evaporite-hosted K				L	M D L
Fe-oxide Cu–Au (IOCG)	M D	M D	D	D	
Broken Hill type Ag–Pb–Zn	M (D)	M	D		
Volcanogenic massive sulphide (VMS) Cu–Pb–Zn	M (D)	M	D	D	
Magmatic Cu, Ni, Cr and Pt-group	M D	M D	D		
Primary diamonds	M	M	(M)		
Uranium	M	M	M	D L	
Porphyry Cu, Mo	M	M D	D	D	
Sedimentary exhalative (SEDEX) Pb–Zn	M	M (D)	D		
Greenstone belt Au	M	M			
Epithermal Au	M	M		M	
Placer deposits	M	(M)	M		M
Sediment-hosted Cu–Pb–Zn	M	M	D		
Skarns	M	M D	(D)		
Heavy mineral sands		M D		M D	
Mineralisation in regolith and cover materials, e.g. Al, U, Ni			D	M D	
Groundwater studies			M D L	L	M
Petroleum exploration and production	(M) L	(M)	(M) L	L	M (D) L

platform, although AEM-derived conductivity measurements are becoming more common. Direct detection of a mineralised environment may depend upon any one or more of density, magnetism, radioactivity, electrical properties and possibly elasticity. Table 1.1 summarises how contrasts in physical properties are exploited in exploration and mining of various types of mineral deposits, and in groundwater and petroleum studies.

1.2.3 The cost of geophysics

The effectiveness and cost of applying any 'tool' to the exploration and mining process, be it geological, geochemical, geophysical, or drilling, are key considerations when formulating exploration strategies. After all, the ultimate aim of the exploration process is to discover ore within the constraints of time and cost, which are usually determined outside the realms of the exploration programme. In both exploration and production the cost of drilling accounts for a large portion of expenditure. An important purpose of geophysical surveying is to help minimise the amount of drilling required.

The cost of a geophysical survey includes a fixed mobilisation cost and a variable cost dependent upon the volume of data collected, with large surveys attracting

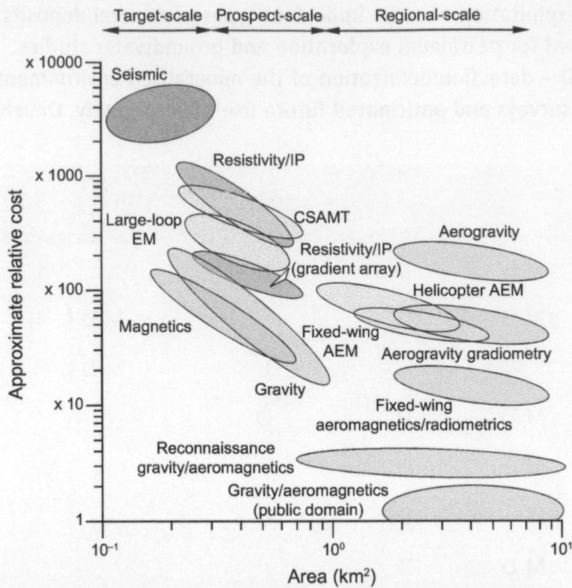

Figure 1.2 Approximate relative costs per square kilometre of different kinds of geophysical surveys and the approximate variation with size of the survey area. AEM – airborne electromagnetics, CSAMT – controlled source audio-frequency magnetotellurics, EM – electromagnetics, IP – induced polarisation. Redrawn with additions, with permission, from Fritz (2000).

favourable economies-of-scale. Additional costs can be incurred through 'lost time' related to factors such as adverse weather and access restrictions to the survey area, all preventing progress of the survey. Local conditions are widely variable, so it is impossible to state here the costs of different kinds of geophysical surveys. Nevertheless, it is useful to have an appreciation for the approximate relative costs of various geophysical methods compared with the cost of drilling. Drilling is not only a major, and often the largest, cost in most exploration and mining programmes, it is often the only alternative to geophysics for investigating the subsurface.

Following the approach of Fritz (2000), Fig. 1.2 shows the approximate relative cost of different geophysical methods. Of course the costs on which these diagrams are based can be highly variable owing to such factors as the prevailing economic conditions and whether the surveys are in remote and rugged areas. They should be treated as indicative only. The seismic method is by far the most expensive, which is one reason why it is little used by the mining industry, the least expensive methods being airborne magnetics and radiometrics. The areas over which information is gathered for each method are compared in Fig. 1.3, noting that cost estimates are equated to the estimated total cost of a single 300 m drillhole, including logging, assaying, remediation etc. The drillhole provides reliable geological information to a certain depth, but only from a very small area. Drilling on a grid pattern at 25 m intervals over an area of 1 km^2 would cost a few tens of millions of dollars, but would only sample 3 ppm of the volume. Geophysical methods provide information from vastly greater areas and volumes, albeit in a form that is not necessarily geologically explicit and will not necessarily directly identify mineralisation. Despite this, appropriately designed geophysical surveys and appropriately chosen data analysis are highly effective for optimally targeting expensive drillholes.

1.3 About this book

Our focus is an explanation of the principles and modern practice of geophysics in the search for mineral deposits. The explanations are presented from a perspective relevant to a mining industry geologist.

Throughout the text we emphasise the key aspects of mineral exploration geophysics, in particular those aspects that affect the interpretation of geophysical data. These include petrophysics, the foundational science of geophysics; numerical processing of the data; the creation and interpretation of raster imagery; problems presented by deeply weathered environments; geophysical characteristics of geologically complex basement terrains; and the inability to remove noise completely from the measurements. We introduce the term '*geophysical paradox*', where to fully understand the geophysical signal (the information of interest) and the noise (the interference producing uncertainty in the signal) requires information about the subsurface, but the purpose of the geophysical survey is to acquire this very information. We emphasise the need to understand this fundamental aspect of geophysics when working with geophysical data.

There have been many developments in geophysics in recent years. We have deliberately avoided presenting older techniques and practices not used widely today, leaving descriptions of these to earlier texts.

The text is structured around the main geophysical methods with each described in its own separate chapter. General aspects of the nature of geophysical data, their

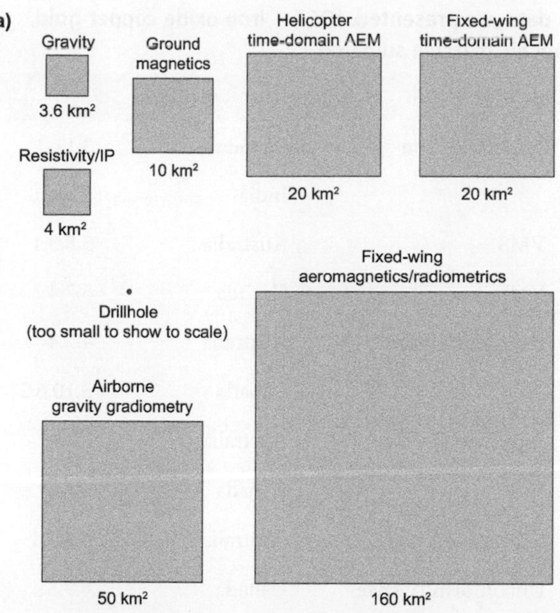

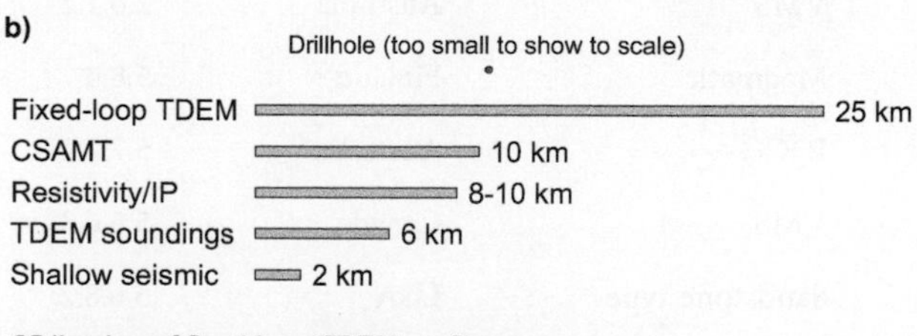

Figure 1.3 Approximate relative (a) areas and (b) line lengths sampled by geophysical surveys costing the equivalent of a single 300 m deep diamond drillhole. The area of the drillhole is shown for comparison. AEM – airborne electromagnetics, CSAMT – controlled source audio-frequency magnetotellurics, GPS – global positioning system, IP – induced polarisation, TDEM – time domain electromagnetics. Redrawn with additions, with permission, from Fritz (2000).

acquisition, processing, display and interpretation, common to all methods, are described first in a general chapter, Chapter 2. Essential, and generally applicable, details of vectors and waves are described in the online Appendices 1 and 2, respectively. The other chapters are designed to be largely self-contained, but with extensive cross-referencing to other chapters, in particular to Chapter 2. We have responded to the widespread complementary use of gravity and magnetics by describing them in a single combined chapter, Chapter 3. Geophysical methods less commonly used by the mining industry are described in online Appendices 3 to 6. Appendix 7 lists sources of information about mineral exploration geophysics, especially case histories. The principles described are demonstrated by examples of geophysical data and case studies from a wide variety of mineral deposit types from around the world. All deposits referred to are listed in Table 1.2 and their locations shown on Fig. 1.4.

At the conclusion of each chapter we provide a short list of appropriate resource material for further reading on the topic. The references cited throughout the text emphasise those we believe suit the requirements of the exploration geoscientist.

Table 1.2 **Locations of deposits and mineralised areas from which geophysical data are presented. IOCG – iron oxide copper gold, MVT – Mississippi Valley-type, SEDEX – sedimentary exhalative, VMS – volcanogenic massive sulphide.**

Number	Deposit name	Commodities	Deposit style/type	Country	Section
1	Adams	Fe	Iron formation	Canada	3.11.3
2	Almora	Graphite		India	5.5.4.1
3	Balcooma	Cu–Ag–Au	VMS	Australia	5.8.3.1
4	Bell Allard	Zn–Cu–Ag–Au	VMS	Canada	6.7.4.2
5	Blinman	Cu	Sediment hosted	Australia	4.7.4
6	Bonnet Plume Basin	Coal		Canada	3.10.6.2
7	Broken Hill area	Pb–Zn–Ag	Broken Hill type	Australia	3.7
8	Buchans	Zn–Pb–Cu	VMS	Canada	4.7.5
9	Butcherbird	Mn	Supergene	Australia	5.9.5.1
10	Cluff Lake area	U	Unconformity style	Canada	4.7.5
11	Cripple Creek district	Ag–Au–Te	Epithermal	USA	3.4.7
12	Cuyuna Iron Range	Fe	Iron formation	USA	5.5.3.2
13	Dugald river	Zn–Pb–Ag	SEDEX	Australia	4.7.5
14	Eloise	Cu–Au	SEDEX	Australia	5.7.7.1
15	Elura	Zn–Pb–Ag	VMS	Australia	2.6.1.2
16	Enonkoski (Laukunkangas)	Ni	Magmatic	Finland	5.8.4
17	Ernest Henry	Cu–Au	IOCG	Australia	5.7.7.1
18	Estrades	Cu–Zn–Au	VMS	Canada	5.6.6.3
19	Franklin	U	Sandstone type	USA	5.6.8.2
20	Gölalana	Cr	Magmatic	Turkey	3.11.5
21	Golden Cross/Waihi-Waitekauri epithermal area	Au–Ag	Epithermal	New Zealand	3.9.7 4.6.6 4.7.3.2 A4.7.2
22	Goongewa/Twelve Mile Bore	Pb–Zn	MVT	Australia	5.6.7
23	Goonumbla/North Parkes area	Cu–Au	Porphyry	Australia	3.11.4 4.6.6
24	Iron King	Pb–Zn–Cu–Au–Ag	VMS	USA	4.6.6
25	Jharia Coalfield	Coal		India	3.11.5 5.5.3.2
26	Jimblebar	Fe	Iron formation	Australia	4.7.5
27	Joma	Fe–S	Massive pyrite	Norway	2.9.2 5.5.3.1
28	Kabanga	Ni	Magmatic	Tanzania	3.9.8.2
29	Kerr Addison	Au	Orogenic	Canada	3.11.3
30	Kimheden	Cu	VMS	Sweden	5.5.3.2

Table 1.2 (*cont.*)

Number	Deposit name	Commodities	Deposit style/type	Country	Section
31	Kirkland Lake	Au	Orogenic	Canada	2.8.1.1 3.11.3
32	Las Cruces	Cu–Au	VMS	Spain	3.7
33	Lisheen	Zn–Pb–Ag	Carbonate-hosted	Eire	5.7.4.2 5.7.4.3
34	London Victoria	Au	Lode	Australia	3.11.4.1 A6.3.5
35	Maple Creek	Au	Placer	Guyana	A5.3.4.1
36	Marmora	Fe	Skarn	Canada	2.6.4 3.11.5
37	Mirdita Zone	Cu	VMS	Albania	5.5.3.1
38	Mount Isa	Pb–Zn–Cu	SEDEX	Australia	5.8.2 A5.4.1
39	Mount Keith area	Ni	Magmatic	Australia	A3.3.1.1
40	Mount Polley	Cu-Au	Porphyry	Canada	2.8.2
41	Murray Brook	Cu–Pb–Zn	VMS	Canada	2.9.2
42	New Insco	Cu	VMS	Canada	5.5.3.1
43	Olympic Dam	Cu–U–Au–Ag–REE	IOCG	Australia	2.7.2.3 5.6.6.3
44	Pajingo epithermal system (Scott Lode, Cindy, Nancy and Vera)	Au	Epithermal	Australia	5.6.6.4
45	Palmietfontein	Diamond	Kimberlite-hosted	South Africa	5.6.6.1 5.6.6.2
46	Pine Point	Pb–Zn	MVT	Canada	2.9.2 5.6.6.4
47	Port Wine area	Au	Placer	USA	3.11.1
48	Poseidon	Ni	Magmatic	Australia	A3.4.1
49	Prairie Evaporite	K	Evaporite	Canada	4.7.5 6.5.2.5
50	Pyhäsalmi	Ni	Magmatic	Finland	2.10.2.3
51	Qian'an District	Fe	Iron Formation	China	3.10.1.1
52	Red Dog	Zn–Pb	SEDEX	USA	5.6.6.3
53	Regis Kimberlite	Diamond	Kimberlite-hosted	Brazil	A4.7.1
54	Rocky's Reward	Ni	Magmatic	Australia	A5.3.4.2
55	Safford	Cu	Porphyry	USA	5.5.4.2
56	Sargipalli	Graphite		India	5.5.3.1
57	Silvermines	Zn–Pb–Ag	Carbonate-hosted	Eire	5.6.6.2

Table 1.2 *(cont.)*

Number	Deposit name	Commodities	Deposit style/type	Country	Section
58	Singhblum	Cu	Disputed	India	5.5.3.2
59	South Illinois Coalfield	Coal		USA	6.7.4.1
60	Sulawesi Island	Ni	Lateritic	Indonesia	A5.3.4.1
61	Telkkälä Taipalsaari	Ni	Magmatic	Finland	2.10.2.3
62	Thalanga	Zn–Pb–Cu–Ag	SEDEX	Australia	2.8.1
63	Thompson	Ni	Magmatic	Canada	3.11.5
64	Trilogy	Cu–Au–Ag–Pb–Zn	VMS	Australia	5.7.7.1
65	Tripod	Ni	Magmatic	Canada	5.7.7.1
66	Uley	Graphite		Australia	5.6.8.1
67	Uranium City area	U	Unconformity style	Canada	4.7.3.1
68	Victoria	Graphite		Canada	5.6.9.5
69	Voisey Bay	Ni	Magmatic	Canada	6.8.2
70	Wallaby	Au	Orogenic	Australia	3.11.2
71	Witwatersrand Goldfield	Au	Palaeoplacer	South Africa	6.7
72	Woodlawn	Cu–Pb–Zn	VMS	Australia	5.6.9.4
73	Yankee Fork Mining District	Ag–Au	Epithermal	USA	3.8.6 3.9.7
74	Yeelirrie	U	Calcrete-hosted	Australia	4.7.3.1

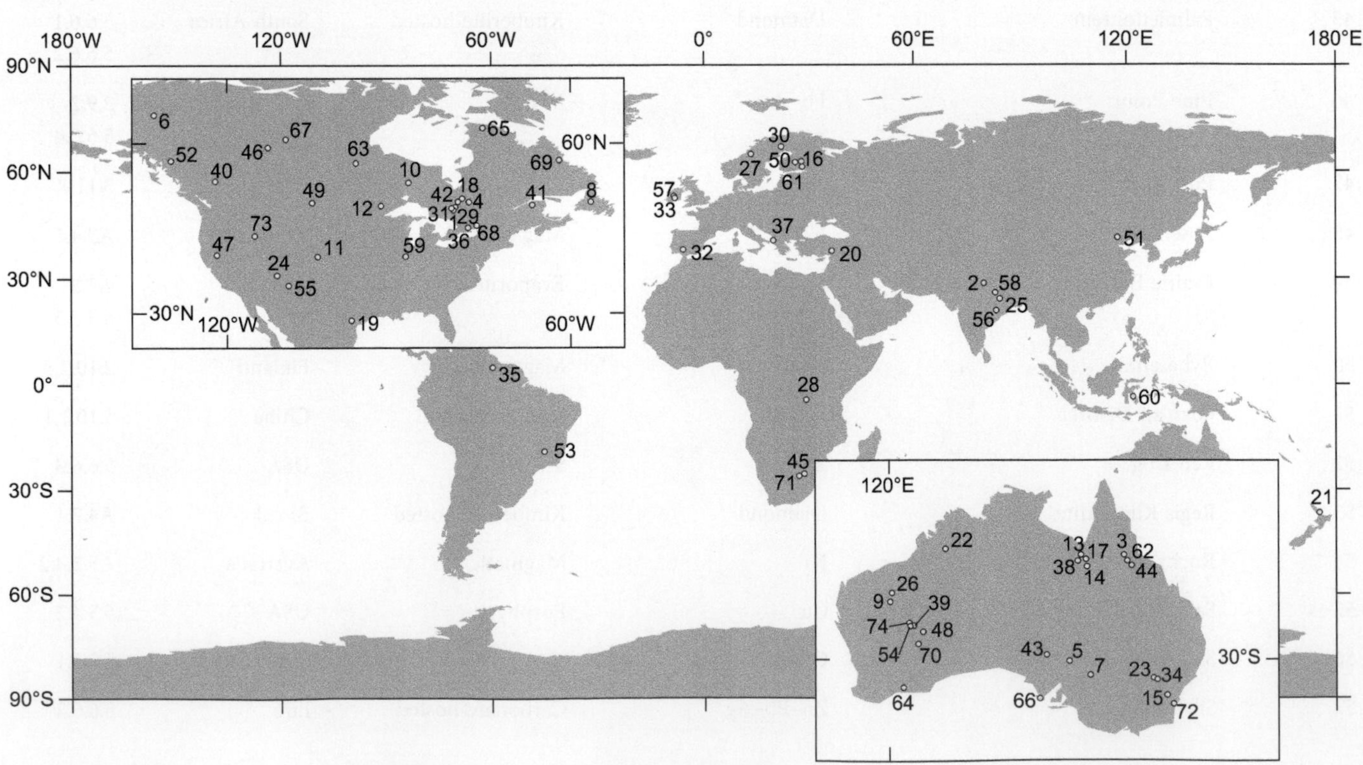

Figure 1.4 Locations of deposits and mineralised areas from which geophysical data are presented.

FURTHER READING

Blain, C., 2000. Fifty year trends in minerals discovery – commodity and ore types. *Exploration and Mining Geology*, 9, 1–11.

Nabighian, M.N. and Asten, M.W., 2002. Metalliferous mining geophysics – state of the art in the last decade of the 20th century and the beginning of the new millennium. *Geophysics*, 67, 964–978.

A summary of the state of the art of mining geophysics, still relevant even though more than 10 years old.

Paterson, N.R., 2003. Geophysical developments and mine discoveries in the 20th century. *The Leading Edge*, 22, 558–561.

This and the first paper provide data on mineral deposit discovery rates and costs, and their relationships with the use and development of geophysical methods.

ZS-PWS

CHAPTER

2 Geophysical data acquisition, processing and interpretation

2.1 Introduction

The use of geophysical methods in an exploration programme or during mining is a multi-stage and iterative process (Fig. 2.1). The main stages in their order of application are: definition of the survey objectives, data acquisition, data processing, data display and then interpretation of different forms of the data. The geologist should help to define the objectives of the survey and should have a significant contribution during interpretation of the survey data, but to ensure an optimum outcome, an understanding of all the other stages highlighted by Fig. 2.1 is required. Survey objectives dictate the geophysical method(s) to be used and the types of surveys that are appropriate, e.g. ground, airborne etc. Data acquisition involves the two distinct tasks of designing the survey and making the required measurements in the field. Data processing involves reduction (i.e. correcting the survey data for a variety of distorting effects), enhancement and display of the data, all designed to highlight what is perceived to be the most geologically relevant information in the data. The processed data can be displayed in a variety of ways to suit the nature of the dataset and the interpreter's requirements in using the data. Data interpretation is the analysis of the geophysical data and the creation of a plausible geological model of the study area. This is an indeterminate process; an interpretation evolves through the iterative process as different geological concepts are tested with the data. It is often necessary to revise aspects of the data enhancement as different characteristics of the data assume greater significance, or as increased geological understanding allows more accurate reduction.

The interpreter needs to have a good understanding of the exploration strategy which was the basis for defining the survey objectives. Ideally the interpreter should also have a working knowledge of the geophysical acquisition–processing sequence since this impinges on the evolving interpretation of the data. The type of survey and the nature of the data acquisition affect the type and resolution of the geological information obtainable, whilst the interpretation of geophysical data is dependent on the numerical methods applied to enhance and display the data. Analysis of the data involves their processing and interpretation. We emphasise that interpretation is not a task to be undertaken in isolation; it is an inextricable part of the iterative and multi-stage analysis shown in Fig. 2.1.

Figure 2.1 illustrates the framework for this chapter. We discuss in turn the various stages in using a geophysical method, but before doing so we discuss some general aspects of geophysical measurements, geophysical responses, and the important concepts of signal and noise.

◂ Airborne magnetic survey aircraft. Image provided by New Resolution Geophysics.

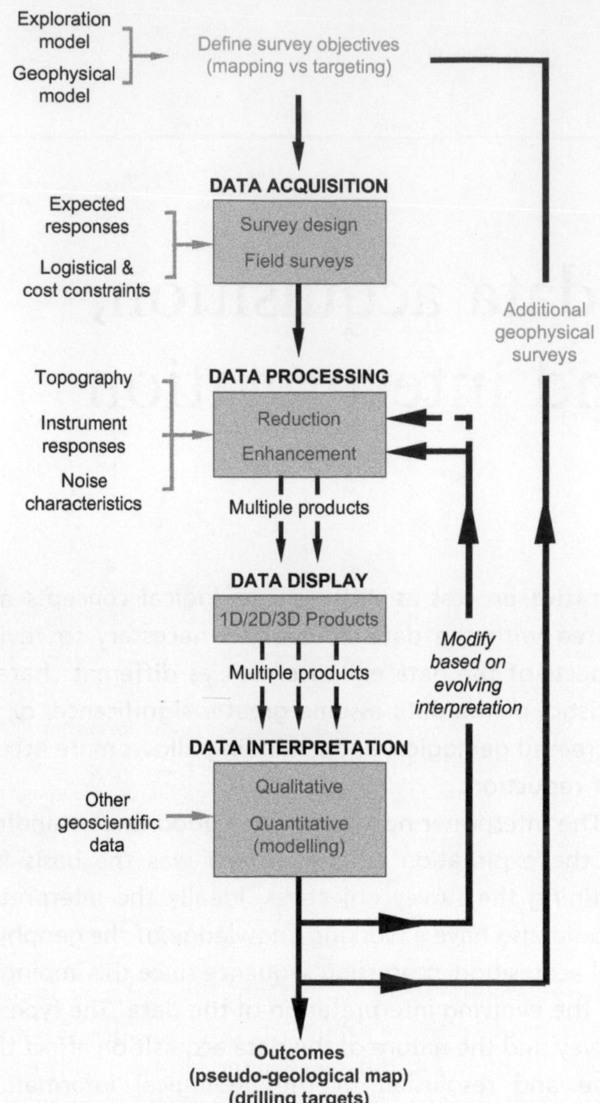

Figure 2.1 The principal stages of a geophysical programme in mineral exploration: from identifying the objectives of the geophysical survey(s) through to providing an interpretation of the subsurface geology.

2.2 Types of geophysical measurement

The parameters measured in the various types of geophysical surveys described in Section 1.2 are continuous, i.e. they vary in time or space and without gaps or end. The variations are an analogue representation of the physical property variations that occur in the subsurface. Measuring or sampling an analogue signal at discrete times or at discrete locations is known as *digitisation*. The continuous variation is then represented by a series of data samples forming a digital series, a form most convenient for storage and processing by a computer.

A geophysical survey consists of a series of measurements made at different locations; usually different geographic locations, or different depths in a drillhole. The location assigned to the measurement is usually the sensor location but may be some point between the transmitter and the sensor. The resultant measurements, i.e. the *dataset*, comprise a *spatial series* in the *spatial domain.* Each of the measurements may comprise a single reading, or may be a series of readings made over an interval of time to form a *time series* in the *time domain*, or over a range of frequencies to form a *frequency series* in the *frequency domain*. In some geophysical methods (e.g. electrical measurements), time- and frequency-series data provide the information about the nature of the rocks at the measurement location; and in other methods (e.g. seismic and some kinds of electromagnetic measurements) they are used to infer variations in the geology with distance from the measurement location. This might be lateral distance from a drillhole, but is most commonly depth below a surface reading. The latter are then known as *soundings*.

Series of all types of geophysical data can be conveniently treated as waves, and we use wave terminology throughout the text. It is strongly recommended that those readers unfamiliar with waves and their properties consult online Appendix 2 for details.

2.2.1 Absolute and relative measurements

Most kinds of geophysical surveys make absolute measurements of the parameter of interest. This is not always necessary; for some kinds of survey, notably gravity and magnetic surveys, relative measurements provide sufficient information. In general, relative measurements have the advantage of being cheaper and easier to make than absolute measurements.

A survey comprising relative measurements requires one or more reference locations, called *base stations*, and the measurements are said to be ‘tied’ to the base stations. The absolute value of the parameter at the base stations may be known, in which case making comparative measurements at other locations allows the absolute values to be determined elsewhere. For example, when we say that the strength of Earth’s magnetic field at a base station is 50,000 nanoteslas (nT), we are referring to the absolute value of the field. If the field strength at a second station is 51,000 nT, then its relative value with respect to the base station is +1000 nT (and the base station has a relative value of −1000 nT with respect to the second station). If the magnetic field at a third station has a relative strength of +2000 nT with respect to the base station, then it has an absolute value of 52,000 nT. In terms of relative values, the base station is assigned a value of zero. In large surveys there may be a master base station from which a series of subsidiary base stations are established. This facilitates surveying by reducing the distance that needs to be travelled to the nearest base station. Note that the accuracy of the absolute value of a parameter obtained by relative measurement from a base station is dependent on the accuracy of the absolute value at the base station and the accuracy of the relative measurement itself.

2.2.2 Scalars and vectors

Physical quantities are classified into two classes. Those that have magnitude only are known as *scalar quantities* or simply scalars. Some examples include mass, time, density and speed. Scalar quantities are described by multiples of their unit of measure. For example, the mass of a body is described by the unit of kilogram and a particular mass is described by the number of kilograms. Scalar quantities are manipulated by applying the rules of ordinary algebra, i.e. addition, subtraction, multiplication and division. For example, the sum of two masses is simply the addition of the individual masses.

Some physical quantities have both magnitude and direction and are known as *vector quantities* or simply *vectors*. Some examples are velocity, acceleration and magnetism. They are described by multiples of their unit of measure and by a statement of their direction. For example, to describe the magnetism of a bar magnet requires a statement of how strong the magnet is (magnitude) and its orientation (direction). The graphical presentation and algebraic manipulation of vectors are described in online Appendix 1.

Measuring vector parameters in geophysics implies that the sensor must be aligned in a particular direction. Often components of the vector are measured. Measurements in perpendicular horizontal directions are designated as the *X* and *Y* directions, which may correspond with east and north; or with directions defined in some other reference frame, for example, relative to the survey traverse along which measurements are taken. Usually the *X* direction is parallel to the traverse. Measurements in the vertical are designated as *Z*, although either up or down may be taken as the positive direction depending upon accepted standards for that particular measurement. We denote the components of a vector parameter (**P**) in these directions as P_X, P_Y and P_Z, respectively.

a)

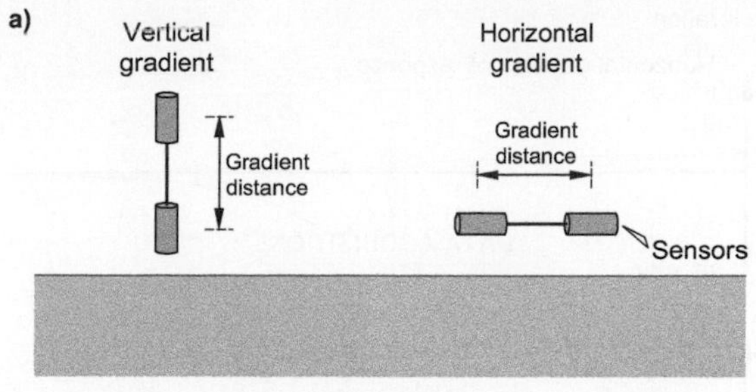

b)

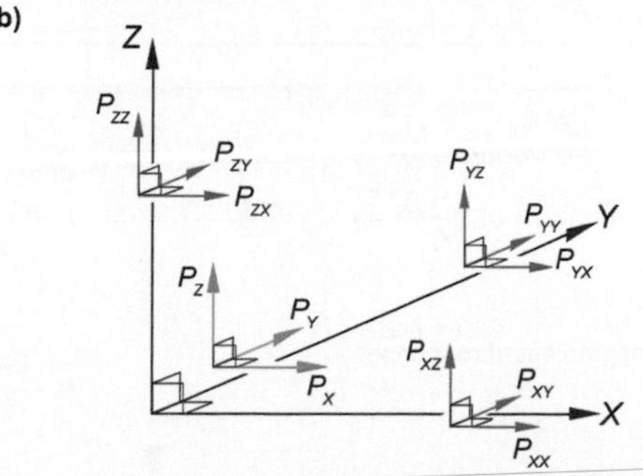

Figure 2.2 Gradient measurements. (a) Vertical and horizontal gradiometers. (b) The three perpendicular gradients of each of the three perpendicular components of a vector parameter **P** forming the gradient tensor of **P**, shown using tensor notation; see text for details.

2.2.3 Gradients

Sometimes it is useful to measure the variation in the amplitude of a physical parameter (**P**) over a small distance at each location. The difference in the measurements from two sensors separated by a fixed distance and oriented in a particular direction is known as the *spatial gradient* of the parameter. It is specified as units/distance in the measurement direction, and so it is a vector quantity. As the measurement distance decreases, the gradient converges to the exact value of the *derivative* of the parameter, as would be obtained from calculus applied to a function describing the parameter field. For the three perpendicular directions *X*, *Y* and *Z*, we refer to the gradient in the *X* direction as the *X*-derivative and, using the notation of calculus, denote it as $\partial\mathbf{P}/\partial x$. Similarly, we denote the *Y*-derivative as $\partial\mathbf{P}/\partial y$ and the *Z*-derivative as $\partial\mathbf{P}/\partial z$.

Gradients may be measured directly using a *gradiometer*, which comprises two sensors positioned a short distance apart (Fig. 2.2a). Alternatively, it is usually possible to compute gradients, commonly referred to as derivatives, directly from the non-gradient survey measurements of the field (see *Gradients and curvature* in Section 2.7.4.4).

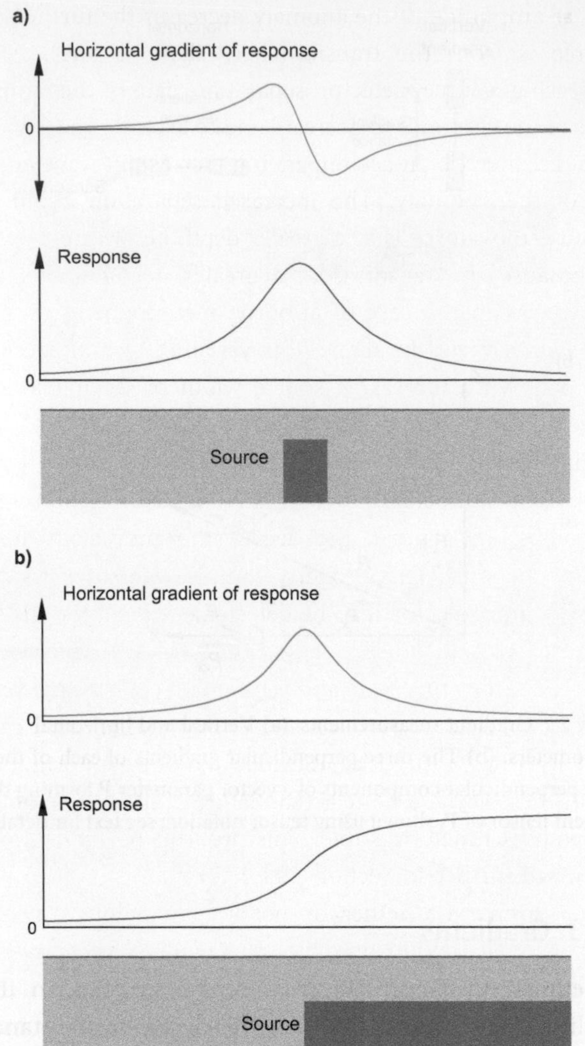

Figure 2.3 Horizontal gradient data across (a) a localised source, and (b) a contact. Note how the gradient response is localised near the source edges.

Gradient measurements have the advantage of not being affected by temporal changes in the parameter being measured; the changes affect both sensors in the same way so any difference in the parameter at each sensor is maintained. Gradient data are very sensitive to the 'edges' of sources. They comprise variations that are more spatially localised than non-gradient data and so have an inherently greater spatial resolution (Fig. 2.3). The main disadvantage of gradient measurements is that they are very sensitive to variations in the orientation of the sensor. Also, long-wavelength variations in the parameter, which produce very small gradients, are often not large enough to be detected.

The derivatives in the three perpendicular directions of each of the three components of a vector parameter (**P**) (Fig. 2.2b) completely describe the parameter at the measurement point. We denote the derivative of the X component of **P** (P_X) in the X direction as $\partial P_X/\partial x$, and the derivatives of the same component in the Y and Z directions are $\partial P_X/\partial y$ and $\partial P_X/\partial z$, respectively; and similarly for the Y and Z components. They form a *tensor* and are displayed and manipulated in matrix form:

$$\begin{pmatrix} \frac{\partial P_X}{\partial x} & \frac{\partial P_X}{\partial y} & \frac{\partial P_X}{\partial z} \\ \frac{\partial P_Y}{\partial x} & \frac{\partial P_Y}{\partial y} & \frac{\partial P_Y}{\partial z} \\ \frac{\partial P_Z}{\partial x} & \frac{\partial P_Z}{\partial y} & \frac{\partial P_Z}{\partial z} \end{pmatrix} \text{ or } \begin{pmatrix} P_{XX} & P_{XY} & P_{XZ} \\ P_{YX} & P_{YY} & P_{YZ} \\ P_{ZX} & P_{ZY} & P_{ZZ} \end{pmatrix} \quad (2.1)$$

Several components of the tensor are related as follows: $P_{XY} = P_{YX}$, $P_{XZ} = P_{ZX}$ and $P_{YZ} = P_{ZY}$, so it is not necessary to measure all of them. This means that less complex sensors are needed and measurements can be made more quickly.

The full-gradient tensor of nine components, i.e. the gradients in the three components in all three directions, provides diagnostic information about the nature of the source of a geophysical anomaly. Tensor measurements are made in airborne gravity surveying (see Section 3.3.2) but are otherwise comparatively rare in other geophysical surveys at present. It seems likely that they will become more common in the future because of the extra information they provide.

2.3 The nature of geophysical responses

As described in Section 1.1 and shown schematically in Fig. 1.1, geophysical surveys respond to physical property contrasts, so changes in the local geology can produce changes in the geophysical response of the subsurface. When the measured property of a target zone is greater than that of the host rocks, the contrast is positive; when lower, it is negative. Typically the changes are localised, arising perhaps from a body of mineralisation or a contact of some kind. These deviations from background values are called *anomalies*. The simplest form of anomaly is an increase or decrease of the measured parameter as the survey traverses the source of the anomaly. Often, though, peaks in the anomaly are offset from their source and/or may be more complex in form; for example, the response from magnetic sources may comprise both an increase and an adjacent decrease in response, forming a dipole anomaly.

Although the underlying physics of each geophysical method is different, some important aspects of the measured responses are the same. Figure 2.4 shows some general

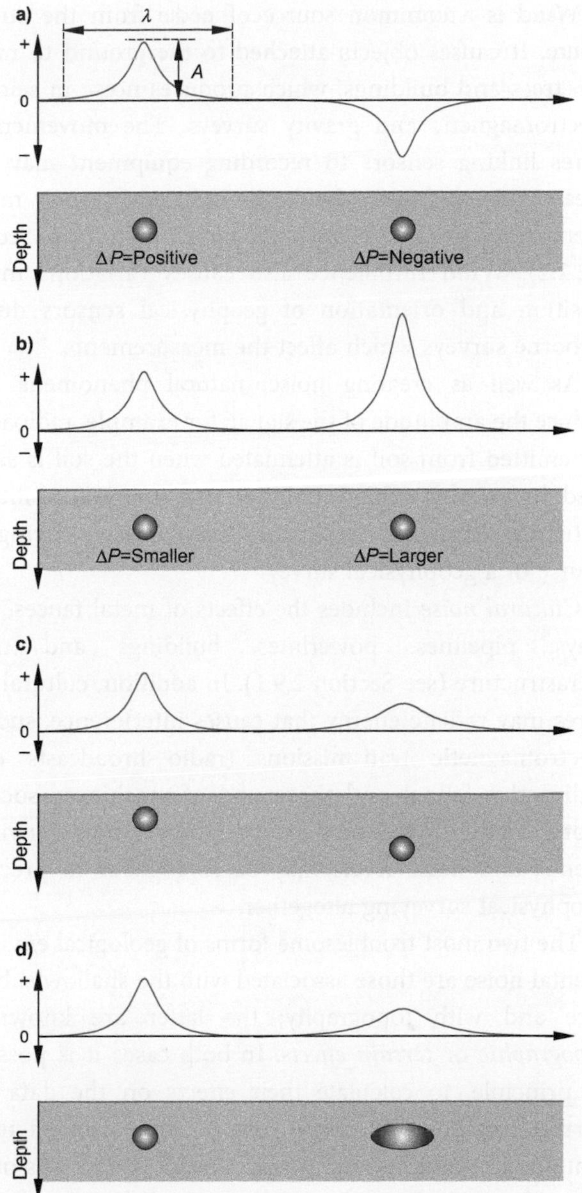

Figure 2.4 The general characteristics of a geophysical response and how these change with variations in (a) sign of the physical property contrast (ΔP), (b) magnitude of the contrast, (c) depth of the source and (d) shape of the source.

characteristics of (non-seismic) geophysical responses using a simple shaped anomaly. A negative contrast produces an anomaly that is an inverted image of that for an identical positive contrast of the same source geometry (Fig. 2.4a). The amplitude (A) of an anomaly depends on the magnitude of the physical property contrast and the physical size of the anomalous distribution. In general, increasing the property contrast increases the amplitude of the anomaly proportionally. Figure 2.4b shows that for two bodies of the same size the anomaly has larger amplitude when there is a larger contrast.

The amplitude of the anomaly decreases the further the source is from the transmitter (if there is one). Also, increasing source–detector separation causes the amplitude of the response to decrease and to extend over a wider area, i.e. there is an accompanying increase in wavelength (λ) of the anomaly. The increased separation could be because the source is at a greater depth below the surface or because the sensor is at a greater height above the surface, as in the case of airborne measurements. When the source varies in shape, this variation also affects the anomaly, with increasing source width producing longer wavelength responses (Fig. 2.4d).

Figures 2.4 and 2.49a (see Section 2.11.4) illustrate some important characteristics of many kinds of geophysical responses. In general, the deeper the anomalous body and/or the smaller its property contrast, the larger its size must be in order for it to be detectable against the inevitable background noise (see Section 2.4). Also, anomalies with the same amplitude and wavelength can be caused by various combinations of source depth, geometry and contrast with the host rocks. Without additional information about these variables, the actual nature of the source of the anomaly is indeterminable. This problem of *ambiguity* is discussed further in Section 2.11.4.

In summary, whether or not an anomalous physical property distribution produces a recognisable geophysical response depends on its size and the magnitude of the contrast between it and the surrounding rocks. In addition, the physical property contrast of a geological feature can change markedly as the properties of the surrounding rocks and/or those of the target feature change, both laterally and with depth. This can significantly change the nature of the geophysical response; it can form multiple geophysical targets related to different parts of the same geological feature and even change the type of geophysical measurements needed to detect it.

2.4 Signal and noise

A measurement of any kind, but especially one made in a setting as complex and unpredictable as the natural environment, will be contaminated with unwanted information. This unwanted information is known as *noise* and is a source of error in a measurement, whilst the information being sought in the measurement is known as *signal.* The relative amounts of signal and noise in a measurement are quantified by the signal-to-noise ratio (SNR). Ideally, one hopes that the amplitude of the signal, the signal level, is as high as possible and that the amplitude of the noise, the noise level, is as low as possible in order to obtain an accurate measurement of the parameter of interest. As a general rule, if SNR is less than one it will be very difficult to extract useful information from the measurement, although data processing techniques are available to improve the situation (see Section 2.7.4).

Suppression of noise is of utmost importance and must be considered at every stage of the geophysical programme, from data acquisition through to presentation of the data for interpretation. Active geophysical methods usually allow the SNR to be improved by changing the nature of the output from the transmitter, e.g. increasing its amplitude or changing its frequency. This advantage is lost with passive methods, where the geophysicist has no control over the natural 'transmitter'.

The signal depends solely on the objective of the survey, and geological responses not associated with the objective of the survey constitute noise. Of course, any response of non-geological origin will always be considered noise. As data are revisited the information required from them may vary, in which case so too do the representations of signal and noise in the data. A useful definition of signal is then 'what is of interest at the time', whilst noise would then be 'everything else'; just as the saying goes, 'One man's trash is another man's treasure', so also one geoscientist's signal is another geoscientist's noise.

Two basic types of noise affect geophysical measurements. Firstly, there are effects originating from the local environment, i.e. *environmental noise*. Secondly, there is *methodological noise*, which includes unwanted consequences of the geophysical survey itself and of the processing of the geophysical data. A feature in the data that is caused by noise is referred to as an *artefact*. It goes without saying that identification and ignoring of artefacts is critical if the data are to be correctly interpreted.

2.4.1 Environmental noise

The main types of environmental noise affecting the different types of geophysical survey are summarised in Table 2.1. Environmental noise can be categorised by its origin; as either geological or non-geological. Geological environmental noise is produced by the geological environment, including topography. Non-geological environmental noise includes sources in the atmosphere and outer space, plus cultural responses associated with human activities.

Wind is a common source of noise from the atmosphere. It causes objects attached to the ground to move, e.g. trees and buildings, which produces noise in seismic, electromagnetic and gravity surveys. The movement of wires linking sensors to recording equipment may also create noise because of voltages induced by their movement through the Earth's magnetic field (see Section 5.2.2.2). Wind turbulence also causes variations in the position and orientation of geophysical sensors during airborne surveys which affect the measurements.

As well as creating noise, natural phenomena may reduce the amplitude of the signal: for example, radioactivity emitted from soil is attenuated when the soil is saturated by rainfall. The variability and unpredictability of natural phenomena cause noise levels to vary during the course of a geophysical survey.

Cultural noise includes the effects of metal fences, railways, pipelines, powerlines, buildings and other infrastructure (see Section 2.9.1). In addition, cultural features may radiate energy that causes interference, such as electromagnetic transmissions (radio broadcasts etc.), radioactive fallout and the sound of machinery such as motor traffic. Mine sites are particularly noisy environments, and noise levels may be so high as to preclude geophysical surveying altogether.

The two most troublesome forms of geological environmental noise are those associated with the shallow subsurface and with topography; the latter are known as *topographic* or *terrain effects*. In both cases it is possible, in principle, to calculate their effects on the data and correct for them. To do so requires very detailed information about the terrain and/or physical properties of the subsurface, which is often lacking. This is an example of the *geophysical paradox* (see Section 1.3). To fully understand the geophysical signal, and the noise, requires information about the subsurface. However, it was to acquire such information that the geophysical survey was undertaken.

2.4.1.1 Topography-related effects

Some examples of topography-related noise are shown schematically in Fig. 2.5a. In rugged terrains, topography creates noise by causing variations in the distance between geophysical transmitters and/or sensors and features in the subsurface. This changes the amplitude and wavelength of the responses (see Section 2.3). These effects can sometimes be accounted for by modifying the measurements, during data reduction. The accuracy with which this can be

Table 2.1 Common sources of environmental noise and the forms in which they manifest themselves for the various geophysical methods. Specific details are included in the relevant chapters on each geophysical method.

Source of noise	Gravity	Magnetics	Radiometrics	Electrical and electromagnetics	Seismic
Regolith	Changes in thickness and internal variations in density causing spurious anomalies	Oxidation of magnetic mineral species Formation of maghaemite causing spurious anomalies	Concealment of bedrock responses Mobilisation of radioactive materials causing responses that are not indicative of bedrock	High conductivity leading to poor signal penetration and electromagnetic coupling with measurement array Internal changes in conductivity (groundwater, clays) causing spurious anomalies Superparamagnetic behaviour (maghaemite)	Changes in thickness and internal changes in velocity affecting responses (statics) Reduction in the energy transmitted from source
Glacial sediments	Changes in thickness and internal variations in density causing spurious anomalies	Magnetic detritus causing spurious anomalies	Concealment of bedrock responses Mobilisation of radioactive materials causing responses that are not indicative of bedrock	Internal changes in conductivity causing spurious anomalies	Changes in thickness and internal changes in velocity affecting responses from below (statics)
Permafrost and snow cover	Changes in ice content causing spurious anomalies		Concealment of bedrock responses	Internal changes in conductivity causing spurious anomalies	Changes in thickness and internal changes in velocity affecting responses from below (statics)
Hydrological	Formation of low density dissolution features in carbonate rocks causing spurious anomalies		Dissolution/precipitation (i.e. transportation) of soluble radioactive elements causing responses that are not indicative of the presence of K, U and Th	Changes in groundwater salinity causing changes in conductivity Movement of electrolytes creating spurious anomalies	Formation of dissolution features in carbonate rocks that scatter seismic waves
Atmospheric phenomena	Turbulence during airborne and ground surveys	Turbulence during airborne surveys	Turbulence during airborne surveys Uneven distribution of radioactive aerosols Movement of gaseous radioactive species Moisture suppressing responses from the ground	Turbulence during airborne and some ground surveys Noise spikes due to lightning strikes (sferics)	Wind (water waves, movement of tree roots etc.) and rain-related noise
Extraterrestrial phenomena	Temporal changes in gravity due to the positions of the Sun and Moon	Magnetic fields associated with processes occurring in the ionosphere	Radioactivity of cosmic origin		

Table 2.1 (*cont.*)

Source of noise	Gravity	Magnetics	Radiometrics	Electrical and electromagnetics	Seismic
Topography	Terrain-related responses Variable relative positions of sensors and anomalous bodies	Terrain-related responses Variable relative positions of sensors and anomalous bodies	Terrain-related responses Variable relative positions of sensors and anomalous bodies	Terrain-related responses Variable relative positions of transmitters, sensors and anomalous bodies	Variable relative positions of sensors and anomalous bodies
Man-made	Responses due to large buildings and excavations, such as open pits	Responses from ferrous objects, roads, pipelines etc. constructed of magnetic materials and electrical powerlines	Responses from materials created by radioactive fallout from nuclear explosions and reactors	Responses from metallic objects, electrical powerlines and electromagnetic transmissions	Sound of motor vehicles and heavy machinery

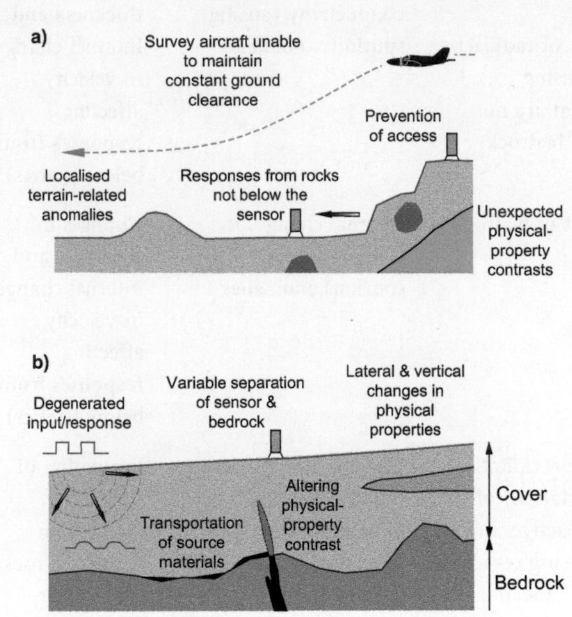

Figure 2.5 Environmental noise associated with (a) topography and (b) the near-surface environment. Cover could be unconsolidated sediments, regolith, glacial till or ice.

achieved depends on a number of factors, notably how well the local geology (and associated physical property variations) and topography are known.

Historically, airborne surveys were usually conducted at a constant barometric altitude above the undulating terrain. Modern airborne surveys are *draped* over the terrain as the aircraft attempts to maintain constant height above the undulating ground surface, i.e. they are flown at constant terrain clearance. Helicopters are better adapted to this type of surveying in very steep terrain than the less-manoeuvrable fixed-wing aircraft. Whether drape is maintained or not depends upon the width (wavelength) and slope of the topographic variation, the speed of the aircraft and its climb capability. Small topographic features tend to be ignored (Figs. 2.6 and 2.7a) producing spurious noise in the measured geophysical response. Another problem for airborne surveys is that an aircraft can descend faster than it can climb. When the terrain begins to rise rapidly in front of the aircraft, the survey pilot must anticipate the need to gain height so as to pass safely over the hill, and so begins ascending before the feature is reached. The result is an increase in terrain clearance adjacent to the hill. The easier descent on the other side of the hill often means it is possible to re-establish the specified terrain clearance relatively quickly. The same occurs when surveying across a valley. The result is illustrated by the actual flight paths from an aeromagnetic survey across part of the Hamersley iron-ore province in northern Western Australia (Fig. 2.6). The terrain clearance not only varies across the hill, but also depends on the direction the aircraft is travelling with respect to the topography. For surveys where adjacent traverses are flown in opposite directions (see Section 2.6.3.3), the terrain clearance is fairly consistent for alternate traverses and different for adjacent traverses resulting in *corrugations*, i.e. traverse-parallel artefacts caused by readings being anomalously high and then low on alternating traverses.

When measurements are made on the ground the geology may be adjacent to, or even above, the sensor (Fig. 2.7b). This is also a common occurrence for

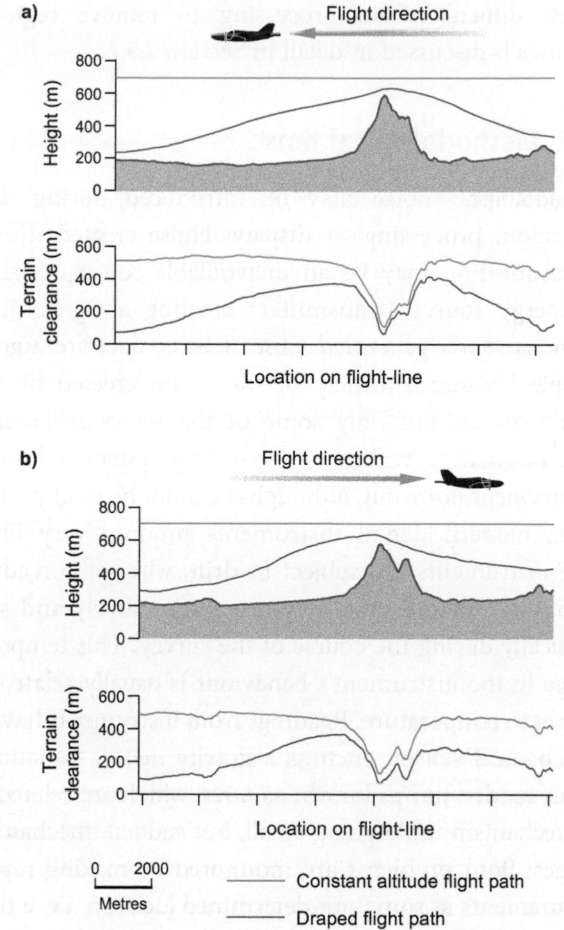

Figure 2.6 Survey height and associated variations in terrain clearance for constant barometric height and draped surveys flown in opposite directions across a ridge. The draped paths are actual flight paths from an aeromagnetic survey in Western Australia. Based on diagrams in Flis and Cowan (2000).

low-level helicopter-borne surveys conducted in rugged terrains. Both situations can cause anomalous responses which might be misinterpreted (Mudge, 1998). Making a measurement in a gully or adjacent to a cliff means more of the 'geology' is closer to the sensor and can create anomalously high readings, with the opposite occurring when on the top of ridges etc. Note that this is exactly the opposite of the effects on imperfectly draped airborne surveys in Fig. 2.7a.

Another problem caused by topography is the distortion of the geophysical response. This is particularly a problem for electrical and electromagnetic surveys where the near-surface flow of electrical current is strongly influenced by the shape of the (conductive) terrain (see Section 5.6.7.3). Also, surveys conducted on sloping terrain need to account for the terrain slope in the analysis of the data, because the

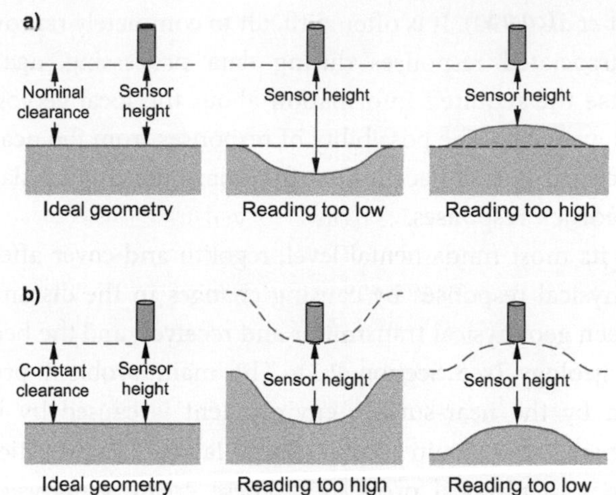

Figure 2.7 Influence of topography on geophysical measurements. (a) Effects when sensor height varies, as with *loose-draped* airborne surveys. (b) Effects when the sensor is maintained at constant terrain clearance, as in ground and *close-draped* airborne surveys in rugged terrain. Note how the effects of the topography on the reading are opposite in (a) and (b).

measured geophysical response is a distortion of that for a horizontal survey surface.

Other features of the terrain producing similar effects to topography include open-pit quarries and mines, tall obstructions such as buildings and other infrastructure, and trees and thick vegetation. The availability of high-resolution digital terrain information from airborne and satellite sensors, and stereo photography, is an important development in terms of compensating for terrain effects. The height of the terrain above sea level can be obtained from airborne geophysical surveys by combining the aircraft height above the terrain, measured with a radio altimeter, with the GPS-derived height. A digital elevation model (DEM), also known as a digital terrain model (DTM), for the survey area can be created in this way. A useful source of radar-derived terrain information is the Shuttle Radar Topography Mission (SRTM) dataset (Cowan and Cooper, 2005). As a general rule, it is good practice to have a terrain model available when interpreting geophysical data.

2.4.1.2 Near-surface and deep-seated responses

A significant source of geological noise is the near-surface environment (Fig. 2.5b). Regolith can present major problems for geophysical surveying (Doyle and Lindeman, 1985), as can cover such as sand dunes, glacial deposits and snow (Smee and Sinha, 1979), and permafrost

(Scott *et al.*, 1990). It is often difficult to completely remove the associated responses during data processing, again because the required information about the local geology is not available. The possibility of responses from the near-surface needs to be accounted for when interpreting data for bedrock responses.

At its most fundamental level, regolith and cover affect geophysical responses by causing changes in the distance between geophysical transmitter and receiver, and the bedrock geology (see Section 2.3). The main problem produced by the near-surface environment is caused by its tendency to contain complicated lateral and vertical changes in physical properties. These create geophysical responses, often of high amplitude, that interfere with those originating from the bedrock. Moreover, they distort the responses from bedrock sources measured at the surface and, for active survey methods, the energy from the transmitter before it penetrates the bedrock. Electrical and electromagnetic methods are particularly susceptible to the effects of the near-surface environment.

The degree and depth of weathering can be an important variable even when there is not a thick regolith present. Some physical properties are altered by the weathering process. Oxidation of magnetic iron-rich minerals can destroy their magnetism but, on the other hand, physical property contrasts can be enhanced by weathering. For example, weathering of the shallower regions of kimberlites can preferentially enhance their electrical conductivity. Mobilisation of materials in the near-surface is another problem that can sometimes result in geophysical responses that do not coincide with the source of the materials. Radioactive elements are especially prone to mobilisation. The depth of erosion can also be an important influence on a target's geophysical response. For example, kimberlites exhibit vertical changes in physical properties associated with crater facies, blue and yellow ground etc. (Macnae, 1995) and will appear different depending on the depth of erosion. Also, porphyry style mineralisation and epithermal deposits have zoned alteration haloes and appear different according to which zones coincide with the ground surface (Irvine and Smith, 1990).

At the other end of a spectrum are responses originating from depths beyond those of immediate interest, or from regional-scale geological features. These forms of geological noise are known as the *regional* response. It is characterised by smooth, long-wavelength variations which often make accurate definition of the shorter-wavelength variations of interest difficult. Data processing to remove regional responses is discussed in detail in Section 2.9.2.

2.4.2 Methodological noise

Methodological noise may be introduced during data acquisition, processing or display. Noise created during data acquisition may be an unavoidable consequence of the energy source (transmitter) creating a signal; it is known as *source-generated noise.* Seismic data are a good example because a variety of waves are created by the seismic source, but only some of the waves are useful. Noise originating within the recording system is known as *instrument noise* but, although it cannot be totally eliminated, modern digital instruments produce very little. Some instruments are subject to drift, where the reading associated with the same response varies slowly and systematically during the course of the survey. This temporal change in the instrument's behaviour is usually related to changes in temperature. Readings from instruments having a mechanical sensor, such as a gravity meter, sometimes exhibit sudden jumps known as *tares*, which are related to the mechanism undergoing small, but sudden, mechanical changes. Both problems are monitored by making repeat measurements at some pre-determined location, i.e. a base station, during the course of the survey. Noise is also caused by the inherent nature of the parameter being measured, as in the measurement of radioactivity in the radiometric method. Here the randomness in time of the radioactive decay process produces statistical noise which can only be reduced by observing and averaging more radioactive 'events' over a longer measurement period (see Section 4.3.1).

Methodological noise can also be caused by errors in orientation and position of the sensor. Accurate orientation of the sensor is fundamental to making many types of geophysical measurements. For example, gravity measurements require the sensor to be accurately aligned in the vertical, and, for all methods, directionally dependent gradient measurements (gradiometry; see Section 2.2.3) require the gradiometer to be orientated consistently in the same direction for the entire survey. Positioning requirements depend on the type of geophysical survey. For example, gravity measurements require the sensor elevation to be measured to within a few centimetres, and to less than 1 cm for high-precision work; whereas measurements to within about a metre are acceptable for magnetic and radiometric measurements. Processing of survey data usually assumes that the location of each reading is adequate for the survey objectives. If this is not the case the results will be erroneous, or at best lack resolution. With the advent of satellite-based positioning, errors in location have been greatly reduced, although elevation is often of more importance than lateral position. Obtaining sufficiently accurate heights from satellite-based data requires specialised equipment and processing of the location data. Another potential source of positioning-related noise is the incorrect and/or inconsistent use of map projections and height datum. It goes without saying that any treatment of data that combines information defined in different positional reference frames will result in errors unless the various data are transformed to a single reference frame. The problem is particularly common when satellite-based positioning information, which is based on a global coordinate system, is combined with data positioned on a ground-based local coordinate system.

Most geophysical surveys also record the time of the measurements to allow temporal corrections to be applied to some types of data, and to enable the integration of different datasets. Minimising errors in time is, therefore, crucial. Satellite navigation systems provide a convenient and highly accurate time signal for this purpose.

Data processing and display techniques may inadvertently create noise. A common example is short-wavelength variations, or ripples, appearing in the processed data through an effect known as *ringing* (see online Appendix 2), which is caused by inappropriate design of the processing algorithm. Sometimes geophysical surveys can be designed to measure responses that predominantly originate from a particular depth. By varying the depth of penetration and the location of the surface measurements a 'cross-section' of data can be constructed. However, the measured responses are not entirely due to features at a particular depth, and determining the depth that has the predominant control on the response is fraught with difficulty. This is a further source of error as some data processing techniques require the depth of particular features to be known in order to apply them. Furthermore, the shape of the measured response profile/cross-section usually does not mimic the true shape of features in the subsurface. In all cases, the result is a potentially misleading display.

2.5 Survey objectives

A geophysical survey should be undertaken with a clear objective in mind since this critically affects both the geophysical method that should be used and also the type of survey, for example airborne or downhole. The two most common objectives of mineral geophysical surveys are to map the local geology and/or to measure responses originating from the mineralised environment. Mapping surveys provide essential geological context, and in areas of poor outcrop the data may comprise the only useful form of 'geological' map available. Surveys designed to target the mineralised environment may be intended to detect or to define the geological/geophysical features of potential significance. Detection simply involves ascertaining whether 'something' is there or not, and obviously the 'something' has to produce a detectable geophysical response so that its presence can be detected. Surveys designed to characterise the source of a response are required when information about the nature of the source is required, usually to design a drilling programme to sample it.

Putting aside the inevitable influence of costs and budgets (see Section 1.2.3), the decision over whether to use geophysics in an exploration or mining programme, and if so which method to use, depends on the geophysical detectability of the features of interest. If significant physical property contrasts do not occur in the survey area then the chances of a successful outcome are much reduced. Clearly, some understanding of the likely physical properties of the geological environment being surveyed is paramount. This might be based on petrophysical data from the survey area; more probably, especially in the early stages of exploration, it will be based on data from other areas or from published data compilations. Successful outcomes are more likely when there is a good understanding of petrophysics, one reason that we emphasise this subject throughout this text.

2.5.1 Geological mapping

Mapping the local geology seeks to identify geological settings conducive to the formation of orebodies. In poorly explored areas the primary intention may be simply to create a geological 'base map', which will form a basis for assessing the area's prospectivity. In better known areas particular deposit types may be sought and the features of primary interest will depend on the exploration model being applied. A common example is seeking to map major faults which may have acted as conduits for mineralising fluids. Alternatively, if the mineralisation being sought is strata-bound then mapping the prospective lithotypes or lithological contacts may be the aim of the survey.

We reiterate the important point made in Section 1.1 that the contrasts in physical properties that control geophysical responses do not necessarily have a one-to-one correlation with contrasts in lithotype. This is because it is the rock-forming minerals that are the basis for assigning lithological names, and these in turn are controlled by rock chemistry. In contrast, the physical properties of relevance to geophysics are not entirely, or sometimes even slightly, controlled by the rock-forming minerals. For this reason, a 'geological' map created using geophysical measurements should be referred to as a *pseudo-geological* map. Compilation of the pseudo-geological map involves identifying the near-surface and deeper responses and classifying the remaining responses according to their possible sources (see Section 2.11). The integrated analysis of multiple data types, i.e. magnetics, radiometrics, conductivity etc., helps to produce a more reliable and accurate model of the subsurface geology.

Surveys designed for geological mapping should provide a uniform coverage of geophysical data across the area of interest. With large areas to cover this will probably require an airborne survey and survey specifications typical of reconnaissance objectives (see Section 2.6.3). As exploration focuses on areas considered to be most prospective, more detailed surveys may be undertaken and ground surveys may be used. The most common types of geophysical surveys for geological mapping are airborne magnetic and radiometric surveys. Airborne gravity surveys are becoming more common as surveying technology improves. These geophysical methods produce responses that distinguish a wide range of lithotypes and are favoured in most geological environments. The more complex, difficult to interpret and expensive electrical and electromagnetic methods tend to be used less, although airborne electromagnetics is increasingly being used in a mapping role.

2.5.2 Anomaly detection

The exploration strategy can involve surveys intended to detect localised responses distinctly different from their surroundings, i.e. anomalies. This approach is sometimes referred to as 'searching for bumps', i.e. looking for localised anomalously 'low' or anomalously 'high' values in various presentations of the survey data. This is a simple and effective form of targeting and is a valid approach when exploration targets give rise to distinct anomalies that are easily distinguishable from the responses of unwanted features, such as the surrounding rock formations and mineralisation of no economic significance. The strategy can be applied at regional scale to select areas for detailed work, and at prospect scale to detect a target anomaly. It is also applicable when a deposit is being mined, with adjacent orebodies being the target. Also, surveys designed to detect faults or dykes ahead of coal mining are basically aimed at identifying anomalous parts of the geological environment.

Anomalous responses associated with the mineralised environment may be caused by the mineralisation itself, although not necessarily the actual ore minerals. Deposits comprising massive or disseminated metal sulphides and oxides are commonly targeted in this manner. Also targeted are alteration zones caused by mineralising fluids, which have the advantage of usually being much larger in area than the target mineralisation, so the geophysical response from the alteration covers a correspondingly larger area, which may help to facilitate its detection. Porphyry style copper deposits are an example of deposits with extensive, geophysically distinctive, alteration haloes. Another form of anomaly targeting seeks to locate specific lithotypes in which mineralisation occurs, e.g. potentially diamondiferous kimberlitic and lamproitic intrusions, or ultramafic intrusions that might contain platinum group elements, or palaeochannels hosting placer deposits.

A survey designed to detect the responses from the mineralised environment requires a survey strategy based on the probability of making a measurement in the right place, i.e. within the bounds of its geophysical response (see Section 2.6.4). This ensures that the anomalous responses are both recorded and recognised as significant (see Section 2.5.3). Needless to say, some knowledge of the physical properties of the targets is required to ensure that responses are anticipated in the chosen form of geophysical data. The depth and volume of the source, and the magnitude of the physical property contrast with its host, are also important since the amplitude of the anomaly depends on this (see Section 2.3). In this context an appreciation of noise levels is required, especially geological noise. For example it may be comparatively easy to identify the magnetic response of a kimberlite intruding a weakly magnetised sedimentary sequence, but the response may be unrecognisable in, say, a terrain comprising a variably magnetised succession of basalts.

2.5.3 Anomaly definition

Surveys designed to improve the definition of an anomalous response are aimed at obtaining more information about the source of the anomaly, and are often used for designing drilling programmes. The surveys are conducted at prospect scale and during exploration in the mine environment. It may be a detailed ground survey to follow up an anomaly detected by an airborne survey or possibly a wider-ranging lower resolution ground survey. Information about the source such as its extent, shape, dip and depth can be obtained, usually by modelling the anomaly (see Section 2.11). Accurately characterising a response requires careful consideration of the survey configuration and the distribution of the measurements within the area of interest (see Section 2.6).

2.6 Data acquisition

We describe the acquisition of geophysical survey data in terms of survey design based on the fundamental concepts of data sampling and feature detection. Throughout this section we encourage the use of computer modelling as an aid to survey design.

Modern geophysical survey equipment stores the digital measurements for later download. Airborne survey systems, and some ground systems, acquire high-precision positional data directly from satellite-based positioning systems, simultaneously with the geophysical measurements. Survey systems often measure 'secondary' data acquired for the purpose of post-survey compensation of survey-induced errors, removal of external sources of noise appearing in the 'primary' data, and use in enhancing the data (see Section 2.7). Examples include: time of each measurement; the orientations of sensors and transmitters; relative positions between sensors on the survey platform; sensor height; air temperature and pressure; and, for moving platforms, velocity.

2.6.1 Sampling and aliasing

As described in Section 2.2, a geophysical measurement may consist of one reading or a series of readings made at the same location. Measurements made over a period of time or range of frequencies form time and spectral series, respectively, for that location. Whether a single reading or a series of readings is made at each location, a survey will consist of measurements at a number of stations to form a spatial series. These series constitute a set of samples of what is a continuous variation in the parameter being measured. In order for the samples to accurately represent the true variation, these series must be appropriately sampled, i.e. readings must be taken at an appropriate spacing or interval.

The rate at which the sampling occurs is known as the *sampling frequency*, which for time series is measured in units of 1/time (frequency, in units of hertz (Hz)); for frequency series, 1/frequency (period, in units of seconds (s)); and for spatial series, 1/distance (units of reciprocal metres (m^{-1})). The time, frequency or distance between samples is the *sampling interval*. A little confusingly, spatial sampling for a moving sensor is usually defined in terms of temporal sampling. For example, detectors used in aeromagnetic surveys measure (or sample) the Earth's magnetic field typically every 0.1 s, i.e. 10 times per second, referred to as 10 Hz sampling. A fixed-wing aircraft acquiring magnetic data flies at about 70 m/s, so at '10 Hz sampling' the spatial sampling interval is 7 m.

If the measurements are not spaced closely enough (in time, frequency or distance) to properly sample the parameter being measured, a phenomenon known as *aliasing* occurs. It is respectively called *temporal, spectral* or *spatial aliasing*. Consider the situation where a time-varying signal being measured varies sinusoidally. The effect of sampling at different sampling frequencies to produce a time series is demonstrated in Fig. 2.8a. The true variations in the signal are properly represented only when the sampling frequency is high enough to represent those variations; if not, the signal is under-sampled or aliased. When under-sampling occurs, the frequency of the sine wave is incorrectly represented by being transformed to spurious longer wavelength (lower frequency) variations. Clearly then, the very act of sampling a signal can produce artefacts in the sampled data series that are indistinguishable from legitimate responses. Strategies can be adopted to combat the problem of aliasing, but if data are aliased it is impossible to reconstruct the original waveform from them.

2.6.1.1 Sampling interval

It is shown in online Appendix 2 that a complex waveform can be represented as a series of superimposed sine waves, each with a different frequency. To avoid aliasing a waveform containing a range of frequencies, it is necessary to sample the waveform at a sampling frequency greater than twice the highest frequency component of the waveform. This is the *Nyquist criterion* of sampling (Fig. 2.8b).

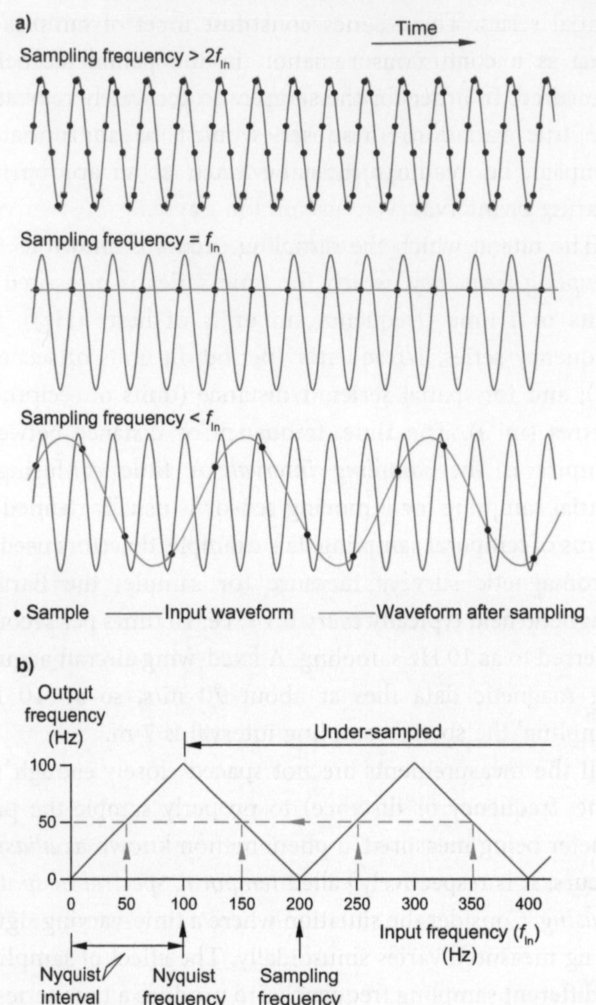

Figure 2.8 Aliasing of a periodic signal of frequency f_{In}. (a) Signal and various sampling frequencies; see text for explanation. (b) The relationship between the frequency of the input signal and the frequency of the sampled output signal for a sampling frequency of, say, 200 Hz. Note how under-sampling causes the output to 'fold-back' into the Nyquist interval (0–100 Hz), e.g. an input signal of 250 Hz produces a sampled output signal of 50 Hz. Redrawn, with permission, from Galybin *et al.* (2007).

In spatial terms, this means that the interval between measurements must be less than half the wavelength of the shortest wavelength (highest frequency) component. Conversely, the maximum component frequency of the signal that can be accurately defined, known as the *Nyquist frequency* (f_N), is equal to half the sampling frequency (f_s). The interval between zero frequency and the Nyquist frequency is known as the *Nyquist interval*. Frequency variations in the input waveform occurring in the Nyquist interval are properly represented in the sampled data series, but frequencies higher than the

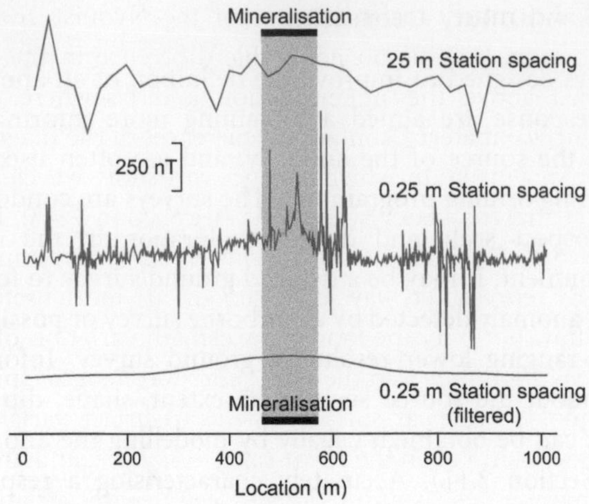

Figure 2.9 Aliasing in total magnetic intensity data across the Elura Zn–Pb–Ag massive sulphide deposit. See text for details. Redrawn from Smith and Pridmore (1989), with permission of J. M. Stanley, formerly Director, Geophysical Research Institute, University of New England, Australia.

Nyquist frequency are under-sampled and converted (aliased) to spurious lower-frequency signals, i.e. they are 'folded back' into the Nyquist interval. The aliased responses mix with the responses of interest and the two are indistinguishable. The aliased responses are artefacts, being purely a product of the interaction between the sampling scheme and the waveform being sampled.

The sampling interval required to avoid aliasing can be established with computer modelling (see Section 2.11), reconnaissance surveys or field tests conducted prior to the actual survey. In practice, economic and logistical considerations mean that aliased data are often, and unavoidably, acquired. This is not necessarily a problem for qualitative interpretations such as geological mapping or target detection. Regions with similar characteristics will usually retain apparently similar appearance even if aliasing has occurred, but different geology will give rise to different geophysical responses. Extreme caution is required when the data are to be quantitatively analysed, i.e. modelled (see Section 2.11.3), because working with an aliased dataset will result in an erroneous interpretation.

2.6.1.2 Example of aliasing in geophysical data

Figure 2.9 shows an example of spatial aliasing in magnetic data collected along a traverse across the Elura Zn–Pb–Ag volcanogenic (pyrrhotite-rich) massive sulphide deposit located in New South Wales, Australia. Data collected at a station spacing of 25 m (minimum properly represented wavelength 50 m) show variations with wavelengths of

50 m and more, a consequence of the Nyquist criteria. Importantly, there is no noticeable difference in the data acquired above the mineralisation and elsewhere. The near-surface material contains occurrences of the magnetic mineral maghaemite, which produce very short-wavelength (high spatial frequency) variations (see Section 3.9.6). Data acquired at a smaller station spacing of 0.25 m (minimum properly represented wavelength 0.5 m) are more useful as they show not only the longer wavelength variation of the mineralisation, but also the very short-wavelength 'spikey' variations of the near-surface. The near-surface variations are under-sampled in the 25-m-sampled dataset and, therefore, have created spurious longer wavelength variations; the true signature of the near-surface response has not been resolved by the survey. The short-wavelength near-surface response is properly defined in the 0.25-m-dataset and, therefore, can be accurately removed using data processing techniques (see Section 2.7.4), to reveal the longer wavelength response of the mineralisation (Fig. 2.9). This example illustrates the need to consider the characteristics of both the signal and the noise when setting the data sampling interval, and the implications that the sampling interval has for the ability to separate the various responses using data processing techniques.

2.6.2 System footprint

The measurement footprint of a geophysical survey system is the volume of the subsurface contributing to an individual measurement. Consistent with the source-to-detector proximity effect on responses (see Section 2.3), the materials exerting the most influence on the measurement are those closest to the sensor, i.e. the point on the ground surface immediately below the sensor, with influence progressively decreasing away from this point. The footprint is often arbitrarily taken as the volume of materials that contribute 90% of the measured response. For measurements made with a stationary sensor in a geologically homogeneous area, the surface projection of the footprint is circular. When measurements are made from a moving platform (an aircraft, ground vehicle or a boat), the system will travel some distance during the time taken to make a measurement, creating a measurement footprint that is elongated in the survey direction. How far the footprint extends into the subsurface varies for different types of geophysical measurement.

The system footprint is a fundamental consideration in survey design, and it increases with survey height. There is minimal benefit (and increased cost for a given survey area) in making measurements closer together than the size of the footprint, because then a large portion of the same geology, in the footprint, would also contribute to the neighbouring measurement. More closely spaced measurements are, however, useful for applying post-survey signal-enhancement processing (see Section 2.7.4).

2.6.3 Survey design

An important consideration in designing a geophysical survey is the optimum number and distribution of the measurements to be made. Too many measurements are a waste of time and money, too few and the survey's objective may not be achieved. In deciding the spacing between measurements it is important to consider the wavelength of the expected responses and the footprint of the measurement, i.e. the size of the area influencing the measurement. The distance between individual measurements should be sufficiently close to satisfy the requirements of the sampling theorem (see Section 2.6.1). It is important that the survey extends across a large enough area to define the longest wavelength response of interest (see online Appendix 2), and that the survey exceeds the limits of the area of interest in order to determine the regional response (see Section 2.9.2), so as to facilitate its removal from the survey data. Survey design should also account for the need to minimise all sources of noise (see Section 2.4). An example of a data acquisition variable directly related to improving SNR is the number of repeat measurements (if any) to make at each location. These allow suppression of random noise (see Section 2.7.4.1), but also increase the time (and cost) to acquire the data. Ultimately, the survey budget will dictate how many measurements may be taken and their accuracy, and geological and logistical factors will also constrain their distribution.

Note that it is common for several parameters to be measured in a single geophysical survey, e.g. airborne magnetics and radiometrics. Characteristics of all the parameters being measured need to be considered when designing the survey, but usually the parameter of principal interest will control the setting of survey parameters.

2.6.3.1 Modelling as an aid to survey design

Whether the aim of a survey is mapping, detection or characterisation, the geological characteristics of the target and its geophysical response, with respect to the response of the surrounding geology, need to be considered during survey design. Geophysical modelling, whereby the geophysical response of a numerical model of the subsurface is computed (see Section 2.11), provides information about the nature of the geophysical response of the expected subsurface geology useful for survey design.

For a specific subsurface feature, including topography, the survey system and configuration most capable of detecting the feature and yielding the required information about it can be determined from modelling. This is particularly important for the electrical and electromagnetic methods where a great diversity of survey systems is available and a great diversity of survey configurations is possible. For a particular survey system and configuration, a suitable sampling interval can be determined; the effects of changes in target parameters on the response can be assessed and those having the greatest or the least influence identified. The amplitude of the response can be used to estimate the acceptable level of noise.

Modelling is an important aid in the design of cost-effective and efficient surveys, and its role should not be underestimated. However, it is only as good as the assumptions that are necessarily made about the geological environment to be surveyed, which in the early stages of exploration may be highly speculative.

2.6.3.2 Survey height

The height of the survey system above the ground is an important parameter in all types of airborne geophysical surveys. Survey aircraft are equipped with radio altimeters to record the aircraft's ground clearance. This is the *flight height* or *survey height* above the ground, which usually varies according to the topography, so it is specified as the mean terrain clearance (MTC) of the survey. Lower survey height reduces the system footprint (see Section 2.6.2), improves spatial resolution and increases the strength of the signal (see Section 2.3). Operational constraints, however, dictate the minimum MTC; and lower survey height is not always an advantage, as close proximity to near-surface sources increases their short-wavelength responses, which may be a source of noise (see Section 2.4.1). A small station interval is required to sample these properly (see Section 2.6.1). Even if this is achieved it can be difficult to correlate the responses between the survey lines, making gridding difficult (see Section 2.7.2) and possibly masking responses from the underlying geology.

2.6.3.3 Survey configurations

Data acquired along a single survey traverse or in a drillhole form a one-dimensional (1D) (spatial) data series, because variations in the measured parameter are depicted in only one direction. A time series is also a 1D data series. A group of 1D spatial series or a random distribution of measurements across a surface can be combined to form a two-dimensional (2D) data series. In this case, the data are presented as a map in terms of two spatial coordinates (easting and northing, or latitude and longitude). Most above-ground surveys are 2D and provide a map showing the spatial variation of the parameter measured.

Some common 2D geophysical survey configurations are shown in Fig. 2.10. The measurement locations are

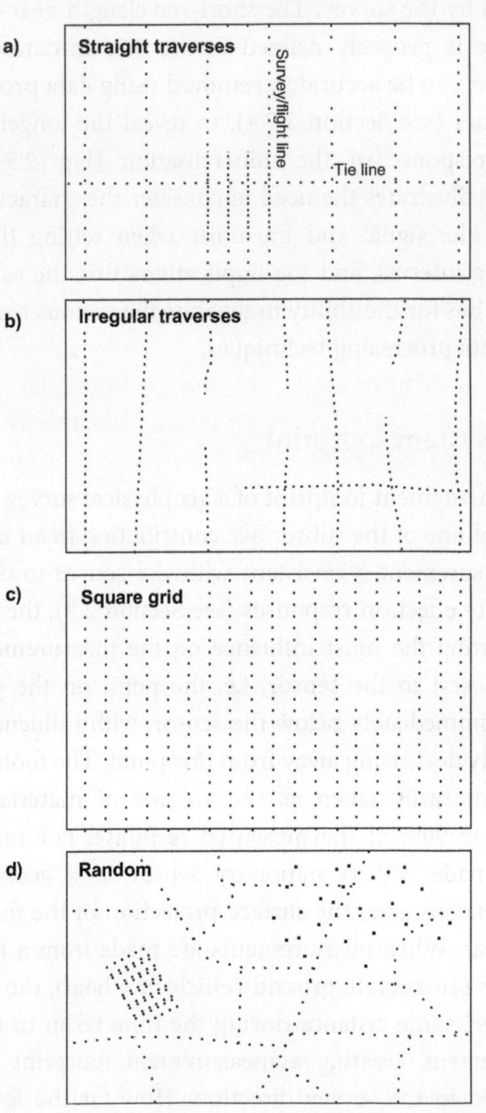

Figure 2.10 Some common configurations for geophysical surveys. The dots represent data points. (a), (b), Various traverse or line configurations; (c) grid network; and (d) configuration typical of ground-based regional surveys.

usually referred to as *stations*, or simply *data points*, and the distance between them is the *station spacing, station interval* or *data interval*. For airborne surveys and downhole logs, the data are acquired using a moving sensor. Measurements are made at a constant time interval so the data interval is equal to the time interval multiplied by the speed of the aircraft/downhole-probe.

When the data are acquired along parallel survey lines or traverses, as is most commonly the case, the distance between the lines, the *line spacing* or *line interval*, has a major influence on the resolution of the resultant map of the measured response. Line spacing is normally much larger than the station spacing, but if they are equal then a regular grid network of measurements is obtained; compare Figs. 2.10a and c. Survey lines are orientated so as to be perpendicular to strike. Note that increasing the survey height increases the system footprint (see Section 2.6.2) and also reduces the short-wavelength component of the measured response. In principle, this means that station/measurement spacing and line spacing can be increased.

In airborne surveying the *survey lines* or *flight lines* are usually flown in alternate directions or headings (Fig. 2.11a). Where the topography is severe, they are flown as groups of lines with the same heading, referred to as *racetrack flying* (Fig. 2.11b). Racetrack flying tends to reduce the problems associated with differences in ground clearance (see Section 2.4.1), so instead of artefacts occurring between each survey line they mainly occur between the line-groups flown in opposite directions. Data are also acquired along a series of *tie lines* oriented perpendicular to the overall survey line orientation, and at the same survey height. The line intersections represent repeat measurements that are theoretically made at the same point in space, and are used to monitor noise and correct errors in the survey data. For the same reasons, ground surveys may include tie lines and/or repeat readings at selected stations.

Ideally, survey lines should be straight and parallel, with line spacing and station spacing kept constant for the entire survey (Figs. 2.10a and 2.12), but in practice any number of factors may prevent this (Fig. 2.10b). There may be gaps in the data caused by, for example, equipment problems, bodies of water, open pits, buildings, severe terrain, areas of denied access or areas where there are high noise levels. Logistical constraints on station distribution are more severe for ground surveys than in airborne operations. There are few access limitations for fixed-wing aircraft and helicopters, unless flying so low that the

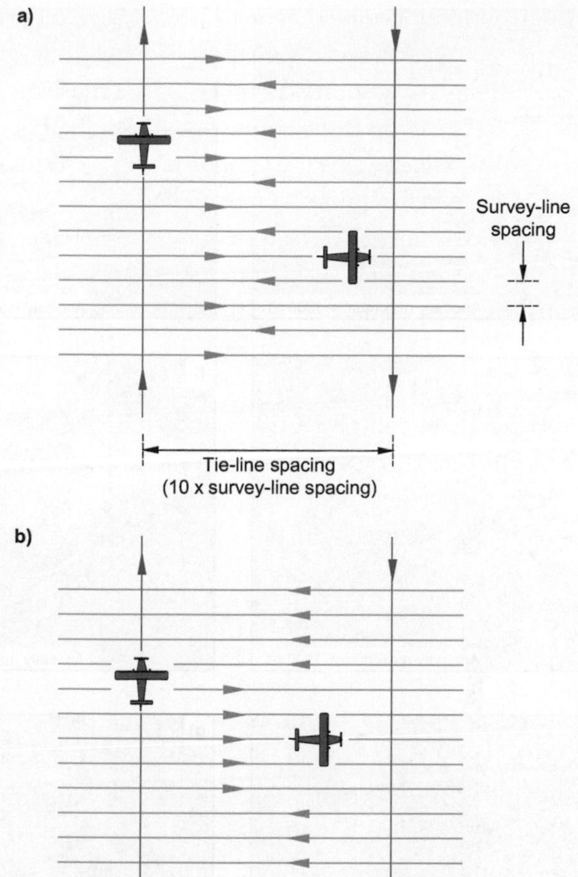

Figure 2.11 Typical flight paths of airborne surveys. (a) Survey with adjacent survey lines flown in alternate directions, and (b) survey flown in racetrack formation. In both cases, perpendicular tie lines are flown with a spacing of usually 10 times the survey line spacing.

topography appears extreme or tall vegetation and man-made structures present a problem.

For ground surveying, there may be a requirement to clear access routes, which can add significantly to the cost of data acquisition. In densely vegetated areas and in very rugged terrains, it may actually be easier to take advantage of lakes and rivers, although of course measurements may then be limited to shorelines. Figure 2.10d illustrates a typical configuration of survey traverses where measurements are made at relatively small spacing along roads (possibly widely spaced and of different and variable orientations), but time and access considerations dictate that there are fewer measurements between the traverses.

The tendency to 'home-in' on features of interest, as exploration progresses, produces an evolving dataset with

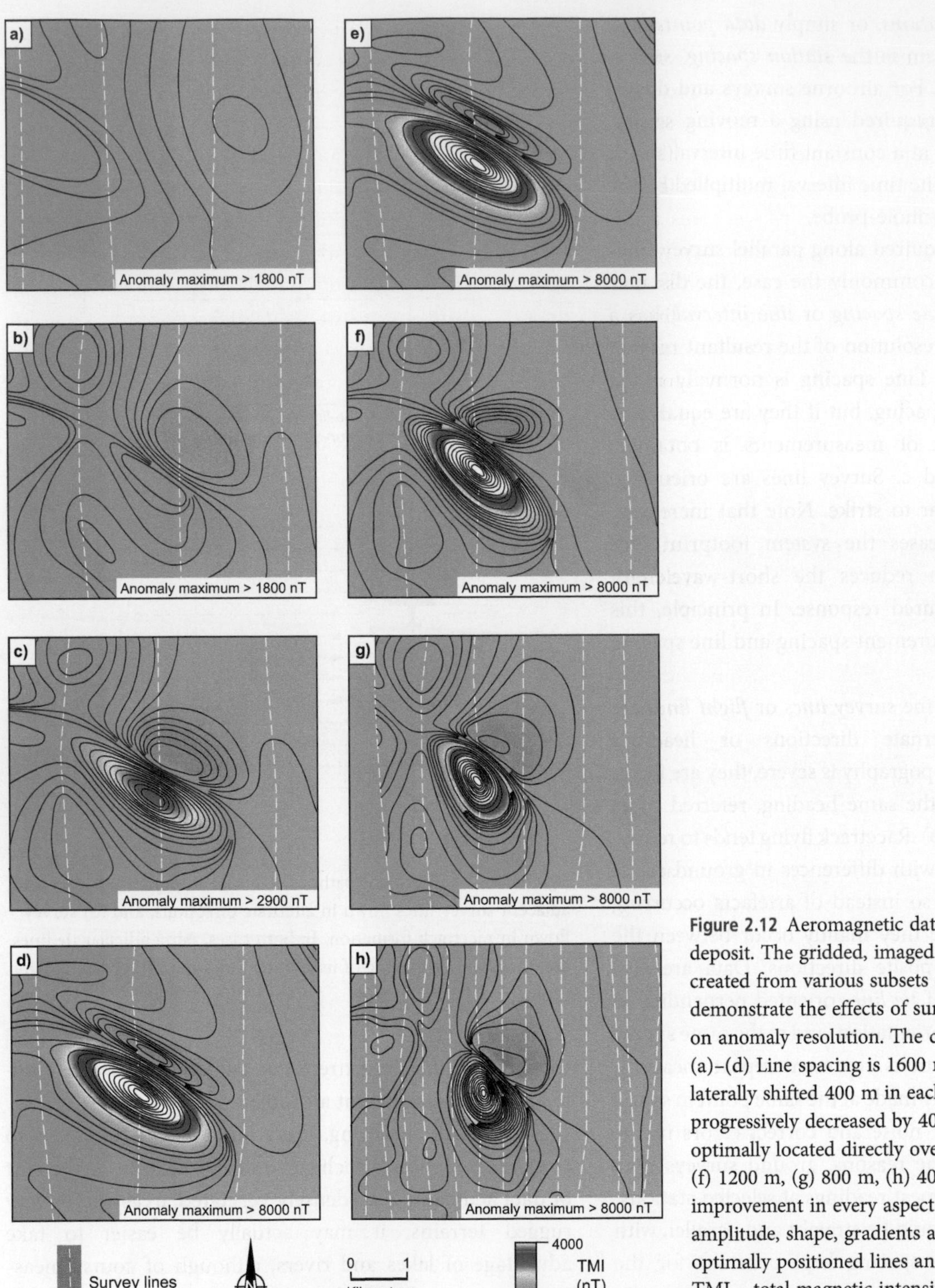

Figure 2.12 Aeromagnetic data from the Marmora Fe deposit. The gridded, imaged and contoured data were created from various subsets of the full dataset to demonstrate the effects of survey line spacing and position on anomaly resolution. The contour interval is variable. (a)–(d) Line spacing is 1600 m, with the lines progressively laterally shifted 400 m in each case. (e)–(h) Line spacing progressively decreased by 400 m, with the central line optimally located directly over the deposit: (e) 1600 m, (f) 1200 m, (g) 800 m, (h) 400 m. Note the spectacular improvement in every aspect of the anomaly, i.e. its amplitude, shape, gradients and trend direction, with optimally positioned lines and decreasing line spacing. TMI – total magnetic intensity. Based on diagrams in Agocs (1955).

a range of station and line intervals (Fig. 2.10d). Here a small area has been surveyed with greater detail, as would be required to characterise a target (see Section 2.6.1). Multiple surveys of differing specifications and extent are a common occurrence and can provide useful information for designing subsequent surveys.

An uneven spatial distribution of samples can lead to aliasing (see Section 2.6.1). Figure 2.10 shows several survey configurations with uneven sampling. In these cases the degree of aliasing will vary with location and, in line-based configurations, with survey line orientation. The smaller along-line sampling interval allows short-wavelength variations to be properly represented (at least up to the Nyquist frequency of the sampling interval), but the much greater distance between lines, i.e. greater across-line sampling interval, means similar variations in the across-line direction are likely to be under-sampled (aliased). This is often less of a problem that it might seem because orientating the survey lines perpendicular to strike ensures there will be a greater degree of variation in the geophysical response along-line than in the strike-parallel across-line direction. Of course, for responses that are equidimensional, the sampling interval should ideally be the same in both directions in order to produce a properly sampled anomaly. In reality, it is the (wider) line spacing that determines overall survey resolution.

For airborne surveys, the nature of the topography in relation to the survey parameters will determine the survey platform and the cost of the survey. Terrains of moderate relief can be surveyed with small fixed-wing aircraft, whilst helicopters are appropriate for low-level surveying of mountainous areas (Mudge, 1996). Cost and logistical considerations of course will ultimately determine the final survey parameters; and to that extent, the line spacing should be increased first and, if necessary, followed by limited increase in survey height.

Unmanned airborne survey platforms are likely to become available for commercial use in the near future (McBarnet, 2005). The advantages of using an unmanned aircraft include reduction in cost, the ability to fly lower and in poorer visibility than is possible with manned aircraft, and the possibility of surveying at night when environmental noise levels are lower.

2.6.4 Feature detection

When the aim of the survey is to identify anomalous responses, the survey parameters should be set to maximise the chance of these responses being measured, and preferably on several survey lines to confirm the validity of the response.

Agocs (1955) shows how the measured anomaly varies with station locations, using aeromagnetic data from the Marmora Fe deposit, located in Ontario, Canada. The deposit is a magnetite skarn which produces a very distinct magnetic anomaly against the quiet background response of the host carbonate sequence. The data shown in Fig. 2.12 were acquired along parallel survey lines spaced approximately ¼ mile (400 m) apart at a nominal terrain clearance of 500 ft (152 m), and were responsible for the discovery of the deposit. The line spacing is typical of a modern reconnaissance aeromagnetic survey, although the survey height is significantly greater than that of current practice. The data have been gridded (see Section 2.7.2) and then imaged and contoured (see Section 2.8). The important aspects of Fig. 2.12 are that the observed response of the mineralisation varies in amplitude and that the locations of the maximum and minimum responses change with line spacing and the location of the survey lines. On using subsets of the data by line spacing and line location, quite significant changes can be seen in the amplitude of the measured anomaly, the apparent location of the source, and its strike length and direction. When the line spacing is large compared with the lateral extent of the actual anomaly (Fig. 2.12a and b), the peak responses are located on the nearest line(s) and the amplitude of the anomaly is underestimated. They are only correctly located when the line fortuitously passes over the centre of the actual anomaly (Fig. 2.12c to h), and then measured amplitudes more closely reflect the actual amplitudes. Every aspect of the anomaly (i.e. its amplitude, shape, gradients and trend direction) improves with optimally positioned lines and decreasing line spacing. The worse-case situation occurs when the anomaly lies entirely between two lines, in which case no anomalous response will be detected.

The Marmora example demonstrates clearly the importance of setting the line spacing to suit the across-line width of the anomaly. It is useful during survey design to determine the probability of detecting a target of particular dimensions with a particular survey configuration. The probability of detecting an anomaly depends on the distribution of the stations (or data points) and the orientation of the survey lines, assuming that the line spacing and the station interval are different, with respect to the dimensions and shape of the target's response. The area of detectable response should not be confused with the surface projection of the target itself because the geophysical response normally extends over an area larger than the target itself (Fig. 2.4), which helps enormously in aiding detection.

Of course, the target's response can be complicated and needs to be recognisable above the noise to ensure its detection. This raises the issue of what constitutes detection. A single anomalous measurement could well be mistaken for noise, so at least two readings, but preferably more, need to be made in the area of a detectable response.

Galybin *et al.* (2007) investigate the problem of recognising an arbitrarily orientated elliptical anomaly with a survey of given line interval and station spacing. The

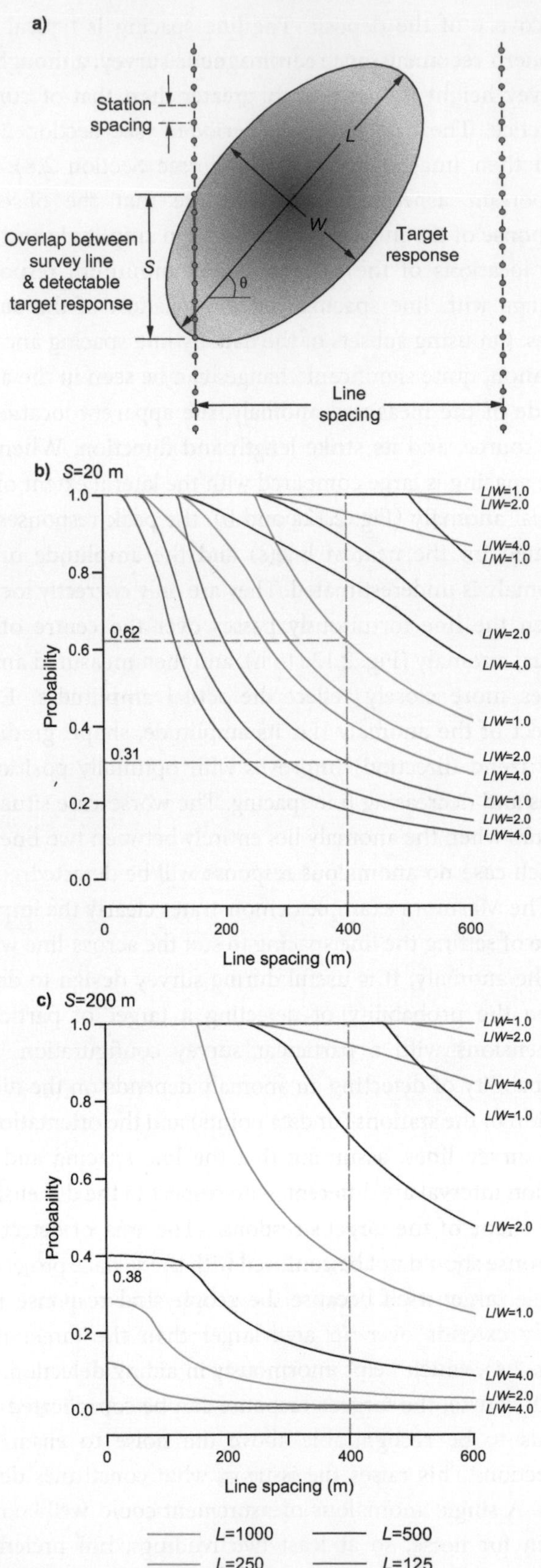

Figure 2.13 Probability of detecting an elliptical anomaly. (a) Parameters used to determine the probability of a line-based survey crossing an elliptical anomaly, with major axis length L and minor axis length W, with an overlap of at least the specified length S. (b) Probability versus line spacing for an overlap of 20 m. (c) Probability of detection for an overlap of 200 m. Redrawn, with permission, from Galybin *et al.* (2007).

probability of detection is based on the requirement to make a specified number of measurements within the bounds of the anomalous region (Fig. 2.13a). Figure 2.13b shows the likelihood of detecting elliptical anomalies with major axes (L) of length 1000, 500, 250 and 125 m, and major to minor axis (W) ratios ranging from 1 to 4. The data in Fig. 2.13b represent aeromagnetic data acquired with a fixed-wing aircraft and a data interval of 5 m. It is assumed that at least five readings are required to produce a recognisable response, corresponding to a survey line overlap interval (S) of 20 m with the anomalous region. For example, a line spacing of 400 m is shown to have about a 31% chance of resulting in a recognisable response from a circular (L/W=1.0) target 125 m in diameter (L), and a 62% chance for a target with twice these dimensions; and as expected, detecting circular features of 500 and 1000 m diameter is a certainty. Figure 2.13c assumes that a 200 m overlap is required for detection, consistent with three readings in a gravity survey with stations spaced 100 m apart. For this case, the circular feature of 125 m diameter is undetectable since the required overlap is larger than the longest dimension of the target, whereas the 250 m feature has a 38% chance of being detected. Again, the largest features will definitely be detected. This type of analysis is applicable to all geophysical methods, provided the characteristics of the target anomaly can be accurately predicted.

2.7 Data processing

We describe data processing as a two-stage process: data reduction and data enhancement (Fig. 2.1), although they have aspects in common. Integral to these is the interpolation of the survey data transforming them into a regularly spaced distribution of sample points, suitable for enhancement using numerical methods, for merging with data from other surveys and for displaying the data.

2.7.1 Reduction of data

For most geophysical methods it is necessary to apply a variety of corrections to the data obtained 'raw' from the field acquisition system in order to 'reduce', or prepare, the data for enhancement and display. It is essential that, as far as is possible, all acquisition-related errors, and errors and noise emanating from external sources, be corrected or compensated as a first stage of data processing. Otherwise, the errors will propagate and increase through the various stages of data processing. The situation is stated succinctly by the maxim: 'Errors don't go away, they just get bigger'.

Reduction compensates for various sources of error and noise, sometimes using secondary data acquired during the survey (see Section 2.6). Some reductions may be carried out automatically by the data acquisition instrumentation, provided that the instrument records the necessary secondary data and that the magnitude of the correction is easily determined. More complicated reduction processes are applied post-survey.

In its simplest form, reduction involves a manual assessment of the data to remove readings that are obviously unsuitable, for example those dominated by noise. Corrections applied during reduction include compensation for the geophysical response of the survey platform; variable sensor orientation/alignment; and various temperature, pressure and instrumental effects. In addition, corrections are applied to suppress environmental noise (see Table 2.1), which may require measurements obtained from a secondary sensor specifically deployed to monitor the noise. The corrections applied during reduction of the data are only as good as the secondary data on which they are based. For example, information about the local topography may be insufficient to remove its effects completely.

Generally, reduction processes are parameter specific and survey type specific, and are described in detail later in our descriptions of each geophysical method. Here we describe some more generally applicable operations.

2.7.1.1 De-spiking

Spikes, or impulses, are abrupt changes in the data which have short spatial or temporal extent, and usually occur as a single data point. These may be caused by instrumental problems or they may originate in the natural environment. Spikes can be removed using various numerical data processing techniques (see *Smoothing* in Section 2.7.4.4) or by manual editing. Erroneous data are then replaced by interpolating from adjacent readings. It is important to de-spike early in the processing sequence since some processing operations will produce artefacts if spikes are not removed, notably those that involve a Fourier transform (see online Appendix 2).

2.7.1.2 Levelling

The success of the data reduction in eliminating errors can be assessed by comparing corrected repeat measurements made at the same location. Ideally the reduced (repeat) values should be identical. Any remaining differences, known as *residual errors*, can be used to further reduce the data to remove these errors. This process is not based on measurements from the natural environment; instead it is a pragmatic attempt, based on statistical methods, to lessen the influence of remaining errors by redistributing them across the entire dataset. The process is known as *levelling* because the adjustment made to the amplitude of each measurement changes the overall amplitude level of the measurements.

Tie lines provide the necessary repeat readings where they intersect the survey lines (see Section 2.6.3.3). The basic idea is to use the tie-line data to 'tie' adjacent survey lines together at regular intervals, with the residual errors at the line intersections used to adjust the data. For airborne surveys this is an imperfect process because of the inevitable errors due to differences in survey heights at the line intersections, i.e. the two readings are not from exactly the same location. Tie lines may also be part of a ground survey; but if measurements have been made at one or more base stations, then these provide the necessary repeat readings. It is common practice to make repeat readings at selected stations for this specific reason.

There are various means of redistributing the residual errors; see Luyendyk (1997), Mauring *et al.* (2002) and Saul and Pearson (1998) for detailed descriptions regarding airborne datasets. One simple approach when tie lines are available is based on using the residual values at the intersections to interpolate a 2D error function across the survey area. By subtracting the relevant error function values from the readings along the survey lines, the residual errors are reduced.

A quick and easy way to assess the quality of the data reduction is to view it with shaded relief (see Section 2.8.2.3), with the illumination perpendicular to the survey line direction, optimum for highlighting across-line levelling errors. Residual errors in the levelled data are revealed as corrugations or ripples between the survey lines (see Fig. 3.24). Also, derivative images, being sensitive to gradients, are usually effective in revealing residual errors in a levelled dataset. Any remaining errors indicate either that the various corrections applied are imperfect or that other sources of survey error have not been accounted for.

2.7.1.3 Microlevelling

Small residual errors can be dealt with through a process known as *microlevelling*. These algorithms operate on the gridded data (see Section 2.7.2); a method based on directional filtering (see *Trend* in Section 2.7.4.4) is described by Minty (1991). Microlevelling aims to remove residual errors remaining after the levelling process and new errors introduced by the gridding process. The computed corrections can be applied to the levelled line data to produce 'microlevelled' line data.

The key aspect of microlevelling is that it is a cosmetic process designed to make the data look good, i.e. to make it look as it is expected to look! If not applied carefully, these methods can remove significant amounts of signal, especially higher-frequency components, and may even introduce unreal features. An example of microlevelled magnetic data is shown in Fig. 3.24d.

2.7.2 Interpolation of data

Most data processing and data display methods require the data points to be regularly distributed, i.e. to be equally spaced. As noted in Section 2.6.3.3, data are rarely acquired in this way, so there is a need to interpolate the data into an evenly spaced network. For example, a 1D unevenly spaced dataset can be interpolated into an evenly spaced series of measurements in terms of time or distance. The interpolated data points are often called *nodes*. Similarly, an uneven 2D distribution of data points, acquired either randomly or along a series of approximately parallel survey lines, is usually interpolated into a regular grid network. Logically, the process is known as *gridding* (Fig. 2.14a) and the distance between the nodes is the *cell size* or *grid interval*.

Gridding is a very common operation in geophysical data processing, and there are various ways in which it can be done. Normally, it is assumed that spatial variations in geophysical parameters will be continuous. Somewhat counter-intuitively, interpolation schemes whose results honour the data points exactly do not usually produce the best results. This is because the data contain both signal and noise, so allowing the gridding algorithm to fit the data to within some prescribed limit helps to reduce the influence of the noise component.

Interpolation is based on an analysis of a window of data points in the vicinity of the node. The window is centred at the node, and when gridding a 2D dataset its shape must

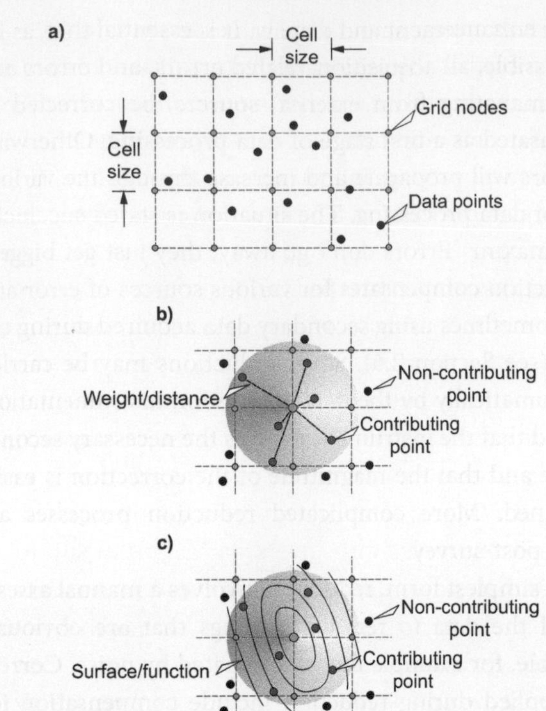

Figure 2.14 Gridding a 2D dataset. (a) Data points in the vicinity of the grid nodes are used to determine an interpolated value at each node. (b) Node-to-station distance-based weighting. (c) A smooth 2D function is fitted to the data and the interpolated value computed from that. In (b) and (c) the grey area represents the region that influences the interpolated value.

be defined. Normally it is circular, but if the data exhibit a well-developed trend direction the window may be elongated parallel to that direction, since this will ensure the trend is preserved in the interpolated data. The size of the window needs to be large enough to enclose a representative sample of measurements, although if it is too large potentially important short-wavelength variations will be lost in the 'averaging' of the data within it.

There are two main ways of establishing the value of the parameter at a node: either statistically or using a simple mathematical function. Both methods can be applied to 1D and 2D datasets; the process as applied to gridding is shown in Figs. 2.14b and c. A key concept in gridding is the concept of minimum curvature. The human vision system perceives smoothness if the first and second derivatives of the parameter being visualised are continuous. Put simply, if the curvature (the gradient of the gradient) of a line or surface varies gradually, it is perceived as smooth. The spatial variation of the parameter being gridded can be thought of as defining a 'topographic' surface. To make this surface appear smooth the values of the grid nodes are adjusted so that the second derivative of the gridded surface varies smoothly, i.e. it has *minimum curvature*. Some types of geophysical data, e.g. gravity and magnetic fields, are smoothly varying, but there is no direct physical basis for using smoothness as a basis for interpolation.

2.7.2.1 Statistical interpolation

The statistical approach involves calculating some form of average of the measurements in the window. The median value has the advantage of being immune to the effects of outliners (extreme values) in the data series, presumed to be noise. Arithmetic or geometric means can incorporate a system of weights, one for each data point scanned and which, for example, vary inversely proportionally to the distance to the data points. Therefore points closer to the grid node exert a greater influence on the interpolated value than those further away. An extreme form of weighted averaging is to assign a value to the node that is equal to that of the closest data point. This is known as *nearest neighbour gridding* and it can be effective if the data are already very nearly regularly spaced. If this is not the case the resultant dataset can have an unacceptably 'blocky' appearance. All of the gridding algorithms described above are suitable for both randomly distributed and line-based data, and minimum curvature adjustment can be applied.

One of the more sophisticated statistical interpolation methods is *kriging*, which is widely used in mineral-resource calculations (Davis, 1986). It is a method using weighted moving averages, where low-valued data points are increased and high values are decreased using smoothing factors or kriging coefficients (weights) dependent on both the lateral dispersion of the data points and their values. The method is not commonly used in geophysics, although it can be very useful for small datasets having an uneven distribution of data points of large dynamic range.

2.7.2.2 Function-based interpolation

An advantage of function-based interpolation methods is that particular behaviour of the measured parameter can be incorporated into the interpolation process, most commonly smoothness. By far the most common function-based interpolation methods use splines. A spline, in its original sense, is a thin strip of flexible material used pre-computer drafting to draw smooth curves. A physical model applicable to 1D data is the drafting spline held in position with weights and distorted so that it passes smoothly through the points to be connected or interpolated. The flexible strip naturally assumes a form having minimum curvature, since elasticity tries to restore its original straightness but is prevented from doing so by the constraining weights. The curve formed is a cubic polynomial and is known as a *cubic spline.*

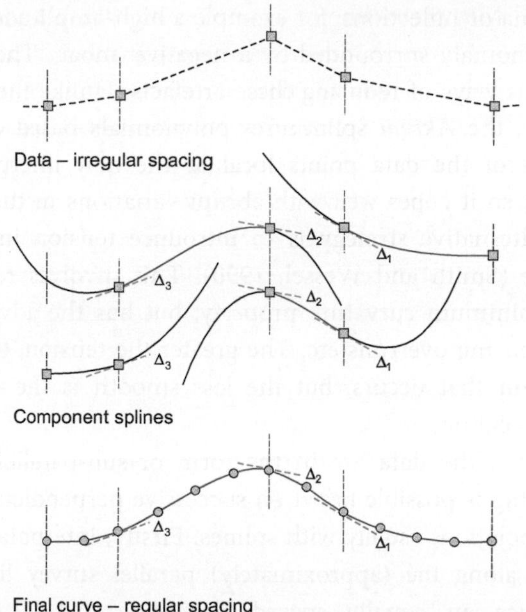

Figure 2.15 1D splining. Cubic splines are fitted to each pair of the irregularly spaced data points so that the gradients of connecting splines are the same at the joining point (Δ_1 etc.). The new values, interpolated at regularly spaced intervals, are obtained from the splines.

Splining in the numerical sense is a line-fitting method that produces a smooth curve (De Boer, 2001). Many types of polynomial functions can be used as splines, but cubic polynomials have less possibility of producing spurious oscillations between the data points, a characteristic of some other functions. The cubic spline consists of a series of cubic functions each fitted to pairs of neighbouring data points. They join smoothly at their common points, where the functions have the same gradients and curvature (i.e. the same first and second derivatives). New data values are calculated in the data intervals using the respective function. 1D interpolation using cubic splines is illustrated in Fig. 2.15.

The motivation for splining is a 'pleasingly' smooth curve. The smoothness of splines may actually be a disadvantage, since if a parameter varies abruptly the requirement for smoothness may result in spurious features infiltrating the interpolated data. Normally there is some kind of *overrun* which creates non-existent maxima, minima or inflections: for example a high-amplitude positive anomaly surrounded by a negative 'moat'. There are various ways of reducing these artefacts. Unlike the cubic spline, the *Akima* spline uses polynomials based on the slopes of the data points local to the new interpolated point, so it copes well with abrupt variations in the data. An alternative strategy is to introduce tension into the spline (Smith and Wessel, 1990). This involves relaxing the minimum curvature property, but has the advantage of reducing overruns etc. The greater the tension, the less overrun that occurs, but the less smooth is the overall interpolation.

When the data are in the form of sub-parallel lines, gridding is possible based on successive perpendicular 1D interpolation, usually with splines. Firstly, interpolation is done along the (approximately) parallel survey lines to produce an equally spaced along-line distribution of samples. The new samples are then used in a second interpolation perpendicular to the interpolated lines, to compute new samples between adjacent lines. This is known as *bi-directional gridding* (Fig. 2.16). The physical analogy would be bending a sheet of flexible material so it approximates the form of the variation in the data with a smooth surface. Two-dimensional spline gridding is often used when the data points are irregularly distributed along a series of approximately parallel survey lines. It is not suitable for randomly distributed data, or line-based data where the lines have random directions.

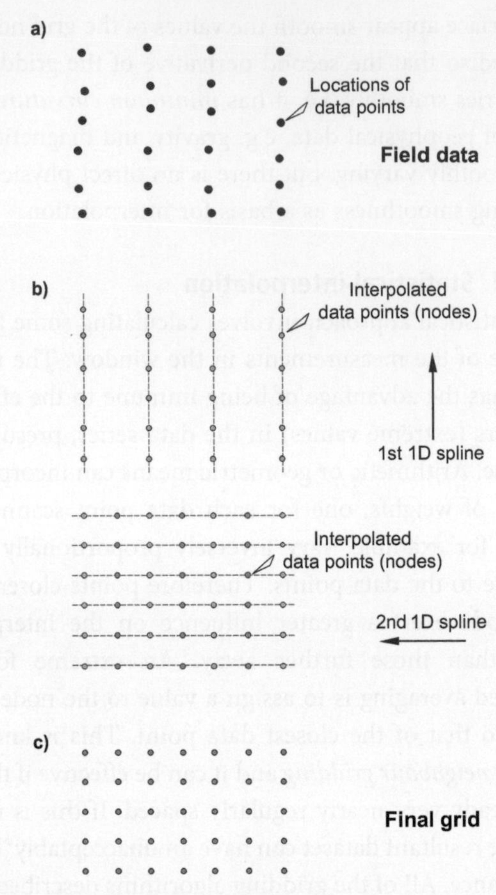

Figure 2.16 Bi-directional gridding of a dataset, comprising a series of approximately parallel survey lines, using splines.

2.7.2.3 Interpolation parameters and artefacts

Setting the appropriate spacing between the interpolated values, i.e. the grid cell size, when interpolating data is fundamental in producing a grid that depicts the survey data with a high degree of accuracy (Fig. 2.14a). If the chosen cell size is too small, instability may occur in the algorithm resulting in artefacts (see below). On the other hand, a very large cell size will result in the loss of useful short-wavelength information and introduce spurious long wavelengths because of spatial aliasing (see Section 2.6.1). In practice, the uneven distribution of samples inevitably leads to variable degrees of spatial aliasing within the interpolated dataset.

When the data are random or form a regular grid network, the cell size is usually set at about half the nominal distance between the data points. Calculating minimum or average spacing from the data is rarely useful since the results may be affected by clusters; see Fig. 2.10c. Comparing the final grid with the distribution of data points is an effective way of determining whether features of interest are properly represented, distorted or significantly aliased. The distribution of the points will strongly influence the gridded data, with most gridding-induced artefacts occurring in areas where there are fewer data points to control the gridding process. Ideally the gridding algorithm should automatically not interpolate beyond some specified distance, assigning 'dummy' values to nodes that the data do not adequately constrain. If this is not the case then features that occur in gaps in the data, or near its edges, should be viewed with suspicion. Furthermore, features centred on a single data point, referred to as *single-point anomalies*, must be considered highly unreliable. For these reasons it is good practice to have a map of survey station/point locations available when analysing the data (see Fig. 3.18). This is also useful for recognising changes in survey specifications, as inevitably occurs when datasets have been merged to form a single compilation (see Section 2.7.3). This can cause changes in the wavelengths

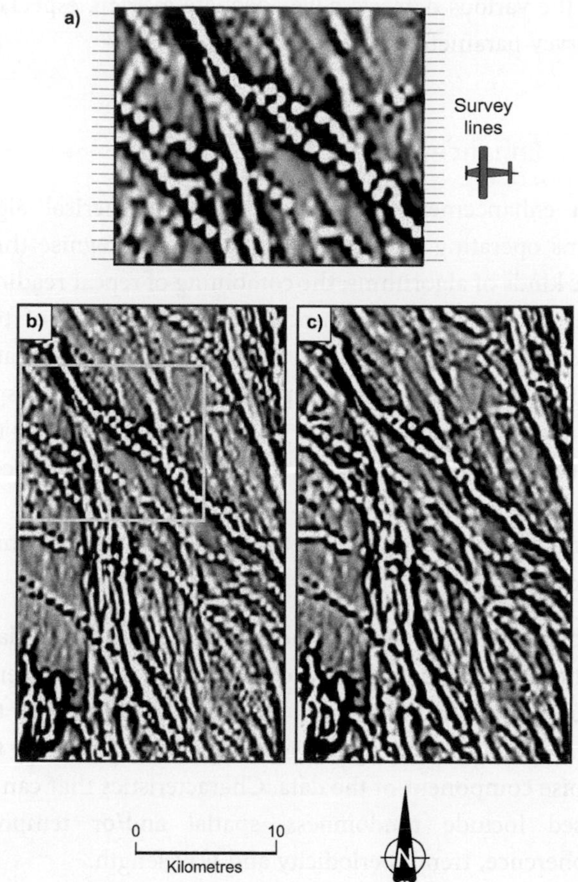

Figure 2.17 Beading in gridded data containing elongate anomalies. The example is aeromagnetic data from South Australia. (a) Close-up of beaded data showing how each bead is centred on a survey line. (b) Data gridded using the inverse-square technique with minimum curvature applied. The rectangle is the area shown in (a). (c) The same data gridded using a trend-enhancement algorithm. Data reproduced courtesy of Department of Manufacturing, Innovation, Trade, Resources and Energy, South Australia.

that make up the gridded data and may cause artefacts to appear along the join of the different datasets.

When the data to be gridded consist of parallel lines, with station spacing much smaller than the line spacing, the cell size must account for the anisotropic distribution of the measurements. It is not uncommon for the across-line sampling interval to be greater than the along-line interval by a factor of 50 or more in reconnaissance surveys, with 1:10 or 1:20 common for detailed prospect-scale surveys. A cell size based on the along-line sampling interval will create major difficulties for interpolation in the perpendicular direction, potentially creating artefacts (see below). On the other hand, choosing a cell size based on the line spacing will result in the loss of a lot of valuable information contained in the line direction. The normal

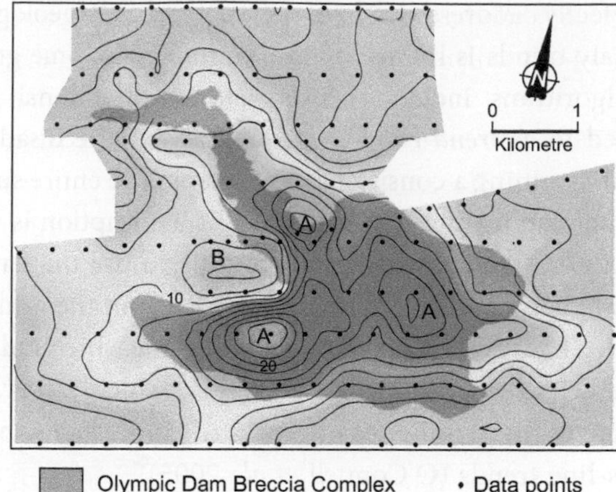

Figure 2.18 Circular anomalies produced by the minimum curvature gridding illustrated in contours of IP-phase response (IP – induced polarisation), at a constant pseudo-depth, from the Olympic Dam IOCG deposit in South Australia. Positive (A) and negative (B) circular anomalies are caused by inadequate sampling of the anomalous areas. Based on a diagram and data from Esdale *et al.* (2003).

compromise is to select a cell size of between 1/5 and 1/3 the line spacing. A smaller cell size can only be justified by having closer sampled data, and in particular closer survey lines.

Even with these cell sizes, interpolation in the across-line direction can be a challenge for gridding algorithms. Laterally continuous short-wavelength anomalies, as might be associated with a steeply dipping stratigraphic horizon or a dyke, can cause particular problems. The phenomenon is variously referred to as *beading, boudinage, steps, step ladders, string of beads* etc. The aeromagnetic data in Fig. 2.17 illustrate the effect. Note how the individual beads have dimensions in the across-line direction equal to the line spacing, allowing this kind of artefact to be easily recognised. As the anomaly trend approaches the survey line direction, individual beads become more elongated towards this direction.

A problem with minimum curvature algorithms (see Section 2.7.2) is their tendency to produce round anomalies, often referred to as *bulls-eye* anomalies. This occurs because an unconstrained minimum curvature surface is a sphere. Where data points are sparse and the true shapes of features not properly defined by the data sampling, the gridding algorithm makes them circular (Fig. 2.18).

Both beading and bulls-eyes are the result of inadequate sampling by the survey. The solution is more measurements, but this may be impractical. The problem can be imperfectly addressed if information about geological/anomaly trends is known or can be inferred. Some gridding algorithms include a user-defined directional bias, referred to as *trend enhancement*. This has the disadvantage of assuming a consistent trend across the entire survey area, but can be highly effective if this assumption is valid (Fig. 2.17c). An alternative approach is to use the data to estimate between-line values and use these in the gridding process. The most effective remedy is to measure gradients of the field in the across-line direction, which act as constraints on the permissible interpolated values and enhance across-line trends (O'Connell *et al.*, 2005).

There is an extensive literature on gridding methods, for example Braile (1978) and Li and Göetze (1999). Different methods perform better in different circumstances; but both inverse-square distance with minimum curvature, and spline algorithms are widely used for geophysical data.

2.7.3 Merging of datasets

When several datasets are available from an area it can be useful to merge or stitch these into a single dataset for processing and analysis. The requirement is to form a coherent and 'seamless' composite dataset from individual surveys that often have quite different survey parameters. This may not be achievable in practice because the different survey parameters will produce datasets with different wavelength content (see Section 2.6.1), and similar geology will not produce identical responses in the different datasets.

Data are usually merged in their gridded form. A description of one kind of grid stitching algorithm is provided by Cheesman *et al.* (1998). Accounting for the range of wavelengths in the datasets is a fundamental aspect of the process. These may vary for each survey in the overlap zone owing to differences in survey size, shape, acquisition parameters and noise levels. Also important is the extent of the overlap between the surveys, with the amount of overlap and the nature of the geophysical field in the overlap zone and at the joins strongly influencing the result. The more similar the datasets in the overlap zone, and the larger their overlap, the better the results are likely to be (particularly in terms of longer wavelengths). When there is no overlap it will be necessary to acquire additional data to link the surveys being merged. For this reason, it is important that any new survey extends far enough into the area of the existing data coverage to ensure that the various datasets have adequate overlap, especially if survey parameters are different.

2.7.4 Enhancement of data

Data enhancement techniques involve numerical algorithms operating on the survey data. We recognise three basic kinds of algorithms: the combining of repeat readings in a process known as *stacking*; the comparison of two measurements by computing their ratio; and more mathematically sophisticated manipulations of the data for specific purposes which we refer to as filtering. Note that the term filtering is often loosely used to describe any process that alters the data in some way.

Enhancements are applied to geophysical data for three basic purposes:

- To enhance the signal-to-noise ratio (SNR) of the data. This involves identifying a characteristic, or characteristics, which are not shared by both the signal and the noise, and then using this as a basis for attenuating the noise component of the data. Characteristics that can be used include randomness, spatial and/or temporal coherence, trend, periodicity and wavelength.
- Most types of geophysical data can also be processed to enhance characteristics of the signal that are considered particularly significant. For example, filtering may enhance short-wavelength responses originating in the near-surface, or longer wavelengths related to deeper features, or may compute the various derivatives (gradients), which are greatest near physical-property contacts etc. (see Fig. 2.3). Ratioing may allow regions with particular combinations of responses to be recognised, for example anomalously high values in one dataset and low values in another.
- For completeness we mention a third form of enhancement that is based on the principle that some types of data can be transformed to produce a dataset that would have been measured with another type of sensor, or with different survey parameters. For example, magnetic and gravity data can be transformed into the equivalent data that would have been obtained if the survey were conducted higher (or lower) above the ground. This not only emphasises features of interest, it also enables data acquired with different survey parameters to be integrated and merged (see Section 3.7.3.2). These kinds of enhancement are described in our descriptions of each geophysical method.

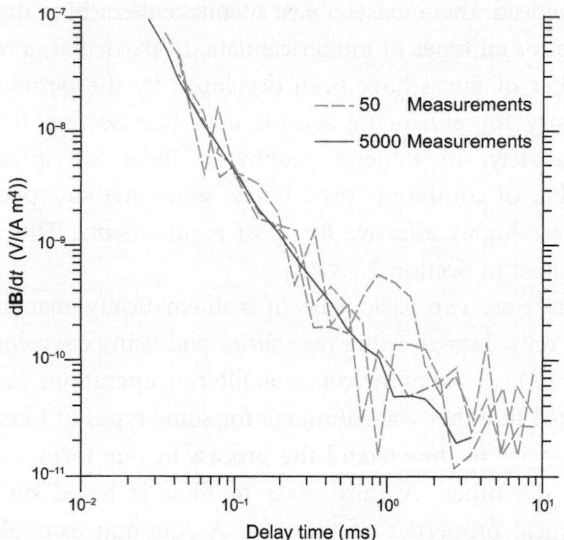

Figure 2.19 Stacking or summation (averaging) of repeat recordings of a repetitive EM signal; the time-varying magnetic field (**B**) is measured as a voltage induced in a coil (see Section 5.7.1). Note the significant reduction in noise in the time series comprising 5000 stacked measurements. Redrawn, with permission, from Sorenson *et al.* (2006).

It is important to appreciate that a variety of enhancement methods, applied sequentially, may be required in order to obtain the desired enhancement of a dataset. Examples are presented in Sections 2.7.4.5 and 6.5.2.3.

2.7.4.1 Stacking

Noise that is random in nature, i.e. unpredictable with no discernible pattern in time or space, is suppressed by stacking (one of several meanings for this term). Random noise is usually a form of environmental noise (see Section 2.4.1), for example caused by the wind (vibrating the sensors and causing ground movement due to tree motion etc.), electrical discharge from airborne sensors moving through moist air, magnetic storms etc.

Stacking involves averaging repeat measurements (Fig. 2.19). Any variations between the repeat measurements should be entirely due to random noise since, ideally, the signal remains constant between readings. The random noise component is likely to be positive or negative in equal proportions and, given that the sum of a large series of random numbers tends to zero, the random component will sum to zero; although for low numbers of readings this is unlikely to be the case with some low-level noise remaining. This means that by simply adding the repeat readings together, and if required dividing by the number of readings so the final value is correct in absolute terms, the noise components will cancel leaving only the signal. The signal to (random) noise ratio is improved by a factor of $\sqrt{N}$, where N is the number of repeat measurements.

Stacking is implemented in many geophysical instruments. In these cases, the instruments are capable of automatically making repeat readings with the measurement recorded being some kind of average of these. In some instruments the values are explicitly compared and recording continues until some level of consistency is achieved, usually based on the standard deviation of the measurements. Alternatively, stacking may be undertaken during post-survey data processing.

Stacking is highly effective where a measurement is made at a fixed location, i.e. during ground and downhole surveying, and over a time period of sufficient duration so as to observe enough repetitions of the signal. A disadvantage of stacking is that it increases the time required to make a measurement. For moving survey platforms, stacking is not as effective because the signal is continually changing as the survey system passes over the ground, so only a smaller number of stacks are possible before distorting the signal too much.

2.7.4.2 Ratios

When measurements are multichannel, i.e. comprising a series of individual readings, it can be useful to display the ratios of pairs of readings. The ratio of two channels A and B is simply A divided by B, and its reciprocal ratio is B divided by A. Multichannel geophysical datasets include radiometrics (K, U and Th channels; see Section 4.3.2.2), and electromagnetic data (different channels in time domain measurements; see Section 5.7.1.6).

Ratios provide information about the relative variation of one channel with respect to another. For example, in the radiometric context, a high uranium channel response combined with a low potassium channel response. The problem with ratios is that they are inherently ambiguous. A high ratio of A/B may be caused either by high values of A or low values of B. The ratio data should always be analysed in association with displays of the channels involved.

Channel ratios can be highly variable and can have large amplitude ranges, which can be attenuated by displaying their logarithm, or by treating them as an angle and applying the inverse tangent function.

2.7.4.3 Filtering

Mathematical operations that change a dataset in some way are known, in signal-processing parlance, as *filters*, and the mathematical representation of the filter, and its parameters, is the *operator*. The dataset is the input to the filter which alters or transforms the data, and the output is the filtered dataset (Fig. 2.20a). The difference between the input and the output depends on characteristics of the filter and is known as the *filter's response*. Some examples of elementary filters are: change a dataset's polarity by simply multiplying every data point in the dataset by −1; slightly more complex, convert the data to an equivalent series having a mean value of zero by subtracting the overall mean of the dataset from each data point.

Filters can be applied to 1D and 2D data. In principle, one can design a filter to do almost anything to the data,

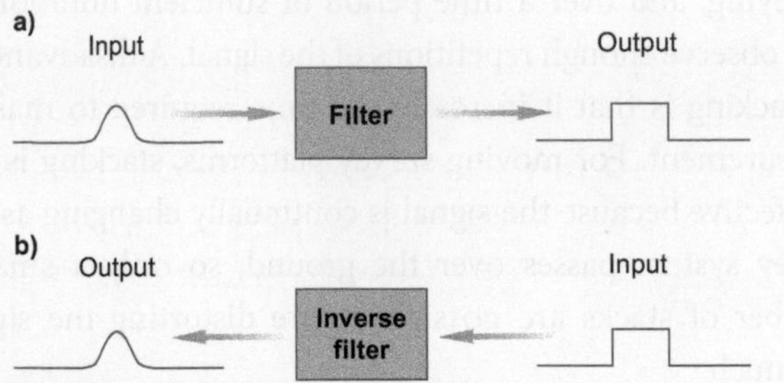

Figure 2.20 Schematic illustration of a filter. (a) A filter with a blocking function, i.e. it converts the data to a step-like form. (b) The inverse filter which reverses the action of the filter in (a).

and indeed, there exists a vast literature describing digital filters for all types of numerical data. In particular, a large number of filters have been developed by the petroleum industry for enhancing seismic data (see Section 6.5.2). Fortunately, in mineral geophysics there are a small number of commonly used filters, some method specific, that are highly effective for most requirements. They are described in Section 2.7.4.4.

There are two basic ways of mathematically enacting a filter on a dataset, using *transforms* and using *convolution* (Fig. 2.21). Most of the common filtering operations can be enacted in either way, although for some types of filtering it is easier to understand the process in one form rather than the other. A third class of filter is based on the statistical properties of the data. A common example is the median filter (see *Smoothing* in Section 2.7.4.4), which simply determines the median value of a moving window of data points and uses this as the filtered value.

Transforms

A data series exists in a particular domain (see Section 2.2) and it is often convenient to transform it into a different domain to facilitate its analysis and filtering. There are several commonly used transforms, but the most important one is the *Fourier transform* (see Appendix 2). Filtering

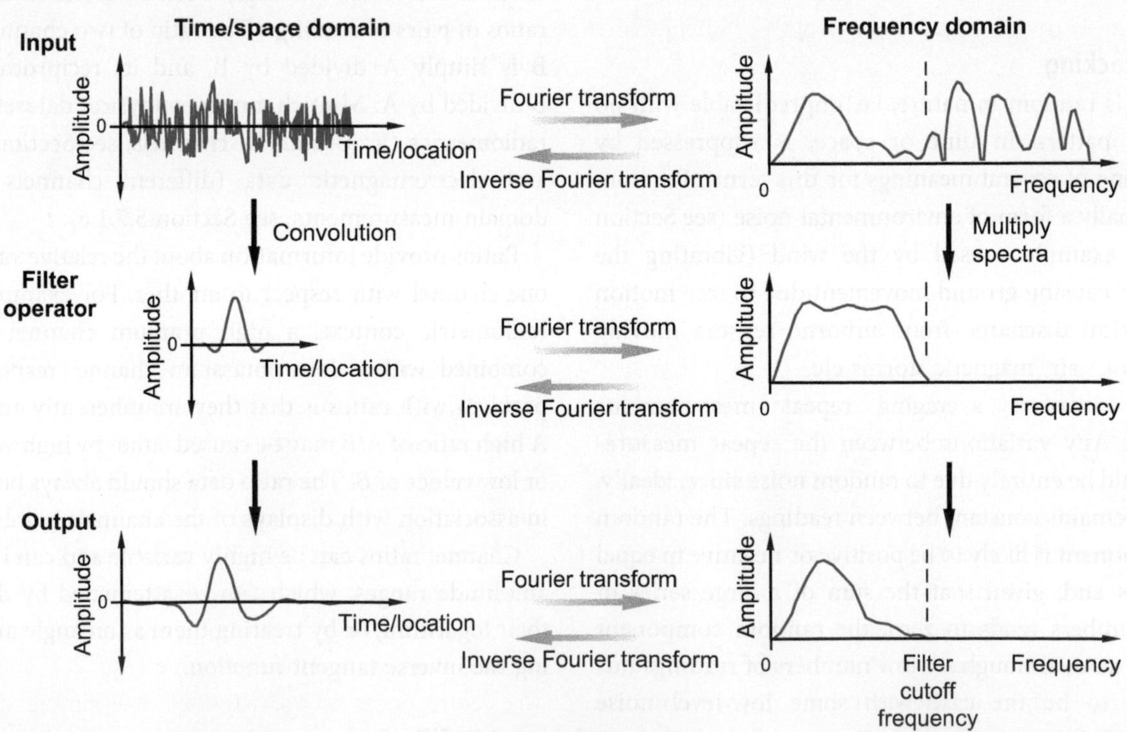

Figure 2.21 Filtering of a 1D dataset in the space/time domain and the frequency domain. The filter is a low-pass frequency filter. Redrawn, with permission, from Ashcroft (2011).

using transforms involves applying a forward transform to the dataset, transforming it into the new domain. Next, the filtering operation is enacted by modifying the transformed data in some desired way. Applying the inverse transform returns the modified transformed data series to its original domain as the filtered data series.

In the case of the Fourier transform, the data are transformed to a series of component sine waves of different frequencies, each represented in terms of its amplitude and phase, i.e. the amplitude and phase spectra; these depend upon the nature of the original data series (see online Appendix 2). This is known as the *frequency domain* (also referred to as the *Fourier domain*), and represents an alternative to the *time* and *spatial domains* in which most data are recorded. The filter operator is defined by its frequency spectrum, in the same way as the transformed dataset, and the two amplitude spectra are multiplied and their phase spectra added to obtain the spectra of the filtered output. Filtering attenuates or amplifies particular frequencies, and/or changes their phases. Recombining the component sine waves transforms the filtered spectra back to the original domain to form the filtered data series. Frequency domain filtering of both 1D and 2D data is possible. Filtering of a 1D dataset is illustrated schematically in Fig. 2.21. In this case the filter removes all frequencies above a defined *cut-off frequency* but does not affect phase, i.e. it is a form of wavelength/frequency filter (see *Frequency/wavelength* in Section 2.7.4.4).

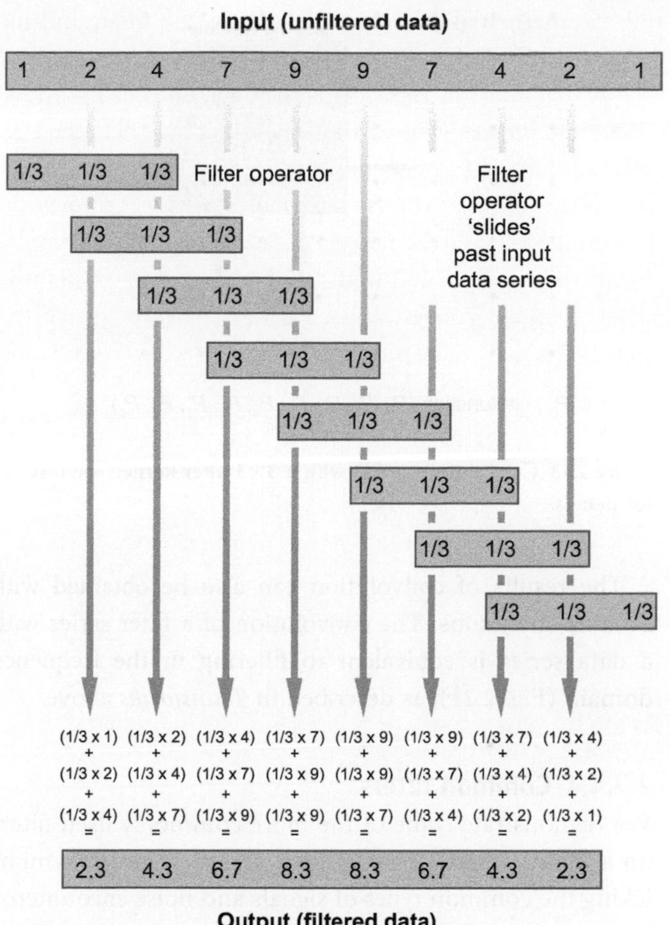

Figure 2.22 Convolution in 1D with a three-point filter kernel. See text for details.

Convolution

A common form of filtering, familiar to most readers, involves computing the running (moving) average of a group of data points. Consider a 1D data series. Obtaining the three-point running average of the data involves summing three consecutive data points and dividing by the number of points (3). This is a filter operator defined by a series of coefficients of equal value, in this case 1/3, and denoted as the series {1/3, 1/3, 1/3}. The set of filter coefficients is also known as the *filter kernel*. It is applied to a subset or window of three data points in the data series by multiplying each filter coefficient with its respective data point, and summing the three multiplied points to obtain the new filtered (averaged) output value. The output is assigned to the point/location at the centre of the window, the operator moved to an adjacent position to window the next consecutive group of data points, and the process repeated until a new, filtered version of the dataset is created. The process is known as *convolution* and is illustrated for 1D data in Fig. 2.22. It is also referred to as filtering in the time/spatial domain.

The example shown above has coefficients of equal value, but an unlimited variety of filters can be produced by varying the number of filter coefficients and their individual values. For example, to change the polarity of every data point (mentioned earlier) involves the very simple, one coefficient operator {–1}; the gradient (first derivative) filter has the coefficients {–1, 1}; and the curvature (second derivative) filter has the coefficients {1, –2, 1}.

Convolution can also be implemented on 2D data (Fig. 2.23), in which case the filter operator has the form of a matrix which is progressively moved through the dataset and the new filtered value (P_{Output}) is assigned to the centre point of the window. The simple three-point running average filter above would then have the form of a 3×3 matrix of nine coefficients each equal to the inverse of their average value, i.e. 1/9.

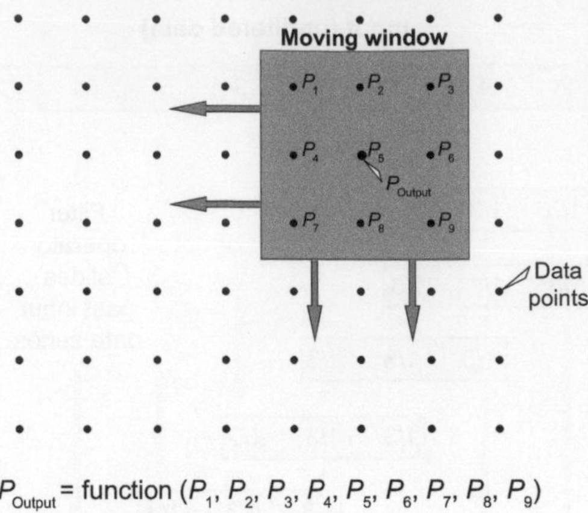

Figure 2.23 Convolution in 2D with a 3×3 filter kernel. See text for details.

The results of convolution can also be obtained with Fourier operations. The convolution of a filter series with a data series is equivalent to filtering in the frequency domain (Fig. 2.21), as described in *Transforms* above.

2.7.4.4 Common filters

We demonstrate some of the more commonly used filters on a 1D data series containing a variety of features mimicking the common types of signals and noise encountered in mineral geophysics (Fig. 2.25). The input series includes random variations, abrupt discontinuities and spikes (common types of noise) and smooth variations with various wavelengths. Depending on the objectives of the filtering, the variations of different wavelength may be considered as either signal or noise.

Gradients and curvature

In Section 2.2.3 we described the measurement of gradients of a physical property field. Some of the gradients can also be computed from the survey data, and enhancements obtained in this way are the most common form of filtering used to assist interpretation of geophysical data. Gradient or derivative filters resolve variations in the geophysical response with distance, with time, or with frequency. The results are usually referred to as derivatives. The term derivative (from calculus) refers to the gradient of continuous functions whose data samples are infinitesimally close, so the gradient of discretely sampled data approximates the derivative; but in geophysics the terms are used synonymously.

For a parameter (*P*) acquired along parallel lines, gradients of the measured field determined in the along-line direction are referred to as the *X-derivative* ($\partial P/\partial x$), and those in the across-line direction as the *Y-derivative* ($\partial P/\partial y$). They are the horizontal gradients of the measured parameter. For some types of geophysical data, it is also possible to compute the vertical gradient, or *Z-derivative* ($\partial P/\partial z$), which shows how the measured parameter changes as the distance to the source (survey height) changes. The vertical gradient, unlike the horizontal gradients, has the advantage of responding to changes irrespective of the orientation of their source with respect to the survey line. These three derivatives are the *first derivatives* in their respective directions.

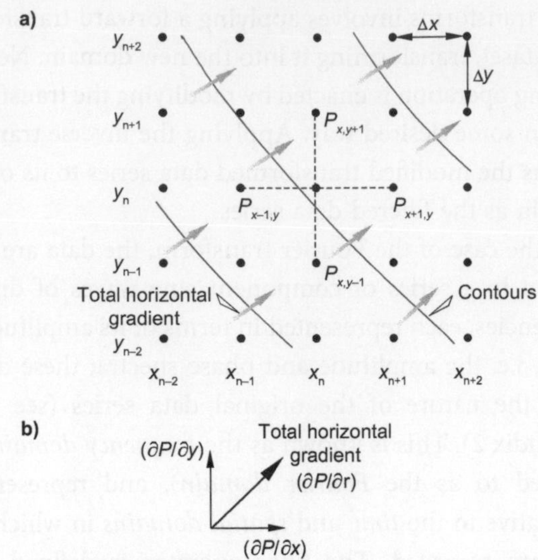

Figure 2.24 Calculation of horizontal gradients. (a) Calculation of horizontal gradients (derivatives) of gridded data based on differences between adjacent grid nodes. See text for details. (b) The total horizontal gradient is the vector sum of the horizontal gradients in the *X* and *Y* directions and is everywhere perpendicular to contours of the data.

Derivative filters can be enacted in the spatial and frequency domains. A 2D spatial domain derivative filter is illustrated in Fig. 2.24a; the derivative of the gridded dataset of *P* is approximated as the gradient at the location (x_n, y_n). The gradients in the *X* and *Y* directions are obtained from the differences in values at adjacent grid nodes, spaced Δx in the *X* direction and Δy in the *Y* direction, using the expressions:

$$\frac{\partial P}{\partial x}(x_n, y_n) \approx \frac{P_{(x_{n+1}, y_n)} - P_{(x_{n-1}, y_n)}}{2\Delta x} \tag{2.2}$$

$$\frac{\partial P}{\partial y}(x_n, y_n) \approx \frac{P_{(x_n, y_{n+1})} - P_{(x_n, y_{n-1})}}{2\Delta y} \tag{2.3}$$

They can be implemented as a convolution (see *Convolution* in Section 2.7.4.3), each gradient kernel being {1/2, 0, –1/2}.

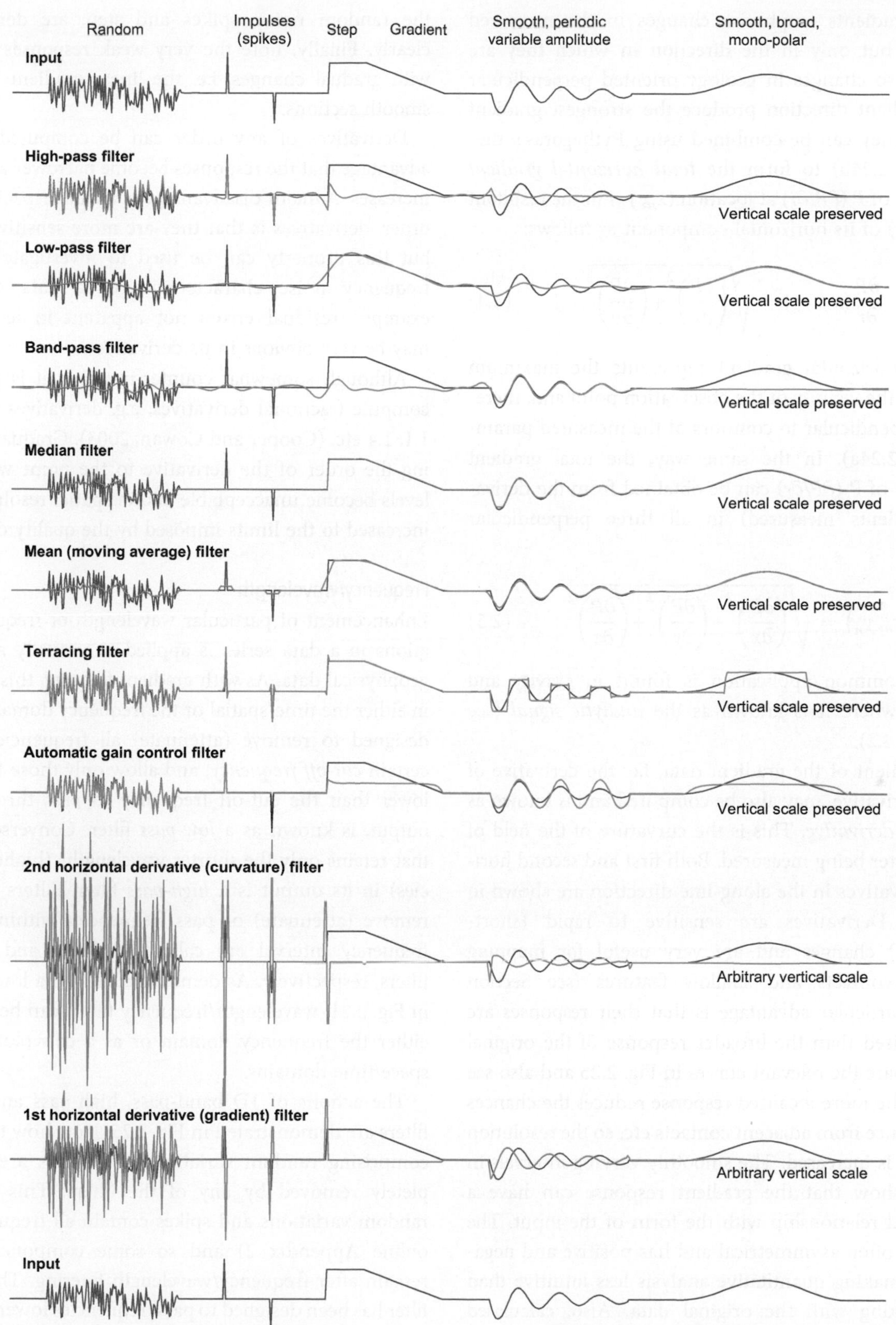

Figure 2.25 Illustration of how data with different characteristics are altered by some of the commonly used filtering operations. The data contain commonly encountered types of signal and noise. For each filter, the red line is the unfiltered input and the blue line is the filtered output.

These gradients emphasise changes in the measured parameter, but only in the direction in which they are calculated, so changes in geology oriented perpendicular to the gradient direction produce the strongest gradient response. They can be combined using Pythagoras's theorem (Fig. 2.24b) to form the *total horizontal gradient* (derivative) of P ($\partial P/\partial r$) at location (x_n, y_n) in the resultant direction (r) of its horizontal component as follows:

$$\frac{\partial P}{\partial r}(x_n, y_n) = \sqrt{\left(\frac{\partial P}{\partial x}\right)^2 + \left(\frac{\partial P}{\partial y}\right)^2} \qquad (2.4)$$

The total horizontal gradient represents the maximum gradient in the vicinity of the observation point and, therefore, is perpendicular to contours of the measured parameter (Fig. 2.24a). In the same way, the total gradient (derivative) of P ($\partial P/\partial r$) can be obtained from the derivatives (gradients measured) in all three perpendicular directions:

$$\frac{\partial P}{\partial r}(x_n, y_n) = \sqrt{\left(\frac{\partial P}{\partial x}\right)^2 + \left(\frac{\partial P}{\partial y}\right)^2 + \left(\frac{\partial P}{\partial z}\right)^2} \qquad (2.5)$$

Its most common application is found in gravity and magnetics where it is known as the *analytic signal* (see Section 3.7.4.2).

The gradient of the gradient data, i.e. the derivative of the first derivative, may also be computed and is known as the *second derivative*. This is the curvature of the field of the parameter being measured. Both first and second horizontal derivatives in the along-line direction are shown in Fig. 2.25. Derivatives are sensitive to rapid (short-wavelength) changes and are very useful for mapping geological contacts and shallow features (see Section 2.2.3). A particular advantage is that their responses are more localised than the broader response of the original data (compare the relevant curves in Fig. 2.25 and also see Fig. 2.3). The more localised response reduces the chances of interference from adjacent contacts etc. so the resolution of the data is increased. The smoothly varying intervals in Fig. 2.25 show that the gradient response can have a complicated relationship with the form of the input. The gradient is often asymmetrical and has positive and negative parts making quantitative analysis less intuitive than when working with the original data. Also, calculated derivatives are very sensitive to noise, especially the Y-derivatives as they are measured across-line where the samples are more widely spaced. Amplification of the abruptly changing sections of the test dataset, such as the random noise, spikes and step, are demonstrated clearly. Finally, note the very weak responses associated with gradual changes, i.e. the linear gradient and broad smooth sections.

Derivatives of any order can be computed, with the advantage that the responses become narrower as the order increases. A major disadvantage of all, but especially higher order, derivatives is that they are more sensitive to noise; but this property can be used to investigate the high-frequency noise characteristics of a data series. For example, residual errors not apparent in levelled data may be very obvious in its derivatives.

Although somewhat counter-intuitive, it is possible to compute fractional derivatives, e.g. derivatives with order 1.1, 1.4 etc. (Cooper and Cowan, 2003). Gradually increasing the order of the derivative to the point where noise levels become unacceptable allows spatial resolution to be increased to the limits imposed by the quality of the data.

Frequency/wavelength

Enhancement of particular wavelength or frequency variations in a data series is applied to virtually all types of geophysical data. As with gradient filtering, this is possible in either the time/spatial or the frequency domains. A filter designed to remove (attenuate) all frequencies above a certain *cut-off frequency*, and allow only those frequencies lower than the cut-off frequency to pass through to its output, is known as a *low-pass* filter. Conversely, a filter that retains only the shorter wavelengths (higher frequencies) in its output is a *high-pass* filter. Filters that either remove (attenuate) or pass frequencies within a defined frequency interval are called *band-stop* and *band-pass* filters, respectively. As demonstrated with a low-pass filter in Fig. 2.21, wavelength/frequency filters can be enacted in either the frequency domain or as a convolution in the space/time domains.

The actions of 1D band-pass, high-pass and low-pass filters are demonstrated in Fig. 2.25. Note how the sections comprising random variations and spikes are not completely removed by any of the filters. This is because random variations and spikes contain all frequencies (see online Appendix 2) and so some components always remain after frequency/wavelength filtering. The low-pass filter has been designed to pass frequencies lower than those comprising the central sinusoidal section of the dataset, and the cut-off frequency of the high-pass filter is greater than the highest frequency in this section. The band-pass filter passes frequencies between the cut-off frequencies of the

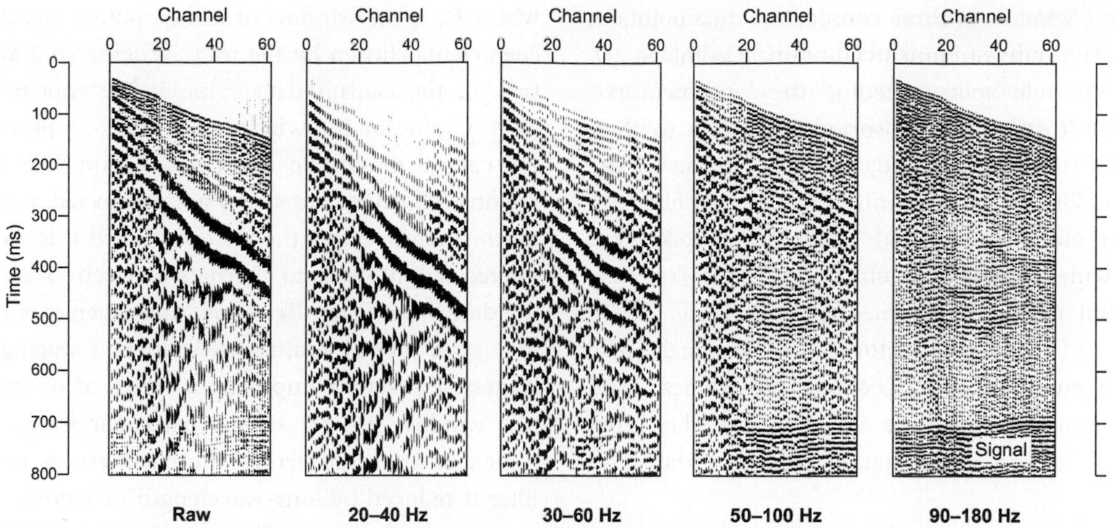

Figure 2.26 Seismic data filtered to pass different frequency bands. Redrawn, with permission, from Juhlin and Palm (2003).

two filters. Consequently, the central sinusoidal section is unaffected by the band-pass filter, but is attenuated by both the high- and low-pass filters. The smooth broad section, being of longer wavelength/low frequency, is unaffected by the low-pass filter, but removed (attenuated) by the high-pass filter. The same is true for the linear gradient, which has a comparatively long wavelength (low frequency), although the abrupt changes are smoothed out.

One-dimensional frequency filtering of seismic data is shown in Fig. 2.26 where the distinct difference in frequency between the signal and noise allows the latter to be largely removed from the data. The signal, barely visible in the unfiltered data, contains higher frequencies (90–180 Hz) than the noise. Therefore, the signal can be preserved and the noise attenuated with a band-pass filter, even though the noise has much higher amplitude than the signal. The results of low-pass filtering of magnetic data are shown in Fig. 4.25d.

Terracing/blocking

Terracing modifies the data so that the variations have a more rectilinear form, i.e. more like a stepped function (Cordell and McCafferty, 1989). Using a topographic analogy, a terrain comprising smooth hills and valleys is converted to one of plains and plateaux with intervening cliffs. The terracing operation is demonstrated in Fig. 2.25. The most obvious result of the filter is that the smoothly varying sections are converted to a step-like form.

Terracing of gridded data produces a result that is akin to a geology map, in that homogenous regions are defined with sharp boundaries between them. The other major application is to downhole logs, where the blocked appearance helps identify lithological boundaries (Lanning and Johnson, 1983).

Smoothing

Smoothing is used to reduce the shorter-wavelength variations in a dataset. There are various forms of smoothing filter (Hall, 2007), of which the low-pass frequency filter (see *Frequency/wavelength* above) is one example. The simplest smoothing filter involves applying a running average (mean) window of the data (see *Convolution* in Section 2.7.4.3 above). Figure 2.25 shows how smoothing using a running-average filter alters the rapidly varying parts of the input data, such as the random variation, the spikes and the abrupt change or step. Note how the spikes are reduced in amplitude and broadened over a distance equal to the width of the filter window. They cannot be completely removed with filters based on window-averaging. There are also some subtle changes to the smoothly varying areas, notably a broadening and reduction in amplitude of the smooth sinusoidal sections. For a filter to be effective in removing a higher frequency component from a data series, the component being removed must be properly defined in the original data series, i.e. the sampling interval needs to satisfy the conditions of the sampling theorem for the higher frequency component being removed (Section 2.6.1).

Another commonly used smoothing filter is the median filter (Evans, 1982; Wessel, 1998). Instead of calculating a mean value from the window, the median value of the windowed data points is used. The median filter is most commonly used to remove impulses or spikes (see Section

2.7.1.1) from a window of three consecutive data points in 1D, or a square window of nine neighbouring points in 2D. The high-amplitude spikes become the 'end-members' when the windowed data are sorted according to their values (to find the median) and, therefore, are never the filter's output. Unlike the mean filter, the median filter can completely remove an isolated single-point spike; but removal of wider 'spikes' and closely spaced spikes may require several passes of the median filter (Hall, 2007), which can lead to unwanted distortion of the signal. Spike removal is a common pre-processing requirement, and they can be removed with little effect on the other types of variations in the input data, as demonstrated in Fig. 2.25.

Amplitude scaling

When a data series exhibit a large amplitude range (as commonly occurs with ratios, gradient data and also electrical property data) or when low-amplitude signals are of interest, it is sometimes helpful to rescale the data. The basic idea is to amplify small-amplitude features and attenuate large amplitudes based on the amplitude of the values and/or the wavelength of a window of the data. To some extent the problem can be addressed during its display (see Section 2.8), the simplest and often highly effective way of amplitude scaling being to work with the logarithm of the amplitude of the parameter P (Morris *et al.*, 2001), which attenuates larger amplitudes more than the lower ones. The square root of the value ($\sqrt{P}$) is a highly effective scaling for magnetic derivatives (Mudge, 1991) and ratios; values less than 1 are amplified and those larger are attenuated. Conversely, squaring the value (P^2) amplifies large values more than smaller values and is often applied to low-count radiometric data (see Section 4.5.3).

Amplitude scaling using automatic gain control (AGC) is routinely applied to seismic data, and increasingly to other kinds of geophysical data, especially magnetic data. An AGC based on both amplitude and wavelength is described by Rajagopalan and Milligan (1995). This involves averaging a window of data points and replacing the centre value of the window (P_{central}) with the rescaled value (P_{AGC}). Varying the window length changes the wavelength dependency of the AGC. A commonly used AGC filter is given by:

$$P_{\text{AGC}} = P_{\text{central}} \Big/ \left[\frac{1}{n}\left(\sum_{i=1}^{n} P_n^p\right)\right]^r \tag{2.6}$$

where P_n is the window of n data points and p and r are coefficients chosen by the user. When $p = 2$ and $r = 0.5$ ($=1/p$), the central data value is the root-mean-square (RMS) value of the window. In lower-amplitude regions the central amplitude is increased, and vice versa; 'flat' regions (all values the same) are unaffected. Care must be taken when selecting the window size; if it is too small the filtered output tends to alternate between −1 and +1 creating data with a step-like form, and when it is too large it may span several anomalous features, reducing its effectiveness. For data comprising a range of wavelengths the best window size for scaling particular features is determined by trial and error. The effectiveness of this AGC filter is reduced by long-wavelength variations in the data, so high-pass frequency filtering prior to applying the AGC filter is often desirable.

The equalising of amplitudes using wavelength-dependent AGC is seen in the sinusoidal variations and the smoothly varying gradient in Fig. 2.25. Note how the random variation and the spikes are largely unaffected by the filter.

Trend

Directional filters respond to the orientation of elongate features and directional trends in the data and therefore are only applicable to 2D data.

A common application for directional filtering is the suppression of coherent noise from seismic data. Another application is the removal of linear spatial features: examples include responses from dykes that may conceal features in the country rock, and the corrugations caused by variable survey height in airborne geophysical datasets (Minty, 1991). Figure 2.27 illustrates directional filtering of aeromagnetic data. The original dataset contains numerous approximately parallel dykes. The linear anomalies associated with the west-northwest trending dykes are greatly attenuated in the filtered data.

Directional filters can also be used outside the spatial domain. For example, seismic reflection data are not displayed as a conventional map with dimensional axes, e.g. easting and northing, but as a group of time-series (traces) plotted side by side in accordance with their relative recording positions along a survey traverse (see Section 6.4.3.2). Different types of seismic waves with different frequencies and velocities cause responses with different trends to appear across the data plotted in this form. Trend filtering based on time-frequency (f) in one direction and spatial frequency (k) in the other (see online Appendix 2) can be

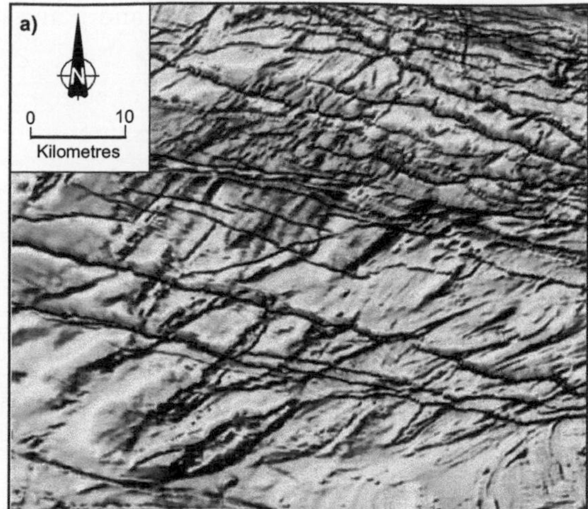

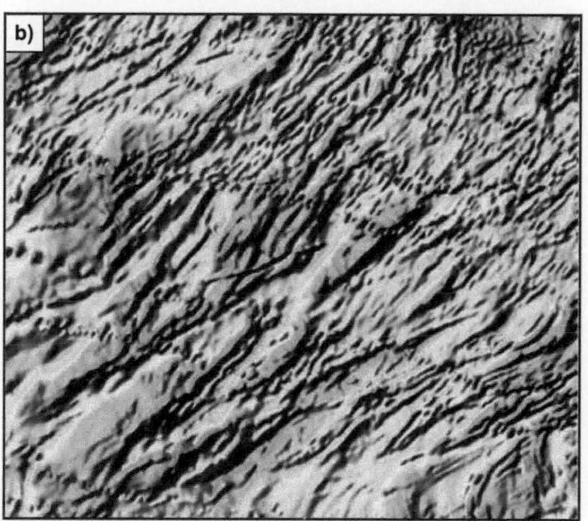

Figure 2.27 Directional filtering of aeromagnetic data to remove linear anomalies associated with dykes. (a) Raw data, (b) filtered data. Original figure based on data supplied courtesy of the Geological Survey of Botswana.

applied and, mathematically, the filtering process is identical to a 'normal' directional filter. Known as *f–k filtering*, and also known as *dip, fan, pie-slice* and *velocity filtering*, it too is based on the 2D Fourier transform. This filtering is a powerful tool for removing noise from seismic data.

Deconvolution

The action of a previous filtering operation can be reversed through a process known as *deconvolution*, also known as *inverse filtering* (Fig. 2.20b). The origin of the term stems from the fact that a filter is defined in the time or spatial domain in terms of a convolution (see *Convolution* in Section 2.7.4.3 above). Confusingly, deconvolution is also a filtering operation, and so can be enacted as a convolution. Less formally, the term deconvolution may be used to mean simplifying or 'unravelling' a complicated response reducing it to a more interpretable form.

The filtering operation to be reversed by deconvolution is normally associated with some undesired aspect of a signal – often modification by its passage through the subsurface, or its imperfect representation by the geophysical detector and recording system. Removing unwanted aspects of a filter's response requires knowledge of the filter's characteristics. This is a realistic expectation when the problem is associated with the data acquisition equipment. However, when the problem is due to the Earth filter it is usually not possible to describe the filter characteristics completely, since again the geophysical paradox applies: the optimal understanding of the geophysical response requires the very knowledge the geophysical response is intended to provide. Deconvolution is very commonly used in the processing of seismic data and is described further in Section 6.5.2.2.

2.7.4.5 Example of enhancing geophysical data

In order to illustrate the enhancement of geophysical data using the techniques described above, an example of removing noise from airborne EM data (see Section 5.9.1) to resolve the signal is presented here.

The EM data are described by Leggatt *et al.* (2000). The signal is a (1D) time series comprising a decaying voltage which is repeating 10 times, with alternate polarity, over a period of 200 ms. The data are sampled every 0.025 ms, resulting in each decay cycle being defined by 768 data points (Fig. 2.28). Note that for clarity only 7 of the 10 decays are shown in the figure.

The unprocessed (raw) data are contaminated by an approximately sinusoidal variation caused by turbulence of the survey aircraft affecting the position and orientation of the detector system, and short-wavelength 'spikes' caused by electrical activity in the atmosphere. The aim is to remove these two kinds of noise from the data. The relatively long-wavelength sinusoidal noise was removed first, with a high-pass filter (25 Hz cut-off), and then the spikes were removed. This required a combination of median filters and stacking of adjacent readings as follows. Firstly, the negative decays were inverted to obtain consistency in polarity. Secondly, the median values of equivalent samples in the ten decays were obtained. The six samples furthest from the median were discarded and the mean obtained from the remaining four. The mean values were used to construct the 'noise-free' decay, which is reproduced (seven times) at the bottom of Fig. 2.28.

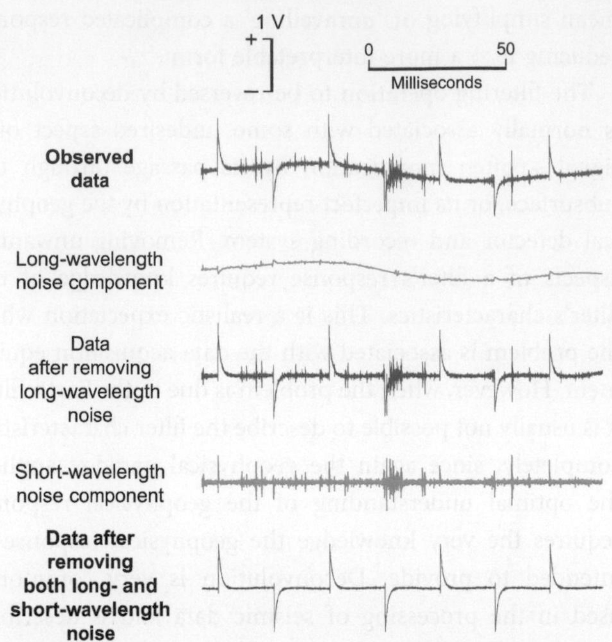

Figure 2.28 Removal of noise from SPECTREM airborne electromagnetic data to resolve the transient decays using a variety of filtering steps. See text for details. Based on diagrams in Leggatt et al. (2000).

The enhanced data are obtained from a series of filtering operations, each treating the data in a different way. This is a common strategy when working with geophysical datasets. Some other common combinations of filters are gradient filtering of gravity and magnetic data followed by low-pass filtering to remove the short-wavelength noise enhanced by the gradient computation, and migration of seismic data (see Section 6.5.2.5) again followed by low-pass filtering to remove the high-frequency noise that may result.

2.8 Data display

A variety of techniques are available for displaying 1D and 2D geophysical data, with the different forms of display emphasising different characteristics of the data. The choice of display should be made according to the objective of the interpretation, in particular targeting versus geological mapping. The quality of the data is also important, as some forms of display can be an effective means of suppressing noise.

2.8.1 Types of data presentation

The simplest form of data presentation is a 1D profile plot displaying the variation of a parameter as a function of distance along a survey traverse, or as a function of time or frequency for a time- or frequency-based parameter,

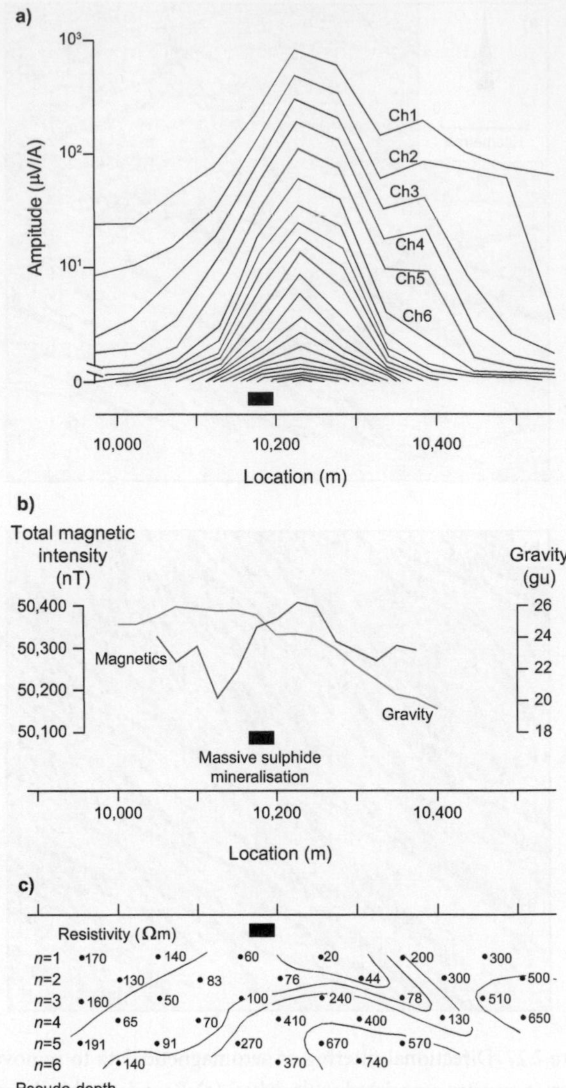

Figure 2.29 Various profile plots of geophysical data from the Thalanga massive Zn–Pb–Cu–Ag sulphide deposit, Charters Towers, Queensland, Australia. (a) Electromagnetic data. The various curves represent the multichannel measurements of the electromagnetic response at increasing times (channels) after transmitter turn-off. (b) Gravity and magnetic data. (c) Electrical resistivity pseudosection. The vertical scale n represents the separation between the source and receiver located at the surface (see Section 5.6.6.3) and is an indication of the depth influencing the measurement. Based on diagrams in Irvine et al. (1985).

respectively. Vertical scales are typically linear or logarithmic, the latter providing amplitude scaling (see *Amplitude scaling* in Section 2.7.4.4). Often several types of data are available at each survey location; these may be multi-parameter measurements or data from different surveys, and they are plotted together to allow integrated interpretation. Profiles are a simple, accurate and effective way to display 1D data (Figs. 2.29a and b). The most common presentation of 2D datasets, when both the spatial coordinates are ground locations, is a map. The various types of map displays are described in Section 2.8.1.1.

Sometimes one of the spatial coordinates in a 2D dataset is the estimated depth at which the response originates, so the presentation comprises a cross-section. Often, a parameter related to the signal's depth penetration is used as a proxy for depth, e.g. time for time-series data such as seismic data (see Chapter 6); or frequency for frequency-series data (see online Appendix 4), or transmitter-to-receiver spacing (distance) for some types of electromagnetic and electrical data (see Chapter 5). Data presented in this way are called a *pseudosection*: see for example the resistivity data in Fig. 2.29c. In this case, an integer multiple (n) of the spacing between transmitter and receiver is used as a proxy depth parameter (see Section 5.6.4.1). A variable other than depth is used because of the uncertain relationship with true depth. It is emphasised that these are pseudo-depths and not true depths and that the discrepancy between the two will vary with depth and across the dataset as the physical properties of the subsurface vary, resulting in a distorted representation of the subsurface.

A more sophisticated presentation displaying depth involves modelling (see Section 2.11) the survey data to infer the subsurface distribution of the physical properties. This invokes a model of the subsurface with associated simplifying assumptions. The simplest approach is based on modelling the data from each survey station/point individually with the assumption that the subsurface physical property distribution below each station comprises a 1D model (see *One-dimensional model* in Section 2.11.1.3). Neighbouring 1D models along the survey line are merged to display a cross-section of the computed true value of the parameter as a function of depth. This is known as a *parasection*, and computed depth is used as the vertical axis, but the simplification of the subsurface in the modelling means that the depths may not be accurate. Parasections are a common form of display for electrical and electromagnetic data.

When the constraints of the 1D model are relaxed so that the physical property is allowed to vary in one lateral direction (depth and along-line direction – but constant in the across-line direction), it is known as a *2D model*, or a 3D model when the property is free to vary in all directions (depth, along-line and across-line directions) (see Section 2.11.1.3). Although more difficult to calculate than the 1D model, they are more realistic representations of the subsurface so the results are referred to as *cross-sections*.

2.8.1.1 Types of 2D display

Regardless of whether the plotting axes are geographic coordinates, or depth/pseudo-depth versus location on the survey line, the same forms of display are used (Fig. 2.30).

The simplest presentation for data comprising parallel survey lines is a series of profiles (Fig. 2.30a), one for each survey line and plotted so that each profile has the same scale and direction for the independent variable (magnetic field strength in the example). The dependent variable of the profile, i.e. location, coincides with the spatial position of the measurement. These are known as *stacked profile*, or a *stack plot*. A distinct advantage of these over other types of display is that every data point is presented, unlike gridded data where an averaged value at a larger data interval is presented. Furthermore, short-wavelength variations in the traverse direction are preserved and can potentially be correlated between adjacent traverses. The disadvantage of this form of display is that it can be confusing when the geology is highly variable, making correlations between the traverses difficult. Also, a large range in amplitude can cause adjacent profiles to overlap, obscuring low-amplitude features. Locating the actual position of features along the survey traverses can sometimes be difficult, especially where gradients are large or anomalies are superimposed on longer-wavelength features, or if the survey lines are not straight. The data may benefit from amplitude scaling (see *Amplitude scaling* in Section 2.7.4.4) and the application of some form of high-pass filter to remove unwanted longer wavelengths (see Section 2.9.2).

Stack plots are commonly used to display aeromagnetic data to reveal the responses of potentially diamondiferous kimberlite or lamproite pipes. These exploration targets are comparatively small and may occur on one traverse only, depending upon survey line spacing, resulting in a subdued expression in the gridded data (Jenke and Cowan, 1994). Stacked profiles are also essential for displaying the weak, short-wavelength, magnetic responses sometimes associated with ilmenite-rich heavy-mineral sand strandline deposits (Mudge and Teakle, 2003).

A classic form of 2D display is the *contour* plot. Here the contour lines 'trace' levels of constant amplitude across the display area. The contour lines are usually black (Fig. 2.30b), but contours colour-coded according to amplitude (or some other parameter) add information to the display that allows the eye to quickly identify and correlate areas with similar characteristics. Contour plots are especially effective for defining gradients and are the only form

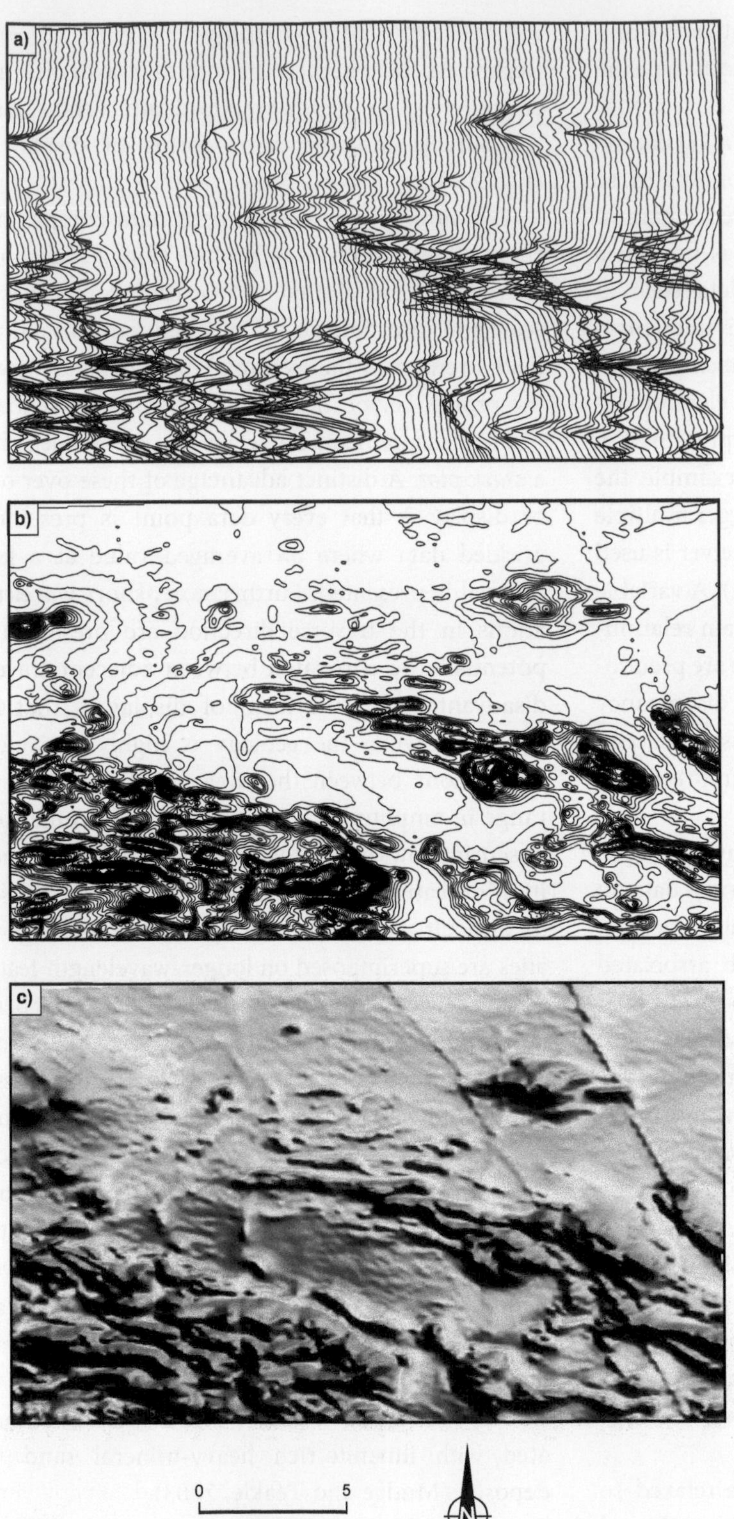

Figure 2.30 Examples of various forms of display for 2D datasets. (a) Stacked profiles, (b) contours, (c) image. Data are aeromagnetic data from the Kirkland Lake area, Ontario, Canada. Source: Ontario Geological Survey © Queen's Printer for Ontario 2014.

of 2D display having the important characteristic of accurately and conveniently depicting the actual amplitude at a point on the geophysical map. They are an effective form of display when the attitude of sources needs to be estimated (see Section 2.10.2.3).

The display of 2D and 3D data as pixelated images (Fig. 2.30c) is ubiquitous in modern geophysics. The technique is based on either varying the intensity or brightness of white light, or applying a range of colours, or applying both simultaneously, over the amplitude range of the data.

An almost infinite array of images can be created from a single dataset. Furthermore, different filtered forms of the same data, or even other types of data, can be integrated into a single image to allow more reliable analysis of the data. It is clear from the various 2D displays shown in Fig. 2.30 that an image allows easy recognition of regions having different characteristics, and also linear features. These are fundamental to the interpretation of geophysical datasets, the principal reason for their popularity. If actual amplitudes are required, or gradients are important, then these can be provided most effectively by plotting contours of amplitude onto the pixel image: for examples see Figs. 2.12, 3.15, 3.21 and 3.70a. Numerous examples of pixel images of different kinds of geophysical data are presented throughout the following chapters. We describe the details of image processing applied to the display of geophysical data in Section 2.8.2.

2.8.1.2 3D display

When data are coordinated in a 3D space and are distributed throughout that space, they are referred to as 3D data, and pseudovolumes, paravolumes or 3D models may be created. They are usually presented with a voxel-based display. The need to integrate, display and analyse many different types of 3D data in unison has led to the development of computer-based 3D visualisation systems. Typically, 3D geophysical models, surface and downhole petrophysical values and vector measurements can be integrated and displayed with many other types of 3D data, such as geochemical, geological, mining and geographical information. Polzer (2007) demonstrates the application of 3D data visualisation in nickel exploration. These systems allow the integrated 3D database to be rotated and viewed from any direction, from within the ground or from above it, and allow multiple views of the data simultaneously. In addition, visual enhancements and filters can be applied to any of the parameters contained in the database, such as data-type, feature, text values, numerical values, voxels etc., and applied to the whole database or parts of it (for example, selected by geological criteria). It is beyond our scope to investigate further this developing area of information technology; but it is likely that the use of 3D visualisation systems for presenting and working with multiple datasets will become the norm.

2.8.2 Image processing

We use the term image processing to mean enhancements applied to pixel displays of gridded data, primarily to emphasise particular features within in the data and create a more easily interpretable product. It is distinct from data enhancement where the emphasis is on numerical processing to highlight particular characteristics of the data or to improve the signal-to-noise ratio. Our description applies equally to the voxel displays used to present data volumes.

At its most fundamental level, digital image processing involves the representation of each grid (data) value as a coloured pixel. The resultant image is known as a *pseudo-colour* display, because the colours are not the true colours of the parameter displayed. The physical size of each pixel and the number of the pixels in the display depends on the hardware used to display the image. Depending on the resolution of the display device and the amount of data to be displayed, each node in the data grid is assigned to a pixel in the display. The amplitude of the data value is used to control the colour of the pixel. This raises two fundamental issues: how the display colours are selected, and how these colours are assigned to the (grid) data values.

2.8.2.1 RGB colour model

The number of different colours available, and therefore the ability to accurately represent the amplitude variations of the data, depends on both hardware and software considerations. Digital display devices use the primary colours of light: red (R), green (G) and blue (B) (Fig. 2.31a), which can be mixed to create a pixel of any colour. For example, mixing equal amounts of two of the primary colours creates the secondary colours yellow, cyan and magenta. Every colour is represented by its location in the RGB colour space which can be visualised as a Cartesian coordinate system (Fig 2.31b). In Fig. 2.31b, black occurs at the origin (R=0, G=0, B=0), with shades of grey and white plotting on the 'grey' or 'intensity axis', along which all three primary colours are mixed in equal amounts (R=G=B). White is produced by mixing the maximum amounts of the three primaries.

Typically an eight-bit hardware architecture is used to represent the amount of each primary colour in a three-colour RGB display, so each primary colour has 2^8 (256) discrete intensity levels (0 to 255) and each pixel carries 24 (3×8) bits. By mixing the three colours, up to 256^3 ($16{,}777{,}216 = 2^{24}$) different colours can be specified, referred to as 24-bit colour. Several techniques are used for assigning colours to pixels.

Hue–saturation–intensity

Colours in the RGB colour space can also be defined in terms of hue (H), saturation (S) and intensity (I)

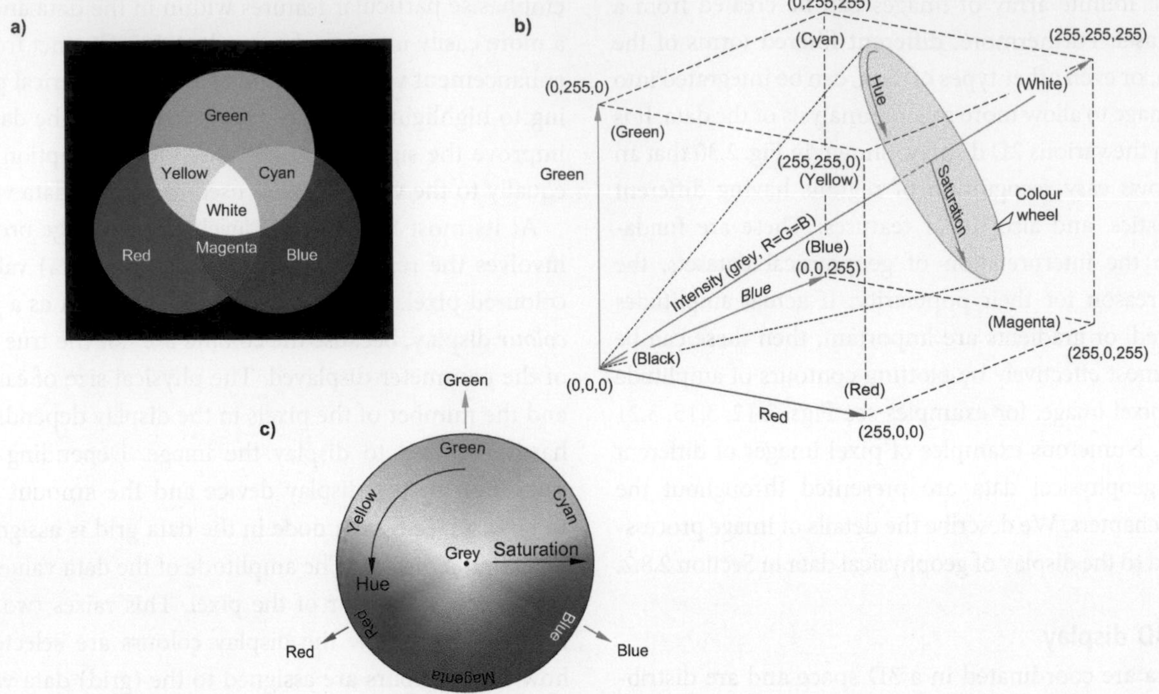

Figure 2.31 Colour models. (a) The primary and secondary colours of light. (b) The red–green–blue (RGB) and hue–saturation–intensity (HSI) colour models. Redrawn, with permission, from Drury (1987). (c) Colour wheel showing changes in tone across a circle whose axis is the intensity axis. Hue changes around the circumference, saturation varies from pure spectral colour at the circumference through pastel colours to grey at the centre, and intensity changes along the central axis.

(Fig. 2.31a). The HSI model is based on the polar coordinate system whose origin is that of the RGB colour space. Colours are represented by hue and saturation, which are defined in terms of a colour wheel centred on the intensity axis of the RGB colour model. Hue varies around the circumference with the primary and secondary colours occurring at equally spaced intervals (Fig. 2.31c). Saturation is the radial distance from the axis, so lines of equal saturation form concentric circles about the intensity axis. Zero saturation occurs on the intensity axis. Hue can be thought of as the dominant colour itself – orange, red, purple etc. – whilst saturation represents the 'whiteness' of the colour. For example, a pure red colour has high saturation, but pastel shades or lavender or pink have a lower saturation. Moving the colour wheel down the axis reduces intensity, so the colours darken but are otherwise unaltered, whilst moving up the axis produces brighter colours. Note that although a pixel's colour may be defined in terms of HSI, it has to be converted to the equivalent coordinates in the RGB space for display, since RGB is the basis on which the hardware creates the colours.

2.8.2.2 Look-up tables

The human eye can perceive many thousands of different colours; however, the enormous number of colours available in the 24-bit RGB colour model is unnecessary in practice as the human eye is unable to resolve them all. Instead, this can be reduced to 256 selected colours, as there is normally only minor visible difference between an image displayed using the full 24-bit colour spectrum and one using 256 colours selected from across the full spectrum. This also significantly reduces the computational resources needed to produce the image. The 256 colours together constitute a colour map or look-up table (LUT).

The LUT is simply a table containing the colours defined by their coordinates in the RGB colour space, with each colour being identified by its position in the table (0 to 255). Numerous LUTs are in common use, with many image processing systems allowing the user to design their own. A simple example is a grey-scale look-up table, where the colours vary from black to white via increasingly lighter shades of grey (Fig. 2.32a). In terms of the RGB colour space, the 256 grey-scale values all lie on the intensity axis.

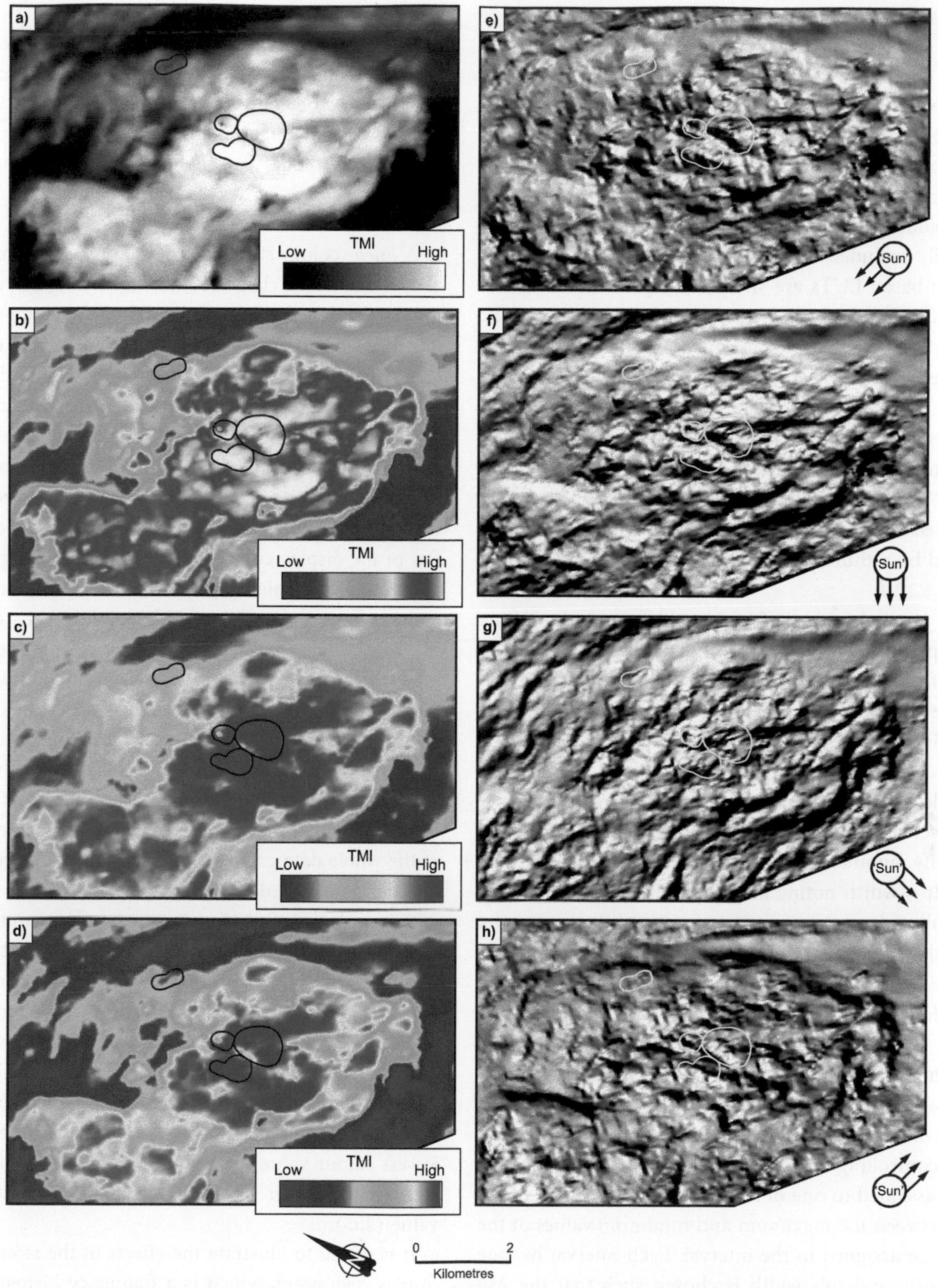

Figure 2.32 Images illustrating the use of different look-up tables applied to the same dataset, and the improvements in resolution achievable with shaded relief. (a) Grey-scale, (b)–(d) various rainbow-style displays, (e) to (h) shaded-relief grey-scale displays illuminated from different directions. Data are aeromagnetic data from the vicinity of the Mount Polley alkalic Cu–Au deposit, British Columbia, Canada. Pit outlines are shown. TMI – total magnetic intensity. Data are reproduced with the permission of the Minister of Public Works and Government Services Canada, 2006, and courtesy of Natural Resources Canada, Geological Survey of Canada.

The human eye can only differentiate about 30 shades of grey, so if more are used in the display the variation appears continuous. Grey-scale displays are useful for some types of data, but generally more detail is revealed when colour is used.

The most common and probably most effective LUTs comprise a rainbow-like spectrum of colours, i.e. the order of the colours in the visible spectrum, although not all are necessarily included. Images created with three different spectrum-based LUTs are shown in Figs. 2.32b, c and d. Lower values in the LUT are assigned to purples and dark blues; intermediate values assigned to shades of lighter blue, green and yellow; and higher values assigned to oranges, reds and finally white. This kind of colour map associates cold colours such as blues with low data values, and warm colours such as oranges and reds with high data values. When displaying data where polarity are important, such as seismic data, a colour map comprising two colours separated by a thin central band of white is very effective (see Fig. 6.12).

There is no firm basis for a particular colour scheme being superior to another; the choice really depends on the nature of the data being displayed, the nature of the interpretation and the interpreter's personal preference (and colour vision). In general, schemes producing greater variation in colour allow more subtle detail to be seen and are preferable for geological mapping. On the other hand, simpler schemes can be effective when the primary aim is to identify anomalous responses which may constitute targets. It is worth noting that the human visual system is not equally adept at seeing variations in different colours (Welland *et al*., 2006). It is worst at recognising different shades of red and blue, so LUTs comprising predominantly these colours should be avoided.

Colour stretch

Once the LUT has been selected, the 256 colours are then assigned or mapped to variations in the data. An important tool for controlling this is a frequency histogram. The data are first assigned to one of 256 class intervals. Data values falling between the maximum and minimum values of the interval are assigned to the interval. Each interval has the same width, and this width is chosen such that the 256 intervals span the entire range of the data, or at least most of the range. To reduce the influence of outlier data values the class interval width may be scaled so as to extend across a fraction of the entire data range, 95% of the range for example. The data histogram shows the number of data values assigned to each class interval (Fig. 2.33). The variations in the data are now no longer represented by their true values; instead they are represented by the class interval in which they fall, 0 representing the lowest value class interval, 255 the highest. The 256 data classes are mapped to the 256 colours comprising the colour map using a stretch function, i.e. the available colours are 'stretched' across the data histogram.

A *linear* colour stretch across the full range of 256 data classes maps each class to its corresponding colour value in the LUT, with the lowest data class (0) mapping to the lowest colour value (0) through to the highest data value (255) mapping to the highest colour value. In other words, there is a linear relationship between the data classes and the colour values. When the data histogram shows that the majority of the data fall within a comparatively small number of classes, which is often the case for geophysical data, a large proportion of the data are assigned to just a few of the display colours. The result is an image which is composed predominantly of just a few colours. In this case, a few extreme data values, the outliers, are exerting a disproportionate influence on the image producing a display which prevents variations across the full range of the data from being recognised. In other words, the data values do not make full use of the available colour values, so the display is said to lack *contrast*. This is not a problem when anomalous readings are the principal interest, as is the case for anomaly detection, but for more general interpretations of the whole data grid, e.g. for mapping, a different stretch function is required.

A simple solution for improving the contrast is to redefine the linear stretch so that it only extends over that part of the data range where most values occur. Figure 2.33a shows a linear function chosen to span the main range of variation in the data. In this case, an arbitrarily chosen data value (G) maps to colour value 209, a mid-range red. The few low-valued data points below the lower limit of the linear stretch all map to the lowest colour value (0) and, similarly, high-valued points beyond the upper limit all map to the highest colour value (255).

It is useful to illustrate the effects of the stretch using a display histogram, which is a frequency histogram of the colour values. It can be plotted next to the data histogram. Both have 256 class intervals; in the data histogram they are populated by the data values and in the display histogram by the colour values. For the case of the localised linear stretch shown in Fig. 2.33a, the display histogram is

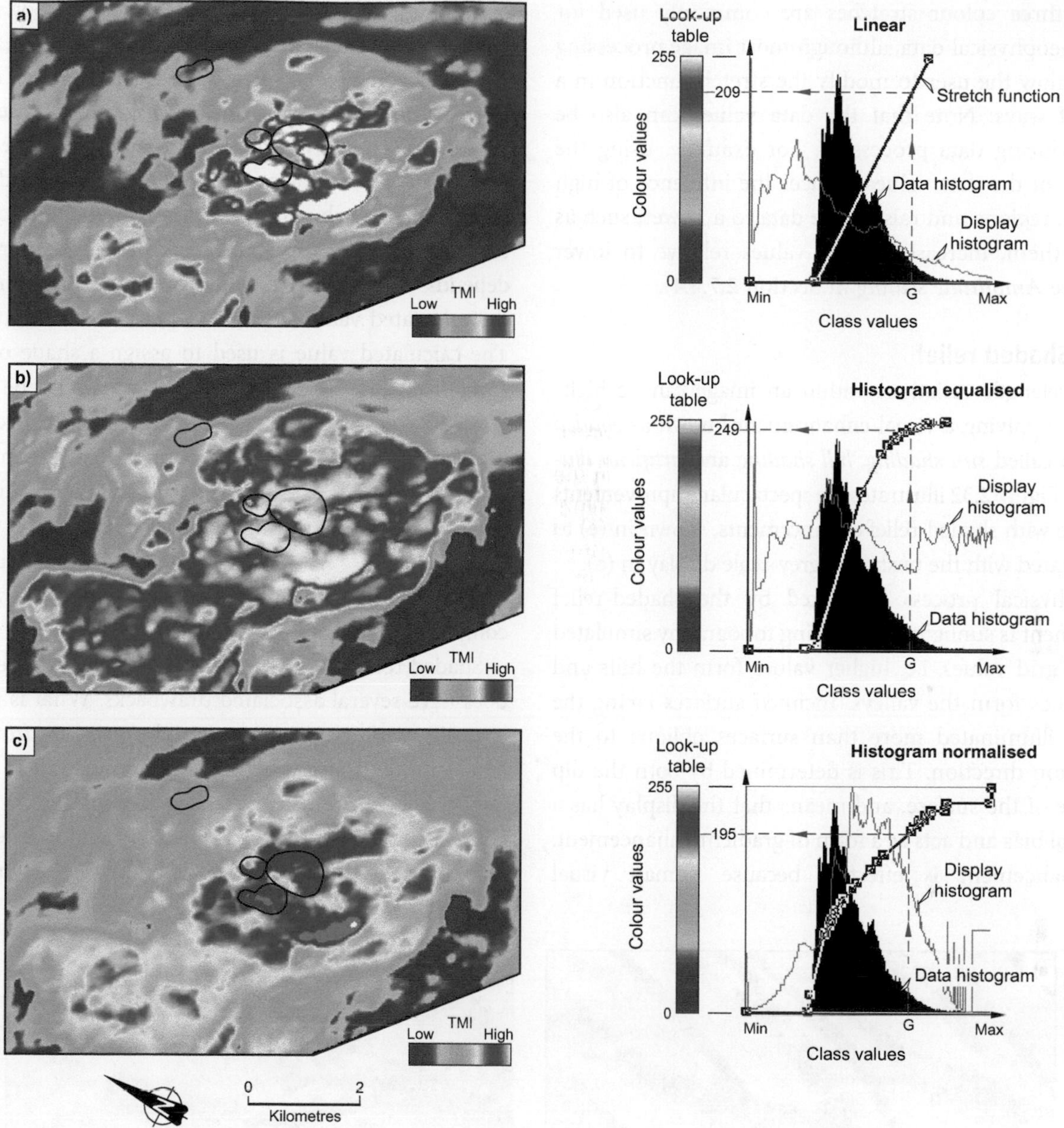

Figure 2.33 Images of the data shown in Fig. 2.32 created using the same look-up table, but with different colour stretches. (a) Linear stretch, (b) histogram equalised, (c) histogram normalised (Gaussian stretch). The frequency histogram of the data and the display histogram are shown for each image. Note how class value 'G' is mapped to different colour values with the various stretch functions. Data are aeromagnetic data from the vicinity of the Mount Polley alkalic Cu–Au deposit, British Columbia, Canada. Pit outlines are shown. Data are reproduced with the permission of the Minister of Public Works and Government Services Canada, 2006, and courtesy of Natural Resources Canada, Geological Survey of Canada.

broadened relative to the data histogram. This means there is a greater range of colours comprising the image.

Alternatively, the *histogram equalisation* or *histogram linearisation* stretch adjusts the size of the data class intervals so that the display histogram has approximately equal number of data points in each class, i.e. the data points are more uniformly distributed across the classes (Fig. 2.33b). The flattened display histogram shows that the image contains approximately equal numbers of pixels of each colour. Value G now maps to display value 249, a pale red. The *histogram normalisation* stretch adjusts the size of the data class intervals so that the display histogram resembles a normal (Gaussian) distribution (Fig. 2.33c). This increases the number of colours assigned to the central part of the data range, where usually most variation occurs. Value G now maps to colour value 195, a shade of orange.

These three colour stretches are commonly used for imaging geophysical data, although most image processing systems allow the user to modify the stretch function in a variety of ways. Note that the data values can also be adjusted during data processing. For example, using the logarithm of the data values reduces the influence of high amplitude regions and raising the data to a power, such as squaring them, increases higher values relative to lower values (see *Amplitude scaling* in Section 2.7.4.4).

2.8.2.3 Shaded relief

Short-wavelength variations within an image can be highlighted by applying a visual enhancement known as *shaded relief*, also called *sun shading, hill shading* and *artificial illumination.* Figure 2.32 illustrates the spectacular improvements obtainable with shaded-relief enhancements, shown in (e) to (h), compared with the unshaded grey-scale display in (a).

The physical process mimicked by the shaded-relief enhancement is sunlight illuminating topography simulated from the grid values, i.e. higher values form the hills and lower values form the valleys. Inclined surfaces facing the 'sun' are illuminated more than surfaces oblique to the illumination direction. This is determined by both the dip and strike of the surface, and means that the display has a directional bias and acts as a form of gradient enhancement. The enhancement is effective because human visual perception of shapes and objects relies heavily on resolving areas of light and shadow (Ramachandran, 1988).

The reflectivity of a surface in the shaded-relief enhancement can be quantified. The most common type is *Lambertian* reflectivity, where the incident light is reflected, and equally in all directions. The reflectivity depends on the orientation of each part of the 'topographic' surface relative to the position of the sun. Since the surface's orientation depends on the relative values of the neighbouring pixels, the calculated value does not depend on individual pixels. The calculated value is used to assign a shade of grey to each pixel (grey-scale displays should always be used for shaded-relief displays). A popular type of shaded relief is the 'wet-look'. It reduces the contrast between mid-range values and emphasises the brightness of highly illuminated areas, making the illuminated surface appear as if it were 'wet' or 'glossy'. As shown in Fig. 2.36b, it is particularly effective in highlighting subtle detail when used as part of a composite display (see Section 2.8.2.4).

Shaded relief is a very useful image enhancement, but it does have several associated drawbacks. What is perceived as positive and negative 'topography' can vary from person to person, and also whether the illumination is from the 'top' or the 'bottom' of the image as displayed on the page; compare Fig. 2.32h with Figs. 2.32e, f and g. The shaded-relief image is also subtly distorted because the illumination

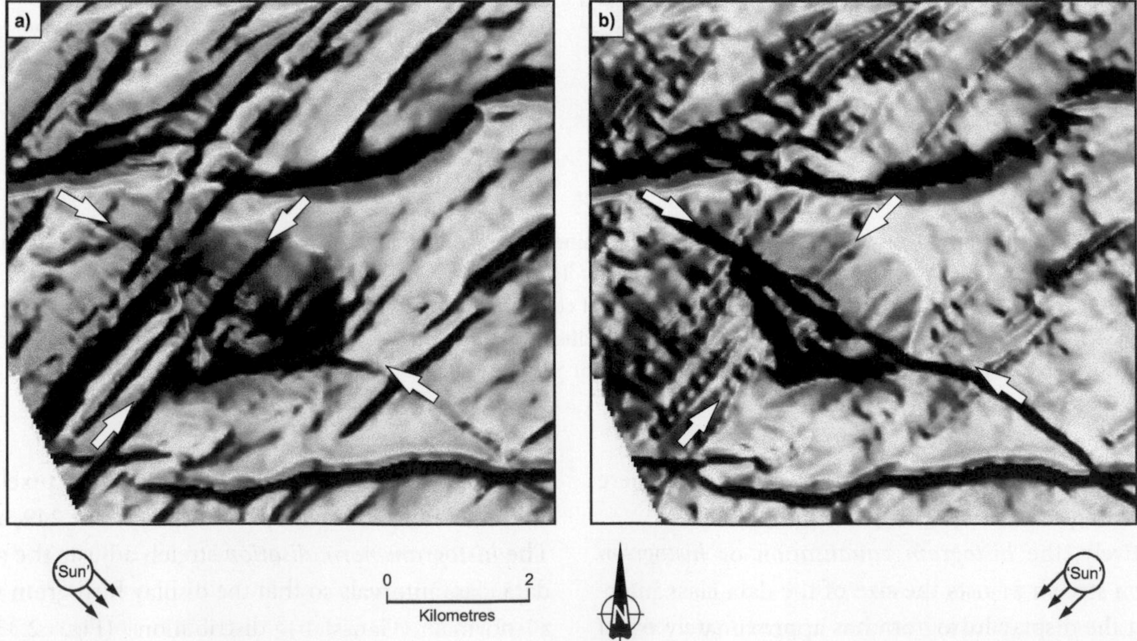

Figure 2.34 Illustration of the change in apparent cross-cutting relationships between anomalies for shaded-relief displays with different illumination directions. This is a composite image combining a pseudocolour display of amplitude with grey-scale shaded relief (see Section 2.8.2.3). Image courtesy of the Geological Survey of Western Australia, Department of Mines and Petroleum. © State of Western Australia 2013.

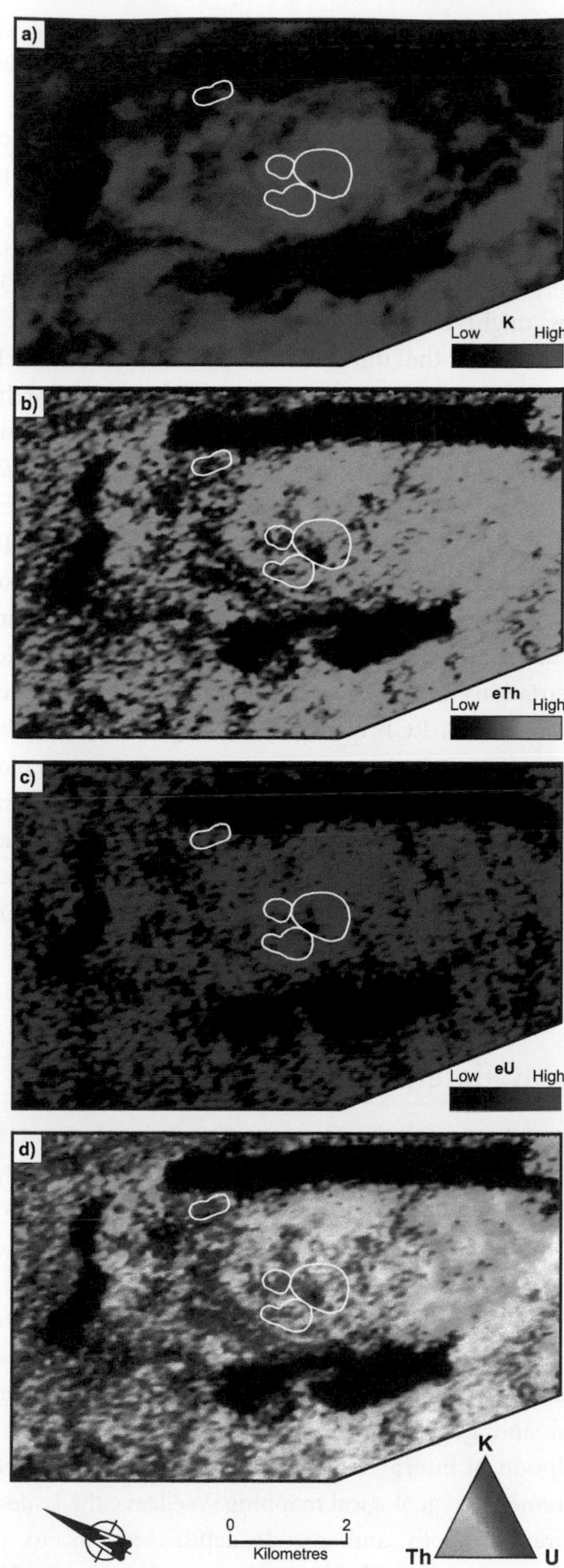

Figure 2.35 Creation of a ternary display (d) by combining different associated datasets represented by variations in (a) red, (b) green and (c) blue. The data are radioelement concentrations (see Section 4.4.7) in the vicinity of the Mount Polley alkalic Cu–Au deposit, British Columbia, Canada. Pit outlines are shown. Note that the black regions are lakes, the water masking the signal. Data are reproduced with the permission of the Minister of Public Works and Government Services Canada, 2006, and courtesy of Natural Resources Canada, Geological Survey of Canada.

position determines where topographic peaks and troughs are perceived to occur. It also has a strong orientation bias, with features orientated at a high angle to the direction of illumination being enhanced more than those trending towards the illumination direction, i.e. shaded relief is a powerful directional filter. Note the suppression of northerly trending features in Fig. 2.32f, which is illuminated from the north. The directional bias of shaded relief means it is important to vary the illumination azimuth when analysing images. It is also important to alter the elevation of the 'sun'. With the 'sun' 'lower in the sky', shadows are more pronounced and more detail is evident, but the directional bias is increased. When the 'sun' is 'too high in the sky', subtle detail may not be evident. An elevation from about 25° to 45° usually produces good results. Figure 2.34 shows a geologically important consequence of the directional bias associated with shaded-relief displays. The apparent cross-cutting relation, i.e. relative age, of the source of the intersecting linear anomalies changes with illumination direction, with anomalies striking perpendicular to the illumination direction appearing to cross-cut those that are more parallel. This is an example of how geophysical responses must be treated differently from geological data.

Most image processing systems allow the position of the 'sun' to be adjusted in real time, with the display changing accordingly. This allows the optimum illuminations to be identified, although it is stressed that at least two (approximately) perpendicular illuminations of the data should always be analysed and that different directions may be required for different parts of a large dataset.

2.8.2.4 Composite displays

It is often convenient, and sometimes necessary, to display several datasets together, allowing an integrated interpretation. Also, different presentations of the same dataset, selected to emphasis different characteristics of the data, can be combined. This can be achieved by integrating the datasets in colour space to produce a single composite image.

For the RGB model, the most common applications are the display of multichannel data such as radiometrics, electromagnetics and the different spectral bands in remote sensing data. Up to three datasets can be displayed: one dataset (or channel) is displayed as 256 variations in red, the second in green and the third in blue. This is known as a *ternary image* (Fig. 2.35) and it makes direct use of the

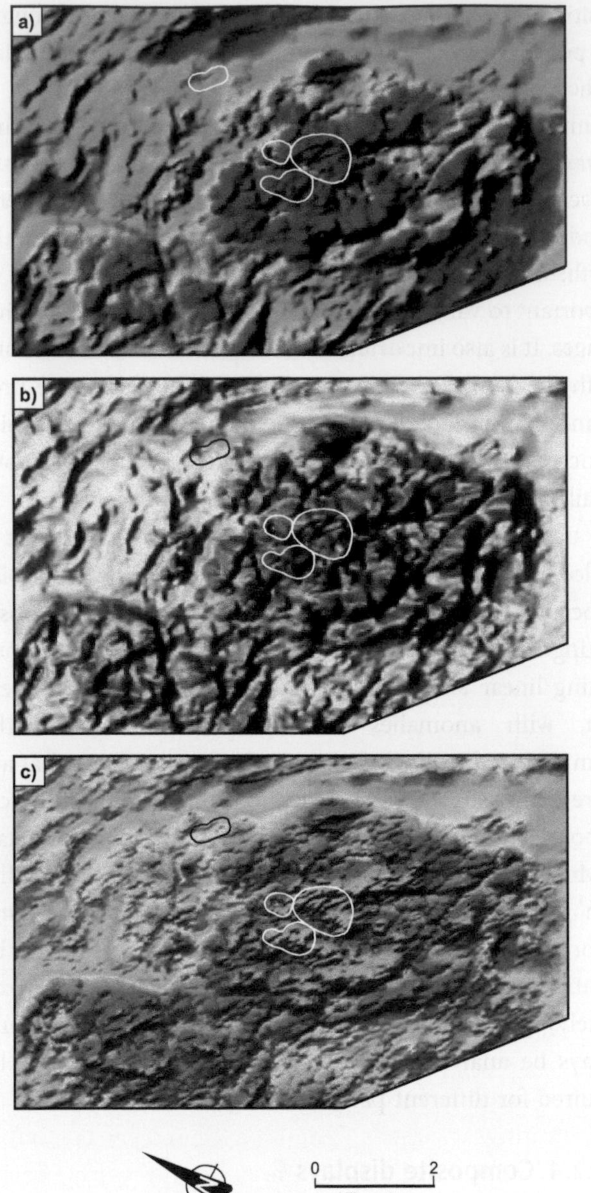

Figure 2.36 Various composite displays of the same aeromagnetic dataset. The pseudocolour emphasises amplitudes whilst the textural information is contained in the grey-scale shaded-relief component as follows: (a) grey-scale shaded relief and (b) wet-look grey-scale shaded relief. (c) High-pass filtered data in pseudocolour with grey-scale shaded relief. All images are illuminated from the north. Data are aeromagnetic data from the vicinity of the Mount Polley alkalic Cu–Au deposit, located in British Columbia, Canada. Pit outlines shown by the solid lines. Data are reproduced with the permission of the Minister of Public Works and Government Services Canada, 2006, and courtesy of Natural Resources Canada, Geological Survey of Canada.

full spectrum of 16.7 million colours. Areas where all three datasets have coincident high values appear white in the ternary image, whilst low levels of all colours appear dark. Where one colour is at low levels, the remaining two combine to create a shade close to the relevant secondary colour (Fig. 2.31a). Each component dataset can be stretched independently to improve resolution and colour balance. An example of a ternary display of radiometric data is shown in Figs. 4.26 and 4.27.

The HSI colour model allows the component datasets to control the colour parameters in a different way. One dataset might be assigned to hue (colour), another to saturation and the third to intensity. A common HSI display uses intensity variation to display shaded relief, hue to display amplitude, and the colours are usually fully saturated. This is sometimes called a *colour drape*, and several example displays combining different data processing and display methods are shown in Fig. 2.36. The simultaneous display of amplitude and textural variations (see Section 2.10.1) is an extremely powerful combination, particularly for mapping-orientated interpretations. Note that variations in any parameter of the HSI model produce a new colour in RGB colour space (Fig. 2.31b), so when working with multichannel data, changes in image colour cannot be easily related to variations in a particular data channel. This is unlike the RGB colour model where variations in the intensity of a primary colour are directly related to variations in its associated data channel, making interpretation easier.

2.9 Data interpretation – general

Interpretation begins with a qualitative analysis of the geophysical data. If the aim is simply to identify and target individual anomalies, i.e. regions with anomalous responses, then this is, in principle, a straightforward task provided the data are appropriately processed and presented (see Sections 2.7 and 2.8). If the aim is to create a pseudo-geological map, then this is usually a more demanding task, not least because it requires both geological and geophysical expertise. For this reason, our description of interpretation is focused more towards the requirements of geological mapping. We leave the issues of interpreting depth and pseudo-depth sections to our descriptions of the individual geophysical methods. Also, we focus on the interpretation of data displayed in the near ubiquitous form of pixel images.

When a target is identified or when difficulties are encountered in making a pseudo-geological map, the observed data can be analysed with the aid of the computed response of a *model*, i.e. a (simplified) numerical representation of the subsurface geology. This is a form of *quantitative analysis*. Obtaining a match between the model's geophysical response and actual observations places constraints on the geology of the subsurface, e.g. the depth of the source of an anomaly, the dip of a stratigraphic contact etc. In most cases, a match is not a definitive result, since more than one model can be made to fit the data. In other words, there is no unique arrangement of the geological elements that explains the observed geophysical data, so the interpretation is ambiguous. This fundamental and important aspect of geophysics is known as *non-uniqueness*, and is discussed in more detail in Section 2.11.4.

Before discussing the general aspects of interpretation, we consider a number of issues affecting interpretation and describe the important matter of accounting for long-wavelength variations, i.e. the regional response.

2.9.1 Interpretation fundamentals

Aside from the difficulties in reconciling the geological and geophysical perspectives of the geological environment (see Section 1.1), an interpretation must consider the following aspects of geophysical data and geophysical responses:

- Petrophysics is the link between the geophysical response and the geology. It is a complex subject. Changes in lithology, mineralogy, alteration, fabrics, weathering and structure can cause great variability in physical properties, so the available data may not be representative of the rocks in the ground. Also, some physical properties are commonly anisotropic, notably electrical conductivity, magnetic susceptibility and seismic velocity, often being different when measured parallel and perpendicular to planar fabrics such as bedding or metamorphic foliations. This will affect the geophysical response.
- The scale of resolvable features needs to be understood: in particular, the diminishing resolution of features with depth and the concept of footprints, i.e. the volume of the subsurface contributing to an individual measurement (see Section 2.6.2).
- Since a geophysical map contains responses from features at a range of depths, the interpreter needs to think in terms of three dimensions. For example, the fact that one anomaly apparently cuts across another does not mean the sources of the anomalies are actually in contact. They may be located at different depths. Shallow and deep responses need to be recognised, a subject we describe later for each geophysical method.
- There is a strong possibility of responses originating in the near-surface cover, regolith, permafrost etc., and that bedrock responses may be distorted and/or attenuated by the effects of the cover.
- Distortions caused by topography may be present. We stress the importance of integrating topographic data into the interpretation of all types of geophysical data; because for most geophysical methods, the shape of the land surface affects the geophysical measurements to produce terrain-induced responses. Nevertheless, it is important to be aware that a correlation between geophysical responses and topography is not necessarily indicative of artefacts in the dataset, since both the topography and the geophysical responses are controlled by the local geology. More likely to be artefacts in the geophysical data are responses not directly corresponding with topographic features, but occurring adjacent to them. For example, a 'moat' surrounding a hill or a linear anomaly adjacent to a ridge should be treated with scepticism. Even then the responses may be reflecting the local geology: a hill due to an intrusion may have a contact aureole, the cliff may be indicative of a fault etc.
- A map or image of the survey stations, or the survey lines and the terrain clearance for airborne surveys, should always be available for the interpretation. Integrating these with the survey data and its various transforms can reveal correlations with responses, possibly indicating artefacts in the data caused by the station distribution itself, or by variations in survey height. Features occurring in areas of sparse data or orientated parallel to the survey lines should be carefully analysed to ensure they are not survey artefacts.
- The survey line direction and spacing have effects on the resolution of particular geological features.
- Geophysical responses may have a complicated form which is not indicative of their source geometry and may be offset from their source location or extend beyond the edges of their source.
- Artefacts caused by data processing and display may be present. Levelling errors may introduce artefacts trending parallel to survey lines. Gridding may introduce beading and bulls-eyes (see Section 2.7.2.3). Apparent 'cross-cutting relationships' are not necessarily indicative of relative ages as gridding may result in datasets where higher-amplitude anomalies appear to cross-cut

low-amplitude ones. Shaded relief also creates spurious apparent cross-cutting relationships (Fig. 2.34).

- Information about infrastructure that might produce artefacts in the geophysical data, e.g. powerlines, pipelines, railways, buildings or open pits, needs to be available. This can usually be obtained from cadastral maps and aerial photography. Responses from these features reflect their form: powerlines create linear artefacts, buildings localised responses etc.
- When several kinds of geophysical data are being interpreted together, which is generally a good strategy, it is very important to consider the geological controls on their individual responses. For example, a geological feature may have a wide gravity response, but only a particular region of the feature may contain minerals that produce a magnetic response, electrically conductive minerals may be confined to other parts of the body, and radioactivity would only be observed where radioactive minerals are exposed at the surface. Often there is partial correspondence between gravity and magnetic datasets, both primarily reflecting bedrock lithological variations. However, there is no requirement for their responses to correspond exactly with, say, the distributions of radioelements in the near-surface as derived from radiometric data, which are much more likely to correlate with spectral remote sensing data. Sometimes the different depths 'probed' by the different geophysical methods cause lateral offsets between the responses of the different data types if the source is dipping. For example, electrical measurements may respond well to the conductive (weathered) near-surface geology, whilst gravity and magnetic data may contain responses from the unweathered deeper regions.

To create an interpretation which accounts for the above requires knowledge that is method specific, together with a good understanding of the nature of geophysical responses, and also how the data were acquired and processed.

2.9.2 Removing the regional response

The measured variations in the geophysical parameter of interest almost invariably consist of a series of superimposed responses from geologic features of different dimensions and/or depths in the vicinity of the measurement. A common manifestation of this is the interference (see online Appendix 2) between an anomaly of interest and the longer-wavelength variations associated with deeper and/or larger features of less interest, i.e. the background or regional response. The regional response is a form of geological environmental noise (see Section 2.4). The target anomaly and the regional field can be separated in a process known as *regional removal* (Fig. 2.37). The variation remaining after the regional has been removed is known as the *residual response*. This is the shorter-wavelength variation correlating with the shallower geology, and is usually the response of interest. In Fig. 2.37 it is the response originating from the mineralised environment. Removing the regional response is important for quantitative analysis (see Section 2.11) and can also greatly facilitate qualitative interpretation (see Section 2.10), as shown in Section 3.11.1.

As with any consideration of signal and noise, the variations of interest depend on the interpreter's requirements of the data. Consider the hypothetical example of a massive nickel sulphide deposit located at the contact between mafic and ultramafic rocks in a greenstone belt, e.g. the komatiitic peridotite-hosted deposits that occur in the Kambalda area in Western Australia (Fig. 2.38). In the early stages of exploration, prospective contacts may be the target being sought. In this case, the contact response is a residual (signal) superimposed on the regional (noise) response of, say, the base of the greenstone succession. When the contact has been located and exploration focuses along it, responses from nickel sulphide mineralisation become the target. The contact response is now the regional (noise), and the response of the mineralisation is the residual (signal).

Ideally, the responses due to the deeper and/or larger geological features are computed (modelled) and subtracted from the data. However, accurately defining the regional response requires detailed knowledge of the local geology, which depends on the geological interpretation of the area (an example of the geophysical paradox; see Section 1.3). Non-uniqueness (see Section 2.11.4) also contributes to the problem because interference between adjacent local anomalies may produce a longer-wavelength composite response which could be mistaken for part of the regional response. Also, the local and the regional responses must both be properly defined in the original data series, i.e. the sampling interval needs to satisfy the conditions of the sampling theorem for the higher-frequency local response (Section 2.6.1), and the data series must extend far enough in distance to define the longer-wavelength regional response. Unfortunately, then, the regional variation cannot be accurately calculated or

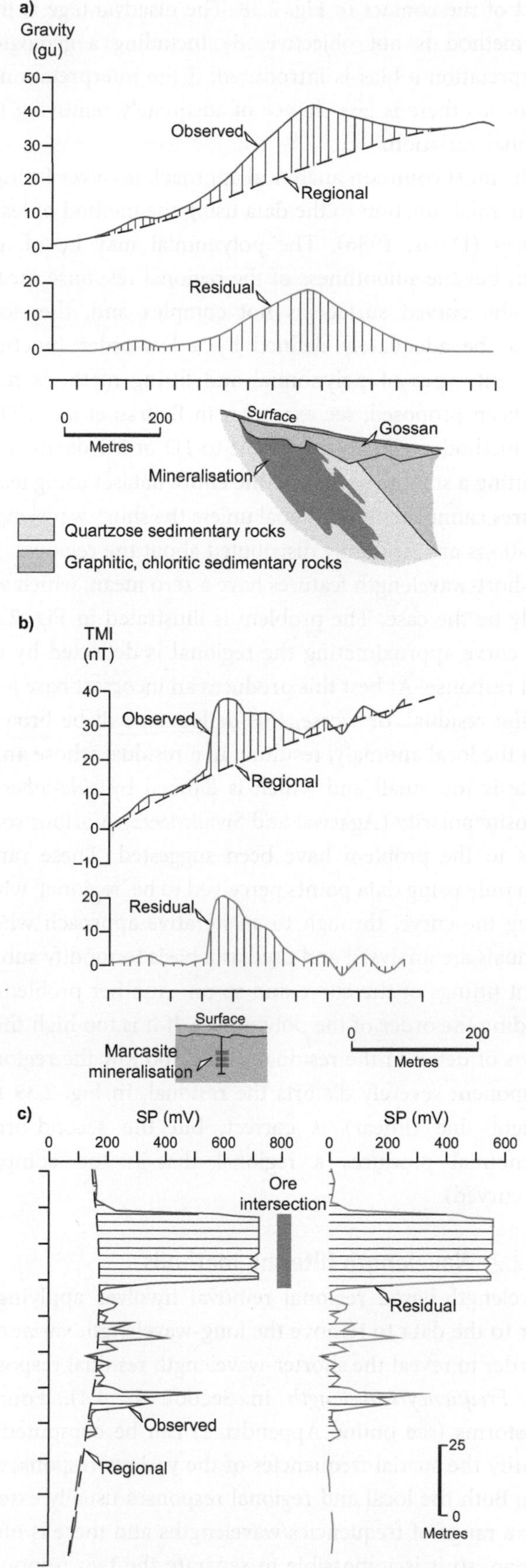

modelled so it cannot be properly removed but, nevertheless, the requirement for removal remains. There exists an extensive literature on regional removal, often referred to as *'regional–residual separation'*, and various techniques are used. Hearst and Morris (2001) show how subjective regional removal can give rise to conflicting interpretations of gravity data from the well-known Sudbury structure in Ontario, Canada.

The two most common approaches to removing the regional response involve either using a smooth mathematical function to approximate the regional or filtering the data to remove the longer-wavelength variations (see *Frequency/wavelength* in Section 2.7.4.4). Neither approach is universally applicable, notwithstanding the fact that determining whether a particular approach is 'correct' is a purely subjective decision. It is often the case that the process needs to be repeated with different interpolation/filter parameters until an 'acceptable' residual response is obtained. Ultimately, the task for the interpreter then is to find a residual response, from an infinite number of possibilities, which is the most geologically plausible. The target anomaly will inevitably be distorted in some way.

We describe here some aspects of regional removal, emphasising the practicalities of the process. At this stage it is worth saying that avoiding the problem altogether may, in some situations, be the best approach. Sometimes it may be better to account for all scales of variation in the observed data, even if this necessitates making gross assumptions about the deep geology of the area, rather than remove some arbitrary regional component. At least, in this case, the assumptions about the local geology are known to the interpreter. On the other hand, not to remove some form of the regional variation can severely limit the interpretation of important local responses.

2.9.2.1 Interpolation methods

Surface and curve fitting techniques involve interpolating the broader regional response into areas dominated by the

Figure 2.37 Examples of regional removal. (a) Gravity data and geology from the Murray Brook massive Cu–Pb–Zn sulphide deposit, New Brunswick, Canada. Redrawn, with permission, from Thomas (1999). (b) Magnetic data across a Mississippi Valley-type Pb–Zn deposit, Pine Point area, Northwest Territories, Canada. Redrawn, with permission, from Lajoie and Klein (1979). (c) Downhole self-potential data from a drillhole intersecting the Joma pyrite deposit, Trøndelag, Norway. Redrawn, with permission, from Skianis and Papadopoulis (1993).

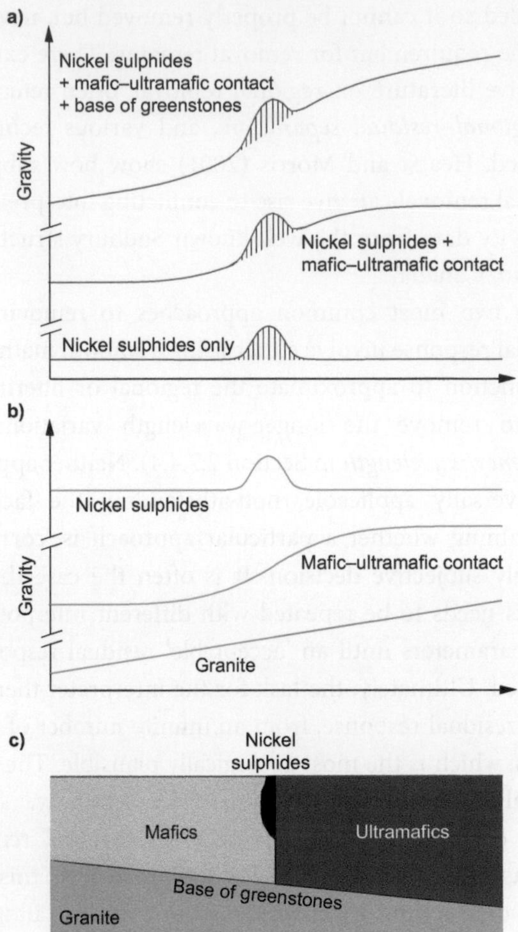

Figure 2.38 A model showing how the various components of the geology contribute to the overall geophysical response; represented with gravity data. A hypothetical massive nickel sulphide body occurs at a mafic–ultramafic contact in a greenstone belt overlying lower density granitic basement; see geological section (c). (a) The different gravity responses of the mineralisation with various component responses of the surrounding geology included. (b) The three component gravity responses producing the resultant measured response.

shorter-wavelength variations. A smooth mathematical function is used to describe the regional variation, a curve for 1D data and a surface for 2D data, and the computed curve/surface is subtracted from the observed data. Often the regional variation over a small area can be adequately represented by a straight line (1D) or a sloping plane (2D).

In its simplest form, the interpolation can be done by manually estimating the form of the regional; the process is given the rather grand name of *graphical*. The basic idea is to extrapolate the field from areas of the data perceived to be free of shorter-wavelength responses into the area containing the local response of interest. The great advantage of the graphical approach is that the effects of the known geological features can be factored in, for example the effect of the contact in Fig. 2.38. The disadvantage is that the method is not objective. By including a geological interpretation a bias is introduced; if the interpretation is erroneous there is less chance of adequately removing the regional variation.

The most common analytical approach involves fitting a polynomial function to the data using the method of least-squares (Davis, 1986). The polynomial may be of any order, but the smoothness of the regional response means that the curved surface is not complex and, therefore, should be adequately defined by a low-order function. Different types of polynomial and fitting methods have also been proposed; see examples in Beltrão *et al.* (1991). The method is equally applicable to 1D or 2D datasets.

Fitting a smooth surface to the entire dataset using least-squares cannot achieve its goal unless the short-wavelength variations are randomly distributed about the regional, i.e. the short-wavelength features have a zero mean, which will rarely be the case. The problem is illustrated in Fig. 2.39. The curve approximating the regional is deflected by the local response. At best this produces an incorrect base level for the residual, or worse, this deflection will be broader than the local anomaly, resulting in a residual whose amplitude is too small and which is flanked by *side-lobes* of opposite polarity (Agarwal and Sivaji, 1992). Various solutions to the problem have been suggested. These range from only using data points perceived to be 'regional' when fitting the curve, through to an iterative approach where residuals are analysed and used as a basis to modify subsequent fittings of the curve and so on. Another problem is deciding the order of the polynomial. If it is too high there is loss of detail in the residual, and if too low the regional component severely distorts the residual. In Fig. 2.39 the straight line (linear) is correct, but the second-order polynomial produces a regional that is too complex (i.e. curved).

2.9.2.2 Wavelength filtering methods

Wavelength-based regional removal involves applying a filter to the data to remove the long-wavelength variations in order to reveal the shorter-wavelength residual response (see *Frequency/wavelength* in Section 2.7.4.4). Fourier transforms (see online Appendix 2) can be computed to identify the spatial frequencies of the various responses in data. Both the local and regional responses usually extend over a range of frequencies/wavelengths and there is often overlap, so it is impossible to separate the two responses completely. The wavelength filter in Fig. 2.39 produces a

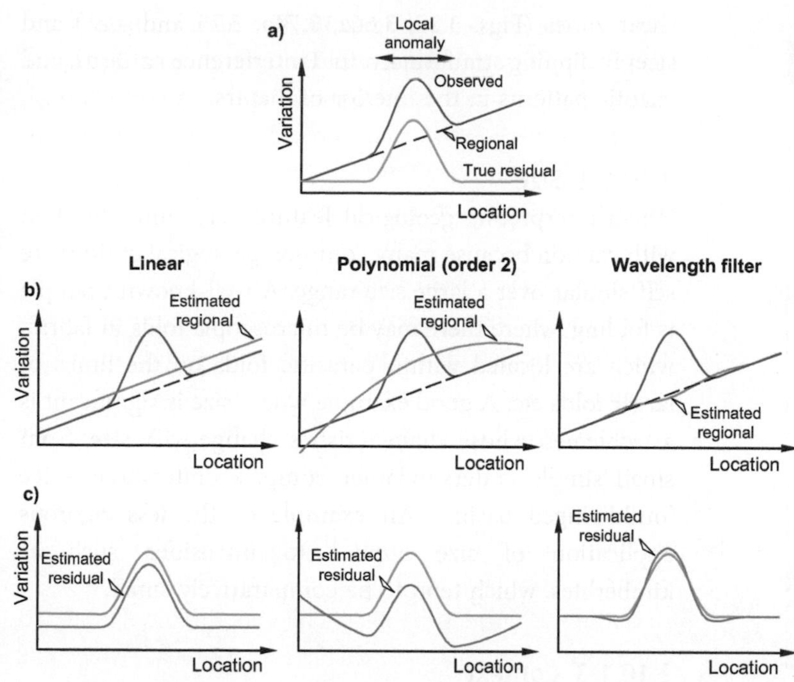

Figure 2.39 Some common consequences of incorrectly estimating the regional response. (a) Synthetic example of observed data comprising a local anomaly (residual) superimposed on a linear regional response. (b) Estimated regional fields. (c) The incorrect estimated residual anomalies resulting from the various regional responses shown in (b). See text for details. Based on a diagram in Leaman (1994).

residual with the correct base level and showing the local anomaly with approximately the correct overall shape, but the responses are distorted owing to the overlap in their frequency content and the imperfect nature of the filter. Cowan and Cowan (1993) describe various wavelength-based filtering methods for obtaining the residual component of aeromagnetic data.

2.10 Data interpretation – qualitative analysis

The qualitative analysis of geophysical data seeks to produce an equivalent (pseudo-)geological map of the survey area. The analysis is based on a map of variations in some geophysical parameter, so a direct correspondence with lithology cannot be expected everywhere. The term 'map' is used here for convenience, but the techniques of analysis are equally applicable to cross-sections and 1D and 3D datasets, after obvious adaptations.

Interpreting a 2D geophysical dataset in terms of geology is, in many ways, similar to interpreting geology from an aerial photograph; both involve pattern recognition and geological inference based on limited information. In both cases, the resultant map is an approximation of the actual geology. The geological data are primary, but they are limited to areas of outcrop and perhaps drilling. Although the geophysical data are secondary, they usually have the distinct advantage of being continuously sampled across the area and are, therefore, a valuable aid when interpolating the geology between outcrops. Of course they also contain information about the geology at depth.

Although semi-automated methods are being developed for geological interpretation of geophysical data (Buckingham *et al.*, 2003), it chiefly remains a manual operation. We describe here some of the generic aspects of 2D data interpretation, with more specific details described later for each geophysical method.

2.10.1 Spatial analysis of 2D data

The spatial analysis of a 2D geophysical dataset presented in pixel form is similar to the analysis applied to any other kind of image and is based on various characteristics, the most fundamental being texture and tone. Tone is the relative colour or brightness of a region and in geophysical terms is usually equated to amplitude. Areas with similar tonal characteristics are best identified in pseudocolour displays (see Section 2.8.2). Texture is the pattern created by short-wavelength variations in tone. The nature of a particular texture can be difficult to quantify, but qualitative descriptions such as 'rippled' and 'granular' adequately describe textural characteristics for geophysical purposes. Textural variations are best seen

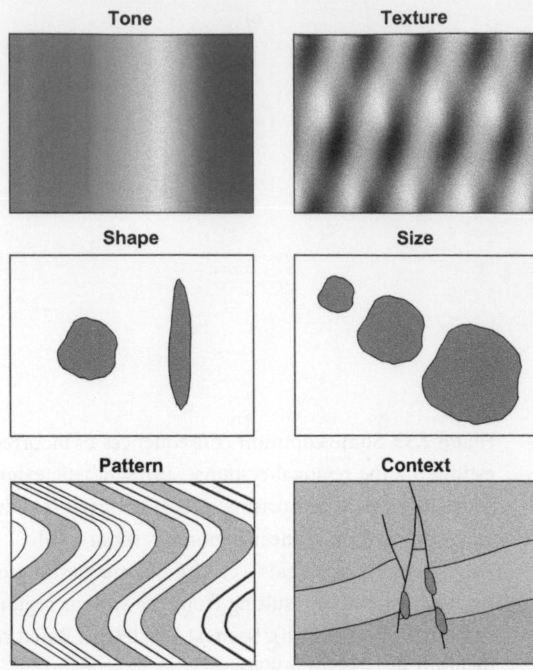

Figure 2.40 Some basic elements of image analysis. Based on diagrams in Lipton (1997).

in grey-scale displays, especially when displayed as shaded relief (see Section 2.8.2.3).

Texture and tone are the usual criteria for delineating (mapping) regions with similar characteristics in an image. This may involve working with a suite of images emphasising these characteristics in complementary ways. The most useful are the colour-draped displays, like that shown in Fig. 2.36 and described in Section 2.8.2.4, because they combine textural and tonal information very effectively. Regions with consistent textural and/or tonal characteristics are interpreted in terms of their *shape, pattern, size (scale)* and (geological) *context (association)* (Fig. 2.40).

2.10.1.1 Shape and pattern

Shapes and patterns of the responses are very important aspects of the interpretation of 2D geophysical data. The inclined lithological layering common in many geological environments is translated into responses that are parallel and extend over a significant area, whether stratigraphic or igneous in origin (see Figs. 3.28 and 4.26). Features that are more equidimensional are often indicative of intrusions (sedimentary or igneous). Examples of some other geological patterns common in geophysical data include dendritic drainage networks (see Figs. 3.75 and 4.24), 'veeing' caused by dipping layers intersecting valleys or by plunging folds (see Figs. 3.28 and 4.26), linears and dykes (Fig. 2.27), shear zones (Figs. 3.27, 3.66a, 3.74a, 3.75 and 4.26) and steeply dipping stratigraphy, fold interference patterns, and chaotic patterns in the interior of diapirs.

2.10.1.2 Size

When interpreting geological features, size must be used with caution because many common geological entities are self-similar over a large size range. A well-known example is folding, where there may be microscopic folds in fabrics which are located within parasitic folds on the limbs of larger folds etc. A good example where size is significant is astroblemes, whose characteristics change with size, from small 'simple' craters to larger 'complex' craters to massive 'multi-ringed basins'. An example of the less rigorous application of size applies to intrusions such as kimberlites, which tend to be comparatively small.

2.10.1.3 Context

Context is ultimately the most important clue to the geological significance of a pattern of geophysical responses. Some examples include the following:

- A response that cross-cuts the dominant anomaly pattern in a region is consistent with an intrusion (Figs. 3.74 and 4.26); if it is linear and extensive it is almost certainly a dyke (e.g. Schwarz *et al.* (1987), (Figs. 3.66a and 3.74).
- The boundary between two regions with different anomaly patterns may be an unconformity (Fig. 4.26).
- A linear feature associated with the lateral displacement or truncation of other responses may be a fault or shear zone (Figs. 3.74 and 4.26).
- Concentric anomaly patterns may be due to zoned intrusive complexes and/or contact aureoles (Hart, 2007; Schwarz, 1991).
- Distinctly curvilinear features are likely to be faults or dykes associated with volcanic centres, plutons, diapirs or impact structures.
- Sets of linear responses at a consistent angle to each other are probably associated with jointing or conjugate faulting.
- Anomalous response(s) at a site geologically favourable for mineralisation may be a potential target. What constitutes a favourable site depends upon deposit type and the exploration model. It could be a zone of structural or stratigraphic complexity, the margin of an intrusion etc.
- Integrating the responses from different geophysical data types also provides context. For example, coincidence

between aeromagnetic and airborne electromagnetic anomalies was very successfully used as a target selection criterion for nickel sulphide mineralisation in the Canadian Shield in the 1950s (Dowsett, 1967).

When the principal object is anomaly detection rather than geological mapping, context can be the only means of ranking the anomalies as exploration targets. However, discriminating a target anomaly from unwanted target-like anomalies is a fundamental challenge for the interpreter. Often geophysical data are unable to differentiate between responses from mineralised environments and geologically similar, but unmineralised, settings.

2.10.1.4 Human perceptions of spatial data

The interpretation process relies on the human mind processing information collected by the human eye to perceive patterns in spatial data, and then interpreting these patterns in terms of the geology. Despite the expression 'seeing is believing', this is far from an infallible process. This aspect of being human is poorly understood and almost completely unexplored from a geophysical context, so only a few general observations are possible. Importantly, we emphasise and demonstrate the real possibility of making erroneous interpretations of patterns in a dataset.

The reader will be familiar with optical illusions, where the eye/brain system misinterprets one or more characteristics of an object in a picture. The most relevant examples in the context of geophysical image analysis are linear and circular features, since these are fundamental components of the geological environment. Figures 2.41a and b demonstrate how obliquely intersecting lines produce illusionary offsets and misjudgement of parallelism. The former is a commonly occurring line-geometry in geophysical interpretations, with the potential result being incorrect interpretations of faulting. The next two examples, Figs. 2.41c and d, illustrate how the human vision system may perceive non-existent lines. The potential exists in geophysical data for the interpretation of spurious linear features. Figure 2.41e shows the strong human tendency to see circles, even when there are none present, or even when there are no curved lines in the area. Figure 2.41f shows how abrupt changes in image intensity create the illusion of deviations in straight lines. Although all three arcs in Fig. 2.41g have the same radius, they are perceived to be different, which may cause problems when interpreting data containing curvilinear features. Finally, the two central squares in Fig. 2.41h are the same shade of grey, but the different shades of the surrounding areas disguise this fact. The same illusion occurs with coloured regions.

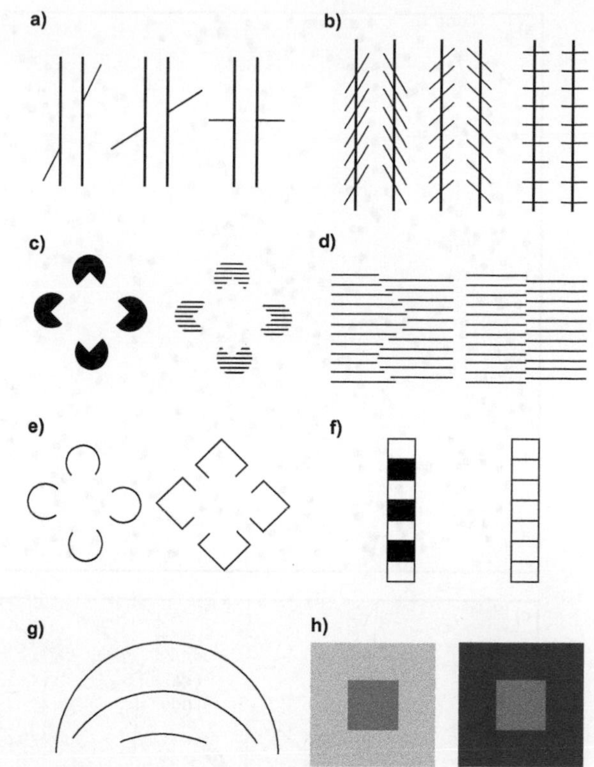

Figure 2.41 Examples of some optical illusions that can be misleading in the interpretation of geophysical data. See text for explanation.

Another common form of optical illusion is the inversion of apparent topography associated with the illumination direction in a shaded-relief display (see Section 2.8.2.3 and Fig. 2.32). Some forms of data enhancement specifically rely on the identification of positive features, for example the analytic signal transformation applied to potential field data (see Section 3.7.4.2). The possibility of making erroneous interpretations is obvious.

2.10.1.5 A cautionary note

Figure 2.42a shows the locations of sources of geophysical responses, perhaps magnetic sources derived from Euler deconvolution (see Section 3.10.4.1) or anomalous radiometric responses (see Section 4.5.1). From the various patterns that can be seen in Fig. 2.42a, in particular linear and circular alignments, and variations in their clustering, a plausible geological interpretation has been made. Linear alignments were identified first and then regions with similar internal distributions of sources were delineated. The interpretation recognised a faulted and folded layered

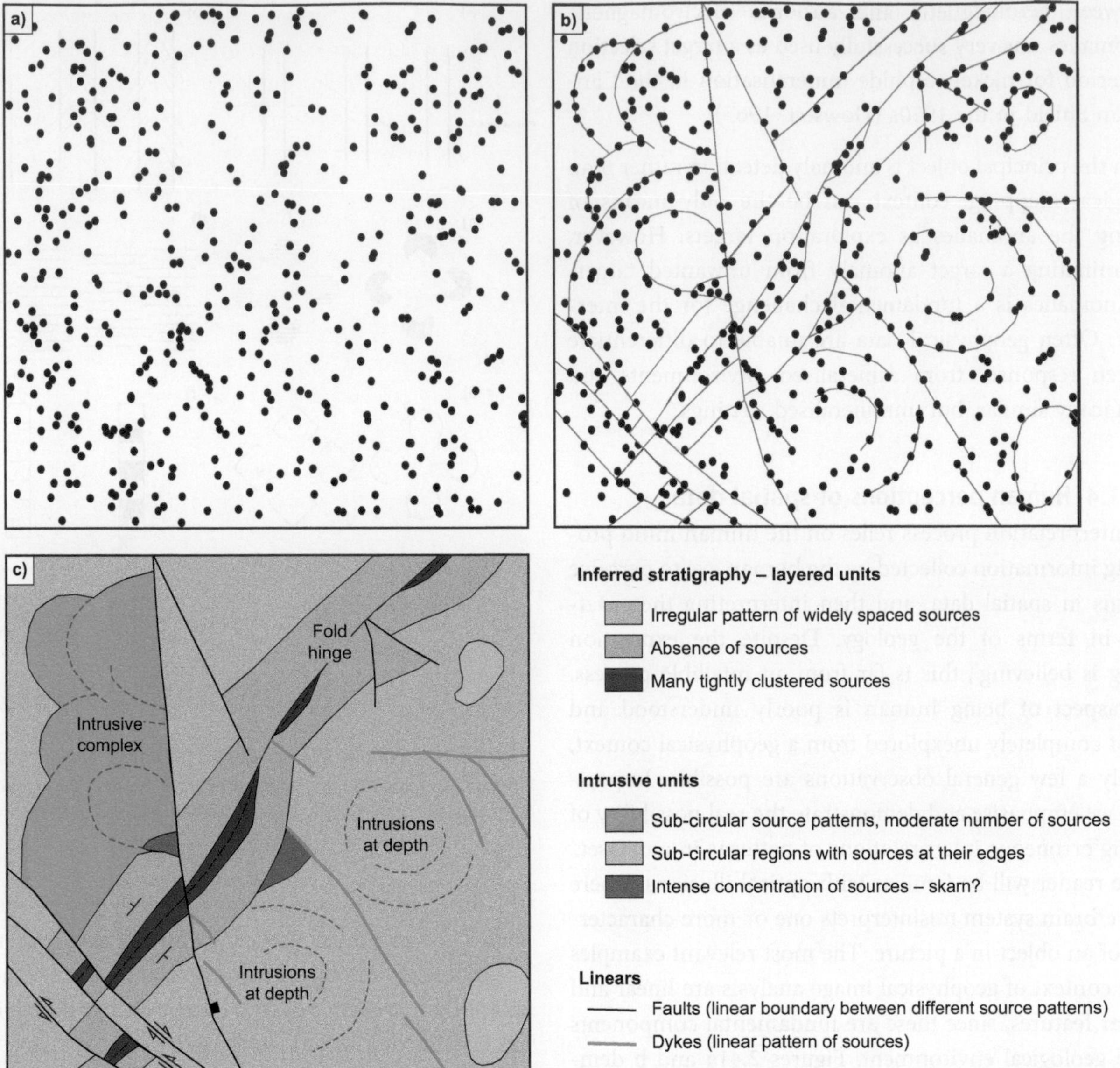

Figure 2.42 A cautionary note. (a) Map of source locations as might be obtained from a geophysical dataset. (b) Linear alignments identified within the distribution, and (c) a possible geological interpretation of the data. Although apparently plausible, the source distribution is in fact random. The figure illustrates the human propensity for seeing patterns where none exist.

sequence intruded by various igneous rocks. Depending on the local geology, sites considered favourable for mineralisation might be the intersection of the major fold and the largest fault in the survey area, or where faults are associated with the intrusive rocks. Although apparently geologically plausible, there is an underlying problem: the dataset comprises 500 randomly distributed points. There is a high probability that any distribution of points will show spatial alignments that form linear and curvilinear features, and also clusters, many or even all of which have no significance. Interpreters can easily fall into this trap when other data are not available to guide the interpretation. In fact, it is very common for humans to mistakenly find structure in random datasets, as described by Taleb (2001) in the context of the analysis of stock markets and more generally by Shermer (2011).

How can traps like those described be avoided? Unfortunately, there is no panacea, but the following cautionary strategies are generally applicable and can help the interpreter to avoid common mistakes:

- One reason that it was possible to create geological 'sense' from a random pattern was that only one form of presentation of the data was used for the interpretation. By using both the data in its basic form and its various transformations, such as derivatives, and by making use of different kinds of display, for example grey-scale, pseudocolour etc., more information can be displayed, making for a much more reliable interpretation.

- Although various forms of the data may emphasise particular characteristics, other important characteristics may be obscured; so the interpreter should continually make reference to the most fundamental form of the geophysical data, e.g. the magnetic or gravity field strength, the calculated electrical conductivity/resistivity etc., described in a simple form such as a colour-draped display (Fig. 2.36).
- As demonstrated by Figs. 2.41 and 2.42, the human vision system is not infallible. Many of the potential traps can be addressed in two ways. For reasons not fully understood, the same interpreters will make different interpretations of the same data if presented in a different orientation (Sivarajah *et al.*, 2013). Why this is the case is the subject of current research, but the problem is easily addressed by rotating the data during interpretation.
- The use of several illumination directions in shaded relief is important because of its inherent directional bias, which can lead to incorrect or incomplete identification of linear features.
- Many optical illusions are a form of incorrect 'mental' extrapolation or interpolation. This can be overcome by checking the interpretation whilst viewing a small window or subarea of the dataset.
- The need for the interpreter to continually evaluate the geological credibility of features interpreted from the data cannot be over-emphasised. The integration of other types of geophysical data, and geological, geochemical and topographic data, is essential for developing an accurate interpretation of the data. The interpreter is more likely to be fooled if they treat their interpretation as 'factual' without making the essential credibility checks.

2.10.2 Geophysical image to geological map

Creating a pseudo-geological map of a large area from geophysical data can be a complex and time-consuming task, so often only areas of particular exploration interest are mapped in detail. Geophysical datasets are as variable as the geological environments that give rise to them, and different datasets must be considered in different ways. Defining a set of rigorous interpretation rules applicable to all situations, and satisfying the requirements of every interpreter, is simply not possible; instead we describe a set of somewhat idealised guidelines. We use the familiar exercise of making a geological map to provide context.

All available relevant factual information from the survey area needs to be compiled prior to beginning the interpretation of the geophysical data. Geological data are fundamental to the process and include outcrop information, orientation measurements (bedding, fold axes etc.) and drillhole intersections. The interpreter needs to consider the possibility of inconsistent lithological identifications if the work of several individuals is combined, and be mindful that a geological map is itself an interpretation. Ideally, a geological fact (outcrop) map should be integrated with the geophysical data being interpreted in a computer-based Geographical Information System (GIS) environment. Examples of this kind of data integration are provided by Haren *et al.* (1997).

Petrophysical measurements, if available, are useful for predicting the responses of the local geology. In the absence of data from the survey area, the generalities about rock physical properties given in our descriptions of each geophysical method can be used, albeit with caution. Particularly important is the identification of the principal physical-property contrasts in the area, since these will correspond with the most prominent features in the geophysical data. Variability within a set of petrophysical measurements can have important implications. For example, a rock unit with a large range in a physical property will exhibit greater textural variation in its geophysical response. Also, a bimodal distribution may suggest that two 'types' of the same lithological unit can be mapped, even though they appear identical when examined in the field. It is also important to bear in mind that the petrophysical sampling may not be representative. The petrophysical data may be biased towards those parts of a unit that are resistant to weathering, and no data at all may be available from units that weather easily.

An interpretation procedure effective for most situations involves, firstly, developing a framework by identifying linear and curvilinear features and, secondly, classifying regions based on their textural and tonal characteristics, and the wavelength or width of the measured responses. Normally the procedure is iterative, i.e. the results from any one stage of the interpretation are re-evaluated in terms of the evolving interpretation. It may be necessary to model some of the observed responses (see Section 2.11.2) in order to help understand both the sources of the responses and the surrounding geology.

The interpretation may commence by delineating the large-scale features, which probably extend across large areas of the survey, and then move on to progressively smaller features. On the other hand, sometimes the data in the vicinity of geological features of interest, or geologically well-known areas, are analysed in detail first, in order to resolve the 'geophysical signature' of the area. The interpretation is then extended into the surroundings, possibly with the intention of identifying repeated occurrences of the same signature. Sometimes only shallow responses or only the deeper features may be all that is of interest, so the interpretation will focus on the appropriate responses, albeit with a 'feeling' for the implications on the interpretation of the other responses in the data.

2.10.2.1 Framework of linear features

The first step in making a pseudo-geological map is the creation of a structural framework from the linear and curvilinear features identified in the data. These may be linear and curvilinear anomalies, lineations interpreted from the alignment or truncation of other features, zones of greater structural complexity or areas of pronounced gradients, and usually coincide with contacts and/or faults/shear zones. Edge-enhancement forms of the data, i.e. some gradient-based transform, are most useful for this aspect of the interpretation.

When identifying linear features on the basis of alignment and truncation of responses, it is important to remember that the geophysical response, especially when presented in its fundamental (untransformed) form, will often extend beyond the surface projection of the source, especially if the source is at depth (Fig. 2.4). This means an interpreted linear can legitimately 'cut off' the end of an anomaly. However, this also means it is easy to mistakenly interpret non-existent features, so caution is required.

2.10.2.2 Classification of regions

The next step is to map the' lithological' units onto the framework of linear features. This involves defining polygons in the GIS. It is essential when analysing images not to assign too much significance to actual colours since these depend on the colour map and the stretch function (see Section 2.8.2) used for the display. Instead, consistent textural and tonal characteristics are the basis for mapping lithological subdivisions. The same unit will change character if it extends under cover owing to the increasing separation between source and sensor (anomalies will decrease in amplitude and increase in wavelength). This must be accounted for in the interpretation. Some examples of this effect are shown in Figs. 3.74 and 3.76.

As in field geological mapping, an early and crucial decision is to determine what constitutes a 'mappable unit'. In both cases this is based on characteristics that can be easily and reliably recognised. Whether working with geological or geophysical data, it is often necessary for different lithotypes to be grouped together into a single mappable unit, especially if individual units comprising a layered sequence are thin, small, or discontinuous, or the inter-relationships between them are too complex to illustrate at the scale of the mapping. Sometimes a single distinctive 'marker' horizon may be recognised and mapped, conveniently revealing the overall structure and structural style. In geological mapping this might be a particularly well-exposed horizon, or one with a distinctive mineralogy or fossil assemblage. In geophysical terms, it might be a unit with anomalously high or low density, magnetism, conductivity or radioactivity etc.

The outcrop map provides important reference information during this part of the interpretation. Nevertheless, the interpreter must be aware of apparent contradictions between the geology and the geophysics caused by the fundamental differences between geological and geophysical responses (see Section 1.1) and due to the depth information in the geophysical signal. A common example of the latter is where a relatively thin sequence of non-magnetic sediment overlies magnetic bedrock. The outcrop map will show the sediment and the magnetic dataset will show responses from the bedrock.

As the interpretation progresses, it is important to check continually for geological credibility. In an area with a layered succession, each unit must be assigned a place within a local stratigraphy, although this will be a pseudo-lithostratigraphy since it will reflect physical property variations rather than explicitly lithotype. A rigorous adherence to stratigraphic rules and the implied outcrop patterns is as necessary for a pseudo-geological map based on geophysical data as it is for one based on geological data. Of course physical properties may vary within a unit just as lithotypes may vary laterally, potentially confusing the mapper.

2.10.2.3 Estimating dip and plunge

During the mapping process it can be very useful to have some idea of the orientation of the sources of the geophysical responses. For example, a fold may be recognised in the geophysical data from its outcrop pattern, but this does indicate whether it is a synform or an antiform. Orientation information can be most accurately determined using modelling (see Section 2.11.2). However, much quicker qualitative methods can be usefully applied during mapping. Some interpreters annotate their maps with dip and plunge symbols equivalent to the plotting of these when geological mapping. How source orientation is estimated depends on the nature of the geophysical response, and it is more easily achieved with some kinds of geophysical data than others. We describe this kind of interpretation in our descriptions of the individual geophysical methods and restrict ourselves to some general statements here.

The key property is anomaly asymmetry which is manifest as the variation in gradients at the margins of a geophysical anomaly (Fig. 2.43). The shallower (low)

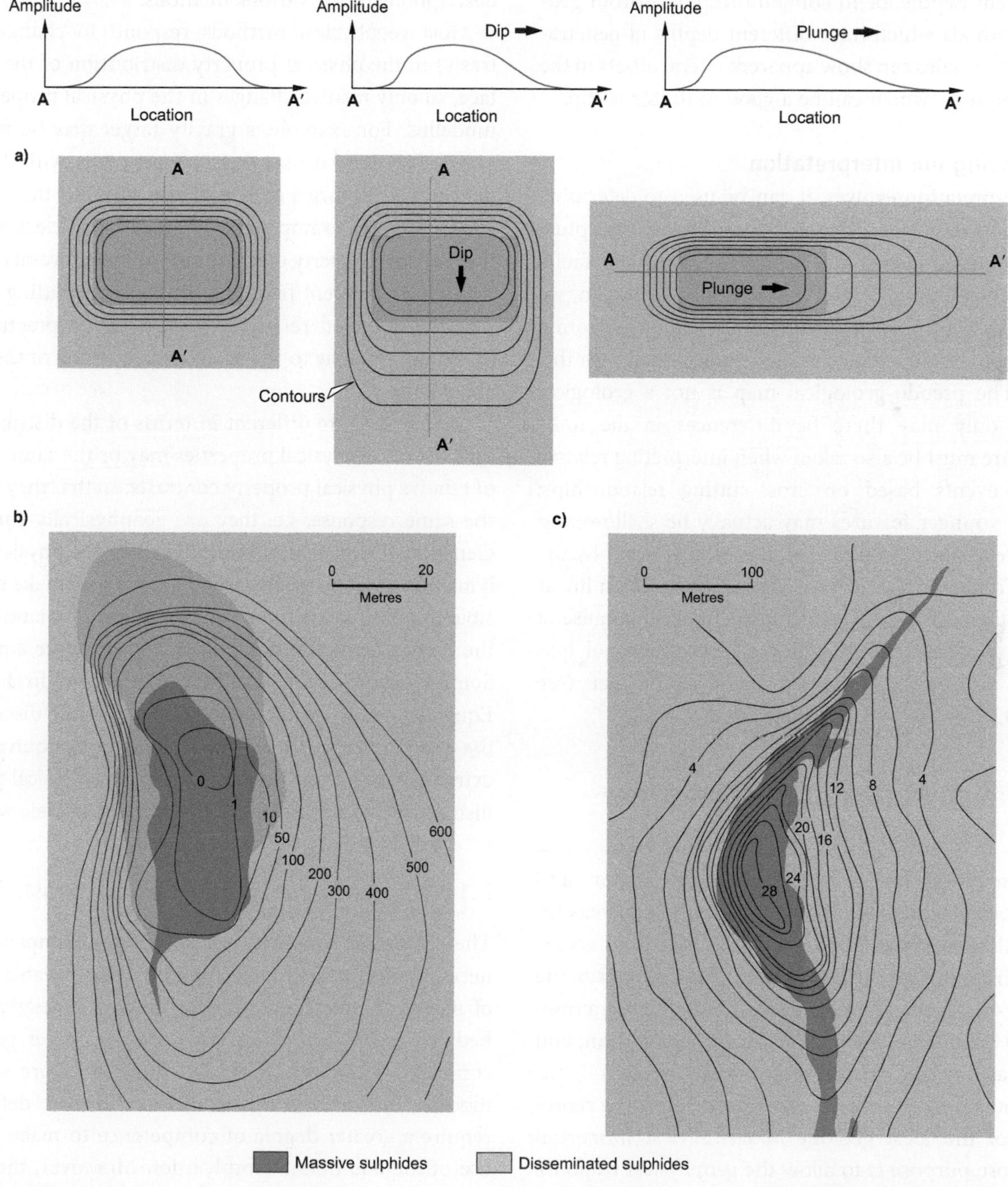

Figure 2.43 Assessing dip and plunge from contour spacing. (a) Schematic illustration of variations in gradients associated with dipping and plunging sources. The increase in contour spacing is the dip/plunge direction. (b) Applied potential data (mV/A) from the Telkkälä Taipalsaari nickel sulphide deposit in southeastern Finland. Redrawn, with permission, from Ketola (1972). (c) Gravity response (gu) of the Pyhäsalmi massive nickel sulphide deposit, Oulu, Finland. Redrawn, with permission, from Ketola *et al.* (1976, Figs. 10a,b, p.231).

gradients usually occur on the down-dip side of the source and, similarly, for a plunging source the gradients are less in the down-plunge direction. This general statement assumes uniform and isotropic physical properties within the source.

Another way to infer dip is to compare the geophysical response in processing products emphasising responses from different depths, or to compare responses from geophysical methods which have different depths of penetration. Dipping bodies can show apparent lateral offsets in the various responses, which can be a good indicator of dip.

2.10.2.4 Using the interpretation

As the interpretation evolves, it can be used to determine the geology of the area, and also for identifying conceptual exploration targets, e.g. dilational zones where faults change their orientation, intersecting faults etc. Again, we emphasise that geophysical data contain responses from a range of depths, so the data should be interpreted with this in mind. The pseudo-geological map is not a geological map. Not only may there be differences in the units mapped, care must be also taken when interpreting relative timing of events based on cross-cutting relationships; apparently younger features may actually be shallower or simply have a much stronger geophysical response obscuring that of others in the vicinity. Determining which linear offsets another can be very difficult to establish, because of smoothing associated with gridding and directional bias associated with data presented as shaded relief (see Section 2.8.2.3).

2.11 Data interpretation – quantitative analysis

When a pseudo-geological map of the desired extent and detail has been created, the interpreter may select areas for more detailed analysis. This is most likely done to accurately establish the depth and geometry of the sources of the anomalies – information required for constructing cross-sections to complement the qualitative interpretation, and in particular for drill-testing the anomaly sources.

Fundamental to quantitative interpretation is the representation of the local geology in terms of a numerical *model*, whose purpose is to allow the geophysical response of geology to be computed. The model is defined by a series of *parameters*, which define the geometry and physical properties of regions within the subsurface. The geophysical response of the model is computed and compared with the observed data and, in an iterative process, the model is adjusted and its response recomputed until there is a satisfactory correspondence. This process is known as *modelling*. We describe here the fundamental aspects of geophysical models and modelling techniques common to the analysis of all types of geophysical data, whilst details specific to particular geophysical methods are given in our descriptions of the various methods.

Most geophysical methods respond to changes (contrasts) in the physical property distribution of the subsurface, so only relative changes in the physical properties are modelled. For example, a gravity target may be modelled as, say, 0.2 g/cm^3 denser than its host rocks, with the actual densities of the target and host unresolved by the model. In some cases, for example with magnetic and electrical data, the physical property of the source may be several orders of magnitude different from that of the surrounding rock, so that the observed relative response is, for practical purposes, entirely due to the absolute properties of the anomalous body.

Models that are different in terms of the distribution of their absolute physical properties may be the same in terms of relative physical property contrasts, so that they produce the same response, i.e. they are geophysically *equivalent*. Generally the relative model is simpler. The physics underlying the different geophysical methods may make it impossible to differentiate the physical property variation within the source, so many different models may be equivalent. Some examples of equivalence are shown in Fig. 3.68. Equivalent models are method-specific and discussed in the chapters on each geophysical method. Equivalence is extremely useful for reducing a complex physical property distribution to a mathematically simpler model.

2.11.1 Geophysical models of the subsurface

The geological environment can be very complex, but by necessity it must be represented by a manageable number of model parameters. This can result in a hugely simplified representation of the subsurface. Greater geological complexity can be represented with more complex models, but these require more effort to define and require a greater degree of competence to make effective use of the additional complexities. Moreover, there is no advantage in using a geophysical model whose complexity is greater than the resolving capabilities of the data being modelled, which may be quite limited, especially for deeper regions.

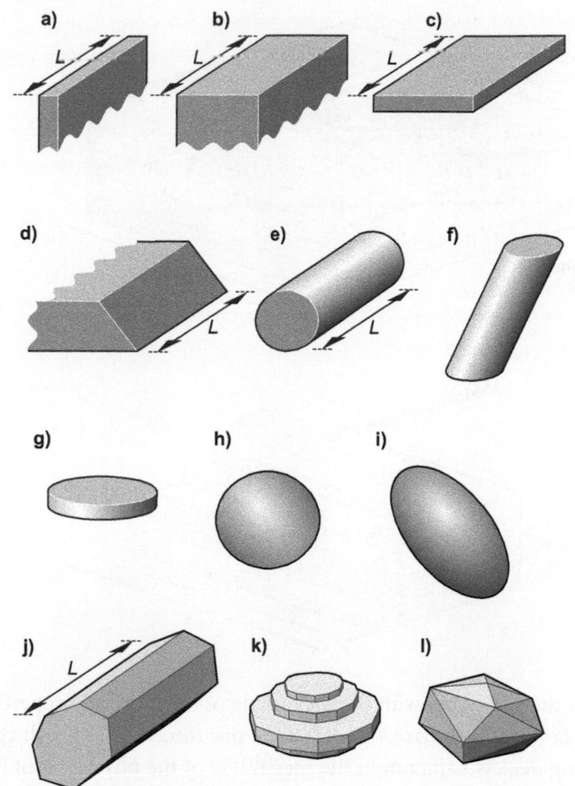

Figure 2.44 Shape-based models. (a) Thin vertical sheet, (b) thick vertical sheet, (c) thin horizontal sheet, (d) step, fault or dipping contact model, (e) horizontal cylinder, (f) plunging cylinder, (g) disc, (h) sphere, (i) ellipsoid, (j) irregular prism, (k) and (l) facet models.

To model a set of geophysical data the interpreter must select an appropriate model to represent the subsurface geology. There are two classes of geophysical model: *shape-based* models and *cell-based* models. In most cases, notably modelling of gravity and magnetic data, the model response is obtained by summing the responses of the ensemble of constituent shapes or cells. The situation is not as simple for electrical and electromagnetic methods where complex interaction between the different components of the model, and with the host rocks, strongly influences the model's overall response.

2.11.1.1 Shape-based models

Models consisting of geometrically simple bodies (Fig. 2.44a to i), each assigned a particular physical property value, are appropriate when a simple homogeneous form is considered to be a reasonable approximation of the shape of the anomaly source, or when too little is known about the subsurface geology and its physical properties to justify a more complex model. Models (a) to (e) and (j) in Fig. 2.44 may be used for 2D and 2.5D modelling, depending on the length L (see Section 2.11.1.3). Multiple bodies can be arranged as an ensemble to simulate a wider variety of realistic 3D geological forms.

During modelling, a body's parameters, i.e. its dimensions, position and physical property, are varied. These models are also known as *parametric models* and they make for quick and easy modelling, but their simplicity means a perfect match with the observed data may not be achievable. This is not necessarily a problem, because the results obtained are a good starting point for building more complex models.

Parametric models include sheets of different thickness (flat, dipping or vertical), dipping contact (sometimes called a step or fault) and a cylinder (vertical, horizontal or plunging) which when thin becomes a disc. The ellipsoid is a particularly useful model. It is an easy matter to stretch and rotate its axes to approximate a range of geological forms such as a plunging pipe-like ore shoot, a flat-lying stratiform orebody, or its axes extended equally to form a sphere to represent equidimensional bodies.

More complex models can be built using bodies whose surfaces are defined by a series of interconnecting planar elements, i.e. facets (Fig. 2.44j to l). Each body is defined by specifying the coordinates of the vertices of the individual facets, which are adjusted during modelling. This provides more flexibility for creating complex shapes, but manipulation of the model is then more complicated. By constructing a model with multiple bodies of this type, complex and realistic geology can be modelled: see McGaughey (2007) and Oldenburgh and Pratt (2007).

2.11.1.2 Cell-based models

Complex physical property models can be created by representing the entire subsurface as a series of discrete cells or elements whose properties are homogeneous throughout the cell/element (Fig. 2.45a and b). A physical property value is assigned to each cell and the data are modelled by adjusting only these values; the cell geometry and positions remain unchanged. The interpreter can define complex geological features on sections and on layers of cells. It is important that models of this type extend well beyond the area of interest to prevent 'edge effects' due to the abrupt change in physical properties at the edge of the model. The challenge of manually manipulating the large number of cells repetitively generally restricts these types of models to inverse modelling (see Section 2.11.2.1).

A variant of this model type represents the subsurface as a series of cells whose dimensions are varied in only one

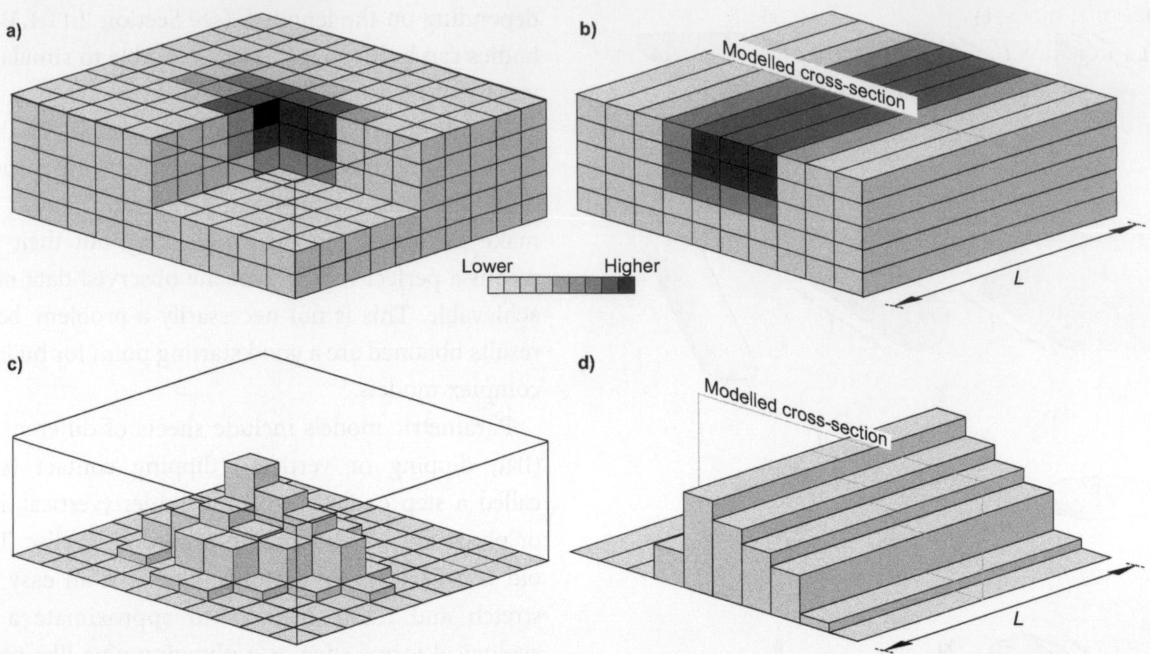

Figure 2.45 Cell-based models. (a) The model comprises a series of voxels of the same size, with the magnitude of the physical property specified for each voxel; (b) 2D or 2.5D version of (a) depending on length *L*; (c) 3D cell-based model where one dimension of each cell is varied; and (d) 2D or 2.5D version of (c) depending on length *L*. The shading depicts variation in the magnitude of the physical property.

direction, usually the vertical (Fig. 2.45 c and d). This kind of model is commonly used to simulate basement topography or sedimentary basins where all the cells are ascribed the same physical property value and the data modelled by adjusting the variable dimension.

Cell size and the number of cells are of fundamental importance in cell-based models. They determine the resolution of the model and the computational effort required to obtain the response. A sufficient number of sufficiently small cells are required to model the shorter-wavelength variations in the observed geophysical response, and to adequately represent areas where physical property variations are more complex. However, too many cells results in a large number of model parameters and demands greater computational resources. Often smaller cells are used in the near-surface, becoming larger with depth where less resolution is acceptable. The cell size may also vary laterally, being smaller where more detail is required in the central part of the model or where the data suggest more complex structure.

2.11.1.3 1D, 2D and 3D models

In addition to selecting the type of model to use, the interpreter must also decide the number of directions, or dimensions, in which the model is to be defined. Both shape- and cell-based models can represent a homogeneous subsurface, or can be arranged into 1-, 2- and 3D forms. The dimensionality should reflect the geological complexity being modelled, which must be justified in terms of the available knowledge of the local geology, and the distribution and quality of the geophysical data.

The model response is usually computed above the surface, and topography may be included in the model. It is also possible to compute responses at locations below the surface, which is the requirement when modelling downhole data.

Half-space model

The simplest representation of the subsurface is a homogeneous volume with a flat upper surface, known as a *half-space* (the other half of the space being the air which, for practical purposes, is a medium with homogeneous physical properties) (Fig. 2.46a). This model depicts those situations where the ground's physical properties are invariant in all directions, including to great depth, i.e. there are no physical property contrasts except at the ground–air interface. It is an important model for calculating the background response of the host rocks of a potential target, and forms part of the response of discrete bodies in electrical and electromagnetic data.

One-dimensional model

An overburden layer on a homogenous basement, and multi-layered geological environments, are represented as a series of flat-lying layers. Each layer has constant thickness and extends laterally to infinity in all directions, and each has its own (constant) physical property (Fig. 2.46b and c). The model parameters are the number of layers and their thicknesses, and the physical properties of each layer and the underlying basement. Variation is only possible in one direction, i.e. vertically, so it is known as a *one-dimensional* model. There is no change in the geophysical response laterally across the surface of the 1D model.

The 1D model is useful when geological features extend laterally well beyond the footprint (see Section 2.6.2) of the geophysical survey. It is used to model (depth) soundings; where the variation of the response with depth is investigated (see Sections 5.6.6.1 and 5.7.4.3). As described in Section 2.8.1, a series of 1D models from adjacent measurements may be used to create parasections, or paravolumes.

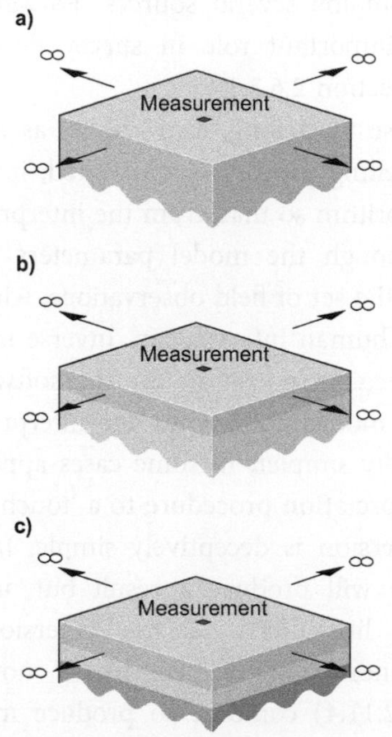

Figure 2.46 One-dimensional models of the subsurface. (a) Half-space, (b) and (c) multiple layers of constant thicknesses and physical properties. All models extend laterally to infinity.

Two-dimensional model

For geological features having a long strike length, a model consisting of a cross-section can be specified and the strike extent of the model set to infinity (Fig. 2.47). What constitutes 'long' depends on the geophysical method and the depth of the source but at least 5 times would be typical. The physical property distribution is specified in only two dimensions, i.e. depth and distance along the survey profile, so it is known as a *two-dimensional* model. When the geophysical data (and geology) are known to be constant along strike, or can reasonably assumed to be, a representative profile perpendicular to the regional strike is selected and can be modelled using a 2D model.

Two-dimensional models have their longest (strike) axis horizontal and the cross-section modelled is vertical. For shape-based models the bodies in the cross-section can be of any shape, such as a rectangle (Fig. 2.44c) or a circle to represent a cylindrical source (Fig. 2.44e), or an arbitrary shape can be defined by specifying the coordinates of the nodes of straight-line segments defining the shape (Fig. 2.44j). For cell-based models the subsurface is defined

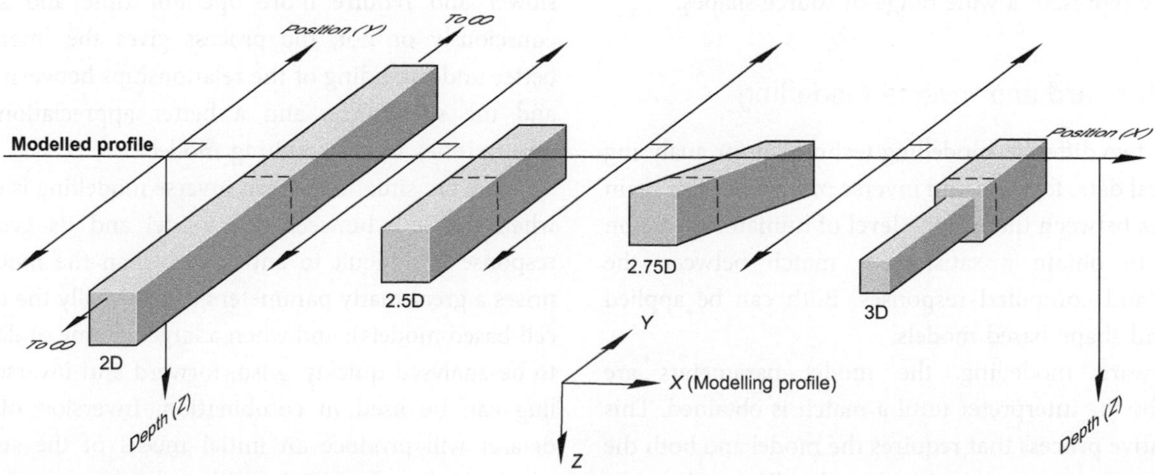

Figure 2.47 Schematic illustration of prism models of different dimensionality.

by cells with their long axes horizontal and perpendicular to the section being modelled (Fig. 2.45b and d).

Two-and-a-half-dimensional model

A very useful variation on the 2D model which removes the restriction of infinite strike length, and is easier to define than the more complex 3D model, is a model with constant cross-section extending over a finite strike length (Fig. 2.47). This is known as a *2½ or 2.5D* model. When the source can have different strike extents on either side of the modelled profile, or the strike or plunge of the body is not perpendicular to the profile, this is sometimes called a 2.75D model. The 2.5D model gives the interpreter control of the third (strike) dimension without the complexity of defining and manipulating a full 3D model, so they are by far the most used models for analysing all types of geophysical data in three dimensions.

Three-dimensional model

When the model of the subsurface can be varied in all three directions, it is known as a *three-dimensional* model. Shaped-based 3D models can be specified in a number of ways, but usually as a network of interconnecting facets (Fig. 2.44k and l). Cell-based models comprise a 3D distribution of uniform cells (Fig. 2.45a and c).

Three-dimensional models can take considerably more effort to define than 2D and 2.5D models and require computer systems to view and manipulate the geometry in three dimensions. The observed and modelled responses are usually displayed as a series of profiles across sections of interest. The simplest, and still very useful, form of 3D model is the ellipsoid (and the sphere – a special case of the ellipsoid). The model is very easy to manipulate and can adequately represent a wide range of source shapes.

2.11.2 Forward and inverse modelling

There are two different modelling techniques for analysing geophysical data: forward and inverse modelling. The main differences between these is the level of human interaction required to obtain a satisfactory match between the observed and computed responses. Both can be applied to cell- and shape-based models.

In forward modelling, the model parameters are adjusted by the interpreter until a match is obtained. This is an iterative process that requires the model and both the observed and computed responses to be displayed graphically so that the result can be assessed by the interpreter. The model parameters are adjusted interactively and the model response recomputed. The number of iterations required correlates closely with the expertise of the interpreter, the complexity of the model and the knowledge of the subsurface geology. Obtaining a suitable match with the observed data can be quite straightforward, provided the computed response is easy to anticipate and that the observed anomaly has a fairly simple form. Responses from gravity and SP models are usually quite easy to predict, but magnetic and electrical and electromagnetic responses can be very difficult to predict, especially when the model contains several sources. Forward modelling also has an important role in survey design which is discussed in Section 2.6.3.1.

With inverse modelling, also known as *inversion*, the iterative modelling process is automated; it is done by a computer algorithm so that, from the interpreter's view, it appears as though the model parameters are obtained directly from the set of field observations, with or without some level of human intervention. Inverse modelling is a far more difficult proposition for the software engineer than forward modelling, but for the interpreter the process is normally simpler, in some cases apparently reducing the interpretation procedure to a 'touch of a button'. However, inversion is deceptively simple. In most cases the algorithm will produce a result but, unfortunately, mathematical limitations of the inversion algorithm and the phenomenon of geophysical non-uniqueness (see Section 2.11.4) combine to produce many possible models that will fit the data. The result from the inversion is one of what may be an infinite number of possibilities, and choosing the best one is often a major challenge in itself for the interpreter. Forward modelling may be slower and require more operator time and skill but, consciously or not, the process gives the interpreter a better understanding of the relationships between the data and the subsurface, and a better appreciation of the uncertainties in the resulting model.

There are situations when inverse modelling is essential: when the link between the model and its geophysical response is difficult to anticipate; when the model comprises a great many parameters (as is usually the case with cell-based models); and when a large volume of data needs to be analysed quickly. Also, forward and inverse modelling can be used in combination. Inversion of a large dataset will produce an initial model of the subsurface which can be refined using either closely controlled inversion or detailed forward modelling. Conversely, forward modelling may be used to produce an approximate match to the data and inversion used to refine the fit.

2.11.2.1 Inverse modelling methods

The mathematics of geophysical inversion is complex and beyond our scope. Only the basic principles are described here to provide the reader with some insight into the source of possible problems when using inversion techniques. Oldenburg and Pratt (2007) provide a comprehensive description of inversion and its applications to mineral exploration. Additional examples are presented by Oldenburg *et al.* (1998).

Inverse modelling requires the interpreter to specify a starting model (although this may be no more complex than a half space) which is systematically refined by the inversion process. The difference between the calculated and observed responses at each data point is called the *residual.* Fundamental to the inversion process is the overall match, or the 'degree of fit', between the two responses, which is mathematically represented by an *objective function* that describes the degree of match as a function of the model parameters. The task for the inversion process is to adjust the model parameters so as to minimise the objective function, i.e. to minimise the residuals.

There are various mathematical methods for minimising the residuals, but the most common involves determining how the residuals vary as each model parameter is altered, which in turn indicates how the parameters should be adjusted to minimise the objective function. Normally in geophysical inversion, it is not possible to directly predict the optimal value for a particular parameter; the problem is said to be *non-linear.* Non-linear inverse problems are solved using an iterative strategy that progressively alters the model parameters, calculates the objective function and then, if necessary, adjusts the parameters again until a satisfactory match between the observed and calculated responses is obtained. The inversion algorithm is described as *converging* on a solution, i.e. a model that produces a satisfactory match to the observed data.

Figure 2.48 illustrates how a gradient-based inversion algorithm, a commonly used strategy, models a set of gravity observations. The source is spherical, and the parameters that can be varied are its depth, lateral position and density contrast with its surrounds. A useful analogy is to visualise the objective function as a terrain, with the algorithm seeking to 'roll downhill' until it reaches the lowest point in the topography, i.e. the place where the objective function and the residuals are minimised. This is a reasonable approach provided the form of the objective function is not overly complex. The problem is that non-linear inverse problems normally have objective functions which have extremely complicated forms and, consequently, it can be very hard to find the overall minima. An algorithm that goes 'down-hill' is likely to be caught in one of many 'valleys' or 'basins', which are called *local minima*, and has little chance of 'rolling to' the absolutely lowest point in the terrain, i.e. the *global minimum*, which is the ultimate objective. Figure 2.48 also demonstrates the reason for non-uniqueness in geophysical modelling, i.e. more than one model will fit the data (see Section 2.11.4). In Fig. 2.48d, where the depth and physical property contrast are allowed to vary, the lowest part of the topography representing the objective function is not localised; rather than being a 'basin', it is instead a 'valley'. Once the inversion algorithm has found the valley floor, moving across the floor creates little change in the value of the objective function. The example shows that multiple combinations of depth and property contrast values can produce the same fit to the observed data, i.e. the result is not unique.

The terrain analogy is a simplification, since mathematically the objective function will be defined in more than three variables (dimensions), i.e. it is a *hyperspace.* Mathematically exploring the hyperspace in an efficient and effective manner can be exceptionally difficult because the gradient-based search algorithm can get hopelessly confused between local and global minima and will not converge at all. Even when a minima is found it may be impossible to tell if this is the global minimum.

If local minima are expected then *global search* methods are required; these 'see through' local features in the function to seek the overall minima. A well-known global search algorithm is the *Monte Carlo* method, which randomly assigns values to model parameters, usually within defined bounds. The match between the computed and observed responses is then determined. If the match is acceptable, according to some defined criteria, the model is accepted as a possible solution and becomes one of a family of solutions. The process is then repeated to find more possible solutions. Monte Carlo methods require a large number of tests, hundreds to millions depending on the number and range of model parameters being varied, and so are computationally demanding. Another disadvantage is that it is still not possible to determine whether all possible models have been identified, a particular problem when very different solutions can exist. Global search

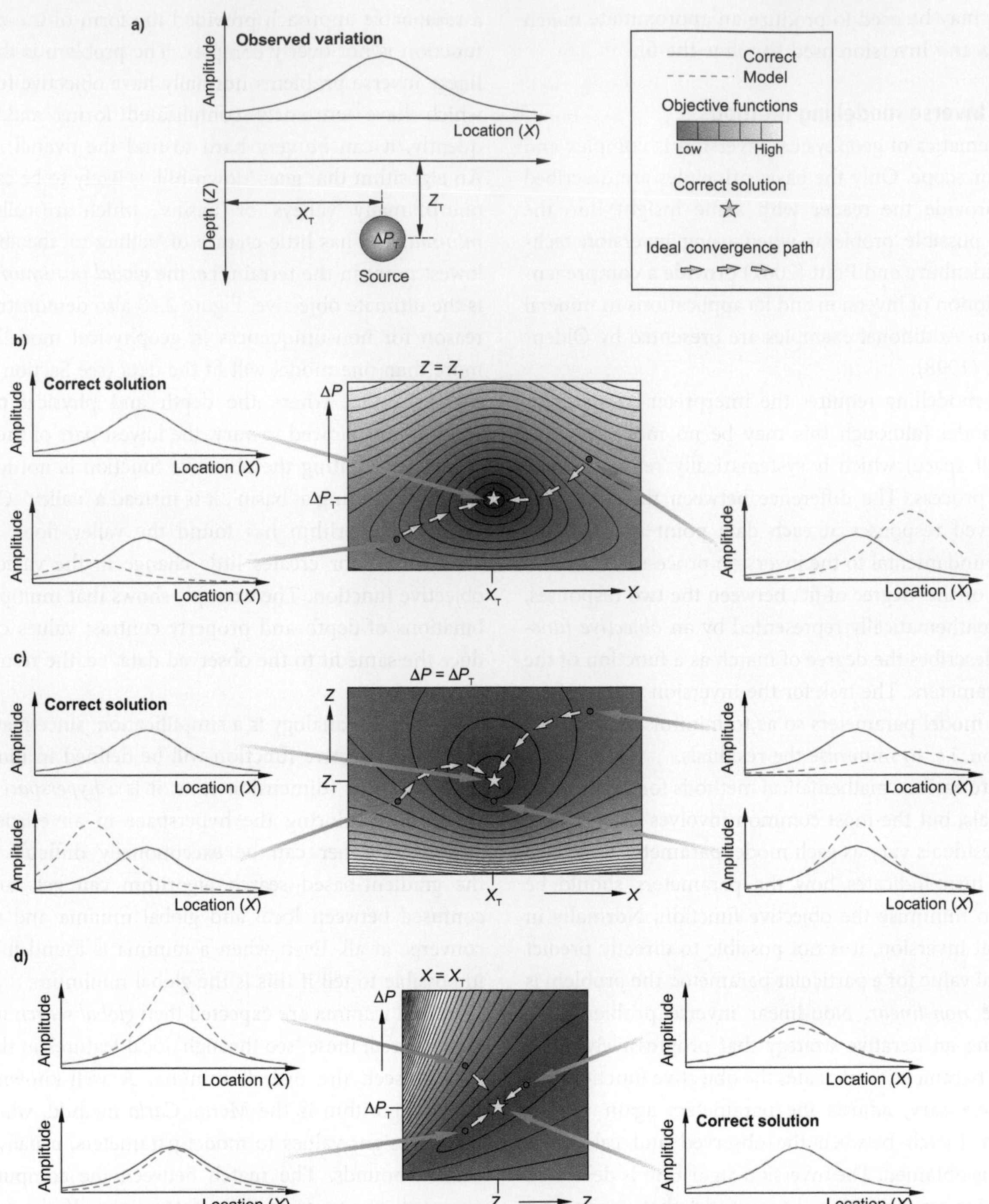

Figure 2.48 Illustration of geophysical inversion based on the gradient of the objective function. (a) Data from a traverse across a body with a positive contrast in some physical property. The inversion seeks the true location of the source (X_T, Z_T) and the true physical property contrast (ΔP_T). (b) Inversion constrained by setting depth (Z) equal to Z_T whilst the lateral position (X) and physical property contrast (ΔP) are allowed to vary. The objective function shows a single minimum coinciding with the correct values X_T and ΔP_T. Also shown are the observed and computed responses for selected pairs of X and ΔP. (c) Inversion constrained by setting ΔP equal to ΔP_T whilst X and Z are allowed to vary. Again, the objective function has a simple form with a single minimum. (d) Inversion constrained by setting X equal to X_T whilst Z and ΔP are allowed to vary. In this case, the ability to balance variations in ΔP and Z between each other produces a broad valley of low values in the objective function. Note that the inverse problem is greatly simplified when there are only three variables, one of which is held constant and set to the correct value, whilst the other two are allowed to vary.

methods that retain and use information about the objective function when searching for the global minima are usually more efficient than Monte Carlo methods. Algorithms of this type include *simulated annealing* and *genetic algorithms*; see Smith *et al.* (1992).

Constrained inversion

Inversion can be directed towards a plausible solution by including known or inferred information about the area being modelled into the inversion process. This is known as *constrained* inversion and the information helps direct the inversion algorithm to that part of the objective function hyperspace where the global minimum exists. Provided the information is accurate, the result will usually be a more useful solution than that from an unconstrained inversion (see Section 3.11.2). One way to constrain the inversion is to set bounds, or limits, on the values of selected parameters (with or without probability components). For example, the variation of physical property may be constrained; non-negativity being an obvious constraint for a property such as density, and one that greatly improves the likelihood of obtaining a geologically realistic result from the inversion. The possible locations of parts of the model may also be constrained: forcing them to below the ground being an obvious constraint; or to honour a drilling intersection. Particularly important in many instances are distance weighting parameters which counter the tendency of inversion algorithms to place regions with anomalous physical properties close to the receiver, which equates to close to the surface for airborne and ground surveys. Other common constraints include geometric controls, for example the source should be of minimal possible volume; it should be elongated in a particular direction (useful in layered sequences); it should not contain interior holes; and adjacent parts of the model should be similar, i.e. a smoothing criterion. Smooth inversion tends to define zones with gradational or 'fuzzy' boundaries and with properties that are less accurately resolved, rather than as compact sharp-boundary zones whose properties are more accurately determined, but with greater probability of uncertainty in their locations and shapes. The inversion process can also be restricted to adjust one or a few of the possible model parameters whilst fixing others, e.g. to invert for dip, in other words to find the 'best-fitting' dip if the source is assumed to be a sheet.

It is preferable to direct the inversion towards more likely solutions through the use of an approximate, but appropriate, forward solution as a starting model; see for example Pratt and Witherly (2003) and the modelling example described in *Inverse modelling* in Section 3.11.2. In practice, there are often limited data available to constrain the nature and distribution of the geological units in the model. Starting models are invariably built from a combination of geological observations and inferences. Petrophysical data, if available, may be incorporated. A relatively small number of constraints can make a big difference to the result (Farquharson *et al.*, 2008; Fullagar and Pears, 2007).

Joint inversion

The reliability of a solution may be improved by modelling two or more types of data simultaneously in a process known as *joint* inversion, which is becoming increasingly common. These data have properties containing common or complementary information about the subsurface. The incremental results for one data type guide the changes made to the model during the inversion of the other data type, and vice versa. Early forms of joint inversion tended to assume a correlation between, say, the density and magnetism in equivalent parts of the model. This assumption is rarely justified, and more recent work has concentrated on correlating regions where the physical properties are changing rather than an explicit correlation of their magnitudes (Gallardo, 2007).

A joint inversion model may be more accurate as it is the 'best fit' to two disparate data types; and it has the additional advantage of reducing the time and effort needed to analyse two different datasets independently. A disadvantage is that more computational effort is required and that the results critically depend on the assumed relationship between the physical properties being inverted for. If the assumption is good then the results will probably be more reliable. If it is not, then the results could be less reliable than individual inversions of each dataset. When jointly inverting different types of geophysical data, it is important to account for the different physics of the methods. For example, and as shown in Fig. 3.65, magnetic data are more influenced by the shallow subsurface than are gravity data. A joint inversion of these two data types will be mostly influenced by the gravity data in the deeper parts of the model, so in this sense the inversion is not 'joint'.

Tomography

An important, but specialised, type of inverse modelling is based on tomography (Dyer and Fawcett, 1994;

Wong *et al.*, 1987). In tomographic surveys, measurements are made 'across' a volume of rock in a range of directions. Usually measurements are restricted to a plane or near plane. A signal is generated by a transmitter, and some property of the signal, usually strength, is measured on the other side of the volume. Any change to the signal, after accounting for the source–receiver separation, depends on the properties of the rock through which it has passed. The method is the geophysical equivalent of computer-axial tomography (CAT scans) which uses X-rays to create images of 'slices' through the interior of the human body.

The volume-slice is represented by a cellular model, and by combining many measurements it is possible, using inversion, to determine the physical property within each cell. The model is usually 2D, i.e. it is only one cell 'deep', so the results represent a 'slice' through the volume of interest. Good results require the signal to cross each cell over a wide range of directions, otherwise the results tend to show anomalies elongated in the measurement direction. This restricts this kind of survey to areas with a suitable number of drillholes and underground access. Even then the directions may be limited to a less than optimal range.

Tomographic surveys are relatively specialised, being mostly restricted to in-mine investigations. Surveys of this type using seismic and electromagnetic waves are described in Section 6.8.2 and online Appendix 5.

2.11.3 Modelling strategy

In modelling all types of geophysical data, model accuracy is determined not just by the match with the observed data, but also in deciding what aspects of the observed response to actually model, the type of model to use to represent the possible source and the degree of accuracy required. Perhaps most important is the skill in recognising the limitations of the result, which in turn dictates the appropriate use of the outcomes. The fundamental phenomenon of non-uniqueness is described in Section 2.11.4.

2.11.3.1 Accounting for noise

It is important to remember that the observed variations will contain a noise component (see Section 2.4). There is no advantage in modelling the data to obtain a match better than the noise level permits. In addition to noise inherent in the actual measurements, noise may be created by incorrect removal of, say, terrain effects and, in particular, the regional response (see Section 2.9.2). Sometimes it is better not to remove these variations and to include them in the model itself, a strategy worth considering if the terrain is severe or the regional response difficult to define. The disadvantage with this approach is that the model must be more complicated if it is to account for these extraneous effects. Often the constraints that can be reasonably applied to the deeper and distant parts of the model, to account for a regional field, are minor, requiring assumptions that are no more justified than those associated with removal of the regional. However, the interpreter needs to establish the ability of the inversion algorithm either to accommodate a background as part of the model or to automatically remove it, and also needs to establish the (mathematical) nature of the removal. It may be preferable, and indeed necessary, for the interpreter to remove a regional field first and apply the inversion to the residual dataset.

Variations in the properties of the near-surface are another source of noise. In general, these effects appear as short-wavelength variations in the data and should be identified for what they are and either filtered from the data before modelling deeper sources, or included as part of the model. The interpreter should be aware that aliasing (see Section 2.6.1) of these responses can introduce non-existent longer-wavelength variations into the data.

2.11.3.2 Choosing a model type

The choice of model type and its dimensionality are of fundamental importance in geophysical modelling. The choice requires consideration of the complexity of setting up the model and adjusting its parameters, and the computation effort and time required to obtain the response. For example, for simulating a dipping thin body or a layered feature, a thin-plate parametric model is simpler to define and faster to compute than a cell-based model. Conversely, a cell-based model may provide the only means of accurately simulating a complex 3D physical property distribution, but with the added complexity of defining and computing the responses of a very large number of model parameters.

2.11.3.3 Source resolution

The further away a geophysical sensor is from an anomalous source the more like a point the source appears to be, and the less is the information obtainable about the source from the measured response. This means that simpler model shapes can be used to obtain valid estimates of only the fundamental body parameters, i.e. depth and overall

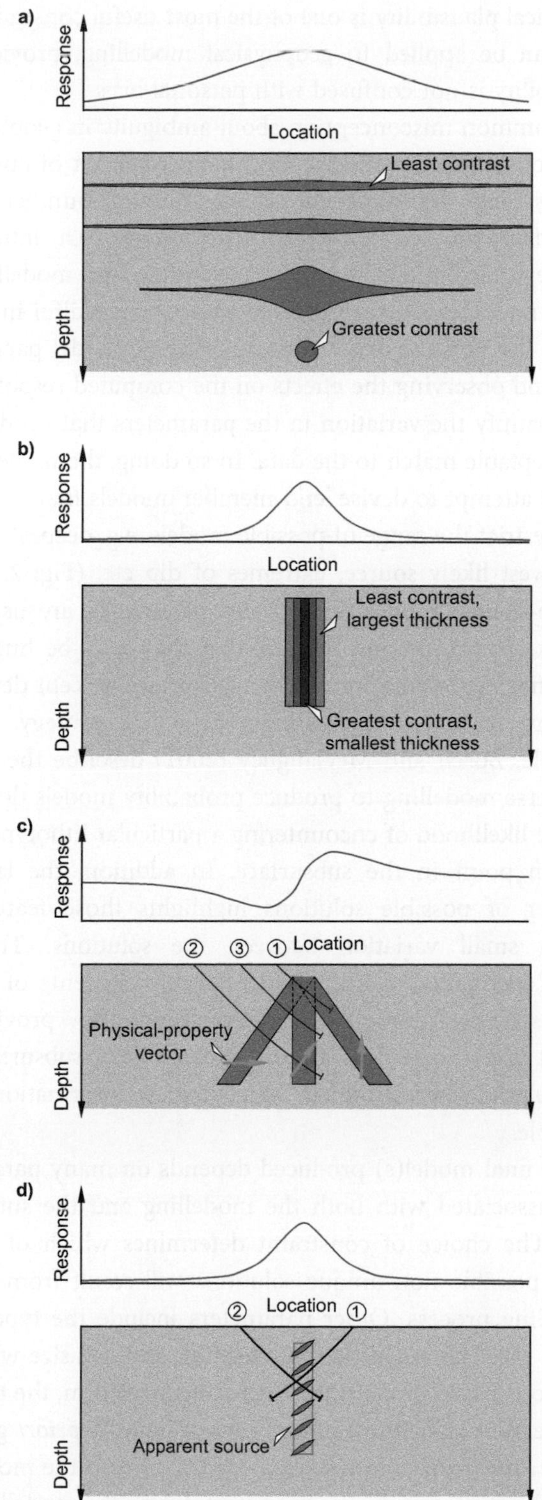

Figure 2.49 Forms of source ambiguity: (a) depth versus width/volume. Redrawn, with permission, from Griffiths and King (1981). (b) Thickness versus property contrast and (c) dip versus directionally dependent physical property (such as magnetism). Note the various drillhole lengths needed to test the various dip directions. (d) Misinterpretation due to lack of resolution. Note how drillhole (1) tests the modelled (apparent) source but misses the actual source.

size. So for very deep sources, a simple sphere, having only depth and diameter, can be an appropriate model to use.

Philosophically it is very important to apply Occam's razor when modelling geophysical data. The simplest possible source geometry should be used despite being sometimes criticised for being 'geologically unrealistic'. Simple geometry is geophysically appropriate when it conveys the actual level of information that the data are capable of providing.

2.11.4 Non-uniqueness

It is crucial to understand when modelling most kinds of geophysical data that there is no single mathematical solution, i.e. no unique model, for an observed response. Many models can be found with computed responses matching the observed data; a particular model is not necessarily the only possibility and not necessarily the correct one. A geological analogy is cross-section balancing as used to test structural interpretations. If the section does not balance, this is equivalent to the computed responses not matching the observed data. In this case the interpretation is definitely wrong. Successfully balancing the section is equivalent to creating a geophysical model that honours the observations. In either case, the interpretation could be right, but is not necessarily so, i.e. it is subject to ambiguity or non-uniqueness.

Non-uniqueness occurs because the fundamental characteristics of an anomaly, i.e. amplitude and wavelength which together control the gradients of the flanks of the anomaly, are reproduced by different combinations of the source's location, geometry/shape and its physical property contrast. As shown in Fig. 2.4, amplitude increases when the body is shallower or its physical property contrast increased, whilst wavelength increases as the body is made deeper or wider. This ability to 'trade off' different model characteristics occurs in several ways:

- Both body width (thickness) and depth affect the wavelength of the geophysical response; so adjusting either parameter allows a match to be found to the data (Fig. 2.49a).
- For sheet-like sources, only the product of the physical property contrast and the thickness can be resolved from a model (Fig. 2.49b). This means that it is impossible to resolve one parameter without knowing the other, e.g. increasing the thickness has the same effect as increasing the physical property contrast. This effect is quite

pronounced when the body is at a depth of about five or more times its horizontal width or thickness, and is then referred to as a *thin body*. When the body's depth is shallower, it appears wider to the survey and is then referred to as a *thick body*. In this case its thickness and physical property contrast influence the response more independently, so there is greater prospect of resolving each parameter.

- When the physical property of a body is a vector quantity such as magnetism, or directionally dependent (anisotropic), changes in the shape of the geophysical response can occur because of changes either in the orientation of the body or the direction of the physical property vector/variations. This is particularly a problem when analysing magnetic data (Fig. 2.49c).

Another source of non-uniqueness is associated with resolution. Obtaining a single coherent property distribution as the solution model does not obviate the possibility of the anomaly source actually being a collection of smaller close-spaced bodies with different physical property contrasts and depths, their properties being 'averaged' to obtain a single-source (Fig. 2.49d). In this case a source comprising en échelon veins is modelled as a vertical sheet. Drillhole 1, designed to intercept the inferred source, misses the actual source, but drillhole 2 would result in an intersection.

2.11.4.1 Dealing with non-uniqueness

When modelling geophysical data it is important to include all available complementary petrophysical, geophysical and geological information into the interpretation to reduce the ambiguity of the result. In practice this can be difficult to achieve as most geological information is depth-limited, being mostly from the near-surface, and drillhole information is often sparse and shallow. Also, physical properties can vary by large amounts over small distances.

A drillhole intersection providing a 'tie-point' on a source's position will hugely reduce the range of possible models. A set of petrophysical data can help limit the range of the physical property and is preferable to just a single-point measurement. A third constraint, which should be used with caution, is geological expectation. Of course, the fact that an anomaly in another part of the area was found to be caused by, say, a flat-lying body of massive sulphide does not mean that only bodies of this type should be used to explain other apparently similar anomalies. Despite this, geological plausibility is one of the most useful constraints that can be applied to geophysical modelling, provided plausibility is not confused with personal bias.

A common misconception about ambiguity in geophysical modelling concerns the fact that a given set of observations may be matched by an infinite number of subsurface physical property distributions. An infinite number of solutions does not mean that the modelling places no constraints on the subsurface. The skilful interpreter, through the experience of changing model parameters and observing the effects on the computed response, can quantify the variation in the parameters that produce an acceptable match to the data. In so doing, the modeller should attempt to devise 'end-member' models that effectively restrict the range of possible models, e.g. deepest and shallowest likely source, extremes of dip etc. (Fig. 2.49). Simple-shaped models having few parameters are useful in this respect, despite the fact that they may be hugely simplified representations of the subsurface. Recent developments in inverse modelling pursue this strategy. For example, Bosch and McGaughey (2001) describe the use of inverse modelling to produce probability models defining the likelihood of encountering a particular lithotype at a given point in the subsurface. In addition, the large number of possible solutions highlights those features having small variation between the solutions. These 'stable' elements are likely to be reliable elements of the final solution model. Non-uniqueness also provides insight into the geological possibilities of the subsurface, particularly where little or no geological information is available.

The final model(s) produced depends on many parameters associated with both the modelling and the survey data. The choice of constraint determines which of the many possible non-unique solutions will result from the modelling process. Other parameters include the type of model (1D, 2D etc.), the discretisation and cell size when using cell-based models, the modelling algorithm, the type of inversion algorithm, numerical accuracy, *a priori* geological knowledge and how it is used to control the model, and subjective user decisions applied during the modelling. The task for the interpreter then is to find the model(s) that offers the most geologically plausible explanation of the data. Given all the limitations imposed on the interpreter and the assumptions adopted in building a model, it is a reality of geophysical modelling that all models are a simplified and imprecise depiction of the subsurface, but some can be useful.

Summary

- Geophysical measurements respond to changes in the physical properties of the subsurface. Some common causes include a body of mineralisation, a contact of some kind, or the effects of weathering and alteration. These cause deviations from the background geophysical response and are called anomalies.
- Geophysical responses measured at the detector become weaker the further the source is from the transmitter (if there is one) and the detector. Increasing source–detector separation causes the geophysical response to extend over a wider area, i.e. there is an accompanying increase in wavelength of the anomaly.
- Responses not associated with the objective of the survey constitute noise. Environmental noise may be geological, e.g. terrain-related, or man-made, e.g. energy originating from powerlines. Methodological noise comprises unwanted consequences of the geophysical survey itself and of the processing of the geophysical data. A feature in the data caused by noise is referred to as an artefact.
- The two most common objectives of mineral geophysics are to map the local geology and/or to measure responses originating from the mineralised environment. Mapping surveys provide essential geological context, and in areas of poor outcrop may comprise the only useful form of 'geological' map available. Surveys for exploration targeting may be intended to detect or to define the geological/geophysical features of potential significance.
- A geophysical survey consists of a set of discrete measurements of what is a continuously varying parameter. In order for the samples to accurately represent the true variation in the parameter, the measurements must be made at an appropriate interval. Failure to do so results in aliasing, i.e. the creation of spurious long-wavelength variations in the data.
- Geophysical data processing comprises two distinct stages: data reduction and data enhancement. These are followed by data display. Reduction is the application of a variety of corrections to the raw survey data to remove obvious errors and correct for sources of noise. Data enhancement involves modifying the data in specified ways with numerical algorithms. We recognise three basic kinds: the merging of repeat readings (stacking); the comparison of two measurements (ratios); and more mathematically sophisticated manipulations of the data (filtering). Image processing methods are used to display, visually enhance and integrate multiple datasets to create readily interpretable pixel images of the survey data.
- Interpretation of geophysical data begins with a qualitative analysis of the data. This usually involves target identification and/or the creation of a pseudo-geological map. This is often followed by matching the measured response of selected features with the computed response of a model, i.e. a (simplified) representation of the subsurface geology. This is a form of quantitative analysis.
- The first step in making a pseudo-geological map is the creation of a structural framework from the linear and curvilinear features identified in the data. These may be interpreted from the alignment or truncation of other features, or areas of pronounced gradients; and they usually coincide with contacts and/or faults/shear zones. Next, consistent textural and tonal characteristics are used as the basis for mapping pseudo-lithological subdivisions.
- Modelling provides important information about source geometry and its location. However, it is important to recognise the limitations of the result, which is ambiguous (non-unique). Ambiguity can be reduced by incorporating petrophysical and geological data. Varying the model parameters and observing the effects on the computed response can quantify the range in model parameters producing an acceptable match to the data.

Review questions

1. What is the difference between absolute and relative measurements? What is a gradient measurement?
2. Give a practical definition of noise in a geophysical context. Describe some common types of non-geological environmental noise.
3. A data series comprises 1001 evenly sampled data values spaced 10 m apart. (a) What is the longest wavelength (lowest frequency) fully represented in the data series; (b) what is the shortest wavelength (highest frequency) present; and (c) what would it be if the sampling interval were reduced to 1 m?
4. What is levelling? Give examples of typical levelling artefacts.
5. Describe how data values can be manipulated to create coloured pixel images on a computer screen.
6. Describe techniques for separating the longer-wavelength regional field and shorter-wavelength local variations in a data series.
7. How can the strike extent and attitude of an anomaly source be estimated using a contour display of geophysical data? How do changes in depth to the source affect the display?
8. What is meant when a model is described as 1D, and as 2.5D?
9. Explain equivalence and the phenomenon of non-uniqueness. How are they related?
10. What is the difference between forward and inverse modelling? Describe the advantages and disadvantages of both.

FURTHER READING

Gupta, V.K. and Ramani, N., 1980. Some aspects of regional–residual separation of gravity anomalies in a Precambrian terrain. *Geophysics*, 45, 1412–1426.

Good clear demonstration of the how and why of the creation of a residual gravity map of greenstone terrain in the Superior Province, Canada. The authors conclude that analytical methods are inferior to manual graphical interpolation. However, Kannan and Mallick (2003), working with the same data, show that interpolation based on the finite-element methods produces a similar result to the manual methods.

Hamming, R.W., 1989. *Digital Filters.* Prentice-Hall International.

This is an introductory text to the broad field of digital signal processing and provides good plain-language explanations of digital sampling, the Fourier transform, and digital filtering, with only the essential mathematics.

Meju, M.A., 1994. *Geophysical Data Analysis: Understanding Inverse Problem Theory and Practice.* Society of Exploration Geophysicists, Course Notes 6.

A very good description of geophysical inverse methods requiring only a low level of mathematical expertise.

Oldenburg, D.W., Li, Y., Farquharson, C.G. *et al.* 1998. Applications of geophysical inversion in mineral exploration. *The Leading Edge*, 17, 461–465.

Oldenburg, D.W. and Pratt, D.A., 2007. Geophysical inversion for mineral exploration: a decade of progress in theory and practice. In Milkereit, B. (Ed.), *Proceedings of Exploration '07: Fifth Decennial International Conference on Mineral Exploration.* Decennial Mineral Exploration Conferences, 61–99.

Phillips, N., Oldenburg, D., Chen, J., Li, Y. and Routh, P., 2001. Cost effectiveness of geophysical inversion in mineral exploration: Application at San Nichols. *The Leading Edge*, 20, 1351–1360.

These three papers discuss the philosophy of inverse modelling and present examples of inversion of various types of geophysical data using mineral exploration examples.

Shermer, M., 2005. *Science Friction: Where the Known Meets the Unknown.* Times Books.

This book is not about geoscience but about scientific thinking and ideas in general, in particular how personal bias and preconceptions can affect scientific judgements. Highly recommended as an interesting read and a warning about how difficult being a good scientist actually is.

Silva, J.B.C., Medeiros, W.E. and Barbosa, V.C.F., 2001. Potential-field inversion: Choosing the appropriate technique to solve a geological problem. *Geophysics*, 66, 511–520.

Using a simple gravity-based example this paper provides a straightforward, non-mathematical summary of the value of constraints in the inverse modelling of geophysical data.

Ulrych, T.J., Sacchi, M.D. and Graul, J.M., 1999. Signal and noise separation: Art and science. *Geophysics*, 64, 1648–1656.

A seismically orientated discussion posed mainly in mathematical terms. Nevertheless still accessible and thought-provoking. Also notable for the rare introduction of a little dry wit into a scientific paper.

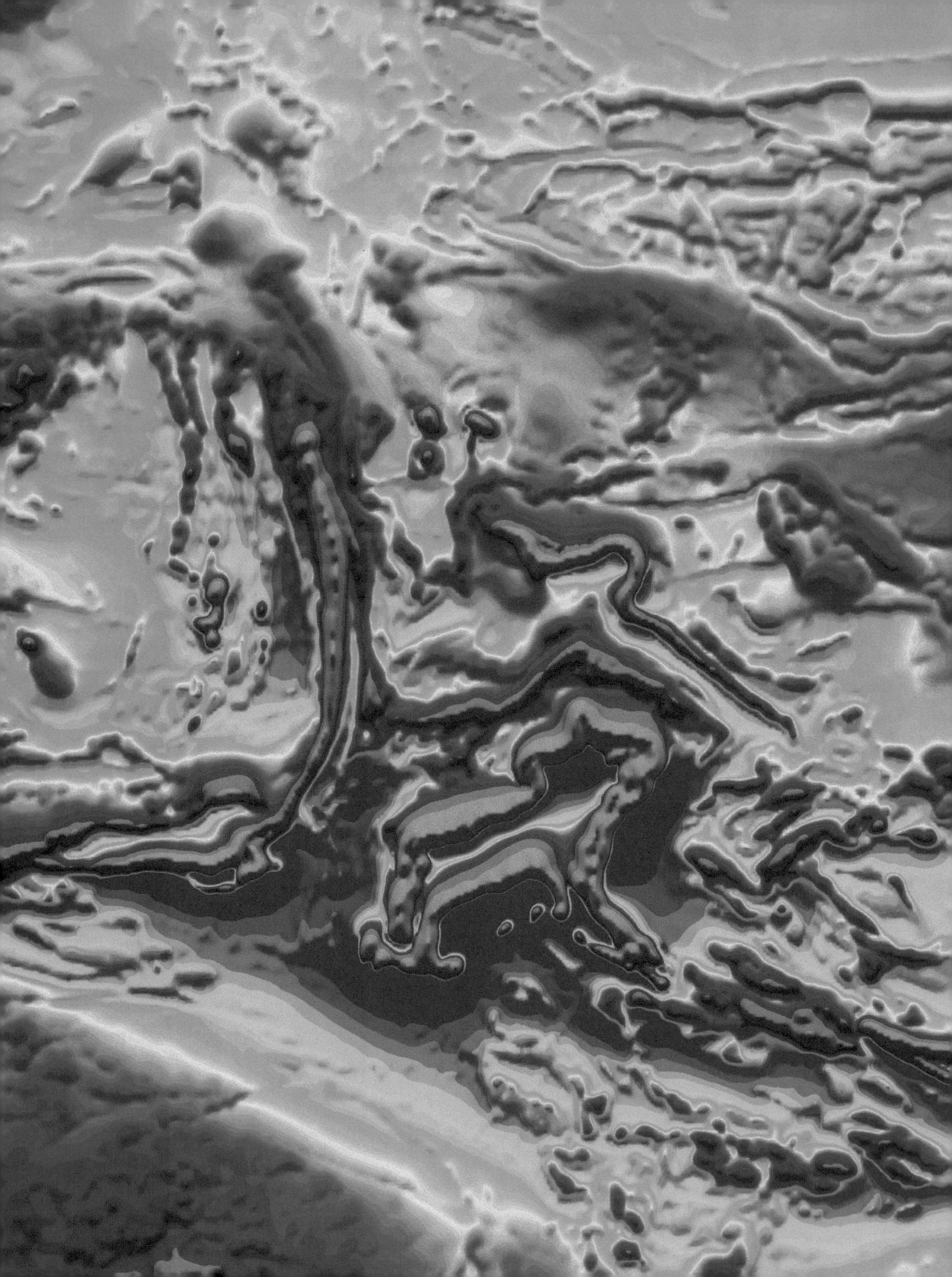

CHAPTER

3 Gravity and magnetic methods

3.1 Introduction

The gravity and magnetic methods measure spatial variations in the Earth's gravity and magnetic fields (Fig. 3.1). Changes in gravity are caused by variations in rock density and those in the magnetic field by variations in rock magnetism, which is mostly controlled by a physical property called magnetic susceptibility. Gravity and magnetic surveys are relatively inexpensive and are widely used for the direct detection of several different types of mineral deposits and for pseudo-geological mapping.

Magnetic measurements made from the air, known as *aeromagnetics*, are virtually ubiquitous in mineral exploration for wide-area regional surveying, for detailed mapping at prospect scale and for target detection. In areas where exposure is poor, aeromagnetics has become an indispensable component of exploration programmes. Gravity measurements are also used for regional and prospect-scale mapping but, historically, measurements of sufficient accuracy and resolution for mineral exploration could only be made on the ground. The development of airborne gravity systems, known as *aerogravity*, with precision suitable for mineral targeting, means that aerogravity in mineral exploration is likely to become as common as aeromagnetics.

Downhole gravity measurements for mineral applications are rare, but the recent development of instruments for use in the small diameter drillholes used by the minerals industry adds a new dimension to geophysical exploration. Downhole magnetic measurements, whilst not common, find application in mining for delineating highly magnetic targets. Downhole logging of rock density and magnetic susceptibility is routinely undertaken.

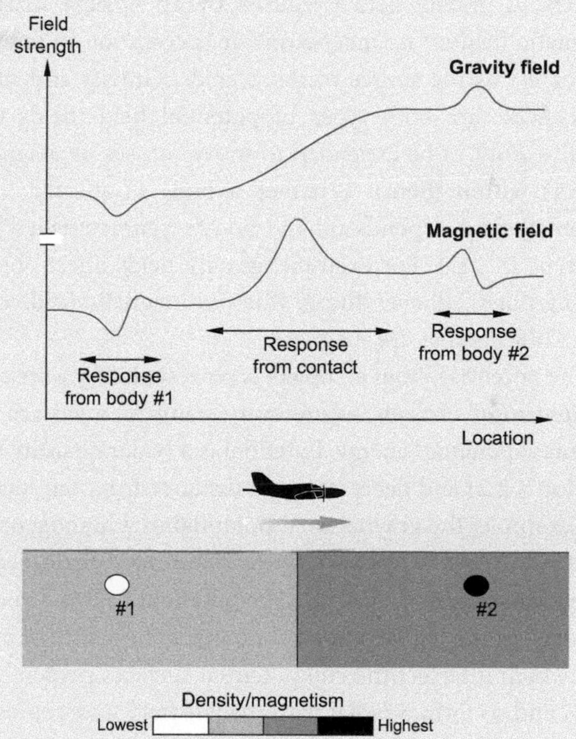

Figure 3.1 Schematic illustration showing variations in the strength of the Earth's gravity and magnetic fields due to variations in crustal density and magnetism, respectively. Note the simpler form of the variations in gravity compared with those in the magnetic field.

The gravity and magnetic methods have much in common, but they also have important aspects that

◀ Enhanced aeromagnetic image of the Padbury and Bryah Basins in Western Australia. The area shown is 75 km wide. Data reproduced with the permission of Geoscience Australia.

are distinctly different. In the first part of this chapter we describe the nature of density and magnetism and the properties of the associated gravity and magnetic fields. The acquisition and reduction of gravity and magnetic data are then described separately, since these are dissimilar. This is followed by a detailed description of aspects common to both methods: namely the processing, display and interpretation of the data. In discussing interpretation we emphasise the geological causes of variations in rock density and magnetism and their relationship to changes in the gravity and magnetic fields. Finally, we present some case studies chosen to demonstrate the application of the gravity and magnetic methods in exploration targeting and geological mapping, and also to illustrate some of the processing and interpretation methods described.

3.2 Gravity and magnetic fields

Gravity and magnetism involve the interaction of objects at a distance through the respective fields surrounding the objects. A gravity field is caused by an object's mass, a magnetic field by its magnetism. It is common to refer to the object as the source of these fields. Gravity and magnetic fields are both types of potential field (fields that require work to be expended to move masses or magnetic objects within them). Whether a body is affected by a potential field depends on the body's characteristics and the type of field. For example, gravity fields affect objects having mass, i.e. everything, whereas magnetic fields only affect objects that are magnetic.

The potential about an object is represented by a series of *equipotential* surfaces, where every point on a surface has the same *potential energy*. Potential is a scalar quantity (see Section 2.2.2) and decreases with distance from the source. For example, the gravitational potential of a homogeneous spherical object is described by surrounding spherical equipotential surfaces (Fig. 3.2a). The potential field is depicted by imaginary *field lines* directed towards the centre of mass and which intersect the equipotential surfaces perpendicularly, and so form a radial pattern. The field lines represent the direction of motion of an object influenced only by the field. Potential fields are vector quantities, i.e. they have both magnitude and direction (see Appendix 1). The strength of the field, also called its magnitude or intensity, decreases with distance from the source. The sources of gravity fields are polar or monopoles, i.e. they have one polarity, and always attract, unlike electrical and magnetic fields, which have poles of opposite polarity that can either attract or repel each other (see Fig. 5.2).

When multiple sources are present the fields of adjacent bodies interact and their equipotential surfaces merge. The field at any point is the vector sum of the fields (see Appendix 1) associated with each body. This is an important characteristic of potential fields for geophysics because sources of gravity and magnetic fields do not exist in isolation; so the ability to determine the combined effect of a group of several adjacent sources is very important. In the case of the Earth's gravity field, the equipotential surfaces approximate the shape of the Earth. The *geoid* (see Section 3.4.4) is the equipotential surface coinciding with mean sea level and everywhere defines the horizontal.

An object's gravitational potential energy changes when it is moved along a path that passes through an equipotential surface, so the potential difference between it and surrounding masses changes. Moving an object from the ground to a higher location opposes the action of gravity and involves crossing equipotential surfaces and increasing

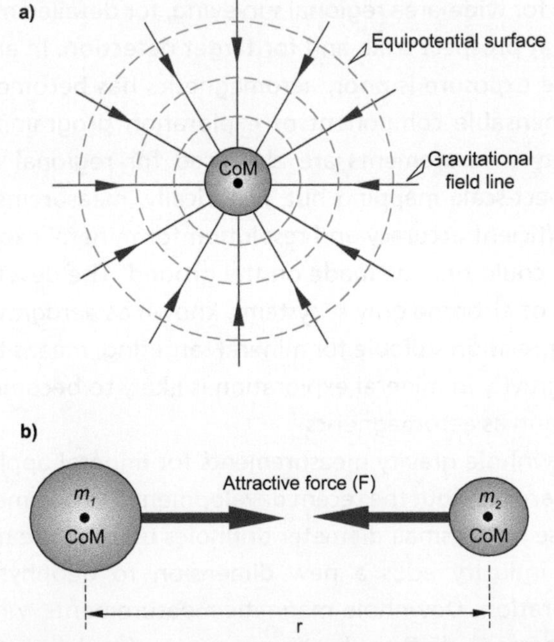

Figure 3.2 Gravitational attraction. (a) The gravitational equipotential surfaces and gravity field of a spherical mass. (b) The gravitational attraction between two masses (m_1 and m_2). CoM – centre of mass.

the object's potential energy. An object falls along the field lines towards the Earth's centre of mass in order to minimise its potential difference with the Earth's mass. Similarly, if a magnetic object is moved across the equipotential surfaces of a magnetic field, the object's magnetic potential energy is changed.

3.2.1 Mass and gravity

The essential characteristics of gravity can be explained in terms of mass, density and the gravity equation. Mass (m) is the amount of the matter contained in an object. Density (ρ) is the mass contained in a unit volume of the matter, i.e. mass per unit volume, and is a measure of the concentration or compactness of a material's mass. It is a fundamental property of all matter and depends on the masses and spacing of the atoms comprising the material. Both quantities are scalars (see Section 2.2.2). Mass is given by:

$$\text{Mass} = \text{density} \times \text{volume} \tag{3.1}$$

and from this

$$\text{Density} = \frac{\text{mass}}{\text{volume}} \tag{3.2}$$

The SI unit of mass is the kilogram (kg) and the unit for volume is cubic metres (m^3), so from Eq. (3.2) density has the units kg/m^3; however, it is common in the geosciences to use grams per cubic centimetre (g/cm^3). The two units differ by a factor of 1000 and, based on 1000 kg being equal to 1 tonne (t), sometimes densities are specified as t/m^3, i.e. 2650 kg/m^3 = 2.65 t/m^3 = 2.65 g/cm^3. The definition of a gram is the mass of 1 cm^3 of pure water at 4 °C. This means that a density is quantified relative to that of an equal volume of water, i.e. a substance with a density of 2.65 g/cm^3 is 2.65 times as dense as water.

The mass distribution and shape of an object are linked by the object's *centre of mass*. It is the mass-weighted average position of the mass distribution and, therefore, the point through which gravity acts on the object. For symmetrical objects of uniform density, like a sphere, cube, sheet etc., the centre of mass coincides with their geometric centres (Fig. 3.2a).

3.2.1.1 The gravity equation

All objects attract one another with a force proportional to their masses and, for spherical masses whose sizes are much smaller than the distance between them, inversely proportionally to the square of the distance between their centres of mass. This is known as the *Universal Law of Gravitation* and is the reason that objects are 'pulled' towards the Earth. The attractive force (F) between the two masses (m_1 and m_2) separated by a distance (r) is given by the gravity equation (Fig. 3.2b):

$$F = G\frac{m_1 m_2}{r^2} \tag{3.3}$$

The constant of proportionality (G) is known as the *universal gravitational constant*. In the SI system of measurement it has an approximate value of 6.6726×10^{-11} m^3 kg^{-1} s^{-2}.

If we suspend an object in a vacuum chamber (to avoid complications associated with air resistance), and then release it so that it falls freely, it will be attracted to the Earth in accordance with the Universal Law of Gravitation. The object's velocity changes from zero, when it was suspended, and increases as it falls, i.e. it accelerates. This is acceleration due to gravity and it can be obtained from Eq. (3.3) as follows. Consider a small object, with mass m_2, located on the Earth's surface. If the mass of the Earth is m_1 and its average radius is r, entering the relevant values into Eq. (3.3) gives:

$$F = 9.81 m_2 \tag{3.4}$$

The attractive force acting on the object, due to the mass of the Earth, is the object's weight, which is proportional to the object's mass. For a body with unit mass ($m_2 = 1$ kg), the average acceleration caused by the mass of the Earth, i.e. gravity, at sea level is approximately equal to 9.81 m/s^2. Acceleration due to gravity is the same for all objects of any mass at the same place on the Earth. However, it does vary over the Earth owing to the Earth's rotation, variations in its radius and variations in its subsurface density; and also varies with height above the Earth's surface (see Section 3.4). Accordingly, a body's weight changes from place to place.

3.2.1.2 Gravity measurement units

Changes in gravitational acceleration associated with density changes due to crustal geological features are minute in comparison with the average strength of the Earth's gravity field. The SI unit of acceleration is metres/second/second (m/s^2) and in the cgs system of measurement the unit is the gal (1 gal = 1 cm/s^2), but they are so large as to be impractical for gravity surveying. Instead, a specific unit of gravity has been defined in the cgs system of measurement and is known as the *milligal* (mgal, where 1 mgal = 10^{-3} gal = 10^{-5} m/s^2). It is still in common use as there is no defined SI unit of gravity. An alternative unit known as the *gravity unit* (gu), which is 1 μm/s^2 (10^{-6} m/s^2), is also used (1 mgal is equal to 10 gu). We use gu throughout our description of the gravity method. Rates of spatial change in gravity, i.e. gravity gradients, are defined in terms of a unit known as the *Eötvös* (Eo), which is a gradient of 10^{-6} mgal/cm, equal to 1 gu/km, 10^{-3} gu/m, 10^{-9} m s^{-2} m^{-1} or 1 ns^{-2}.

3.2.1.3 Excess mass

A buried body with higher density than the surrounding country rocks, i.e. a positive density contrast ($\Delta\rho$), produces an increase in mass above that which would be present if the body had the same density as the country rock. This extra mass is known as *excess mass* (M_e) and is given by:

$$M_e = \Delta\rho V = (\rho_{body} - \rho_{country})V \qquad (3.5)$$

where V is the volume of the body, and ρ_{body} and $\rho_{country}$ are the densities of the body and the country rock, respectively.

The excess mass produces a positive gravity anomaly (see Section 3.2.2). When the body has lower density than the country rock, i.e. a negative density contrast, it exhibits a *mass deficiency* and produces a negative gravity anomaly; see Fig. 2.4a. Note that it is the excess mass (or mass deficiency as the case may be) that gives rise to the gravity anomaly and not the body's absolute mass.

Excess mass may be estimated from modelling (see Section 3.10.3) and the *absolute mass* (M_a) of the source determined as follows:

$$M_a = \frac{\rho_{body}}{\Delta\rho} M_e = \frac{\rho_{body}}{\rho_{body} - \rho_{country}} M_e \qquad (3.6)$$

3.2.2 Gravity anomalies

Figure 3.3 shows the effect on the gravity field of a spherical source in the subsurface that is denser than its surrounds. For the simple model shown, gravity measurements are made on a horizontal surface above the source, and the gravitational attraction due to the Earth is taken as constant over the area of the target response. As shown in Fig. 3.3d, the source's field is radially directed towards its centre of mass. The Earth's field is radially directed toward the centre of the Earth, but the Earth is so large that the field lines are effectively parallel in the area depicted by the figure. In the presence of an excess mass, the field lines deflect towards the anomalous mass; but the effect is negligible because the strength of the Earth's gravity field is substantially larger than that of any excess mass in the crust. The gravity fields of the Earth and the sphere add to form the resultant total gravity field. The gravity instrument is aligned in the direction of the total field (defining the vertical; see Sections 3.2 and 3.4.4), so the measured gravity anomaly (change in gravity) is simply the magnitude of the variation in the total field, which is the vertical component of the field due to the anomalous mass (Fig. 3.3c). Note that because a gravity source is a monopole, its gravity anomaly has single polarity (sign) and has its peak centred directly above the source. This simple model illustrates how a high-density body in the subsurface, such as a mineral deposit, can

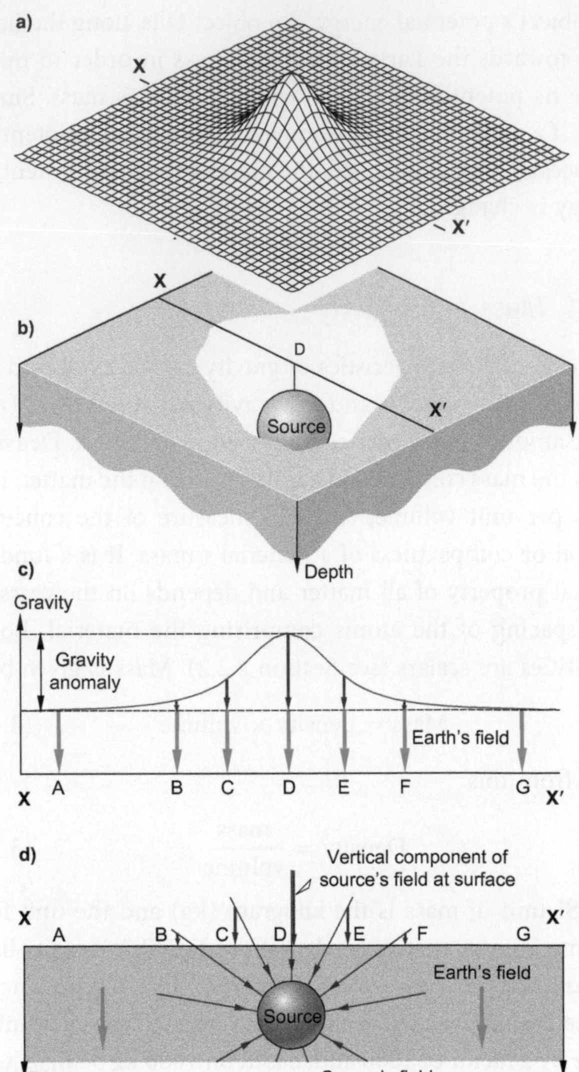

Figure 3.3 Gravitational field of a sphere. (a) Gravity measured on a horizontal surface above a dense spherical source (b). (c) Variation in the vertical component of gravity along the principal profile (X–X′) due to the combined effects of the Earth's field and that of the sphere. (d) The radially directed gravity field of the source and its vertical component. Note that the strength of the Earth's gravity field is many times greater than that of the source, but has been reduced here for clarity.

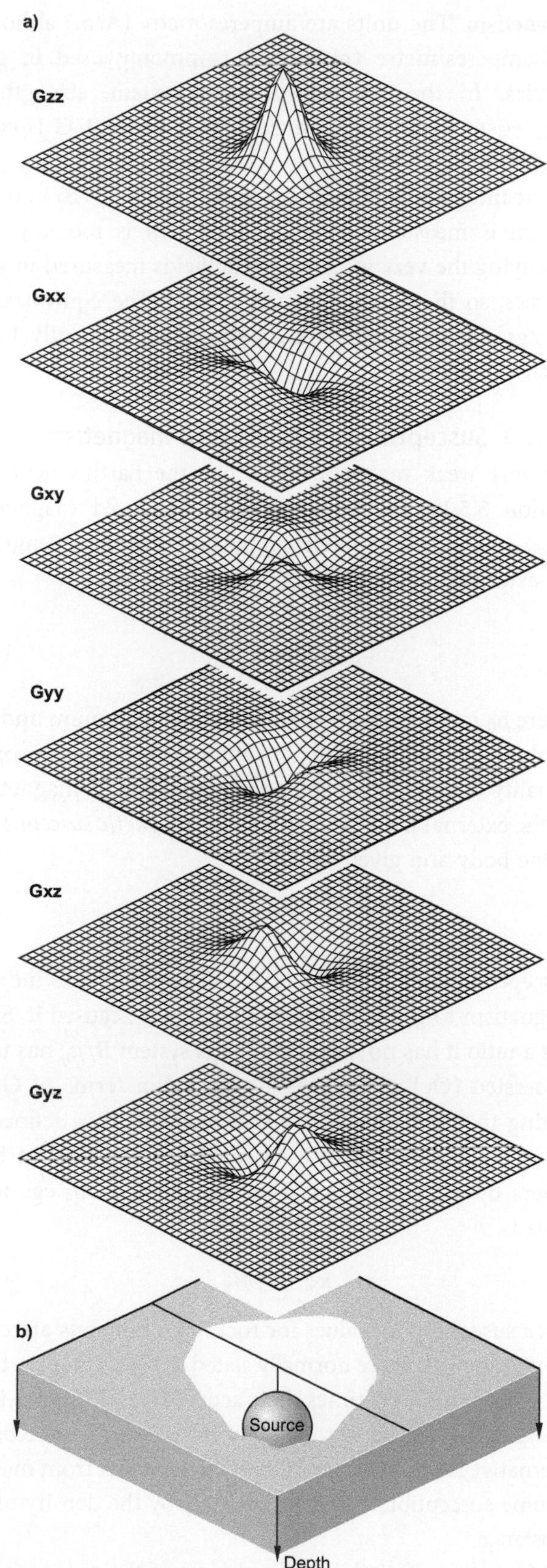

Figure 3.4 The gradients (a) of the gravity field which form the gravity tensor measured on a horizontal surface above a dense spherical source (b).

be remotely detected through its effect on the gravity field. It should be noted that variations in the vertical component of gravity measured beside or below the source, e.g. downhole or with underground measurements, will be different; explanations of these are beyond our scope.

More information about the shape and location of the source of a gravity anomaly can be obtained if, instead of measuring just gravity, tensor measurements are made (see Section 2.2.3). Three-component gradient or full-tensor gradiometers measure the gravity gradient in three perpendicular directions. Data of this kind provide more information about the source than gravity alone. However, and as shown in Fig. 3.4, the individual gradients are not easily correlated with the overall shape of the source, but collectively they are very useful for quantitative interpretation using inverse modelling methods (see Section 2.11.2.1).

3.2.3 Magnetism and magnetic fields

In describing the principles of magnetism and magnetic fields, we will use the term 'magnetism' when referring to the properties of a body that is magnetic and the term 'magnetisation' to describe the process of acquiring magnetism. In the literature it is common to use magnetisation and magnetism as equivalent terms, which can be unnecessarily confusing.

3.2.3.1 Magnets and magnetic dipoles

An important characteristic of magnetism is that it can be intrinsic, i.e. a material can inherently possess magnetism, or the magnetism can be induced through the material's being affected by an external magnetic field, such as the Earth's magnetic field. These two types of magnetism are called *permanent* or *remanent magnetism* and *induced magnetism*, respectively. Some materials exhibit both kinds of magnetism (at the same time); for some their magnetism is predominantly or solely of one type; and some other materials are incapable of becoming magnetic. The overall magnetism of a body is the vector sum (see online Appendix 1) of its induced (J_{Induced}) and remanent (J_{Remanent}) magnetisms.

The ubiquitous bar magnet is a convenient vehicle for describing the principles of magnetism. Figure 3.5a shows such a magnet and its external magnetic field. Magnetism can be described in terms of the *magnetic pole*, the fundamental element of magnetism. This is the basis of the older cgs system of measuring magnetism. Magnetic poles have either positive or negative polarity, referred to as north and south poles, respectively. Unlike poles attract and like poles

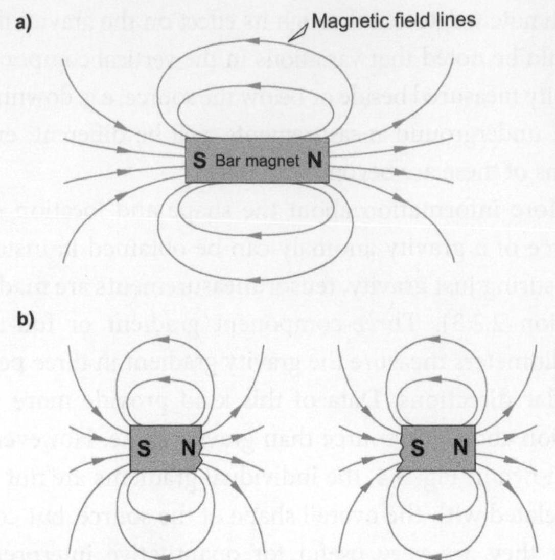

Figure 3.5 Bar magnet. (a) Field lines depicting the magnetic field of a bar magnet. The field is strongest at the poles where the lines converge. (b) Breaking the magnet into two pieces produces two smaller magnets.

repel. Magnetic poles of equal strength but opposite polarity always occur as a pair and are known as a *magnetic dipole*. The field of a magnet (a magnetic dipole) interacts with that of the Earth and the north pole of the magnet is that which is attracted towards geographic north (and similarly the south pole is attracted to geographic south). When a bar magnet is broken into pieces, each forms a smaller dipole (Fig. 3.5b). The process recurs when the pieces are broken into smaller pieces, and so on.

Magnetism is a vector quantity (see Section 2.2.2). The direction of the magnetism internal to a dipole is taken as directed away from the south pole and towards the north pole, the north pole being the positive pole. The external magnetic field of a dipole is the combined effect of the fields from the two magnetic poles. By convention, the direction of the magnetic field is taken as directed away from the north (positive) pole and towards the south (negative) pole. The field lines are closed paths and the strength and orientation of the field varies with location relative to the poles. It is strongest at the poles. The magnetic field of a bar magnet is like that produced by a constant electrical current flowing in a loop of wire (see Section 5.2.2). This electrical model of magnetism is the basis for measurements in the SI system.

3.2.3.2 Magnetic measurement units

In the SI measurement system, the intensity or strength of an object's magnetism is defined in terms of an electric current flowing in a loop of wire producing the same magnetism. The units are amperes/metre (A/m) although milliamperes/metre (mA/m) is commonly used in geophysics. In the cgs measurement system, strength of magnetism is measured in gauss (G), where 1 G is equal to 1000 A/m.

The intensity or strength of a magnetic field (B) in the SI system is measured in tesla (T), but it is too large for specifying the very weak magnetic fields measured in geophysics, so the nanotesla (nT) is used. The equivalent in the cgs system is the gamma (γ), and conveniently 1 γ is equal to 1 nT.

3.2.3.3 Susceptibility and induced magnetism

For very weak magnetic fields like the Earth's field (see Section 3.5.1), the strength of the induced magnetism (J_{induced}) is approximately proportional to the strength of the externally applied field (B) and given by:

$$J_{\text{induced}} = \kappa \frac{B}{\mu_0} \tag{3.7}$$

where μ_0 is the *magnetic permeability* of a vacuum and has a value of $4\pi \times 10^{-7}$ henry/m, and the constant of proportionality (κ), i.e. the degree to which a body is magnetised by the external field, is known as the *magnetic susceptibility* of the body and given by:

$$\kappa = \frac{\mu_0}{B} J_{\text{induced}} \tag{3.8}$$

Susceptibility is the ratio of the strength of the induced magnetism to the strength of the field that caused it. Since it is a ratio it has no units. In the cgs system B/μ_0 has units of oersted (Oe) and κ may be quoted in terms of G/Oe. Owing to the different way that magnetism is defined in the SI and cgs systems, a susceptibility specified in both differs by a factor of 4π. The conversion from cgs to SI units is:

$$\kappa_{\text{SI}} = 4\pi\kappa_{\text{cgs}} \tag{3.9}$$

Since susceptibility values for rocks and minerals are commonly small, they are normally stated as '$\times 10^{-3}$' or '$\times 10^{-5}$', i.e. 1000 or 100,000 times their actual value. Whether in SI or cgs units, *volume susceptibility* (κ) is usually quoted; the alternative is *mass susceptibility* (χ). To convert from mass to volume susceptibility, simply multiply by the density of the substance.

Magnetic susceptibility depends upon the *magnetic permeability* (μ) of the material in the following way:

$$\kappa = \frac{\mu - \mu_0}{\mu_0} = \frac{\mu}{\mu_0} - 1 \tag{3.10}$$

Magnetic permeability is analogous to electrical conductivity (see Section 5.2.1.2) and, put simply, accounts for how easily a magnetic field can exist within a material. For non-magnetic (most) minerals $\mu \approx \mu_0$.The ratio μ/μ_0 is known as the *relative permeability* (μ_r) and is approximately equal to one for non-magnetic materials (Zhdanov and Keller, 1994).

The higher a material's susceptibility, and/or the stronger the external field, the stronger will be the magnetism induced in the body. The induced field is parallel to the field that caused it. However, this description so far neglects the effects of self-demagnetisation and susceptibility anisotropy (see Sections 3.2.3.6 and 3.2.3.7). A magnetic anomaly arises from a contrast in magnetic susceptibility ($\Delta\kappa$), which is given by:

$$\Delta\kappa = \kappa_{body} - \kappa_{country} \quad (3.10.1)$$

where κ_{body} and κ_{country} are the magnetic susceptibilities of the body and the country rock, respectively.

3.2.3.4 Remanent magnetism

For some materials, the external field may cause irreversible changes to the material's magnetic properties (see *Ferromagnetism* in Section 3.2.3.5), so when the external field is removed the material retains a permanent or remanent magnetism. The strength and orientation of remanent magnetism is related to the external magnetic field at the time of its formation and is also affected by the magnetic mineral content of the rock, and by factors such as magnetic grain size and microstructure (see Sections 3.2.3.5 and 3.9.1). From a geological perspective, remanent magnetism is not truly permanent as it does change very slowly with the long-term variations in the Earth's magnetic field; but it may be assumed to be permanent for our purposes. There are numerous processes occurring in the natural environment that affect remanent magnetism. At different times during its existence a rock may be completely or partially remagnetised, and all, or part, of an existing remanent magnetism may be destroyed. Consequently, several phases of remanent magnetism may co-exist in a rock. The overall remanent magnetism is the combined effect, i.e. their vector sum (see online Appendix 1), of the various permanent magnetisms and is called the *natural remanent magnetism* (NRM).

Remanent magnetism is parallel, or very nearly so, to the Earth's magnetic field at the time the magnetism was created. Since the Earth's field changes in polarity and direction with both time and location (see Section 3.5.1), and since a rock unit is very likely to be subsequently rotated by tectonic processes such as faulting and continental drift, only very recently acquired remanent magnetisms are likely to be parallel to the present-day Earth's field. Remanent magnetism acquired at the time of a rock's formation is called *primary*, with *secondary* referring to magnetism acquired subsequently. Prolonged exposure to the Earth's magnetic field can produce a secondary magnetism known as a *viscous remanent magnetism* (VRM). This type of magnetisation accounts for many remanent magnetisms being found parallel to the present-day Earth's field.

The ratio of the strengths of the remanent magnetism (J_{remanent}) and induced magnetism (J_{induced}) is known as the *Königsberger ratio* (Q):

$$Q = \frac{J_{\text{remanent}}}{J_{\text{induced}}} \quad (3.11)$$

Because Q is a ratio it has no units. When it is greater than 1, remanent magnetism is dominant, and vice versa. Although the Königsberger ratio gives an indication as to whether induced or remanent magnetism is dominant, the directions of each component also significantly influence the resultant overall magnetism. When the induced and remanent magnetisms have similar directions, their effects will be mainly additive creating a stronger overall magnetism; and they will be subtractive when they have opposite directions, so the resultant magnetism is less.

3.2.3.5 Types of magnetism

The magnetic properties of a material are determined by the electron spins and their orbital motions in the atoms, the concentration of magnetic atoms or ions, the interaction between the atoms, and the molecular lattice structure. For most atoms and ions the magnetic effects of these cancel so that the atom or ion is non-magnetic. In many other atoms they do not cancel, so overall, the atoms have a magnetic dipole forming the material's intrinsic or spontaneous magnetisation – the magnetisation in the absence of an external magnetic field. The reaction of the atomic structure of matter to an external field can be classified into three distinct types of magnetism: diamagnetism, paramagnetism and ferromagnetism, which are reflected in the material's magnetic susceptibility. Ferromagnetism is about a million times stronger than diamagnetism and paramagnetism.

Diamagnetism and paramagnetism

Materials in which the atomic electron spins align so that their magnetic dipoles oppose an external magnetic field have a characteristic weak negative susceptibility and are diamagnetic. Geological materials with these characteristics can be considered non-magnetic in geophysical surveying. Materials in which the electron spins align so

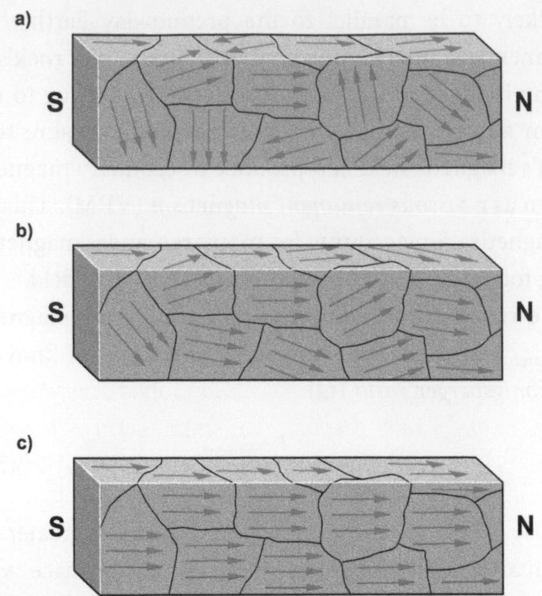

Figure 3.6 Schematic illustration of magnetic domains. In the unmagnetised piece of iron (a), the domains are randomly oriented. Imparting an approximate alignment to the domains creates a weak magnet (b). Magnetic saturation occurs when all the domains are aligned (c). Note that the sizes of the individual domains are greatly exaggerated.

that their magnetic dipoles align with an external field exhibit a weak positive susceptibility and are paramagnetic. They can produce very weak magnetic responses visible in high-resolution geophysical surveys. In both cases only induced magnetism is possible.

Ferromagnetism

Materials in which the atomic dipoles are magnetically coupled are known as *ferromagnetic*; the nature of the coupling determines the material's magnetic properties.

Ferromagnetism can be understood using the concept of *magnetic domains*. Domains are volumes in the lattice within which the magnetic vectors are parallel (Fig. 3.6). A series of domains with different magnetism represent a lower energy state than a single uniform direction of magnetism; their magnetic fields interact and they align to minimise the magnetic forces between adjacent domains. When domains are randomly oriented they cancel each other and the material is non-magnetic (Fig. 3.6a). In the presence of an external magnetic field, normally the Earth's magnetic field in the geological environment, the domains may grow by aligning with the external field leading to a net magnetism of the object as a whole (Fig. 3.6b), i.e. the material becomes magnetic with an induced magnetism.

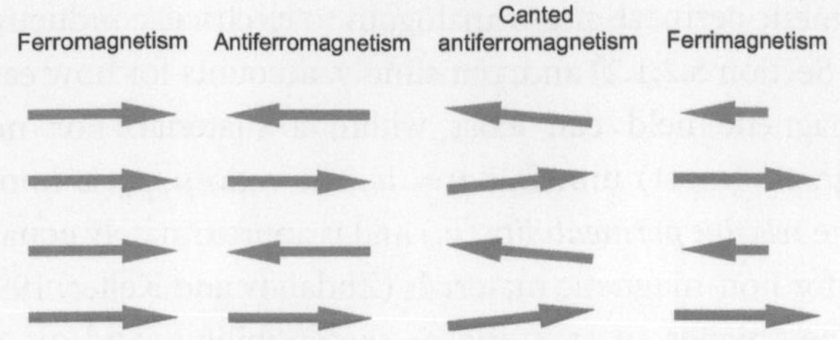

Figure 3.7 Schematic illustration of the alignment of magnetic dipoles in materials with different types of magnetism. Redrawn, with permission, from Harrison and Freiberg (2009).

A material's magnetism, or in the case of a rock the magnetism of an individual mineral grain, can be either *single-domain* or *multidomain*. The greater the number of domains in alignment, the stronger the magnetism, the limit being when all the domains are aligned (Fig. 3.6c). In this case the maximum magnetism possible for the material is reached and it is effectively a single domain. When the external field is removed the domains revert to their original state, but if the external field is sufficiently strong it may cause irreversible changes to the domains. The material will then have a remanent magnetism.

The most important ferromagnetic material is iron, but materials with this kind of magnetism rarely occur in the natural environment. Figure 3.7 shows the various types of ferromagnetism. In a ferromagnet the intra-domain magnetic dipoles are parallel and the material has a strong intrinsic magnetism and high susceptibility. Materials where the magnetic dipoles, of equal strength, are antiparallel with equal numbers of dipoles in each direction are known as *antiferromagnetic*. Imperfect antiparallelism of the dipoles, i.e. canted antiferromagnetism, may cause a small intrinsic magnetism, but the material does not acquire remanence. An example is haematite.

Another form of ferromagnetism occurs either when the antiparallel sub-domains of the lattice have unequal magnetisation or when there is more of one sub-domain type than the other (when the crystal lattice has two types of ions with different electron spins). This is known as *ferrimagnetism*; ferrimagnetic materials have high positive susceptibilities and can acquire remanent magnetism. Nearly all magnetic minerals are ferromagnetic, including monoclinic pyrrhotite, maghaemite, ilmenite and magnetite.

For simplicity we will follow Clark (1997) in referring to all strongly magnetic minerals as ferromagnetic (*sensu lato*). Ferromagnetic materials can have high susceptibility and can produce very strong magnetic responses in geophysical surveys. As the temperature increases, thermal agitation destroys alignment of the dipoles and ferromagnetism decreases. It disappears at a characteristic temperature called the *Curie point* where the material becomes paramagnetic and, therefore, has low susceptibility and no remanent magnetism. The Curie point is an intrinsic property of a ferromagnetic material dependent only on the material's composition. The Curie point of the common magnetic materials is exceeded at mid crustal depth, the exact depth depending on the specific mineral and the local geotherm. Magnetic anomalies originating at greater depths have been detected but their origin is uncertain; see McEnroe *et al.* (2009b).

3.2.3.6 Self-demagnetisation

In addition to the internal magnetisation described in Section 3.2.3.1, there exists a magnetic field internal to a body that, like the external field, extends from the north pole to the south pole. It has the effect of reducing the effective magnetism of the body through a phenomenon known as *self-demagnetisation*. The effect is influenced by the shape of the magnetised body. For example, in tabular bodies self-demagnetisation is greater perpendicular to the plane of the object than in the plane, so the overall magnetism is deflected towards the plane of the object. Self-demagnetisation produces a shape-related anisotropy that operates at a range of scales, from individual grains through to mineralogical layering, to the scale of orebodies and stratigraphic units. It depends on the strength of the external field but is only significant in highly magnetic materials ($\kappa_{SI} > 0.1$). Self-demagnetisation is less than the induced magnetism, but the apparent decrease in magnetism and deflection of the direction of magnetism must be accounted for when interpreting anomalies caused by strongly magnetised rocks. Geological materials that are sufficiently magnetic for self-demagnetisation to be important include banded iron formations and massive magnetite mineralisation.

3.2.3.7 Magnetic anisotropy

A mineral's magnetic properties are affected by its magnetic domain structure, which can differ along the various crystal axes. They are also influenced by the shape and size of the magnetic grain (see Section 3.2.3.6). The result is that individual mineral grains are often magnetically anisotropic, and this affects both their induced and remanent magnetisms. Susceptibility anisotropy deflects the induced magnetism away from the direction of the external inducing field towards the direction of highest susceptibility. The direction of the remanent magnetism is deflected from parallelism with the external magnetic field direction.

Where there is preferential alignment of magnetically anisotropic grains in a rock mass, then an overall magnetic anisotropy forms. Clark and Schmidt (1994) report strong anisotropy in banded iron formations from the Hamersley iron-ore province and the Yilgarn Craton of Western Australia, where susceptibilities parallel to bedding exceed those perpendicular to the bedding in the range of 2 to 4 times. In most cases the consequences of susceptibility anisotropy are insignificant with respect to interpreting magnetic responses for mineral exploration purposes, but anisotropy may be exploited for geological purposes, as reviewed by Hrouda (1982). Measurements of anisotropy of magnetic susceptibility (AMS) are commonly used to detect planar and/or linear *magnetic fabrics* that may be otherwise invisible. Examples of AMS studies with an economic geology context are those of Scott and Spray (1999), who describe work in the nickel sulphide-rich Sudbury Basin of central Canada, and Diot *et al.* (2003) who describe work on the Tellnes ilmenite deposit in southern Norway.

3.2.4 Magnetic anomalies

The strength of the Earth's magnetic field, which includes the main geomagnetic field associated with the Earth's core and the fields associated with the magnetism of the local rocks, is measured in magnetic surveys. Modern instruments measure the total field strength, usually referred as the total magnetic intensity (TMI), which is the resultant, or vector sum, of the vertical and the two horizontal components of the field (see online Appendix 1). Unless otherwise stated, reference to magnetic anomalies invariably means anomalies in TMI, a convention we use throughout our description of the magnetic method.

Figure 3.8 shows the variation in TMI across a horizontal surface above a sphere that is more magnetic than its surrounds. As described in Section 3.5.1, the direction and strength of the geomagnetic field varies around the Earth. Here the source is assumed to have only induced magnetism, and the resulting anomalies for different orientations (inclinations) of the inducing (geomagnetic) field are shown. The inclination is the direction of the field relative to the horizontal (see Section 3.5.1). Clearly, the resulting magnetic anomalies are markedly different, even though the shape of the source is the same. The observed variations in the magnetic field are the combined effects of the

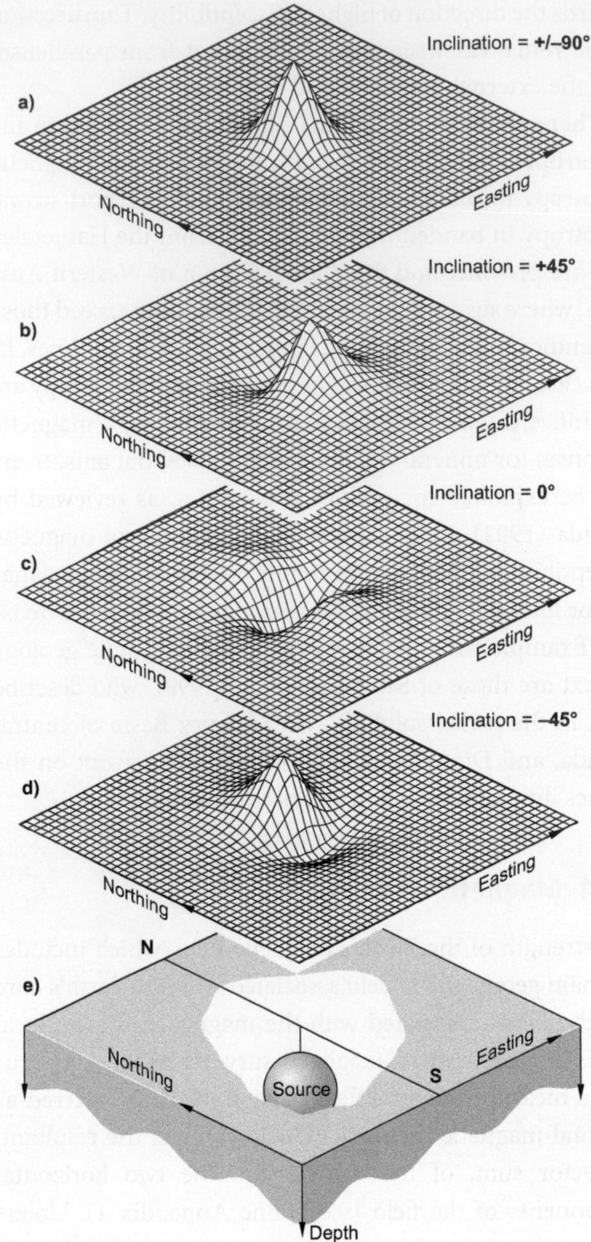

Figure 3.8 Total magnetic intensity measured on a horizontal surface above a sphere (e) magnetised by induction for inducing fields with different inclination. (a) Polar, inclination = 90°. (b) Northern hemisphere, inclination = +45°. (c) Equator, inclination = 0°. (d) Southern hemisphere, inclination = –45°.

magnetic fields of the Earth and the sphere, i.e. the vector addition of the two fields (see online Appendix 1).

Figure 3.9 shows variations in TMI along a traverse across the centre of a source. In some locations the field due to the sphere is in roughly the same direction as the Earth's field, so the strength of the resultant field (TMI) is greater than the Earth's field alone. Elsewhere the two fields are in opposite directions, so the resultant field is weaker than the Earth's field. For the case where the Earth's field is inclined, the body's magnetism is also inclined and the anomaly is asymmetrical and dipolar. Where the Earth's field is horizontal the dominant response is a counter-intuitive decrease in TMI above the body, even though it is more magnetic than its surroundings. This occurs because the two fields are opposed in the region above the body where the TMI is measured. Only when both the Earth's field and the body's magnetism are predominantly vertical does a simple response comparable to gravity occur.

Remanent magnetism, magnetic anisotropy and self-demagnetisation have the effect of changing the strength and direction of the body's total magnetism. The myriad of possible strengths and directions of the total magnetism allows for a great many possible anomaly shapes and amplitudes. Remanent magnetism is often a significantly complicating factor in the analysis of magnetic anomalies. The issue is discussed in detail in Section 3.10.1.2.

Anomalies measured below the surface can be envisaged in a similar manner to those in Fig. 3.9, but they can be more complex. Similar complications may arise in very rugged terrain since the magnetic geology may not all be below the survey point (see Section 2.4.1.1). Measuring the components of the magnetic field in one or more directions provides more information about the source than just a TMI measurement. This is standard practice in downhole surveys. As with gravity data, full tensor measurements fully define the field. Nevertheless, it is currently normal practice just to measure TMI, primarily because this can be done very quickly and easily, and sufficient information is obtained for most needs.

3.3 Measurement of the Earth's gravity field

The vast majority of the Earth's mass is contained within the core and the mantle, the mass of the crust being a tiny fraction of the total (about 0.4%). It is the interior components of the Earth that are mostly responsible for its gravity For this reason variations in the gravity field related to geological features in the crust are very small compared with the absolute value of gravity of about 9,800,000 gu. Note that the Earth's gravity field has no part in producing anomalies of interest; it is an external interference that must be compensated in order to resolve the anomalies. This is unlike the magnetic field whose presence is fundamental in producing magnetic anomalies.

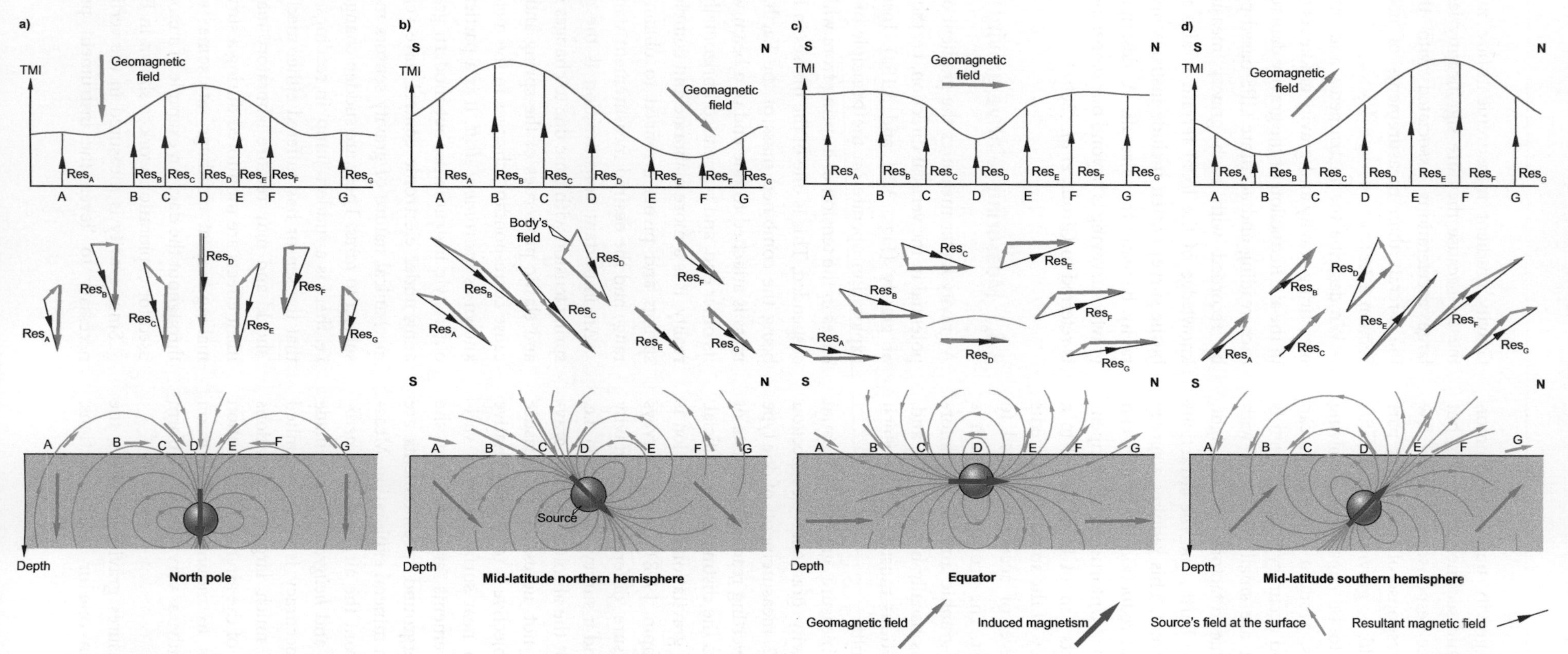

Figure 3.9 Schematic illustrations of the induced magnetic fields of a spherical source. (a) At the magnetic north pole, (b) mid-latitude in the northern hemisphere, (c) the magnetic equator and (d) mid-latitude in the southern hemisphere. Shown are the variations in TMI along the principal profile over the source resulting from vector addition of the Earth's (geomagnetic) field with that of the source, the induced magnetic field. The geomagnetic field strength is many times greater than that of the source, but has been reduced here for clarity.

Anomalies of hundreds of gravity units are common over large igneous intrusions and sedimentary basins, but most mineral deposits produce responses of only a few gravity units. An anomalous response of 0.1 gu is 1 part in 10^8 of the Earth's gravity field, so gravity measurements and survey procedures need to be capable of resolving these extremely tiny responses. Modern instruments can measure gravity to the required accuracy. The problem is that the responses of interest are small compared with variations in the gravity field due to factors such as height, topography, the rotation of the Earth, and the attractions of other bodies in the solar system. This requires a survey strategy designed to ensure that reduction of the data can remove or reduce these effects. Unfortunately, compensation for some of these effects cannot be achieved to a level comparable to the accuracy of the survey instruments and sometimes not to the level of accuracy needed to recognise the signals of interest. The principal problem is that sufficiently accurate topographic and density information from the survey area is usually not available and, inevitably, simplifying assumptions must be made which reduce the accuracy of the results.

Geophysical surveys may measure gravity or spatial variations in gravity, i.e. gravity gradients (see Section 2.2.3). As noted in Section 3.3.2, measurements of one type may be used as a basis for converting readings into data of the other type so in some ways the distinction is artificial. The instrument for measuring gravitational acceleration is known as a *gravity meter* (Chapin, 1998). Gravity surveys for mineral exploration measure differences in gravity between the survey stations and a survey base station, i.e. relative measurements, and not the absolute value of gravity (see Section 2.2.1). It is not necessary to measure absolute gravity because the objective is to identify relative changes in gravity related to near-surface density variations. Most gravity measurements are made on the ground. Downhole and underground measurements are possible, but are uncommon in mineral exploration. Measurements can also be made from the air, known as *aerogravity* for fixed-wing aircraft and *heligravity* when made from a helicopter, but their accuracy is severely limited by the need to remove the much larger accelerations associated with the movement of the aircraft. One solution which has shown promise is to mount the sensor on an airship, which is inherently a very stable platform (Hatch and Pitts, 2010).

A *gravity gradiometer* measures gravity gradients, the reading being the gradient in one or more directions. Gravity gradient measurements are mostly made from the air because they are significantly less affected by the large accelerations associated with the movement of the aircraft than measurements of normal gravity (see Section 3.3.2).

We describe the instruments and procedures used for measuring gravity and gravity gradients on the ground and in the air. Reduction of the gravity data requires the time of each reading and accurate GPS-based positioning data to be recorded with each gravity measurement. Detailed knowledge of the local terrain is also required, and airborne surveys often include instruments to survey topography beneath the aircraft. A description of downhole gravity surveying is beyond our scope; the interested reader is referred to Giroux *et al.* (2007).

3.3.1 Measuring relative gravity

A gravity meter measures the gradient of the gravitational potential in the vertical direction, i.e. the vertical attraction of gravity (Figs. 3.3d and 3.15a). Instruments used in exploration operate on the principle of measuring differences in the tension of a spring from which a small mass is suspended. This is one of the masses in Eq. (3.3), the other being the combined mass of the Earth. In principle, the mass is attached to one end of a beam which is pivoted at its other end, and suspended horizontally by the spring. In reality, it is a more elaborate and complex arrangement of springs and pivots arranged to obtain a large dynamic range and the desired measurement stability.

An important consideration is the gradual change in spring tension with time due to changes in the temperature and elastic properties of the spring and the beam, which cause corresponding changes in the meter reading. This is known as *instrument drift*. It is a particular problem with older-style instruments, but modern gravity meters operating under electronic control have very little drift. The mechanical nature of gravity sensors means they are also subject to *tares*. These are sudden changes in the drift rate, i.e. there is a sudden 'jump' in readings, usually indicating that the sensor has suffered either mechanical or thermal 'shock' and may require laboratory maintenance. Instrument effects are monitored during a survey by periodically making repeat readings at some reference location throughout the day. An example of a modern gravity meter used for exploration work is shown in Figure 3.10.

Since gravity is measured in the vertical direction, it is necessary to 'level' the instrument prior to making a

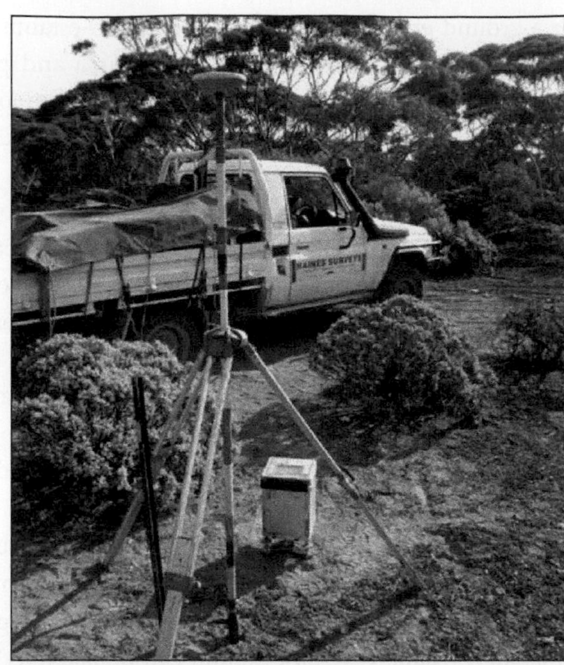

Figure 3.10 Scintrex CG-5 gravity meter in the field along with GPS equipment used for positioning. Reproduced with the permission of Haines Surveys Pty Ltd.

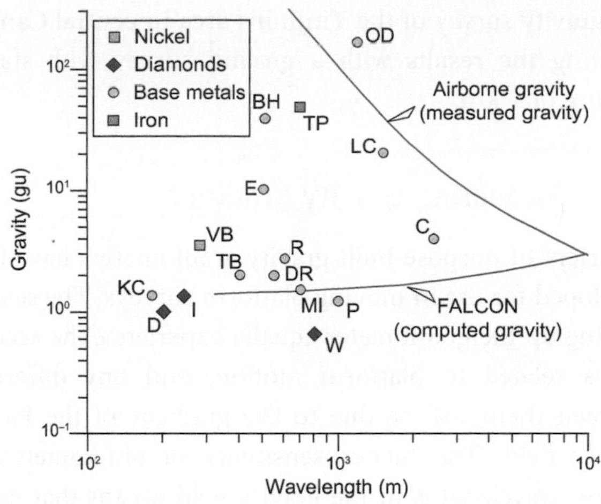

Figure 3.11 Plot of gravity versus wavelength showing the responses of various mineral deposits and the detectability limits of airborne gravity systems and the FALCON gravity gradiometer system. OD – Olympic Dam, TP – Mt Tom Price, BH – Broken Hill, LC – Las Cruces, E – Elura, C– Century, VB – Voisey's Bay, TB – Teutonic Bore, R – Rosebery, DR – Dugald River, MI – Mount Isa, P – Palabosa, KC – Kidd Creek, I – Impala, D – Diavik, W – Warrego. Redrawn, with permission, from Anstie *et al.* (2010).

measurement. This is facilitated by placing it on a tripod or a concave-upwards *base plate*, which provides a stable operating surface on which the instrument can easily be approximately levelled and operated. Modern electronically controlled digital gravity meters require the operator to level them only approximately, after which they automatically self-level. The instruments then automatically make a series of repeat gravity readings to improve the resolution of the reading, an example of stacking (see Section 2.7.4.1). The reading, its location and the time are recorded into an internal memory and the correction for Earth tides (see Section 3.4.2) may also be applied.

3.3.1.1 Airborne measurements

The measurement of gravity on a moving aircraft is a complex process given that the aircraft is undergoing movement around all three axes (roll, pitch and yaw) and experiences very large vertical accelerations up to the order of 1,000,000 gu. The gravity sensor must be mechanically stabilised so that its sensitive vertical axis remains vertical as the aircraft 'moves' around it. This requires the sensor to be mounted on a three-axis gyroscopically stabilised inertial platform. In addition, aircraft motions are measured and recorded with GPS data, and used post-survey to compensate the gravity data for the aircraft's accelerations and resolve the gravity signal from the underlying geology. The changing mass of the aircraft, mainly owing to fuel burn, is also measured. Errors in sensor orientation and stability, and in measurement of aircraft motion, propagate into the gravity signal as noise. Noise levels will vary according to local conditions, i.e. the amount of wind turbulence and the ruggedness of the terrain.

Short-wavelength noise is removed with a low-pass filter, so any signal with comparable wavelength is also removed. The performance of airborne gravity systems is defined by the magnitude and wavelength of the minimum detectable anomaly – parameters which are related to noise levels, the speed of the aircraft and the accuracy of the GPS-derived locations. Their detection limit is shown in Fig. 3.11; the curve represents the limit imposed by the accuracy of error correction based on GPS data. At the time of writing, these systems are capable of detecting the responses of only the largest of mineral deposits. Aerogravity surveying has proven to be effective for reconnaissance geological mapping in terrains with suitably large density contrasts over reasonably long wavelengths, so they are commonly used for petroleum-basin mapping. Recent developments, however, have produced systems useful for regional-scale mineral exploration.

Several aerogravity systems are operational at the time of our writing. Sander (2003) presents data from an aerogravity survey of the Timmins area in central Canada, equating the results with a ground survey with station spacing of 1 km.

3.3.2 Measuring gravity gradients

A variety of purpose-built gravity gradiometers have been developed for use in moving platform surveys. The sensors making up the gradiometer equally experience the accelerations related to platform motion, and any difference between them will be due to the gradient of the Earth's gravity field. The higher sensitivity of gradiometers to shorter wavelengths in the gravity field means that gradiometer data are more suitable for mineral exploration targeting than normal gravity systems measuring gravity directly. Importantly, gradiometer data can be transformed to normal gravity data showing the required shorter-wavelength variations with less error than obtained from normal gravity systems, or that would otherwise be unobtainable (Dransfield, 2007).

Two gravity gradiometer systems in use at the time of our writing are FALCON (Dransfield, 2007) and Air-FTG™ (Murphy, 2004). FALCON produces the vertical gradient of gravity from which normal gravity is computed. Dransfield (2007) provides examples of data from several types of mineral deposits. Air-FTG determines the gradient in each of the three perpendicular directions of all three components of the gravity field to produce full-tensor gravity (FTG) measurements (see Section 2.2.3). Hatch (2004) describes its application to kimberlite exploration. Figure 3.11 shows the detection limit of the FALCON system, which is also representative of other existing gradiometer systems. The ability of the higher-sensitivity gradiometers to detect a wide range of mineral deposits, compared with gravity systems, is clearly demonstrated.

Gradiometer measurements are very sensitive to near-surface features so the data are also very sensitive to variations in survey height and topography. Airborne gradiometer systems carry laser survey instruments for the acquisition of digital terrain data to centimetre accuracy for post-survey data reduction (see Section 3.4.5).

3.3.3 Gravity survey practice

Comprehensive descriptions of all aspects of geophysical data acquisition and survey design are given in Section 2.6. Details pertaining specifically to gravity surveying are described here.

For a ground gravity survey, it is essential to establish a *base station* close to, or within, the survey area and permanently mark it for later reoccupation. This is the location of repeat readings used to monitor instrument drift and detect tares (see Section 3.3.1) and the origin point for determining the latitudinal gradient of normal gravity (Section 3.4.4.1). Furthermore, because gravity meters measure the difference in gravity between survey stations, the base station is the essential common point of reference for all the survey measurements (see Section 2.2.1). For larger surveys, logistical constraints may dictate that a network of base stations be established throughout the survey area. In these cases, a separate survey is needed to establish the relative differences in gravity between the stations comprising the network. Strict adherence to measurement procedures is necessary in order to minimise errors in the network, as these will ultimately propagate into the survey data. One base station can be taken as the master base station for the entire survey, and often this is tied to a national gravity station where absolute gravity and height are accurately known. As an ongoing check for errors, repeat readings at various stations within the survey area is normal practice.

Absolute measurements of gravity have been made at numerous reference stations around the world to form a global network known as the *International Gravity Standardisation Net 1971* (IGSN71). Often the stations coincide with geodetic survey marks. To determine absolute gravity at a local base station, the difference in gravity between it and the reference station is measured. Tying a gravity survey to a national network is not necessary for identifying the types of responses of interest to mineral explorers. However, it does allow different surveys to be compared and merged at a later time.

In addition to the gravity measurement, the time of the reading, its location and height above sea level to centimetre accuracy are required for data reduction. These ancillary data can be obtained using GPS-based survey methods (Fig. 3.10).

Airborne gravity and gravity gradiometer surveys are usually conducted at a terrain clearance ranging from 80 to 150 m with survey lines spaced from 100 m to several kilometres apart depending on the nature of the survey. A magnetometer and sometimes radiometric equipment are carried on the aircraft. Both helicopters and fixed-wing aircraft are used but the gravity sensors are sensitive to 'strong' aircraft manoeuvres, so the tight turns at the end of flight-lines and the rapid elevation changes common in

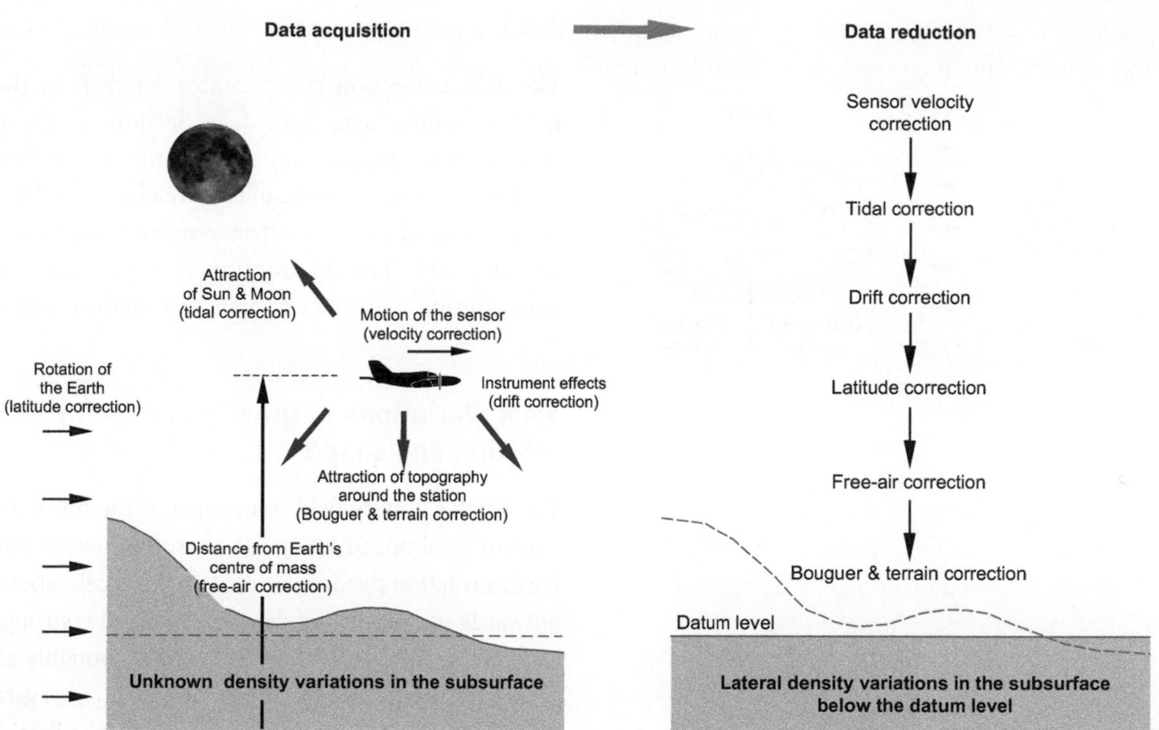

Figure 3.12 Schematic illustration of the non-geological causes of variations in measured gravity, and the sequence of their removal in data reduction.

low-level aeromagnetic surveys are generally to be avoided in gravity work. The survey heights are chosen to suit the gravity data and are usually greater than ideal for the magnetic and radiometric measurements (see Section 2.6).

3.4 Reduction of gravity data

The corrections applied to remove unwanted variations (noise; see Section 2.4) in gravity are known by the parameter that they correct or compensate, and we consider them in the order that they are applied to the observed data. The data reduction sequence is summarised in Fig. 3.12. Derivations of the various correction factors can be found in Lowrie (2007).

3.4.1 Velocity effect

The Earth's rotation produces an outward-directed centrifugal acceleration which acts on all objects and is compensated by the latitude correction (see Section 3.4.4.1). Objects moving across the rotating Earth experience an additional acceleration related chiefly to the east–west component of their velocity. The effect is greatest on the equator and decreases with increasing latitude. A body moving eastward (in the same direction as the Earth's spin) will experience an increase in centrifugal acceleration, whereas a body moving westward (against the Earth's spin) will experience a decrease in centrifugal acceleration. The vertical component of this acceleration, plus a small acceleration related to motion on the curved Earth, is known as the *Eötvös effect*. It needs to be accounted for in gravity measurements made from a continuously moving survey platform, such as an aircraft. The *Eötvös correction* (G_E) is:

$$G_E = 40.40V \cos\phi \sin\alpha + 0.01211V^2 \text{ gu} \tag{3.12}$$

where ϕ is the latitude of the measurement, V the velocity (in kilometres per hour) of the platform and α its heading direction with respect to true north. The correction is added to the gravity reading, although its polarity depends on the direction of motion of the gravity instrument. Error in the Eötvös correction depends chiefly on the accuracy in determining the platform's instantaneous velocity and heading.

3.4.2 Tidal effect

Gravity at every location varies with time owing to the gravitational attraction of the Moon, and to a lesser extent

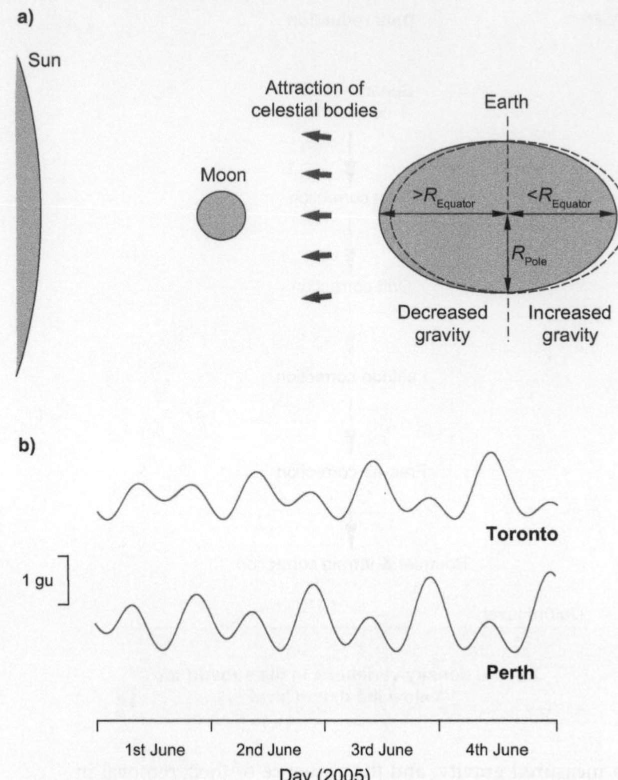

Figure 3.13 Earth tides. (a) Schematic illustration of the gravitational effects of the Sun and the Moon on gravity measurements on the Earth's surface. (b) Earth tides at Toronto (northern hemisphere) and Perth (southern hemisphere) over a 4-day period in June 2005. Note the approximate 12-hour periodicity.

that of the Sun. Even though the Moon has far less mass than the Sun, its closer proximity to the Earth means it exerts a greater influence, cf. Eq. (3.3). The gravitational attraction of extraterrestrial bodies has two components: the actual attraction of the bodies, and the distortions they cause to the Earth's shape known as *solid Earth tides* (Fig. 3.13a). The latter lead to changes in the Earth's radius of a few centimetres, and produce changes in the distance and mass between an observation point and the Earth's centre of mass. They are collectively referred to as *tidal effects* and produce gravitational changes of less than 3 gu. The effect varies with time and latitude owing to variations in the positions of the Moon and the Sun. It has a period of about 12 hours (Fig. 3.13b) and is greatest at low latitudes.

The variation in gravity due to the tidal effect can be accurately calculated for any time at any location on the Earth using the method of Longman (1959). Modern computer-based gravity meters automatically calculate and apply the tidal correction to the measured gravity.

3.4.3 Instrument drift

The drift correction compensates for drift in the instrument's reading over time (see Section 3.3.1). The drift correction is based on repeat readings at the base station (s). By recording the time of each reading, the drift rate can be determined and the effect removed from each gravity measurement. The assumption is made that the drift is linear during the time between base station readings.

3.4.4 Variations in gravity due to the Earth's rotation and shape

The Earth's gravity field is stronger at the poles than at the equator by about 51,000 gu as a result of several effects. The Earth's rotation produces a centrifugal acceleration directed outwards and perpendicular to its axis and is strongest at the equator. As a result the Earth's shape is roughly ellipsoidal (Fig. 3.14a), with the equatorial radius ($R_{equator}$) greater than the polar radius (R_{pole}). As a consequence, the Earth's surface gets progressively closer to its centre of mass at higher latitudes. This proximity effect is greater than the attraction at the equator from the extra mass between the surface and the centre due to the larger equatorial radius, as expected from Eq. (3.3). The centrifugal acceleration opposes the acceleration due to gravity, so gravity is less at the equator (Fig. 3.14a). At progressively higher latitudes, the centrifugal acceleration decreases and becomes increasingly oblique to the direction of gravity. It is zero at the poles.

The surface of equal gravitational potential (see Section 3.2) corresponding with mean sea level is known as the *geoid*. It is an undulating surface mainly influenced by variations in the distribution of mass deep within the Earth. It defines the horizontal everywhere and is an important surface for surveying. Numerous global and local geoids have been computed, a recent example being EGM96 (Lemoine *et al.*, 1998).

The direction of a suspended plumb line is the direction along which gravity acts (the field lines in Fig. 3.2a). It is directed downwards towards the Earth's centre of mass and is perpendicular to the geoid. It defines the vertical (see Fig. 3.15a).

The undulating geoid is not a convenient surface to represent the Earth for geodetic purposes. Instead a smooth ellipsoid is used that is a 'best-fit' to the geoid (Li and Göetze, 2001). It is known as the *reference ellipsoid*, also referred to as the *reference spheroid*, and is the idealised geometric figure to which all geographical locations are referenced in terms of their geographic coordinates,

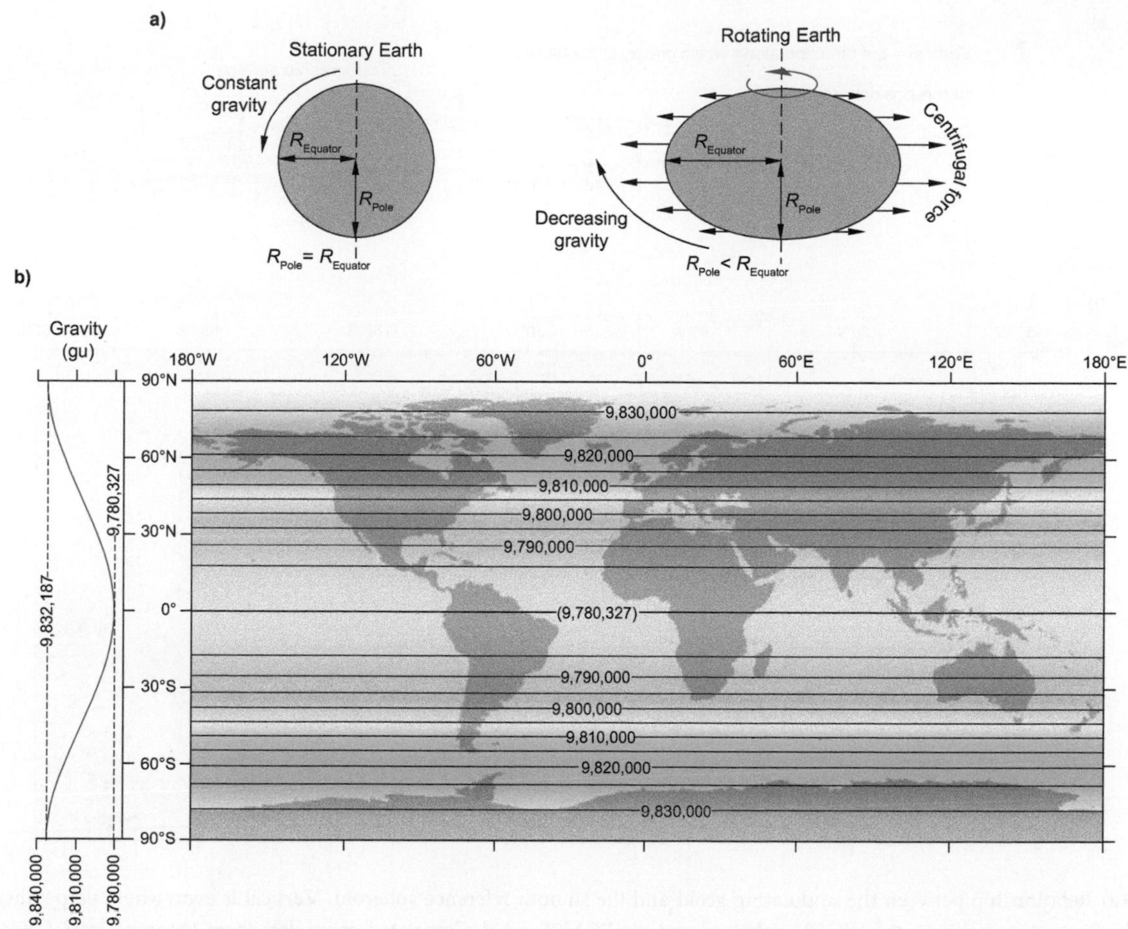

Figure 3.14 Gravity variations due to rotation and shape of the Earth. (a) Schematic illustration of the effects of the Earth's rotation and shape on gravity. (b) Variation of normal gravity with latitude calculated from the International Gravity Formula 1980. Units are gu.

latitude and longitude (Fig. 3.15a). There have been a number of determinations of the best-fitting reference ellipsoid. The geocentric Geodetic Reference System 1980 (GRS80) spheroid (similar to the World Geodetic System 1984 (WGS84) spheroid) is used as the datum for positioning and gravity surveying. The separation between the WGS84 spheroid and the undulating geoid is shown in Fig. 3.15b. Typically the separation is just a few tens of metres.

The variation in gravity with latitude (ϕ) is defined in terms of gravity on the surface of the spheroid, and is known as the *normal* or *theoretical gravity* (g_ϕ). It can be calculated using the International Gravity Formula, which is periodically updated. For the case of the GRS80 spheroid, the Gravity Formula 1980 given by Moritz (1980) is:

$$g_{\phi(1980)} = 9{,}780{,}326.7715 \frac{1 + 0.001\,931\,851\,353 \sin^2\phi}{\sqrt{1 - 0.006\,694\,380\,022\,9 \sin^2\phi}} \text{gu} \tag{3.13}$$

Normal gravity across the Earth is shown in Fig. 3.14b.

3.4.4.1 Latitude correction

The variation in gravity due to the difference in latitude between the survey station and the survey base station is compensated with the *latitude correction*. When latitude and the absolute gravity are known at all the survey stations, the latter obtained by tying the base station to a permanent absolute gravity mark or when making absolute gravity measurements, the latitude correction is the normal gravity calculated for that location (Eq. (3.13)). It is subtracted from the drift, Eötvös and tide-corrected reading. Since latitude is spheroid-dependent it is important that position is determined according to the correct spheroid (see Featherstone and Dentith, 1997).

Alternatively, the latitude correction can be obtained from the latitudinal gradient of normal gravity. At the scale of most mineral exploration activities, the change in normal gravity is sufficiently smooth and gradual that it appears as a very small linear increase in the direction of the nearest geographic pole. The latitude gradient

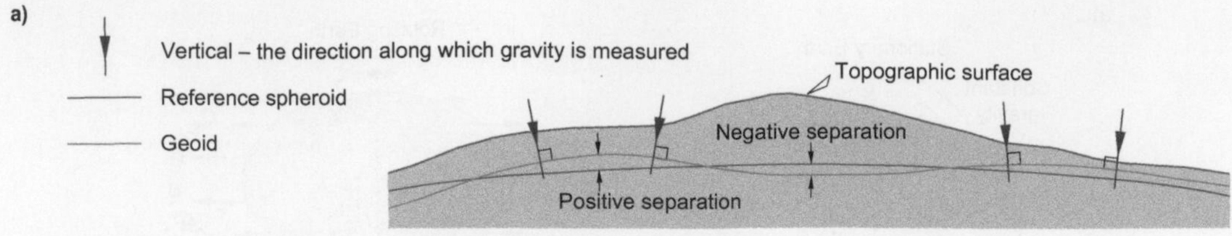

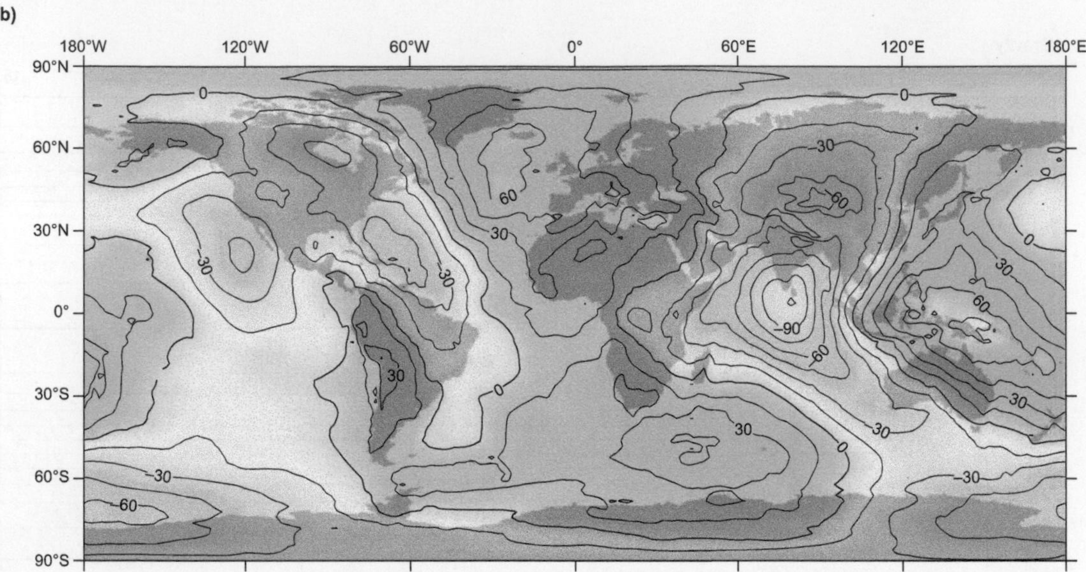

Figure 3.15 (a) Relationship between the undulating geoid and the smooth reference spheroid. Vertical is everywhere perpendicular to the geoid. (b) Separation between the WGS84 spheroid and the EGM96 geoid. Computed using data from Lemoine *et al.* (1998). Units are metres.

represents the change in gravity with north–south distance from a base station and is given by:

$$g_{\phi(\mathrm{N-S})} = 0.00812 \sin 2\phi \, \mathrm{gu/m}_{(\mathrm{N-S})} \tag{3.14}$$

where ϕ is the latitude (negative for southern hemisphere) of the base station. This formula is accurate enough for most exploration applications for distances up to about 20 km north and south of the base station (preferably located central to the survey area). It is useful where the latitudes of the gravity stations are unknown. The latitude correction at a gravity station is simply the north–south latitude gradient multiplied by the north–south distance (in metres) between the station and the base station. Since gravity increases towards the poles (Fig. 3.14b), the correction is subtracted for stations located on the pole side of the base station and added for stations on the equatorial side.

Station location needs to be known to about 10 m north–south in order to calculate the latitude correction to an accuracy of 0.1 gu.

3.4.5 Variations in gravity due to height and topography

Topography influences gravity measurements because it causes variations in station elevation, i.e. the distance between the station and the centre of the Earth's mass; see Eq. (3.3). Also significant are the effects of the materials forming the mass of the topography. They exert their own gravitational attraction which tends to oppose the decrease in gravity caused by increasing elevation, but as predicted by Eq. (3.3), the attraction of the mass has less effect than the distance factor. Terrain effects can be as complicated as the terrain itself; they are particularly strong in gravity gradient measurements.

Compensation of height and topographic effects involves three sequential corrections, known as the *free-air*, *Bouguer* and *terrain* corrections. When the gravity meter is located on the surface of the topography, the usual situation for a ground survey, all three corrections are then based on the height of the topography above some *datum*

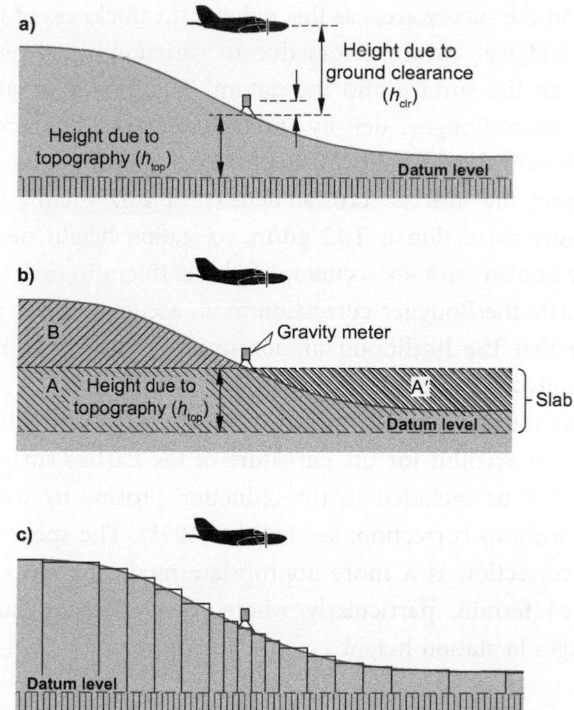

Figure 3.16 (a) The relationship between instrument height, topography and the datum level for the free-air correction. (b) The concept of the Bouguer slab. The significance of the various hatched areas is described in the text. (c) Representation of topographic variation (in 2D) by a series of flat-topped prisms. Note the smaller size of the prisms close to the gravity station.

level, usually the geoid (sea level), so they can be combined into a single *elevation correction*. When the gravity meter is elevated above the topography (Fig. 3.16a), e.g. on a tripod or in an aircraft, it is not possible to apply a single height-dependent elevation correction. Usually the three corrections are applied separately and in the sequence described below.

A common misconception regarding these corrections is that their application produces the equivalent reading that would have been obtained if the gravity meter had been located at the datum level. This is not the case, since such a process must include continuation of the field (see Section 3.7.3.2). Instead, what the reduction process does is correct for height and topographic effects whilst retaining relative changes due to lateral density changes in the subsurface.

Another influence related to topography is the isostatic state of the survey area, in particular whether local or regional isostatic compensation occurs (Lowrie, 2007). For example, the mass deficit of a mountain's root in local Airy-type compensation leads to lower gravity than a setting where the flexural rigidity is sufficient to carry the load of the mountain in a Pratt-type model. This is because the mass deficiency comprising the root zone partly balances the gravitational consequences of the mass excess represented by positive topography. If there is no local compensation, then obviously the gravitational effects of the terrain alone affect the reading. The isostasy-related changes in gravity tend to be of sufficiently long wavelength that they appear as a constant component in most exploration-related gravity surveys. They can be compensated using an *isostatic correction* (Lowrie, 2007), important for very large regional surveys, but not normally required for smaller mineral exploration-scale surveys.

3.4.5.1 Free-air correction

If a measurement of gravity were made on flat ground and then, at the same location, the measurement repeated at the top of a tall step ladder, the value of observed gravity would be less at the top of the ladder owing only to the increase in distance between the measurement location and the Earth's centre of mass; the $1/r^2$ component of Eq. (3.3). The change in gravity with height is compensated with the free-air correction (g_{FA}) given by:

$$g_{FA} = \frac{2G}{R}h \tag{3.15}$$

where h is the height (m) of the gravity station above the datum (usually the geoid), R is the average radius (m) of the Earth (6371 km) and G the universal gravitational constant given in Section 3.2.1.1. This reduces to:

$$g_{FA} = 3.086h \text{ gu} \tag{3.16}$$

which is simply the free-air gradient multiplied by the height. Station height needs to be measured with an accuracy of about 3 cm in order to calculate the free-air correction to an accuracy of 0.1 gu. The free-air correction compensates for height variations of the gravity meter above or below the datum level, where the station height is that of the topography (h_{top}) plus the height above the ground surface of the gravity meter (h_{clr}) (Fig. 3.16a). The free-air correction is added to the gravity reading, noting that height is negative for stations below the datum level so the free-air correction is then negative.

3.4.5.2 Bouguer correction

The free-air correction only accounts for the difference in height between the instrument and the datum level. When the height variation is caused by topographic variations, as is usually the case, the intervening mass of a hill above the datum, or lack of it in the case of the valley below the datum, affects gravity. When the gravity station is located on topography above the survey datum, it is located on a layer of rock that contributes mass between the gravity meter and the datum. The rock layer increases the measured value of gravity. Its effect is compensated by the Bouguer correction (g_{BOU}), which assumes that the Earth is flat, i.e. not curved, and that the rock layer is a flat slab of uniform density extending to infinity in all directions, i.e. the hatched area comprising zones A and A′ in Fig. 3.16b. It is given by:

$$g_{BOU} = 2\pi G\rho h_{top} \tag{3.17}$$

where h_{top} is the height (m) of the topography, or the thickness of the slab (m); ρ is the *Bouguer density* (g/cm^3), the average density of the slab between the gravity station and the datum level; and G the universal gravitational constant given in Section 3.2.1.1. This reduces to:

$$g_{BOU} = 0.4192\rho h_{top} \text{ gu} \tag{3.18}$$

The Bouguer correction is subtracted from the free-air gravity, noting that height is negative for stations below the datum level so the Bouguer correction is then negative.

The Bouguer correction is an approximation as it does not account for variable topography around the station (the top of the slab is flat). Furthermore, selecting the appropriate Bouguer density can be a problem. The average crustal density (2.67 g/cm^3; see Section 3.8) is often used but other values are appropriate for particular geological environments. For example, values as low as 1.8 g/cm^3 may be used in sedimentary basins, 0.917 g/cm^3 for ice-covered areas.

Since it is unlikely that the density of the rocks underlying the entire survey area will be the same, a variable Bouguer density may be more appropriate. The subsurface geology and density distribution needs to be known to determine the appropriate Bouguer density which, obviously, is a problem since determining subsurface density variations is the aim of the gravity survey itself (an example of the geophysical paradox; see Section 1.3). If variable density is chosen based on outcrop geology, an explicit assumption is being made that the geology is consistent vertically downwards to the datum level, which may well not be the case. Selecting a single density is usually the only practical solution, reflecting an acceptance of one's ignorance. The resulting errors are small, except in rugged terrain. If variable density is chosen, the datum level should be as high as possible, for example the lowest topographic level in the survey area, as this reduces the thickness of the slab and will reduce errors due to variations in density between the surface and the datum. The use of variable or constant Bouguer density and the consequences thereof are described in detail by Vajk (1956).

Using the average crustal density of 2.67 g/cm^3, the Bouguer correction is 1.12 gu/m, so station height needs to be known with an accuracy of about 10 cm in order to calculate the Bouguer correction to an accuracy of 0.1 gu. Note that the likelihood of non-uniform density is not accounted for in this calculation.

The infinite flat-slab model of the Bouguer correction does not account for the curvature of the Earth's surface. This can be included in the reduction process by using *spherical cap* correction; see LaFehr (1991). The spherical cap correction is a more appropriate model in areas of rugged terrain, particularly where there are very large changes in station height.

3.4.5.3 Terrain correction

The Bouguer correction assumes that the rock occupying the height interval between the datum level and the station is a uniform slab extending to infinity in all directions. Referring to Fig. 3.16b, the Bouguer correction removes the effect of the mass in zones A and A′. It fails to account for mass above the slab (B), i.e. mass above the gravity station. In contrast, the correction accounts for too much mass in regions where the topographic surface is lower than the station, i.e. the non-existent mass where the slab is above the actual ground surface (A′). The terrain correction explicitly addresses these limitations and therefore must be used in conjunction with the Bouguer correction.

Referring to Fig. 3.16, the vertical component of the gravitational attraction of the mass above the slab (in zone B) acts against the gravitational attraction of the subsurface to 'pull' the gravity sensor up. The observed gravity is then 'too low' and a positive correction is required. In the lower-lying area, zone A′, the Bouguer correction has assumed that mass is present here and in so doing has over-corrected; it has taken mass away, reducing the gravitational attraction. Since the Bouguer correction is subtracted, the corrected gravity is too small and again a positive correction compensates for this.

The terrain correction is the gravitational attraction, at the gravity station, of all the hills above the Bouguer slab and all the valleys occupied by the slab. It is obtained by determining the mass of the hills and the mass deficiencies of the valleys using topographic information and the

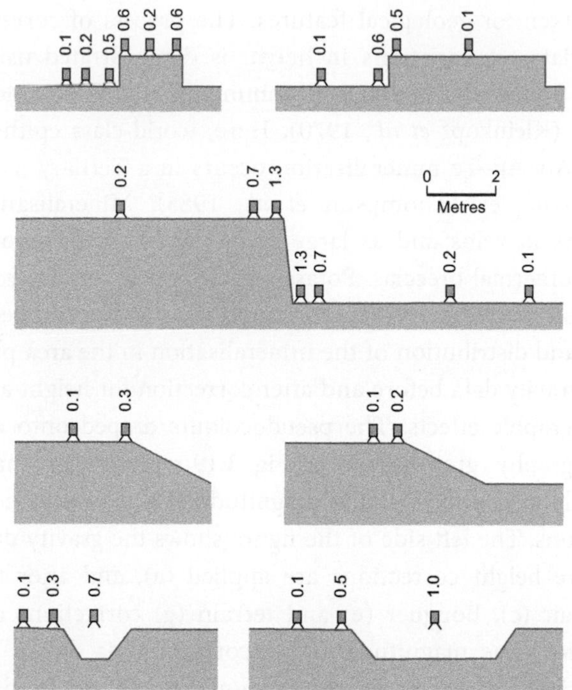

Figure 3.17 Estimated magnitude of the terrain corrections, in gu, due to small-scale topographic features close to the gravity station, calculated for a density of 2.67 g/cm^2. Redrawn, with permission, from Leaman (1998).

Bouguer density. The process is repeated for each gravity station as they will all (or mostly) have different heights so the Bouguer slab at each station has different thickness (Eq. (3.17)), and they all have a different relationship with the topography. Alternatively, the Bouguer correction can be ignored, and the gravitational attraction of the undulating terrain surrounding the gravity station can be computed as a full terrain correction with respect to the datum level.

The gravitational attraction of topography depends on the size of the topographic features, and decreases with their increasing distance from the gravity station. Depending on the desired accuracy of the survey and the ruggedness of the terrain, this means that relatively small features close to the survey station, such as culverts, storage tanks, reservoirs, mine dumps, open-pits and rock tors, can have a significant effect on the measured gravity and can be a major source of error in high-resolution gravity work (Fig. 3.17; Leaman, 1998). In addition, topographic features tens of kilometres away from the station, and even very large mountain ranges more than a hundred kilometres away, may need to be accounted for. The effects of more distant features appear as a regional gradient in the data so their effect could be removed with the regional field (see Section 2.9.2). The terrain correction is by far the most complex correction to implement as it requires the topography and its density distribution be accurately known; but this is usually difficult to achieve, so it is prone to error.

Digital terrain data provide the essential heights and shape of the topography needed for calculating the terrain correction. The ability to mathematically describe the terrain in terms of the digital data depends on the nature of the data and their resolution. Some airborne gravity systems are equipped with a laser scanner which maps the terrain in an across-line swathe below the aircraft, providing terrain information of sufficient resolution and accuracy for the terrain correction. For ground surveys, data may be true point (spot) heights taken from aerial photography or contour maps, or may be average heights of compartments subdividing the photography or contour map. It is important that the actual elevations of the gravity stations match the equivalent points on the DEM as discrepancies in heights and locations are sources of errors. The DEM may lack sufficient resolution to adequately define small topographic features and large abrupt surface irregularities, such as the edges of steep cliffs and the bases of steep hills, in the immediate vicinity of the gravity station, an additional source of error for stations affected in this way. During the gravity survey it is necessary to record details of small local features manually. Leaman (1998) provides practical advice regarding terrain effects close to the gravity station.

Various procedures for geometrically describing the digital terrain, so that its volume above the datum level can be calculated, have been implemented. The most common approach divides the terrain into a large number of volume elements (voxels), usually flat-top juxtaposed prisms, extending down to the datum level. Their gravitational attractions are computed and summed at each gravity station (Fig. 3.16c) and the whole process repeated for each gravity station. Given that topography closer to the gravity station exerts greater influence than more distant terrain features, efficient algorithms implement smaller prisms with more realistic, i.e. topography resembling, upper surfaces close to the station in order to minimise errors. Distant features, having a smaller gravity effect, are described more crudely with larger and fewer voxels, which also significantly reduces computing resources. Specific densities can be assigned to each prism but, as with the Bouguer density, using variable density implies significant understanding of the density variation in the survey area and its (distant) surrounds.

For gravity stations located near or over a mass of water, such as a lake, the terrain correction needs to correct down to the floor of the lake. Also, glaciers (density ranging from 0.79 to 0.88 g/cm^3) need to be accounted for in the terrain correction. A summary of terrain corrections, including unusual situations, is provided by Nowell (1999).

3.4.6 Summary of gravity data reduction

Gravity data may be presented in a number of different forms. Perturbations to the gravity field of non-geological origin are compensated by applying a sequence of corrections to produce observed gravity.

$$\text{Observed gravity} = \text{gravity meter reading} + \text{Eötvös correction (if required)} + \text{tidal correction} + \text{instrument drift correction} \quad (3.19)$$

Observed gravity is reduced to either the *free-air anomaly* (FAA) or the *Bouguer anomaly* (BA).

$$\text{FAA} = \text{observed gravity} - \text{latitude correction} + \text{free-air correction} \quad (3.20)$$

BA data are of two kinds. The Bouguer-corrected but not terrain-corrected Bouguer anomaly is sometimes referred to as the *partial* or *incomplete Bouguer anomaly*, and terrain-corrected Bouguer anomaly data as the *full* or *complete Bouguer anomaly.*

$$\text{Partial BA} = \text{FAA} - \text{Bouguer correction} \quad (3.21)$$

$$\text{Complete BA} = \text{FAA} - \text{Bouguer correction} + \text{terrain correction} \quad (3.22)$$

The greatest source of error in obtaining the complete Bouguer anomaly is uncertainties associated with the terrain correction. In many mineral terrains, topography and density are often highly correlated and highly variable. Flis *et al.* (1998) examine the underlying assumptions of the standard gravity reduction procedures for these environments and assess the errors they introduce in the context of iron ore exploration in the rugged Hamersley iron-ore province of Western Australia. They advocate the use of more accurate corrections based on a 'complete Bouguer correction' incorporating a full terrain correction and a variable density model.

3.4.7 Example of the reduction of ground gravity data

The correction of gravity data for height and topography is the process most likely to create artefacts that might be mistaken for geological features. The process of correcting data for variations in height is demonstrated using data from the Cripple Creek mining district in Colorado, USA (Kleinkopf *et al.*, 1970). Here, world-class epithermal Au–Ag–Te mineralisation occurs in a Tertiary igneous complex (Thompson *et al.*, 1985). Mineralisation occurs in veins and as large bodies within tectonic and hydrothermal breccias. Potassic alteration occurs in association with mineralisation. Figure 3.18 shows the geology and distribution of the mineralisation in the area plus the gravity data before and after correction for height and topographic effects. The pseudocolours draped onto the topography (grey surface) in Fig. 3.19 represent the magnitude of gravity and the magnitude of the various corrections. The left side of the figure shows the gravity data before height corrections are applied (a), and after the free-air (c), Bouguer (e) and terrain (g) corrections are applied. The magnitude of the corrections is shown on the right side of the figure. The uncorrected gravity data show a clear correlation between 'low' gravity and topographic 'highs'. This is due to the dominance of the free-air effect ($1/r^2$ in Eq. (3.3)). Applying the free-air correction reveals 'high' gravity correlating with high ground, which is mainly due to the mass of the hills. The Bouguer correction partially removes the effect of the mass so the correlation between topography and gravity is reduced; see, for example, the high ground in the foreground. Applying the terrain correction completes the correction process. Note how the highest values of the terrain correction correlate with the largest topographic features, i.e. where the slab approximation of the Bouguer correction is least appropriate.

The final reduced data show a negative gravity anomaly which correlates with the extent of the igneous complex (Fig. 3.18d). The mis-matches are mostly due to a lack of gravity stations to constrain the data. The majority of the mineralisation and associated alteration coincides with the lowest values of gravity. Kleinkopf *et al.* (1970) state that density determinations show that the volcanic breccia is about 0.20 g/cm^3 less dense than the surrounding Precambrian rocks, which is sufficient to explain the negative gravity anomaly.

3.5 Measurement of the Earth's magnetic field

Like the gravity field, most of the Earth's magnetic field does not originate in the crust, and it is smaller localised

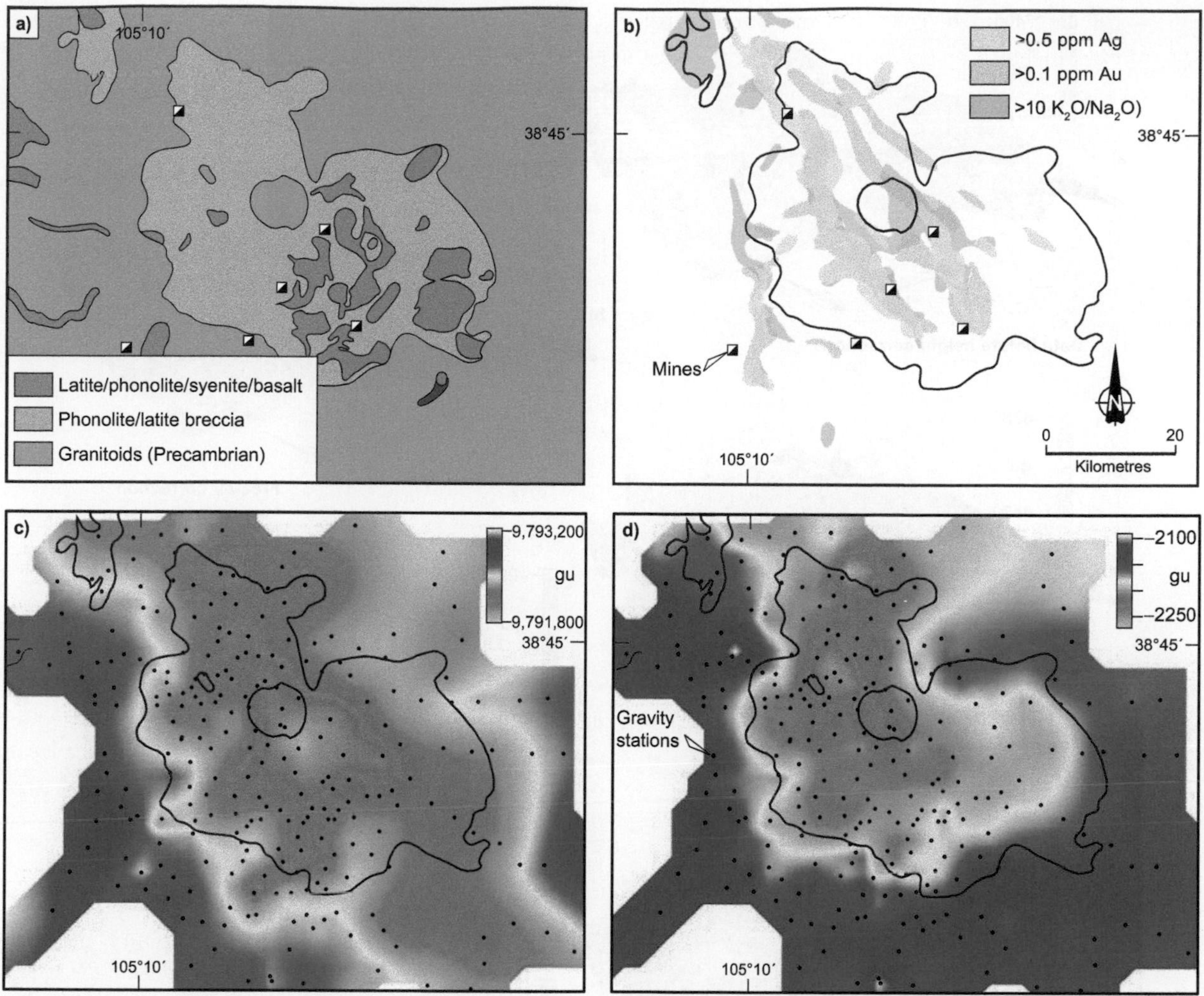

Figure 3.18 Geological and geophysical data from Cripple Creek. (a) Simplified geology, (b) distribution of Au–Ag–Te mineralisation and associated hydrothermal potassic alteration, (c) gravity before correction for height and topography, and (d) complete Bouguer anomaly. The gravity images were created from open-file gravity data. Geological data based on diagrams in Kleinkopf *et al.* (1970).

variations of the field due to magnetic materials in the crust that are of interest in mineral exploration. The average strength of the Earth's magnetic field is about 50,000 nT and variations of geological origin may exceed 10,000 nT, which is about 20% of the field strength, so variations can be extremely large compared with the very small influence that local geological features have on the gravity field. However, these large variations are rare. More common are variations of tens or hundreds of nanoteslas.

Unlike the gravity field, the Earth's magnetic field is fundamental in determining the strength and shape of crustal magnetic anomalies. Without it there would be no magnetic anomalies and no formation of remanent magnetism. The magnetic field changes significantly in both direction and strength over the Earth and at time scales which are significant for exploration surveys. These short-term temporal variations are a source of noise during magnetic surveying, and corrections can be applied to compensate for them.

The instrument used to measure the magnetic field is called a *magnetometer*. Magnetometers used in exploration make absolute measurements, although only relative differences are actually required, and these are usually the scalar strength of the field (TMI). The strength of the field in a particular direction, i.e. a component of the field, is made with a vector magnetometer. The instruments are relatively small and lightweight, and measurements are routinely made from the air, on the ground and downhole. Magnetic surveys conducted in the air with a fixed-wing

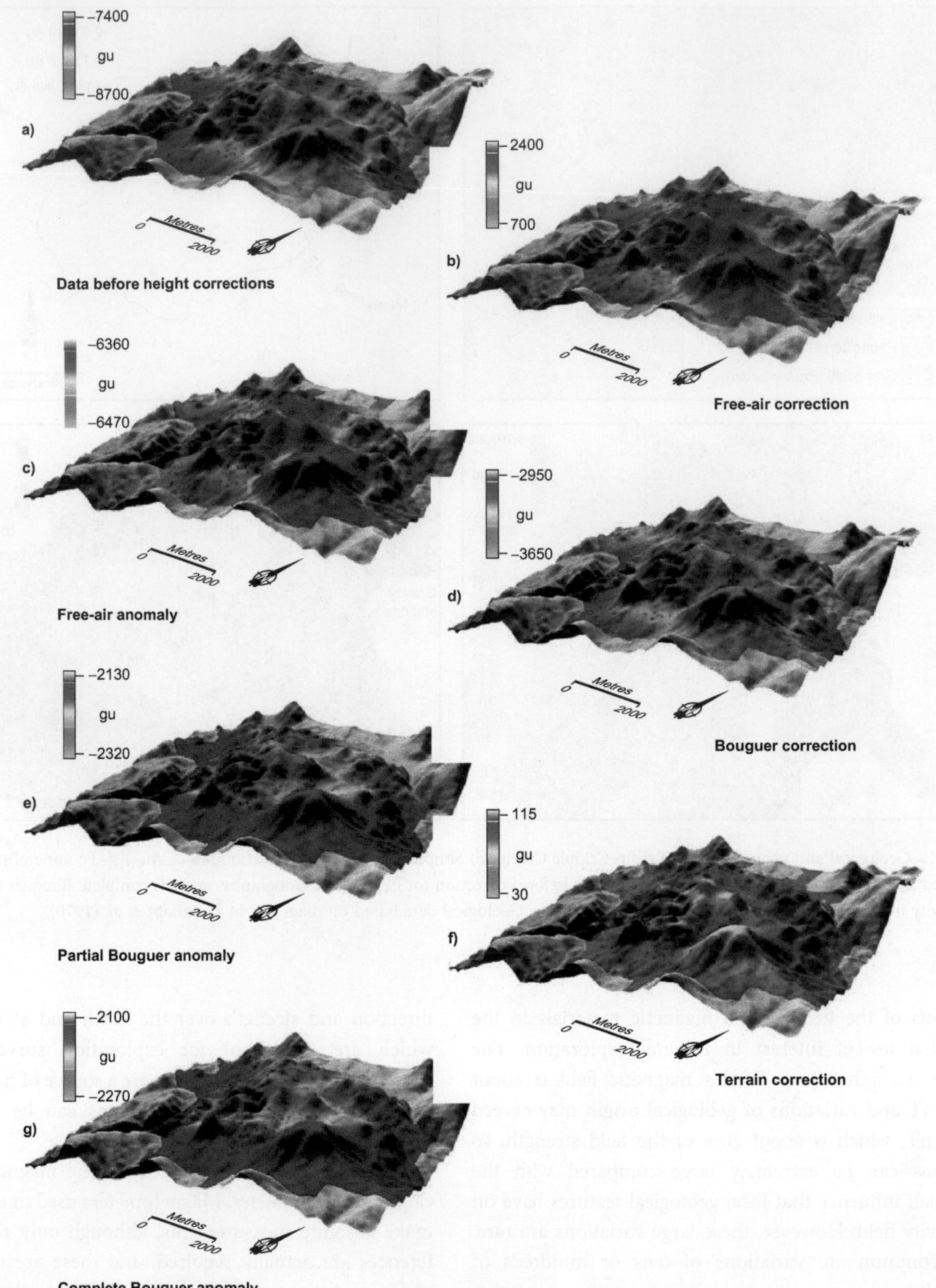

Figure 3.19 Height and terrain corrections applied to gravity data from Cripple Creek. The grey surface is topography, which has an average elevation of around 3000 m and a variation of more than 800 m. The pseudocolours correspond with (a) observed gravity corrected for tidal, drift and latitude effects, (b) free-air corrections, (c) FAA, (d) Bouguer corrections, (e) partial BA, (f) terrain corrections and (g) complete BA. The images were created from open-file gravity and digital terrain data obtained from the United States Geological Survey.

aircraft are known as *aeromagnetics*, and those made from a helicopter as *helimagnetics*. Magnetic gradiometer measurements are also common and have the same advantages and disadvantages as their gravity counter parts, i.e. improved spatial resolution but inferior detection of deeper sources. Tensor measurements (see Section 2.2.3) are sometimes made but are not yet common.

The magnetic field varies less with elevation than the gravity field, so it is not necessary to collect accurate height information during a survey. Also, TMI measurements do not require the sensor orientation to be monitored. Consequently, the process of measuring the magnetic field is logistically simpler than the gravity measurements described in Section 3.3.

Here we describe the characteristics of the Earth's magnetic field, and the instruments and survey procedures used for measuring it on the ground and in the air. Downhole magnetic surveys are undertaken to explore around a drillhole and to assist with target delineation, but are not described here. Hoschke (1985, 1991) provides detailed descriptions of using a downhole magnetometer to investigate Au–Cu–Bi mineralised ironstones at Tennant Creek, Northern Territory, Australia; and Hattula (1986) describes downhole magnetometer surveys in the Otanmäki vanadium mineralised ironstone and the Kotalahti Ni–Cu sulphide vein system in Finland.

3.5.1 The geomagnetic field

Approximately 90% of the Earth's magnetic field can be represented in terms of the field of a very large hypothetical bar magnet located within the Earth (Fig. 3.20a). It is a good representation of the field resulting from, most likely, a complex system of electric currents flowing in the Earth's core and driven by convection-related processes. The north pole of a magnetic compass needle, which is a small bar-magnet, seeks the south magnetic pole of the Earth's field (because unlike poles attract). This is actually located in the vicinity of the north geographic pole, and vice versa. Following the convention described in Section 3.2.3.1, the geomagnetic field at the surface of the Earth is then directed towards the North. The long axis of the hypothetical bar magnet is orientated approximately 10° from the Earth's axis of rotation, so the geomagnetic and geographic poles are not coincident.

The relatively stable main field originating from the core is known as the *internal field*. It is responsible for the induced and remanent magnetism of rocks. Changes in

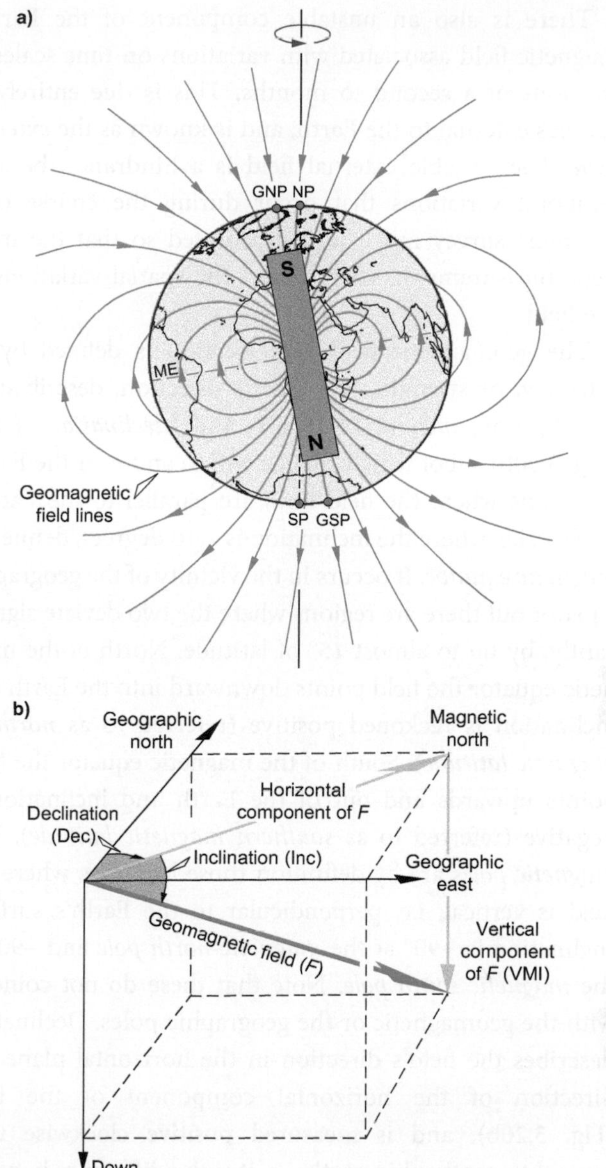

Figure 3.20 (a) Approximation of the geomagnetic field by a dipole inclined to the Earth's axis of rotation. The actual magnetic poles do not coincide with the dipole's geomagnetic pole or the geographic poles. GNP – geomagnetic north pole, GSP – geomagnetic south pole, ME – magnetic equator, NP – north geographic pole, and SP – south geographic pole. VMI – vertical magnetic intensity. (b) Elements defining the geomagnetic field (*F*). The field direction is defined by the angles of inclination (Inc) and declination (Dec).

rock magnetism in the upper crust cause short-wavelength spatial variations in the field, which are superimposed on the long-wavelength variations originating within the core. Mapping the spatial changes in the magnetic field due to crustal features is the principal objective of magnetic surveying.

There is also an unstable component of the Earth's magnetic field associated with variations on time scales of fractions of a second to months. This is due entirely to sources external to the Earth, and is known as the *external field*. The variable external field is a hindrance, because temporal variations that occur during the course of a magnetic survey must be compensated so that the magnetic measurements solely reflect the spatial variations in the field.

The geomagnetic field at a location is defined by its intensity, or strength, (F) and its direction, described by its dip or *inclination* (Inc) and *declination* (Dec) (Fig. 3.20b), all of which change within and over the Earth. Locations where the field lines are parallel to the Earth's surface, i.e. where the inclination is zero degrees, define the *magnetic equator*. It occurs in the vicinity of the geographic equator but there are regions where the two deviate significantly, by up to almost 15° of latitude. North of the magnetic equator the field points downward into the Earth and inclination is reckoned positive (referred to as *northern magnetic latitude*). South of the magnetic equator the field points upwards and out of the Earth and inclination is negative (referred to as *southern magnetic latitude*). The *magnetic poles* are by definition those locations where the field is vertical, i.e. perpendicular to the Earth's surface. Inclination is +90° at the *magnetic north pole* and −90° at the *magnetic south pole*. Note that these do not coincide with the geomagnetic or the geographic poles. Declination describes the field's direction in the horizontal plane, i.e. direction of the horizontal component of the field (Fig. 3.20b), and is measured positive clockwise with respect to geographic north, i.e. it is the difference between *true north* and *magnetic north*. A compass needle aligns itself in this direction, which is known as the *magnetic meridian*.

The geomagnetic field reaches its maximum strength at the magnetic poles, about 61,000 nT at the magnetic north pole and 67,000 nT at the magnetic south pole. Its minimum strength, about 23,000 nT, occurs in Brazil and the southern Atlantic Ocean. This is important because it means that a source's induced magnetism is less where the field is weaker (see Section 3.2.3.3). Magnetic anomalies in South America and Southern Africa have about half the amplitude of the same sources in, say, Australia, where the field inclination is similar (Figs. 3.21a and b).

The relatively stable component of the geomagnetic field is described by a mathematical model developed from observations over many years at a large number of locations around the globe. This is known as the *International Geomagnetic Reference Field* (IGRF). It gives the direction and strength of the field at any location, and attempts to predict temporal changes in these for the 5-year epoch ahead, which is known as the *provisional* field. It is updated every 5 years to account for observed (actual) magnetic activity. In magnetic surveying, the IGRF represents a smooth long-wavelength variation superimposed on the shorter-wavelength crustal features of interest. The 2004 edition of the IGRF is shown in Fig. 3.21. Note how the magnetic poles are not diametrically opposite each other. The importance of the IGRF for magnetic surveying is that it allows the strength and orientation of the geomagnetic field at the time and location of a magnetic survey to be determined, which is important for both enhancing and interpreting the data; see Sections 3.7.2 and 3.2.4.

3.5.1.1 Temporal variations in the geomagnetic field

The longer-period variation of the Earth's magnetic field, of greater than about a year, is known as the *secular variation*. It is thought to be primarily due to changes in the electric currents producing the internal field. There are also variations in the inclination and declination, and the magnetic north pole is also drifting westwards. However, these very small changes are of little significance in magnetic prospecting. Secular variations in the strength of the field have an effect when merging data from magnetic surveys conducted over periods of years or decades, typical of national survey programmes.

Palaeomagnetic studies indicate that the geomagnetic field has reversed a number of times (Collinson, 1983). The present direction of the field is known as the *normal direction*, established from its previously *reverse direction* about 780,000 years ago. Intervals of consistent polarity range from about 50,000 years to about 5 million years. The pole reversals have an obvious impact on the remanent magnetisation of crustal rocks, which strongly influences their magnetic responses.

There are several short-period variations of the Earth's field that affect magnetic surveying; they are a form of environmental noise (see Section 2.4.1), which we now describe.

Diurnal variation

Variations of the Earth's magnetic field of less than about a year are due to changes in the external field, which are related to sources external to the Earth. These are chiefly

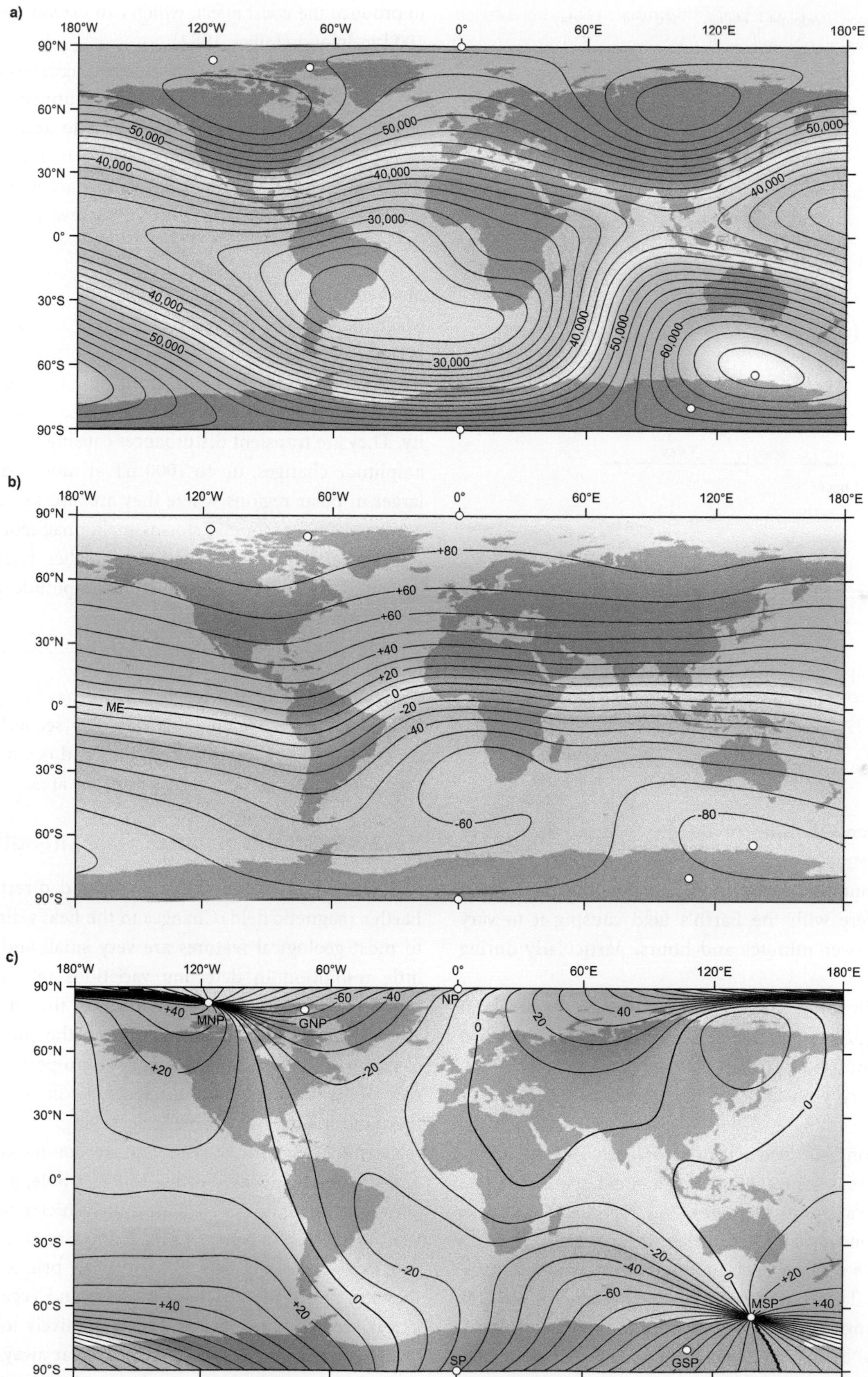

Figure 3.21 The International Geomagnetic Reference Field (2004). (a) Intensity (nT), (b) inclination (degrees), and (c) declination (degrees). GNP – geomagnetic north pole, GSP – geomagnetic south pole, ME – magnetic equator, MNP – magnetic north pole, MSP –magnetic south pole, NP – north geographic pole, and SP – south geographic pole.

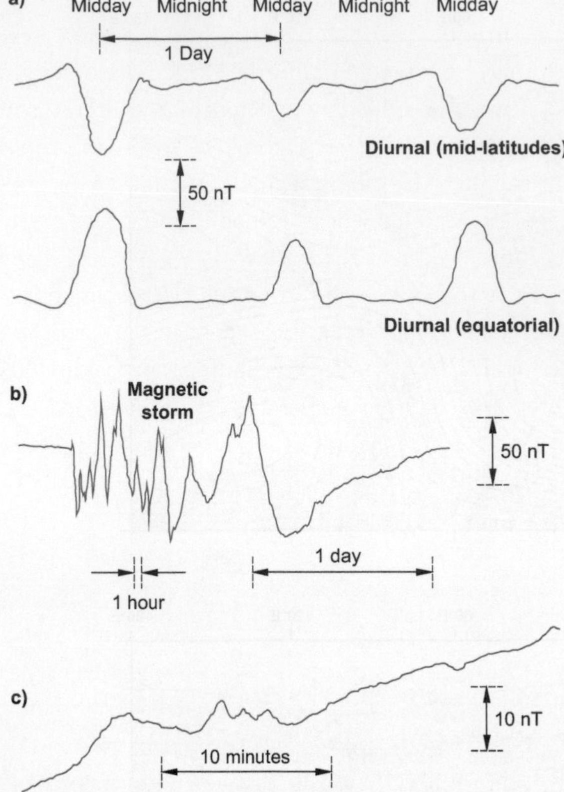

Figure 3.22 Temporal variations in the TMI of the geomagnetic field. (a) Diurnal variations,(b) a magnetic storm and (c) micropulsations. Redrawn from Breiner (1973), with permission of Geometrics Inc., San Jose.

electric currents flowing in the ionosphere, the ionised layer of the upper atmosphere, and are associated with radiation from the Sun. These generate magnetic fields which interfere with the Earth's field causing it to vary and wander over minutes and hours, particularly during daytime.

These daily or *diurnal* variations range up to about 30 nT (Fig. 3.22a) and are at a minimum at night when the night hemisphere is shielded from the Sun's radiation. They also vary with latitude, being greater in equatorial areas.

Observations of field disturbances in oceans, coastal areas and across continents reveal distinct spatial changes in the field variations. The atmospheric electric currents produce secondary currents in electrically conductive areas of the Earth's surface, such as the oceans and conductive areas inland. These, in turn, generate magnetic fields whose strength changes throughout the day. The strength of the disturbances increases toward the coast of continents, where the land mass meets the highly conductive sea water, to produce the *coast effect*, which can extend up to at least 100 km inland (Lilley, 1982).

The diurnal variation of the field strength has a big effect on magnetic survey data, so it is routinely monitored during the survey. It is relatively easy to account for (see Sections 2.7.1 and 3.6.1) since change is gradual both in space and time, although this is harder for surveys extending over large areas and for surveys conducted near the coast.

Magnetic storms

Magnetic storms are rapid variations of the field with periods of milliseconds to minutes, often appearing as irregular bursts and lasting from hours to several days, and are associated with the 11-year cycle of sunspot activity. They are transient disturbances causing a wide range of amplitude changes, up to 1000 nT at most latitudes, and larger in polar regions where they are also associated with aurora events. Magnetic storms make magnetic surveying impractical because the target anomalies related to the crustal rocks are often smaller in amplitude than these erratic variations (Fig. 3.22b).

Micropulsations

Micropulsations of the magnetic field have amplitudes of less than 10 nT and periods of a few seconds to 300 s (Fig. 3.22c). They are very common and occur randomly, their amplitude varying across a survey area.

3.5.2 Measuring magnetic field strength

Magnetic objects alter the strength and direction of the Earth's magnetic field. Changes in the field's direction due to most geological features are very small and offer very little resolution in detecting variations in the magnetic properties of the subsurface. However, the effect on field strength is significant and is by far the most sensitive element to changes in the magnetic properties of crustal rocks. It is the strength of the magnetic field that is measured and mapped in magnetic surveying.

Ground magnetic surveys can be conducted 'on foot' with the sensor located at the top of a pole, a few metres above the ground. Small all-terrain vehicles may also be used where access permits. Low-flying aircraft offer significant advantages over ground surveying primarily because they are unimpeded by terrain access and vegetation, and can cover large areas faster and at relatively lower survey cost. The magnetic sensor is located as far away as possible from the survey platform in order to minimise its magnetic

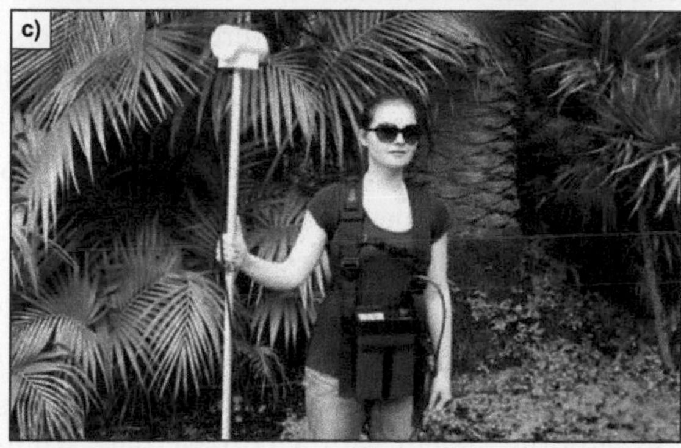

Figure 3.23 (a) Helicopter magnetic survey system. Picture courtesy of GPX Surveys. (b) Airborne magnetic gradiometer survey system. Note the magnetic sensors on each wing-tip. Picture courtesy of Geotech Airborne. (c) Ground magnetic survey system.

effect at the sensor. Usually it is mounted at the end of a non-magnetic support extending from the vehicle/aircraft. Sometimes in airborne work the magnetic sensor is towed, on a long cable, behind and below the aircraft. This is usually when magnetic data are being collected along with other types of geophysical measurements, for example, when making electromagnetic measurements (see Section 5.9). Figure 3.23 shows equipment for ground surveying and some examples of aircraft configured for aeromagnetic surveying.

Sometimes the difference in TMI is measured between one or more pairs of closely spaced sensors forming a *magnetic gradiometer*, the reading being a TMI gradient in a particular direction (see Section 2.2.3). The gradiometer is usually oriented so as to measure either the vertical or horizontal gradient of the magnetic field. For ground surveys, the two sensors are mounted several metres apart on a pole. In airborne work, horizontal gradients are measured by magnetometers on the aircraft's wing tips (horizontal gradient) (Fig. 3.23b), or at the top and base of the tail (vertical gradients). Orientation errors can be large in gradient measurements and they must be closely monitored and recorded.

Hogg (2004) provides an extensive discussion about the advantages and disadvantages of magnetic-gradient measurements. Although there are advantages in making tensor measurements (see Section 2.2.3) of the magnetic field, equipment to do this routinely is not yet available (Schmidt and Clark, 2000).

3.5.2.1 Magnetometers

Three types of magnetic sensor are commonly used in geophysical surveying, namely the proton-precession, the Overhauser and the alkali-vapour sensors. The operation of all three is based on quantum-mechanical properties of atoms. Importantly, they are sensitive to the strength of the Earth's magnetic field, *F* in Fig. 3.20b, and do not measure its direction, i.e. they measure the total magnetic intensity (TMI). Magnetometers using these sensors are equipped with solid-state memory and can store thousands of measurements, and some can also internally record and store satellite-derived location information.

The proton-precession sensor makes a measurement of magnetic field strength over a period of time. Typically, measurements can be made at the rate of about one per second, although smaller cycle times are possible but result in lower accuracy. Modern instruments can measure the TMI to at least 0.1 nT, and to greater precision with longer measurement periods. The Overhauser sensor can make a measurement more quickly, typically to 0.01 nT several times a second. They can also operate in higher field gradients than the proton-precession sensor.

The alkali-vapour sensor is an analogue device producing a continuous signal so field measurements can be made rapidly. Measurements of TMI are typically made to 0.01 nT, and to 1 pT for sensors based on particular types of alkali-metal vapours. They can operate in higher field gradients than the proton-precession and Overhauser instruments. Alkali-vapour magnetometers are routinely used for most survey work. Their high precision and greater tolerance of gradients mean that they are the preferred sensor for high-precision gradiometers.

The strength of the magnetic field in a particular direction can be measured with a *flux-gate* sensor, whose operation depends upon the magnetic properties of iron–nickel alloys. It measures the strength of the magnetic field in the direction of its sensor, the alloy cores, which need to be oriented in the direction of the field component being measured. Orientation errors are a source of measurement noise. The flux-gate magnetometer is a vector magnetometer and finds application where the direction to a magnetic source is required, such as downhole magnetometery and downhole electromagnetics (see Section 5.8).

3.5.3 Magnetic survey practice

Comprehensive descriptions of general aspects of geophysical data acquisition and survey design are given in Section 2.6. Details pertaining specifically to magnetic surveying are described here.

Strategies for conducting magnetic surveys are principally determined by the need to record temporal changes in the geomagnetic field occurring during the survey so that they can be removed during data reduction. The problem is addressed by a combination of repeat readings and continuously monitoring the field. The latter is done by establishing a second, base station, magnetometer at a fixed location close to or central to the survey area to record the changing geomagnetic field, usually at a time interval of several minutes. For large surveys spatial variations in diurnal changes require multiple base stations.

The base station also allows the rate of field change to be monitored, which is the operational basis for identifying a magnetic storm (see *Magnetic storms* in Section 3.5.1.1). The survey data acquired during storm periods will be dominated by storm events, so they are rejected and the survey repeated when the field is deemed to be stable.

3.5.3.1 Ground and airborne surveys

Ground and airborne magnetic surveys usually comprise a series of parallel traverses and tie lines as appropriate (see Fig. 2.10a). The line spacing and line direction, and survey height in the case of airborne surveys, are the main parameters to consider in survey design. When the survey is conducted with a moving vehicle, usually an aircraft, the along-line data interval is determined by the speed of the vehicle and the sampling rate of the magnetometer. Modern instruments sample the field rapidly, resulting in a very small along-line data interval compared with the survey line spacing.

In addition to the TMI measurements, station location, obtained from the GPS, and aircraft height, obtained from a radio altimeter, are measured and recorded, typically, every 0.1 s (10 hertz sampling). As far as is possible, the aircraft maintains constant survey speed, typically 70 m/s, 50 m/s for low-flying crop-duster aircraft and as low as 30 m/s for helicopters. The constantly sampled data are then spaced along-line at intervals of approximately 7, 5 and 3 m respectively. Obviously, the sampling interval is determined by the survey speed, but these very small sampling intervals are usually more than adequate for nearly every geological environment.

For ground magnetic surveys, readings are taken at specified locations. Usually a station interval of 10 m, or even 5 m, is suitable for most purposes, but for environments with near-surface occurrences of highly magnetic materials, a closer interval of 1 m or less is desirable. These highly magnetic and disrupted features produce very large amplitude variations in the magnetic field over small distances that require a very close sampling interval in order to avoid aliasing (see Section 2.6.1). When properly sampled, the high-frequency noise can be removed with low-pass filtering (see *Frequency/wavelength* in Section 2.7.4.4). A better and logistically more convenient solution in these environments is to use a continuously time-sampling magnetometer (see Section 3.5.2.1) configured like an airborne system with a portable data acquisition computer and a positioning system. At normal walking speed, the data interval can be small as 10 cm, producing very dense sampling of the high-frequency noise which allows its removal with filtering.

The key variables when designing magnetic surveys are survey height, line direction and spacing.

Survey height

The sensor height is a key variable in magnetic survey design affecting both the amplitude and wavelength of the anomalies detected (see Sections 3.10.1.1 and 2.3). Reducing the height places the sensor closer to the magnetic sources, increasing the measured amplitude of any anomalies, i.e. survey sensitivity is increased. It also means that there will be more shorter-wavelength, i.e. high-frequency, information in the field, increasing the risk of aliasing (see Section 2.6.1), a problem mainly with ground surveys. For example, in areas where highly magnetic material occurs near-surface, decreasing the survey height causes the measurement to be dominated by the effects of the near-surface material. This may not be a bad thing if the near-surface feature is the target of interest, but it can be a major source of interference where measuring the lower-amplitude, longer-wavelength responses of deeper features is the main interest. In this case, increasing the survey height rapidly attenuates the near-surface high-frequency response, but with smaller effect on the deeper-seated responses.

The amplitude and wavelength of the magnetic response is sensitive to source–sensor separation. Figure 3.66 shows two magnetic datasets recorded at different survey heights. The data from Western Australia were acquired at a height of 40 m and the data from Canada at a height of 70 m. The obviously smoother appearance of the Canadian data reflects the greater attenuation of shorter wavelengths with increasing height compared with the longer wavelengths. The significant effect of source–sensor separation on magnetic responses means that maintaining constant separation across the survey area is important in order to minimise noise (Cowan and Cooper, 2003a).

Ground surveys by their nature are conducted with the magnetic sensor at constant (survey) height. Portable instruments typically have the sensor at a height of 1–2 m, although up to 6 m height may be required where surface magnetic noise is a problem. A wide range of line spacing is used depending on the purpose of the survey. Ground vehicle-borne magnetometer systems operate in a similar way to airborne systems and provide very dense sampling, less than a metre, but the locale must be suitable for vehicles.

Airborne surveys, on the other hand, are conducted at a constant nominal survey height (see Fig. 2.6), but suffer unavoidable height variations due to topography. Where weak magnetic responses are anticipated or where high resolution is required, surveys can be conducted as low as 20 m above the ground with small crop-duster aircraft or helicopters, topography permitting. Regional surveys designed to map the longer-wavelength responses of large-scale geological features are usually conducted at higher terrain clearances, typically 60 to 150 m, which also offers easier survey logistics in undulating terrains. In very rugged terrains it can be logistically simpler to conduct the survey at a constant barometric height above all the topography, in which case the terrain clearance is variable (see Fig. 2.6).

When designing airborne magnetic surveys it is better to lean towards lower survey height since it is always possible to upward-continue the data to a greater height, but the reverse is much more difficult (see Section 3.7.3.2).

Survey line direction and spacing

Survey line direction requires particular attention when designing both ground and airborne magnetic surveys as it is somewhat dependent on the magnetic latitude of the survey area. In middle to high latitudes, i.e. towards the poles, survey lines ought to be orientated perpendicular to the regional strike of the magnetic sources.

In contrast, at low latitudes (less than about 30°), the 'high' and 'low' of the anomaly are displaced in the direction of the magnetic meridian to the northern and southern edges of the source (see Figs. 3.8 and 3.26), respectively, irrespective of the shape and strike of the magnetic body. Survey lines orientated along the magnetic meridian, i.e. magnetic north–south, measure both the 'low' and 'high' of the anomaly and provide more information about the source than lines orientated perpendicular to strike. This is critically important as it ensures that the survey lines actually traverse the anomaly dipole, essential for its analysis. However, this strategy assumes the anomalies are entirely due to induced magnetism. At low latitudes, remanent magnetism can transform the response to that suited to an alternative line orientation. This is a particular advantage for long linear sources striking north–south, and particularly at the equator, where induction alone produces anomalies at its northern and southern edges with little or no measureable response over the central part of the body.

Survey line spacing is fundamental in determining the lateral resolution of the survey and a major control on overall survey cost. The survey lines can be widely spaced when measuring long-wavelength responses and should be closer for measuring shorter wavelengths, but in all cases must be sufficiently close to adequately resolve the features of interest. Choice of line spacing is not independent of survey height. A lower survey height will enhance short-wavelength responses which will require a smaller survey line spacing to reduce across-line aliasing and allow the data to be properly gridded (see Section 2.7.2). Ideally the survey line spacing should not exceed twice the survey height, but a wider spacing is acceptable for areas where magnetic responses are continuous along strike (Cowan and Cooper, 2003a).

Line spacings for magnetic survey are typically a few tens of metres to a few hundred metres, increasing with survey height. This results in between one and two orders of magnitude difference in sampling interval in the along-line and across-line directions. Tie lines (see Section 2.6.3.3) are typically spaced 10 times the survey line spacing. For ground surveys a single tie line could be sufficient for the typically smaller area surveyed.

Survey specifications

Based on the characteristics described above for each parameter, and considering the cost of conducting a survey, a range of line/height combinations has evolved that have, by experience, been found to satisfy the requirements of the various classes of magnetic surveys.

For wide-ranging regional airborne surveys aimed at providing the 'first view' of the large-scale geological environment, and useful for targeting areas for more detailed work, the lines are often spaced 400 m apart and the survey conducted at 80 to 120 m above the terrain. In highly prospective areas, the line spacing and height are often reduced to 200 m and 50 to 60 m, respectively. Surveys for prospect targeting require the lines to be spaced 40 to 100 m apart, and the survey conducted at 20 to 50 m above the ground.

Of course, there are exceptions to these relationships. For example, the very long and thin ilmenite-rich heavy mineral sand strandlines, which typically are many kilometres in length, are surveyed with lines spaced about 400 m apart, but at a height of 20 m above the ground, and even lower when helicopter systems are available (Mudge and Teakle, 2003). The exceptionally low survey height is essential for detecting the very weak responses, typically less than 1 nT, associated with these near-surface mineral deposits. Similarly low survey heights, but with closer lines, are typical in kimberlite exploration where the laterally confined targets can be weakly magnetic. For highly magnetic banded iron formations, low survey height improves resolution of the thin magnetic bands.

3.6 Reduction of magnetic data

The reduction of magnetic survey data is principally aimed at removing the effects of temporal variations in the Earth's magnetic field that occur during the course of the survey (see Section 3.5.1.1). Like gravity reduction, planetary-scale spatial variations in the field are compensated for, but in contrast, elevation-related variations in the magnetic field are minor. Corrections for variations in survey ground clearance may also be required if the terrain is rugged. With moving platform systems, there is a small residual magnetic response related to the platform's magnetism that varies with its attitude and heading with respect to the Earth's magnetic field and known as the *heading error*. In the case of airborne systems, it is continuously measured by a dedicated on-board magnetic sensor and a heading correction is automatically applied, as the data are recorded, to compensate for it.

Note that the various corrections also remove the background value of the magnetic field, the reason that the reduced data sometimes have unexpectedly low absolute values. This is not a problem since the mapping of spatial variations in the field is the object of the survey. Sometimes an average field strength value from the survey area is added to return the data to a realistic absolute value.

We briefly describe the various corrections in the order that they are applied, and after which the corrected data are levelled to remove residual errors (see Section 2.7.1).

3.6.1 Temporal variations in field strength

Very short-period variations in the strength of the magnetic field, such as micropulsations (see *Micropulsations* in Section 3.5.1.1) and noise 'spikes', are identified and excised from both the survey data and the base station record. This is based primarily on manual inspection and editing of the data. The edited base station record is then synchronised in time with the survey data, and diurnal variations removed by simply subtracting the base station record from the observations. It is worth noting here that any noise in the base station record, owing to magnetic disturbances local to the base station, needs to be identified and removed from the recording first; otherwise the noise will propagate via the diurnal correction and contaminate the survey data.

Continuous base station recordings (see Section 3.5.3) are usually quite effective for removing diurnal variations in surveys conducted close to the base station. The process is imperfect, however, because in the case of airborne surveys diurnal variations at the survey level are often not accurately represented by a base station located at ground level. Also, features, of all frequencies, in the diurnal variation will have different amplitudes at different locations. Short-wavelength variations (micropulsations) are a particular problem and the diurnal data must be low-pass filtered (see *Frequency/wavelength* in Section 2.7.4.4) before subtraction. Deploying a network of base stations around a large survey area usually presents major logistical challenges and is rarely undertaken. Errors related to the use of a single base station are the generally largest source of error in modern airborne TMI data. Diurnal variation can often be removed without the base station record, in which case it is treated as a tie-line cross-over error and removed during the levelling process (see Section 3.6.4).

3.6.2 Regional variations in field strength

The geomagnetic field (see Section 3.5.1) appears as a very long-wavelength regional field. Removing this increases the resolution of magnetic anomalies associated with crustal sources. For small surveys, the geomagnetic field varies smoothly and by only a small amount and can be removed as part of a single regional gradient (see Section 2.9.2).

For magnetic surveys that extend over large areas, such as airborne surveys, the gradient of the geomagnetic field can be removed by simply computing the field strength from the IGRF model (Fig. 3.21) for the time and location of each measurement, and subtracting these from the observations. The correction also compensates for the slower secular variations of the geomagnetic field, important for surveys that extend over long periods, such as airborne surveys as these typically take days or weeks to complete, and even months and years for national-scale surveys.

3.6.3 Terrain clearance effects

In the absence of anomalous features, and unlike the gravity field, the magnetic field varies little with distance from the Earth. However, in the presence of crustal magnetic features, variations in the distance between the magnetometer and the Earth's surface can be important (Ugalde and Morris, 2008). This is usually not a problem for ground surveys, but it is for airborne surveys. Also, when processing and interpreting magnetic data, it is generally assumed that the survey plane is horizontal and always above the rocks, but this is often not the case in rugged terrain.

It is common for geological controls on topography to produce strike-parallel ridges and valleys. Aerial surveying across strike causes changes in ground clearance, as shown in Fig. 2.6. For these cases, it is possible to estimate what the measurements would have been had the ideal flight path with constant terrain clearance been achieved. In order to mimic an ideal drape-flown survey (see Section 2.4.1.1), a *drape correction* is applied (Flis and Cowan, 2000). The basis for this is a procedure known as *continuation*, whereby a measurement at one height can be used to determine the equivalent measurement at another height (see Section 3.7.3.2). In practice the correction is not routinely applied because continuation of data to lower heights is inherently unstable and can introduce noise that degrades the quality of the data.

Where the terrain is very rugged and the hills are formed of magnetic material, a magnetic measurement made in a valley will be influenced by magnetic rocks located below, adjacent to and above the magnetometer. In this case, magnetic profiles measured across rugged terrain exhibit complex responses due to the rapidly changing ground clearance in airborne surveys, sloping survey plane and, particularly in valleys, interference of magnetic material contained in the neighbouring hills (Mudge, 1988). The varying ground clearance and the shape of the terrain need to be included in the modelling of target anomalies (see Section 2.11.3). Otherwise, to correct for these effects would require detailed knowledge of the magnetism of the terrain, which is not available (an example of the geophysical paradox; see Section 1.3).

3.6.4 Levelling

The final stage in the reduction of magnetic data is levelling (see Section 2.7.1.2) based on the repeat readings associated with tie lines crossing the survey lines. A detailed description of the procedures for aeromagnetic data is provided by Luyendyk (1997) and Saul and Pearson (1998). Errors in the magnetic levels of each data point, remaining after the application of the various corrections described above, appear in aeromagnetic datasets as corrugations parallel to the survey line direction. Levelling redistributes these around the whole dataset, and any further remaining errors are usually removed with microlevelling (see Section 2.7.1.3).

The quality of the data reduction can be accessed by viewing the levelled data with shaded relief (see Section 2.8.2.3), with the illumination perpendicular to the survey line direction, optimum for highlighting across-line levelling errors. Derivative images, being sensitive to gradients, often reveal residual errors when the levelled TMI data appear to be free of corrugations.

3.6.5 Example of the reduction of aeromagnetic data

Figure 3.24 shows an aeromagnetic survey over an area of moderate terrain in the Northern Territory, Australia. The mean terrain clearance is 80 m, but narrow ridges due to erosion resistant units caused significant height variations in their vicinity. The survey lines are spaced 100 m apart

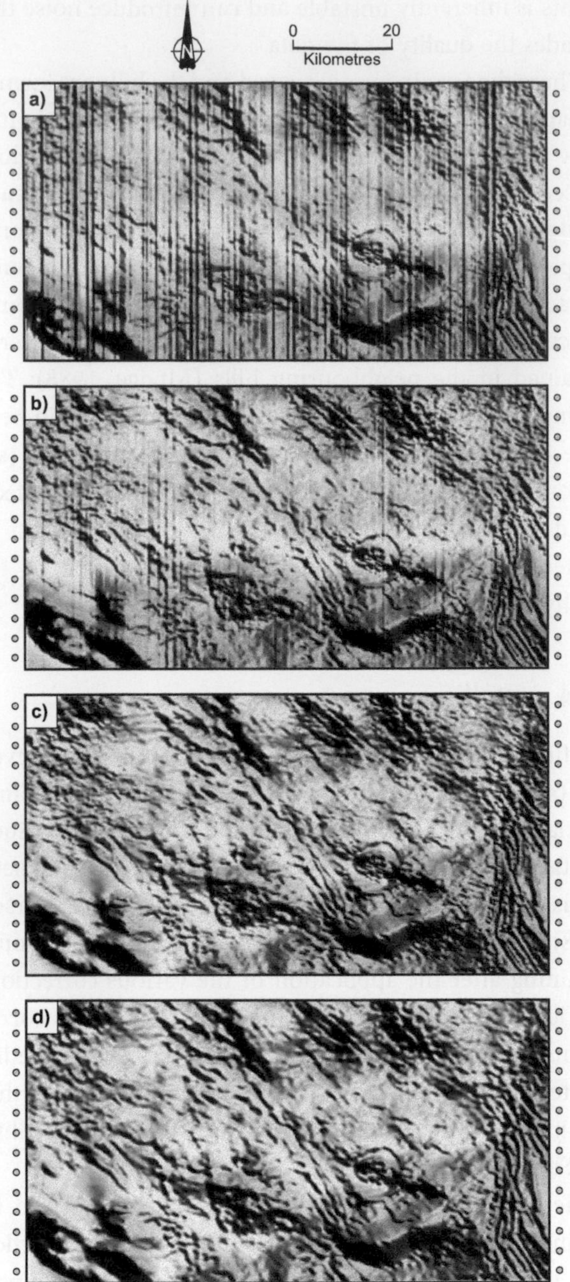

Figure 3.24 Removal/attenuation of unwanted artefacts at different stages of the reduction of aeromagnetic data acquired along north–south survey lines. All images are TMI (in pseudocolour) with shaded relief illuminated from the northeast (as a grey-scale). (a) Data corrected for aircraft orientation and compensated for IGRF; (b) the data in (a) with base-station-recorded diurnal variations removed; (c) the data in (b) after tie-line levelling; and (d) the data in (c) after microlevelling. The grey circles show the ends of the east–west tie lines. The prominent circular feature in the right centre of the images is an astrobleme. Data reproduced with the permission of the Northern Territory Geological Survey.

and flown north–south in a racetrack pattern (see Section 2.6.3.3). The figure shows images of the data at different stages of reduction, Fig. 3.24a being the data after removal of the IGRF and heading effects. Corrugations parallel to the survey line direction, due to errors, are obvious. They are greatly reduced by correcting for diurnal variations measured at the base station (Fig. 3.24b) and are almost totally removed after tie-line levelling (see Section 3.6.4) (Fig. 3.24c). The application of microlevelling attenuates some residual short-wavelength features to improve the overall appearance (Fig. 3.24d), for example by improving the continuity of linear anomalies in the southeast of the dataset, and improving the clarity of detail throughout.

3.7 Enhancement and display of gravity and magnetic data

The interpreter's ability to analyse gravity and magnetic data can be greatly assisted by various filters that emphasise particular characteristics of the data, or suppress undesirable characteristics. The methods used on magnetic and gravity data are similar since the appropriate filtering techniques are based on the physics of potential fields. They may be applied to 1D profiles of data, for example individual flight lines of an aeromagnetic survey; but working with 2D gridded data is more common.

The filters described in Section 2.7.4.4 are routinely applied to gravity and magnetic data. Only those filtering operations specific to these kinds of data are described here, and then we focus only on the most commonly used of a vast number of potential-field filters. A comprehensive summary of potential-field filtering operations, as applied to magnetic data, is provided by Milligan and Gunn (1997). In order to demonstrate the various filters we have computed their effects on the gravity and magnetic responses of a square prism with vertical sides (Figs. 3.25 and 3.26) and have also applied them to actual gravity and TMI datasets.

The gravity survey data (Fig. 3.27) are from the vicinity of the Las Cruces Cu–Au volcanogenetic massive sulphide deposit, located in the Iberian Pyrite Belt near Seville, Spain. Mineralisation occurs below 150 m of Tertiary sediments and comprises massive to semi-massive sulphides underlain by a pyritic stock work. Host rocks are volcaniclastics and black shales. The geophysical characteristics of this deposit are described by McIntosh *et al.* (1999) and the geology by Doyle *et al.* (2003). The deposit is associated with a roughly circular positive gravity anomaly with

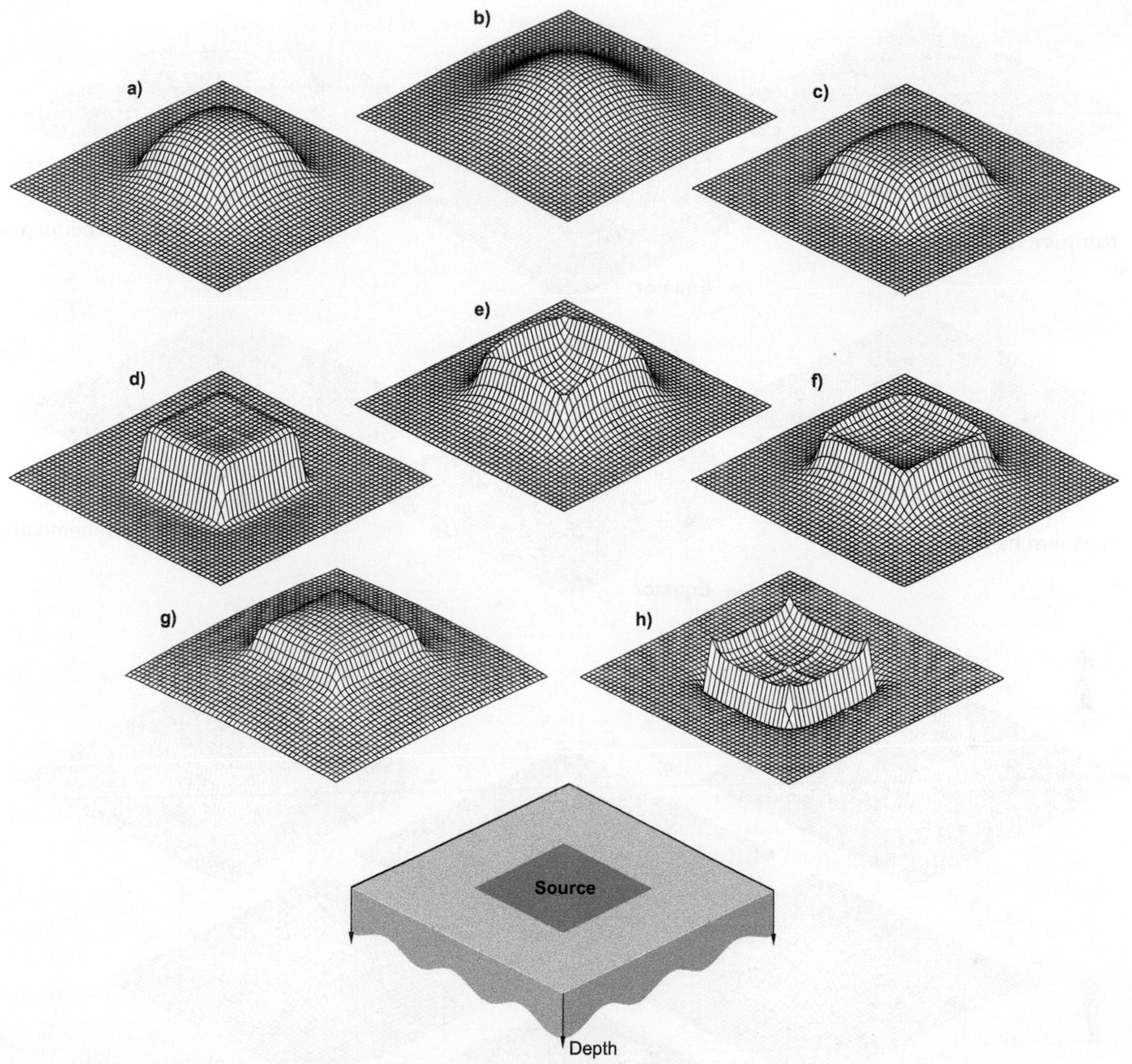

Figure 3.25 The effects of various filter operations on synthetic gravity data. (a) Calculated gravity field due to a vertically sided square prism and (b) after upward continuation. Note that the anomaly now has lower amplitude and longer wavelength. (c) Residual anomaly obtained by subtracting the upward continued response (b) from (a). This produces a sharper short-wavelength response. (d) First vertical derivative of the anomaly in (a). Note the more localised response compared with the original anomaly. (e) Total horizontal gradient of the anomaly in (a). Note how the maxima are coincident with the edges of the source, revealing its geometric form. (f) 3D analytic signal of the anomaly in (a). The crests of the ridges are also coincident with the edges of the source, revealing its geometric form. (g) Tilt derivative of the anomaly in (a). Note the positive response above the source. (h) Second vertical derivative of the anomaly in (a). The response is localised over the source edges.

amplitude of about 25 gu. This dataset is chosen to demonstrate the various enhancements in a situation where anomaly detection is the primary requirement. Stations are on loose grid with a spacing of about 250 m, with more detailed data over the orebody.

The airborne TMI survey data (Fig. 3.28) are from the vicinity of the world-class Broken Hill Zn–Pb–Ag deposit, in New South Wales, Australia. The regional geology comprises high-grade metasediments, amphibolite and gneiss which have experienced multiple phases of deformation and metamorphism. The Broken Hill mineralisation does not give rise to a magnetic response but the local structure and stratigraphy is clearly seen from the pattern of anomalies. This dataset was chosen to demonstrate the use of data enhancements to assist geological mapping. Haren *et al.* (1997) and Maidment *et al.* (2000) describe the interpretation of these data and argue for structural as well as stratigraphic

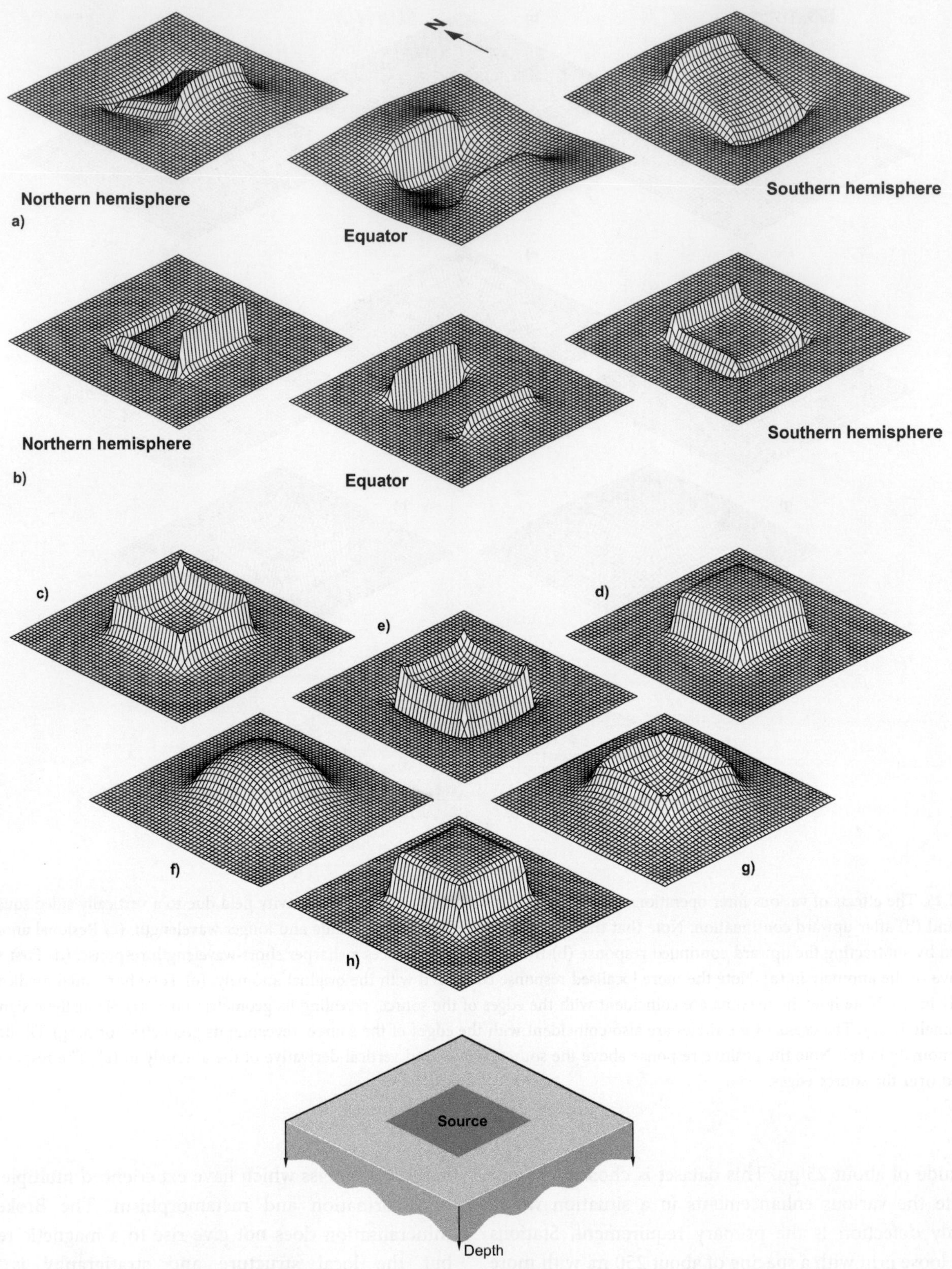

Figure 3.26 The effects of various filter operations on synthetic TMI data. (a) Calculated TMI field due to a vertically sided square prism with induced magnetism directed downwards (northern hemisphere) and towards the North, horizontal (at the equator) and towards the North, and directed upwards (southern hemisphere) and towards the North. Note the markedly different anomaly shapes due entirely to the different magnetisation directions. (b) First vertical derivatives of anomalies in (a). (c) The 3D analytic signal is the same for all directions of magnetisation; note that the crests of the ridges are coincident with the edges of the source prism, revealing its geometric form.

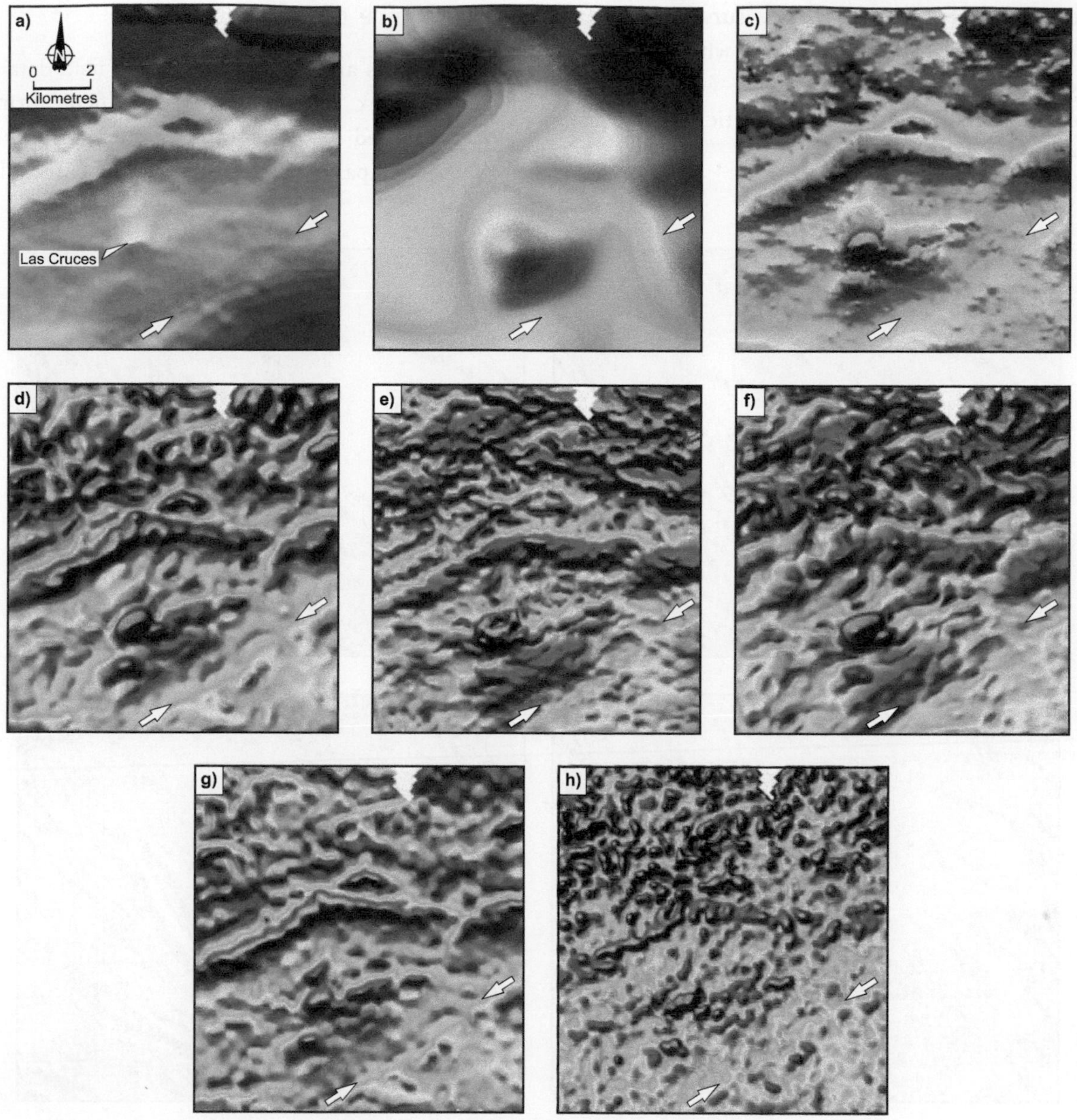

Figure 3.27 Gravity data from the vicinity of the Las Cruces Cu–Au massive sulphide deposit. All images are pseudocolour combined with grey-scale shaded relief illuminated from the northeast. (a) Bouguer gravity, (b) data in (a) upward-continued by 800 m, (c) residual dataset obtained by subtracting (b) from (a), (d) first vertical derivative, (e) total horizontal gradient of data in (a), (f) 3D analytic signal of data in (a), (g) tilt-derivative of data in (a), and (h) second vertical derivative of data in (a). The arrows highlight a prominent linear feature. Note how this is more easily seen in the various derivative-based enhancements. These data are used with the permission of First Quantum Minerals Ltd.

Caption for Figure 3.26 (*cont.*) (d) TMI reduced-to-pole data: note the symmetrical response, revealing the form of the source, unlike the complex TMI anomalies in (a). (e) First vertical derivative of reduced-to-pole data in (d): note the symmetrical response revealing the form of the source more clearly than the complex derivative anomalies in (b). (f) Pseudogravity response; compare with Fig. 3.25a. (g) Total horizontal gradient of pseudogravity response in (f), again revealing the form of the source. (h) Tilt-derivative of reduced-to-pole data in (d). Note the positive response of the source. The responses in (c) to (h) are more easily correlated with the source geometry than the non-polar responses in (a) and (b).

sources of anomalies. The magnetic survey was conducted at a terrain clearance of 60 m with the survey lines spaced 100 m and orientated east–west. The area is located in the southern magnetic hemisphere (Inc = –44°, Dec = –5.5°).

3.7.1 Choice of enhancements

Enhancements are applied to potential field data to transform complex responses to simplified forms more directly related to their sources, and to improve the resolution of particular features in the data. Filters that

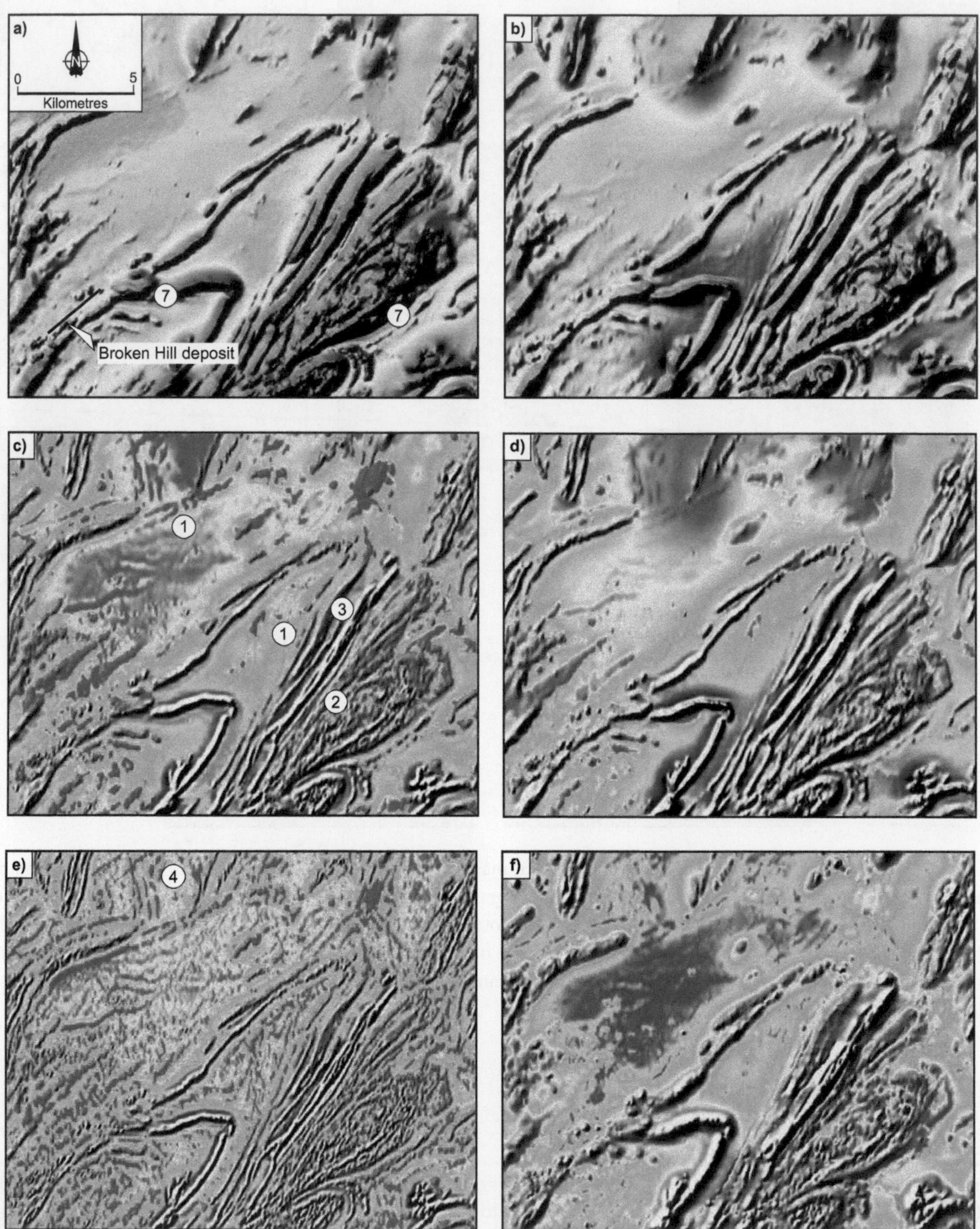

Figure 3.28 TMI data for the vicinity of the Broken Hill Pb–Zn–Ag deposit. (a) TMI, (b) TMI reduced to pole, (c) first vertical derivative of TMI, (d) first vertical derivative of TMI reduced to pole, (e) second vertical derivative, (f) 3D analytic signal, (g) tilt derivative, (h) total horizontal gradient of pseudogravity, and (i) pseudogravity. These data are used with the permission of Geoscience Australia.

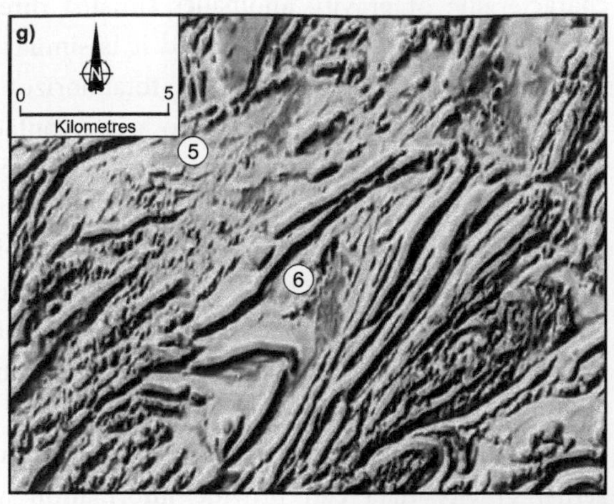

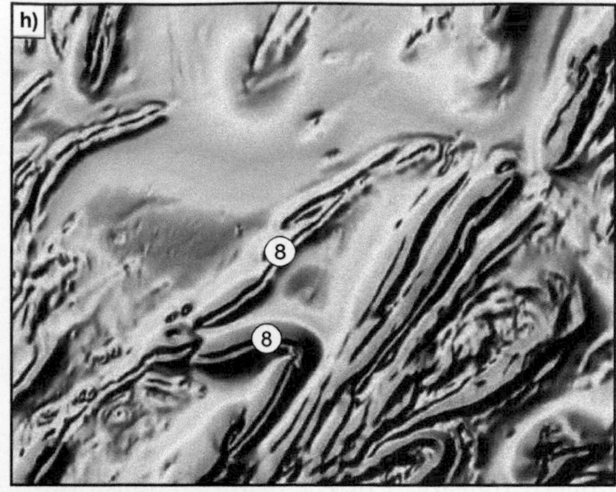

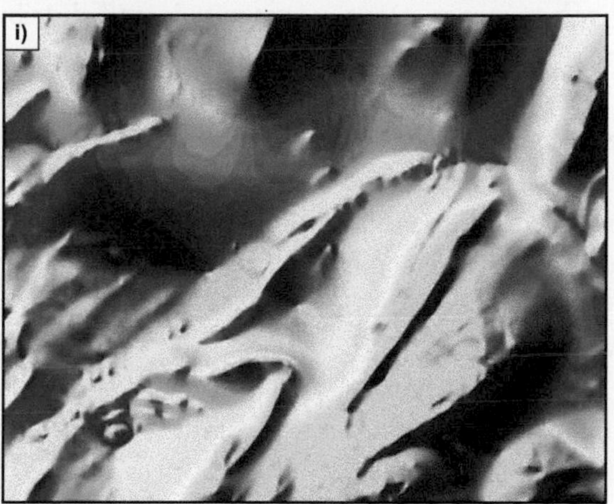

Figure 3.28 (*cont.*)

simplify the response are mostly applicable to magnetic data. The aim is to transform the dipolar response characteristic of mid-latitudes to the simpler monopole response occurring at the poles (Fig. 3.8). Filters enhancing detail emphasise responses from shallow depths by enhancing their short-wavelength responses. Both wavelength and gradient filters are commonly used and allow structure to be mapped in considerably more detail than in the 'raw' data. The long-wavelength responses of large and deep-seated features can also be enhanced using wavelength filters.

3.7.2 Reduction-to-pole and pseudogravity transforms

As shown in Figs. 3.25a and 3.26a, magnetic responses are much more complex than gravity responses, owing to the dipolar nature of magnetism and the possibility that the magnetisation of the body can be in any direction. Unless the body's magnetism is vertical, the resultant anomaly can be difficult to relate to the geometry of its source. Fortunately for the interpreter, the problem is addressed by two processing operations: the reduction-to-pole (RTP) and the pseudogravity (PSG) transforms. Both are forms of anomaly simplifier making magnetic data much easier to analyse and allowing comparisons of data from different magnetic latitudes.

3.7.2.1 Reduction-to-pole

The reduction-to-pole operator transforms magnetic anomalies resulting from the inclined magnetism of non-polar regions into their equivalent polar response where a body's magnetism is vertical; compare Figs. 3.26a and d. Polar magnetic anomalies resemble gravity anomalies in that they are monopolar and occur directly above their source, but with the advantage of being more localised to the source than gravity anomalies. The resolution of close-spaced sources is improved. Derivative enhancements are similarly simpler when calculated from RTP data (Figs. 3.26b and e).

There are, however, some practical difficulties with the RTP transform. The magnetic inclination and declination of the inducing field are required for the transformation (obtained from the IGRF; see Section 3.6.2), and it is normally assumed that the magnetism of all the rocks in the area is parallel to the geomagnetic field and by implication is entirely induced. Remanent magnetism, and to a lesser extent self-demagnetisation and AMS (see Section 3.2.3), change a body's magnetism so it is not parallel to the geomagnetic field everywhere and, therefore, anomalies will not be properly transformed. The distorted transformations are most noticeable where the remanence is strong and in a direction significantly different from that of the Earth's field.

RTP algorithms often perform well when the geomagnetic field inclination is steep, but some algorithms are much less effective when it is shallow, i.e. close to and at the magnetic equator (Li, 2008). Prominent north–south orientated artefacts are introduced into the transformed data, as demonstrated by Cooper and Cowan (2003). An alternate strategy is to reduce the data to the equator instead of the pole; but although this generally places a response over the magnetic source, it is negative (see Fig. 3.26a) even though the body is more strongly magnetised than its surrounds, and the anomaly does not mimic the geometry of the source.

The Broken Hill data are notably different after reducing to the pole (Figs. 3.28a and b). Note how the dipolar anomalies in the TMI data have become monopoles (7). The anomalies in the first vertical derivative data are also more localised to their sources (Figs. 3.28c and d).

3.7.2.2 Pseudogravity

The pseudogravity operator transforms the TMI anomaly into the gravity-like response that would be obtained if the body's magnetism were replaced with the same density distribution, and with the density as a multiple of the magnetisation. By treating the magnetism like density, it aims to simplify the analysis of magnetic data. Like the RTP operator, pseudogravity assumes that the magnetism of the source is parallel to the geomagnetic field, and by implication is entirely induced; so strong remanent magnetism and AMS are sources of error.

The pseudogravity data shown in Fig. 3.26f, obtained from the TMI data shown in Fig. 3.26a, exhibit the smoother, longer-wavelength monopole responses characteristic of gravity anomalies situated directly over the magnetic source. In this regard it is similar (but not identical) to the RTP response. The total horizontal gradient of pseudogravity can be used to locate contacts in the same way as the gradient of normal gravity (see Section 3.7.4). The example shown in Fig. 3.26g confirms the formation of peaks over contacts. The mapping of contacts is well demonstrated by the double-peaked response over stratigraphic sourced anomalies in the Broken Hill data, e.g. (8) in Fig. 3.28h, although there is less detail than in the various magnetic derivative images.

Comparisons between computed pseudogravity and actual gravity data can provide information about the magnitude and distribution of magnetism and density in the magnetic sources and the shape and size of the respective portions of the source causing the gravity and magnetic responses. We emphasise that pseudogravity is not the actual gravity response and is related only to the magnetic rocks in the survey area; it is not the total gravity response of all the (magnetic plus non-magnetic) geology.

3.7.3 Wavelength filters

A common filtering operation applied to potential field data involves directly modifying the frequency content of the data (see online Appendix 2). These are the wavelength/frequency filters described in *Frequency/wavelength* in Section 2.7.4.4.

Separation of local and regional responses is often a very effective enhancement to potential field data. It can be an important precursor in the application of the various derivative filters and transforms described in the next sections, because their ability to accurately resolve features of gravity and magnetic anomalies can be strongly degraded in the presence of a regional field.

3.7.3.1 Spectral filters

Gravity or magnetic datasets contain variations with a range of wavelengths, and usually this is a continuous range. Deep sources and broad near-surface sources contribute the longer wavelengths, whilst smaller near-surface sources produce the shorter wavelengths (see Section 2.3). It is important to appreciate that wavelength filtering to isolate responses from particular depth intervals, so-called *depth slicing*, is impossible. This is due to overlap in the frequency spectra of the responses from different depths. Consequently, it is impossible for wavelength filtering to completely separate the responses of a series of discrete depth intervals. In practice it is possible to partly separate, say, shallow, intermediate and deep responses.

3.7.3.2 Continuation filters

Particularly useful filters applicable to potential field data are the upward and downward continuation filters. In physical terms, the filter transforms the data to what it would have been if the measurements had been made at a different height above the source. Increasing the height 'continues' the potential field upwards, i.e. upward continuation, and the reverse for downward continuation. Upward continuation is equivalent to increasing the survey height, downward continuation to decreasing it. Continuation is possible because the frequency spectrum (see online Appendix 2) of potential fields varies in a predictable manner with distance from the source. The data are usually continued between planar horizontal surfaces, but it is also possible to work with irregular surfaces; this is the basis for the aeromagnetic drape corrections mentioned in Section 3.6.3.

The effects of changing source–detector separation are shown in Fig. 2.4. Increasing the separation causes a decrease in amplitude and increase in wavelength of the response. Upward continuation is a form of low-pass filtering. The amplitude of the whole data spectrum is attenuated with height, but the rate of attenuation is wavelength-dependent. Shorter wavelengths (high frequencies) associated with near-surface sources attenuate more rapidly with height than the longer wavelengths (lower frequencies), so the shallow-sourced responses, plus any short-wavelength noise, are suppressed. Longer-wavelength components dominate the filtered data so they have a smoother appearance (Fig. 3.25b and 3.27b). Conversely, downward continuation amplifies the spectrum with decreasing height, with shorter wavelengths (high frequencies) amplified more than longer wavelengths (lower frequencies) so the near-surface response is enhanced. In practice, data cannot be downward-continued very far because the transform becomes unstable. This is because the shorter wavelengths are inherently low in amplitude, or at worst unavailable owing to inadequate sampling (see Section 2.6.1), in measurements made at greater heights. Also, wavelengths with amplitudes similar to the noise level cannot be accurately downward-continued. Instability in the filter is seen as extreme variations in amplitude over short distances. Downward continuation is only effective when applied to very high-quality datasets.

Downward continuation should be used with caution, but we recommend upward continuation as the filter of choice for most low-pass filtering operations. A high-pass filter (detail enhancer) can be obtained indirectly by subtracting the resultant upward-continued data from the unfiltered data to remove the longer-wavelength component emphasised in the continued data. Figures 3.25c and 3.27c were produced in this way. Note the 'sharper' appearance resulting from suppression of longer-wavelength variations, the removal of the regional gradient that is obvious in Fig. 3.27a and the reduced interference between adjacent anomalies, especially in the northern and eastern parts of the Las Cruces data.

3.7.4 Gradients/derivatives

Gradients or derivatives of magnetic and gravity fields are more sensitive than the measured TMI and normal gravity to changes in the physical properties of the subsurface, so they are a detail-enhancement filter. Derivatives emphasise shallow bodies in preference to the deeper-seated broader features, which produce small changes (gradients) in the fields. Magnetic and gravity derivatives are calculated from the TMI and gravity data (see *Gradients and curvature* in Section 2.7.4.4), if they were not directly measured with a gradiometer as described in Section 2.2.3. The use of the vertical gradient is especially common. It can be visualised as the difference between the upward- and downward-continued (see Section 3.7.3.2) responses at equivalent locations, and normalised (divided) by the difference in height between the two continuation responses.

Vertical and horizontal gradients are very sensitive to the edges of bodies and are 'edge detectors'. Note how the relevant responses in Figs. 3.25 and 3.26 are concentrated at the edges of the prism. For vertical dipping contacts, and with vertical magnetisation in the case of magnetic sources, the derivative responses coincide with the contact. They are displaced slightly from the body when the contact is dipping or the source is narrow, although this displacement is only normally significant for very detailed studies (Grauch and Cordell, 1987).

Figures 3.25d and 3.27d show the first vertical derivatives of the gravity model and the gravity data, respectively, with the equivalent responses for the magnetic data shown in Figs. 3.26b and e, and Figs. 3.28c and d. As shown in Figs. 3.25 and 3.26, the relationship between the derivatives and their respective sources is quite simple for gravity, but can be more complex for magnetics. The form of the magnetic derivatives depends on the direction of the source magnetism and is usually multi-peaked and dipolar unless the magnetism is vertical. The asymmetric responses due to non-vertical magnetisation can be overcome by applying the derivatives to pole-reduced data (see Section 3.7.2.1).

The first vertical derivatives of the Las Cruces and Broken Hill data are dominated by much shorter-wavelength anomalies than the Bouguer anomaly and TMI data. Features in areas which are smooth in the TMI data now contain recognisable anomalies (e.g. (1)) and details of the structure of an intrusive (2) and folded stratigraphy (3) are easier to resolve. The first vertical derivative of the Las Cruces gravity data shows a slightly spotty appearance; this is noise caused by the poor definition of the short-wavelength component (shallow sources) of the gravity signal due to the comparatively large distance between stations. This is nearly always present in ground gravity data. Note the northeast–southwest linear feature traversing the southeast part of the image (1), which is much less obvious in the original data.

A useful enhancement for gravity data (and also pseudogravity data; see Section 3.7.2.2), is the total horizontal gradient which peaks over vertical contacts (Fig. 3.25e), or forms a ridge if the source is narrow. This enhancement creates a circular ridge at the Las Cruces anomaly (Fig. 3.27e), but since the edges of the source are not vertical these do not correspond with the edges of the deposit. The magnetic data from Broken Hill clearly emphasise source edges (Fig. 3.28h).

3.7.4.1 Second-order derivatives

Second-order derivatives (see *Gradients and curvature* in Section 2.7.4.4) can be an effective form of enhancement. The zero values of the second vertical derivative coincide with the edges of sources if their edges are vertical, but more importantly, the response is localised to source edges increasing their resolution. Figures 3.25h and 3.27h show the second vertical derivatives of the gravity model and the Las Cruces gravity data respectively, with the equivalent responses for the Broken Hill magnetic data shown in Fig. 3.28e. The improved resolution is clearly seen in the Broken Hill magnetic data, e.g. (4). However, second- and higher-order derivatives are very susceptible to noise and so are only useful on high-quality datasets. As shown by the 'spotty' appearance of the Las Cruces image, they are rarely useful for ground gravity data unless the station spacing is very small.

3.7.4.2 Analytic signal

Combining the three directional gradients of the gravity or magnetic field to obtain the total gradient (see *Gradients and curvature* in Section 2.7.4.4, and Eq. (2.5)) removes the complexities of derivative responses. When applied to potential field data, the total gradient at a location (x, y) is known as the *analytic signal* (AS) and given by:

$$\mathrm{AS}(x, y) = \sqrt{\left(\frac{\partial f}{\partial x}\right)^2 + \left(\frac{\partial f}{\partial y}\right)^2 + \left(\frac{\partial f}{\partial z}\right)^2} \tag{3.23}$$

where f is either the gravity or the magnetic field. Where the survey line spacing is significantly larger than the station spacing, the across-line Y-derivative is not accurately defined. It is preferable then to assume that the geology is two-dimensional and set the Y-derivative to zero.

The analytic signal has the form of a ridge located above the vertical contact (Figs. 3.25f and 3.26c), and is slightly displaced laterally when the contact is dipping. The form of the source prism is clearly visible in the transformed datasets; the crest of the ridge delineates the edge of the top surface. The magnetic data from Broken Hill show distinct peaks above the relatively narrow sources which define the northeast–southwest trending fold (Fig. 3.28f). Being based on derivatives, the gravity data from Las Cruces are quite noisy (Fig. 3.27f). Another example of the analytic signal of TMI data is shown in Fig. 4.25b, in this case the response is controlled by magnetism-destructive alteration which greatly reduces the short-wavelength component of the signal to which the analytic signal responds.

The analytic signal is effective for delineating geological boundaries and resolving close-spaced bodies. Since the magnetic analytic signal depends upon the strength and not the direction of a body's magnetism, it is particularly useful for analysing data from equatorial regions, where the TMI response provides limited spatial resolution, and when the source carries strong remanent magnetisation (MacLeod *et al.*, 1993).

The gradient measurements are susceptible to noise which can severely contaminate the computed analytic signal. This is especially a problem with comparatively sparsely sampled datasets. Higher-order total gradient signals calculated from higher-order gradients, such as the second derivative, have also been proposed, e.g. Hsu *et al.* (1996). Like the derivatives, the response becomes narrower as the order increases, offering the advantage of higher spatial resolution. However, higher-order gradients amplify noise which leads to a noisier analytic signal.

3.7.4.3 Tilt derivatives

Shallow sources produce large amplitudes in the vertical and horizontal gradients. The large amplitude range presents a problem for display. Ratios of the derivatives of each class of source have similar amplitudes, so the resolution of both classes can be balanced by dividing the vertical derivative by the amplitude of the total horizontal derivative. Furthermore, the ratio can be treated as an angle and the inverse tangent function applied to attenuate high amplitudes (see *Amplitude scaling* in Section 2.7.4.4). This is known as the *tilt derivative* (TDR) (see Miller and Singh, 1994), and at a location (x, y) it is given by:

$$\mathrm{TDR}(x, y) = \tan^{-1}\left[\frac{\partial f}{dz} \Big/ \sqrt{\left(\frac{\partial f}{\partial x}\right)^2 + \left(\frac{\partial f}{\partial y}\right)^2}\right] \qquad (3.24)$$

where f is either the gravity or the magnetic field.

For gravity anomalies and vertical magnetised bodies the tilt derivative is positive over the source and negative outside of it. Its form mimics the 2D shape of the anomaly source with the zero-value contour line delineating the upper boundaries of the source (Figs. 3.25g and 3.26h). Its shape is more complex for non-polar magnetisation. The amplitude scaling properties of the enhancement are well demonstrated in the Broken Hill magnetic data in areas of weak TMI, e.g. (5) and (6) in Fig. 3.28g; compare these with the responses of the other derivative enhancements.

3.8 Density in the geological environment

An understanding of the density of geological materials and geological causes for variations in density is crucial for making geologically realistic interpretations of gravity data.

Rock density affects the Earth's gravity field, but in the wider geophysical context density is also an important control on radiometric and seismic responses (see Sections 4.2.3 and 6.6). Figures 3.29 and 3.30 show the variations in density for materials commonly encountered in the geological environment. By convention, the average density of the continental crust is taken to be 2.67 g/cm^3 (see Hinze, 2003), which is consistent with the data presented here. Densities fall within a very limited range when compared with the other physical properties relevant to geophysics, notably magnetic and electrical properties (see Sections 3.9 and 5.3) which vary by many orders of magnitude. Also, unlike most other physical properties important in geophysics, rock density is a bulk property. It depends on the entire mineral assemblage and, in relevant rock types, their total porosity.

We present here an overview of the density of various rock types and geological environments, at the same time considering geological processes that produce changes in the density of minerals and rocks.

3.8.1 Densities of low-porosity rocks

The bulk density of a rock (ρ_{bulk}), i.e. the entire rock including both matrix and pore contents, can be calculated from the densities and proportions of its constituents as follows:

$$\rho_{\text{bulk}} = \sum_{i=1}^{n}\left(\frac{V_i}{V}\right)\rho_i \qquad (3.25)$$

where the rock comprises n components and V_i/V is the volume fraction of the ith component with density ρ_i. Where a rock has minimal porosity, it is the density and relative amounts of the constituent minerals which control its density.

The density of a particular mineral species depends on the masses of its constituent atoms and how closely they are packed together within its crystal lattice. The majority of rock-forming minerals contain the elements Al, Fe, Mg, Ca, K, Na, C, O and Si which have comparatively low mass numbers. The largest is that of Fe with a mass number of about 56 whilst the others have mass numbers between 12 and about 40. Metal sulphides and oxides contain heavier elements, such as Zn, Ni, Cu and Pb, whose mass numbers range from about 59 to 207 and are therefore expected to be denser than the rock-forming minerals, as is confirmed by Fig. 3.29.

Minerals that have a range of compositions, in particular in their cations, for example the feldspars, olivines etc., will vary in density according to their composition. Figure 3.31a shows density variations in olivine. The Fe-rich varieties have significantly higher densities, reflecting the high mass number of Fe. A further cause of density variations are impurities in the crystal lattice, causing even those minerals that have a specific composition to vary in density. Clearly some variation in the density of a particular mineral species is almost always to be expected.

The influence of crystal structure on density is demonstrated by the carbon allotropes. The tightly packed

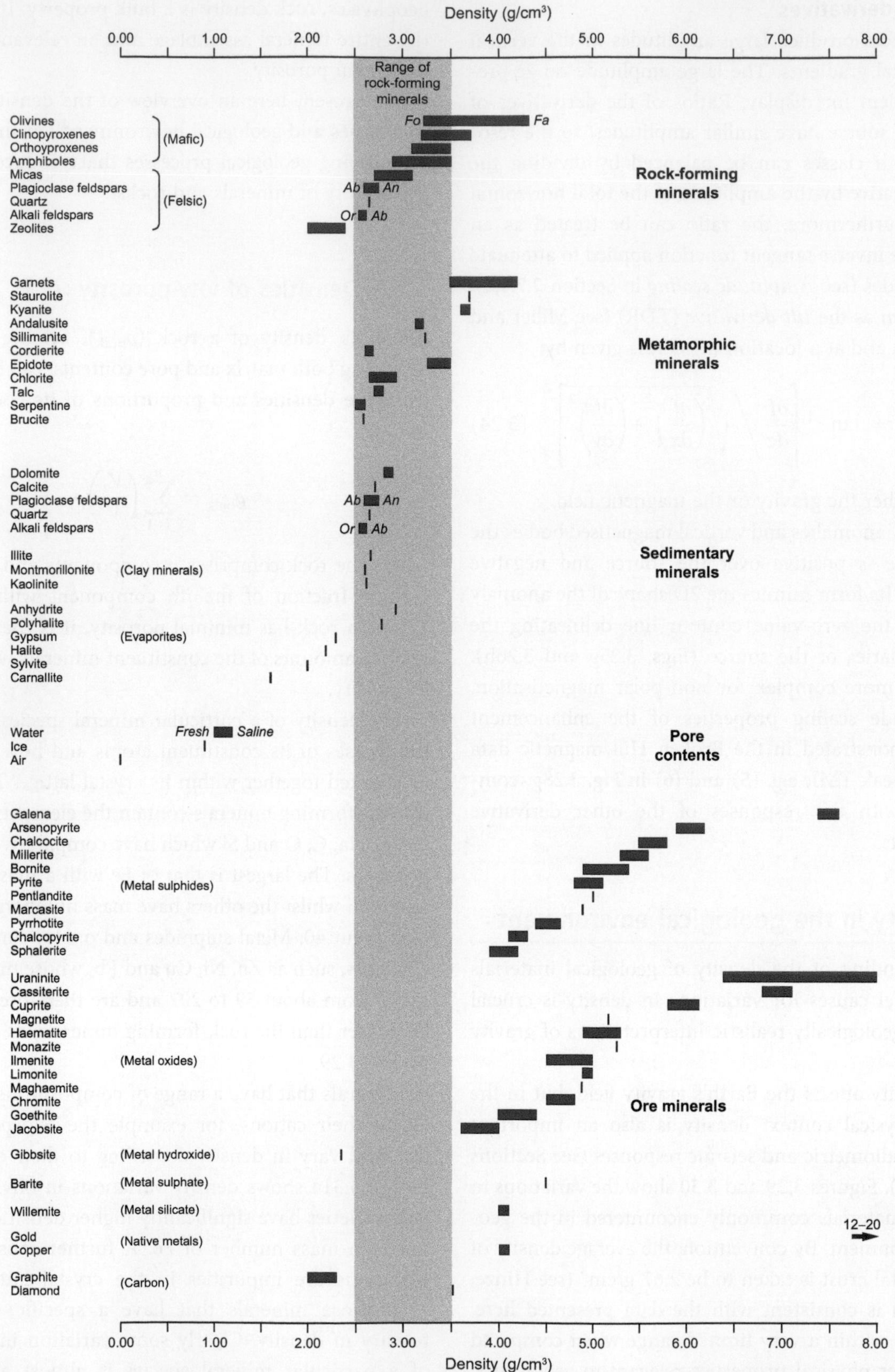

Figure 3.29 Density ranges of common rock-forming and ore minerals, and their likely pore contents. The shaded area corresponds with the density range of the main rock-forming minerals shown in Fig. 3.30. Based mainly on data in Read (1970).

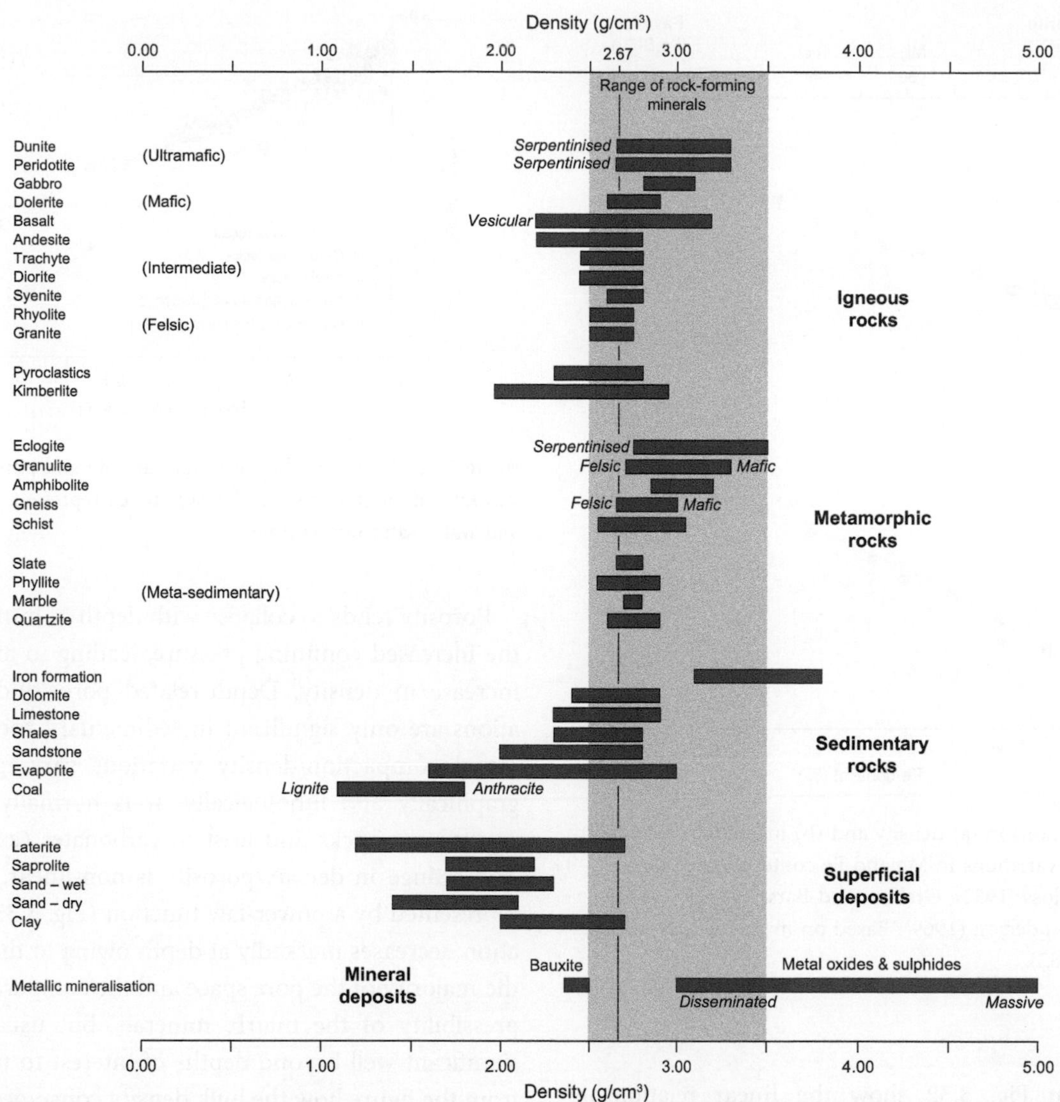

Figure 3.30 Density ranges of various rock types. Based mainly on diagrams and data in Emerson (1990), Schön (1996) and Wohlenberg (1982).

structure of diamond results in a density of 3.52 g/cm^3 whereas the more loosely packed atoms in graphite result in a density of 2–2.3 g/cm^3. The density difference of other relevant dimorphic substances is usually much less, but still significant. For example, the two forms of $CaCO_3$ (aragonite and calcite) and FeS_2 (pyrite and marcasite) both have a density difference of 0.22 g/cm^3.

Figure 3.29 demonstrates that the densities of the main rock-forming minerals lie mostly in the range 2.5 to 3.5 g/cm^3. Felsic minerals, containing lighter cations such as Na and K, are noticeably less dense than mafic minerals whose cations are mainly heavier elements such as Fe, Mg and Ca. Several common metamorphic minerals, e.g. garnet and kyanite, have unusually high densities.

3.8.2 Densities of porous rocks

The pore contents of porous rocks are an important control on the bulk density of the rock because the density of the pore contents is much lower than that of any of the matrix minerals (Fig. 3.29). Consequently, a change in porosity exerts much more influence than a change in the mineralogy of the matrix. The relationship between bulk density (ρ_{bulk}), and the (average) density of the matrix (ρ_{matrix}) and the density of the pore contents ($\rho_{\text{pore contents}}$) is given by:

$$\rho_{\text{bulk}} = \rho_{\text{matrix}}(1-\phi) + \rho_{\text{pore contents}}\phi \qquad (3.26)$$

where ϕ is the *fractional porosity* (0 to 1).

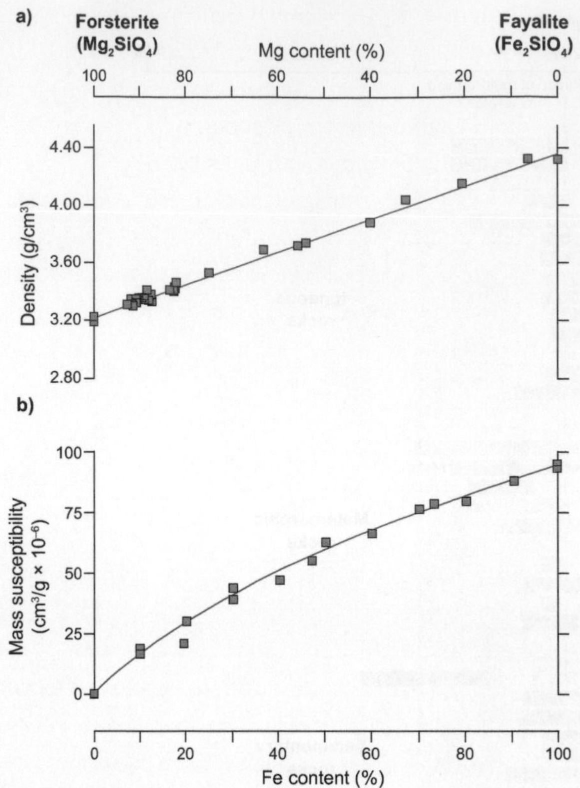

Figure 3.31 Variations in (a) density and (b) magnetic susceptibility of olivine due to variations in Mg and Fe content. Based on density data in Bloss (1952), Graham and Barsch (1969), and Kumazawa and Anderson (1969). Based on magnetic data in Bleil and Petersen (1982).

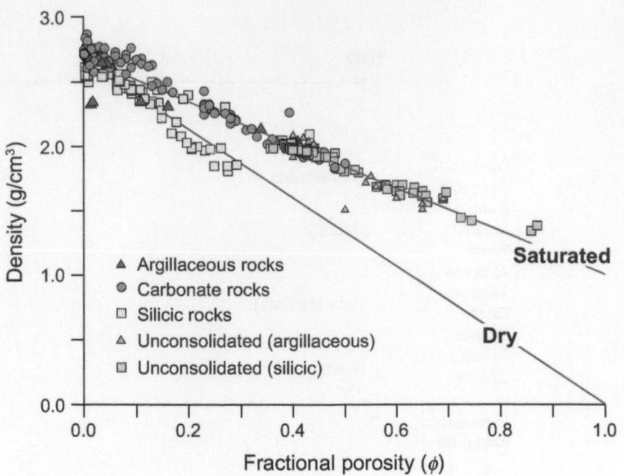

Figure 3.32 Correlation between bulk density and porosity for various sedimentary rocks. The two trends represent dry (air-filled) and water-saturated samples.

The data in Fig. 3.32 show the linear relationship between porosity and bulk density predicted by Eq. (3.26). The data lie on mixing lines between pure matrix materials, usually calcite or quartz (all data are from sedimentary materials), and the pore contents (air or water); and they extend from the density of the matrix (normally 2.65–2.75 g/cm^3) at zero porosity to the densities of air and water at 100% porosity. The significant influence of low-density pore contents on the overall density is clearly evident.

If the pore space is not totally saturated, or contains more than one fluid phase, the density of the pore contents is then the average of the constituents weighted according to their relative abundance. The low density of air means that saturated and unsaturated rocks may have quite different densities, but fresh and salt water have similar densities so salinity makes little difference. Ice may occupy pore space in permafrost areas resulting in a slightly lower density than if saturated with water.

Porosity tends to collapse with depth of burial, owing to the increased confining pressure, leading to an equivalent increase in density. Depth-related porosity/density variations are only significant in sedimentary rocks, with the actual compaction/density variations varying both geographically and lithologically. It is normally greatest in argillaceous rocks and least in carbonates (except chalk). The change in density/porosity is non-linear, often being represented by a power-law function (Fig. 3.33). The variation decreases markedly at depth owing to the collapse of the majority of the pore space and the comparative incompressibility of the matrix minerals, but usually remains significant well beyond depths of interest to miners. Note from the figure how the bulk density converges toward the matrix density. If gravity responses are to be modelled (see Section 3.10.3) in terrains containing Phanerozoic sedimentary rocks, it is usually necessary to account for the depth dependency of density.

3.8.3 Density and lithology

Figures 3.29 and 3.30 show, as expected, that crystalline rock types have densities which lie within the range of the rock-forming minerals. Density is not diagnostic of rock type, but a number of relationships are evident in Fig. 3.30.

- For unweathered and unaltered rocks, felsic rocks are normally less dense than intermediate rocks, which are less dense than mafic rocks, and ultramafic rocks normally have the highest densities. This correlation between mineralogy and chemistry is primarily due to the relative

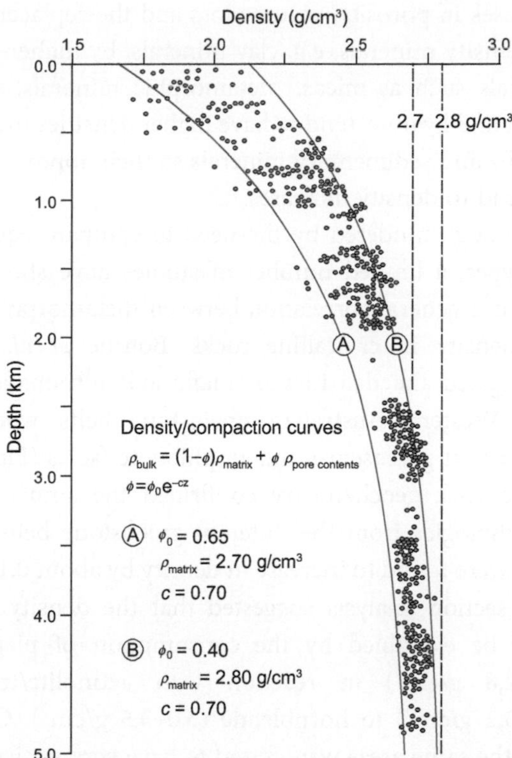

Figure 3.33 Increase in density with depth in sandstones and siltstones owing to decrease in porosity associated with compaction. Redrawn, with permission, from Schön (1996).

proportions of low-density felsic to high-density mafic minerals, e.g. feldspars and quartz versus amphiboles, pyroxenes and olivine. Igneous ultramafic rocks continue this trend, but metamorphic examples tend to be denser, largely owing to the presence of garnet-bearing eclogites.

- Felsic rocks exhibit smaller ranges in density than mafic and ultramafic rocks. This is due to the susceptibility of the latter two to alteration, especially serpentinisation and weathering (see Sections 3.8.4 and 3.8.5). Also vesicles and amygdales may reduce their bulk density.

Figures 3.29 and 3.30 show that the density of sedimentary rocks extends to values that are significantly less than those of the rock-forming minerals. This is primarily due to the dominant influence of porosity over mineralogy. Commonly, porous rocks also have a greater range of density than those not having significant porosity.

Although density is again not diagnostic of lithology, reference to Fig. 3.30 shows some general statements can be made:

- In general, carbonates are denser than siliciclastics.
- Evaporites are some of the lowest density species owing to the low density of their constituents, but the large range in density of their constituent minerals (Fig. 3.29) gives this class a very large density range.
- The presence of iron oxides in iron formations makes these one of the densest rock types.
- Coals are amongst the lowest density rocks.
- Heavy mineral sands are not particularly dense, because the effects of the 'heavy' components are countered by high porosities. Even with 50% titanomagnetite content, Lawton and Hochstein (1993) measured densities of only about 2.8 g/cm^3 for samples from New Zealand deposits.

3.8.4 Changes in density due to metamorphism and alteration

Any alteration or metamorphic reaction involving the replacement of mineral species with others of different density will affect the density of a rock. Metamorphic rocks generally have higher density than the rocks from which they are derived because the process of metamorphism reforms the source rocks into more compact forms. Any changes in porosity will also change its density.

Serpentinisation is arguably the most important alteration/metamorphic process in terms of geophysical responses, significantly affecting density, magnetism and acoustic properties. Put simply, serpentinisation is the low-temperature hydrothermal alteration of olivine and orthopyroxene to serpentine group minerals, brucite and magnetite. A comprehensive list of serpentinisation reactions, in the context of their physical property implications, is provided by Toft *et al.* (1990).

Figure 3.34 shows that bulk density is inversely proportional to the degree of serpentinisation. Any rock containing the requisite minerals may be affected by serpentinisation, but it is especially common in ultramafic species comprising predominantly serpentinisable minerals, e.g. peridotite, dunite, harzburgite and lherzolite. Mafic minerals with densities of around 3.1 to 3.6 g/cm^3 are replaced by serpentine-group minerals whose densities range from 2.55 to 2.61 g/cm^3. In addition, brucite also has a low density (2.39 g/cm^3), and highly serpentinised rocks may have porosities as high as 15%. Even allowing for the creation of magnetite (density about 5.2 g/cm^3), the reduction in density caused by these effects is significant, with bulk density decreasing by up to about 0.9 g/cm^3 demonstrated by Toft *et al.* (1990). The fact that some degree of serpentinisation is common in mafic and ultramafic rocks is the reason for their density range extending to low values, similar to those of felsic rocks etc.

Figure 3.30 shows that metamorphosed sedimentary rocks, e.g. marble, slate and quartzite, have much smaller density ranges than their sedimentary precursors, limestone, shale and sandstone respectively. Their densities coincide with the upper limits of their precursors owing to decreases in porosity, volume loss and the replacement of low-density minerals, e.g. clay minerals, by higher-density minerals such as micas. Metamorphic minerals, such as garnet and kyanite, tend to have higher densities than most igneous and sedimentary minerals so their appearance can also lead to density increases.

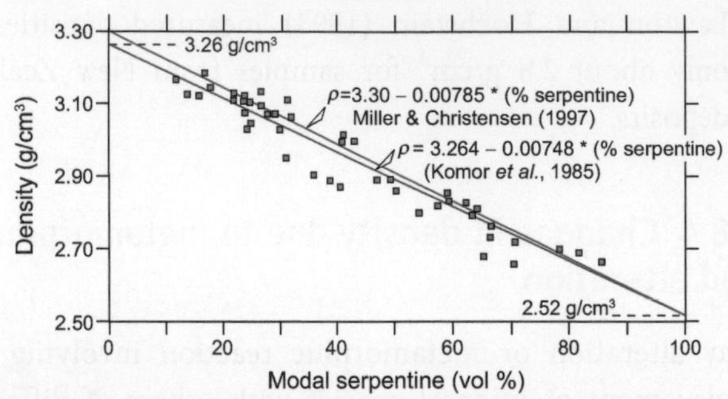

Figure 3.34 Inverse linear relationship between density and degree of serpentinisation as reported by Komor *et al.* (1985) and, similarly, by Miller and Christensen (1997). The data are from a predominantly dunite-wehrlite suite from the Bay of Islands ophiolite in western Newfoundland, Canada. Serpentinisation in this case primarily affects olivine. Deviations from the main trend are caused by inter-sample variations in the amounts of unaltered olivine, orthopyroxene, plus spinel and the magnetite and brucite formed during serpentinisation. Redrawn with additions, with permission, from Komor *et al.* (1985).

Although hindered by the need to compare equivalent lithotypes, a limited number of studies have shown that there is a general correlation between metamorphic grade and density in crystalline rocks. Bourne *et al.* (1993) investigated the densities of mafic and ultramafic rocks from Western Australian greenstone belts where the rocks are at greenschist and amphibolite facies (Fig. 3.35). Whole rock geochemistry confirmed the similarities of the lithologies from the different greenstone belts. Mafic rocks were found to increase in density by about 0.1 g/cm^3. Thin section analysis suggested that the density change could be explained by the consumption of plagioclase (2.6–2.8 g/cm^3) in reaction with actinolite/tremolite (2.9–3.2 g/cm^3) to hornblende (3.0–3.5 g/cm^3). Granites from the same areas were found to have very similar densities, reflecting their more stable felsic mineral assemblages. Olesen *et al.* (1991) describe an increase in average density, from about 2.75 to 2.81 g/cm^3, associated with the change from amphibolite to granulite conditions in a Scandinavian high-grade gneiss terrain. However, the scatter of densities

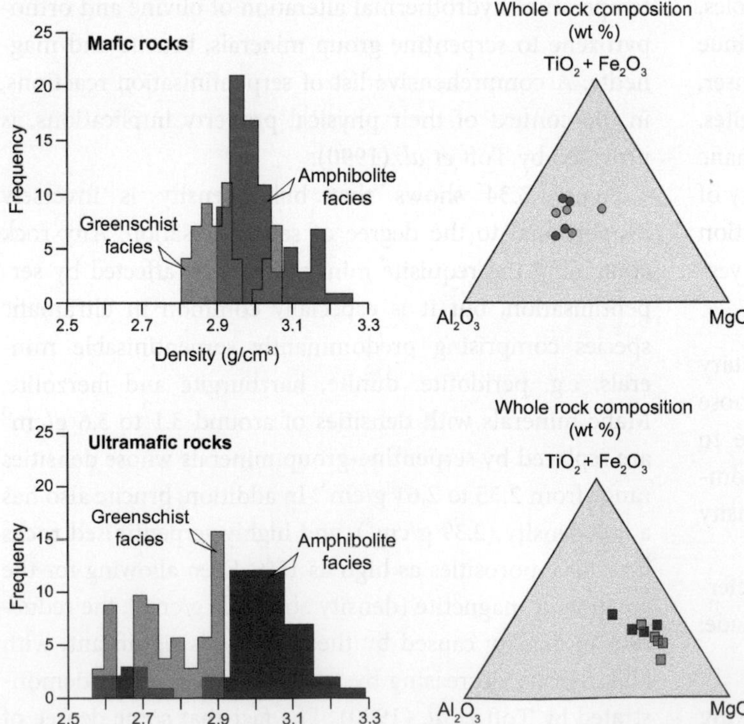

Figure 3.35 Frequency histograms of the density of mafic and ultramafic rocks from two greenstone belts in Western Australia showing increase in density from greenschist to amphibolite facies. Based on diagrams in Bourne *et al.* (1993).

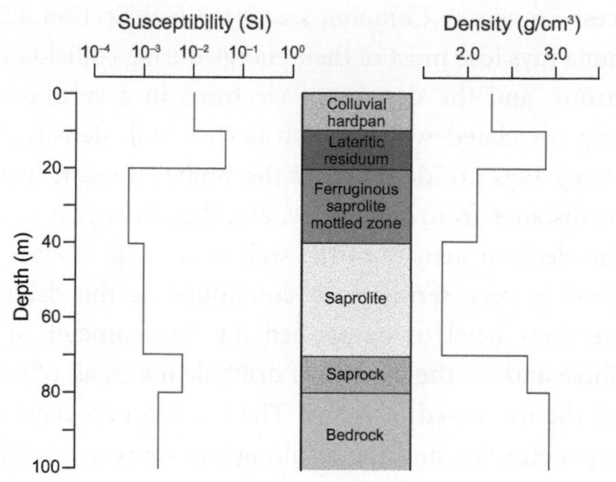

Figure 3.36 Variations in magnetic susceptibility and average density through overburden and regolith in a greenstone terrain in Western Australia. Based on diagrams in Emerson *et al.* (2000).

in each group is significantly larger than the increase, which is attributed to an increase in magnetite content and the alteration of biotite and amphibole to pyroxene.

In terms of interpreting gravity data for mineral exploration purposes, regional metamorphic gradients cause relatively small and gradual changes in density which are unlikely to be significant. The effects of alteration are important, however, since serpentinisation may cause ultramafic rocks to have unexpectedly low densities.

3.8.5 Density of the near-surface

Weathering significantly reduces density, and since weathering is normally greatest near the surface, decreasing progressively with depth, a density gradient is to be expected. The main causes are increased porosity and the creation of lower-density secondary minerals, e.g. feldspars being replaced by clay minerals. Ishikawa *et al.* (1981) describe an inverse correlation between density and the degree of weathering in granites, the data defining a continuous variation between that of the fresh material and that of unconsolidated siliceous sediments (see Section 6.6.5).

As shown in Fig. 3.30, unconsolidated sediments have lower density than most lithified materials, so the bedrock–cover interface coincides with a significant density contrast. Also, density variations within the cover material may be significant, especially where there is a well-developed regolith. Figure 3.36 shows intra-regolith density variations to be quite large with a dense, and magnetic, laterite horizon occurring at the top of the regolith. The regolith in its entirety is of lower density than the protolith and often affects gravity responses, especially through changes in its thickness. The volume of cover and/or regolith material may not be great, but density contrasts may be large compared with those at depth and their influence on the gravity reading can be significant, simply by virtue of the fact that these features are closer to the gravity meter than the underlying geology.

Where there is a cover of snow and ice, and/or permafrost is present, there are associated density contrasts. Ice and snow are less dense than the underlying rocks, so where the former are thick they may affect gravity readings. Permafrost is less dense than identical materials filled with water, although the contrast is normally small and depends on porosity. Sufficiently large areas of greater ice content may cause significant decreases in gravity.

3.8.6 Density of mineralised environments

The distribution and variation in density of a mineralised environment is often complex. Common gangue minerals such as quartz and calcite have low density. The heavy elements within metal oxide and sulphide minerals make these noticeably denser than the rock-forming minerals, at roughly 4.0 to 7.0 g/cm^3. Native metals, although some of the densest substances in the geological environment, are too scarce to affect the bulk density of their host rocks. Another dense form of mineralisation is that containing barite (density = 4.48 g/cm^3). Graphite is a rare example of low-density mineralisation; bauxite is another. A gravity survey will respond to the entire ore–gangue package but normally it is the dense ore minerals that dominate, making the mineralised assemblage denser than its host rocks (Fig. 3.30).

Hydrothermal alteration and preferential weathering of unstable ore minerals, such as sulphides, will usually be associated with a reduction in overall density of the affected zone. The same is true for alteration of feldspars to clay minerals. This is evident in data from the hydrothermal alteration zone associated with the General Custer Mine in the Yankee Fork Ag–Au mining district in Idaho, USA (see Fig. 3.57 in Section 3.9.7) and the lower gravity in the Cripple Creek example described in Section 3.4.7. Conversely, processes such as propylitisation that create epidote (density = 3.5 g/cm^3) may lead to an increase in density, as will deposition of alteration-related minerals in pore space. These competing effects are the reason that both positive and negative gravity anomalies have been reported over zones of hydrothermal alteration (see Irvine and Smith, 1990).

A common occurrence is preferentially deeper weathering of mineral deposits that are in contact with the base of weathering. Weathering into the deposit reduces density and causes a reduction in the overall gravity response of the mineralisation. Similarly, kimberlite pipes extending to the near-surface are sometimes preferentially weathered deeper than the surrounding area to produce a 'low' gravity response over these features.

Surveys designed to detect gravity variations due to the rocks hosting mineralisation include those targeting sulphides in mafic and ultramafic intrusions and also kimberlite pipes. Sulphide-bearing intrusions are invariably denser than their host rocks. Kimberlites have highly variable density but generally it is lower than their host rocks. Any diamonds they may contain will be too scarce to influence the overall density of the intrusion significantly.

These complex controls on density mean that, depending on the area being explored, geologically similar exploration targets may be associated with either positive or negative gravity anomalies.

3.8.7 Measuring density

Density of rock specimens can be measured using Archimedes' principle. The specimen's weight in air and when immersed in water is compared. This gives the specific gravity of the specimen, i.e. its density relative to water, but given that water attains its maximum density of 1 g/cm^3 at 4 °C, and that it changes very little with temperature, specific gravity is taken as the density of the rock specimen ($\rho_{specimen}$) and is given by:

$$\rho_{specimen} = \frac{\text{Weight}_{\text{in air}}}{\text{Weight}_{\text{in air}} - \text{Weight}_{\text{in water}}} \tag{3.27}$$

In order to obtain accurate measurements the specimen should be fully saturated, and should be as large as possible to increase the likelihood of using representative samples of what may be a heterogeneous lithotype. Lipton (2001) describes appropriate laboratory procedures.

3.8.7.1 Density logging

Density can also be measured *in situ* with continuous downhole density (gamma–gamma) logging. The method is based on measuring gamma rays, emitted as a focused beam from a radioactive source on the logging tool, that interact with the electrons in atoms of the wall rocks in a process known as *Compton scattering* (see Section 4.2.3). Gamma rays lose most of their energy during collision with electrons, and the density of electrons in a substance is closely correlated with the substance's bulk density. The scattered rays are detected by the probe's sensor, located some distance from the source, and their energy is related to the electron density of the wall rocks (Fig. 3.37a). The method is very sensitive to conditions in the drillhole. Corrections need to be applied for the diameter of the drillhole and for the density of drillhole fluids, all of which affect the measured intensity. The gamma rays have very little penetration, and the resulting logs represent the bulk density of the rocks within a few centimetres of the drillhole.

The density log can be a very useful lithological indicator, especially for high-density ore minerals or low-density coal seams (Fig. 3.37b). In soft-rock environments it can be useful for hole-to-hole correlation as demonstrated by Mach (1997), who describes the use of density logs for correlation within coal measures in the Czech Republic.

3.8.8 Analysis of density data

It is often necessary to estimate a representative density of some geological entity, for example to correlate lithotypes with variations in gravity or for modelling. As with other petrophysical data, rather than trying to identify a specific value a better approach is to identify a likely density range. This should use measurements made on a large number of samples chosen to represent observed variations in the formation's geological characteristics. The data can then be analysed as a frequency histogram (Fig. 3.38). In practice only a limited number of samples may be available.

Ideally, the density measurements will show a normal distribution, and the arithmetic mean and standard deviation can be determined. Skew toward lower values may occur owing to the greater porosity, and probably also changes in mineralogy, of weathered materials taken from the near-surface. Multimodal distributions may result from inherently heterogeneous lithotypes such as gneisses, layered intrusions and banded iron formation, when the scale of variation is larger than the size of a sample. The number of samples required to characterise the actual distribution correctly depends on the homogeneity of the formation. The more homogenous the rock, the fewer the samples required. As with all geological sampling, the danger exists of taking an unrepresentatively large number

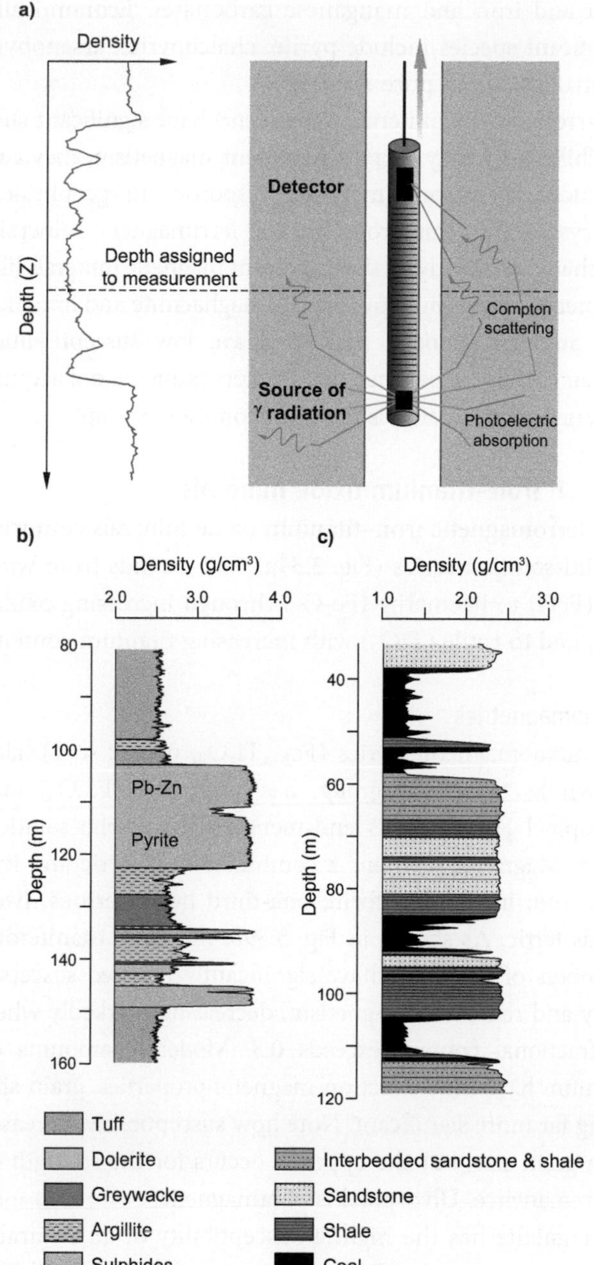

Figure 3.37 Downhole density logs. (a) Schematic illustration of gamma–gamma density logging, redrawn, with permission, from Schmitt *et al.* (2003). (b) Density log from the Stratmat Zn–Pb–Cu–Ag–Au VMS deposit, New Brunswick, Canada. Based on a diagram in Killeen (1997b). (c) Density log from a coal-measure sequence. Based on diagrams in Renwick (1981).

of samples of lithotypes that are resistant to erosion, and so outcrop.

Figure 3.38 shows frequency histograms for various lithologies. Generally, they are roughly symmetrical and have a single mode. The distribution for the granite is typical, being rather narrow. Basalts tend to have broader

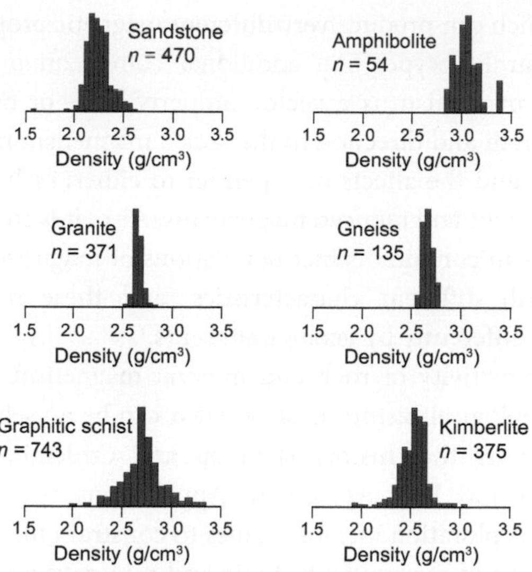

Figure 3.38 Frequency histograms of density for various lithotypes. Based on diagrams in Airo and Loukola-Ruskeeniemi (2004), Mwenifumbo *et al.* (1998), Mwenifumbo *et al.* (2004), and Subrahmanyam and Verma (1981).

distributions, and alteration may lead to several sub-populations (Fig. 3.35). Sedimentary rocks tend to have the most complicated distributions reflecting variations in composition and, in particular, porosity.

3.9 Magnetism in the geological environment

An understanding of the magnetism of the geological environment and geological causes for variations in magnetism is crucial for making geologically realistic interpretations of magnetic data.

The magnetism of rocks depends on the magnetic properties of their constituent minerals. Rock magnetism varies far more than density with susceptibility varying by approximately five orders of magnitude in the common rock types. The most fundamental control on rock magnetism is iron content. Put simply, without iron magnetic minerals cannot form, but not all iron-minerals are strongly magnetic. Magnetic minerals may be formed in the primary igneous environment, others may be created by secondary processes such as metamorphism. They all may be destroyed by secondary processes, notably weathering.

Unlike density, rock magnetism is not a bulk rock property. Instead it depends on mineral species that usually comprise only a few per cent by volume of a rock's mineralogy. Magnetic mineralogy is often strongly affected by subtle changes in geochemistry and secondary processes, all of which can produce very different magnetic properties in similar lithotypes. An additional complication arises because magnetism is a vector property. One or both of the strength and direction of the rock's magnetism may be affected, and the affects may pertain to either, or both, of the remanent and induced magnetisms. Also, it is common for rocks to contain distinct populations of magnetic minerals with different characteristics, and these may be affected differently by geological events.

The sensitivity of rock and mineral magnetism to the rock's geological history means that it can be a useful tool for studying that history (Dunlop and Ozdemir, 1997; Harrison and Freiburg, 2009). Applications relevant to mineral exploration include studies to constrain the nature and timing of magmatic, tectonic and mineralising events (Borradaile and Kukkee, 1996; Lockhart *et al.*, 2004; Symons *et al.*, 2011; Trench *et al.*, 1992).

We present here an overview of the mineralogical controls on rock magnetism followed by a description of the magnetic properties of different rock types and geological environments, at the same time considering geological processes that lead to the creation or destruction of magnetic minerals and/or the modification of the magnetic properties of existing minerals.

3.9.1 Magnetic properties of minerals

As described in Section 3.2.3.5, materials can be categorised according to their magnetic characteristics with diamagnetic, paramagnetic and ferromagnetic behaviour important for materials commonly occurring in the geological environment. It is the ferromagnetic group that have the strongest magnetic properties and that have the greatest influence on the magnetic properties of rocks.

Diamagnetic minerals have very weak negative susceptibility (about 10^{-5} SI) and are incapable of carrying remanent magnetism. Rock-forming minerals of this type include pure phases of quartz, calcite and feldspar. Economically significant examples include pure galena and sphalerite, graphite, halite, gypsum and anhydrite. Many clay minerals and water are diamagnetic.

Paramagnetic minerals have weak, but still potentially significant, positive susceptibilities (about 10^{-3} SI), but again no remanent magnetism. Paramagnetism is associated with the presence of iron and, less importantly, manganese. The susceptibility of paramagnetic species correlates with their total iron content (Fig. 3.31b). Minerals of this type include olivine, pyroxene, amphibole, garnet, mica and iron and manganese carbonates. Economically significant species include pyrite, chalcopyrite, arsenopyrite, marcasite and pure ilmenite.

Ferromagnetic minerals *(sensu lato)* have significant susceptibility and may carry a remanent magnetism; they can produce very strong magnetic responses in geophysical surveys. Within this group are the ferrimagnetic minerals, which include nearly all the important magnetic minerals like magnetite, monoclinic pyrrhotite, maghaemite and ilmenite. The antiferromagnetic minerals have low susceptibilities similar to those of paramagnetic materials and do not acquire remanence. Haematite is the most common example.

3.9.1.1 Iron–titanium oxide minerals

The ferromagnetic iron–titanium oxide minerals comprise a solid-solution series (Fig. 3.39a) that extends from wüstite (FeO) to haematite (Fe_2O_3) through increasing oxidation, and to rutile (TiO_2) with increasing titanium content.

Titanomagnetites

The titanomagnetite series ($Fe_{3-x}Ti_xO_4$; $0 < x < 1$), also known as the *spinel group*, has magnetite (Fe_3O_4) and ulvospinel (Fe_2TiO_4) as end-members of a solid-solution series. Magnetite contains a combination of ferric and ferrous iron: in its pure form, one-third being ferrous, two-thirds ferric. As shown in Fig. 3.39b, the more titaniferous members of the series have significantly reduced susceptibility and remanent magnetism, decreasing markedly when Ti fractional content exceeds 0.8. Moderate amounts of titanium have little effect on magnetic properties, grain size being far more significant. Note how susceptibility increases with grain size, but the opposite occurs for the strength of the remanence. Ulvospinel is paramagnetic.

Magnetite has the highest susceptibility of all naturally occurring minerals. Pure magnetite has susceptibility ranging from 13 SI for fine-grained, poorly crystalline, inhomogeneous or stressed grains, up to about 130 SI for very coarse, well-crystallised magnetite. Magnetite accounts for about 1.5% of crustal minerals, occurring in igneous, metamorphic and sedimentary environments. Its Curie temperature is 578 °C.

An important characteristic of titanomagnetites is the tendency for exsolution to occur during cooling. This is best developed in grains formed in the slower cooling plutonic rocks. At greater than 600 °C, i.e. magmatic temperatures, there is a complete solid solution between ulvospinel and magnetite. At lower temperature there is increasing immiscibility in the central section of the compositional

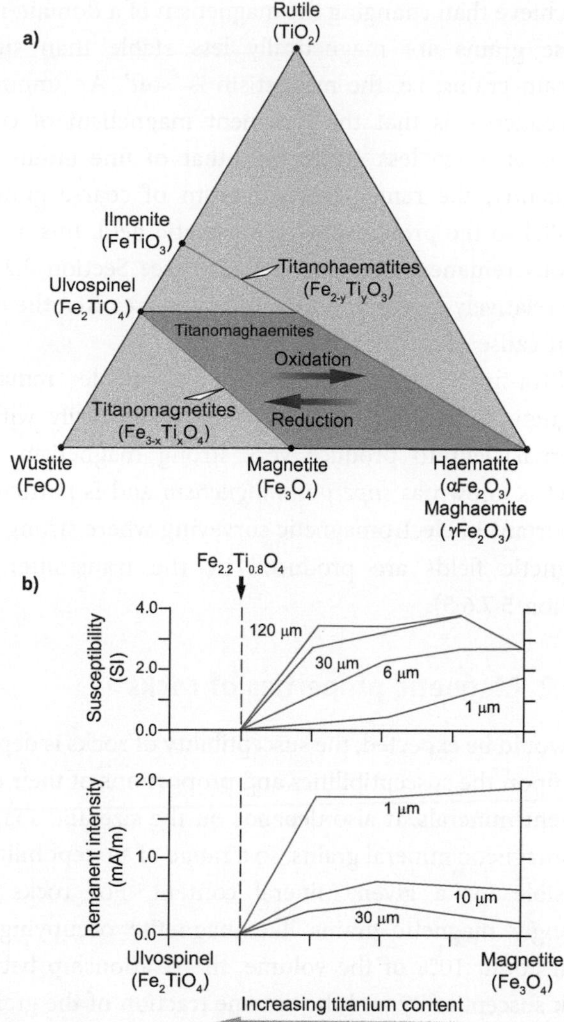

Figure 3.39 (a) Ternary diagram showing the chemistry of Fe–Ti oxides and their change with increasing oxidation. (b) Compositional control of magnetic properties in the titanomagnetite series. The individual curves are for the grain sizes shown. Redrawn, with permission, from Clark (1997).

range. Grains with compositions in this range undergo subsolidus exsolution into magnetite- and ulvospinel-rich, usually lamellar, intergrowths. Photomicrographs illustrating this phenomenon in an ore environment are provided by Alva-Valdivia and Urrutia-Fucugauchi (1998). The exsolved ulvospinel then tends to be oxidised to ilmenite and magnetite, if dissociated water is present, i.e. deuteric oxidation. Both phenomena tend to increase the magnetism of the grain. In more rapidly cooled extrusives, the titanium and iron oxides tend to remain in metastable equilibrium. In addition to producing iron-rich magnetic grains, exsolution may also effectively partition larger grains into smaller ones, whose 'effective' grain size affects their magnetic properties (see Section 3.9.1.3).

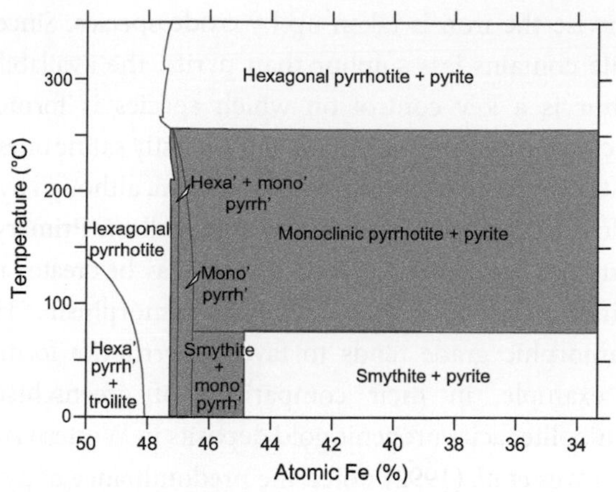

Figure 3.40 Relationship between condensed phases in part of the Fe–S system for temperatures below 350 °C. Redrawn, with permission, from Kissin and Scott (1982).

Titanohaematites

The titanohaematite series ($Fe_{2-y}Ti_yO_3$; $0 < y < 1$) has ilmenite ($FeTiO_3$) and haematite (αFe_2O_3) as end-members. Titanohaematites with 50–80% ilmenite are strongly magnetic and carry remanence. Pure ilmenite is paramagnetic and therefore carries no remanence and has low susceptibility. Some ilmenite compositions exhibit the rare phenomenon (in rocks) of acquiring a self-reversal remanent magnetisation, where the remanence is opposite to the applied magnetic field. Haematite exhibits weak susceptibility and its Curie temperature is 680 °C.

Maghaemite (γFe_2O_3) is ferrimagnetic and strongly magnetic, although its magnetic characteristics are complex and poorly understood.

3.9.1.2 Iron sulphide minerals

The iron sulphide mineral pyrrhotite can be magnetic. The general formula for the pyrrhotites is $Fe_{1-z}S$ ($0 < z < 0.13$) with values of z appearing to correspond with particular ratios of Fe:S. Its crystal structure is temperature-dependent and sensitive to composition (Fig. 3.40). Pyrrhotites occur with various crystal structures, but a key generalisation, from a geophysical perspective, is that monoclinic (4C) pyrrhotite (Fe_7S_8) is the only common pyrrhotite that is ferrimagnetic with high susceptibility. It has a Curie temperature of 320 °C. Note that monoclinic pyrrhotite is primarily a lower-temperature phase. It is unstable above about 250 °C, and higher-temperature forms have hexagonal structure and are antiferromagnetic.

Pyrrhotite often occurs in association with pyrite (FeS_2), which is paramagnetic, especially in ore environments. Both minerals are favoured by strongly reducing conditions, otherwise the iron is taken up by oxide species. Since pyrrhotite contains less sulphur than pyrite, the availability of sulphur is a key control on which species is formed. In reducing sedimentary environments, mostly saline ones, pyrrhotite may form by sulphurisation of iron, although pyrite is produced if sufficient time and sulphur allow. Primary pyrrhotite can occur in igneous rocks and may be created by desulphurisation of pyrite during metamorphism. Higher metamorphic grade tends to favour pyrrhotite formation. For example, in their comparison of greenschist and amphibolite facies orogenic gold deposits in Western Australia, Groves *et al.* (1992) noted the predominance of pyrite in deposits in lower-grade hosts, but pyrrhotite when the grade was higher. Geological descriptions rarely consider the habit of the pyrrhotite. Whether the occurrences described above are of monoclinic pyrrhotite will depend primarily on the temperature/cooling history of the environment.

Monoclinic pyrrhotite often carries strong remanent magnetisation. This can be stable for geologically long periods at low temperatures, particularly in fine-grained materials. However, because of its relatively low Curie temperature, remanent magnetism can be altered relatively easily. Susceptibility is dependent on grain size, decreasing with decreasing grain size. Monoclinic pyrrhotite has strong intrinsic anisotropy within its basal plane which results in significant susceptibility anisotropy in the rocks having a preferred orientation of the pyrrhotite grains.

Less common magnetic iron sulphide minerals include greigite (Fe_3S_4) and smythite ($(Fe, Ni)_9S_{11}$). Greigite occurs in young sediments; smythite, although rare, occurs in some magmatic ores and some sedimentary rocks. In both cases their magnetic properties probably resemble those of monoclinic pyrrhotite.

3.9.1.3 Grain size and rock magnetism

In general, very fine grains of magnetic minerals have a single magnetic domain structure, so the grain itself is uniformly magnetised. The magnetism of these grains is very stable or 'hard', because to alter it requires changing the direction of the magnetism itself. They are the most important carriers of intense remanence in many rocks even though they may constitute a minor proportion of the overall magnetic mineral assemblage.

For larger grains, the shape-dependent, internal self-demagnetisation (see Section 3.2.3.6) interacts with their intrinsic magnetisation, producing multiple domains with different magnetic orientations. The magnetism can be changed by migration of domain boundaries (see *Ferromagnetism* in Section 3.2.3.5), a process that is much easier to achieve than changing the magnetism of a domain itself. These grains are magnetically less stable than single-domain grains, i.e. the magnetism is 'soft'. An important consequence is that the remanent magnetism of coarse grains is much less stable than that of fine grains and, frequently, the remanent magnetism of coarse grains is parallel to the present-day geomagnetic field, this being a viscous remanent magnetism (VRM; see Section 3.2.3.4). The relatively large multidomain grains are often the dominant cause of a rock's susceptibility.

Ultra-fine grains do not retain stable remanent magnetism, but they do realign, or relax, easily with an external field to produce very strong magnetism. This effect is known as *superparamagnetism* and is particularly important in electromagnetic surveying where strong local magnetic fields are produced by the transmitter (see Section 5.7.6.5).

3.9.2 Magnetic properties of rocks

As would be expected, the susceptibility of rocks is dependent upon the susceptibilities and proportions of their component minerals. It also depends on the size and shape of the magnetic mineral grains, so a range of susceptibilities is possible for a given mineral content. For rocks with strongly magnetic grains like magnetite occupying less than about 10% of the volume, the relationship between rock susceptibility and the volume fraction of the grains is approximately linear on a log–log graph (Fig. 3.41). When

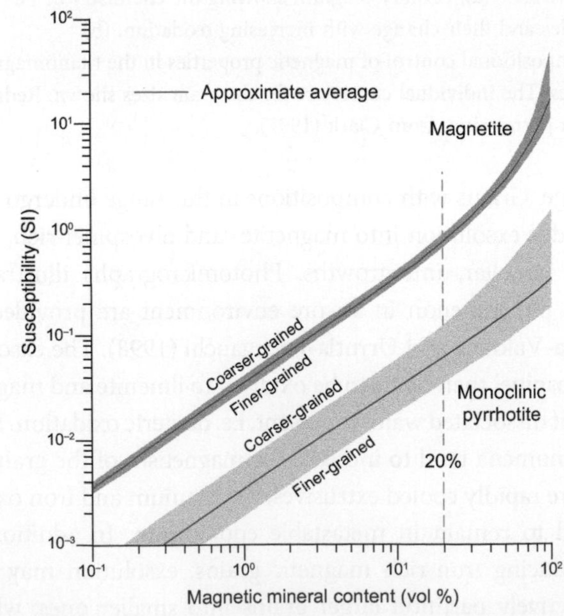

Figure 3.41 Magnetic susceptibility versus the volumetric content of magnetite and pyrrhotite. Fine-grained: <20 μm, coarse-grained: >500 μm. Redrawn, with permission, from Clark (1997).

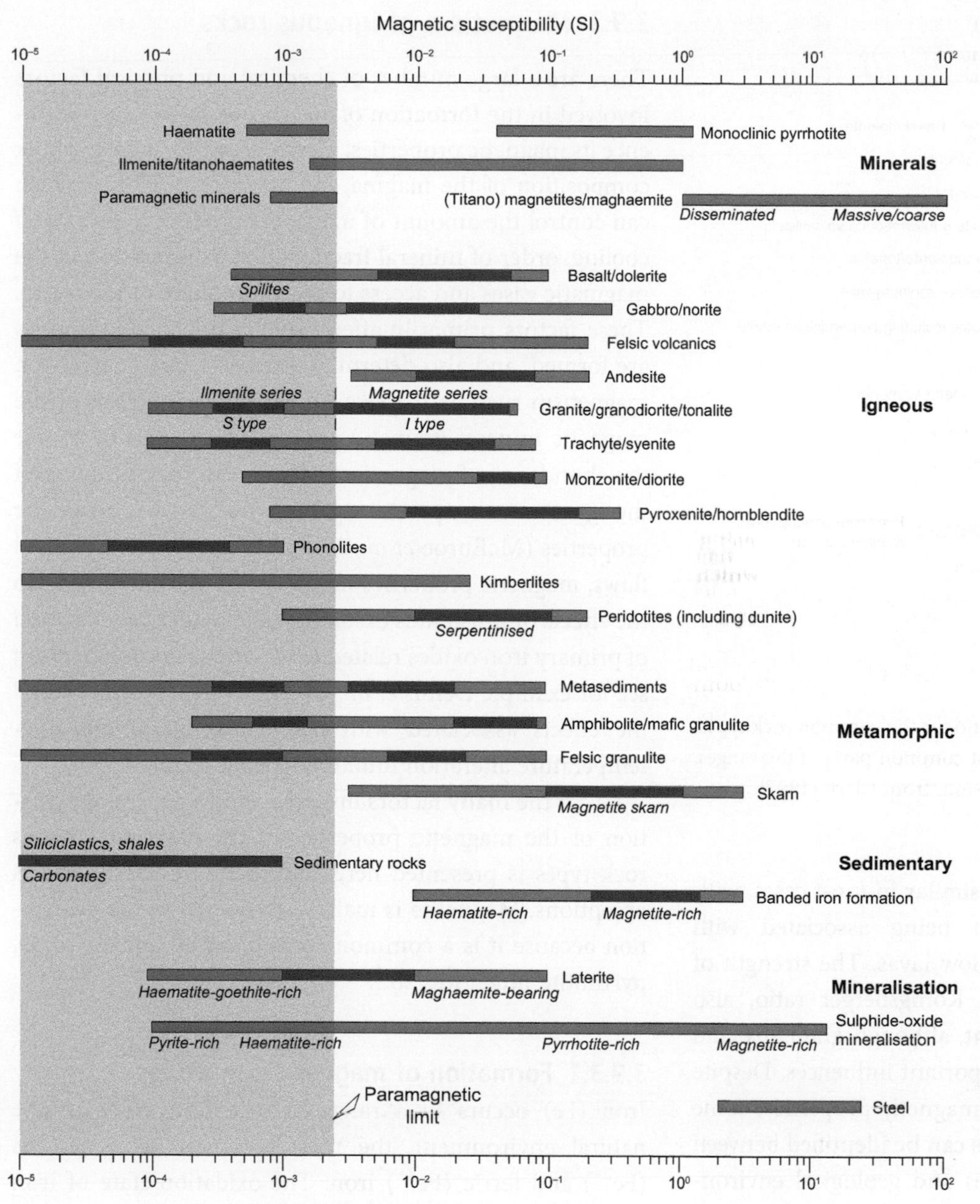

Figure 3.42 Susceptibility ranges for common minerals and rock types. The darker shading indicates the most common parts of the ranges. Redrawn with additions, with permission, from Clark (1997).

the magnetic grains occupy more than about 20% of the rock volume the relationship becomes substantially non-linear and susceptibility increases faster. This is because the strongly magnetic grains are packed closer together, increasing interactions between the grains. For weakly magnetic minerals, such as haematite and paramagnetic species, susceptibility is essentially proportional to the magnetic mineral content, right up to 100% concentration.

Figures 3.42 and 3.43 show the variation in susceptibility and Königsberger ratio (see Section 3.2.3.4) for common rock types. All of the significantly magnetic minerals are accessory minerals, except in certain ore environments. Their presence or absence is largely disregarded when assigning a name to a particular rock or rock unit. Since it is the magnetic minerals that control magnetic responses, there is no reason to expect a one-to-one correlation between magnetic anomalies and geologically mapped lithological boundaries. In most cases the overall magnetism of rocks reflects their magnetite content, because of magnetite's common occurrence and strong magnetism. Consequently, magnetic maps are often described as 'magnetite-distribution maps', which for most types of terrain is a valid statement.

Magnetic susceptibility varies widely in all the major rock groups (Fig. 3.42). Often the distribution is bimodal, having members with high and low susceptibilities reflecting the presence or absence, respectively, of ferromagnetic minerals along with paramagnetic species. The range in

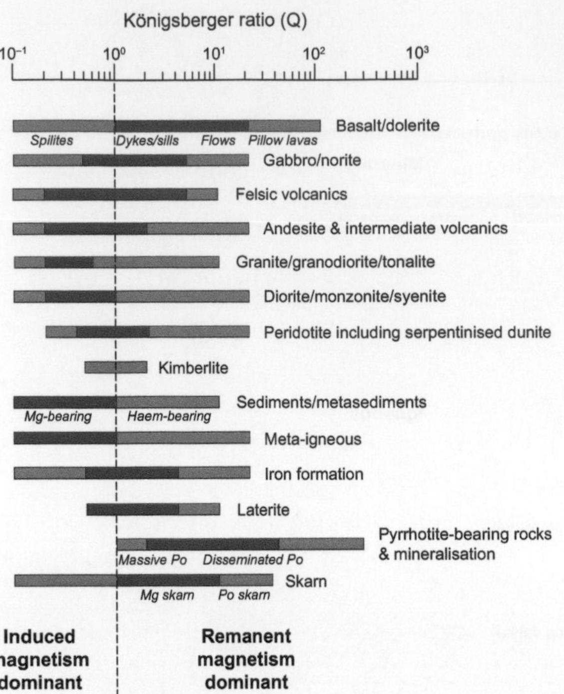

Figure 3.43 Ranges in Königsberger ratio for the common rock types. The darker shading indicates the most common parts of the ranges. Redrawn with additions, with permission, from Clark (1997).

Königsberger ratio (Fig. 3.43) is similar in most cases, with particularly high values often being associated with pyrrhotite-bearing rocks and pillow lavas. The strength of remanent magnetism, and the Königsberger ratio, also reflects magnetic mineral content, although grain size and grain microstructure are also important influences. Despite this and the wide variability of magnetic properties, some important magnetic relationships can be identified between and within different rock classes and geological environments. Nevertheless, the fundamental problem remains that the relationship between rock magnetism and geological processes is not fully understood for all geological processes and environments. When a particular aspect of rock magnetism is studied in detail, the relationship is almost invariably shown to be complex. Much work remains to be done to further our understanding of magnetic petrology.

We discuss the large and complex subject of magnetic petrophysics in terms of the various rock classes and different geological environments and processes. It is impractical to provide a comprehensive description of the subject in the space available to us, and inevitably there will be exceptions made to the generalisations below. Important sources of further information are Clark (1997; 1999) and Grant (1985).

3.9.3 Magnetism of igneous rocks

There are a large number of chemical and physical factors involved in the formation of an igneous mass which influence its magnetic properties. These factors include: the bulk composition of the magma, the presence of biotite which can control the amount of magnetite produced, the rate of cooling, order of mineral fractionation, composition of the magmatic gases and access to the atmosphere or the ocean. These factors primarily affect whether magnetic minerals are formed, and also determine the secondary controls on magnetism such as the size, shape and orientation of the magnetic mineral grains. All these factors, plus of course the abundance of magnetic minerals, can vary throughout the igneous mass, producing large variations in magnetic properties (McEnroe *et al.*, 2009a). Even within individual flows, magnetic properties may vary significantly owing to the effects of subsolidus exsolution and deuteric oxidisation of primary iron oxides related to variations in cooling rates; see for example Delius *et al.* (2003). To this must be added the effects associated with the almost ubiquitous low-temperature alteration found in igneous terrains.

Given the many factors involved, only a general description of the magnetic properties of the different igneous rock types is presented here, and there are nearly always exceptions. Magnetite is mainly considered in the description because it is a common constituent of igneous rocks, pyrrhotite much less so.

3.9.3.1 Formation of magnetic iron oxides

Iron (Fe) occurs in a range of oxidation states in the natural environment, the most common being ferrous (Fe^{2+}) and ferric (Fe^{3+}) iron. The oxidation state of iron depends on oxygen fugacity (fo_2); see Frost (1991). In a chemical system without Ti or Mg, when fo_2 is very low, such as occurs in the Earth's core, iron may occur as a metal (Fe^0). At higher fo_2, in a silicate-bearing system ferrous iron occurs and is mostly incorporated into paramagnetic silicate minerals like fayalite. The reaction (QIF) is (written with the high-entropy side to the right, as are others in this section):

$$\text{QIF}: \quad Fe_2SiO_4 \Rightarrow 2Fe^0 + SiO_2 + O_2$$

Fayalite (F) ⇒ Iron (I) + Quartz (Q) + Oxygen

At increasingly higher fo_2, iron is present in either its ferrous or ferric states, and is mostly incorporated into magnetite. The change is described by the reaction (FMQ):

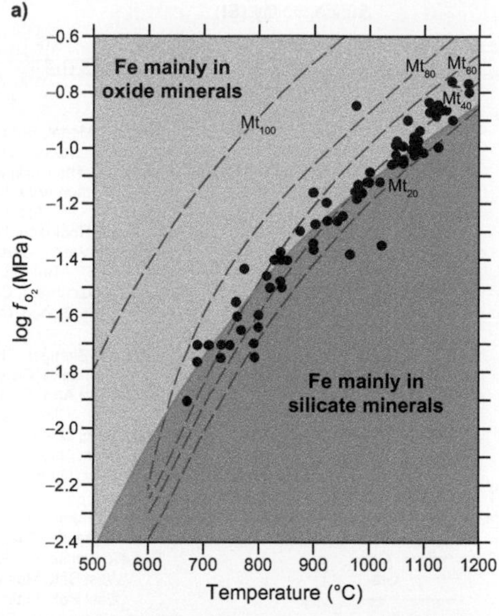

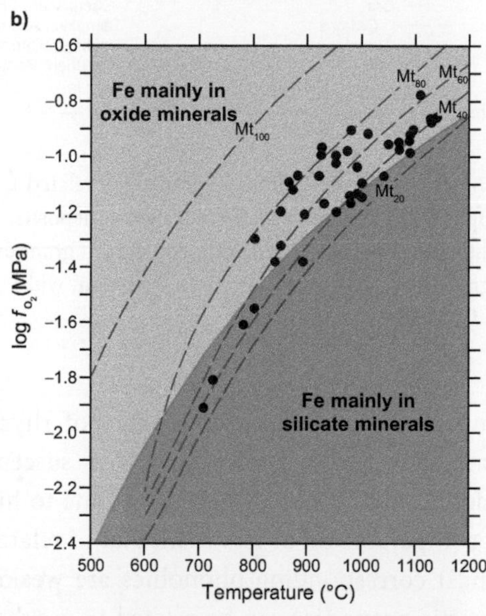

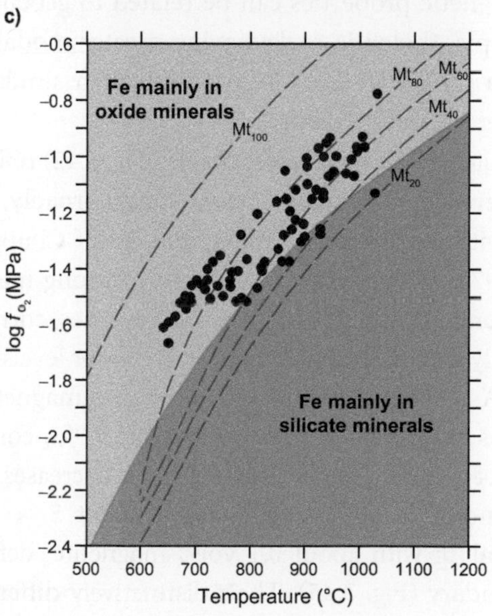

$$\text{FMQ}: \quad 2Fe_3O_4 + 3SiO_2 \Rightarrow 3Fe_2SiO_4 + O_2$$

$$\text{Magnetite (M) + Quartz (Q)} \Rightarrow \text{Fayalite (F) + Oxygen}$$

At very high fo_2, iron occurs in its ferric state, mainly in haematite, as shown by the reaction (MH):

$$\text{MH}: \quad 6Fe_2O_3 \Rightarrow 4Fe_3O_4 + O_2$$

$$\text{Haematite (H)} \Rightarrow \text{Magnetite (M) + Oxygen}$$

For the system Fe-O-SiO_2 the reactions FMQ and QIF mark the fo_2 limits of fayalite, and MH the upper fo_2 limit for the stability of magnetite. Consequently, fo_2 is a useful guide to whether the iron occurs in silicates, and if not, in which iron oxide.

Magnesium and iron can substitute for each other in a variety of silicates. When magnesium substitutes for iron in silicates it stabilises them to higher fo_2, i.e. the FMQ is shifted upwards, decreasing the stability field of titanomagnetites. If they contain enough magnesium, iron silicates, containing ferrous iron, may be stable even in the presence of haematite. The presence of titanium makes these species more stable relative to silicates than their iron-only end-members. Ferrous iron substituting for ferric iron has the same effect.

In summary, in a mineral assemblage containing silicates and oxides, the Fe/Mg ratio of the silicates, the titanium contents, ferrous/ferric ratios of the oxides, and oxygen fugacities are all inter-related. However, fo_2 is not a control imposed upon the system, but rather is governed by the composition of the primary melt (if the rock is igneous) and by the mineral reactions that occurred during the formation of the rock. Oxygen is extremely rare in the geological environment (except at, or very near, the Earth's surface). Therefore, a reaction such as the oxidation of fayalite to magnetite can only take place in association with some other reaction that is a source of oxygen.

Figure 3.44 shows estimates of temperature–fo_2 conditions for various types of igneous rocks, along with selected titanomagnetite and titanohaematite curves. The curves are for solid-solution pairs that are in equilibrium. Of course, this is unlikely to be the case in reality, especially

Figure 3.44 Temperature, oxygen fugacity and silica-content controls on mineralogy (a) basic extrusive suites, (b) intermediate extrusive suites and (c) acid extrusive suites. The heavy line is the fayalite–magnetite–quartz buffer curve at 10^5 Pa. The dashed lines represent experimentally coequilibrated ulvospinel–magnetite solid solutions. Redrawn, with permission, from Haggerty (1979).

if cooling was rapid, but nevertheless the curves give useful indications of the compositions towards which the Fe–Ti oxides will tend. Introducing SiO_2 to the system allows the FMQ curve to be added, representing a boundary below which iron will tend to be mainly in silicate minerals. The curve shown is for a silica-saturated system. If it is not silica-saturated the curve is shifted downwards, encouraging the formation of the Fe–Ti oxides.

The titanohaematites are all very close to the ilmenite end-member (less than 10% haematite), but the titanomagnetites have a much broader range of compositions. More than about 30% magnetite (Mt_{30}) is required for significant magnetism (see Fig. 3.39b). Data from felsic lavas almost entirely plot above the FMQ curve, and magnetic titanomagnetite is predicted. Basic lavas tend to straddle the FMQ curve, and ulvospinel-rich titanomagnetites are expected. Although these grains may be less magnetic than their equivalents in felsic rocks, they are more abundant, and exsolution and oxidation to more magnetic species may occur during cooling. Intermediate lavas plot in an intermediate position, albeit at high temperatures. Data from intrusives show a broadly similar pattern, albeit shifted to lower temperatures parallel to the FMQ curve, a result of slower cooling. Slower cooling permits the titanomagnetite to evolve towards more iron-rich varieties, since the isopleths are oblique to the FMQ curve. Of course these observations are not definitive, since a complex chemical system has been simplified, and the rapid cooling of extrusives may prevent equilibrium conditions being attained. However, most points on Fig. 3.44 are above the Mt_{30} isopleth indicating that, in virtually every case, if a titanomagnetite species occurs it will be ferromagnetic. Even if this is not the case, subsolidus exsolution is likely to create magnetic species.

3.9.3.2 Magnetism and igneous rock types

The multitude of controls on rock magnetism means that it is unwise to use magnetic measurements from igneous rocks in one area to predict responses of similar rock types in another. Within the same area some generalisations may be made because magmatic characteristics, which control the all-important partitioning of iron between silicates and Fe–Ti oxides, are more likely to be similar. For example, magnetic susceptibility generally increases with maficity for rocks from the same area, and andesites generally have lower or similar susceptibilities to their related basalts. Rhyolites show a bimodal susceptibility distribution with the ferromagnetic rhyolites less magnetic than the more basic members of the series. Other types of rhyolite are usually paramagnetic and have very low susceptibility. Trachyandesites and trachytes have moderate to high susceptibility comparable to, or less than, that of related alkali basalts, whilst corresponding phonolites are weakly magnetic. Magnetic properties can be related to geochemistry. For example, tholeiitic rocks having greater modal titanomagnetite exhibit higher susceptibility than similar rocks with lower Fe and Ti contents.

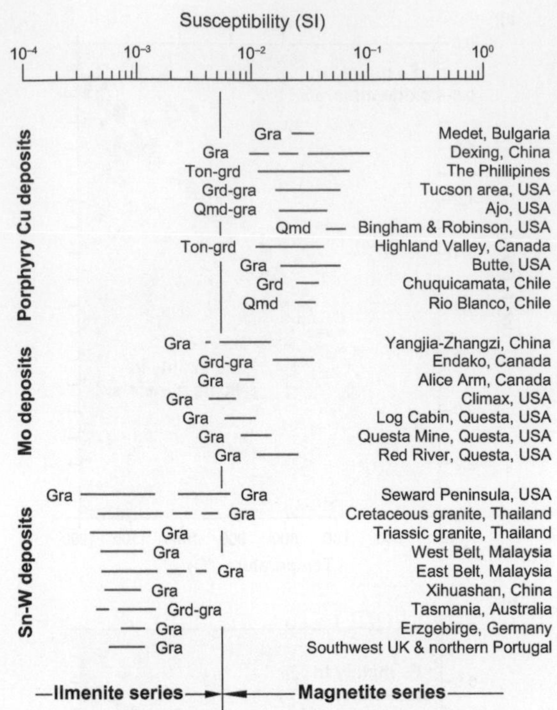

Figure 3.45 Magnetic susceptibilities of granitoids related to porphyry copper, molybdenum and tin–tungsten deposits. Ton – tonalite, Qmd – quartz monzodiorite, Grd – granodiorite, and Gra – granite (mostly monzogranite). Redrawn, with permission, from Ishihara (1981).

In granitoids, bimodal susceptibility (Fig. 3.42) reflects two distinct series of granitoids corresponding broadly, but not exactly, with the S- and I-type granitoids of Chappell and White (1974). The magnetite-series have strong magnetism due to abundant magnetite. The ilmenite-series, corresponding roughly with S-type granitoids, have lower levels of magnetism. A-type granites are also poorly magnetic. The differences are due to different temperature–fo_2 conditions; see Fig. 3.44. In both series, susceptibility decreases as silica content increases. A susceptibility of about 5×10^{-3} SI, corresponding with about 0.1 vol% magnetite, defines the class boundary (Fig. 3.45). Their distinctively different susceptibilities allow the magnetite- and ilmenite-series granites to be distinguished in the field using a portable susceptibility meter (see Section 3.9.8.1). They can also be identified from magnetic maps, but the possibility of significant remanent magnetism needs to be accounted for in analysing anomalies.

The characteristics of the magnetite- and ilmenite series extend over many lithotypes, including gabbroids, syn-, late- and post-orogenic calc-alkaline granitoids and alkaline-series anorogenic granitoids. The difference is economically important since copper and gold are associated with the magnetic magnetite-series intermediate I-type granitoids; molybdenum is associated with more fractionated and oxidised magnetite-series I-type suites; and tin with the paramagnetic, reduced, fractionated I- or S-type suites. The relationship is due to redox conditions in the magma affecting the behaviour of metals such as Cu, Mo, W and Sn within the melt.

Magnetic properties of granitoids can also be predicted from the presence of various minerals, which in turn are indicative of oxidation conditions. For example, hornblende-biotite granodiorites are predominantly ferromagnetic and exhibit moderate susceptibility, whereas the muscovite-biotite granitoids are usually paramagnetic and weakly magnetic.

The zoned Alaskan-type igneous complexes have primary magnetite as the main magnetic mineral. The ultramafic rocks in these intrusions, such as pyroxenites, hornblendites and serpentinised dunites, generally have high susceptibility, and the associated mafic and intermediate rocks, such as gabbro, diorite and monzonite have moderate to high susceptibility. Unaltered komatiitic lavas and alpine-type peridotites have low susceptibility. However, serpentinisation of these rocks creates magnetite, which makes serpentinised ultramafic rocks strongly magnetic (see Section 3.9.5.4).

3.9.3.3 Remanent magnetism of igneous rocks

Remanent magnetism of igneous rocks is acquired when the magnetic mineral grains pass through their Curie temperature, as the lava/magma cools. This primary remanent magnetism is known as *thermoremanent magnetisation* (TRM) (Fig. 3.46a). TRM is a very stable primary magnetisation that can exist for long periods on the geological time scale. It is parallel to the Earth's field and approximately proportional to its strength at the time of cooling through the Curie point. As cooling progresses inward from the outer margins the Earth's field may reverse, so that inner cooling zones may acquire different remanent

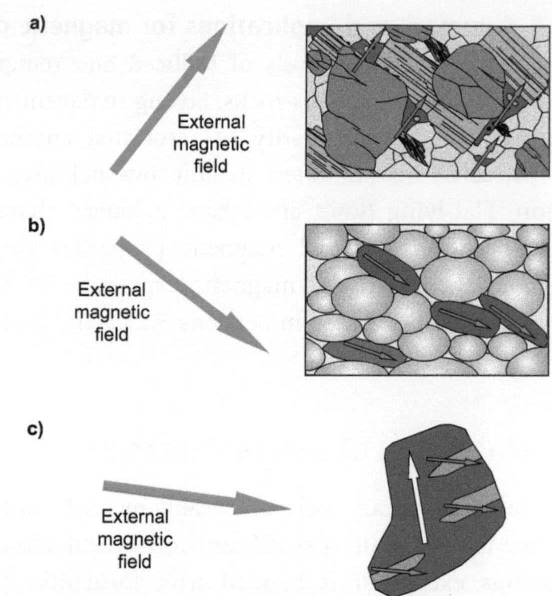

Figure 3.46 Schematic illustrations of common types of remanent magnetism. (a) Thermal remanent magnetism (TRM). (b) Depositional remanent magnetism (DRM); note the difference in inclination between the DRM and the Earth's field. (c) Crystallisation remanent magnetism (CRM); the white arrow represents a pre-existing magnetism.

magnetisms. Large intrusive masses often have internal magnetic zonation related to magnetic mineral fractionation, multiple intrusive phases and changes in remanent magnetisation related to the cooling period extending through multiple reversals in the Earth's magnetic field (McEnroe *et al.*, 2009a).

The cooling rates of igneous rocks influence their remanent magnetism, although this can be overprinted by later thermal or chemical changes. The slowly cooled plutonic rocks usually have coarse-grained multidomain magnetite, causing these rocks to have very small Königsberger ratios (Fig. 3.43). However, some gabbroic and dioritic intrusives, which contain very fine single-domain magnetite within silicate minerals, are notable exceptions and may have significant remanent magnetism. Rapidly chilled igneous rocks contain fine-grained titanomagnetites and consequently can have very large Königsberger ratios. The largest Königsberger ratios are associated with the fastest cooling portions of the melt, such as the sub-aqueous chilled margins and small pillows, decreasing away from the margin. On the other hand, slower cooling, thick dolerite sills and dykes can exhibit high Königsberger ratios. Kimberlites often have ratios greater than one.

3.9.3.4 Summary and implications for magnetic data

Overall, widely variable levels of induced and remanent magnetism occur in igneous rocks. Strong remanent magnetism can affect the polarity of associated anomalies. Large intrusive masses often exhibit internal magnetic zonation. Flat-lying flows often have a 'noisy' character due to the cooling-related magnetic properties varying throughout the flow. The magnetic expression of some igneous rocks is illustrated in Sections 3.11.3 and 3.11.4.

3.9.4 Magnetism of sedimentary rocks

In general, sedimentary rocks are weakly magnetic because magnetite is not usually a significant constituent, although an obvious exception is banded iron formation (BIF) (Fig. 3.42).

The magnetic properties of clastic sediments reflect the mineralogy of their source. Quartz-rich units from a mature source area have lower susceptibilities than sediments sourced from immature volcanic terrains, although it is sometimes uncertain whether the magnetite within these sediments is detrital or due to decomposition of mafic minerals.

Iron is most abundant in fine-grained sedimentary rocks, often associated with clays into which it is absorbed. Waters that are in contact with the atmosphere are oxidising, eventually causing minerals containing ferrous iron (e.g. magnetite) to be oxidised to ferric iron minerals (e.g. haematite). Reducing environments occur in stagnant water-logged environments, particularly those rich in organic matter. Magnetite and pyrrhotite are stable in these environments, but only form when there are low levels of sulphur (otherwise iron is taken up by pyrite) or carbonate (otherwise iron is taken up by siderite).

Variations in susceptibility can be correlated with grain size in clastic sediments, graded bedding being mimicked by decreases in susceptibility (Fig. 3.47). This is interpreted as due to density stratification causing iron-rich grains to occur mainly at the bottom of units, and/or a systematic upward decrease in the number of larger (multidomain) magnetite grains.

The iron-deficient carbonates have very low susceptibility. In contrast, sediments deposited in iron-rich solutions associated with volcanogenic activity or Precambrian chemical precipitates, can contain appreciable magnetite or pyrrhotite. These sediments may be transitional to syngenetic massive mineralisation or BIFs. These rocks are

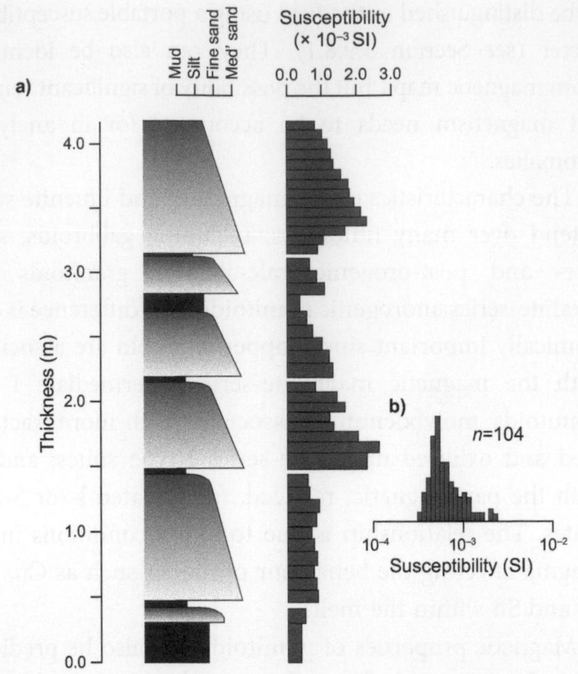

Figure 3.47 Variation in grain size and magnetic susceptibility of a greywacke sequence in southern Scotland. (a) Grain size distribution within lenses of the sequence. (b) Frequency histogram of the susceptibility measurements, which are the arithmetic means of 12 measurements. The correlation between the two parameters is quite clear. Redrawn, with permission, from Floyd and Trench (1989).

highly magnetic, generally remanently magnetised and are characterised by strong anisotropy of susceptibility. In the Hamersley iron-ore province of Western Australia, the susceptibility of BIFs parallel to bedding exceeds the susceptibility normal to bedding, typically by a factor of 2 to 4 (Clark and Schmidt, 1994). Continental red bed sequences are rarely significantly magnetised, since the colouring attests to the dominance of haematite.

3.9.4.1 Remanent magnetism of sedimentary rocks

Detrital magnetic mineral grains acquire a *primary detrital remanent magnetism* when they align themselves in the Earth's magnetic field as they settle through water (Fig. 3.46b). Its direction is sub-parallel to the direction of the Earth's field, owing to gravity causing elongated grains to lie flat on the bottom. Subsequent rotation of magnetic grains can occur in pore spaces. Diagenesis, alteration and weathering of sediments can produce physical and chemical changes to magnetic minerals, below the Curie point, to produce a secondary *crystallisation remanent magnetism* (CRM), which is parallel and proportional to the strength of the ambient field.

Fine-grained sediments may acquire a *post-depositional remanent magnetism* (pDRM) where fine-grained magnetic minerals suspended in the water-filled pore spaces of sediments align with the Earth's magnetic field. This can occur in water-logged slurries that form at the sediment–water interface. The pDRM is fixed in the sediment by compaction and de-watering, which occur at a depth of about 10 cm; so the pDRM is acquired subsequent to sedimentation and so is secondary. It is a common mechanism for remanent magnetisation of pelagic limestones.

3.9.4.2 Summary and implications for magnetic data

The magnetic responses of sedimentary rocks are often featureless owing to their low levels of magnetism; for examples see Figs. 3.74 and 3.75. Moderate levels of induced and remanent magnetism may occur in some sedimentary rocks owing to the presence of magnetite or pyrrhotite. Variations in magnetism in a succession are usually correlated with stratigraphy: see for example Schwarz and Broome (1994).

The most magnetic sedimentary rocks are BIF. Magnetic responses from these units can be extremely strong but complicated by the effects of strong remanent magnetism, self-demagnetisation (see Section 3.2.3.6) and anisotropic magnetic properties (Clark and Schmidt, 1994).

3.9.5 Magnetism of metamorphosed and altered rocks

In a geochemical system as complicated as the Earth's interior, there is an almost infinite range of possibilities in terms of changes in magnetic mineralogy and the size and shape of these mineral grains. In general, conditions become more reducing as depth and metamorphic grade increase. For this reason, magnetite and ilmenite are the main Fe–Ti oxides in metamorphic rocks. Haematite is much less common, although it may survive in metamorphosed iron formation and oxidised metasedimentary rocks, and in low- to medium-grade metabasites.

As would be expected, any metamorphic or alteration process that creates or destroys magnetic minerals is the primary cause of changes in rock magnetism, especially if the minerals are ferromagnetic. Obviously, if these are to occur the rock must contain iron, either primary or introduced by hydrothermal processes, but whether ferromagnetic species are formed depends on the specific physical and chemical conditions (see Section 3.9.3). Metamorphism can significantly change magnetic properties, but, with such enormous diversity in the conditions possible, only generalisations may be made about the magnetic consequences of metamorphism.

In addition to changes in mineralogy, other significant factors in metamorphic environments include heating, which encourages exsolution of iron- and titanium-rich phases of titanomagnetite, and deformation causing recrystallisation into coarser grains or altering grain shape, or causing preferential alignment of grains. The texture of the rock can also be important since it controls grain size. If magnetite is locally very abundant, for example in a cumulus phase, it tends to be more stable than if more widely distributed. This is probably because the large volume of iron released from the altering magnetite saturates the metamorphic fluids to preserve the remaining magnetite. Also, if magnetic minerals occur as very fine grains within stable minerals they may survive under conditions outside their normal stability range.

Metamorphic effects that occur and change over short distances are much more likely to produce recognisable magnetic variations in magnetic data. Situations where this occurs include metamorphic processes that are structurally controlled, for example retrogressive processes concentrated along shear zones; and situations where metamorphic conditions may change rapidly, such as contact aureoles.

3.9.5.1 Regional metamorphism of crystalline rocks

There are a number of published studies investigating the changes in magnetic properties associated with changes in metamorphic grade, but these tend to be specific to particular areas and rock types (mostly mafic rocks) and they cover a relatively small range of metamorphic conditions. Figure 3.48d is an attempt to summarise how the magnetism of mafic rocks changes as metamorphic grade increases. There is a general decrease with increasing grades until the possible creation of secondary magnetite at granulite facies. At the highest grades, magnetism is reduced as ferromagnetic species breakdown to form paramagnetic minerals. Summary conclusions of several studies follow.

Magnetic properties of sea-floor basalts from the Atlantic Ocean ranging from fresh to greenschist facies are shown in Fig. 3.48a. There is much scatter but the fresh samples are the most magnetic. Note the high Königsberger ratios of the fresh samples, the reason for the characteristic striped ocean-floor magnetic anomalies. Hydrothermal alteration at zeolite facies reduces the

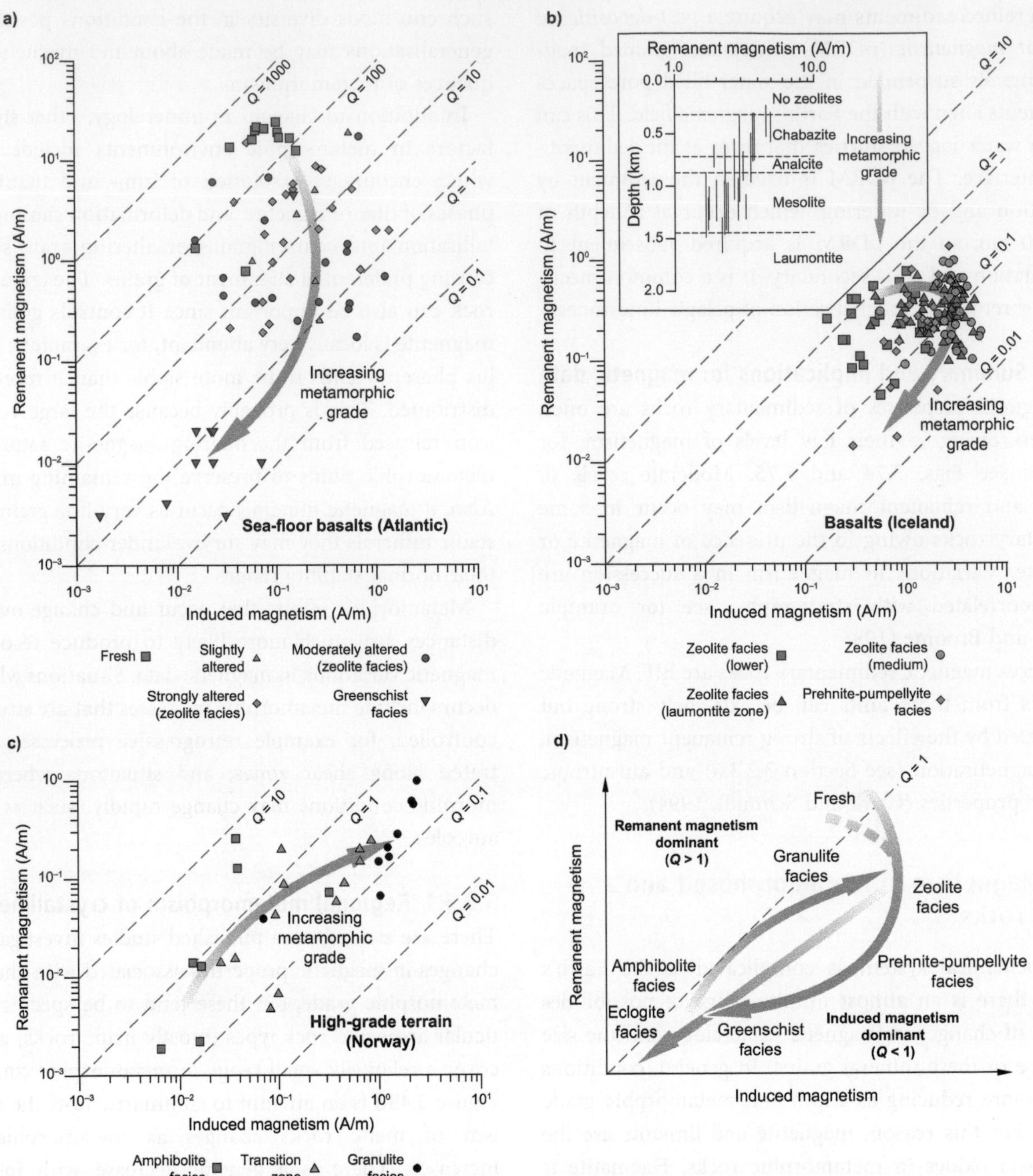

Figure 3.48 Examples of the effects of metamorphism on rock magnetism. (a) Variations in the strength of magnetism for sea-floor basalts from the Atlantic Ocean showing the changes in magnetism as metamorphic grade increases to greenschist facies. Based on data in Fox and Opdyke (1973). (b) Variation in the strength of magnetism as a function of grade in the zeolite to prehnite–pumpellyite facies from a 3 km thick pile of basaltic lavas in eastern Iceland. Based on data in Bleil *et al.* (1982), Mehegan and Robinson (1982), and Robinson *et al.* (1982). Inset: Variation in the strength of remanent magnetism versus depth/metamorphic grade. Metamorphic grade varies from zero through various zones of the zeolite facies. The vertical errors bars reflect uncertainty in estimated depth. Based on diagrams in Wood and Gibson (1976). (c) Changes in magnetism of acid to intermediate gneisses associated with the transition from amphibolite to granulite facies. Based on data in Olesen *et al.* (1991). (d) Schematic illustration of the effects of metamorphism on the magnetism of mafic rocks.

strength of remanent magnetism with possibly a slight increase in induced magnetism. When greenschist facies is attained, both forms of magnetism are reduced compared with the protolith, especially the remanent magnetism. The decrease in strength of remanent magnetism in basalts during low-grade metamorphism is probably due to oxidation of iron-rich titanomagnetites. Once greenschist facies are reached, induced magnetism has also been significantly reduced.

In gabbro the fine magnetic grains within silicates may be protected by their silicate hosts from low- to medium-grade metamorphism, so gabbro can retain its magnetic properties. Magnetite in felsic plutons appears to be more resistant to metamorphic destruction than felsic and mafic volcanic rocks at greenschist and also amphibolite grade.

A 3 km thick sequence of basaltic flows exposed in eastern Iceland provides an opportunity to monitor magnetic properties as a function of hydrothermal alteration in the zeolite and prehnite–pumpellyite facies (Fig. 3.48b). There is a significant scatter in the data but it is clear that an increase in susceptibility occurs into about the middle of the zeolite facies. As metamorphic grade increases, the strengths of both the induced and remanent magnetisms decrease significantly, especially when prehnite–pumpellyite conditions are reached. From the figure inset, a decrease in the strength of remanent magnetism with increasing metamorphic grade is clear.

As demonstrated by Bourne *et al.* (1993) and Hageskov (1984), the transition from greenschist to amphibolite facies has little influence on magnetism, other than slightly reducing already low levels of magnetism. In general, amphibolite tends to be weakly magnetic if it contains chlorite and/or biotite, but hornblende-bearing varieties are much more magnetic.

The effects of the amphibolite–granulite transition are described by Olesen *et al.* (1991) and Schlinger (1985) using data from an acid to intermediate gneissic terrain in northwestern Norway, and shown in Fig. 3.48c. Much of the scatter is due to varying lithotypes, although there is no compositional difference between the lower- and higher-grade equivalents. The transition from amphibolite facies to granulite facies is associated with substantial increase in the strengths of both the induced and remanent magnetisms, owing primarily to the growth of metamorphic magnetite. The metamorphism is interpreted as breaking down hydrous iron-bearing silicates (biotite and amphibole) with the iron taken up by iron–titanium oxides. In this area, retrogressive metamorphism of the granulite-facies mangerites returned both susceptibility and remanent magnetism to amphibolite-facies levels. Petrologically this is due to magnetite, ilmenite and ilmenite–haematite being replaced by silicates, titanite and haematite. A further factor at granulite facies grade is that temperatures are sufficient for stability of the complete titanomagnetite solid-solution series. Exsolved titanomagnetites can therefore recombine, which may result in a loss of magnetism due to the creation of less iron-rich members.

High-pressure granulites and eclogites tend to be paramagnetic, with magnetite breaking down at 1000–2000 MPa (Clark, 1997), and the iron being taken up by garnet and clinopyroxene. Decompression of high-pressure granulite during rapid uplift can produce fine-grained magnetite by breakdown of garnet and clinopyroxene.

3.9.5.2 Metamorphism of sedimentary rocks

The magnetic properties of metamorphosed sediments depend largely on the nature of the original sediment. An iron-poor protolith, such as a mature sandstone or pure carbonate, cannot produce significant quantities of magnetic minerals. This means that only metamorphosed argillaceous rocks are likely to be magnetic. When iron is present, the oxidation state is important, in particular the ratio of ferric to ferrous iron. Low ratios favour the formation of iron silicates, intermediate values favour magnetite, and high ratios favour haematite and ilmenite. Reduced sediments devoid of Fe^{3+} will generally produce a non-magnetic metasediment. This may result in variations in magnetic properties reflecting original facies variations.

The magnetic susceptibility of the shales is due to iron-bearing silicates, magnetite and pyrrhotite. Rochette (1987) reports on data from the Swiss Alps (Fig. 3.49). The overall susceptibility of the rocks depends on all the minerals making up the rocks, but only the ferromagnetic mineralogy is affected by changes in metamorphic grade. As metamorphic grade increases, initially susceptibility decreases owing to destruction of detrital magnetite to create mainly pyrite, then it sharply increases as pyrite is converted into pyrrhotite. These reactions are caused by increasingly reducing conditions resulting from the maturation of organic matter within the shales.

Sediments in which most of the iron is in iron oxides, such as banded iron formations and haematite-bearing sandstones, retain their unaltered assemblages up to the highest metamorphic grades. This occurs because oxygen released from the oxidation of magnetite during

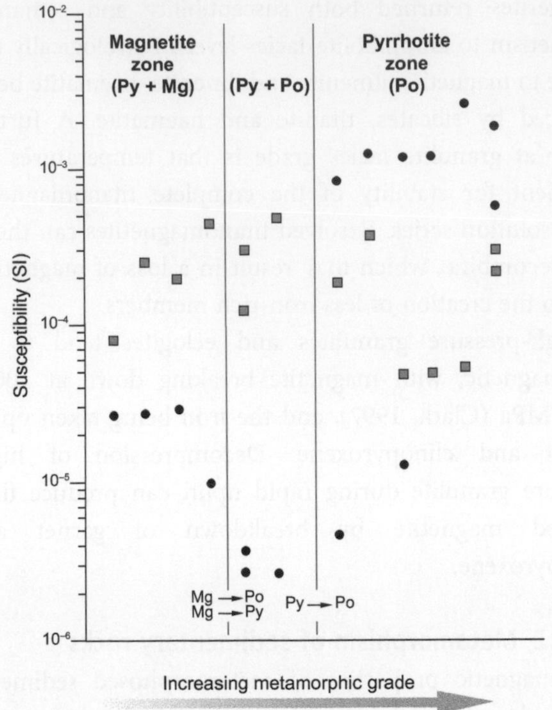

Figure 3.49 Variations in magnetic susceptibility with metamorphic grade in a black shale sequence from the Swiss Alps. The black circles represent the susceptibility due to ferromagnetic minerals; the blue squares represent the contribution from diamagnetic and paramagnetic minerals. Redrawn, with permission, from Rochette (1987).

metamorphism does not diffuse away, unless accessed by mobile reducing fluids, so it buffers further reaction to maintain haematite stability. Consequently, magnetite- and haematite-rich bands are preserved. In general, iron oxides in metasediments are predominantly magnetic, in contrast to the predominance of the more oxidised haematitic forms in unmetamorphosed sediments. This reflects their approach toward chemical equilibrium.

Carbon is a reducing agent with its effect increasing with temperature but limited by pressure, so graphitic schists formed from carbonaceous shales are deficient in magnetite. Several other factors can favour the formation of magnetite during metamorphism, i.e. low silicate content, low titanium content, low pressure and dehydration both affecting oxidation and reduction, excess aluminium favouring the formation of muscovite mica and magnetite rather than biotite, high temperatures in biotite rocks, equilibrium to low temperatures in titaniferous rocks, and absence of carbon. Pyrrhotite is the most common magnetic mineral in reduced graphitic metasediments, e.g. graphitic schists, especially above greenschist facies (Clark and Tonkin, 1994), its high Königsberger ratio and anisotropic magnetic properties (see Sections 3.2.3.4 and 3.2.3.7) being reflected in the magnetic characteristics of the rocks.

3.9.5.3 Contact metamorphism

Magnetic anomalies associated with thermal/contact metamorphism (Schwarz, 1991) are quite common. They are the result of hot igneous fluids penetrating the colder country rock and, where the rocks are chemically reactive, producing usually a zone of reduced magnetism. Examples due to both magnetite and pyrrhotite have been reported (Hart, 2007; Speer, 1981). If metamorphic magnetic minerals are formed, their remanent magnetism may be oriented differently from that of the source intrusion giving the metamorphosed zone a distinctive contrasting magnetic signature. Some examples of the magnetic responses of the rock surrounding intrusions are described in Section 3.11.3.

Figure 3.50 shows the variations in magnetic properties through a doleritic dyke and adjacent metabasalt country rocks from the Abitibi area, Ontario, Canada. The baked-contact zone has higher levels of remanent magnetism and a higher Königsberger ratio than either the dyke or the country rocks, but induced magnetism is highest in the central part of the dyke. The majority of the remanent magnetic vectors are sub-parallel to the present-day field, so the overall magnetism is higher, roughly equal to the combined strengths of J_{Induced} and J_{Remanent}. Depending on the survey height, the responses from the contact zones could appear as a single anomaly, which would then probably be erroneously correlated with the dyke itself.

Kontny and Dietl (2002) describe a detailed study of the magnetic mineralogy in a contact aureole in eastern California. They explain the creation and destruction of magnetite as a consequence of changing physical and chemical conditions and their effects on iron–titanium oxides and iron magnesium silicates. A metasedimentary succession (lower greenschist facies) has experienced prograde contact metamorphism in the aureole of a large pluton, with some areas subsequently experiencing retrograde metamorphism. Three zones are recognised in the aureole based on silicate and magnetic mineralogy (Fig. 3.51a). Susceptibility is low outside the aureole, but a zone characterised by andalusite and cordierite, and a mineralogical transition zone, have high magnetism. The zone closest to the intrusion, characterised by cordierite and K-feldspar, has low susceptibility. The conversion of limonite, rutile +/- haematite to magnetite and ilmenite–haematite causes the

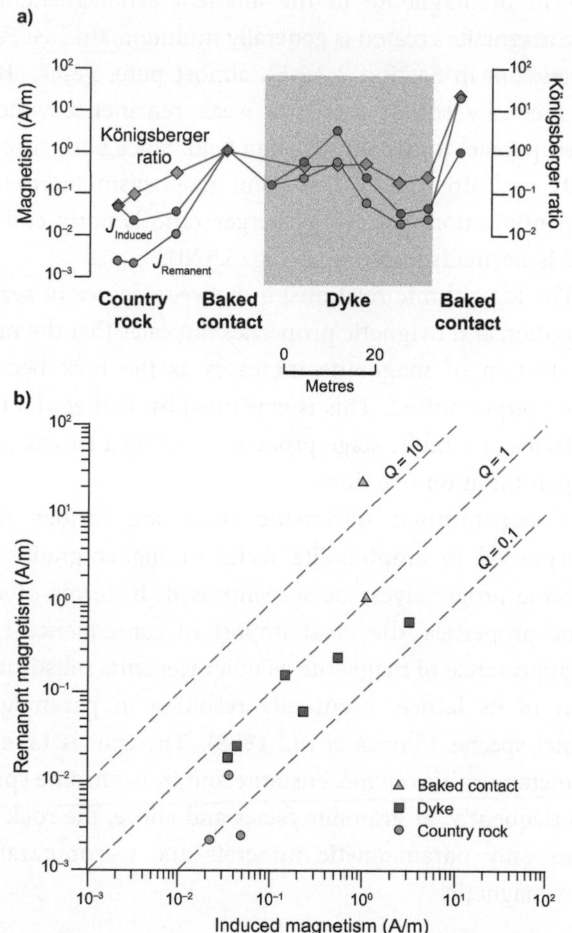

Figure 3.50 Variations in magnetic properties through a doleritic dyke intruding a metabasalt in the Abitibi area. (a) Profile across the dyke, and (b) strengths of the induced and remanent magnetisms. Based on data in Schwarz (1991).

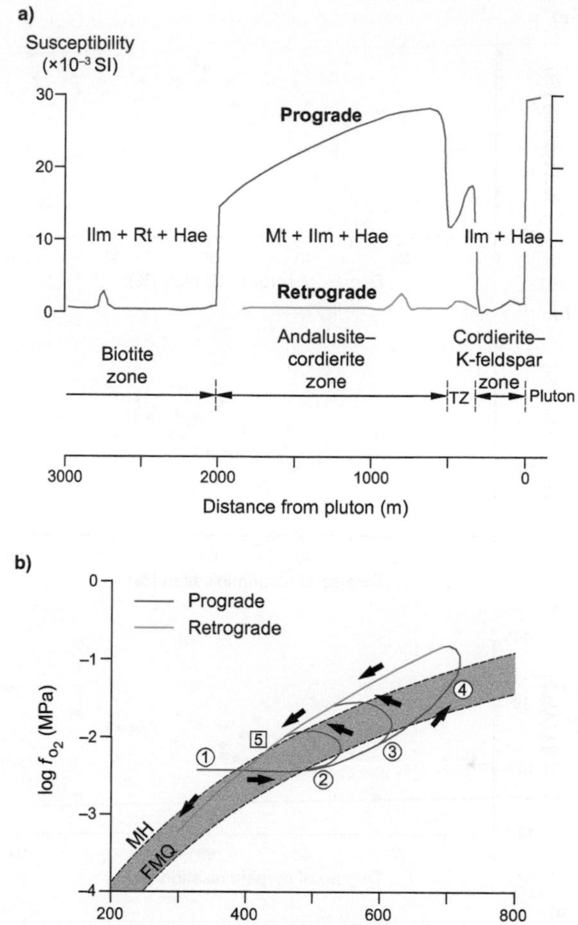

Figure 3.51 (a) Variation in magnetic susceptibility across different zones comprising a contact aureole in California. (b) Schematic representation of changes in temperature and oxygen fugacity within the aureole. Paths 1–4 refer to the biotite, andalusite–cordierite, transition, and cordierite–K-feldspar zones respectively. Path 5 represents retrogressive metamorphism. Hae – haematite, Ilm – ilmenite, Mt – magnetite and Rt – rutile. Redrawn, with permission, from Kontny and Dietl (2002).

increased magnetism during prograde metamorphism. In the inner zone magnetite has been converted to ilmenite and haematite, causing a decrease in magnetism. Retrograde metamorphism is magnetite destructive. Figure 3.51b is a schematic illustration of the variation in temperature and oxygen fugacity within the contact aureole and relates the metamorphism to the stability of magnetite and haematite, as described in Section 3.9.3.1. Using mineralogically derived estimates of physical and chemical conditions, the appearance of magnetite is shown to be associated with an increase in temperature. A further increase in temperature and also oxygen fugacity leads to destruction of magnetite in the inner zone of the aureole. Retrograde metamorphism is associated with decreases in both temperature and oxygen fugacity but conditions remain outside the magnetite stability field. Figure 3.51b shows how only small changes in oxygen fugacity and temperature across the haematite–magnetite boundary cause significant changes in rock magnetism due to the changes in magnetite abundance.

3.9.5.4 Serpentinisation

Serpentinisation is one of the most significant metamorphic processes in terms of its effects on physical properties. Density is inversely proportional to the degree of serpentinisation (see Section 3.8.4), but magnetism increases with serpentinisation owing to reactions that produce magnetite at the expense of the paramagnetic minerals olivine and orthopyroxene.

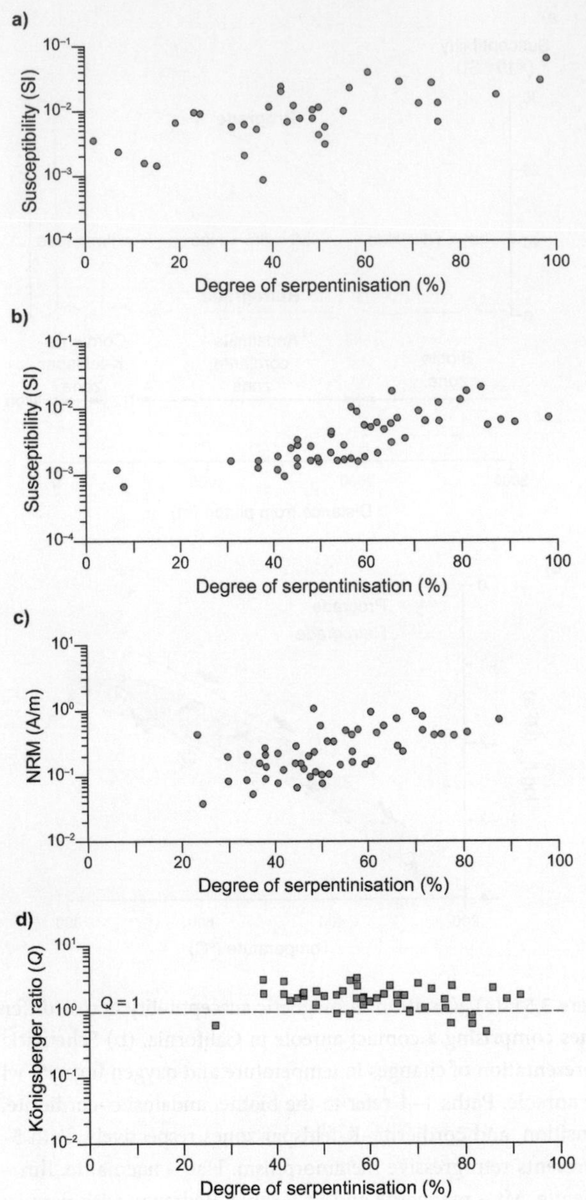

Figure 3.52 Magnetic properties versus degree of serpentinisation for some continental serpentinites. (a) Data from Josephine Creek, Oregon (Toft *et al.*, 1990). Data in (b), (c) and (d) from Red Mountain. California (Saad, 1969).

Figure 3.52 shows changes in magnetic properties with degree of serpentinisation, for which density has been used as a proxy (see Section 3.8.4). Susceptibility data from Josephine Creek, Oregon, USA, and Red Mountain, California, USA (Figs. 3.52a and b) show a broadly linear increase in the logarithm of susceptibility with increasing serpentinisation. The susceptibility of completely serpentinised material is about 10 times that of unserpentinised material. A similar relationship is indicated for remanent magnetism in Fig. 3.52c. This is a CRM caused by the growth of magnetite in the ambient geomagnetic field. The magnetite created is generally multidomain (see *Ferromagnetism* in Section 3.2.3.5), almost pure, Fe_3O_4. Being magnetically soft it carries a weak remanence which is often parallel to the present-day field. Since both susceptibility and strength of remanent magnetism increase on serpentinisation, the Königsberger ratio is fairly constant and is normally about 2–4 (Fig. 3.52d).

The logarithmic relationship between degree of serpentinisation and magnetic properties indicates that the rate of production of magnetite increases as the rock becomes more serpentinised. This is explained by Toft *et al.* (1990) in terms of a multi-stage process involving a succession of serpentinisation reactions.

If serpentinised ultramafic rocks are further metamorphosed to amphibolite facies or higher grades they become progressively de-serpentinised. In terms of magnetic properties, the most important consequence is the disappearance of magnetite as other elements substitute for iron in its lattice, eventually resulting in paramagnetic spinel species (Shives *et al.*, 1988). The iron is taken up in metamorphic olivine, enstatite and iron–chrome spinels. Consequently, at granulite facies and above, the rock contains only paramagnetic minerals and is comparatively non-magnetic.

3.9.5.5 Remanent magnetism of metamorphosed and altered rocks

Remanent magnetism can be reset by the metamorphic process if magnetic minerals are destroyed, or formed and cooled through their Curie temperature. Low- and medium-grade metamorphism can overprint a primary remanence with a partial TRM. High-grade metamorphism can remove the remanent magnetism by taking the magnetic minerals above their Curie point, with the remanence reset as the cooling passes back through the Curie temperature of each magnetic mineral.

Alteration can superimpose a CRM on an earlier remanence (Fig. 3.46c). When coarse-grained secondary magnetite is produced, its multidomain character means its remanent magnetism is easily changed and it will often have a 'soft' viscous remanent magnetism (VRM) parallel to the present-day Earth's magnetic field. Otherwise, remanent magnetism is often parallel to metamorphic lineations and in the plane of metamorphic fabrics, owing to preferred alignment of mineral grains, especially when pyrrhotite is the carrier.

3.9.5.6 Summary and implications for magnetic data

From a magnetic interpreter's perspective, metamorphic grade is important since it affects the relative magnetism of different rock types. Changes in metamorphic grade in a survey area may lead to change in the magnetic character of the same rock types, e.g. Robinson *et al.* (1985), characterised by variations in both the strength and direction of the magnetisation. In some instances it may be possible to identify a magnetic (mineral) isograd. This is most likely in the case of the high metamorphic gradients associated with contact metamorphism. Regional metamorphism in high-grade terrains tends to result in magnetic transitions occurring over hundreds to thousands of metres, and so will not give rise to a discrete anomaly. At lower grades, appearance and/or destruction of magnetic minerals has been used to define 'magnetic isograds' within metasedimentary terrains (Rochette, 1987; Rochette and Lamarche, 1986).

3.9.6 Magnetism of the near-surface

Figure 3.53 shows Eh–pH conditions in a variety of near-surface environments, within the boundaries of water stability. Outside these boundaries, water disassociates into hydrogen and oxygen. Also shown are the stability fields for various iron oxides, sulphides and carbonates for different levels of total dissolved sulphur (ΣS) and total dissolved carbonate (ΣCO_2). Equilibrium conditions are assumed, although they may never be attained in the natural environment. Note how non-magnetic species are stable in nearly all the natural settings. Weathering tends to transform iron to the haematitic Fe^{3+} state because this is the stable state of iron in the atmosphere, so weathered rocks usually have lower magnetism than their fresh equivalents. Other possible consequences of weathering are the creation of a CRM, and the formation of strongly magnetic maghaemite.

Commonly, ferrous iron silicates and crystalline iron oxide minerals break down in carbonate-rich groundwater, in which Fe^{2+} is slightly soluble. This can oxidise to Fe^{3+}, which is much less soluble and precipitates as the hydrated ferric oxides goethite and lepidocrocite, which are weakly and variably magnetic. In a weathered terrain, the low solubility of Fe^{3+}, compared with most other cations, forms *in situ* laterites and iron-rich soils which have significant magnetism. In these environments, strongly magnetic maghaemite can form. It is extremely stable and occurs widely in the soil profile. The saprolites and saprock that comprise the deeper parts of the regolith are normally non-magnetic, with magnetism increasing again at the bedrock interface, depending on the nature of the protolith (Fig. 3.36). Figure 3.54 shows variations in the strength of induced and remanent magnetisms through a regolith profile at Lawlers, in the Yilgarn Craton of Western Australia. Königsberger ratios are as high as 100 in the iron-rich laterites, with the strength of remanent magnetism reaching as much as 100 A/m. This is a very strong magnetism, and magnetic responses from these near-surface layers are often seen in magnetic surveys, often to the detriment of bedrock responses.

Destruction of magnetism, or demagnetisation (not to be confused with self-demagnetisation; see Section 3.2.3.6), by alteration and/or weathering is almost ubiquitous in fault/shear/facture zones. These structural features are easily seen in magnetic maps and in particular where they affect strongly magnetised rocks; several good examples are shown in Sections 3.11.3 and 3.11.4. Figure 3.55 shows the variations in magnetisation through a fracture zone in a Scandinavian granite (Henkel and Guzmán, 1977). The strength of both the induced and remanent magnetisms decreases in the fracture zone, but more so for the induced magnetism, resulting in an increase in the Königsberger ratio. The country rock contains disseminated almost fresh magnetite; but where the magnetism decreases, magnetite has been oxidised to haematite (martite) with very little magnetite remaining. Iron hydroxides are also present, indicating hydration. There is a clear correlation between the loss of magnetism and the degree of alteration which occurred near the ground surface, so it is interpreted to be the cause of the demagnetisation.

Dendritic anomaly patterns associated with present and past drainage systems are often seen in aeromagnetic data, most easily when the bedrock is weakly magnetic (see Section 3.11.4). The source of these anomalies is magnetic detritus. In weathered terrains such as Australia, the source of the anomalies is often maghaemite. In glaciated terrains it may be magnetic boulders in glacial sediments (Parker Gay, 2004).

3.9.7 Magnetism of mineralised environments

The magnetic responses of various types of mineralisation are summarised by Gunn and Dentith (1997). Anomalous magnetism of the mineralisation itself, the alteration zone and the host lithologies is relatively common and routinely exploited during exploration.

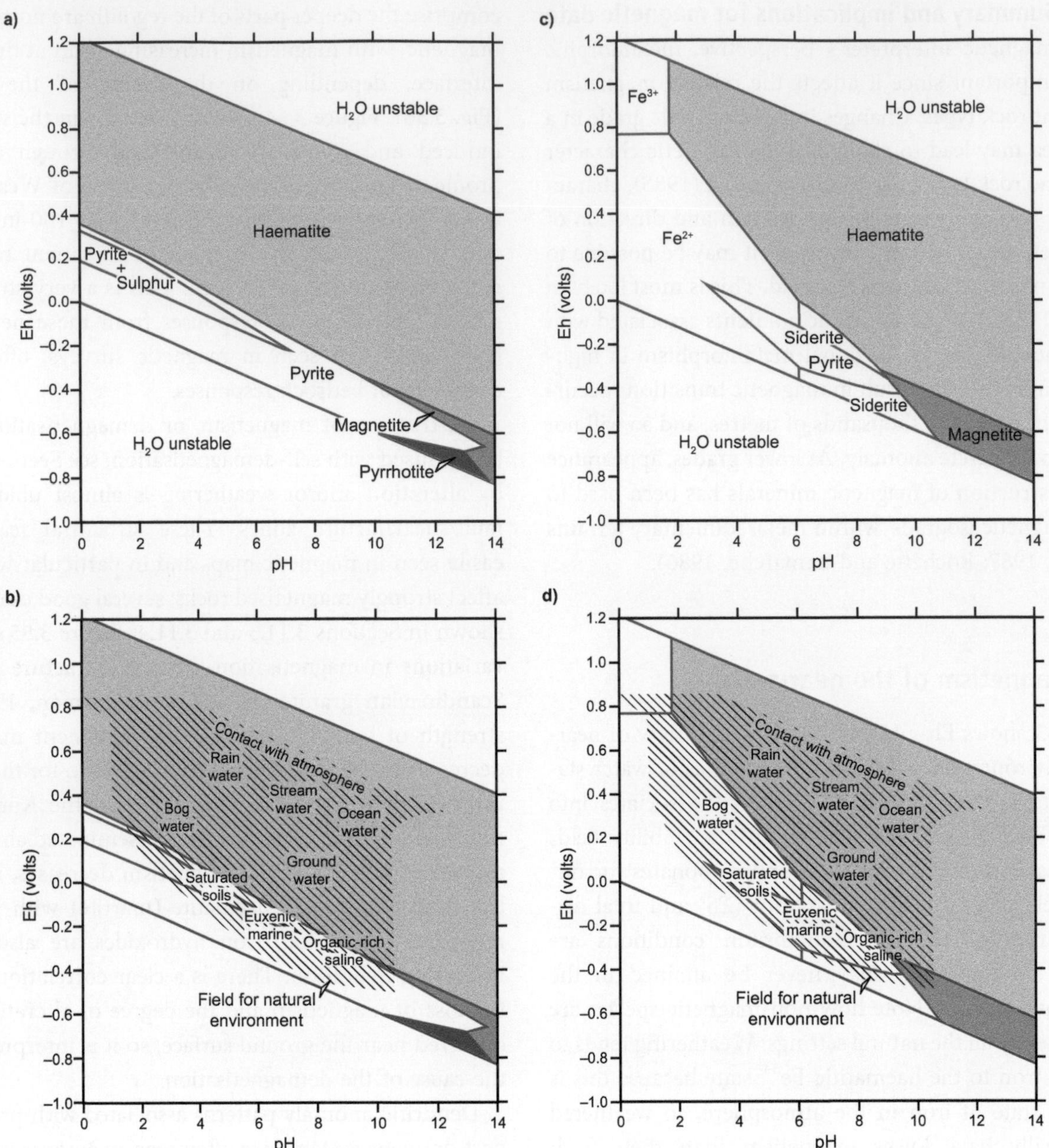

Figure 3.53 Eh–pH diagrams showing the stability fields of various iron oxides, sulphides and carbonates in relation to conditions in near-surface environments in the presence of water. Pressure = 0.1 MPa, temperature = 25 °C. (a) Haematite, pyrite, magnetite and pyrrhotite for $\Sigma S = 10^{-1.5}$, $\Sigma CO_2 = 10^{-2.5}$ M (as in sea water) (b) Comparison of fields in (a) with those in the natural environment. (c) Stability fields after an increase in the availability of CO_2 and decrease in the availability of S; $\Sigma S = 10^{-6}$, $\Sigma CO_2 = 1$M. (d) Comparison of fields in (c) with those in the natural environment. (a) Redrawn, with permission, from Mueller (1978); (c) redrawn, with permission, from Garrels and Christ (1965); (b) and (d) based on diagrams in Baas Becking *et al.* (1960) and Garrels and Christ (1965).

The metal oxide and sulphide minerals are mostly weakly magnetic, although magnetite mineralisation is an obvious exception. Deposits containing massive magnetite are extremely magnetic, for example magnetite-bearing skarns, although self-demagnetisation (see Section 3.2.3.6) may reduce this. Magnetite-bearing heavy mineral sands may have high susceptibilities, and available data suggest that remanent magnetism of the deposits is very weak (Lawton and Hochstein, 1993), probably owing to random orientation of the magnetic grains. Other heavy minerals are weakly magnetic (ilmenite) or non-magnetic. Mineralisation comprising nickel and base metal sulphide and oxide species can be strongly magnetic (see Fig. 3.42). This is due to magnetite and pyrrhotite in the mineralisation,

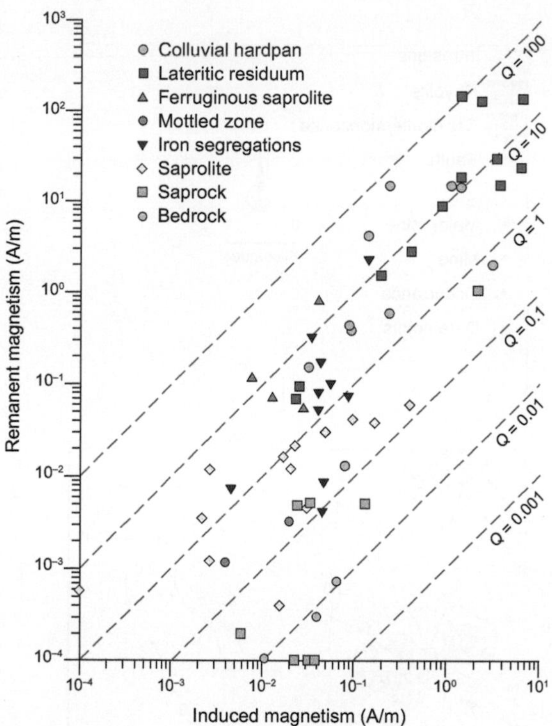

Figure 3.54 Variation in induced and remanent magnetisms through the regolith profile at Lawlers. Redrawn, with permission, from Emerson and Macnae (2001).

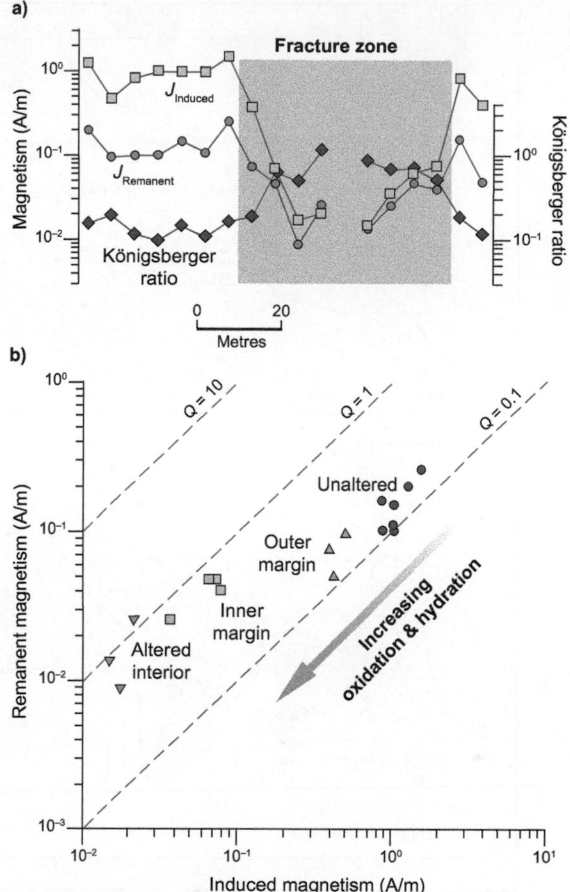

Figure 3.55 Magnetic property variations through a fracture zone in a granite. (a) Profiles across the fracture zone, and (b) remanent versus induced magnetism and the Königsberger ratio (Q). Based on diagrams and data in Henkel and Guzmán (1977).

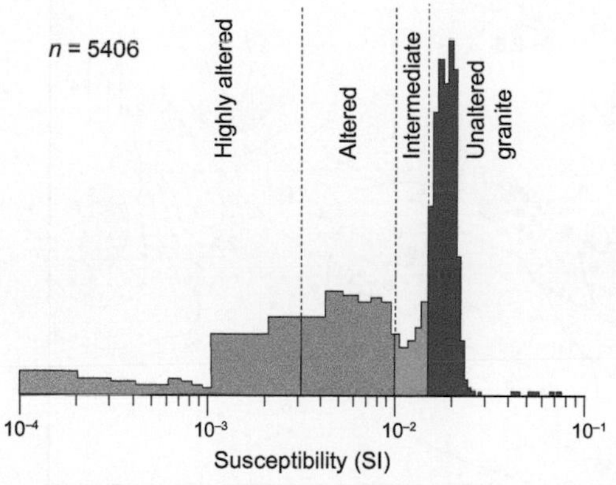

Figure 3.56 Frequency histogram of magnetic susceptibility data from a drillhole in a Canadian granite. Redrawn, with permission, from Lapointe *et al.* (1986).

which are not the economic minerals. Mineral zonation of a deposit may mean that the ore and magnetic minerals occur in different parts of the mineralised system. For example, in VMS deposits magnetite and pyrrhotite tend to concentrate near the base of the massive sulphide mineralisation. An example of reduced magnetism in ore zones is haematitic iron ores formed by enrichment of banded iron formation due to weathering-related destruction of magnetism.

Alteration related to hydrothermal fluids can cause significant changes in rock magnetism, including the creation of CRMs. Hydrothermal temperatures of greater than 150 °C start to break down titanomagnetites to form titanohaematites. For a Canadian granite (Fig. 3.56), mineralogical analysis revealed a correlation between susceptibility and the intensity of hydrothermal alteration within fractures. The 'highly altered' material comprises clay, carbonate and iron hydroxide minerals; the 'altered' material comprises chlorite-group minerals; the 'intermediate' group comprises epidote minerals; and the 'unaltered' material is fresh biotite–hornblende granite. Since alteration associated with mineralisation normally extends over a much larger area than the mineralisation itself, the destruction (or creation) of magnetism can be a useful exploration vector. Another important example is the alteration haloes associated with porphyry style mineralisation. Phyllic,

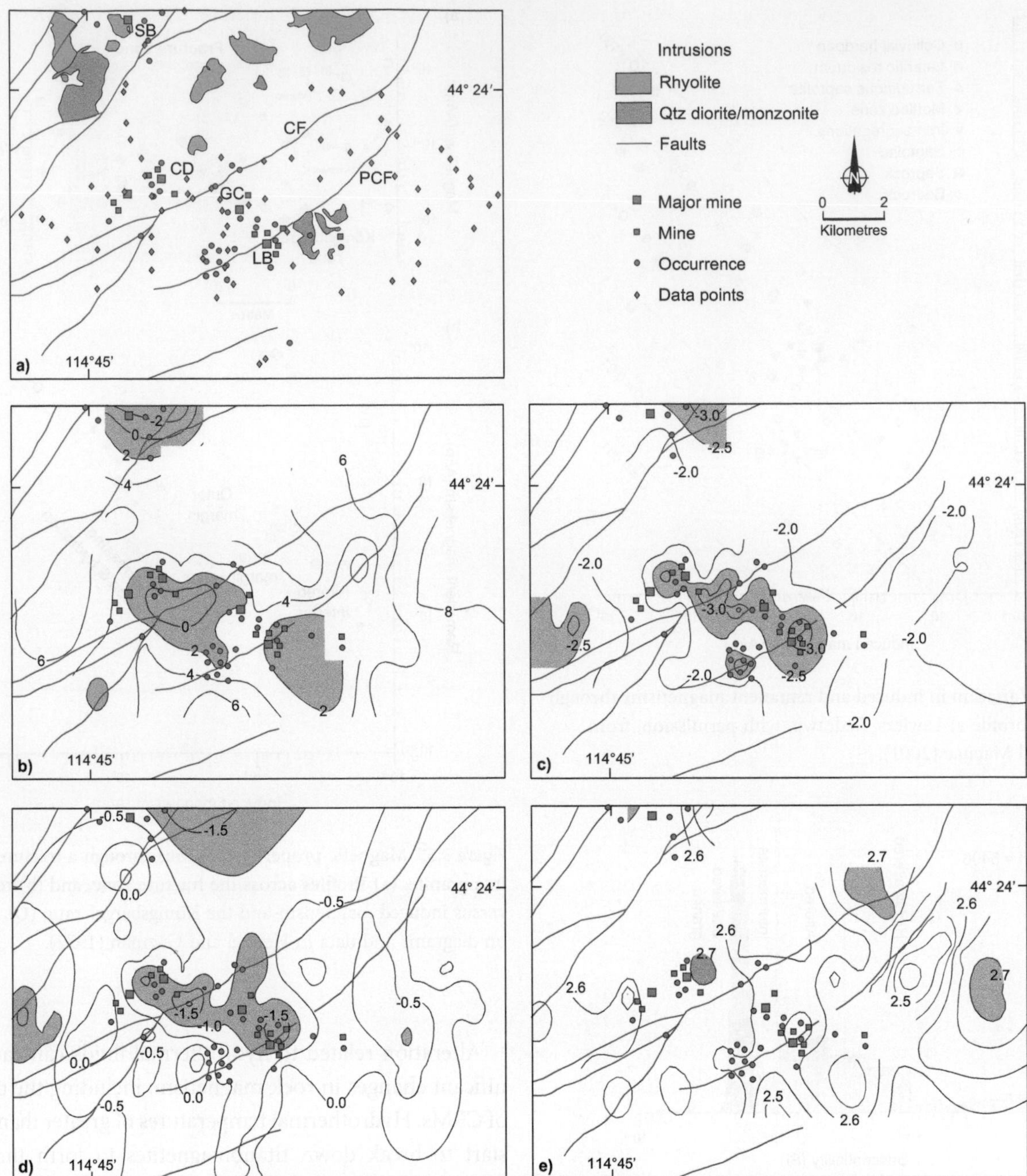

Figure 3.57 The hydrothermal alteration system associated with the Yankee Fork mining district. (a) Simplified geology and occurrences of mineralisation. CD – Charles Dickens Mine, CF – Custer Fault, GC – General Custer Mine, LB – Luck Boy Mine, PCF – Preacher's Cove Fault, SB – Sunbeam Mine. (b) Contours of hydrothermal alteration as indicated by whole-rock $\delta^{18}O$ analyses. (c) Contours of the logarithm of magnetic susceptibility (SI). (d) Contours of the logarithm of strength of remanent magnetism (A/m). (e) Contours of density (g/cm^3). The hydrothermal alteration has produced a general decrease in magnetism and variable density. Note the increase in density near the Charles Dickens Mine and decrease around the General Custer and Lucky Boy mines. Based on diagrams, and created using data, from Criss *et al.* (1985).

argillic and intense propylitic alteration tends to destroy magnetite, although less intense propylitic alteration may allow magnetite to survive. In contrast, the potassic alteration zone is magnetite rich, especial in Au-rich systems (Hoschke, 2011).

Criss *et al.* (1985) describe demagnetisation by a mineralising hydrothermal system in the Yankee Fork mining district, Idaho, USA. The system contains Ag–Au vein mineralisation (Fig. 3.57), which occurs within Eocene calc-alkaline lavas, tuffs and intrusives, and the area is intersected by several large northeast-trending faults. A sinistral displacement of about 2 km on the Custer Fault post-dates the mineralisation, which occurs in two distinct districts a few kilometres apart. The southern area, from where most of the data were collected, occurs around the General Custer Mine and has produced several million ounces of Ag from a vein-system within andesitic lavas. Wall-rock alteration is pronounced and includes silicification, sericitisation and chloritisation. The system is thought to occur above a quartz monzonite intrusion, whose roof zone is exposed in places, which would have been the source of hydrothermal fluids. The second area, which contains the Sunbeam Mine, contains both Ag and Au but production is much less than the southern area. Mineralisation is spatially associated with rhyolitic intrusions and associated with argillic alteration surrounded by wide zones of propylitisation.

Density, magnetic susceptibility and strength of remanent magnetism, and the variation on oxygen isotope ratio ($\delta^{18}O$) data were collected at 78 locations mainly in and around the southern mineralised district. The oxygen isotope data show two zones of very low values interpreted to be due to the effects of meteoric–hydrothermal fluids. These coincide with the two mineralised districts, allowing for the later offset on the Custer Fault which cuts the southern district, and probably define the extent of intense hydrothermal activity. The susceptibilities and strengths of remanent magnetism of intermediate lavas within the low $\delta^{18}O$ zones, indicative of hydrothermal alteration, are distinctly lower than the surrounding areas. The rocks with lower magnetism have undergone intense argillisation with their primary magnetite destroyed by oxidation and hydration. In a few cases, hydrothermal magnetite occurs in highly altered rocks, increasing their magnetisation. The isotopically inferred hydrothermal zone closely correlates with the regions of reduced magnetism. The reduction in magnetic properties is sufficient to be detected by aeromagnetic measurements. An excellent example of modern aeromagnetic data mapping alteration zones associated with epithermal Au–Ag mineralisation at Waihi-Waitekauri, New Zealand, is described in Section 4.7.3.2.

Examples of anomalous magnetism of host lithotypes include large intrusions hosting magmatic deposits, kimberlites and Au–Cu–Bi mineralised ironstones in the Tennant Creek Inlier, Northern Territory, Australia (Hoschke, 1985, 1991).

3.9.8 Magnetic property measurements and their analysis

Magnetic susceptibility is relatively easy to measure in the field on rock samples and on actual outcrops. Remanent magnetism on the other hand is far more difficult to measure. It requires specialist laboratory equipment, and samples must be geographically oriented in order to resolve the direction of the remanent field. An unfortunate consequence of this is that usually more susceptibility data than remanence data are available, so the actual magnetic properties of the various rock formations are only partially defined.

3.9.8.1 Measuring magnetic susceptibility of samples

A range of hand-held susceptibility instruments for use in the field is available, and one kind is shown in Fig. 3.58. They are lightweight, robust and easy to use: this involves placing the instrument next to the rock sample, or outcrop face, and pressing a button to display the measurement. Some models store the data for subsequent downloading. The measurement is influenced chiefly by material within about 20 to 30 mm of the surface, so the shape and size of the specimen influence the reading obtained. Usually the specimen needs to be at least the size of one's fist and have a flat surface for the measurement, and be free of magnetic materials such as steel core-trays. An instrument

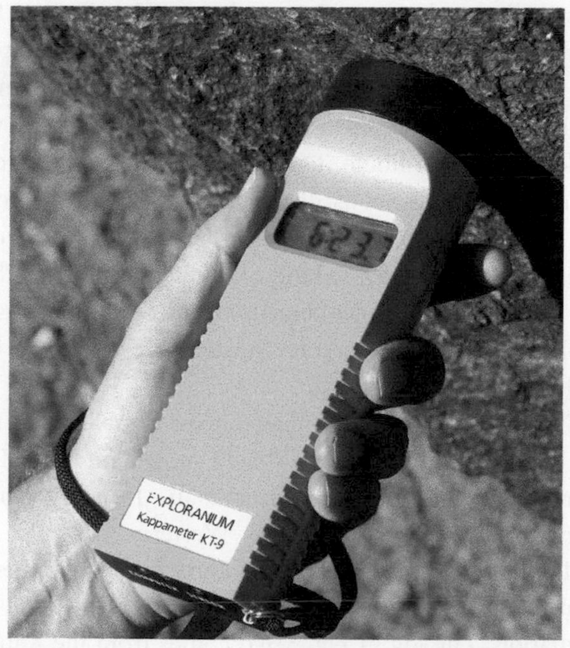

Figure 3.58 A portable hand-held magnetic susceptibility meter for field use.

correction factor can be applied for undersized specimens, rough surfaces and circular drillcores. High electrical conductivity of samples can disturb the operation of the sensor. When susceptibility anisotropy is of interest, oriented rock samples are required and laboratory measurement techniques must be used.

Hand-held susceptibility meters are a means of quickly and cheaply acquiring data, provided high precision is not required. Care is required in selecting samples to ensure that they are representative of the rocks being investigated. Oxidised samples in particular need to be avoided because of the likelihood of magnetite being destroyed. Also, the presence of magnetic veins and the vein/fracture density of the sample can produce large variations in the measurements. The major mistake is to take too few readings. When working with drillcore, it is 'good practice' to take three equally spaced susceptibility readings around the perimeter of the core sample, at the same downhole-depth. For measurements on outcrop, a sufficiently large dataset ought to be acquired in order to accurately resolve the natural spatial variation in susceptibility across the outcrop. The data can then be analysed in terms of their frequency distribution (see Section 3.9.8.3).

3.9.8.2 Magnetic susceptibility logging

Magnetic susceptibility can be measured *in situ* with a probe in a drillhole to quickly obtain a continuous susceptibility log. Measurement interval is typically 10 cm. The logs produce a far larger dataset than can be obtained from manual measurements. The probe senses roughly a spherical volume of rock a few tens of centimetres in diameter, so the log shows the variation in susceptibility of the wall-rock along the drillhole. The small sample volume and the generally heterogeneous distribution of magnetic properties mean that susceptibility logs are often locally quite variable. Normally some kind of mathematical smoothing, e.g. a median filter (see *Smoothing* in Section 2.7.4.4), is applied to the log to resolve relationships with the wall-rock geology.

Figure 3.59a shows a magnetic susceptibility log from the banded iron formation ('shale') succession in the Dales Gorge Member of the Brockman Iron Formation, Hamersley iron-ore province, Western Australia. The shales are in fact ferruginous cherts and are important stratigraphic markers within the thick and invariant iron formation. The log in Fig. 3.59b comes from the Kabanga Ni sulphide deposit, Tanzania. The sulphide mineralisation has significantly higher susceptibility than the host rocks.

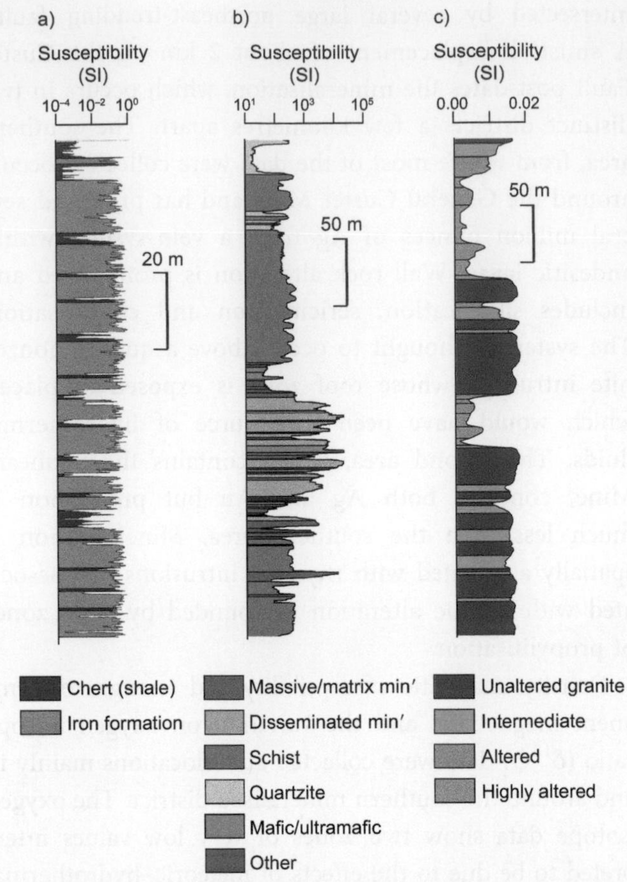

Figure 3.59 Magnetic susceptibility logs from various geological environments: (a) Banded iron formation from the Hamersley iron-ore province. Based on diagram in Cowan and Cooper (2003b), (b) Kabanga nickel sulphide deposit, reproduced with the permission of Barrick (Australia Pacific) Ltd, and (c) alteration zone in a Canadian granite. Based on a diagram in Lapointe *et al.* (1986).

Figure 3.59c is the log from a hydrothermally altered granite in Canada and is the source of the data shown in Fig. 3.56. The decrease in susceptibility with increasing alteration correlates with fracture zones.

Magnetic susceptibility logs are used for lithological discrimination, hole-to-hole correlations and identification of alteration zones, and are commonly used to characterise variations in the local stratigraphy. Where mineralisation has different magnetic properties from the host sequence, there have been some attempts to use downhole magnetic susceptibility logs as a fast, low-cost, basis for ore-grade control, either alone or in association with other physical property logs.

3.9.8.3 Analysis of magnetic susceptibility data

Magnetic susceptibility data can be displayed as a frequency histogram for each rock type sampled. Mostly,

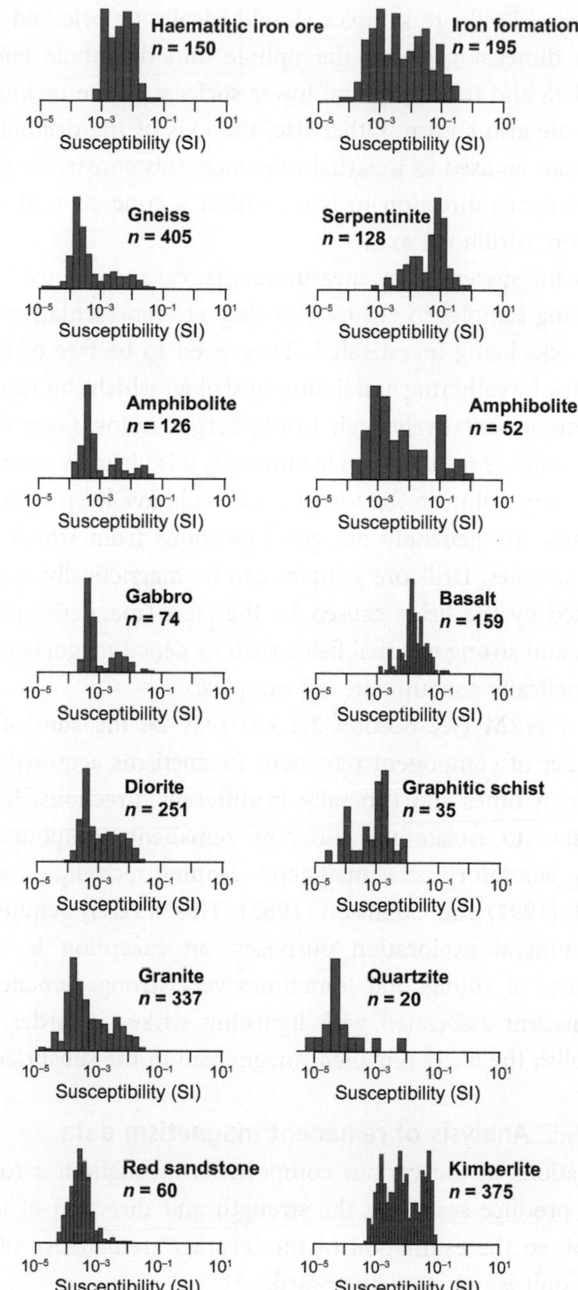

Figure 3.60 Frequency histograms of magnetic susceptibilities for various lithotypes. The data for each lithotype are from the same area and are all measurements made on outcrop. Based on diagrams in Irving *et al.* (1966), Karlsen and Olesen (1996), Mwenifumbo *et al.* (1998), Puranen *et al.* (1968), Tarling (1966) and unpublished data of the authors. Note how bi-modal or multi-modal successions are the norm.

but not invariably, the distribution will be skewed, so it is usual to plot the logarithm of susceptibility to make the frequency distribution more symmetrical (Irving *et al.*, 1966). The data are often multimodal (Figs. 3.47 and 3.60) owing to the presence of different populations of

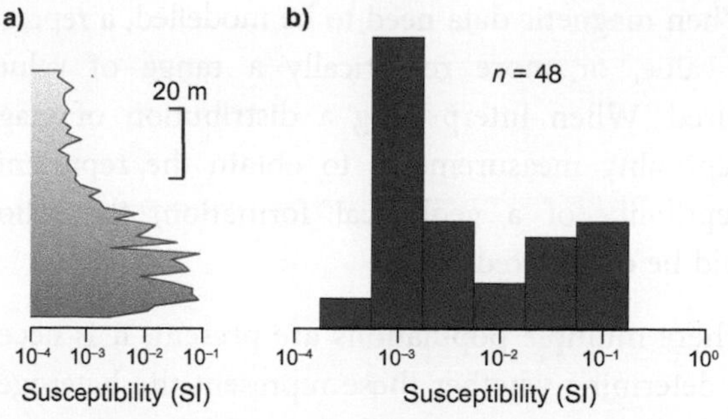

Figure 3.61 Magnetic susceptibility data from a thick komatiite flow in a greenstone belt in Western Australia. Measurements plotted (a) in their locations across the flow and (b) as a histogram. Based on data in Keele (1994).

magnetic minerals within the dataset. A single-mode distribution is comparatively rare, even when the lithology sampled appears homogenous. Figure 3.61 shows susceptibility variations through a thick komatiite flow. The zoning of these flows produces a complicated frequency histogram, but the actual variation through the flow is revealed when the data are displayed in terms of their location within the flow.

It may be possible to distinguish both ferrimagnetic and paramagnetic populations in the data. For this to occur in fresh rocks the minerals of the two types must be segregated within the rock at a scale greater than the dimensions of the volume sampled by each measurement. In outcrop, however, localised weathering may have oxidised titanomagnetites to less magnetic species resulting in a measurement influenced mainly by the paramagnetic constituents of the rock. Statistical methods can be applied to susceptibility data to identify individual populations; for example see Larsson (1977) and Lapointe *et al.* (1986).

Before applying statistical methods to a susceptibility dataset, it is worth considering the purpose for which the data were acquired. When the primary objective is to delineate contrasts in magnetisation and to characterise the magnetic responses of the various geological units, a qualitative assessment of a complex distribution of susceptibility identifying a range of magnetic responses is probably sufficient. For example, the average or range of susceptibility from a unit could be used to establish an informal magnetisation hierarchy from 'highly magnetic' through to 'non-magnetic' and to which the likely geology in concealed areas could be referred. Although statistical analysis may be unnecessary for these applications, adequate sampling to create a representative distribution is essential.

When magnetic data need to be modelled, a representative value, or more realistically a range of values, is required. When interpreting a distribution of magnetic susceptibility measurements to obtain the representative susceptibility of a geological formation, the following should be considered:

- Where multiple populations are present, it is necessary to determine whether these represent the heterogeneity of the geology being sampled, or are a sampling artefact. As noted above, susceptibility readings made with a hand-held instrument will be affected by the surface roughness and volume of the sample.
- It is possible that one or more populations are of local significance and can be ignored: for example chilled margins of intrusions and areas of alteration close to faults and joints etc., which may be either magnetite constructive or destructive. Also, weathering will tend to skew the distribution towards lower susceptibilities or may even result in an identifiable weathered population.
- A fundamental aspect of all geological field mapping is that outcrop is often a biased sample, being overly representative of those units that are more resistant to erosion. This may also lead to bias in susceptibility measurements.

The inherent complexity of magnetic mineralogy and limitations of sampling mean that it is very difficult to assign an accurate 'average' susceptibility value to a geological unit. Selecting a range that encompasses the majority of the data is often the best approach. The aim should be to identify likely susceptibility ranges for individual units and where significant inter-unit contrasts occur.

3.9.8.4 Measuring remanent magnetism

Being a vector quantity, both the strength and direction of remanent magnetism need to be measured. Measurements are made in the laboratory on geographically oriented rock samples. The methodology is described by Clark and Emerson (1991), and Collinson (1983) who also provide detailed descriptions of rock sampling and sample preparation techniques. For accurate measurements a specialised drill is used for taking a mini-core 16 mm in diameter and 10–30 mm in length. For most mineral exploration applications, orientated bulk samples are usually suitable. When using a magnetic compass to determine orientation, care should be taken to ensure that the magnetism of the sample does not deflect the compass. For strongly magnetised rocks, such as BIF, it may be necessary to use a sun compass. Drillcore samples should ideally be oriented in three dimensions: with the uphole and downhole ends labelled, and the upper and lower surfaces of the inclined drillhole also shown. Otherwise, the axis of the drillhole/core can be used as a partial reference; this constrains the magnetism's direction to lying within a cone centred on the core (drillhole) axis.

As for susceptibility measurements, care is required in selecting samples to ensure that they are representative of the rocks being investigated. They need to be free of the effects of weathering and lightning strikes, which can result in unrepresentatively high Königsberger ratios. Over the time scales required to erode outcrops, it is virtually certain that every point on the top of a hill will have been struck, so these are generally not good locations from which to take samples. Drillcore samples can be magnetically overprinted by the fields caused by the steel (magnetic) drill rods, and strong external fields such as pencil magnets can magnetically contaminate the samples.

The NRM (see Section 3.2.3.4) may be the sum of a number of component remanent magnetisms acquired at different times and generally in different directions. It is possible to isolate the different remanent components using laboratory palaeomagnetic cleaning techniques; see Clark (1997) and Collinson (1983). This is rarely required for mineral exploration purposes; an exception is the removal of young, and sometimes very strong, remanent magnetism associated with lightning strikes in order to establish the likely remanent magnetism in the subsurface.

3.9.8.5 Analysis of remanent magnetism data

Variations in the various components throughout a rock unit produce scatter in the strength and direction of the NRM, so the estimation of the 'average' remanence of a rock unit is not straightforward.

Like susceptibilities, a number of measurements of remanent magnetism should be made on each rock unit. Strengths can be plotted as histograms and multimodal distributions are common. Directions can be plotted on a stereographic projection, as used for working with other geological orientation data. However, both the upper and lower hemispheres must be used when plotting magnetic directions. By convention the orientations of fold axes etc. are plotted in the down-plunge direction: that is, the lower hemisphere is used to project onto the stereo net. An up-plunge measurement of, say, 45° upwards towards the north would be converted to 45° downwards to the south before plotting. The convention does not apply to magnetic

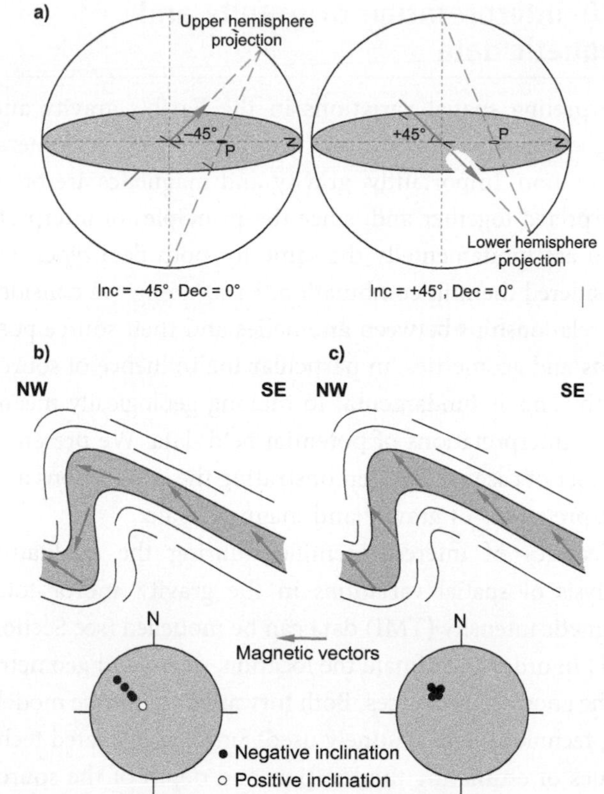

Figure 3.62 Measuring the direction of remanent magnetism. (a) Upper and lower hemisphere stereographic projections, as needed, to account for the possibility of positive or negative magnetic inclinations. Directions of remanent magnetism formed (b) before and (c) after a folding event, and their appearances on a stereographic projection.

directions; negative inclinations must be projected from the upper hemisphere. Magnetic directions of −45° (Inc) and 0° (Dec), and +45° (Inc) and 0° (Dec), will plot at the same point on the stereo net (Fig. 3.62a). They are distinguished by using different symbols for positive and negative inclinations.

The directions of individual components, and therefore the NRM itself, can vary across an area of interest. For example, a remanent magnetism that precedes a folding event will have a consistent direction relative to the bedding plane but will vary across a fold in the same way as the dip of the bedding (Fig. 3.62b). If the direction of the remanent magnetism is post-folding its direction is unaffected by the fold but varies relative to the bedding plane (Fig. 3.62c). The result is that the magnetic anomaly produced by similar geology will be significantly different.

In practice remanent magnetism is either ignored or assumed to be parallel to the induced magnetism, unless magnetic anomalies have characteristics that demonstrate that this is not the case. Even after careful analysis, estimating a representative remanent magnetism for a unit is extremely difficult. More measurements than can be justified economically are virtually always required. As with susceptibilities, the practical solution is to identify likely ranges in directions and intensities, which may vary from area to area, and use these as a basis for interpreting magnetic responses.

3.9.9 Correlations between density and magnetism

Gravity and magnetic data are often used together for geological mapping and target identification. However, it must be stressed that there is no physical reason why the density and magnetism of a rock formation, and indeed variations in these properties, must be related. Recall that magnetism depends primarily on the content of what is usually an accessory mineral; whereas, in contrast, density is a bulk property of the rocks reflecting their matrix mineralogy and, in the presence of significant porosity, the pore volume and contents as well. Consequently, gravity and magnetic maps reflect fundamentally different physical characteristics of the rocks, so they may differ significantly.

For some rock types, density and magnetism roughly correlate because the variation in magnetite content correlates with felsicity, which in turn controls the proportion of the denser mafic minerals. So correlations may occur in crystalline rocks, but are less likely in sedimentary rocks where magnetism is generally very weak and density is controlled primarily by porosity. Henkel (1991, 1994) describes the correlation between magnetism and density of crystalline lithotypes in the Baltic Shield. In a plot of density versus magnetic susceptibility, most of the data plot in one of two fields, termed the *paramagnetic* and *magnetite trends* by Henkel (1991) (Fig. 3.63a). Within each field there is some correlation between density and susceptibility, although it is weak compared with the effects of geological processes that tend to change (usually) magnetite content and cause lithotypes to move from one field to the other, or to some intermediate position as with metamorphism and serpentinisation. Serpentinisation is a very significant factor because of the large associated reduction in density and increase in magnetism (Figs. 3.34 and 3.52), and the breaking of the general observation that density and magnetism often increase together.

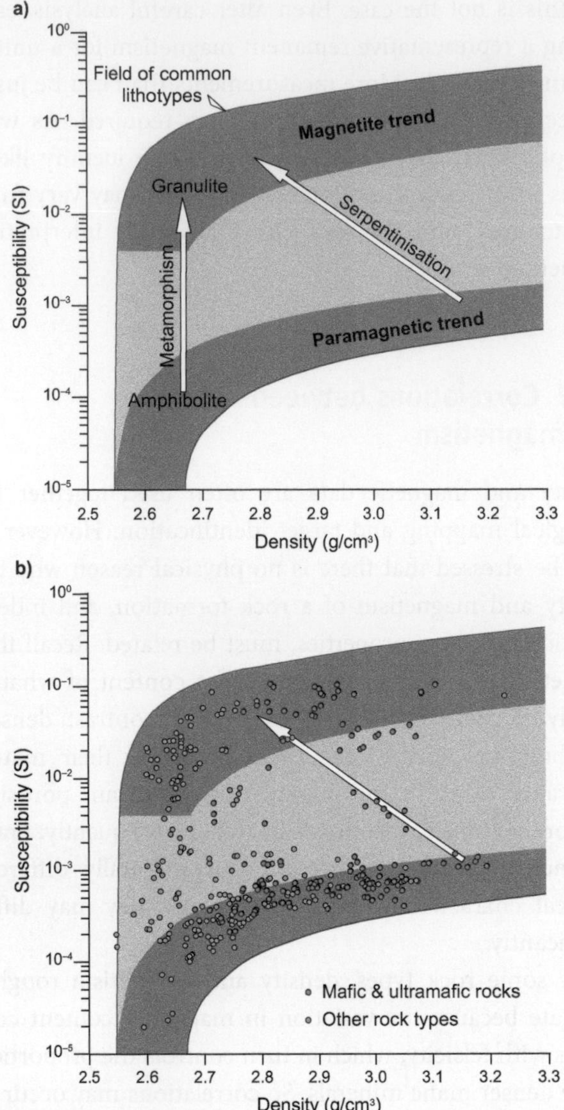

Figure 3.63 Correlation between density and magnetic susceptibility. (a) Correlation fields for most lithotypes; and some common geological processes, schematically indicated, that move the property values from one field to the other. Redrawn, with permission, from Henkel (1991). (b) Data for various lithotypes from a granitoid–greenstone terrain in the Abitibi Subprovince. Data from Ontario Geological Survey (2001).

Figure 3.63b shows data from a granitoid–greenstone terrain in the Abitibi Subprovince, Ontario, Canada (see Section 3.11.3). Data plot in both the paramagnetic and magnetite fields. The metamorphic grade in the region is greenschist facies so mafic rocks tend to have low susceptibilities and, therefore, occupy the paramagnetic field, although the data falling on the serpentinisation trend suggest this process has occurred.

3.10 Interpretation of gravity and magnetic data

Interpreting spatial variations in the Earth's gravity and magnetic fields is a common 'geophysical' task in mineral exploration. Importantly, gravity and magnetics are often interpreted together and, since the principles of interpretation are fundamentally the same for both data types, we considered them in combination. In so doing, we consider the relationships between anomalies and their source positions and geometries, in particular the influence of source depth. This is fundamental to making geologically meaningful interpretations of potential field data. We present a number of case studies demonstrating the applications and interpretations of gravity and magnetic data.

Features of interest identified during the qualitative analysis of spatial variations in the gravity and/or total magnetic intensity (TMI) data can be modelled (see Section 2.11) in order to estimate the location, depth and geometry of the anomalous sources. Both forward and inverse modelling techniques are routinely used. Simple automated techniques of estimating the location and depth of the source offer the advantage of being able to analyse large volumes of data rapidly, the results being used as the basis for more sophisticated analysis of targeted features. Parametric models (see Section 2.11.1.1) provide additional information for a particular source shape, such as the depth, width and dip. Property distributions of greater complexity (see Section 2.11.1.2) can be resolved using inversion techniques, but require considerably more time and effort to apply.

Removal of the regional response is often required in order to resolve the target anomaly and can be done with any of the techniques described in Sections 2.9.2 and 3.7.3; but obtaining satisfactory resolution of the local anomaly is not always easy. Non-uniqueness or ambiguity (see Section 2.11.4) is always prevalent, being reduced with greater geological control. These two factors represent the greatest sources of uncertainty in the modelling of gravity and magnetic data.

3.10.1 Gravity and magnetic anomalies and their sources

The fundamental features of geophysical responses described in Section 2.3 apply to potential field responses. These are confirmed here by showing variations in total magnetic intensity and gravity across sources with simple geometry embedded in a 'background' volume.

The amplitude of a gravity anomaly is proportional to both the volume of the buried body and its density contrast (see Section 3.2.1.3). This is because the force experienced by a body is proportional to its mass; see Eq. (3.3). In other words, larger and denser features produce larger-amplitude anomalies. The same is generally true for contrasts in magnetic susceptibility (see Section 3.2.3.3), although the source's self-demagnetisation (see Section 3.2.3.6) counteracts this relationship, and quite significantly so at high susceptibility. As shown in Fig. 2.4, increasing the contrast simply linearly increases the amplitude of the anomaly. The wavelength of the response does not change. Reversing the sign of the physical property contrast reverses the polarity of the response, i.e. an anomaly simply 'flips over'.

3.10.1.1 Effects of source depth

As shown in Fig. 2.4, increasing the distance between source and the detector, whether by moving the source deeper into the subsurface or making measurements at greater height above the ground (increased survey height), increases the wavelength and decreases the amplitude. The change in response with source–detector separation is a very important issue because it controls the depth to which geophysical data can 'see' into the subsurface. As the amplitude of an anomaly decreases it eventually falls to below the noise level, making its detection impossible.

The magnetic survey data shown in Fig. 3.64 illustrate changes in wavelength and amplitude with source depth. In the Qian'an area of China, strata-bound iron ore deposits occur in banded iron formation (BIF). Country rocks are high-grade metamorphic rocks. The BIF and associated ores are much more magnetic than the country rocks. Near-surface mineralisation gives rise to distinctive short-wavelength anomalies (S) whilst the deeper mineralisation (D) is associated with longer-wavelength responses; which in this case led to their discovery.

For a spherical gravity source (a monopole), the strength of the gravity field decreases as the square of the distance, i.e. as $1/\text{distance}^2$. For larger and more complex source geometries, it reduces more slowly and can be proportional to distance, i.e. as 1/distance, or even less than this (Dobrin and Savit, 1988). The decrease in the strength of a magnetic anomaly away from a magnetic source depends on source geometry and the orientation of the body's magnetism. Specifically, it depends on the distance between the sensor and the two magnetic poles of the source's magnetism. The effects of both poles are significant in sources with small depth extent, and the field strength decreases faster. For

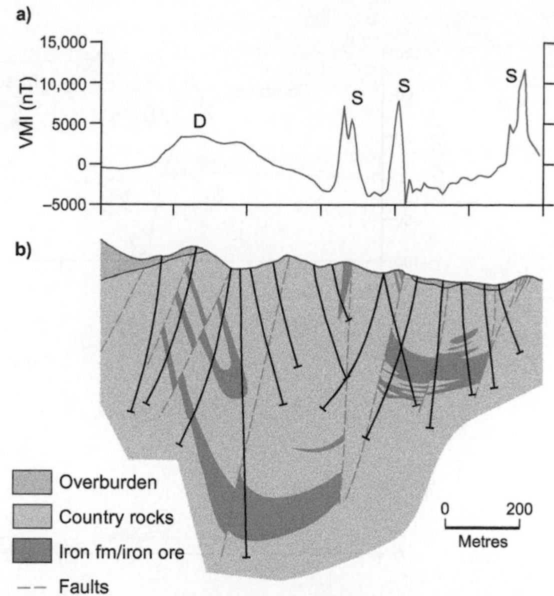

Figure 3.64 (a) Magnetic data and (b) geological cross-section of iron ores in the Qian'an district. Responses due to shallow sources (S) and deep sources (D) are highlighted. VMI – vertical magnetic intensity. Redrawn, with permission, from Qinfan (1988).

example, for compact sources like a sphere (a magnetic dipole) it decreases as the cube of the distance, i.e. as $1/\text{distance}^3$, and as $1/\text{distance}^2$ as the strike length increases (a line of dipoles). At the other extreme, when the source's depth extent is so large that the reduction in field strength with depth is due almost entirely to the effects of the closest pole to the sensor, it decreases as $1/\text{distance}^2$ for pipe-like sources to 1/distance for a source having large strike extent, such as a dyke. More complex source geometries have more complex pole distributions and produce decreases in the range of $1/\text{distance}^2$ to about $1/\text{distance}^{0.5}$ (Breiner, 1973). Note that for a (semi-) infinite horizontal distribution of poles or dipoles, as would occur on the surface of a rock body of very large horizontal extent, the field does not change with height above the surface so the variation can be expressed as $1/\text{distance}^0$, or 1. The Bouguer gravity slab described in Section 3.4.5.2 has this property.

The contributions of the various depth portions of a source are demonstrated in Fig. 3.65 for a vertical prism and a vertical step contact. Both models have higher density and susceptibility than the background and are subdivided into five depth portions. The gravity and TMI responses (shown for mid latitude) of the complete structures are the sum of the responses of their individual portions. In both cases the shallowest portion, being closest

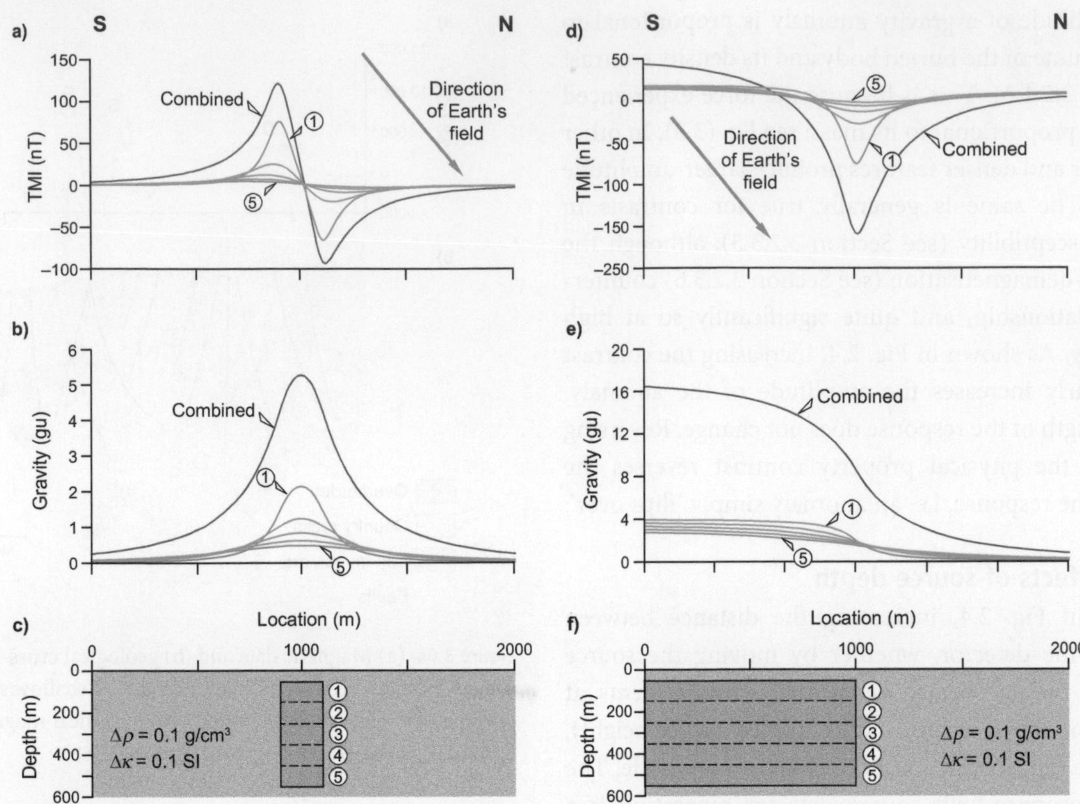

Figure 3.65 Depth-dependent characteristics of the TMI and gravity responses of a vertical prism and a vertical step; both are magnetic and have high density. In both cases the source is divided into five depth-portions. See text for details.

to the sensor, makes the greatest contribution to the amplitude of the responses, with the contributions from the deeper portions decreasing with depth. As expected, the amplitude of the magnetic response decreases faster with depth than the gravity response.

Figure 3.65 confirms the increase in wavelength of responses from deeper sources. Note that the gravity response has the greater wavelength, which increases more rapidly with source depth than the magnetic response. Two other important differences between gravity and magnetic data are illustrated in the figure. Firstly, magnetic anomalies are more localised to their source than gravity anomalies and, secondly, gravity data provide information to greater depth than magnetic data (survey parameters and noise notwithstanding).

In general, and importantly, the lesser rates of amplitude decrease for gravity responses means there is more information about the geology at depth in gravity data than magnetic data. In terms of target detectability, it means gravity surveys are usually better at detecting deeper targets than magnetic surveys. In terms of mapping, the gravity data will be more prone to interference between responses from different depths than the magnetic data.

3.10.1.2 Effects of remanent magnetism

If an observed variation in TMI does not have the general form of the anomaly, e.g. dipole with low to the north etc., expected for the local orientation of the geomagnetic field (Figs. 3.8 and 3.26a), then remanent magnetism (see Section 3.2.3.4) carried by the source is strong and orientated in a direction different from that of the present-day Earth's field. The form of the magnetic response depends on the source geometry and the strength and direction of its magnetism. In principle a remanent magnetism can have any direction, so there is clearly an infinite range of variations possible for the resultant response. However, the following observations about anomaly character provide some information about the nature of the remanent magnetism carried by the source, essential for interpretation and modelling.

Where the remanent magnetism's direction is roughly the same as that of the present-day Earth's field, the anomaly amplitude will be stronger and the source will appear to have higher susceptibility than it actually has. Where the remanent magnetism is in roughly in the opposite direction, the anomaly amplitude will be weaker and the source appears to have lower susceptibility. The anomaly will have opposite polarity (to that expected owing to induced magnetism) only

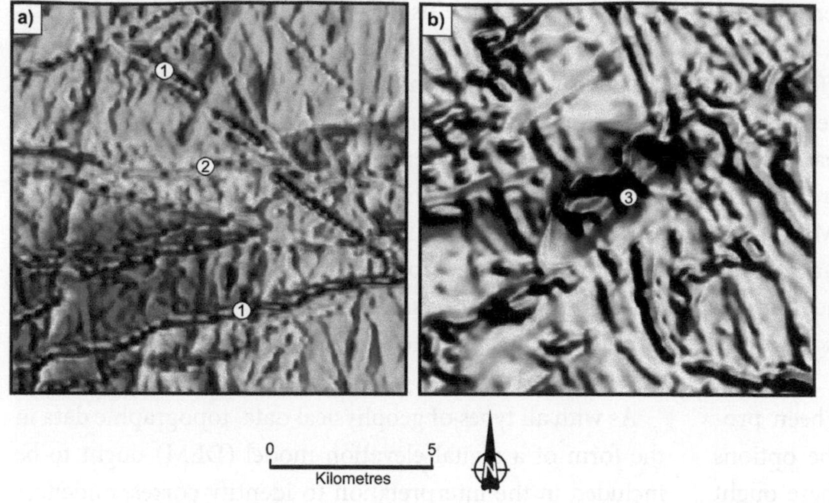

Figure 3.66 Examples of airborne TMI data with anomalies caused by dominant remanent magnetism. (a) Data from the central Yilgarn Craton, Western Australia, acquired at a survey terrain clearance of 40 m. The local geomagnetic field has a declination of 0° and an inclination of –66°. (b) Data from the Wawa area, Ontario, Canada, acquired at a survey terrain clearance of 70 m. The local geomagnetic field has a declination of –8° and an inclination of +74°. The survey line separation for both areas is 200 m. Source (b): Ontario Geological Survey © Queen's Printer for Ontario 2014.

when the remanent magnetism is significantly stronger than the induced magnetism, i.e. a high Königsberger ratio (Q). In certain circumstances a magnetic 'low' may result from a highly magnetic source. Lockhart *et al.* (2004) describe kimberlites with minimal magnetic responses due to near-cancelling induced and remanent magnetisms.

The TMI data from Western Australia in Fig. 3.66a show the prominent responses of several dyke swarms. The dyke labelled (1) has a dipolar response with the positive component on the northern (equatorial) side, consistent with the direction of induced magnetism in the area. The dyke labelled (2) has a symmetrical negative anomaly inconsistent with an induced magnetism, implying dominant remanent magnetism. The northerly trending linear features of lower amplitude are caused by shear zones. The data in Fig. 3.66b from Ontario show a negative anomaly (3) consistent with a source dominated by remanent magnetism. It is likely to be caused by gabbro and diorite mapped in the area (D. Rainsford, pers comm.). Intrusions with responses consistent with a dominant remanent magnetism are described in Section 3.11.4.

The change in the direction of the resultant magnetism due to remanence strongly affects the response in the same way as changing the dip and strike of the source with respect to the Earth's field. Dip is the most strongly affected source parameter and is generally difficult to resolve for a source carrying significant remanent magnetism.

3.10.2 Analysis of gravity and magnetic maps

The guidelines described in Section 2.10.2 for the analysis of geophysical data in terms of geological mapping and anomaly identification, apply equally to gravity and magnetic data. Interpretation involves delineating linears, which will be due to contacts and structures, and interpreting mappable lithological units, with repeated checks made of the evolving interpretation for geological credibility. It is essential to be aware of the fact that potential field data are inherently smooth and that responses due to variations in density or magnetism extend beyond the edges of their sources. The magnetic responses from two very geologically different areas are described in detail in Sections 3.11.3 and 3.11.4. In these two examples the magnetic responses of some commonly encountered geological features are presented.

When working with magnetic data, it is essential to know the magnetic inclination and declination of the inducing field in the survey area in order to understand the expected forms of the magnetic anomalies. This also assists in the recognition of responses that are 'out of character' for the area, and possibly attributable to the presence of remanent magnetism. Many interpreters prefer to work with data that have been reduced-to-pole (RTP) (see Section 3.7.2.1), but this may not be possible if the local magnetic inclination is too low. Also, RTP can in some circumstance introduce artefacts into the data.

3.10.2.1 Data selection and integration

The foundations of the interpretation of potential field data are the fundamental datasets of TMI and Bouguer or free-air gravity. However, interpretation is invariably based on more than one representation of the data. Vertical derivatives (see Section 3.7.4) resolve contacts, edges and boundaries, and emphasise features in the near-surface; but the fact that some of these may be in the cover material rather than the bedrock should borne in mind. The analytic signal (see Section 3.7.4.2) is also very effective for resolving contacts. It has the distinct advantage of being independent of the magnetic inclination and is particularly useful at low magnetic latitudes where the nature of magnetic sources can be difficult to resolve from TMI images, but the polarity of the physical property contrast across the contact is lost. The merits of these and other methods for locating contacts using magnetic data are discussed by Pilkington and Keating (2004).

A vast number of enhancement filters have been proposed for potential field data in addition to the options available regarding display (see Section 2.8). These ought to be selected to provide complementary information useful in terms of the geological information being sought. For example, when the aim is to map the geology, a product that emphasises contacts and one emphasising linear features would complement each other. Where the geology changes significantly, for example strongly magnetised basalt at the surface in some places, it may be necessary to create different enhancement products applicable to particular areas. Some filters are more susceptible to noise than others reducing their effectiveness on lower quality datasets. In aeromagnetic data, noise commonly manifests itself as corrugations due to mis-levelling (see Section 3.6.4), and in gravity data as a speckled appearance related to incomplete sampling of the variations due to non-ideal station spacing (see Section 2.6.1.1).

The removal of long wavelengths with high-pass filters (see Section 3.7.3) or using upward continuation (see Section 3.7.3.2), or by simply removing a regional gradient (see Section 2.9.2), improves the resolution of short-wavelength shallow features. When the intent is to resolve specific target anomalies from the response of the local geology this is particularly important. A little judicious experimentation with the various parameters is usually required to resolve anomalous responses in the local geology. An example of regional removal to facilitate interpretation of a gravity map is presented in Section 3.11.1. In many cases it is also worthwhile to produce a version of the data containing the highest frequencies (shortest wavelengths) to identify density or magnetisation contrasts in the cover material; second horizontal derivatives computed on the line data can be very effective in this regard.

Various transformations of the data can be integrated into single images using imaging techniques (see Section 2.8). Examples include grey-scale shaded relief with a pseudocolour drape useful for displaying multiple aspects of the data (see Fig. 2.36); shaded relief (see Section 2.8.2.3) applied as a directional enhancement; shaded-relief grey-scale of first vertical derivative with the gravity or TMI data as a pseudocolour overlain; or contours of one particular parameter overlaid on an image of another parameter. Ternary images (see Section 2.8.2.4) can be created by integrating different forms of gravity and magnetic data. Stacked profiles are useful for displaying detail on survey line data that may otherwise be distorted by the gridding process (see Fig. 2.30).

As with all types of geophysical data, topographic data in the form of a digital elevation model (DEM) ought to be included in the interpretation to identify correspondences between the terrain and possible anomalies, and also to identify effects due to changes in ground clearance (see Section 2.4.1.1).

3.10.3 Interpretation pitfalls

Gravity and magnetic data are subject to a variety of spurious responses from a number of sources. Their responses distort genuine target anomalies and can masquerade as target anomalies. They are usually fairly easy to recognise in the data, particularly after removal of the regional response.

3.10.3.1 Cultural effects

Cultural features can produce anomalous responses in gravity and magnetic data. Features such as dams, reservoirs, buildings, open pits and mine dumps influence the gravity terrain correction and are a source of error. The iron content of buildings, roads, railway lines, pipelines and powerlines produces responses in magnetic data. Because they are at, or near, the surface, the responses are usually localised. Aerial photographs can be used to confirm the presence of cultural features.

3.10.3.2 Overburden effects

The near-surface environment can also be a source of noise. For the case of gravity, variations in the thickness of the regolith or cover produce responses that superimpose geological noise on the responses of underlying sources, which can be mistaken as bedrock responses. In lateritic terrain, the near-surface may be highly magnetic, giving rise to large-amplitude, short-wavelength anomalies, which when gridded create a mottled appearance (Dentith *et al.*, 1994). Magnetic detritus and maghaemite in drainage networks can also be a problem (see Section 3.11.4), although these are easy to recognise by their characteristic dendritic form. Glacial deposits can be a significant source of magnetism owing to the presence of magnetic materials plucked from the bedrock and redeposited by glacial processes. Magnetic responses due to glacial features are described by Parker Gay (2004). Both gravity and magnetic responses can be caused by lateral changes in the thickness of the glacial deposits, which may be due to channels; but particularly important are the effects of glacial landforms whose relief affects the distance between sensor and source, such as drumlins, moraines and younger river channels that may have removed large sections of laterally extensive deposits. Effects of the near-surface diminish in downhole work.

3.10.3.3 Topographic effects

Topographic effects can be expected in both gravity and magnetic data acquired in rugged terrains (see Section 2.4.1.1). Variations in terrain clearance are common in airborne data. Gravity data are highly susceptible to errors in the terrain correction (see Section 3.4.5.3), and particularly for topography local to the station for which accurate definition of features is essential. Care is required where responses coincide with 'step-like' changes in the topography, such as a scarp; they may be poorly described by the DTM data. In high-resolution surveys, even sand dunes may affect the data, owing to their topographic effect in gravity and their possible magnetic mineral content in magnetics. Topographic effects can be identified by observing correlations between gravity and magnetic images and topographic data. There is no practical way of removing topographic effects other than to include the topography in computer models and account for its effect in the interpretation of the survey data.

3.10.3.4 Survey and data processing effects

An irregular distribution of survey stations can produce distortions in the observed anomalies (see Fig. 2.18) and in the various transforms of the gridded data, particularly where the stations are sparse. This can be easily confirmed by overlaying the station locations on the data and its various transforms. In the case of airborne surveys, the survey line direction and spacing has an important effect on the dataset because, as shown in Fig. 3.24, inter-line variations can be difficult to remove completely (see Section 3.6.4). Features orientated parallel to the survey lines should be carefully considered to ensure they are not survey artefacts.

The interpreter should always be aware of the possible presence of processing artefacts in the transformed datasets and avoid erroneously interpreting them as genuine geological features. The possibility of artefacts being produced by the reduction-to-pole and pseudogravity transforms is a particular problem for magnetic data (see Section 3.7.2).

3.10.4 Estimating depth-to-source

The large contribution that the upper region of a source (Fig. 3.65) makes to the gravity and magnetic responses above the ground surface means that the data are sensitive to the depth to the top of the source. Depth-to-source is most accurately determined by modelling the anomaly (see Section 2.11). Often though, it is necessary to make an estimate of the depth of the source of an anomaly quickly and easily, or depth estimates of a large number of anomalies in a dataset may be required. Depth-to-source estimation is the simplest form of inverse modelling (see Section 2.11.2).

Manual depth-to-source techniques, now largely obsolete, involve identifying certain characteristics (zero cross-over points, gradients, distance between inflection points etc.) of the central or principal profile of the anomaly which, ideally, passes through its maximum and minimum peaks, and usually trends perpendicular to its gradients. Details for various source geometries are given by Am (1972), Atchuta Rao and Ram Babu (1984), and Blakely (1995) provides a compilation of the various methods; see also Salem *et al.* (2007). Modelling techniques that work with all the data points to find a 'best-fit' to the whole profile provide more reliable results.

3.10.4.1 Euler deconvolution

Euler (pronounced 'oiler') deconvolution (Reid *et al.*, 1990; Zhang *et al.*, 2000) is a commonly used semi-automated depth-to-source method useful for quickly analysing a large number of responses in a dataset. The method is based on anomaly gradients for selected source geometry and is sequentially applied to all the points along the anomaly profile.

Euler's equation represents the strength (f) of the potential field at a point (x, y, z) in space, due to a source located at (x_0, y_0, z_0), in terms of the first-order derivatives $(\partial f/\partial x$ etc.) of the field in the following form:

$$(x - x_0)\frac{\partial f}{\partial x} + (y - y_0)\frac{\partial f}{\partial y} + (z - z_0)\frac{\partial f}{\partial z} = N(B - f) \quad (3.28)$$

which includes a background (regional) component (B). Note that for magnetic data, information about the direction of the magnetism is not required, so remanent magnetism does not present a problem.

The *structural index* (N) accounts for the rate of decrease in the amplitude of the response with distance from the source (see Section 3.10.1.1). This affects the measured gradients and depends on the source geometry. For the case of a spherical source, N is equal to 2 for gravity data and 3 for magnetics. Indices have been derived for a variety of source types, and they fall in the range 0 to 3 (for 0 the equation has to be modified slightly).

The source position (x_0, y_0, z_0) and the background field are obtained by solving the Euler equation (Eq. (3.28)). If N is too low the depth estimate (z_0) will be too shallow, and if N is too high, the depth will be overestimated. The horizontal coordinates are much less affected. An effective strategy is to work with all values of N between 0 and 3, in increments of, say, 0.5. This will account for the geology not being properly represented by any one of the idealised model shapes, and also it has been shown that for more realistic models N varies with depth and location.

The derivatives/gradients are usually calculated but, as discussed in *Gradients and curvature* in Section 2.7.4.4, gradients are susceptible to noise, especially as their order increases, so the quality of the results will be affected accordingly. Note that in magnetic data the gradient in the across-line can be poorly constrained owing to spatial asymmetry in the sampling, and is potentially a major cause of error in Euler deconvolution.

Figure 3.67 shows the implementation for profile gravity (g) data. In this case the across-line (Y) component of the field is assumed to be symmetric about the profile and a 2D result is obtained (x_0, y_0). A window of predefined length (n) is progressively moved along the gravity profile and the profiles of its vertical and horizontal derivatives, and the background field and source coordinates are obtained for each measurement location. Where there are three unknowns (x_0, z_0 and B), the window must span a minimum of three points; in practice at least double this number are used, which allows the reliability of the result to be estimated. For simplicity, Fig. 3.67 shows a five-point window ($n = 5$, centred at x_0). An expression for the unknowns is set up for each location in the window, forming a set of n simultaneous equations of the form:

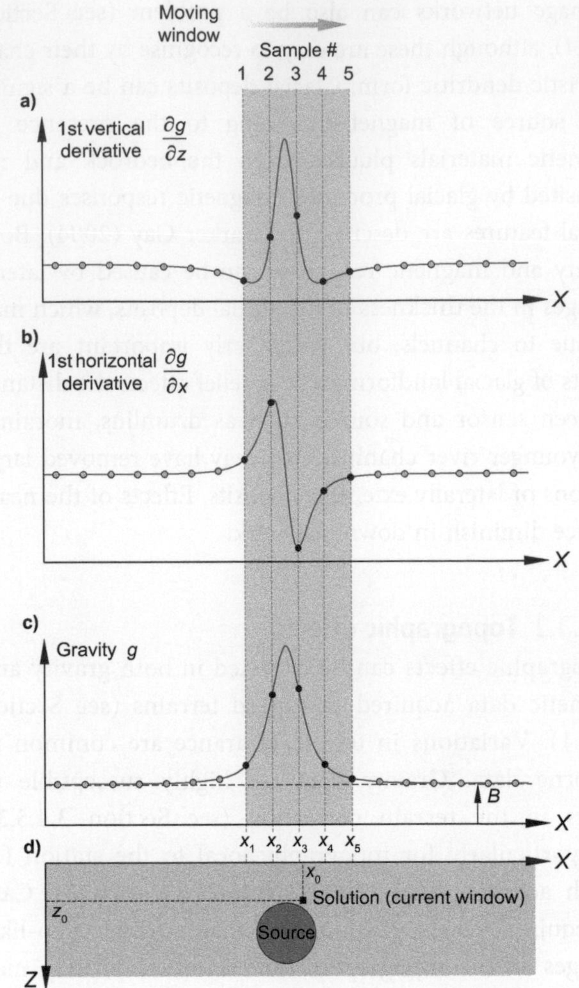

Figure 3.67 Schematic illustration of 2D Euler deconvolution using a window of five data points applied to gravity data. See text for details.

$$(x_i - x_0)\frac{\partial g}{\partial x_i} + (z_i - z_0)\frac{\partial g}{\partial z_i} = N(B - g_i) \quad (3.29)$$

where i identifies the data points in the window $i = 1$ to n. The equations are solved for the three unknowns.

The window is then moved to the next point along the profile and the process repeated. Each window position provides an estimate of source location. For 3D implementation, the window is moved across the gridded data and a 3D (x_0, y_0, z_0) solution for the source position is obtained. Window size in either case represents a trade-off between resolution and reliability; increasing the size reduces the former and increases the latter. The window size must be large enough to include significant variations in the field being analysed, and is usually set to the wavelength of the anomalies of interest. It can be used as a crude means of focusing the process on sources at different depths. However, the larger the window the more likely that more than one source will influence the data within it, which will create spurious results. The quality of the data, i.e. sampling interval and noise, affects the result.

Euler deconvolution produces many solutions for an anomaly, most of which are spurious. Various techniques are used to identify the best solutions. These include assessing how well the solutions fit the data in each window; assessing the clustering of solutions from different windows; and comparing results for various values of N and only providing the solutions for the best-fit value. Various improvements have been proposed such as solving for several sources simultaneously (to address anomaly overlap), extracting additional information about source characteristics, and more sophisticated means of choosing the best solution from the large number produced. See, for example, Mushayandebvu *et al.* (2001).

Euler solutions are usefully analysed in terms of depths and horizontal positions. Three-dimensional results are presented in map form with different source depths and geometries (different N) distinguished using variations in colour and/or symbol type (Fig. 3.71). The edges of source bodies can be mapped in this way. Two-dimensional data are presented as points on a cross-section, with different N represented by different symbols.

Euler deconvolution can be applied to gradient data, in which case second-order derivatives are required. The advantage of using gradients is that they are more localised to the source and have greater immunity to neighbouring sources, resulting in better spatial resolution in the Euler solutions. A disadvantage is that gradients have poorer signal-to-noise ratios than normal field data, especially higher-order derivatives, so the quality of the results may be significantly degraded.

3.10.5 Modelling source geometry

The principles of modelling and the types of models available for the quantitative analysis of geophysical data are described in Section 2.11. Here we discuss some aspects specific to modelling potential field data. A practical description of how to go about modelling gravity and magnetic data is provided by Leaman (1994). A demonstration of modelling the magnetic anomaly associated with the Wallaby Au deposit is given in Section 3.11.2. Note that when working with magnetic data it is essential that the strength and orientation of the inducing field in the survey area be known as they are fundamental in determining the response (see Section 3.2.4).

Modelling a targeted anomaly usually requires that it be isolated from the background response, achieved by removing the background as a regional field (see Section 2.9.2). The broader gravity responses almost always contain superimposed responses from neighbouring sources, and separating the target anomaly from these can be difficult. This is less of a problem with magnetic data. Modelling the background variation with the target anomaly can often provide a satisfactory result (see Section 2.11.3.1).

When modelling Bouguer gravity anomalies, it can be confusing to reconcile subsurface depths and depths of model sources which occur below the surface, but above the level of the reduction datum. In this case it may be easier to include the topography in the model and work with free-air gravity anomalies. The Bouguer slab and terrain effects (see Section 3.4.5) are accounted for by including the topography in the model, and depths of features in the model then correspond to depths below the topographic surface. Also, the model can then directly account for possible lateral and vertical changes in density above the datum level, and so avoids the problem of estimating an average single density for the Bouguer correction. The modelling of magnetic data should also include topography.

3.10.6 Modelling pitfalls

Modelling of any geophysical data is only as good as the assumptions made when simplifying the complex real-world variations in physical properties into a model that is defined by a manageable number of parameters. We reiterate the points made in Section 2.11.3 on the importance of accounting for noise levels, choosing an appropriate type of model and creating the simplest possible model that matches the observations so as not to imply information that is not supported by the data. In particular, when modelling potential field data, account must also be taken of ambiguity: the unavoidable fact that without constraints provided by other sources of data, the effects of equivalence and non-uniqueness mean that an infinite number of density or magnetism distributions in the subsurface can reproduce the observed anomalies.

3.10.6.1 Equivalence

Recall that only relative changes in gravity and magnetic strength are of significance in mineral exploration

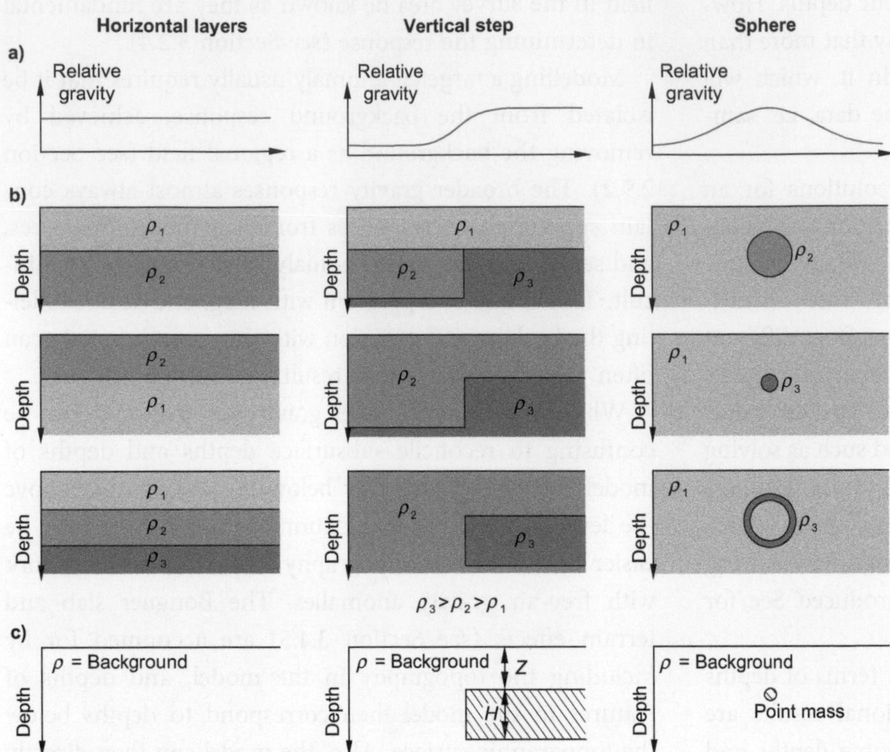

Figure 3.68 Equivalent gravity models for three common geological sections: (a) gravity anomaly, (b) some possible density cross-sections that produce the gravity anomaly, and (c) equivalent model.

geophysics. Referring to Fig. 3.68, laterally continuous horizontal layers do not cause relative changes in gravity on a surface above them, so the structure of a horizontally layered sequence cannot be determined from measurements made from above. The response is the same as a homogeneous half-space. This is a simpler geometrical form and is, therefore, the equivalent gravity model.

Only lateral changes in density will cause variation in the gravity response. Careful examination of the 'vertical step' model in Fig. 3.68 shows that in terms of lateral density change the three density cross-sections are equivalent. In all cases there is a lateral density contrast ($\rho_3 - \rho_2$) which has a vertical thickness H and depth to top Z.

For the case of spherical density distributions, where the mass is distributed radially and evenly around the centre of the sphere, the gravity effect is the same as if its mass was concentrated at a point at its centre. The same gravity effect is obtained when the mass is distributed homogeneously as a sphere of larger volume and smaller density contrast, or as a spherical shell, or as a multiple spherical density zonation. In all cases the total excess mass (see Section 3.2.1.3) and depth to centre are the same, but the densities and radii vary accordingly. The solid homogeneous sphere, being the simplest form, is the equivalent model for all spherical distributions.

In each of these three examples the gravity response is identical even though the subsurface density models are different. It goes without saying that the gravity response cannot be used to tell them apart.

3.10.6.2 Non-uniqueness

In Section 2.11.4 we described the phenomenon of non-uniqueness in geophysical modelling, whereby many different physical property distributions can produce the same geophysical response, and strategies for dealing with it.

Physical property/geometry ambiguity, as it relates to gravity data, is illustrated in Fig. 2.49a. Here, a broad low-density feature at shallow depth produces the same gravity response as a range of more compact bodies, of increasing density, at progressively greater depths. They all have the same excess mass (see Section 3.2.1.3). The shallowest, widest and least dense body, and the deepest, smallest and most dense compact body, represent end-members of an infinite number of solutions to the measured response. As shown in Section 3.10.1, amplitude increases when the body is shallower or its density contrast increased, whilst wavelength increases as the body is made deeper or wider. So long as the excess mass of the solutions is the same, the same gravity response will be obtained.

For magnetic modelling, there is the additional complication of the direction of the body's magnetism (see Sections 3.2.4 and 3.10.1.2), which can be traded off against changes in its geometry, usually its dip. Different combinations of source dip and magnetism direction can produce identical magnetic responses; this type of ambiguity is illustrated in Fig. 2.49c. Failure to recognise the presence of remanent magnetism which is not parallel to the present-day Earth's field will cause the model to be in error, since the assumption of only induced magnetism implies the source's magnetism to be parallel to the Earth's field. Even when there is no significant remanent magnetism, problems may occur owing to anisotropy of magnetic susceptibility (see Section 3.2.3.7) and, for the case of highly magnetic sources, self-demagnetisation (see Section 3.2.3.6). Both deflect the induced field away from parallelism with the Earth's field.

The strategies described in Section 2.11.4.1, of using knowledge of the nature of non-uniqueness to guide the interpretation, can be applied in the analysis of gravity and magnetic data. Of course, some prior knowledge of the geology, and the target, is required in order to identify realistic source geometries. Ideally there are also petrophysical data available, but physical properties and especially magnetic properties can vary by large amounts over small distances (Fig. 3.61), so identifying the correct susceptibility etc. to use can be equivocal. For these reasons, the interpreter should create several models designed to represent 'end-member' models of the range of possibilities: for example the deepest possible source, the shallowest possible source etc. In practice, time constraints often mean only one 'preferred' interpretation is produced.

Figure 3.69 shows three gravity models across the Bonnet Plume Basin, Yukon Territory, Canada. All are geologically plausible and their responses (not shown) fit the data to an acceptable degree. The particular problem was to determine the extent and thickness of the coal-bearing upper and lower members of the Bonnet Plume Formation. The models represent the maximum and minimum possible formation thickness, and also what is considered to be the mostly likely form of the underlying geology.

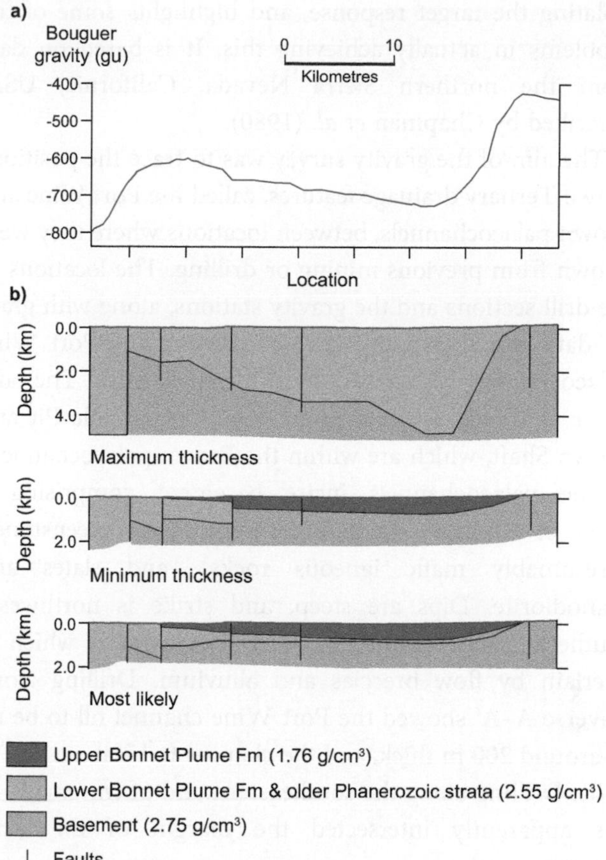

Figure 3.69 Ambiguity in modelling of gravity data from the Bonnet Plume Basin. (a) Observed gravity profile, and (b) cross-sections comprising end-member density models: maximum and minimum thicknesses, and what is considered to be the most likely model. Redrawn, with permission, from Sobczak and Long (1980).

3.11 Examples of gravity and magnetic data from mineralised terrains

We present here a selection of examples to show the types of gravity and magnetic responses produced by various types of mineral deposits, as would be sought during exploration targeting, and to demonstrate various aspects of their interpretation. Included are aeromagnetic data from an Archaean granitoid–greenstone terrain and a low-grade metamorphosed Palaeozoic orogenic belt to show the responses typical of these environments, and to demonstrate the integrated use of potential field data for regional mapping, i.e. lithotype identification, structural mapping and target identification.

3.11.1 Regional removal and gravity mapping of palaeochannels hosting placer gold

This case study demonstrates the use of gravity to map prospective stratigraphy, which in this case is low-density palaeochannel fill containing placer gold deposits. It illustrates the importance of regional gradient removal for isolating the target response, and highlights some of the problems in actually achieving this. It is based on data from the northern Sierra Nevada, California, USA, described by Chapman *et al.* (1980).

The aim of the gravity survey was to trace the positions of two Tertiary drainage features, called the Port Wine and Brown palaeochannels, between locations where they were known from previous mining or drilling. The locations of the drill sections and the gravity stations, along with gravity data, are shown in Fig. 3.70. Part of the Port Wine palaeochannel was worked from the Iowa Mine. The only other workings in the area are the Hardscrabble Pit and Brown Shaft, which are within the Brown palaeochannel.

The palaeochannels incise basement comprising a range of lithotypes, including amphibolites, 'greenstones' (presumably mafic igneous rocks), and slates and granodiorite. Dips are steep, and strike is northwest–southeast. The channel fill is Tertiary gravel which is overlain by flow breccias and alluvium. Drilling along traverse A–A′ showed the Port Wine channel fill to be up to around 200 m thick, and the channel to be about 500 m wide. Drilling along the southeastern end of traverse B–B′ has apparently intersected the margin of the same palaeochannel.

Gravity measurements were made along a series of traverses oriented roughly perpendicular to the palaeodrainage, but access restrictions produced an uneven station distribution. The resultant station locations are shown in Fig. 3.70. As noted previously, for surveys with irregular station distribution it is good practice to display the stations so that features can be assessed in terms of artefacts produced by the uneven data sampling and poorly constrained interpolation by the gridding algorithm. The data were reduced to Bouguer anomaly using a density of 1.85 g/cm^3, and terrain corrections were applied to compensate for the steep terrain.

The Bouguer anomaly (BA) is dominated by a northeasterly negative gradient (Fig. 3.70a), culminating in a prominent gravity 'low' anomaly in the eastern part of the survey area. Chapman *et al.* (1980) suggest that this is due to a shear zone within greenstones. This is important to the interpretation of the data and will be referred to as the 'shear-zone anomaly'. There is evidence of a decrease in gravity related to the Port Wine channel along traverse B–B′, but elsewhere any response from the channel fill is disguised by the regional gradient.

The first vertical derivative of the gravity emphasises shallow features. It is reasonably effective in defining the course of the Port Wine channel as a northeasterly trending gravity 'low' extending from the drillhole intersections on traverse A–A′ through the Iowa Mine to traverse B–B′. Note that the map of gravity stations shows that the northwestern bank of the channel is not defined as the survey did not extend far enough to the northwest. This allows the gridding algorithm to extend the low channel-related values to the northwest, and possibly erroneously. A subsidiary northwest-trending channel appears to intersect the main channel near the Iowa Mine. Continuation of the Port Wine channel to the northeast, and its relationship with the Brown channel, are not clearly defined owing to the effects of the shear-zone anomaly. The Brown palaeochannel appears to become wider to the south, with the Hardscrabble Pit and Brown Shaft lying on its western edge.

The shear-zone anomaly presents a significant problem for regional removal. Its wavelength is similar to the response of the palaeochannels. Moreover, it is geologically quite likely that a basement shear zone would be preferentially eroded and, consequently, influence the drainage system; so they are likely to be spatially coincident. It is, therefore, unlikely that the shear zone and palaeochannel anomalies can be completely separated based on their wavelengths. It is uncertain whether the shear-zone anomaly should be treated entirely as part of the regional field, or whether all or part should be considered residual in nature.

Three methods of regional removal have been applied to the BA data. In all cases the residual is obtained by subtracting the regional field from the Bouguer anomaly data. Figures 3.70c and d show the data after low-pass wavelength filtering with the cut-off wavelength (500 m) chosen to include as much of the shear-zone anomaly as possible in the regional variation, so that it is attenuated in the residual data. The residual data confirm the course of the Port Wine channel suggested by the first vertical derivative data, and the tributary near the Iowa Mine is also confirmed. In addition, evidence is revealed of two palaeochannels east of the area. The Hardscrabble Pit and Brown Shaft are located in the middle of the western one of these, which trends north–south. The other gravity 'low' trends parallel to the shear-zone anomaly and may be due to this and/or palaeodrainage.

Figures 3.70e and f show the results for a regional field obtained by upward continuation to a height of 400 m so as to remove as much of the shear-zone anomaly as possible (compare Figs. 3.70e and c). In the residual response the Brown palaeochannel appears to be a single drainage

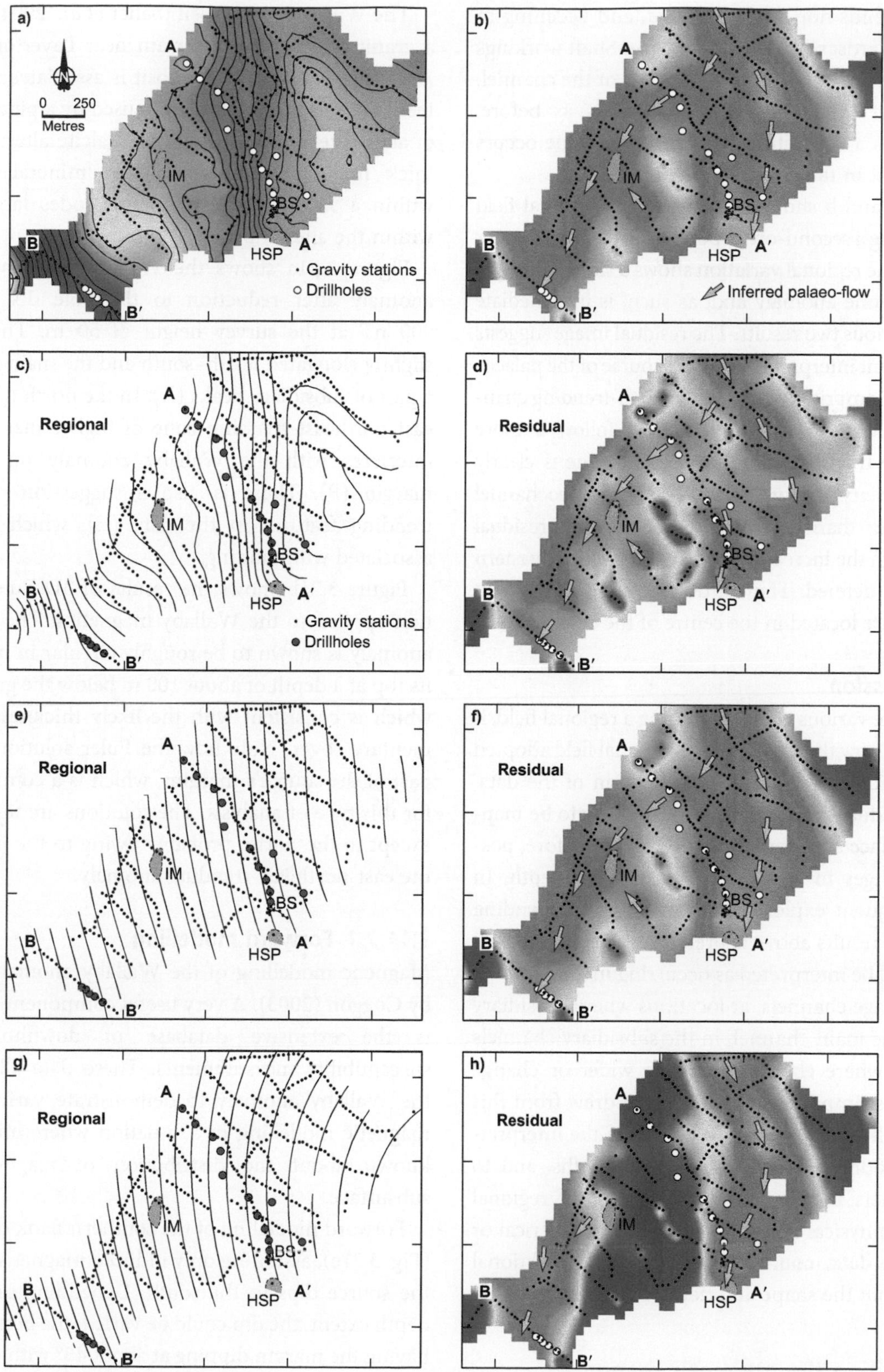

Figure 3.70 Regional removal in gravity data from the Port Wine area. (a) Contours and image of the Bouguer anomaly (BA). Regional fields obtained from the BA data shown as contours and the residuals as images as follows: (b) first vertical derivative; (c) regional obtained through low-pass filtering; (d) residual response from (c); (e) regional obtained from upward continuation; (f) residual response from (e); (g) regional obtained by fitting second-order polynomial; and (h) residual response from (g). Contour interval in all figures is 2 gu. IM – Iowa Mine, BS – Brown Shaft, HSP – Hardscrabble Pit. Based on data in Chapman *et al.* (1980).

feature which trends north of northwest, and widening to the south. The Hardscrabble Pit and Brown Shaft workings lie on the western margin of the wider part of the channel. Interpretation of the Port Wine channel is as before, although there is a suggestion that the Iowa Mine occurs at a distinct bend in the channel.

Figures 3.70g and h show the results for a regional field obtained by fitting a second-order polynomial to the Bouguer anomaly data. The regional variation shows a slight influence from the shear-zone anomaly and, as such, is intermediate between the previous two results. The residual image suggests a radically different interpretation of the course of the palaeo-drainage, which comprises two roughly north-trending channels. The Port Wine channel appears to follow a more northerly course than before and the Iowa Mine is clearly within a subsidiary feature. The Brown palaeochannel appears narrower than depicted in the previous residual datasets, although the increase in gravity on the southeastern bank is not well defined. The Hardscrabble Pit and Brown Shaft workings are located in the centre of the channel.

3.11.1.1 Discussion

In illustrating the various ways of defining a regional field, it is demonstrated how the nature of the regional field adopted has major implications for the interpretation of the data. The various residual gravity data are also likely to be mapping the subsurface at different depths and, therefore, possibly show changes in channel geometry with depth. In terms of subsequent exploration activity, and depending on which of the results above is preferred, the known mineralisation could be interpreted as occurring in the centre of the major drainage channels, at locations where subsidiary channels join the main channel, in the subsidiary channels themselves, or where channels become wider or change course. The most important conclusion to draw from this is that more drilling is required to constrain the interpretation of the geophysical data for various depths, and to eliminate artefacts introduced by the choice of regional field. Other geophysical methods, for example electrical or electromagnetic data, could probably provide additional information about the shape and depth of the channels.

3.11.2 Modelling the magnetic response associated with the Wallaby gold deposit

This case study demonstrates forward and inverse modelling of a magnetic anomaly and shows the inherent ambiguity in modelling potential field data.

The Wallaby Au deposit (Salier *et al.*, 2004) is located in a granitoid–greenstone terrain near Laverton in Western Australia. The Wallaby deposit is associated with a prominent aeromagnetic anomaly caused by a pipe-shaped zone of actinolite–magnetite–epidote–calcite alteration within a thick mafic conglomerate. Gold mineralisation occurs within a series of sub-horizontal lodes largely confined within the altered pipe.

Figure 3.71a shows the Wallaby aeromagnetic (TMI) anomaly after reduction to the pole. Its amplitude is 900 nT at the survey height of 50 m. The anomaly is slightly elongated north–south and the main peak contains a pair of subsidiary peaks (A). In the north there is a broad east-northeast-trending zone of higher magnetism which interferes with the Wallaby anomaly on its northern margin (B). There is also a suggestion of northwest-trending features in the data (C), which are probably associated with faulting.

Figure 3.71b shows the results of 3D Euler deconvolution applied to the Wallaby magnetics. The source of the anomaly is shown to be roughly circular in map view with its top at a depth of about 100 m below the ground surface, which is consistent with the likely thickness of the sedimentary cover. Note how the Euler solutions only define part of the source's margins, which is a common outcome for this type of analysis. The solutions are tightly clustered except to the north, probably owing to the interference of the east-northeast-trending anomaly.

3.11.2.1 Forward modelling

Magnetic modelling of the Wallaby anomaly is described by Coggon (2003). A very useful component of this project is the extensive database of downhole magnetic-susceptibility measurements. These data allow us to use the Wallaby anomaly to demonstrate various aspects of magnetic modelling in a situation where much is already known about the distribution of magnetism in the subsurface.

Forward modelling of the southern flank of the anomaly (Fig. 3.71c), assuming only induced magnetism, shows that the source dips to the south. Depending on the source's depth extent, the dip could be varied, the best-fitting model having the margin dipping at about 45° with the base of the source at a depth of 700 m. An alternative model, producing a slightly poorer fit to the observations, comprises a source with lower susceptibility, steeper dip and much greater depth extent. Modelling the northern flank of the anomaly is hampered by interference from other anomalies

(B in Fig. 3.71c). It was not considered justifiable to model this part of the profile using the model representing the Wallaby alteration pipe, but a simple and geologically plausible interpretation is that the northern margin is parallel to the southern margin.

The top of the prism model, shown in Fig. 3.71b, agrees quite well with the Euler solutions except on the northern margin, which is to be expected since this part of the data was not modelled. The modelling was undertaken without reference to the known susceptibility distribution, but these data from beneath the modelled profile are in excellent agreement with the depth to the top of the source and its dip. The absolute and relative susceptibilities are also correctly represented, notably the lower-susceptibility core of the alteration zone. The depth to the base of the anomaly source was poorly predicted; but the geometry of the deeper parts of an anomaly source is usually not well constrained because the observed response is dominated by the response of the shallower parts of the source (see Section 3.10.1.1).

3.11.2.2 Inverse modelling

Inverse modelling was also applied to the Wallaby magnetic anomaly using a 3D voxel-based algorithm (see Section 2.11.1.2) to explore the range of possible source geometries. Firstly, a half-space model was used and the inversion applied unconstrained. It was then reapplied but constrained by the results of the 2.5D modelling, with the inversion adjusting the body parameters only as necessary to fit the data. The computed responses are shown as contours in Fig. 3.71d; note the excellent match between these and the actual data. The subsurface susceptibility distributions produced by the modelling are shown in Figs. 3.71e and f for the principal profile and two depth slices, along with equivalent displays of the downhole (observed) susceptibility data.

The inversion results from the shallower parts of the subsurface, i.e. the 200 m depth slice and the upper part of the cross-section, agree quite well and are a good representation of the observed susceptibility distribution. The spatial extent of the anomalous susceptibility is correctly defined, including the lower-susceptibility core, although the absolute value of susceptibility was less well predicted by the unconstrained inversion. At greater depths, the correspondence between the modelled and observed distributions deteriorates, especially for the unconstrained inversion which has incorrectly predicted a steep northerly dip for the source. The constrained inversion has performed better, but this is partly due to being constrained by a forward modelling result which accurately predicted the susceptibility distribution of the subsurface.

3.11.2.3 Discussion

The modelling of the Wallaby anomaly demonstrates that it is good practice to begin by forward modelling the data, incorporating all available information in order to understand the range of possible source geometries, and then use 3D inversion to refine the model. Proceeding directly with inverse modelling is a very high-risk strategy, especially when it is unconstrained. In common with all magnetic modelling, the results are more likely to be correct in terms of source depth and extent for shallow regions of the subsurface, but information about deeper regions is likely to be less accurate. This also applies to source dip, which is a critical parameter in designing a drilling programme to test the source of an anomaly (see Section 2.11.4). This example also shows that although the Wallaby anomaly is a relatively isolated feature in the aeromagnetic data, even here modelling was hindered by interference from other anomalies.

3.11.3 Magnetic responses from an Archaean granitoid–greenstone terrain: Kirkland Lake area.

Archaean granitoid–greenstone terrains contain a wide variety of lithotypes ranging from very weakly magnetised sedimentary and felsic igneous rocks through to ultra-magnetic iron formations. Moreover, metamorphic grade is generally quite low and, although brittle and ductile deformation is ubiquitous, the overall structure is normally not too complicated at the scale of hundreds of metres to kilometres. Aeromagnetics is a useful aid to geological mapping in these environments at both the regional and prospect scale.

The most magnetic lithotypes found in greenstone belts, in addition to iron formations, are ultramafic rocks such as komatiites and some types of granitoids. Metamorphic grade is normally greenschist to amphibolite facies, so mafic rocks tend to be weakly magnetic (see Fig. 3.48). Other usually weakly magnetised lithotypes include felsic and intermediate igneous rocks, and clastic sedimentary rocks. Structures such as faults and shear zones are important targets because of their association with mineral deposits. They are often recognised as offsets and truncations to stratigraphic anomalies, and may appear as zones

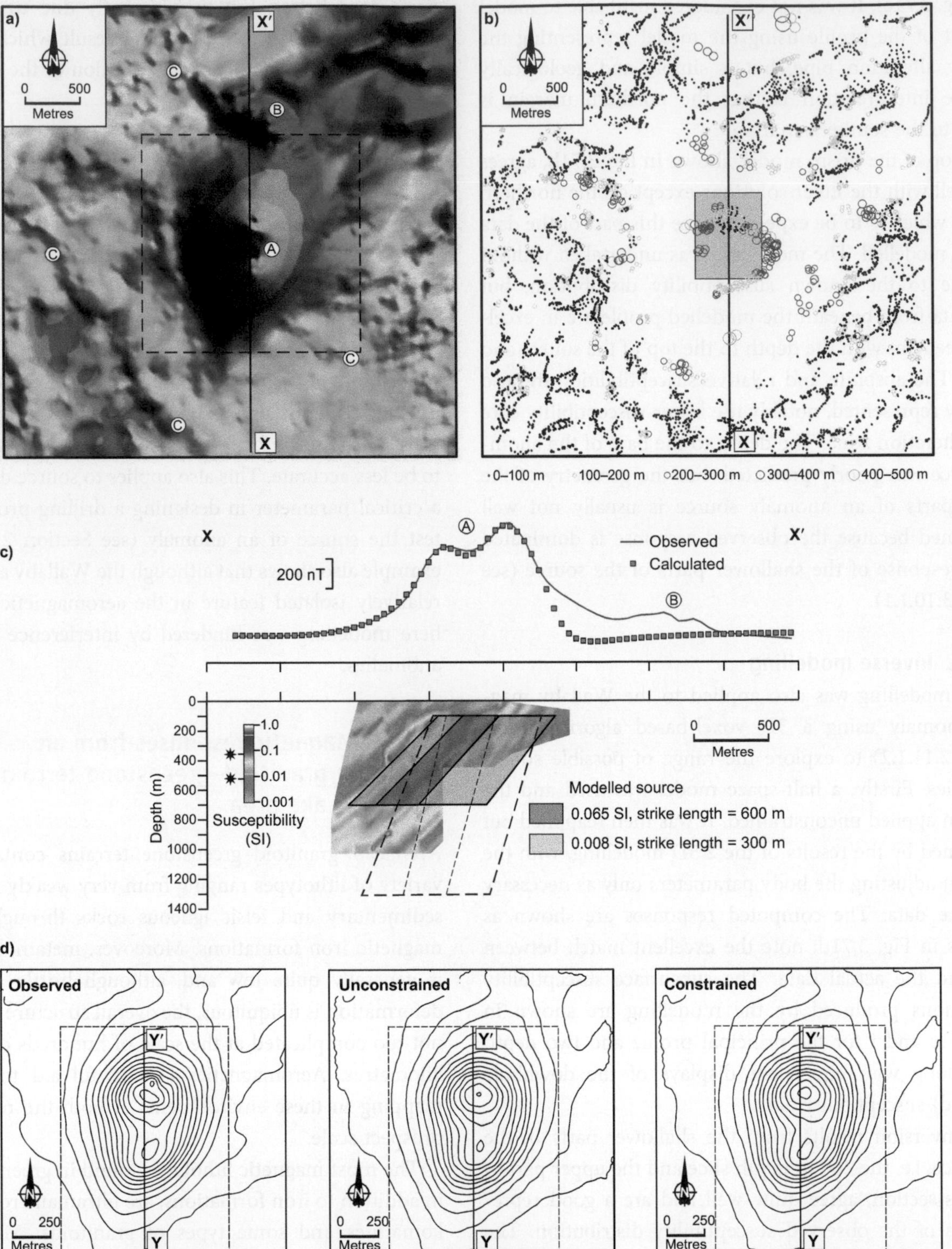

Figure 3.71 Aeromagnetic data from the Wallaby Au deposit, Western Australia, and the results of forward and inverse modelling. (a) Reduced-to-pole TMI data. The letters refer to features discussed in the text. X–X′ is the profile used for 2.5D modelling, and the dashed-line rectangle defines the region used for 3D modelling. (b) Results from 3D Euler deconvolution applied to the data in (a) using an 11-point window and structural index (N) = 1.5. The grey rectangle is the surface projection of the 2.5D model shown in (c). (c) 2.5D dipping prism model of the source of the magnetic anomaly drawn on the known subsurface susceptibility distribution. The dashed outline corresponds to an alternative model compatible with the data. The starred values on the susceptibility scale correspond to the susceptibilities of the two models. (d) Contours of the observed TMI and the TMI anomalies calculated from subsurface susceptibility distributions created by 3D inverse modelling. Contour interval is 50 nT in all cases. Profile Y–Y′ (a subprofile of profile X–X′) and the dashed outline show the locations of

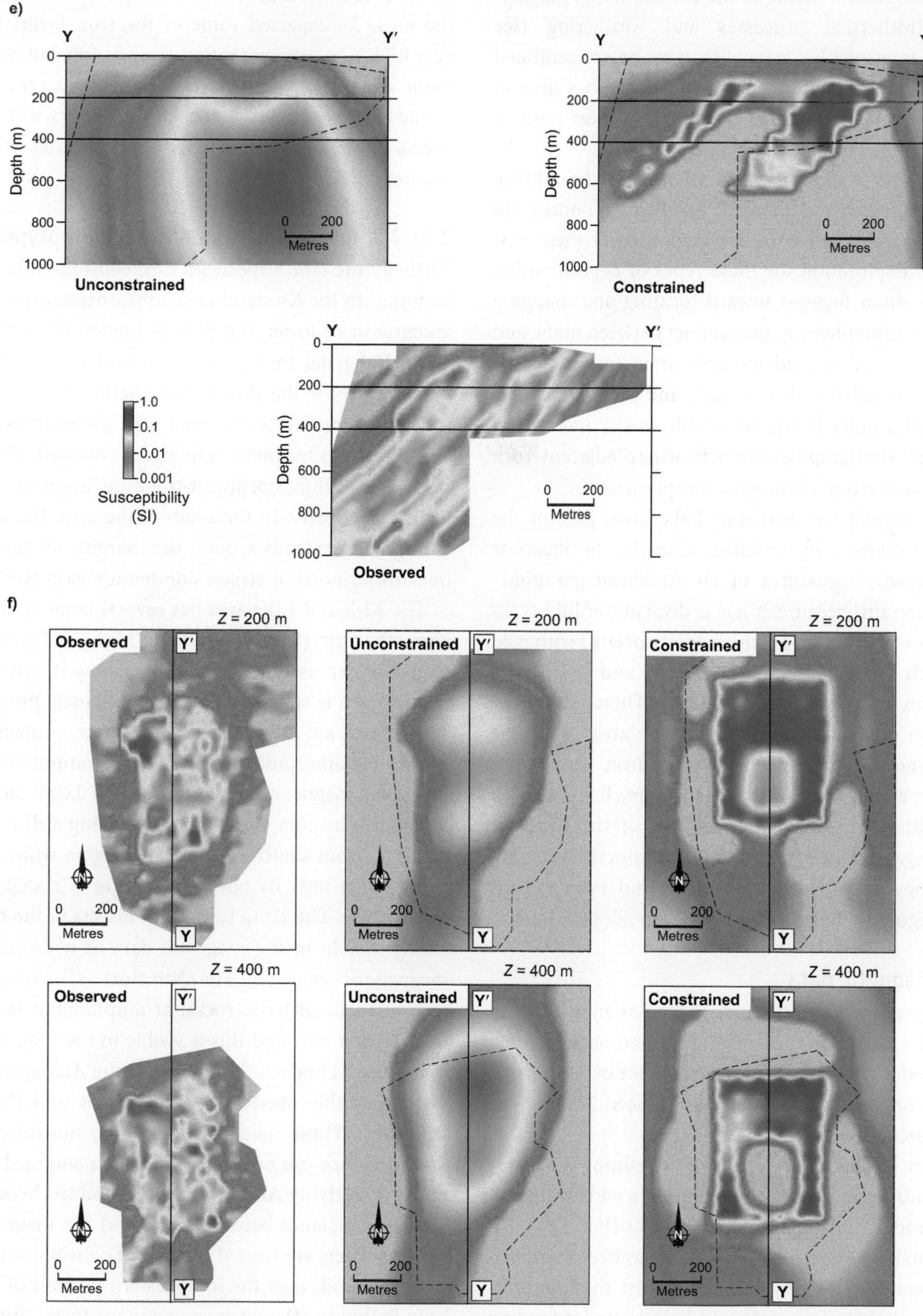

Figure 3.71 (*cont.*) the cross-sections and depth slices, respectively, shown in (e) and (f). (e) Observed and modelled susceptibility distributions along cross-sections below Y–Y′. (f) Observed and modelled susceptibility distributions along depth slices 200 and 400 m below the surface. The dashed lines in (e) and (f) represent the limits of reliable subsurface susceptibility data. Magnetic and susceptibility data reproduced with the permission of Barrick (Australia Pacific) Ltd. Susceptibility model created and supplied by Mines Geophysical Services. Euler deconvolution by Cowan Geodata Services.

of reduced magnetism owing to the destruction of magnetite by hydrothermal processes and weathering (see Fig. 3.55). Occasionally, magnetite may be concentrated more in structures such as faults and shear zones than in the surrounding rocks, so that they exhibit linear positive responses.

The magnetic responses from nickel and base-metal sulphides can rarely be defined, as their responses are usually weak compared with the responses of their host stratigraphy. Exploration for these types of deposits using magnetics is often focused toward locating and mapping favourable stratigraphy, e.g. the contact between mafic and ultramafic units. Also, gold mineralisation may be stratigraphically controlled with iron-rich, and therefore potentially magnetic, units being favourable hosts; and it may also occur at stratigraphic contacts where adjacent rock types have contrasting rheological properties.

We have chosen the Kirkland Lake area, part of the Abitibi Subprovince in Ontario, Canada, to illustrate the aeromagnetic signatures of an Archaean granitoid–greenstone terrain, because it has a diversity of lithotypes and several major structures which are associated with gold deposits such as the giant Kirkland Lake and world-class Kerr Addison and Chesterfield deposits. There is also significant iron-ore mineralisation in the area, with the Adams Mine producing magnetite iron ore from Algoman-type iron formation. In addition, there exists a significant literature about the geology of the Kirkland Lake area, e.g. Ispolatov *et al.* (2008). A concise summary of the geology is provided by Jackson and Fyon (1991), whose stratigraphic terminology has been adopted here.

3.11.3.1 Magnetic data

The geomagnetic field in the survey area has an inclination of +74° and a declination of −12°. The aeromagnetic data were acquired at a nominal terrain clearance of 70 m along survey lines oriented north–south and spaced 200 m apart (Ontario Geological Survey, 2003).

The survey has been very effective in defining the major geological entities in the region, mapping a wide variety of lithotypes, and delineating major structures (Fig. 3.74). An extensive database of petrophysical data has been compiled (Ontario Geological Survey, 2001) to assist the interpretation of the aeromagnetic data and help in understanding of the geology. Frequency histograms of magnetic susceptibility and strength of remanent magnetism for selected lithotypes are shown in Fig. 3.72 and the strength of remanent and induced magnetism compared in Fig. 3.73. These data show that remanent magnetism is significant in the area. As expected some of the iron formations have very high Königsberger ratios, close to 100, but many felsic, mafic and ultramafic rocks also have ratios greater than 1, in most cases between 1 and 10. The rocks with values in excess of 10 may be affected by lightning strikes (see Section 3.9.8.4).

3.11.3.2 Responses from different lithotypes

Virtually the entire spectrum of greenstone lithotypes can be found in the Kirkland Lake area; metamorphic grade is greenschist or lower, but may be higher adjacent to intrusions. Referring to Figures 3.74a and b, the largest geological entity is the Round Lake batholith (A), a gneissic tonalite–granodiorite intrusion with dimensions of tens of kilometres. Its magnetic response is subdued, allowing the responses of thin, northwest-trending lamproite dykes (B) to be seen clearly. In the south of the area, the greenstone stratigraphy extends around the margins of the batholith, but further north it strikes dominantly east–west.

The Kirkland Lake area has several large syenitic intrusions: the Otto (C), Lebel (D) and McElroy (E) stocks. The magnetic expression of these intrusions is generally subdued, which is consistent with the magnetic property data. The stocks are surrounded by narrow contact aureoles where metamorphic grade reaches amphibolite facies. Based on magnetic fabric studies, the Lebel stock (D) is thought to be comparatively thin, having a disc-like form, resulting from southward flow of magma which ascended along faults near its northern margin (Cruden and Launeau, 1994). The Long Lake Fault bisects it, the fault being clearly visible in the magnetic data as a decrease in the magnetic anomaly (F). The Otto stock (C) contains a large roof pendant of basic rocks, at amphibolite facies, in its west-central area and this is visible in the magnetics (G).

In the east and south of the area, the Archaean rocks are unconformably overlain by Proterozoic and Phanerozoic strata (H). These cover units are fairly non-magnetic, but their presence can be inferred from the subdued responses of the underlying Archaean rocks (subdued because of the increased distance between them and the magnetometer). Similar effects are caused by lakes. The western part of the survey extends over the southeastern portion of the Watabeag Batholith (I), which is weakly magnetic. Intermediate to felsic igneous rocks of the Watabeag Assemblage occur within the batholith (J) but, also being poorly magnetic, are not well resolved in the magnetic data. More obvious are the north–south trending dolerite dykes of the

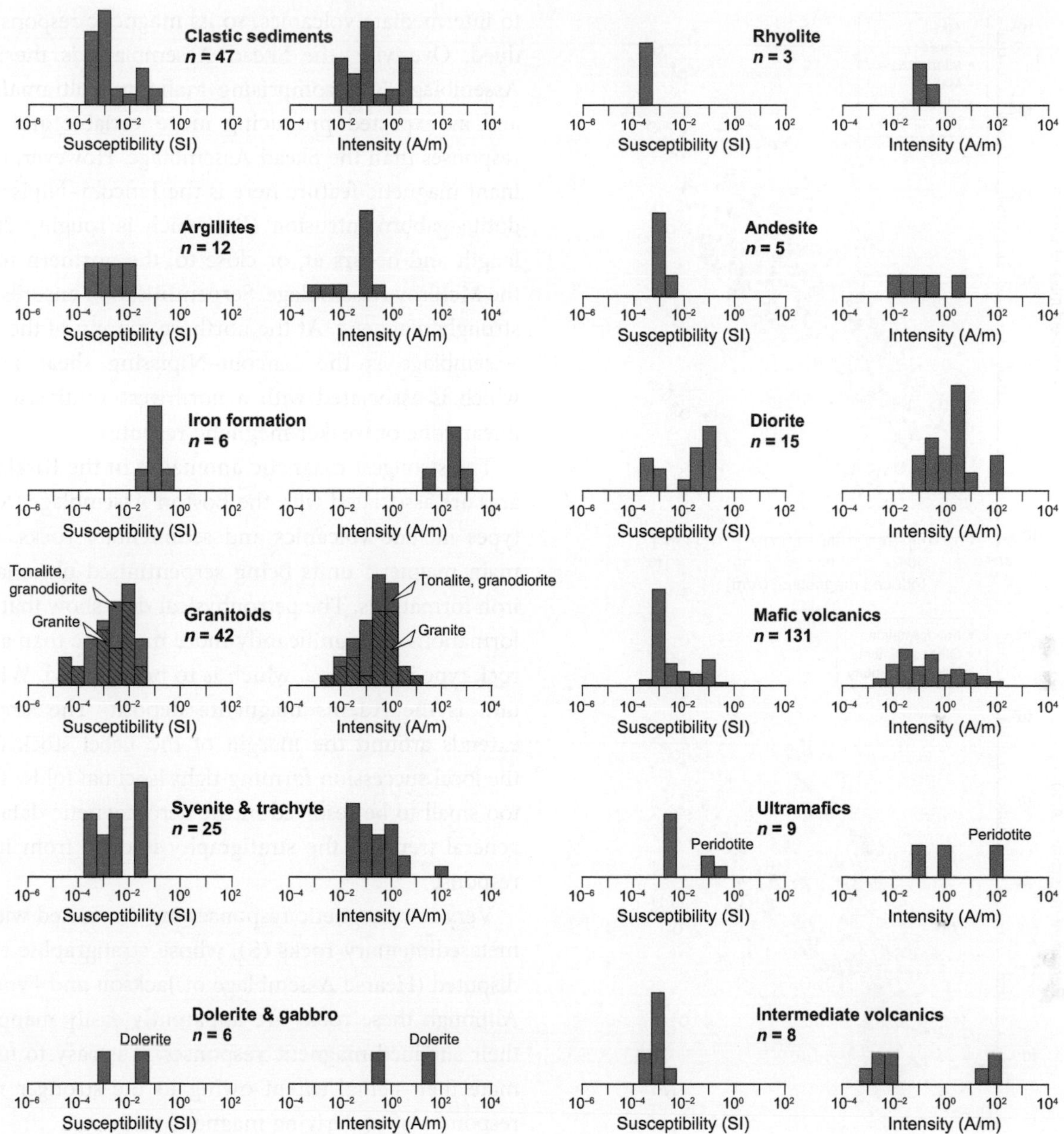

Figure 3.72 Frequency histograms of magnetic susceptibility and intensity (strength) of remanent magnetism for various lithotypes in the Kirkland Lake area. Based on data in Ontario Geological Survey (2001).

Matachewan swarm (K). Note that the dyke anomalies probably appear wider in the image than their true width owing to their strike being parallel to the survey lines. Interpolation between survey lines to create the gridded data (see Section 2.7.2) tends to 'smear' the anomalies making their widths appear bigger. Another intrusion (L), occurring at the southern end of the Watabeag Batholith (I), exhibits the rugose texture often associated with granitoids, but it is significantly more magnetic than the large granitoids in the survey area. The variable magnetic responses of the local granitoids are reflected in the multimodal distribution of their magnetic properties.

Good examples of stratigraphically controlled linear magnetic responses occur in the greenstone succession to the east of the Round Lake batholith (A). Immediately east of the batholith is the Catharine–Pacaud assemblage (M), which includes komatiites, intermediate and basaltic intrusive and extrusives, and close to the batholith a sulphide-facies iron formation. There is no significant contact aureole associated with the intrusion, so primary lithologies are

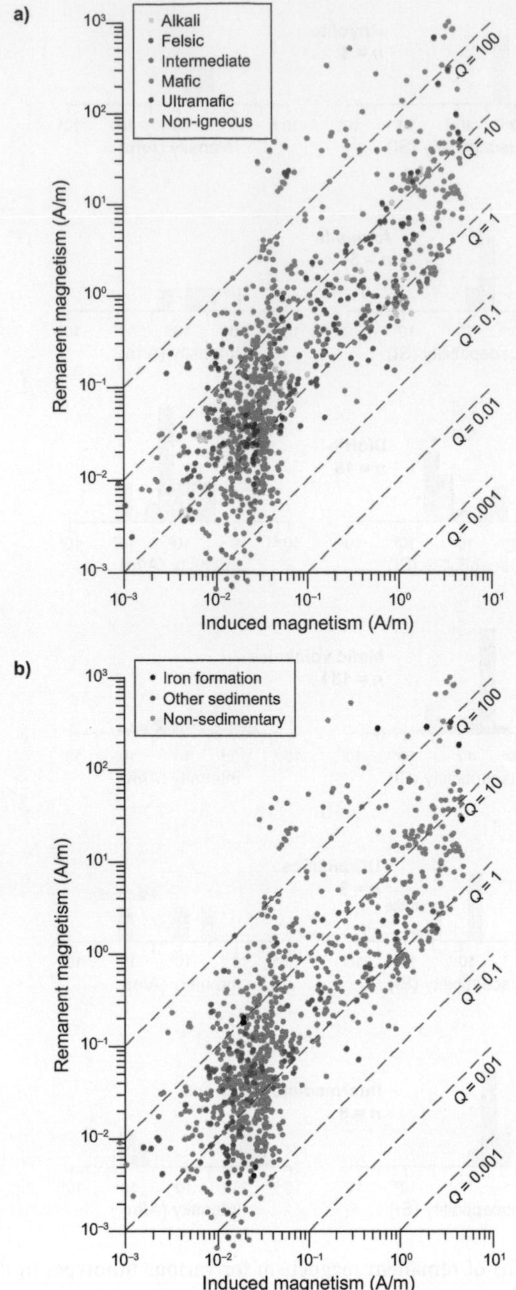

Figure 3.73 Plots comparing the strength of the induced and remanent magnetisms of various lithotypes in the Kirkland Lake area. Based on data in Ontario Geological Survey (2001).

the principal control on the magnetic responses, with the linear anomaly pattern reflecting the interbedded lithotypes, the more mafic rocks being associated with the stronger magnetic responses. The contrasts in the magnetic properties of these lithotypes are confirmed by the petrophysical data. The contact with the overlying Skead Assemblage (N) is a sheared primary contact. The Skead Assemblage is composed predominantly of calc-alkali felsic to intermediate volcanics, so its magnetic response is subdued. Overlying the Skead Assemblage is the McElroy Assemblage (O), comprising mafic and ultramafic rocks, and as expected producing more variable and stronger responses than the Skead Assemblage. However, the dominant magnetic feature here is the Lincoln–Nipissing peridotite–gabbro intrusion (P), which is roughly 20 km in length and occurs at, or close to, the northern margin of the McElroy Assemblage. Serpentinisation ensures that it is strongly magnetic. At the northern margin of the McElroy Assemblage is the Lincoln–Nipissing shear zone (Q), which is associated with a northwest–southeast trending linear zone of weaker magnetic response.

The strongest magnetic anomalies in the Kirkland Lake area are associated with the Boston Assemblage (R). Lithotypes include volcanics and sedimentary rocks, with the main magnetic units being serpentinised ultramafics and iron formations. The petrophysical data show that the iron formations are significantly more magnetic than any other rock type in the area, which is to be expected. Within this unit is the Adams magnetite deposit. The stratigraphy extends around the margin of the Lebel stock (D) with the local succession forming tight isoclinal folds. These are too small to be resolved in the aeromagnetic data, but the general trend of the stratigraphy is clear from its strong response.

Very low magnetic responses are associated with clastic metasedimentary rocks (S), whose stratigraphic affinity is disputed (Hearse Assemblage of Jackson and Fyon, 1991). Although these rocks are apparently easily mapped from their subdued magnetic responses, it is easy to underestimate their actual extent owing to the stronger magnetic responses of underlying magnetic units.

The Timiskaming Assemblage (T and U) is the youngest rock formation in the area and has an unconformity at its base. It comprises alluvial–fluvial and turbidite sedimentary rocks, consisting mainly of conglomerate, plus sandstones and argillites. Associated with the sedimentary rocks are alkalic flows and pyroclastics. The succession faces mainly south and has a moderate southerly dip. It is cut by numerous faults and shear zones, and despite the consistent facing does not consistently young southwards because of strike faults and folds. The magnetic response is variable, the metasedimentary components being mostly weakly magnetised (T), but the igneous components give rise to moderate positive anomalies (U). It is the truncation and displacement of anomalies related to the igneous rocks that allow faults to be delineated.

Kimberlites (Y), some diamondiferous, occur in the north of the survey area. These are associated with subtle magnetic anomalies which are only likely to be recognised in magnetically inactive areas. The aeromagnetic and ground magnetic responses of kimberlites of the Kirkland Lake swarm are described in more detail by Brummer *et al.* (1992). The various ways of presenting geophysical data described in Section 2.8 are illustrated using magnetic data from the Kirkland Lake area which include responses from the local kimberlites (see Fig. 2.30). Note the significantly different anomaly shapes, probably owing to variations in remanent magnetism carried by the kimberlites.

3.11.3.3 Responses from major fault zones

The two major fault structures in Kirkland Lake area are the Larder–Cadillac deformation (fault) zone and the Kirkland Lake main break (fault). Both are associated with important gold deposits.

The Larder–Cadillac deformation zone is a major structure extending for hundreds of kilometres (Wilkinson *et al.*, 1999). It dips to the south and has a major reverse-movement component, and is one of several high-strain zones in the Abitibi Subprovince. It is characterised by intense hydrothermal alteration, metasomatism and veining, which is 500 m wide in places. This major deformation zone has a complex history of reactivation with different senses of movement in different areas, depending on their strike, and is generally accepted to mark a suture zone between major crustal blocks. The deformation zone is easily recognised in the magnetic data from the associated linear zones of lower magnetic responses (V), and the truncation of discordant anomaly trends across it (W). Mapping its exact position can be difficult when it parallels the stratigraphy (X).

The Kirkland Lake main break is a system of several closely spaced faults within the Timiskaming Assemblage (T and U). It trends east-northeast and has vertical to steep southerly dip with southerly movement of a few hundred metres, but no significant lateral component. Since its strike direction is parallel to the local stratigraphic trend, it is not especially well defined in the magnetic data.

3.11.3.4 Discussion

Although the Kirkland Lake area dataset would not be considered of particularly high resolution by modern standards, it does show just how effective aeromagnetics can be as a mapping tool. A wealth of stratigraphic and structural information is evident in the images making up Fig. 3.74. In addition, geological environments which exploration models predict to be prospective can be identified relatively easily.

3.11.4 Magnetic responses in a Phanerozoic Orogenic terrain: Lachlan Foldbelt

The Palaeozoic Lachlan Foldbelt extends along much of the eastern seaboard of Australia. We have chosen a section of it from central New South Wales (NSW) to illustrate the magnetic signatures of a Palaeozoic orogenic belt. The tectonic history of the region is generally held to involve the interaction of suspect terranes in an arc/convergent margin environment experiencing oblique-compression. Degeling *et al.* (1986) describe one perspective on its tectonic history, including its implications for metallogeny. A more recent interpretation emphasising thin-skinned tectonics is that of Glen (1992). Gray and Foster (2004) provide the most recent synthesis.

The region was chosen because its geology is well documented (e.g. Scheibner and Basden, 1998), it is only moderately deformed and metamorphosed, and it contains a variety of sedimentary rocks and both intrusive and extrusive igneous rocks. There are also major fault structures and important porphyry style Cu–Au deposits, e.g. Cadia–Ridgeway and Goonumbla–North Parkes.

3.11.4.1 Magnetic data

The aeromagnetic data were acquired at a terrain clearance of 80 m along survey lines orientated east–west and spaced mainly 250 m apart. Figures 3.75a and b show images of the TMI and its first vertical derivative for the study area. The geomagnetic field in the area has an inclination of –63° and declination of +11°. The survey has been very effective in defining the major geological entities, mapping a wide variety of lithotypes, and delineating major structures.

3.11.4.2 Regional geology

The Lachlan Foldbelt comprises a series of north–south-trending structural domains separated by major fault structures (which can be seen in the magnetic images in Figs. 3.75a and b) along which deformation has been concentrated, e.g. (E) and the Gilmore (A), Tullamore (B) and Coodalc–Narromine (D) sutures and the Parkes Thrust (C). These structures often have similar trend to the local stratigraphy, so truncated anomalies are not always obvious. However, they can be recognised as either a linear

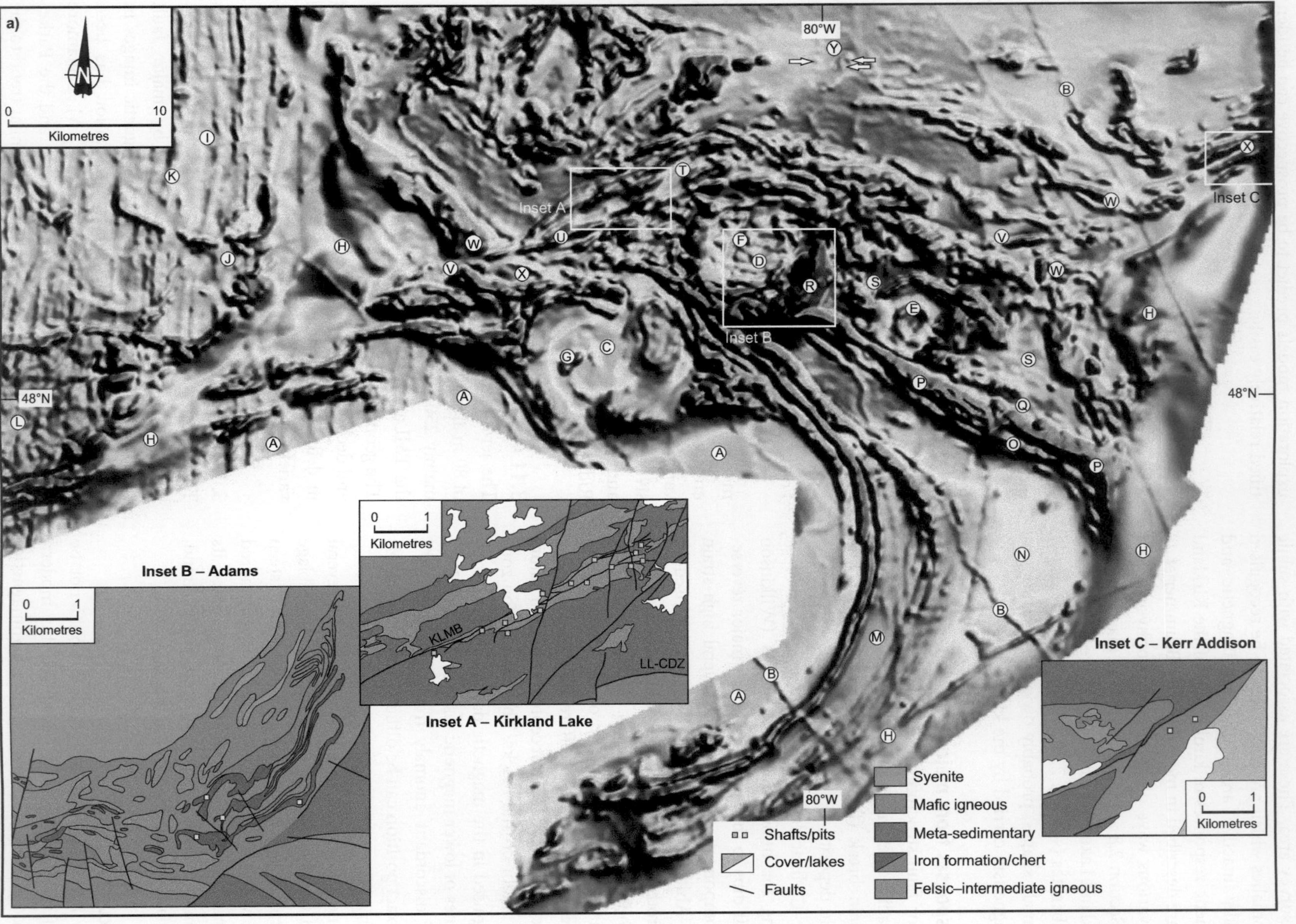

Figure 3.74 Aeromagnetic data and selected geology from the Kirkland Lake area. (a) Pseudocolour TMI, and (b) grey-scale first vertical derivative. Both images are illuminated from the northeast. The various labelled features are discussed in the text. KLMB – Kirkland Lake Main Break, and LL-CDZ – Larder Lake-Cadillac Deformation Zone. Geological maps redrawn, with permission, from Dubuc (1966) and from Robert and Poulsen (1997; Taylor & Francis Ltd, http://www.informaworld.com). Data source: Ontario Geological Survey © Queen's Printer for Ontario 2014.

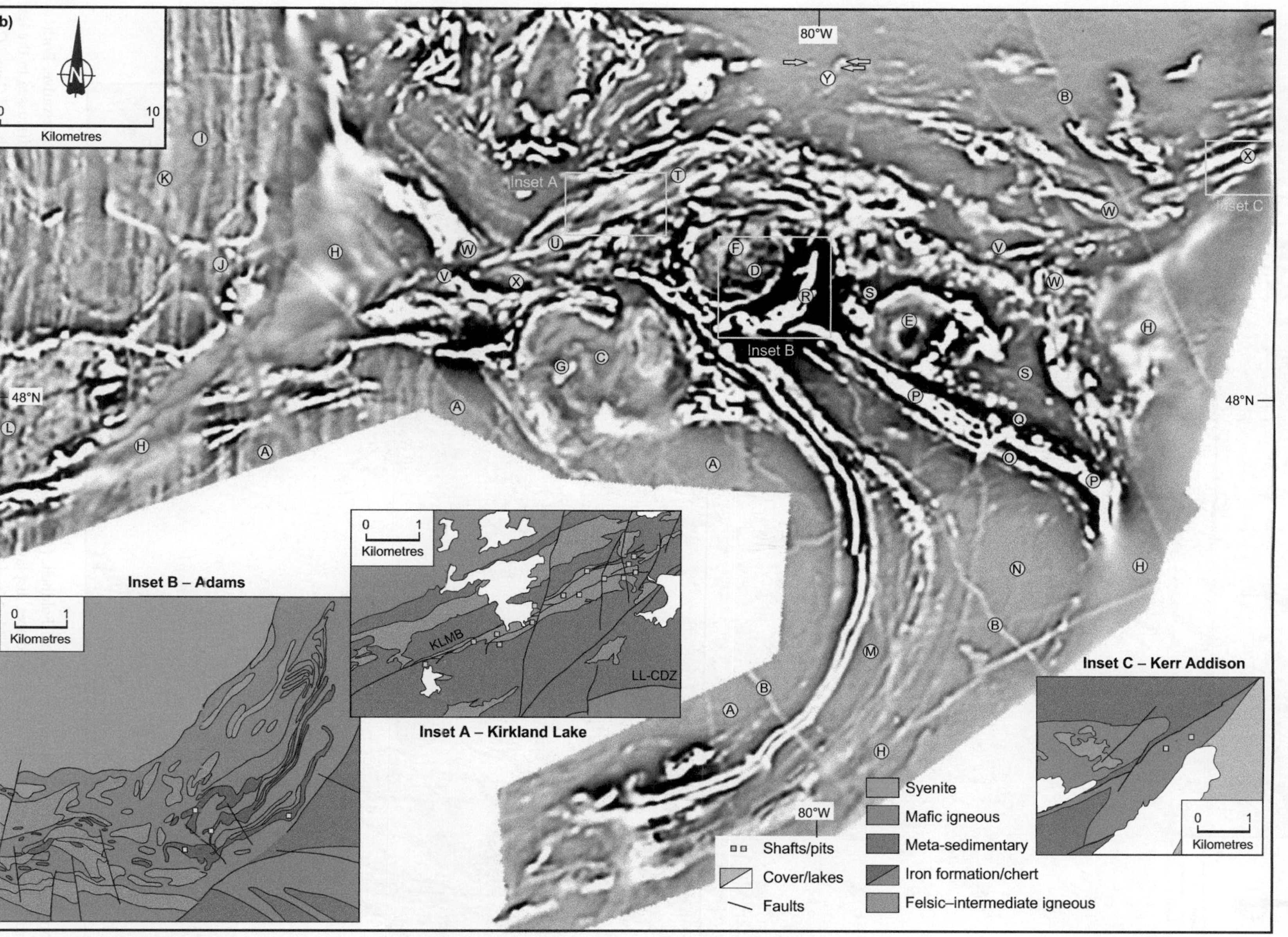
b)
0
10
Kilometres
80°W
48°N
Inset A
Inset B
Inset C
Inset B – Adams
Inset A – Kirkland Lake
Inset C – Kerr Addison
KLMB
LL-CDZ
Shafts/pits
Cover/lakes
Faults
Syenite
Mafic igneous
Meta-sedimentary
Iron formation/chert
Felsic–intermediate igneous

Figure 3.74 (*cont.*)

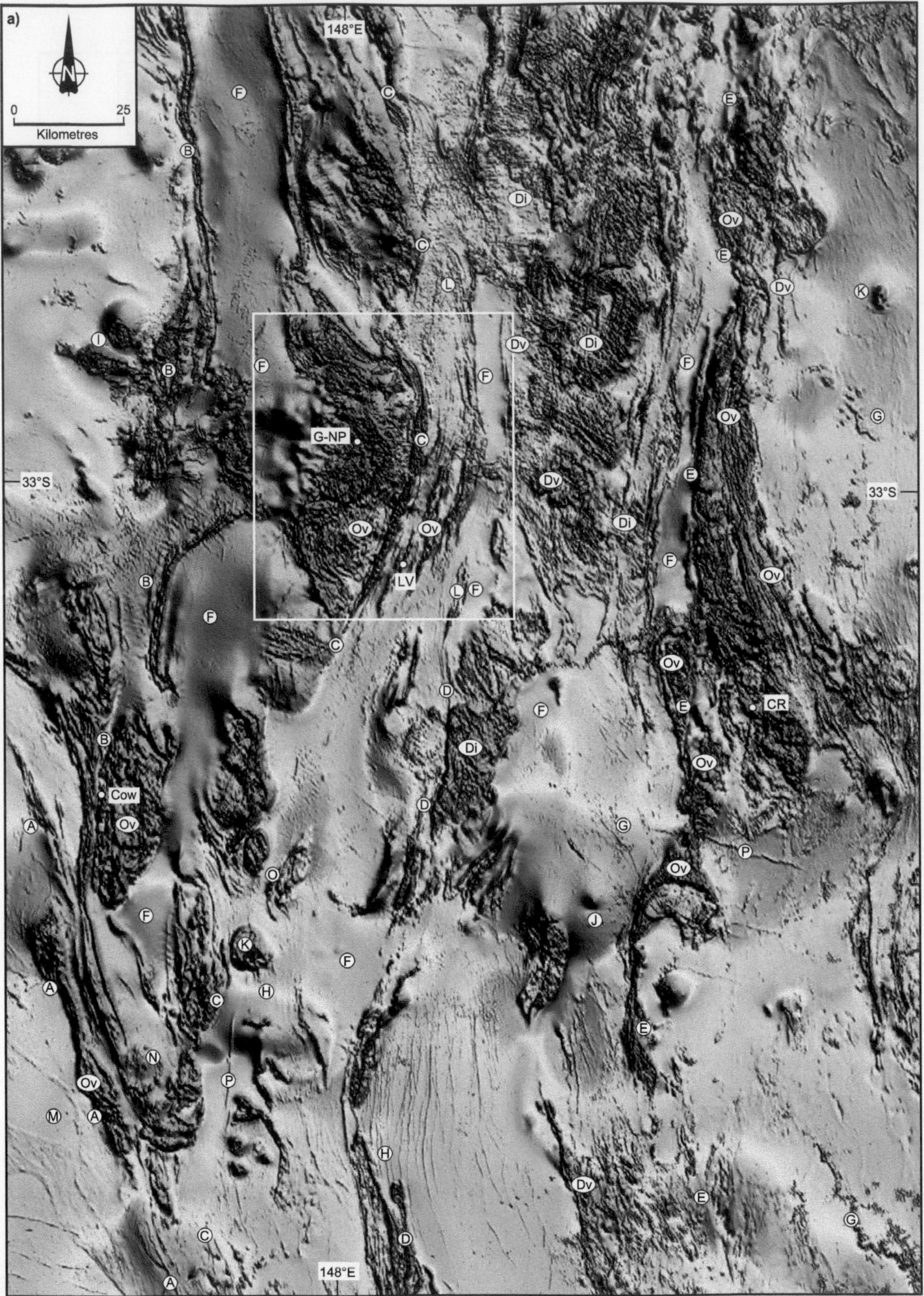

Figure 3.75 Aeromagnetic data from the central Lachlan Foldbelt. (a) Pseudocolour TMI, and (b) grey-scale first vertical derivative. Both images are illuminated from the northeast. Highlighted area displayed in Figure 3.76 and the various labelled features are discussed in the text. CR – Cadia-Ridgeway Cu–Au deposit, G-NP – Goonumbla-North Parkes Cu–Au deposit, LV – London Victoria Au deposit, and Cow – Cowal Au deposit. Data provided courtesy of the Geological Survey of NSW, NSW Trade & Investment.

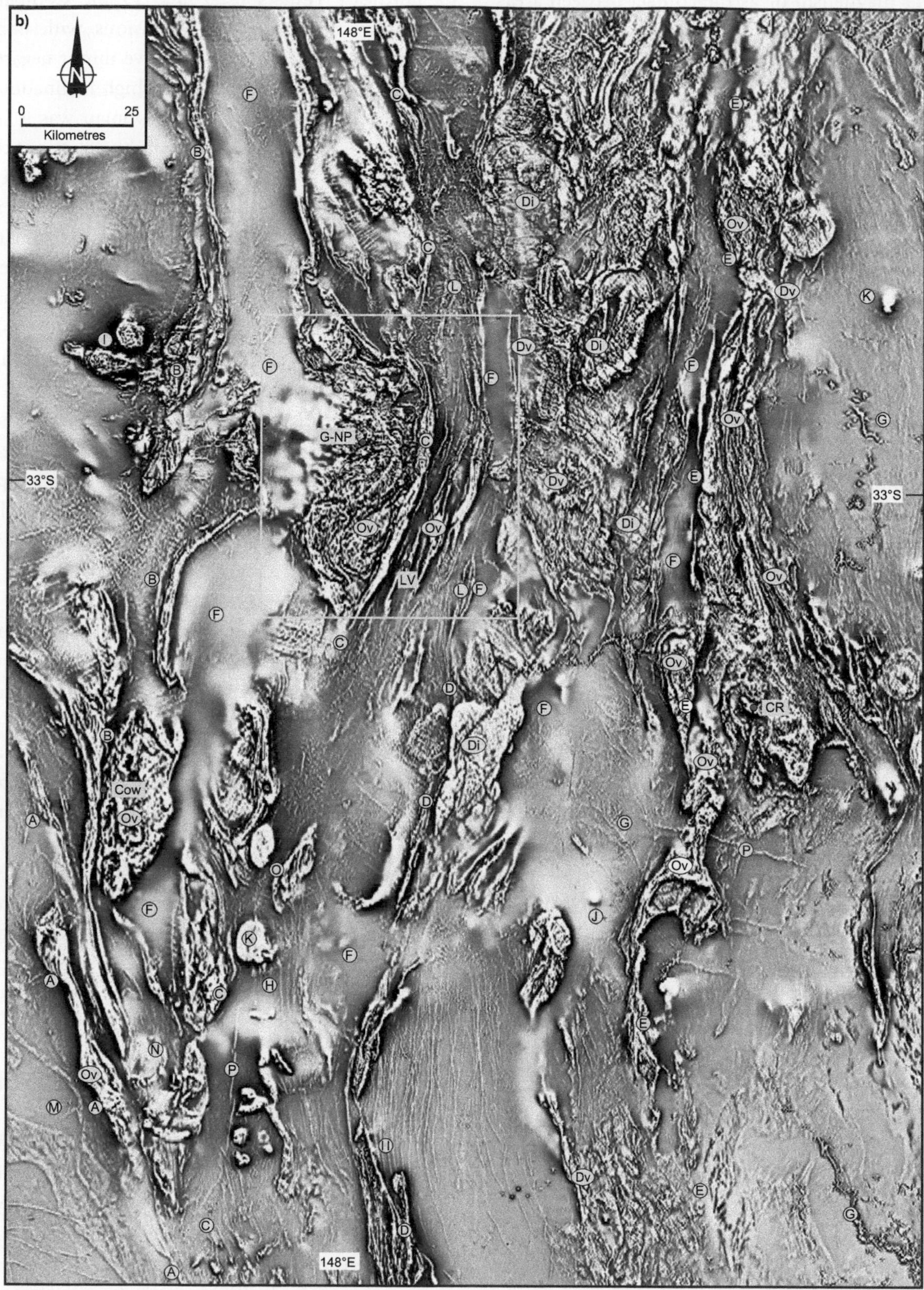

Figure 3.75 (*cont.*)

zone of lower magnetism or as the contact between areas with different magnetic textures.

Belts of deformed volcanic, volcaniclastic and intrusive igneous rocks are separated by 'troughs' containing sedimentary and volcanic sequences. From the Ordovician to the early Devonian there was a series of tectonic and magmatic episodes, often of comparatively localised areal extent. There is a dominant north–south trend from folds and thrust faults, plus some conjugate strike-slip faulting with northeast and northwest trends. Deformation was predominantly thin-skinned, above detachments in the upper and middle crust.

3.11.4.3 Magnetic responses from lithotypes and structures

Ordovician quartz-rich turbidites, black shales and chert, plus mainly shoshonitic igneous rocks (Ov), exhibit the highly textured magnetic response typical of igneous rocks. In the Silurian at, and to the east of, the Gilmore Suture (A), thick sequences of turbidites accumulated in mainly narrow graben or graben-like troughs, e.g. the Tumult, Cowra–Yass and Hill End troughs. These rocks have a subdued magnetic signature although, in places, linear anomalies delineate structures. The regions between the troughs were sites of shallow-water sedimentation, mainly clastics, but also some carbonates. These rocks are very weakly magnetised and so tend to exhibit lower magnetic intensity.

In the Late Silurian to Early Devonian, the Tumult Trough was inverted resulting in a zone of folding, thrusting and metamorphism mainly in the region between the Gilmore (A) and Coolac-Narromine sutures (D). Linear magnetic anomalies in these areas allow some folds to be recognised (L). Further east, e.g. the Hill End and Cowra–Yass Zones, deposition continued uninterrupted. During the Devonian there was also significant magmatism of mafic to felsic composition (Di, Dv), e.g. I-type granitoids and volcanics. Their magnetic responses are similar to the Ordovician volcanics. Intrusion of gabbro–peridotite or Alaskan-type intrusions occurred in the west, for example the Fifield Complex (I). Magnetic anomalies associated with these rocks show evidence for dominance of remanent magnetism, i.e. they are dipolar anomalies with the negative component to the southwest of the positive component. Without remanent magnetism, a response like that seen in Fig. 3.8d would be expected, i.e. a 'low' to the south of the source with a 'high' north of it. This interpretation is supported by petrophysical data which show magnetic vectors with shallow inclinations directed northwest to northeast (Emerson *et al.*, 1979). Compare these responses with those from intrusions with dominantly induced magnetism (K), which have minor negative components due to the comparatively high inclination.

Deformation in the Middle Devonian was followed by molasse sedimentation, comprising the clastic rocks of the Hervey Group (F) in the study area. These sequences are non-magnetic and create a characteristic subdued response in the magnetic data. This is partly due to their lack of magnetism, but also due to underlying magnetic sources being at great depth. In the area to the east of the Goonumbla–North Parkes porphyry deposits (G-NP), the change in magnetic character with source depth is quite clear. Outcropping volcanic rocks have a highly textured response with many short-wavelength variations. Where the same rocks underlie sediments of the Hervey Group (F), the responses coalesce creating a smoother appearance. Cratonisation in the Early Carboniferous was followed by erosion which transported magnetic detritus into the drainage network, delineated by its prominent sinuous and dendritic anomaly shapes (G).

The magnetic images show several other features of general interest. The cross-cutting nature of intrusions is revealed by the truncation of stratigraphical anomalies (H). Anomaly (J) has long wavelength and does not correlate with the surface geology. It is probably related to an intrusive mass at depth. Most granitoids have comparatively weak magnetism (M) and there is evidence for a magnetic halo (N), possibly due to magnetite created by contact metamorphism, or maybe remanently magnetised zonation related to cooling of the mass. Small intrusive bodies have a characteristic textured appearance similar to the Ordovician volcanics they intrude. Linear positive anomalies are caused by dykes (P). There are also good examples of anomalies truncated by minor faults (O).

3.11.4.4 Gravity and magnetic responses in the North Parkes area

The North Parkes porphyry copper district (G-NP) contains several Cu–Au deposits, e.g. Endeavour 26 North, Endeavour 22 and Endeavour 27 (Heithersay *et al.*, 1990). Terrain is subdued and outcrop sparse, with cover material tens of metres thick. The base of oxidation varies from zero to 80 m depth. Magnetic surveys have proven to be an effective means of mapping the bedrock geology in the district. A geological map and images of the magnetic and Bouguer gravity data from the North Parkes area are shown in Fig. 3.76.

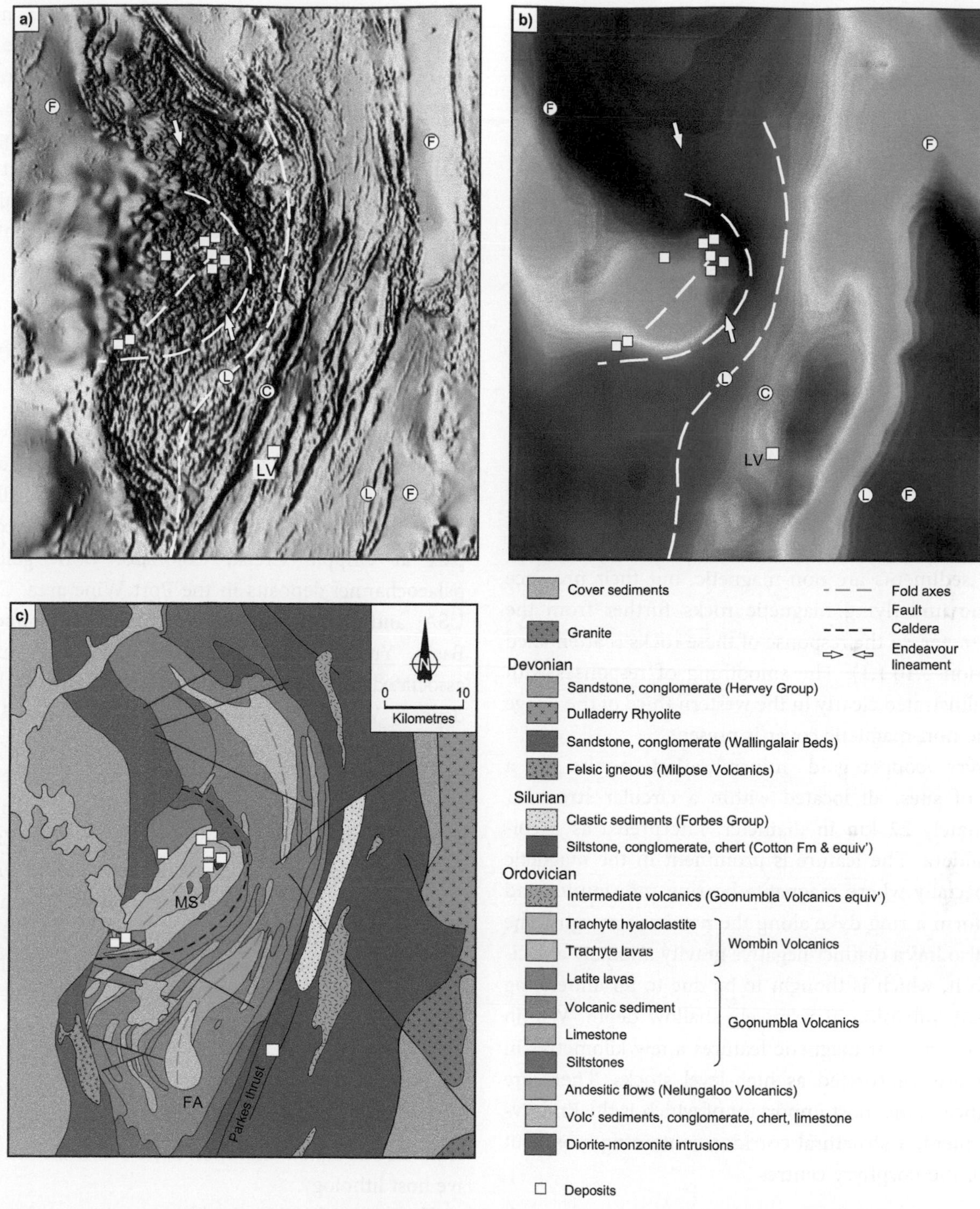

Figure 3.76 (a) TMI data from the Goonumbla–Parkes subarea of Figs. 3.75a and b. (b) Bouguer gravity. Data provided courtesy of the Geological Survey of NSW, NSW Trade & Investment. (c) Geological map of the subarea. Redrawn, with permission, from Heithersay and Walshe (1995); and from Heithersay *et al.* (1990) with the permission of The Australasian Institute of Mining and Metallurgy. The various labelled features are discussed in the text. FA – Forbes anticline, LV – London Victoria Au deposit and MS – Milpose syncline.

Referring to Fig. 3.76, the Parkes Thrust (C) is the major structure in the area and is defined in both the gravity and magnetic data. There is a distinct gravity gradient and change in magnetic character across the structure because it marks the boundary between a volcanic area to the west and a sedimentary/volcanic area to the east. The sediments and volcanics to the east have been metamorphosed to lower greenschist facies and tightly folded. Although subtle, the folds are defined by the magnetic data, often as linear anomalies forming elongated closed loops (L). Felsic rocks in the extreme east of the image give rise to subdued responses. The volcanic terrain to the west of the Parkes thrust has the textured appearance typical of a succession of variably magnetic igneous rocks. Volcanic lithologies are predominantly andesitic, but tend towards basaltic in places. The succession is unmetamorphosed with dips of 30–50°. Two major folds, the Forbes anticline (FA) and Milpose syncline (MS), have curvilinear traces trending northeasterly. The sedimentary units are significantly less magnetic than the volcanics, allowing the broad structure to be delineated from the magnetic data. The comparatively wide survey line spacing of the data in Fig. 3.76a causes aliasing of the short-wavelength responses which produces a spurious north–south 'fabric' in the image. The clastic sediments of the Hervey Group are again associated with subdued magnetic responses and lower gravity (F). The sediments are non-magnetic, but their presence places the underlying magnetic rocks further from the magnetometer, so the response of these rocks is attenuated (see Section 3.10.1.1). The smoothing of responses with depth is illustrated clearly in the western third of the image where the non-magnetic cover is present.

Porphyry copper–gold mineralisation occurs at a number of sites, all located within a circular structure, approximately 22 km in diameter, interpreted as a collapsed caldera. The feature is prominent in the magnetic data, especially where magnetite-bearing monzonites and diorites form a ring dyke along the northern margin. The caldera also has a distinct negative gravity anomaly associated with it, which is thought to be due to an underlying low-density intrusion at relatively shallow depth. Within the caldera, circular magnetic features a few kilometres in diameter are interpreted as high-level stocks. There are several linears, the most important of which is the Endeavour lineament, a structural corridor containing most, but not all, of the porphyry centres.

3.11.4.5 Discussion

Although a very different geological environment to the granitoid–greenstone terrain of the Kirkland Lake example (see Section 3.11.3), the aeromagnetic data are again extremely effective at mapping the geology. The data are characterised by responses of very different character: the 'busy' response of the volcanics versus the smooth response from the sediments. The ability of magnetics to map different aspects of the geology is demonstrated by the data containing responses from Palaeozoic volcanic centres and modern-day drainage. Completely different exploration models could be applied to these data.

3.11.5 Magnetic and gravity responses from mineralised environments

Data presented elsewhere in this chapter show gravity responses associated with: the Las Cruces Cu–Au massive sulphide deposit in the Iberian Pyrite Belt, Seville, Spain; epithermal Au–Ag–Te mineralisation in an igneous complex at Cripple Creek, Colorado, USA; gold-bearing palaeochannel deposits in the Port Wine area, California, USA; and coal-bearing sequences in the Bonnet Plume Basin, Yukon, Canada. Also presented are alteration-associated magnetic responses from the Wallaby Au deposit, Western Australia, and an iron-ore deposit in Qian'an District, China. In Section 4.7.3.2 we present the magnetic responses associated with Au–Ag mineralisation in the Waihi-Waitekauri region in New Zealand, in Section 2.6.1.2 we showed the response from the Elura Zn–Pb–Ag volcanogenic massive sulphide deposit in eastern Australia, and in Section 2.6.4 the response from the Marmora magnetite skarn in Ontario, Canada. In Section 2.10.2.3, we showed the gravity response from massive nickel sulphide mineralisation at Pyhäsalmi, Oulu, Finland. Further examples are shown in Fig. 3.77. Together these examples have been chosen to illustrate the three basic types of potential field responses that can be used for direct detection of mineral deposits, i.e. responses from mineralisation, from an associated alteration zone and from the prospective host lithology.

The mineralisation itself gives rise to the magnetic response from the magnetite skarn in Figs. 2.12 and 3.77a, the iron-formation hosted iron ore in Fig. 3.64 and the cumulate chromite deposits in Fig. 3.77b. The

Figure 3.77 Gravity and magnetic responses associated with various types of mineralisation. (a) Aeromagnetic response of the Marmora magnetite skarn, Ontario, Canada. Redrawn, with permission, from Wahl and Lake (1957); (b) the positive gravity anomaly associated with the Gölalan Cr deposit, Elaziğ, Turkey. Redrawn, with permission, from Yüngül (1956); (c) magnetic response of the Thompson Ni deposit, Manitoba, Canada. Redrawn, with permission, from Dowsett (1967); (d) gravity response of the sedimentary basin hosting the Jharia Coalfield, Jharkhand, India. Redrawn, with permission, from Verma *et al.* (1979).

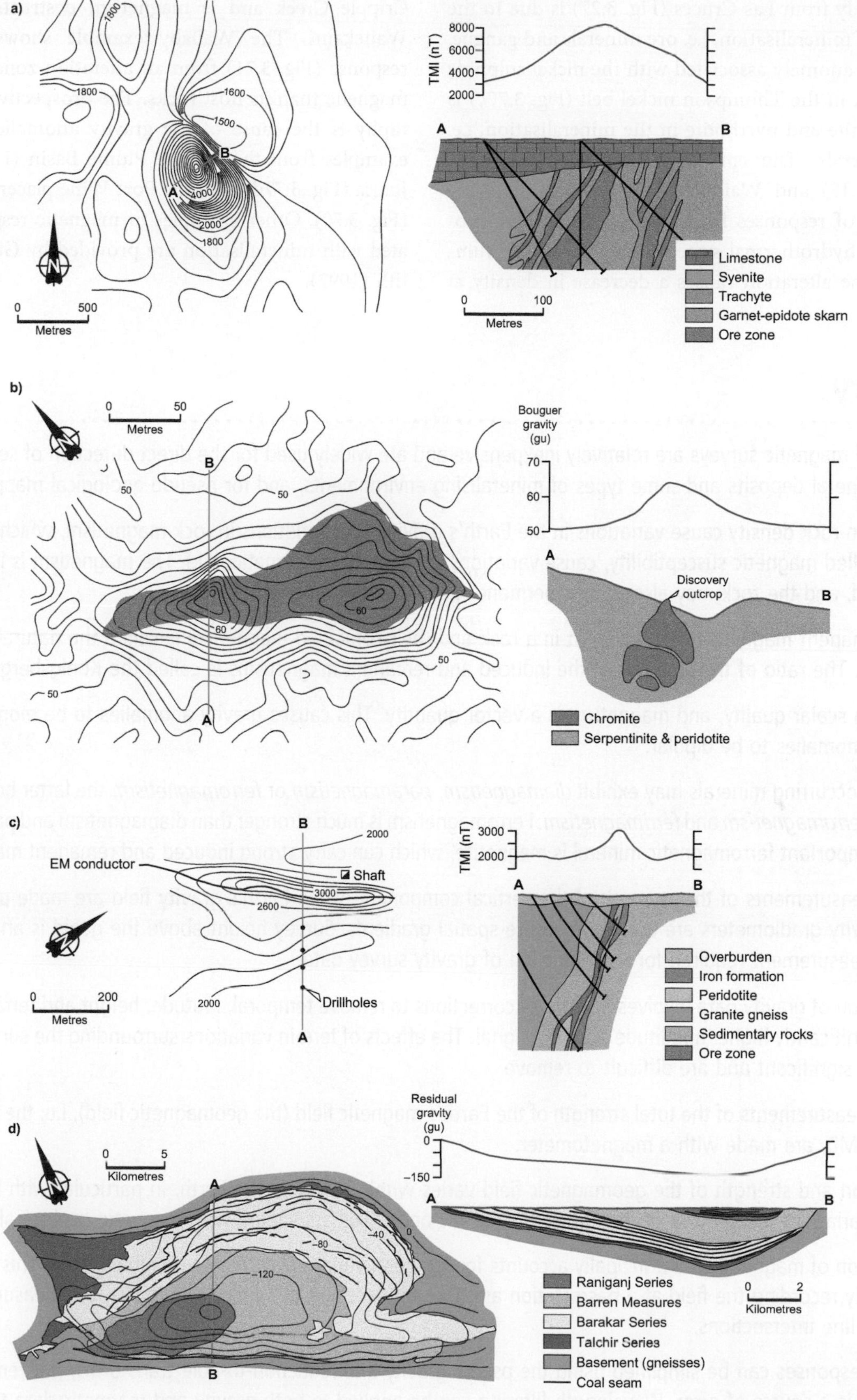

a)
TMI (nT)
8000
6000
4000
2000
A
B
1800
1600
1500
4000
2000
Limestone
Syenite
Trachyte
Garnet-epidote skarn
Ore zone
0 500
Metres
0 100
Metres
b)
0 50
Metres
Bouguer gravity (gu)
70
60
50
Discovery outcrop
Chromite
Serpentinite & peridotite
c)
EM conductor
Shaft
3000
2600
2000
Drillholes
0 200
Metres
Overburden
Iron formation
Peridotite
Granite gneiss
Sedimentary rocks
Ore zone
d)
Residual gravity (gu)
0
–150
0 5
Kilometres
–40
–80
–120
Raniganj Series
Barren Measures
Barakar Series
Talchir Series
Basement (gneisses)
Coal seam
0 2
Kilometres

gravity anomaly from Las Cruces (Fig. 3.27) is due to the entire body of mineralisation, i.e. ore minerals and gangue. The magnetic anomaly associated with the nickel sulphide mineralisation in the Thompson nickel belt (Fig. 3.77c) is due to magnetite and pyrrhotite in the mineralisation, i.e. non-ore minerals. The epithermal deposits at Cripple Creek (Fig. 3.18) and Waihi-Waitekauri (see Fig. 4.25) are examples of responses from the alteration zone produced by the hydrothermal system that gave rise to mineralisation. The alteration causes a decrease in density at Cripple Creek and is magnetism destructive at Waihi-Waitekauri. The Wallaby example shows a magnetic response (Fig. 3.71) from an alteration zone that is more magnetic than its host rocks. The prospective host stratigraphy is the cause of the gravity anomalies in the coal examples from the Bonnet Plume Basin (Figs. 3.69) and Jharia (Fig. 3.77d), and the Port Wine placer gold example (Fig. 3.70). Other examples of magnetic responses associated with mineralisation are provided by Gunn and Dentith (1997).

Summary

- Gravity and magnetic surveys are relatively inexpensive and are widely used for the direct detection of several different types of mineral deposits and some types of mineralising environments, and for pseudo-geological mapping.
- Variations in rock density cause variations in the Earth's gravity, and variations in rock magnetism, which depends on a property called magnetic susceptibility, cause variations in the Earth's magnetic field. The magnetism is induced by the Earth's field, and the rock may also carry a permanent or remanent magnetism.
- Several remanent magnetisms may coexist in a rock and their combined magnetism is called the natural remanent magnetism. The ratio of the strength of the induced and remanent magnetisms is called the Königsberger ratio.
- Density is a scalar quality, and magnetism is a vector quantity. This causes gravity anomalies to be monopolar and magnetic anomalies to be dipolar.
- Commonly occurring minerals may exhibit *diamagnetism*, *paramagnetism* or *ferromagnetism*, the latter being of two types, *antiferromagnetism* and *ferrimagnetism*. Ferromagnetism is much stronger than diamagnetism and paramagnetism. The most important ferromagnetic mineral is magnetite, which can carry strong induced and remanent magnetisms.
- Relative measurements of the strength of the vertical component of the Earth's gravity field are made using a gravity meter. Gravity gradiometers are used to measure spatial gradients. Survey height above the geoid is an essential ancillary measurement required for the reduction of gravity survey data.
- The reduction of gravity data involves a series of corrections to remove temporal, latitude, height and terrain effects that all have significantly higher amplitude than the signal. The effects of terrain variations surrounding the survey station are particularly significant and are difficult to remove.
- Absolute measurements of the total strength of the Earth's magnetic field (the geomagnetic field), i.e. the total magnetic intensity (TMI), are made with a magnetometer.
- The direction and strength of the geomagnetic field varies within and over the Earth, in particular with latitude. This produces variations in the form of the induced magnetic anomalies associated with magnetic geological features.
- The reduction of magnetic data principally accounts for temporal variations in the geomagnetic field. This is achieved by continuously recording the field at a base station and through levelling the data based on repeat measurements at tie-line/survey line intersections.
- Magnetic responses can be simplified using the pseudogravity and reduction-to-pole transforms, but remanent magnetism is a source of error. Wavelength filtering can be applied to both gravity and magnetic data to separate the

responses of large, deep and shallow sources. Enhancements to increase resolution of detail are most commonly based on spatial gradients (derivatives).

- Density variations in crystalline rocks are controlled by the densities and proportions of the constituent minerals. When the rocks have significant porosity, as is the case for most sedimentary rocks, then the amount of porosity and the pore fluid become the dominant control on density.
- The magnetism of rocks is roughly proportional to their magnetite content. Other, sometimes significantly, magnetic minerals are haematite and pyrrhotite. Whether an igneous or metamorphic rock contains magnetite depends on various factors, but a key control is whether the right chemical conditions exist for iron in the system to form oxides (which may be highly magnetic) or silicates (which are weakly magnetic).
- There is no physical reason that the density and magnetism of a rock formation, and indeed variations in these properties, must be related. Correlations may occur in crystalline rocks, but are less likely in sedimentary rocks. Gravity and magnetic maps reflect fundamentally different physical characteristics of the rocks, so they may differ significantly.
- Interpretation of gravity and magnetic data presented as pseudocolour images is usually based on a combination of the 'raw' data, i.e. Bouguer or free-air gravity and TMI, and some form of derivative-based enhancement of which the first vertical derivative is most common. Care must be taken to avoid misinterpreting responses due to levelling errors, effects of terrain and remanent magnetism.
- Quantitative interpretation of gravity and magnetic data is hindered by non-uniqueness. The most reliable approach is to create several models representing 'end-member' solutions. In its simplest form, modelling may involve automated estimates of source depth based on anomaly amplitude and gradient relationships. Sophisticated forward and inverse modelling methods are also available, but the results remain inherently ambiguous, although this may be reduced with petrophysical data and other geological information.

Review questions

1. Explain why gravity responses are monopoles and magnetic responses dipoles.
2. Explain induced magnetism, remanent magnetism, total magnetisation and the Königsberger ratio. How does magnetic susceptibility affect all of these parameters?
3. Explain how a change in survey height (or depth to the source) affects gravity and magnetic responses. Sketch the various anomalies.
4. What are the benefits of gravity and magnetic gradiometer measurements over conventional field measurements?
5. Contrast the sequence of reductions applied to gravity and magnetic survey data.
6. Sketch in the principal meridian plane the magnetic field of a sphere and the induced magnetic anomaly above it for magnetic inclinations of +90°, +45° and 0°. How do these differ in the opposite magnetic hemisphere?
7. Describe four data processing techniques you would use to assist in the analysis of low-latitude TMI data.
8. What are horizontal and vertical derivatives, the tilt derivative and the analytic signal? What benefits do they provide over the TMI and gravity field measurements?
9. Describe the magnetic properties of magnetite. How does grain size influence magnetic properties?
10. Serpentinisation is an important control on rock magnetism and density – discuss.
11. Describe an appropriate workflow for the analysis of a set of magnetic susceptibility measurements.
12. What is Euler deconvolution?
13. How might magnetic data be used during greenfields exploration for gold deposits in Archaean greenstone belts?

FURTHER READING

Blakely, R.J., 1995. *Potential Theory in Gravity and Magnetic Applications*. Cambridge University Press.

This provides a thorough description of the principles behind the methods, and includes comprehensive mathematical descriptions.

Gunn, P.J., 1997. Airborne magnetic and radiometric surveys. *AGSO Journal of Australian Geology & Geophysics*, 17, (2).

This is a special edition of the AGSO Journal of Australian Geology and Geophysics *comprising 17 papers on virtually every aspect of the acquisition, processing and interpretation of aeromagnetic (and radiometric) data. The collection includes a paper specifically describing the magnetic responses of different kinds of mineral deposits.*

Isles, D.J. and Rankin, L.R., 2013. *Geological Interpretation of Aeromagnetic Data*. CSIRO Publishing.

This is an e-book published by the Australian Society of Exploration Geophysicists and the Society of Exploration Geophysicists. It is an excellent resource for interpreters of magnetic datasets.

Reeves, C., 2005. *Aeromagnetic Surveys: Principles, Practice and Interpretation*. Geosoft.

This is an e-book obtainable from the website of Geosoft®, Canada. It is a comprehensive description of all aspects of the aeromagnetic method.

Thomas, M.D., 1999. Application of the gravity method in mineral exploration: case histories of exploration and discovery. In Lowe, C., Thomas, M.D. and Morris, W.A. (Eds.), *Geophysics in Mineral Exploration: Fundamentals and Case Histories*. Geological Association of Canada, Short Course Notes Volume 14, 101–113.

A useful compilation of gravity responses from a variety of deposit types.

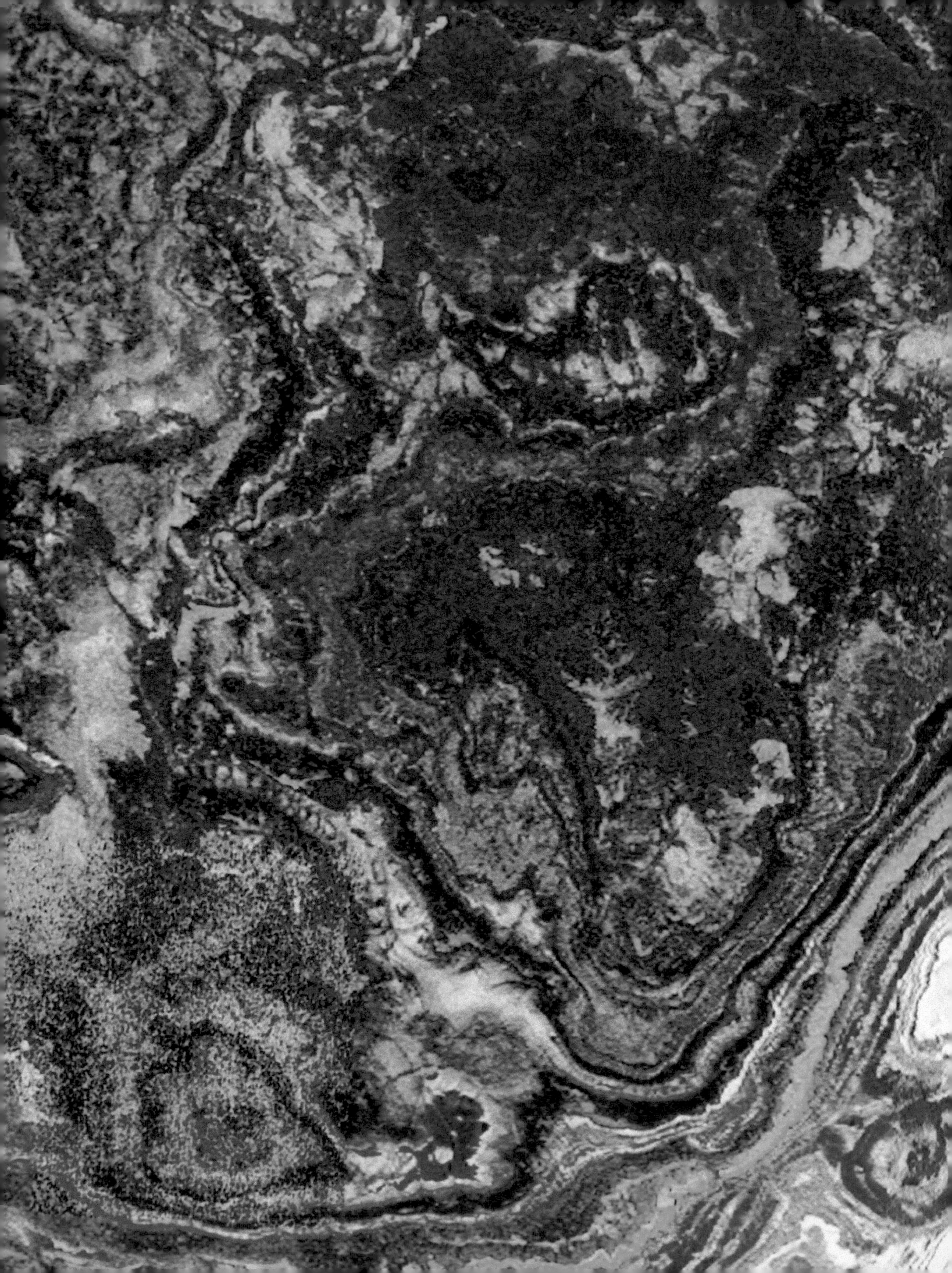

CHAPTER

4 Radiometric method

4.1 Introduction

The radiometric method, or radiometrics, measures naturally occurring radioactivity in the form of gamma-rays (γ-rays) (Fig. 4.1). Most of this radiation originates from mineral species containing radioactive isotopes of potassium (K), uranium (U) and thorium (Th). Radiometrics is a passive geophysical method because it measures a natural source of energy. Radiometric surveys for mineral exploration are routinely made from the air, on the ground and within drillholes. Airborne radiometrics is particularly common in mineral exploration where the radiometric data are acquired simultaneously with magnetics during airborne surveying, although it would be unusual for a solely radiometric survey to be conducted. Ground radiometric surveys are usually conducted with hand-held instruments, but γ-ray detectors are sometimes mounted on a moving vehicle for larger-scale surveys. Downhole radiometric measurements are limited to γ-logging; downhole surveying as defined in Section 1.2.1 is not possible in radiometrics. Gamma-logging is a common component of multiparameter logging for rapidly and economically measuring *in situ* physical properties, correlating stratigraphy between drillholes and assisting in the evaluation of uranium deposits. It is also used extensively by the petroleum industry in well-logging.

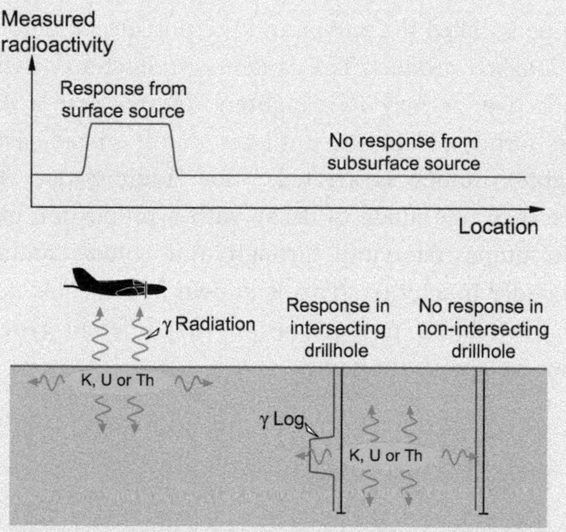

Figure 4.1 Schematic illustration of the radiometric method, which measures natural radioactivity from K, U and Th. The emissions are absorbed by even minor amounts of rock, so surface occurrences or drillhole intersections are required to obtain recognisable responses.

Historically, the main use of radiometrics in mineral prospecting was detection of anomalies caused by outcropping, highly radioactive, uranium deposits. With improvements in sensor technology and data-processing algorithms, downhole radiometric techniques were developed for estimating the grade of uranium ores intersected by drillholes. The advent of multichannel detectors for airborne surveying, capable of distinguishing radiation from different radioactive elements, the increased sensitivity and resolution of airborne surveying techniques, and the development of new data reduction algorithms have focused airborne radiometrics more toward geological mapping than purely anomaly detection. Newer applications include detecting and mapping areas of hydrothermal alteration and weakly radioactive mineral deposits,

◀ Ternary radiometric image of the Kimberley Basin in north-west Western Australia. The area shown is 140 km wide. Data reproduced with the permission of Geoscience Australia.

e.g. heavy-mineral sands. Renewed interest in the method has also been driven by the requirement to map the materials making up the near-surface. In this context, the radiometric data are usually interpreted in combination with topographic and satellite-borne remote-sensing data. Some applications include regolith mapping for mineral exploration and the mapping of soil types for environmental studies.

The radiometric method has several characteristics that make it unique amongst the geophysical methods. Firstly, the measured radioactivity originates from only the top few centimetres of the Earth's crust so, unlike other geophysical methods, radiometrics has only a very limited ability to see into the subsurface. Secondly, because it is possible to identify the elemental source of the radiation from the energy of the γ-rays emitted, radiometric data are used to map variations in the chemical rather than the physical characteristics of the survey area. Interpretation of radiometric data straddles the boundary between geochemistry and geophysics. This may explain why the methodologies for interpreting radiometrics are less well-developed than those of other geophysical methods.

There exists a large literature describing the radiometric method, although much of it is focused on applications in the glaciated terrains of the northern hemisphere. Significant developments in the past 10 years, particularly in data processing and reduction methods, and progress in understanding the behaviour of the radioelements in arid terrains such as Australia, mean that much of the published material has now been superseded.

The process of converting measured radioactivity to elemental concentrations of the ground involves a multi-stage reduction of the survey data to remove responses of non-geological origin. It is based largely on *calibration* measurements made specifically for this purpose. Much of this reduction process is empirical, and understanding its basis and limitations requires a reasonable knowledge of the characteristics of radioactivity. This, therefore, is the starting point for our description of the radiometric method, with emphasis on aspects relevant to geophysical surveying. The interpretation of radiometric data and the computed elemental maps requires an understanding of the behaviour of radioactive elements in the geological environment, which is emphasised in the latter part of the chapter.

4.2 Radioactivity

An atom that contains equal numbers of protons and electrons will be electrically neutral. The number of protons or electrons in the atom is the element's atomic number (Z), which defines its chemical properties and its place in the periodic table of the elements. The number of neutrons in the nucleus of the atom is the atom's neutron number (N), and the total number of protons and neutrons ($Z + N$) in the nucleus is the atom's mass number (A). Atoms of the same element having different neutron numbers are called *isotopes* and are identified by their mass number. For example, uranium (U) contains 92 protons, but can have 142, 143 or 146 neutrons, forming the isotopes ^{234}U, ^{235}U and ^{238}U respectively.

Radioactive materials, also called *radioisotopes*, are unstable and spontaneously emit radiation as part of their transformation into a more stable, non-radioactive, state. The radiation occurs in three forms, known as *alpha-* (α), *beta-* (β) and *gamma-* (γ) radiation, with the emission of the different kinds of radiation referred to as α-, β- and γ-decay, respectively. The original (pre-decay) isotope is called the *parent* and the post-decay isotope is the *daughter* product. The daughter product may itself be radioactive, as may its daughters. In this case a *decay series* forms, continuing until a stable (non-radioactive) daughter isotope is created. Some radioisotopes have more than one mode of decay with a proportion of the radioisotope decaying through one mode and the remainder in another. This is known as *branched decay*, and the relative proportions of the different types of decay are always the same.

4.2.1 Radioactive decay

Alpha-decay involves emission of a helium (^{4}He) nucleus, β-decay the emission of an electron (β^-) or positron (β^+), and γ-decay the emission of a photon. Alpha- and β-decays are the emission of particles whereas γ-decay is the emission of high-frequency electromagnetic radiation (see Fig. 5.1). The particles originate from within atomic nuclei and, since their emission causes a change in atomic number, a different element is created.

Gamma-decay follows after either α- or β-decay and does not change the atomic number or the mass number. Emission of γ-radiation is associated with a reduction in energy of the daughter nucleus from an *excited state* to either a *ground state*, or to an excited state of lower energy and in which case further γ-emissions occur. The change in energy, and hence the energy of the γ-ray, depends on the particular parent–daughter combination. This is important from a geophysical perspective since the energy of a γ-ray can be measured and used to infer the type of emitting isotope present.

Gamma-radiation is also produced by another mechanism called K-capture. This occurs when an inner-orbital atomic electron is captured by the nucleus and a proton is converted to a neutron. A new element is created and, again, γ-emission allows the daughter to lose energy to attain its ground state.

The changes in atomic number (Z) and mass number (A) associated with the various forms of radioactive decay are as follows:

$$^{A}_{Z}\text{Parent} \rightarrow {}^{4}_{2}\text{He} + {}^{A-4}_{Z-2}\text{daughter} \qquad \alpha\text{-emission} \qquad (4.1)$$

$$^{A}_{Z}\text{Parent} \rightarrow e^{-} + {}_{Z+1}^{A}\text{daughter} \qquad \beta\text{-emission} \qquad (4.2)$$

$$^{A}_{Z}\text{Parent} + e^{-}_{\text{orbital}} \rightarrow {}_{Z-1}^{A}\text{daughter} \qquad \text{K-capture} \qquad (4.3)$$

It is the ratio of neutrons to protons in the nucleus that controls the type of decay that occurs. When the ratio is high, β-emission is favoured, because it tends to reduce the ratio. When the ratio is low the other types of emission occur. Radioactive decay ceases when a stable combination of protons and neutrons is achieved.

4.2.2 Half-life and equilibrium

The radioactive decay of an individual nucleus is spontaneous, completely random and independent of external conditions such as changes in temperature and pressure. In any period of time it is not possible to identify which particular nuclei will disintegrate. It is a statistical process, so on average half of the original population of radioactive nuclei will undergo radioactive decay in a time period known as the *half-life* ($T_{1/2}$). Half of the remaining nuclei will then decay in the next half-life and so on. The result is an exponential decrease in both the number of parent nuclei and the rate of decay, and an increase in the amount of daughter nuclei produced (Fig. 4.2). The process is described by the expression:

$$N = N_0 e^{-\lambda t} \qquad (4.4)$$

where N_0 is the original number of atoms, N is the number of atoms at time t and λ is the *decay constant* for the radioelement. The half-life of the radioisotope, i.e. when $N = N_0/2$ at $t = T_{1/2}$, is:

$$T_{1/2} = \frac{\ln(2)}{\lambda} = \frac{0.693}{\lambda} \qquad (4.5)$$

The various radioactive nuclei have vastly different half-lives, ranging from seconds to billions of years, and can be significant on geological time scales (allowing radioactive decay to be used for dating of geological events).

The time required for a complete decay series to be established depends on the various half-lives of its components. When the series is fully established, the relative amount of each component remains constant and the decay series is then said to be in *secular equilibrium*. This means that if the abundance of one component is known, the abundance of another component may be determined. If any member of the decay series is disturbed, i.e. material is removed or added, the relative amounts of each component are altered and a state of *disequilibrium* occurs. The time required for equilibrium to re-establish depends on the half-lives of the radioisotopes involved. Consider the situation where the decay of a parent results in a radioactive daughter having a half-life shorter than its parents. Depending on the half-lives involved, after a

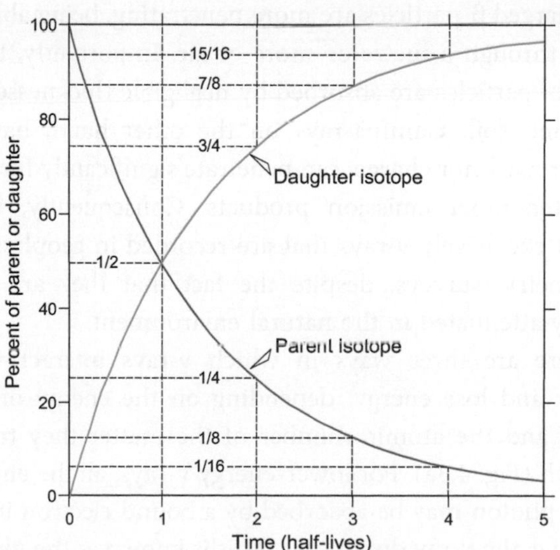

Figure 4.2 Exponential decay process. Sum of the parent and daughter nuclei at any time is 100%. Redrawn, with permission, from Lowrie (2007).

period of time the daughter product will be decaying as rapidly as it is being produced, so its concentration remains constant. The time taken to approximately restore equilibrium, if a member of a decay series is disturbed, is about 5 times the half-life of the disturbed member (the time required to form 97% of the daughter isotope volume; see Fig. 4.2). The time for the whole decay series to reach equilibrium is governed by the longest half-life in the series; so it would be at least 7 times the half-life (to form 99%) of the longest-living daughter element.

4.2.3 Interaction of radiation and matter

Radiation produced by radioactive decay can pass through various materials. As the radiation passes through, it is attenuated as emission products lose energy to the atoms of the material through scattering, collision and absorption. The ability to pass through a material is basically a function of the energy, size and charge of the different emission products, and the density of the material through which they pass. From a geophysical perspective this is important because it controls the depth beneath the ground surface from which detectable decay products may originate. It also has a significant influence on how close the detector must be to the radioactive source for the radiation to be detected.

Alpha-particles are comparatively large and highly charged and, with the energy levels that occur in the natural environment, can pass through a few centimetres of air before being completely absorbed. The smaller and less charged β-particles are more penetrating, being able to travel through a metre or more of air. Importantly, both types of particles are absorbed by negligible thicknesses of rock and soil. Gamma-rays on the other hand, having neither mass nor charge, can penetrate significantly further than the other emission products. Consequently, it is almost exclusively γ-rays that are recorded in geophysical radiometric surveys, despite the fact that they are still rapidly attenuated in the natural environment.

There are three ways in which γ-rays interact with matter and lose energy, depending on the energy of the γ-rays and the atomic number of the matter they travel through (Fig. 4.3a). For lower-energy γ-rays, all the energy of the photon may be absorbed by a bound electron in an atom, i.e. the γ-ray disappears. This is known as the *photoelectric effect*. High-energy photons are absorbed, creating an electron–positron pair in a process known as *pair production*. For γ-radiation of intermediate energy, and

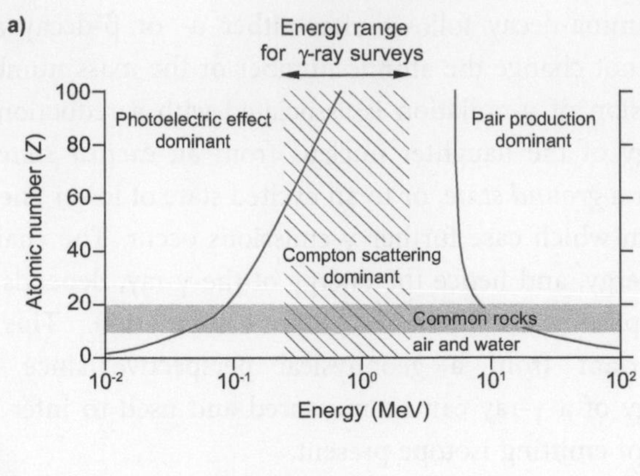

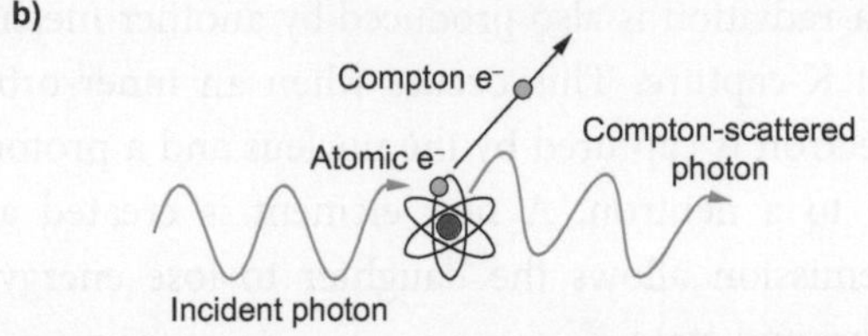

Figure 4.3 Interaction of γ-rays with matter. (a) Mechanisms of energy loss in γ-rays of different energies. Redrawn, with permission, from Minty (1997). (b) Compton scattering due to collision of a photon with an electron.

most important in geophysical exploration, the incident photon transfers some of its energy to an electron and the photon continues its motion in a different direction (Fig. 4.3b). The process is known as *Compton scattering*; as it occurs, the γ-ray loses energy and eventually is completely absorbed.

The attenuation of γ-rays of three different initial energies by air, water, overburden and rock is shown in Fig. 4.4. An exponential decrease in intensity with thickness is observed, with the denser materials causing greater attenuation, and the lower-energy γ-rays being attenuated more than those of higher energy. It is clear that only γ-rays originating in rock and overburden located a few tens of centimetres from the surface can escape into the atmosphere and potentially be detected by a radiometric survey. A similar thickness of overburden will absorb the radiation from an underlying bedrock source. This is a demonstration of the earlier statement that the radiometric method provides very little information about the subsurface. Attenuation in water is less than in the bedrock and overburden, but is still important. Any significant body of water will prevent the detection of γ-emissions from the underlying material; for example, the dark regions in Fig. 2.35 are due to the masking of bedrock responses by lakes. In particular, moisture in the overburden will significantly increase its

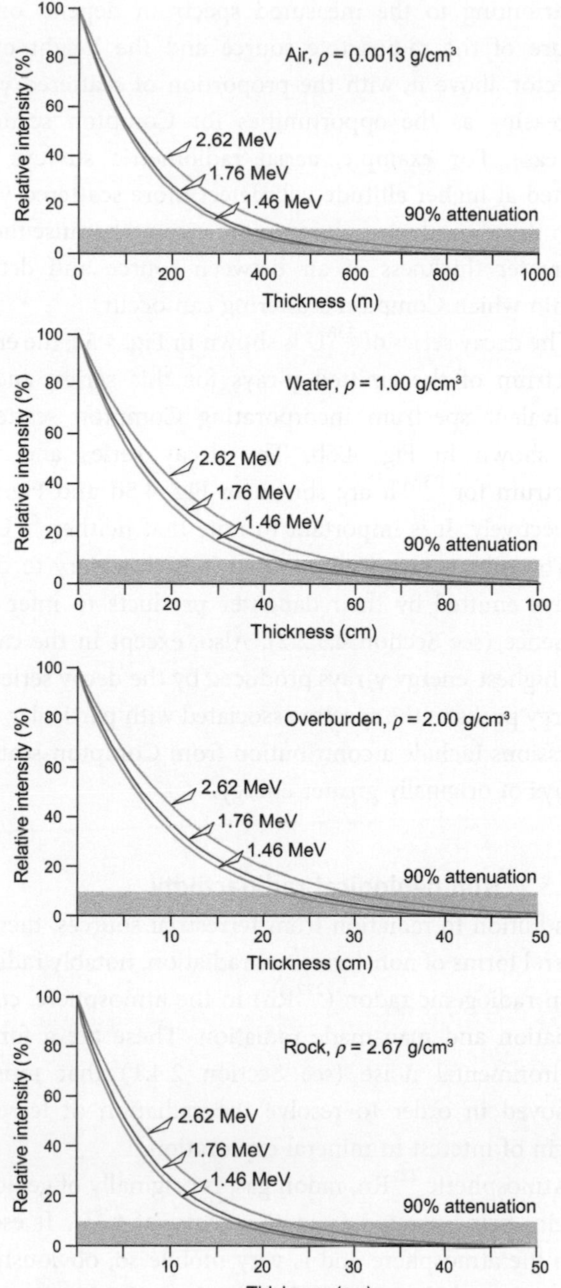

Figure 4.4 Attenuation of γ-rays of different initial energies by various materials encountered in the geological environment. The curves show the intensity of the γ-radiation after passing through a given thickness of the materials, relative to their initial intensity. Based on data in Grasty (1979). Note the different horizontal scales.

attenuation over that of a dry overburden. Upon entering the atmosphere the γ-rays are subject to much less attenuation, although radiometric measurements must still be made within a few hundred metres of the ground surface in order to detect radiation of terrestrial origin.

4.2.4 Measurement units

A variety of units are used for measuring radioactivity. The Curie (Ci) is defined as the disintegration rate of one gram of radium, equal to 3.7×10^{10} radioactive disintegrations per second. In the SI system of measurements, one disintegration per second is called a becquerel (Bq). Intensity of radiation can also be measured in terms of the quantity of radiation absorbed per unit volume or mass of a material, the measurement unit being the roentgen (R). Although these units are officially recognised by the scientific community, they are rarely used in exploration geophysics, where it is usual to measure count rate. This is the number of emissions measured by the detector in a specified time interval known as the *integration period* or *integration time*. The integration period is usually one second (giving the number of counts per second – cps) or one minute (giving counts per minute – cpm). Count rates measured during a radiometric survey depend on the integration period, the instrument used, survey variables such as the elevation of the detector, and even the weather; so the resulting dataset is a series of relative measurements.

Modern survey instruments measure not only the number of radioactive emissions, but also their energies. The energy of the radioactive emissions is measured in electron volts (eV), this being the energy acquired by an electron when accelerated through an electrical potential of one volt. However, this unit is too small for practical purposes so energies are normally expressed as millions of electron volts (MeV), i.e. 10^6 eV. As described later, instruments are calibrated to enable the conversion of the measured counts, in different energy intervals, to concentrations of the radioactive elements comprising the radioactive source. The units for radioelement concentrations are percentages or parts-per-million (ppm). This presents the data in a more geologically meaningful form. It also allows the results of different surveys to be directly compared. The data from the older uncalibrated surveys are typically presented as count rates, with the unfortunate consequence that it is difficult to compare data between different surveys.

Downhole radiometric data may be presented as cps, or if the logging tool is calibrated, in American Petroleum Institute (API) units. API units have no particular significance, merely representing an arbitrary standard to enable comparison of different datasets. API units are normally used in the petroleum industry, but count rates are more commonly used by the mineral industry.

4.2.5 Sources of radioactivity in the natural environment

There are over 50 naturally occurring radioactive elements, but terrestrial radiation is dominated by the emission products from just three elements: potassium (K), uranium (U) and thorium (Th). The half-lives of their radioactive isotopes are of the same order as the age of the Earth (5×10^9 years) and are sufficiently long that they remain comparatively abundant. The other naturally occurring radioactive elements are too rare and/or too weakly radioactive to be of significance.

Several estimates of the abundances of these elements in the continental crust are presented in the literature, but they differ owing to the assumptions upon which they are based. The figures quoted here are from Krauskopf and Bird (1995). Potassium is a volumetrically significant component of the continental crust, averaging about 25,900 ppm, but only the ^{40}K isotope is radioactive, with a half-life of 1.31×10^9 years. It comprises just 0.012% of K in the natural environment, representing an average crustal abundance of about 3.1 ppm. Thorium-232, the only naturally occurring isotope of Th, has an average crustal abundance of about 7.2 ppm. Uranium, having an average crust abundance of 1.8 ppm, has two naturally occurring isotopes, ^{238}U and ^{235}U, both of which decay via series to isotopes of lead (Pb). Uranium-238 decays to ^{206}Pb and accounts for 99.275% of naturally occurring uranium. We will not further discuss ^{235}U since it represents only 0.72% of naturally occurring uranium, and the γ-rays associated with its decay series are of low energy and not useful in radiometric surveying.

Changes in atomic number and neutron number associated with radioactive decay are described using the standard graphical presentation shown in Fig. 4.5a. Potassium-40 (^{40}K) undergoes branched decay (Fig. 4.5b). In 89% of cases, decay involves β-emission and the creation of radiogenic calcium (^{40}Ca). The remaining 11% is more significant in that it involves electron capture (K-capture) and emission of a γ-ray, with energy of 1.46 MeV. The daughter product is radiogenic argon (^{40}Ar). In practice, the γ-rays due to K detected in the natural environment do not have a single energy, but instead have a range of energies with the energy spectrum having a peak, known as a *photopeak*, at 1.46 MeV. The spectrum is a result of unscattered rays and a continuum of γ-rays of lower energy caused by Compton scattering (Fig. 4.6a). The relative amounts of scattered and unscattered γ-rays contributing to the measured spectrum depend on the nature of the radioactive source and the height of the detector above it, with the proportion of scattered γ-rays increasing as the opportunities for Compton scattering increase. For example, aerial radiometric surveys conducted at higher altitude will detect more scattered γ-rays than those conducted closer to the ground because there is a greater thickness of air between source and detector within which Compton scattering can occur.

The decay series of ^{238}U is shown in Fig. 4.5c; the energy spectrum of the emitted γ-rays for this series, and the equivalent spectrum incorporating Compton scattering, are shown in Fig. 4.6b. The decay series and γ-ray spectrum for ^{232}Th are shown in Fig. 4.5d and Fig. 4.6c, respectively. It is important to note that neither ^{238}U nor ^{232}Th emits γ-rays itself, and so it is necessary to detect γ-rays emitted by their daughter products to infer their presence (see Section 4.3.2.2). Also, except in the case of the highest-energy γ-rays produced by the decay series, the energy peaks in the spectra associated with particular γ-ray emissions include a contribution from Compton-scattered γ-rays of originally greater energy.

4.2.5.1 Non-geological radioactivity

In addition to radiation from terrestrial sources, there are several forms of non-terrestrial radiation, notably radiation from radiogenic radon (^{222}Rn) in the atmosphere, cosmic radiation and man-made radiation. These are a form of environmental noise (see Section 2.4.1) that must be removed in order to resolve the radiation of terrestrial origin of interest in mineral exploration.

Atmospheric ^{222}Rn, radon gas, is originally of geological origin, being created from the decay of ^{238}U. It escapes into the atmosphere and is very mobile so, obviously, the presence and abundance of atmospheric Rn bears no relationship to radioactive materials in the ground below. Since mapping the ground is the ultimate objective of a radiometric survey, atmospheric Rn is considered here to be of non-geological origin.

Gamma-radiation of cosmic origin is produced by primary cosmic radiation reacting with the atmosphere. Man-made radiation is the fallout from nuclear accidents and explosions. For reasons outlined in Section 4.4.4, the most significant product is caesium (^{137}Cs), which emits γ-rays with energy of 0.662 MeV. It has a half-life of 30 years and is present in much of the Earth's northern hemisphere.

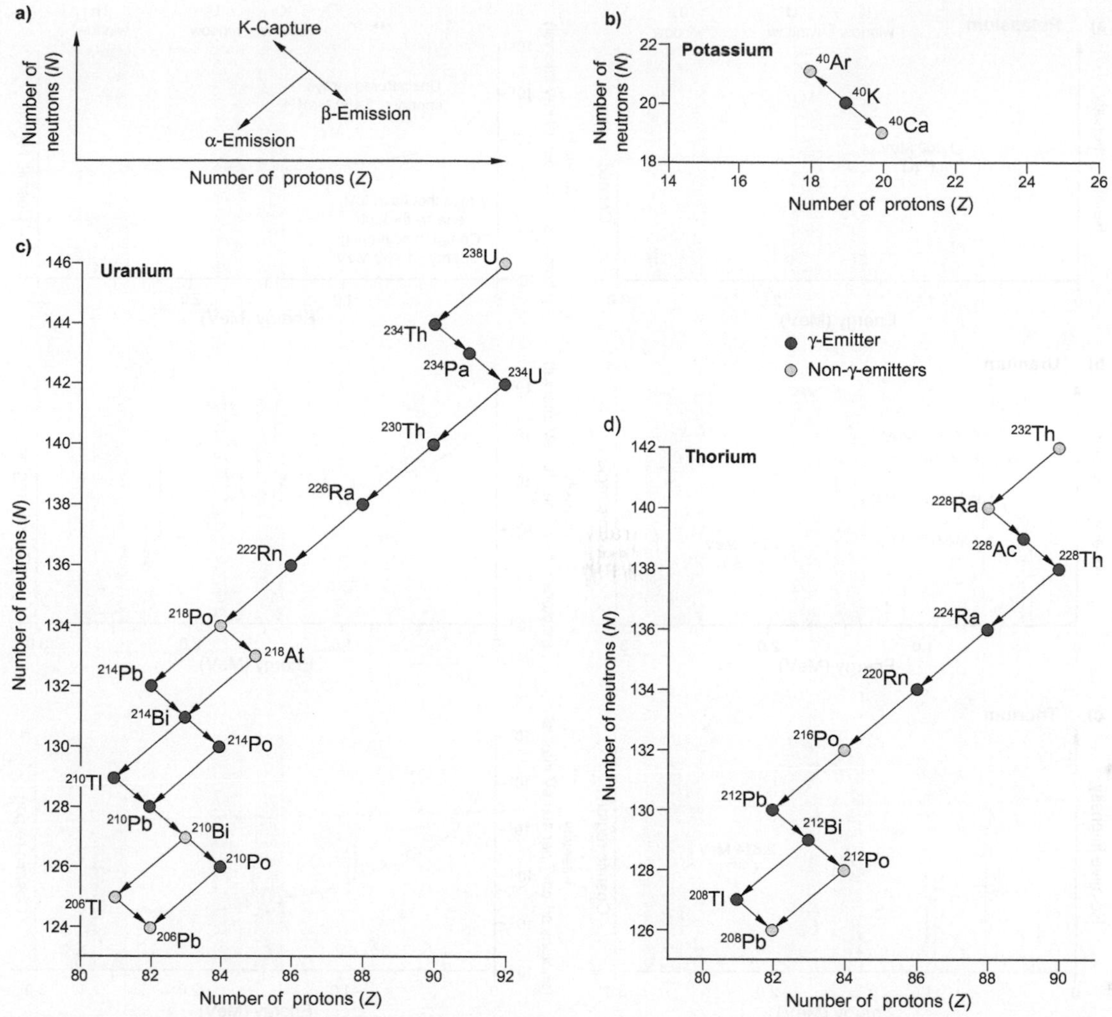

Figure 4.5 Changes in atomic number (Z) and neutron number (N) associated with radioactive decay. (a) The scheme for showing the changes associated with α- and β-emissions and K-capture (there are no changes in Z and N with γ-emission); (b) ^{40}K series, (c) ^{238}U series, and (d) ^{232}Th series.

4.3 Measurement of radioactivity in the field

The intensity of the radiation measured, from either a stationary or moving survey platform, depends on a number of parameters. These include the actual intensity of the radiation emitted by the radioactive sources, the spatial size of the radioactive sources with respect to the spatial sampling interval, the type and nature of the detector, and the time available to make individual measurements. These parameters also affect the accuracy of the measurement.

4.3.1 Statistical noise

The random nature of the radioactive decay process requires measurements to be made over comparatively long integration periods. The number of emission products measured over a series of integration periods of the same duration will generally show a statistical scatter, or deviation, about a mean value. This is known as *statistical noise*, and the standard deviation (SD) is given by

$$\mathrm{SD} = \sqrt{N} \tag{4.6}$$

where N is the number of emissions counted during the integration period. This can be expressed as a percentage statistical error of N and given by

$$\text{Statistical error } (\%) = 100/\sqrt{N} \tag{4.7}$$

This shows that the error in a measurement is inversely proportional to the square root of the total number of emissions recorded (Fig. 4.7). A small count has an inherently higher error, so it is preferable to measure a large number of emissions in order to obtain an accurate

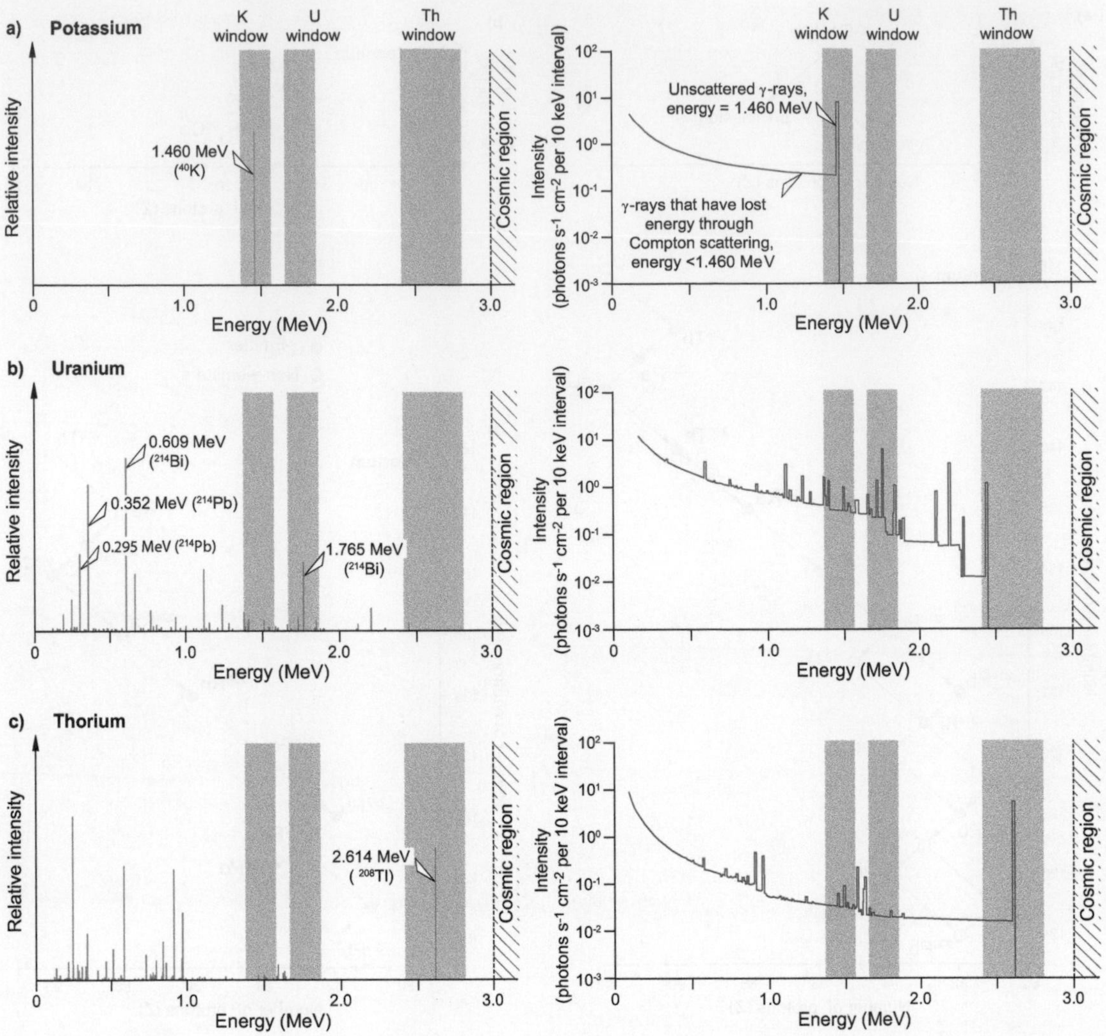

Figure 4.6 Energy spectra of γ-rays produced by the decay of the three main naturally occurring radioelements. Line spectra are shown on the left, whilst on the right are shown simulated spectra incorporating the effects of Compton scattering for a source and receiver separated by 300 m of air, (a) ^{40}K series, (b) ^{238}U series, and (c) ^{232}Th series. Also shown are the three standard energy windows used to measure γ-rays originating from these radioelements. Redrawn, with permission, from Minty (1997).

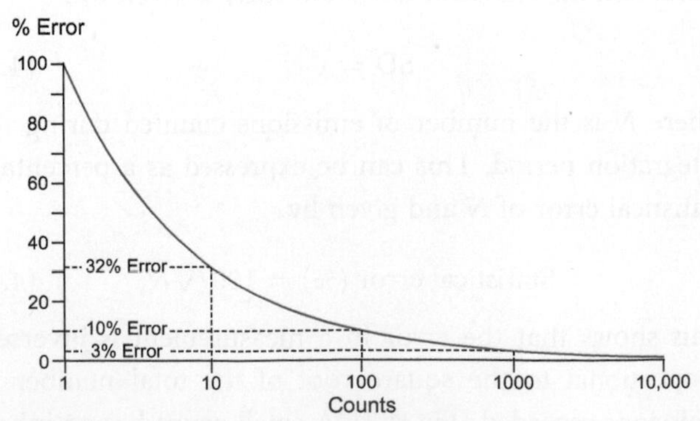

Figure 4.7 Statistical measurement error versus the number of emission counts recorded, calculated from Eq. (4.7).

measurement of radioactivity. For example, if 10 emissions are measured over an integration period of one second the error is 32%. However, if the same rate of emission is detected over a longer integration period of, say, 10 seconds to measure approximately 100 counts, the error reduces to 10%; and 1000 counts results in an error of 3%.

Measurements made over an integration period of one second can range from just a few counts to many tens of thousands of counts for highly radioactive uranium deposits. Ideally, the counting (integration) period should be longer in areas of low radioactivity where count rates are small. In practice it is kept constant, since determining levels of radioactivity is the purpose of the survey. This is an example of the geophysical paradox (see Section 1.3).

The result is that the reliability of the individual measurements may vary significantly across the survey area.

4.3.2 Radiation detectors

The popular image of radiation detectors is a Geiger–Müller counter emitting a buzzing sound that varies in intensity according to the amount of radiation present. These instruments were used in the early era of radiometric surveying, but they respond primarily to β-radiation and so could only be used for ground surveying. Furthermore, they are an extremely inefficient detector of γ-rays and are therefore no longer used for geophysical exploration.

The instruments used in modern radiometric surveys are *scintillometers* and *spectrometers* that detect γ-rays by their interaction with matter. They contain crystals of thallium-activated sodium iodide (NaI), which is quite dense and therefore effective in absorbing γ-radiation. When the crystal encounters γ-radiation, and also β-radiation, it luminesces (this property is enhanced by the thallium), and a pulse of ultraviolet light is generated in a process known as *scintillation*. The light pulse is detected by a photomultiplier tube, optically connected to the crystals, which converts it to an electrical pulse for counting by ancillary electronics. The intensity of the light pulse and, hence, the amplitude of the electrical pulse is proportional to the energy of the incident radiation. The resolution of the detector, or its ability to resolve incident γ-rays of different energies, is measured in terms of the half-width of a photopeak as a fraction of its maximum energy. Resolution is about 7 to 8% for a scintillometer, and varies slightly with the energy of the γ-ray; so rather than distinct lines in the spectra corresponding with each γ-ray emission (Fig. 4.6) there are broader peaks. Thermal drift of the detector severely degrades resolution, i.e. it causes a change in the relationship between the intensity of the light pulse and the energy of the incident γ-ray.

Increasing the volume of the NaI crystals increases the efficiency and sensitivity of the detector to γ-radiation. A detector with a large crystal volume will measure more counts over a given integration period than one containing a smaller volume, resulting in lower statistical error in the measurement. Because the integration period affects the rate at which measurements can be made, it should be as short as possible whilst achieving sufficient count levels to maintain acceptable measurement error. The higher sensitivity of larger crystal volumes allows a smaller integration period to be used for counting the pulses. However, the benefits of using larger crystals come at the expense of increased size and weight of the sensor, an important consideration for weight-sensitive airborne and size-sensitive downhole instruments.

Hand-held and downhole instruments typically have crystal volumes ranging from about 0.1 to 0.3 litres, allowing the instrument to be of an acceptable size and weight. Instruments for airborne surveying require significantly larger volumes in order to obtain the sensitivity needed to measure the lower radiation levels distant from the source. Fixed-wing airborne surveys typically use crystals of at least 16 litres volume, often 33.6 litres and sometimes larger for particularly high-sensitivity applications. Helicopter surveys may utilise smaller crystals to comply with aircraft weight limitations, although often the lower sensitivity of the sensor can be compensated by lower survey height (measurements made closer to the radioactive sources).

4.3.2.1 Gamma-ray scintillometer

Scintillation meters or scintillometers are the simplest form of instrument used for detecting γ-rays. Those for hand-held and downhole use are compact, weigh a few kilograms and are comparatively inexpensive. The total number of counts from a broad range of energy is recorded and known as the *broadband*, or more commonly, the *total-count* (TC) response. Hand-held instruments often produce an audio signal whose frequency is proportional to the count rate. A threshold capability is incorporated whereby sound is emitted only if the intensity of the radiation (the count rate) exceeds a particular (background) level. They are easy to use and are convenient for locating and mapping any source of radioactivity.

4.3.2.2 Gamma-ray spectrometer

A more complex type of instrument is the γ-ray spectrometer which can measure the number of γ-rays in discrete, narrow, pre-defined energy bands, also referred to as *windows* or *channels*. Spectrometers provide information about the geochemical nature of the radioactive source, a distinctive advantage over the limited information obtainable from single-channel broad-band scintillometers.

Spectrometers are usually 'tuned' to the energies of the radioelements of geological interest. The names and the internationally agreed limits of these energy windows are shown in Table 4.1.

The K, U and Th energy windows are chosen so as to detect energetic γ-rays emitted from relevant elements in parts of the energy spectrum where emissions from the

Table 4.1 Gamma-ray energy windows monitored during radiometric surveying.

Window	Target nuclide	Energy (MeV)
Total count		0.410–2.810
Potassium	^{40}K (1.460 MeV)	1.370–1.570
Uranium	^{214}Bi (1.765 MeV) in the ^{238}U series	1.660–1.860
Thorium	^{208}Tl (2.614 MeV) in the ^{232}Th series	2.410–2.810
Cosmic		3.0 to infinity

IAEA, 1991

other elements are weak or rare (Fig. 4.6). Since neither U nor Th emits γ-rays, emissions from daughter products are used and it is emissions from ^{214}Bi and ^{208}Tl that are measured in the U and Th windows, respectively. A fundamental assumption in measuring the emission products from a daughter element, for indirectly inferring the concentration of the parent, is that the concentration of the γ-emitting element must be in proportion to the concentration of its radioactive source, i.e. the decay series must be in equilibrium (see Section 4.2.2).

It is also worth noting from Fig. 4.6b that the intensity of the radiation in the U window is significantly less than that in the other windows, resulting in greater noise levels in this window (see Section 4.3.1). The cosmic window is measured so that γ-rays of cosmic origin can be removed from the data (see Section 4.4.3).

Small hand-held and larger portable spectrometers for ground surveying have internal memories to store the large quantity of data acquired, which is generally restricted to measurements in the K, U and Th energy windows, and the total count. Some examples are shown in Fig. 4.8. Vehicle and airborne radiometric survey systems use larger and more sophisticated spectrometers. Older-generation instruments monitored the same four energy windows. Modern instruments usually measure counts in a minimum of 256 energy windows, although 512-channel recording is also available. The large number of additional channels allows more sophisticated data processing techniques to be applied in the reduction of the survey data to obtain more accurate K, U, Th and TC data.

4.3.2.3 Fields of view

An important parameter in radiometric survey planning and data interpretation is the field of view of the detector,

Figure 4.8 Examples of modern γ-ray spectrometers. Exploranium GR-320 enviSPEC with large external sensor (rear); Exploranium GR-135 miniSPEC (front left); Radiation Solutions RS-125 Super-Spec (front right).

an example of the measurement or system footprint (see Section 2.6.2). Consider a stationary γ-ray detector located some distance above a flat ground extending laterally to infinity. Let us assume that the radioactive materials in the ground are uniformly distributed and extend to sufficient depth that the source region appears to be infinitely thick. If there were no attenuation of the γ-rays within the ground and in the intervening air, then radiation from any point on the ground could be contributing to the measured counts. This measurement is known as the *infinite-source yield*. In reality, the closer the detector is to the source the less opportunity there is for the γ-rays to be attenuated by the air. As a consequence, the radiometric measurement is biased towards radioactive sources closer to the detector, so the maximum contribution comes from immediately below it. Since attenuation is less for higher-energy γ-rays, the measurement is also influenced more by higher-energy radiation.

Points on the ground equidistant from the detector, i.e. separated by equal thickness of air and source material, define a *circle-of-investigation* in the plane of the ground surface, with the detector located above its centre (Fig. 4.9a). The amount of radiation originating from within the circle, as a proportion of that from a theoretical infinite source, can be calculated for a range of attenuation and a range of detector height above ground. Figure 4.9b shows the relationship between the diameter of the circle-of-investigation and the detector height for various percentages of the infinite-source yield, indicating the bias toward radioactive material located near the centre of the

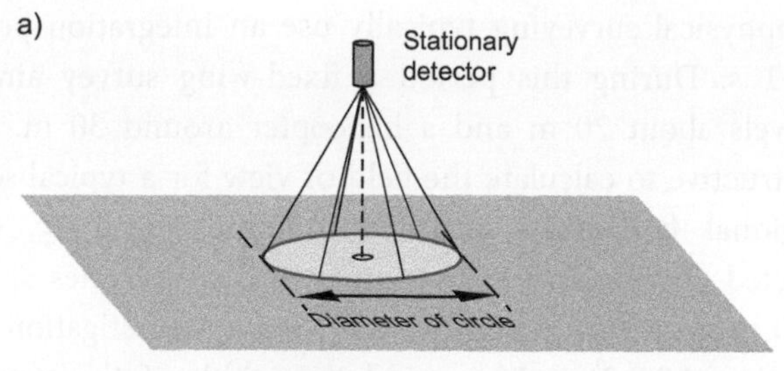

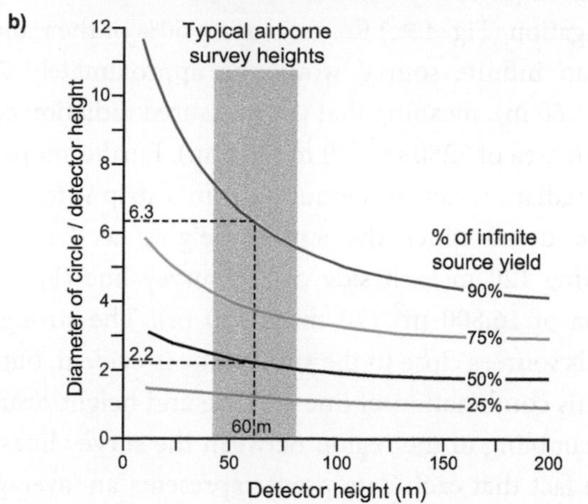

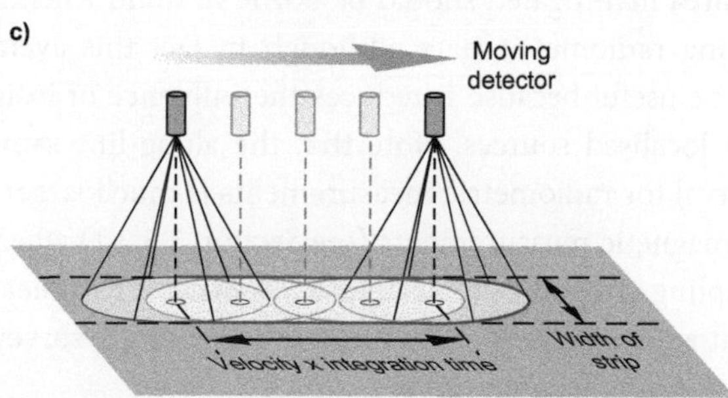

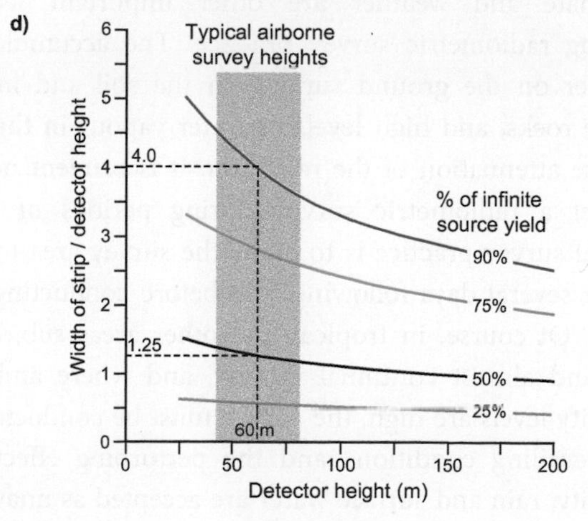

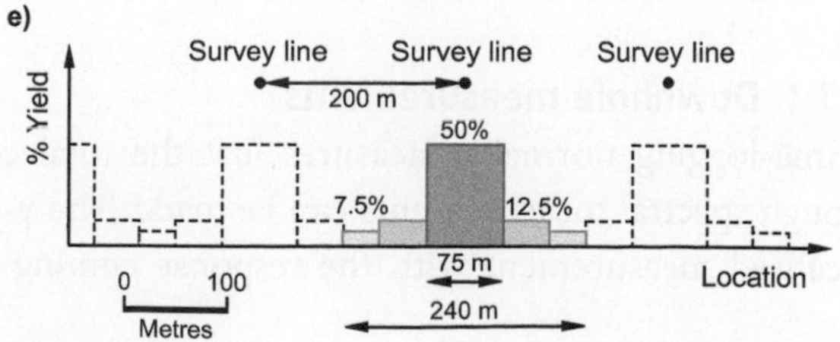

circle, i.e. within smaller circles. Note that, for a particular percentage of the infinite-source yield, the size of the circle-of-investigation increases with increasing detector height; so reducing the height means that the individual survey readings need to be closer together to avoid under-sampling the distribution of radioactivity.

For the case of a moving detector, the circle-of-investigation is replaced by a *strip-of-investigation* whose centre is directly beneath the detector (Fig. 4.9c). The length of the strip in the survey line direction depends upon the integration period and the speed at which the detector is moving, plus a contribution from the semicircular ends of the strip. Radiation emanating from parallel strips on both sides of the survey line will contribute equally to the measurements, with the greatest contribution from the strip immediately beneath the detector, and progressively less from those offset to either side, neglecting the end effects of the strips. Figure 4.9d shows the relationship between the width of the strip-of-investigation and the detector height for various percentages of the infinite-source yield, confirming the bias toward radioactive material located near the centre of the strip. For a given detector height, the width of the strip-of-investigation is considerably smaller than the diameter of the circle-of-investigation because the moving detector produces overlapping sample-areas along-line. To understand why, it is useful to think of the strip as being composed of a series of overlapping circles-of-investigation (Fig. 4.9c). Each radioactive source will lie within more than one circle and will be closer to the centre as the detector moves past them, so those sources closer to the centre line contribute most to the measurement. This results in a greater bias from sources closer to the detector (the survey line) than for the case of a stationary detector.

It is clear from Figs. 4.9b and d that detector height is the most important parameter in determining the area of the circle-of-investigation, and the strip-of-investigation

Figure 4.9 Fields of view. (a) The field of view for a stationary γ-ray detector. (b) The relationship between the diameter of the circle-of-investigation (expressed as a multiple of detector height) and height of a stationary detector for selected percentages of the infinite-source yield. (c) The field of view for a moving γ-ray detector. (d) The relationship between the width of the strip-of-investigation (expressed as a multiple of detector height) and height of a moving detector for selected percentages of the infinite-source yield. (e) The relative contributions of strips of ground parallel to the survey line for a moving detector located 60 m above the ground on survey lines spaced at 200 m apart. (b) and (d) redrawn, with permission, from Pitkin and Duval (1980).

for a moving sensor. Topographic variations within the circle or strip could change the detector height significantly, particularly if they form a large portion of the circle/strip. For an airborne survey, terrain rising up to the survey platform (hills) will reduce the investigation area, and falling terrain (valleys) will increase the area.

These definitions of the field of view are based on the geometry of the detector and sampling system, and assume that the radioactive source is uniformly distributed. However, the relative contributions of different areas of the ground also depend on the distribution of the radioactive sources. A strong source of radioactivity located outside the nominal field of view can be detected. The detection of smaller and weaker sources requires closer-spaced measurements at lower survey height. It is possible to calculate the theoretical response of zones of increased or decreased radioactivity of a given shape, size and position relative to the survey line, and use these for survey design. The presence of other variable factors, such as overburden thickness and moisture content, reduces the usefulness of these calculations. Often, in an exploration programme survey, specifications are optimised for the acquisition of other types of data, such as magnetics or EM, with space and weight limitations within the survey aircraft also a constraint on detector (crystal) size. Consequently, the specifications may not be optimal for the radiometric data.

4.3.3 Survey practice

The size of the detector crystal has a significant influence on radiometric survey practice. For measurements made on, or in, the ground, a stationary measurement can be made over a large integration period, possibly several minutes, in order to minimise measurement error, albeit at the cost of longer data acquisition time. This is particularly advantageous when the crystal volume is small and of low sensitivity, or when high-sensitivity measurements are required. For reconnaissance work, it is often convenient to select a small integration period, say 1 s, and traverse the area in search of 'hot spots' of elevated radioactivity worthy of further, more detailed, investigation.

For the case of mobile platforms, the along-line sampling is determined by the instrument's integration period and the platform's speed over the ground; the moving platform travels a significant distance during the integration period, and this reduces measurement precision and resolution of ground features. Instruments for airborne geophysical surveying typically use an integration period of 1 s. During this period a fixed-wing survey aircraft travels about 70 m and a helicopter around 30 m. It is instructive to calculate the field of view for a typical semi-regional fixed-wing airborne radiometric survey, conducted 60 m above the ground with survey lines spaced 200 m apart. The length of the strip-of-investigation (see Section 4.3.2.3) is 70 m and the width of the strip-of-investigation (Fig. 4.9c) for measuring 50% of the radiation from an infinite source would be approximately 75 m (1.25 × 60 m), meaning that the measured radiation comes from an area of 5250 m^2 (70 m × 75 m). Furthermore, 90% of the radiation can be measured from a strip whose width is around 4.0 times the survey height, i.e. 240 m, or extending 120 m each side of the survey line (Fig. 4.9e), an area of 16,800 m^2 (70 m × 240 m). The strong bias towards sources close to the survey line is evident, but note how this combination of line spacing and height results in poor sampling of the region between the survey lines.

The fact that each data point represents an 'average' of the area investigated should be borne in mind when interpreting radiometric data, although in fact this averaging can be useful because it reduces the influence of insignificant localised sources. Note that the along-line sampling interval for radiometric measurements is much larger than for magnetic measurements (see Section 3.5.3.1) where the sampling interval of typically 0.1 s translates to a measurement approximately every 7 m for a fixed-wing survey and 3 m for a helicopter survey.

Climate and weather are other important factors affecting radiometric survey practice. The accumulation of water on the ground surface, in the soil and in the surface rocks, and high levels of water vapour in the air, increase attenuation of the radiation. It is prudent not to conduct a radiometric survey during periods of rain. Normal survey practice is to allow the survey area to dry out for several days following rain before conducting the survey. Of course, in tropical and other areas subject to large and almost continual rainfall, and where ambient humidity levels are high, the survey must be conducted in the prevailing conditions and the perturbing effects of humidity, rain and surface water are accepted as unavoidable form of environmental noise.

4.3.3.1 Downhole measurements

Gamma-logging normally measures just the total count, although spectral measurements can be made. The γ-log is a localised measurement with the response coming from materials within a few centimetres of the drillhole wall. The denser the surrounding rock the smaller is the contributing volume. This is not necessarily a major problem, however, since it means that the log has high spatial resolution, allowing centimetre-scale downhole variations to be identified. This is also dependent on the logging speed, the rate at which the tool is raised in the drillhole. Typical speeds are 1 to 6 m/min with spatial resolution of 'thin' formational layers and contacts increasing with decreasing speed. See Section 4.7.5 for details of the interpretation of γ-logs.

4.4 Reduction of radiometric data

The γ-ray spectrum recorded in geophysical surveying comprises radiation from a number of different sources (see Section 4.2.5) in varying proportions. Of these, radiation originating from one or more of the three common radioelements present in the ground, i.e. potassium (K), uranium (U) and thorium (Th), is the signal of interest, and all other sources of radiation form unwanted noise.

The survey data are reduced to remove various sources of noise, both environmental and methodological (see Section 2.4), to reveal the radiometric response of geological significance, which is used to calculate elemental ground concentrations of the three radiometric elements. Significant non-geological factors affecting the measurements include: the detector's inability to perfectly count all the γ-rays; the position, size, shape and chemical composition of the radioactive sources; survey height; nature of the overburden and its thickness; vegetation; air temperature, pressure and humidity; the presence of moisture; and the presence of atmospheric radon. The effects of some of these can be partly corrected by making appropriate secondary measurements, for example air temperature and pressure. Other variables require knowledge of the local geology, some to the extent that would make a radiometric survey superfluous (an example of the geophysical paradox; see Section 1.3). In the absence of the necessary data or theoretical models from which to calculate their affect, many of the required corrections are empirical. Consequently, the reduction of radiometric survey data depends upon complex instrument calibration procedures and the adoption of some simplifying assumptions about the local environment. Errors associated with each stage of the reduction process are additive, so adherence to calibration procedures and constant monitoring of ancillary parameters throughout the survey are essential in order to minimise uncertainty in the final result.

Since the 1990s, the reduction of radiometric data uses entire recorded γ-ray energy spectra, generally 256 energy channels, whereas previously it was based primarily on the counts recorded in just the K, U and Th energy windows. Current developments in data reduction focus on improved methods of noise reduction based on statistical smoothing of the measured spectra. The various corrections applied are known by the parameters that they compensate, and we consider each one in the order that it is applied to the survey data (Fig. 4.10), as this can be critical. A detailed description of these procedures applied to airborne surveying is provided by Minty *et al.* (1997).

The corrections described here are pertinent for spectrometer measurements made above ground. Downhole measurements require less correction: for example, those related to survey height, and to atmospheric and cosmic effects, are not applicable. Total-count data are not usually corrected at all.

4.4.1 Instrument effects

The first stage in the reduction process is to correct for the effects of the measuring equipment. Thermal drift of the spectrometer reduces its ability to accurately discriminate the energy of the incident γ-rays, i.e. it degrades the spectrometer's resolution. Stability of the system gain and loss of sensitivity due to crystal or photomultiplier damage are monitored by checking the resolution pre- and post-survey. This is done by calibrating the instrument's measured energy response against a reference source of known energy, either ^{137}Cs or ^{133}Ba as these elements emit γ-rays at prominent energy levels of 0.662 and 0.352 MeV respectively. Modern spectrometers automatically monitor and correct for drift during operation and contain heaters which maintain the crystals at a constant temperature.

The γ-ray spectrometer takes a finite time to record and process the radiation detected during the integration period. During the recording time it is unable to process incoming γ-rays, so these are lost. This is known as the *dead time* and is typically less than 15 μs. It only becomes important for count rates above about 1000 cps when the number of counts occurring during the dead time, and therefore lost by the instrument, becomes significant. Modern instruments correct for this automatically.

Other problems related to the instrument include the fact that the NaI sensor is more efficient at detecting lower-energy γ-rays, and to the possibility of accidental

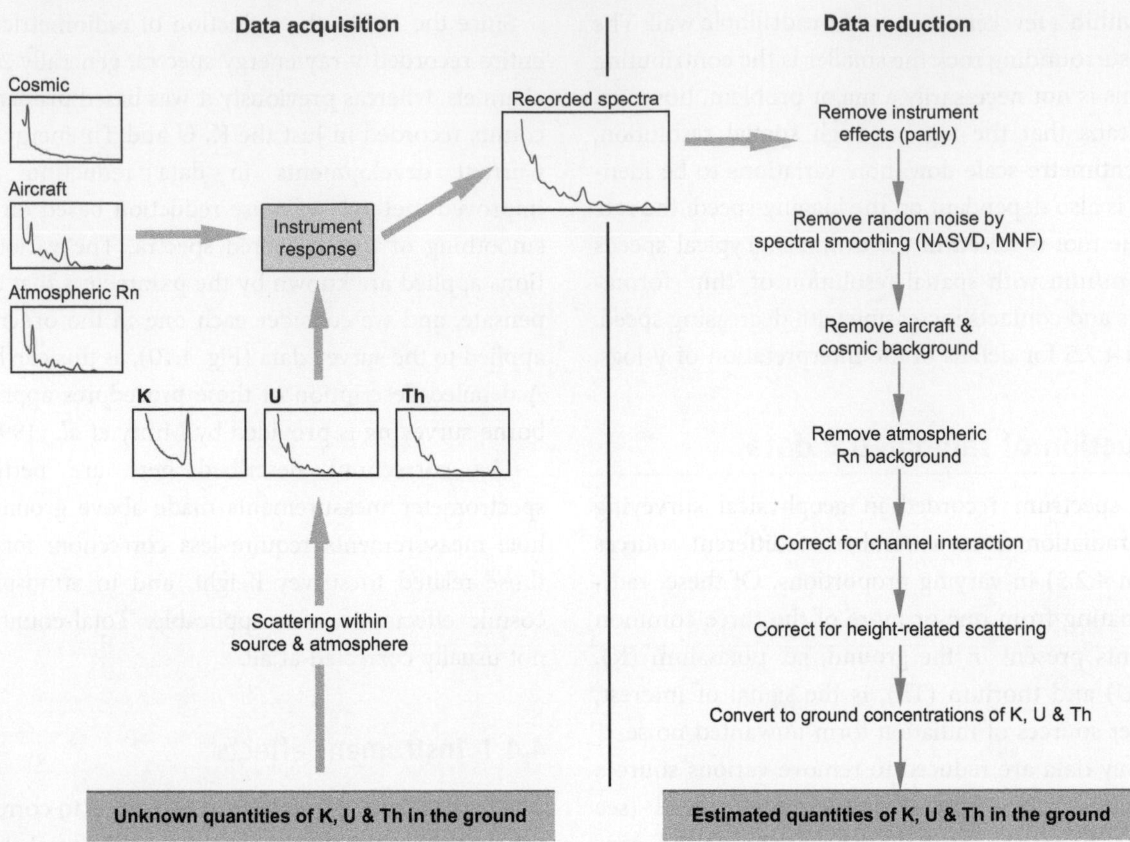

Figure 4.10 Factors contributing to the γ-ray spectra measured in the field and the corrections applied to reduce the survey data. The reduction process is designed to resolve and count γ-rays emitted by radioelements in the ground beneath the sensor, and determine the local concentrations of these elements.

summing – where the photomultiplier combines the effects of two nearly coincident radiation events and erroneously records them as a single higher-energy event. These reduce the instrument's ability to count scintillation events accurately across the energy spectrum. There is no correction available for these effects.

4.4.2 Random noise

A significant part of the noise component of a radiometric measurement is random, caused by the random nature of the radioactive decay process. This means there is no correlation between the noise recorded in different energy channels and, therefore, it can be suppressed using channel correlation-based smoothing techniques, also referred to as *spectral smoothing*. One approach is to treat the measured spectra as though they were composed of several component spectra. The challenge then is to identify these components and reconstruct the measured spectra from each of them, and in so doing identify the noise component. A more widely utilised approach is based on statistical analysis of the data. The two most common methods are maximum noise fraction (MNF) and noise-adjusted singular value deconvolution (NASVD), both fairly recent developments.

The NASVD and MNF methods are both quite complex, as are the arguments about their relative merits (Minty, 1998). Both utilise *principal components analysis* (PCA), which in turn requires pre-conditioning of the survey data so that its statistical properties meet fundamental requirements of the PCA method (see Davis, 1986). The main differences between the two methods relate to how this is achieved, because both are based on the fact that the signal is concentrated in the lower-order principal components, whilst the uncorrelated (random) noise is concentrated in the higher-order components. The smoothed spectra are reconstructed using the lower-order components only, thus removing the random noise. It appears that in many cases similar results are obtained, but when they differ it is a function of the survey data, as a high correlation between U and Th affects the result. The important fact is that both methods can significantly reduce errors in channel counts compared with conventional processing: typically by 42%

in U, 37% in Th and 14% in the K channel (B. Minty, pers comm).

There is no doubt that these reduction algorithms significantly increase the resolution of survey data. The improvements exhibited in the corrected survey data, as judged from the clarity of geology-related patterns and textures, are often dramatic. Good examples are presented by Minty (1998) and Dickson and Taylor (2000).

4.4.3 Background radiation

Gamma-radiation from the aircraft and from space is background radioactivity that contributes to the measurements. The former is constant and due to radioactive elements in the materials from which the aircraft and recording equipment are made. Cosmic radiation has a fairly constant energy spectrum everywhere, although its amplitude reduces with decreasing altitude.

Removal of the cosmic background is based on the fact that any radiation greater than 3.0 MeV must be of cosmic origin, since terrestrial γ-rays have lower energies (see Fig. 4.6). Airborne spectrometers include a cosmic-energy channel to measure the cosmic radiation, usually in the range 3 to 6 MeV (Table 4.1). The cosmic contribution in any energy window is proportional to the radioactivity in the cosmic window and, because its energy spectrum is fairly constant, its contribution can be determined. Correction parameters for the cosmic background are determined by acquiring survey data at a number of survey heights (typically 3000 to 4000 m) over a large body of water, such as a lake or the ocean, where ground radiation is minimal owing to the water's ability to absorb γ-rays (see Section 4.2.3). For all energy windows, there is a linear relationship between the intensity of the radiation and survey height, with the aircraft contribution appearing as a constant background. The cosmic contribution in each channel of normal survey data can be calculated from the slope of the linearity, for that channel, and the counts measured in the cosmic channel. This is combined with the constant aircraft response for the energy window and removed as a single *background correction*.

4.4.4 Atmospheric radon

One of the most difficult aspects of radiometric data reduction is compensating the effects of radon gas (^{222}Rn) and its daughters in the atmosphere. Radon-222 occurs above ^{214}Bi in the ^{238}U decay series (Fig. 4.5c) so it is a source of uranium-channel γ-rays. It is constantly present in the atmosphere because its daughter products attach themselves to airborne aerosols and dust particles. Its distribution is not uniform; it is affected by such factors as wind, moisture and temperature, resulting in significant variations in concentration with location (especially height) and time. Furthermore, increasing barometric pressure forces Rn into the pores of the soil, from which it is then released into the atmosphere during periods of lower pressure. The problem can be severe, with Rn sometimes accounting for a significant portion of the counts in the U channel.

One method of removing atmospheric Rn requires the use of an upward-looking radiation detector shielded from the terrestrial radiation. Calibration surveys are conducted over water on days having different amounts of ^{222}Rn in the air. The ratio of counts measured by the upward-looking detector to those measured by the main downward-looking detector in each energy window provides channel correction factors that are applied to the downward-looking measurements. The main disadvantage of this approach is the additional space and weight associated with the shielded upward-looking detector, and that Rn is not necessarily uniformly distributed within the air.

Spectral signature correction methods are based on the differential attenuation of γ-rays of different energies. As described in Section 4.2.3, lower-energy γ-rays experience greater attenuation than higher-energy rays. This means that a decay series spectrum changes as the distance between source and detector changes, with the greatest spectral modification occurring in the lower energy part of the spectrum. The relative amplitudes of the various photopeaks will be different for a source close to the detector, such as atmospheric Rn, and for more distant sources in the ground. Spectral signature methods compare peaks in the observed spectra and estimate the relative contributions from terrestrial and atmospheric sources. Those used are the ^{214}Bi peaks at 0.609 and 1.765 MeV, and the ^{214}Pb peaks at 0.295 and 0.352 MeV (see Fig. 4.6b). Detailed descriptions of these methods and examples of their effectiveness in removing Rn-related noise are provided by Minty (1992) and Jurza *et al.* (2005).

Note that methods based on the ^{214}Bi 0.609 MeV photopeak are only effective in areas devoid of ^{137}Cs (see Section 4.2.5.1), because its single photopeak at 0.662 MeV contaminates the 0.609 MeV emissions from U and Rn. Consequently, these methods are likely to be ineffective in much of the northern hemisphere.

4.4.5 Channel interaction

The spectrometer energy windows (Table 4.1) are chosen so that they contain unscattered γ-rays whose energies are diagnostic of their source element. However, Compton scattering of the higher-energy γ-rays causes them to appear in the energy windows used to infer the presence of particular elements. The highest energy window, Th, is unaffected (Fig. 4.6), but γ-rays scattered from the Th decay series will 'contaminate' the lower-energy U and K windows. The lowest energy window, K, is further contaminated by γ-rays scattered from the U decay series. Errors due to these interactions between the energy windows are compensated by the Compton correction coefficients, or channel stripping ratios (to strip the contamination from the various channels). The stripping ratios are an estimate of the ratio of counts due to scattered γ-rays in the relevant lower-energy channel(s) relative to the counts of unscattered γ-rays in the relevant elemental channel(s). They predict the number of counts expected in the lower-energy channel(s) for every count in the higher-energy channel(s).

The stripping ratios can be obtained from pure U and Th spectra. They are determined by making radiometric measurements with man-made sources of known elemental concentration, and depend on detector height and the intervening material. The effect of source geometry on the stripping ratios is minor, although sensor height is important. Commonly, for airborne systems these are concrete slabs or pads, with dimensions of several metres, located at airfields. Smaller hand-size samples are used for portable instruments. Spectra are obtained by locating the survey spectrometer (the survey aircraft) over the synthetic sources (the pads), and the effects of different survey heights simulated by partial shielding of the detector. The spectra so obtained are converted to pure K, U and Th spectra based on the known elemental concentrations of the samples (the pads), and the stripping ratios obtained for the various simulated survey heights.

The reduced γ-ray counts obtained by applying the dead-time correction, spectral smoothing, cosmic background correction, atmospheric Rn removal, and the channel interaction correction to each measurement channel are known as *stripped data*.

4.4.6 Height attenuation

The intensity of the γ-radiation measured at the detector decreases with increasing height above the radioactive source. The rate of attenuation with increasing height depends on the energy of the radiation and source geometry, with greater attenuation from narrower sources than from broader ones. In addition, the varying amount of air between the source and the detector causes variations in the rate of attenuation (see Section 4.2.3). Overall, height attenuation is an approximately exponential decrease in measured counts with increasing height. It is usually assumed that topography is subdued: that is, the sources of radiation are broad and flat.

A correction for attenuation with height in each energy window is obtained from calibration surveys conducted at a number of heights over a calibration range, ideally an area where the elemental distributions in the ground are known and constant with time. The *height attenuation coefficient* for each energy window is applied to the measured counts using the actual measurement height corrected for the effects of changes in temperature and pressure, since these affect the density of the air and its attenuation.

4.4.7 Analytical calibration

The height-corrected stripped counts in each of the three energy channels, K, U and Th, are linearly related to the ground concentrations of their respective elements. The relationship between counts and ground concentration is a function of source type and the detector, and is established for each radioelement through analytical calibration of the survey system.

As with the attenuation-with-height correction, surveys are conducted over a calibration range where radioelement concentrations of the ground are known. Ideally, ground radiometric measurements should be made simultaneously to allow correction for environmental variables such as moisture content, which is not possible if only geochemical assay data are available. The relationships between counts and element concentrations are known as the *sensitivity coefficients* and are obtained by comparing the corrected counts in each measurement channel with the known elemental concentrations of the calibration range.

The reduced survey data obtained by applying the sensitivity coefficients to the height-corrected stripped data are the estimated ground concentrations of the radioelements, expressed as a percentage for K and as parts per million for U and Th. Since the U and Th concentrations are actually inferred from emissions from Bi and Tl, respectively (see Section 4.3.2.2), they will only be correct if the relevant decay series are in equilibrium, so they are referred to as *equivalent concentrations*, eU and eTh.

4.5 Enhancement and display of radiometric data

Reduction of radiometric data to elemental concentrations transforms the data into a readily interpretable form; there is limited scope for enhancement of the individual channel data, cf. the methods described in Section 2.7.4. The most important forms of numerical enhancements for radiometric data are filtering to remove short-wavelength noise, amplifying the variations in a channel, e.g. working with eU^2 instead of eU (see *Amplitude scaling* in Section 2.7.4.4), and computing ratios of the channels (see Section 2.7.4.2). Several techniques are used to visually enhance and integrate the different channels to assist with their analysis, which we describe later. The individual elemental concentrations and integrated channel data are usually displayed as maps and images, with total-count data also commonly used.

The integration of the three elemental concentration channels, potassium (K), uranium (eU) and thorium (eTh), with the total-count (TC) channel and other data-sets, such as terrain and multispectral scanner data, is an important aspect of the geochemical/geological interpretation of radiometric data. The need to integrate multiple data-types has led to method-specific forms of display, and the use of multivariant statistical methods to quantitatively analyse the integrated data.

In addition to geochemical/geological mapping, radiometric data may be used to search for anomalies indicative of radioactive mineralisation. In these cases the mode of display is chosen to emphasise target areas (see Section 2.8).

4.5.1 Single-channel displays

A form of display that is still occasionally encountered when working with old total-count datasets is shown in Fig. 4.11. It is simply a map showing various symbols plotted on the survey lines to indicate when particular levels of radiation of a single channel, relative to a predetermined background level, are exceeded. It is important to note that the anomaly is offset, in the survey direction, from its source owing to the finite integration time of the sensor – a fact that must be accounted for when locating the anomaly on the ground. This type of display was common during early exploration for uranium when

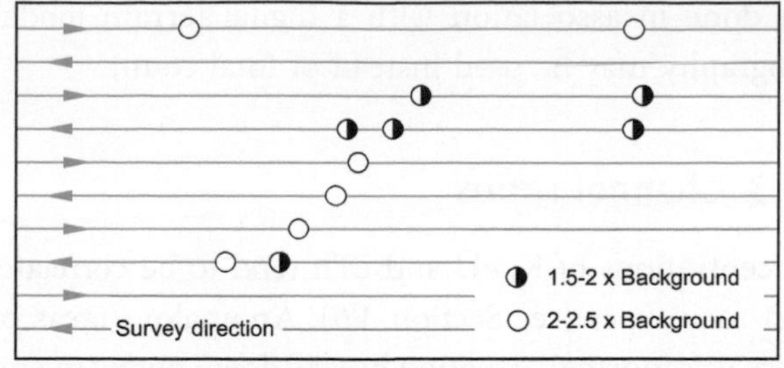

Figure 4.11 Radiometric data presented to highlight only anomalous responses. The symbols show the locations of anomalous readings along the survey lines.

exploration strategies focused on the identification of ‘radioactive hot-spot’ anomalies.

Downhole measurements are displayed as logs showing the measured level of radioactivity versus depth. Occasionally, equivalent displays are used for surface measurements, i.e. single-channel stacked-profile plots (see Cowan *et al.*, 1985), but grey-scale or pseudocolour images are the normal form of presentation. The larger along-line data interval and larger footprint of radiometrics, compared with the smaller sampling dimensions of aeromagnetics, need to be considered when gridding the line-data (see Section 2.7.2) to ensure that the resultant images depict the actual (relatively lower) resolution of the radiometric survey. A cell size of 1/3 of the line spacing usually accounts for the larger sampling dimensions. Note that the gridded values may span a large amplitude range and form a complicated multimodal distribution, so care must be taken when creating an image from the gridded data.

4.5.2 Multichannel ternary displays

Ternary images (see Section 2.8.2.4) are a very effective way of displaying and visually enhancing multichannel radiometric data in a single image. By convention, K is assigned to red, eTh to green and eU to blue. When all three elements are abundant the display tends towards white, and when all three are sparse it tends towards black. They allow (semi-)quantitative classification of the ground’s chemical signature and lithological discrimination based on colour. Consequently, they are by far the most common form of display for radiometric data.

Sometimes it is useful to integrate and display total count with the element concentrations displayed in the ternary image by adjusting the brightness of the colour at each point with the total count values at the point. As discussed below, the interpretation of radiometric data is best done in association with a digital terrain model, so topography may be used instead of total count.

4.5.3 Channel ratios

Concentrations of K, eU and eTh tend to be correlated in most rock types (see Section 4.6). Anomalous areas occur where this 'normal' situation breaks down owing to enrichment and/or depletion of one or more of the three radioelements, which can often be identified in the elemental channel data. The individual channels show the absolute concentrations of the elements, whereas their ratios, i.e. eTh/K, eU/K, eU/eTh and their reciprocals, show their relative concentrations. Using amplified channel values increases the influence that variations in those channels have on the ratio and reduces the effects of variations in the denominator term. They are particularly effective for channels having low count rates with respect to the other channels, typically eU and eTh. The ratios eU^2/eTh and eTh^2/K often provide higher resolution than their corresponding elemental ratios.

Another strategy is to combine various elements into a single term. The ratios K/(eU + eTh), eU/(K + eTh) and eTh/(K + eU) show the variation of one element with respect to the other two; and (eTh × eU)/K can be used to target areas where both U and Th vary with respect to K. Amplified values can be substituted where appropriate.

Note that the elements forming the various types of ratios should be specified in the same measurement units, i.e. as counts per second or as concentrations in ppm (multiply K% concentration by 10,000 to obtain ppm). Channel ratios can be highly variable and large amplitudes can be attenuated by displaying the logarithm or the inverse tangent of the ratio.

Channel amplification and channel ratios are very important enhancements for radiometric data because they reduce the effects of variable outcrop and soil water-content etc. For example, the absolute level of radiation from a particular rock outcrop will be reduced in areas where it is weathered deeper or subject to surface moisture, but the elemental ratios can be expected to remain (relatively) unaffected, indicating the same lithotype. They are also less sensitive to errors in the height correction and to non-planar source geometry etc. Ratios of elemental concentrations provide higher resolution of variations in elemental concentrations than the elemental channels themselves, so they correlate better with lithological units and are useful for characterising different lithotypes. They are also useful for highlighting zones of preferential radioelement enrichment and alteration.

Caution must be exercised when working with channel ratios, as a high ratio value is obtained when the level of the radioelement in the denominator (lower term) is anomalously low. This can be a spurious indicator of concentrations of the radioelement occupying the numerator (upper term); amplifying (squaring) the numerator helps to minimise this effect.

The various ratio groups can be displayed as ternary images, e.g. (K/eTh, eTh/eU, K/eU) etc. Moreover, a particular radioelement can be combined with its respective ratios to form a composite ternary image of that radioelement, for example (K, K/eTh, K/eU), (eTh, eTh/K, eTh/eU) and (eU, eU/eTh, eU/K), and amplified values can be substituted where appropriate. These images emphasise the abundance of the principal element relative to its ratios with the other two elements. Areas where the principal element has high concentration will appear white when the other two elements both have lower concentrations, and will appear red where the concentrations of both increase.

4.5.4 Multivariant methods

Multivariant techniques can be applied to assist integration and interpretation of radiometric and complementary datasets. These include factor, cluster and principal-component analysis (PCA) (Davis, 1986). Their common goal is to reduce the complexity of interpretation by identifying fewer, or particularly significant, parameters that most effectively represent variations in the various datasets. A discussion of these quite complex statistical methods is beyond our scope. Some examples are described by Pirkle *et al.* (1980), Lanne (1986), Pires and Harthill (1989), and Ranjbar *et al.* (2001).

4.6 Radioelements in the geological environment

Radiometric responses ultimately depend on the presence of mineral species that contain one or more of the radioelements K, U and Th. These may be primary constituents of the mineral or may occur in trace quantities as impurities in the crystal lattice.

Potassium, which is by far the most abundant of the three radioactive elements in the geological environment,

Table 4.2 **Chemistry of various potassium, uranium and thorium minerals.**

Mineral	Formula	Wt% K	Wt% U	Wt% Th
Feldpars				
Orthoclase	$KAlSi_3O_8$	14.05		
Microcline	$KAlSi_3O_8$	14.05		
Sanidine	$(K,Na)(Si,Al)_4O_8$	10.69		
Anorthoclase	$(Na,K)AlSi_3O_8$	3.67		
Micas				
Biotite	$K(Mg,Fe^{2+})_3[AlSi_3]O_{10}(OH,F)_2$	9.02		
Muscovite	$KAl_2(AlSi_3)O_{10}(OH,F)_2$	9.81		
Phlogopite	$KMg_3(AlSi_3)O_{10}(F,OH)_2$	9.33		
Glauconite	$K(Fe^{3+},Al)_2(Si,Al)_4O_{10}(OH)_2$	5.49		
Clay minerals				
Illite	$K_yAl_4(Si_{8-y}Al_y)O_{20}(OH)_2$	6.03		
Evaporites				
Sylvite	KCl	52.45		
Carnallite	$KMgCl_3{\cdot}6H_2O$	14.07		
U and Th minerals				
Brannerite	$(U,Ca,Ce)(Ti,Fe)_2O_2$		33.54	
Carnotite	$K_2(UO_2)_2(VO_4)_2{\cdot}1\text{–}3H_2O$	8.67	52.77	
Coffinite	$U(SiO_4)_{1-x}(OH)_{4x}$		72.63	
Monazite	$(Ce,La,Nd,Th,Y)PO_4$			4.83
Thorite	$(Th)SiO_4$			71.59
Uraninite (pitchblende)	UO_2		88.15	
Uranophane	$Ca(UO_2)_2Si_2O_7{\cdot}6H_2O$		40.59	

is an alkali element and occurs mainly in alkali feldspars, micas and illite (Table 4.2). It also occurs in evaporite minerals, which may constitute ore horizons for mining of potash.

Uranium and thorium have large ionic radii and are comparatively highly charged, so they rarely substitute for other elements in common silicate minerals, but instead tend to form minerals of their own (Table 4.2). Uranium can form either an oxide or a silicate and in these forms may occur as a film along grain boundaries. Quartz, feldspar and other rock-forming minerals may contain trace quantities of U and Th, but in the main these occur in accessory minerals such as zircon, monazite, sphene, rutile and xenotime. Uranium can substitute for a variety of cations of similar size and charge, for example thorium, zirconium, titanium and calcium. The U-rich varieties of thorite (urano-thorite) may contain more than 20% U. The degree of substitution of U and Th can vary widely, but the common rock-forming minerals, such as feldspars, pyroxene etc., may contain up to a few 10s of ppm of either (Table 4.3). The relative abundance of these minerals means they may represent a significant component of a rock's U and Th content.

Table 4.3 Radioelement contents of selected minerals in which the radioelements substitute for other elements.

Mineral	K (%)	U (ppm)	Th (ppm)
Amphibole		0–0.3	0–0.5
Ilmenite		0–50	
Olivine		0–1.5	0–4
Plagioclase	0–0.5	0–5	0–3
Pyroxene		0–40	0–25
Quartz		0–5	0–6
Rutile		0–194	
Sphene		0–700	0–1000
Zircon		0–6000	0–4000

The behaviour of U and Th in the geological environment is described in detail by Gascoyne (1992). Under chemically reducing conditions, both U and Th exist in a tetravalent state but, importantly, under oxidising conditions U occurs in a hexavalent state. Tetravalent U and Th have similar ionic radii, equal coordination number with respect to oxygen (8) and complete outermost electron shells. Consequently, they tend to remain together in geological processes occurring in a reducing environment. Both tetravalent ions are relatively insoluble, but the hexavalent uranyl ion (UO_2^{2+}) is soluble in water. Uranium can form complexes with a wide variety of ions in aqueous environments. Organic compounds may enhance Th solubility in neutral conditions, but normally Th has very low solubility in natural waters and is largely transported in particulate matter. The mobility of U in its hexavalent state is also affected by adsorption on hydrous iron oxides, clay minerals, zeolites and colloids. A commonly cited mechanism for uranium concentration involves precipitation from oxidised groundwater when reducing environments are encountered.

Concentrations of K, U and Th for various lithotypes and mineral species are plotted in Figs. 4.12, 4.13 and 4.14. An important characteristic of the three radioelements is that their concentrations tend to be correlated within most lithotypes: see, for example, Galbraith and Saunders (1983) and references therein. The data from the different rocks types are mostly explainable in terms of the radioelement content of the common rock-forming minerals, with micas and feldspars being the dominant influences. Data that plot outside the region bounded by these minerals are usually due to the presence of U and Th, explainable in terms of the substitution of these elements in accessory minerals and other rock-forming species.

4.6.1 Disequilibrium in the geological environment

A significant feature of both the ^{238}U and ^{232}Th decay series is that neither of these isotopes decays by γ-emission (see Section 4.2.5) and so it is the emission products of daughter elements that are detected by radiometric surveys and used to indirectly infer concentrations of the parent. However, and importantly, an assumption is made that daughter isotopes are neither added to nor removed from the system, i.e. the decay series is closed and in equilibrium. As explained in Section 4.2.2, the time for the whole decay series to reach equilibrium is governed by the longest half-life in the series. Members of ^{232}Th decay series have half-lives ranging from nanoseconds to a few years, with the exception of ^{232}Th itself, so equilibrium can be re-established instantaneously in terms of the geological time scale. The ^{232}Th series is virtually always in equilibrium.

The half-lives of members of the ^{238}U decay series range from fractions of a second to thousands of years, and a variety of physical and chemical mechanisms can remove isotopes from the series, or introduce isotopes created elsewhere (Fig. 4.15). Consequently, disequilibrium is common for the ^{238}U decay series. Uranium and radium (Ra) are both soluble so they can be removed by groundwater. Radium can be mobilised by most groundwater, with its mobility restricted by co-precipitation with barium sulphates, iron-manganese oxides or sulphates, or through adsorption by organic matter. The presence of radon (Rn), a gas albeit an inert one, also encourages disequilibrium since there are ample opportunities for movement of gas in the geological environment. In terms of the 'theoretically closed' decay series, the loss of ^{222}Rn will require about 27 days for equilibrium to be re-established. However, if ^{234}U is leached relative to Ra, equilibrium would take as long as 1.74 million years to re-establish. The long half-life of ^{230}Th (next in the series) ensures that the effects are not felt further down the decay series for a considerable time. If ^{230}Th were lost it would require about 530,000 years for equilibrium to re-establish. Clearly then, the age of a U occurrence is a critical factor in determining the

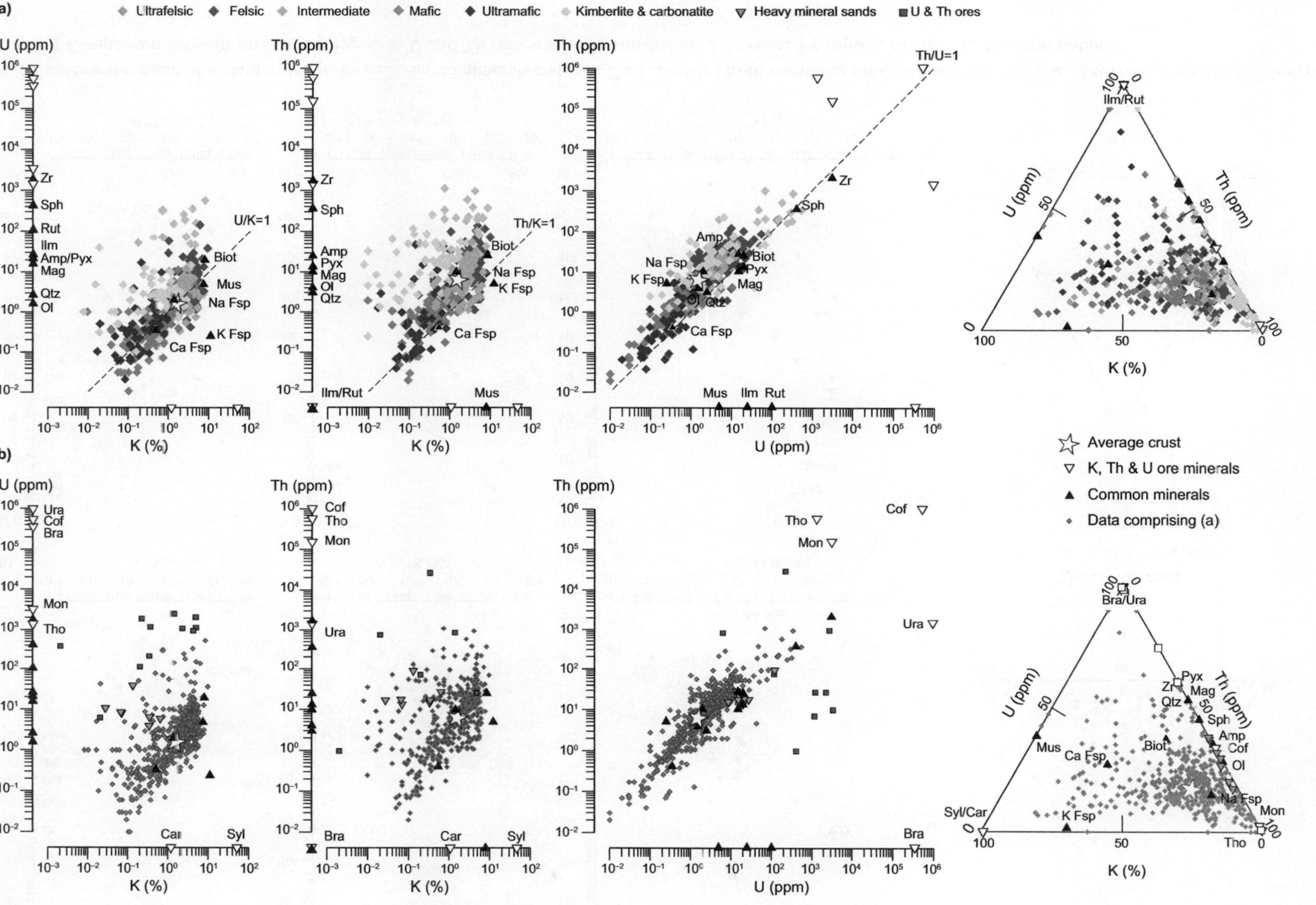

Figure 4.12 Radioelement content of igneous rocks and selected minerals and ores. Data compiled from numerous published sources. (a) Data subdivided according to lithotype and geochemistry. (b) Comparison between various lithotypes, K, U and Th ores and heavy mineral sand deposits. Amp – amphibole, Biot – biotite, Bra – brannerite, Ca Fsp – plagioclase, Car – carnallite, Cof – coffinite, Ilm – ilmenite, K Fsp – potassic feldspar, Mag – magnetite, Mon – monazite, Mus – muscovite, Na Fsp – sodic feldspar, Ol – olivine, Pbl – pitchblende, Pyx – pyroxene, Qtz – quartz, Rut – rutile, Sph – sphene, Syl – sylvite, Tho – thorite, Ura – uraninite, Zr – zircon.

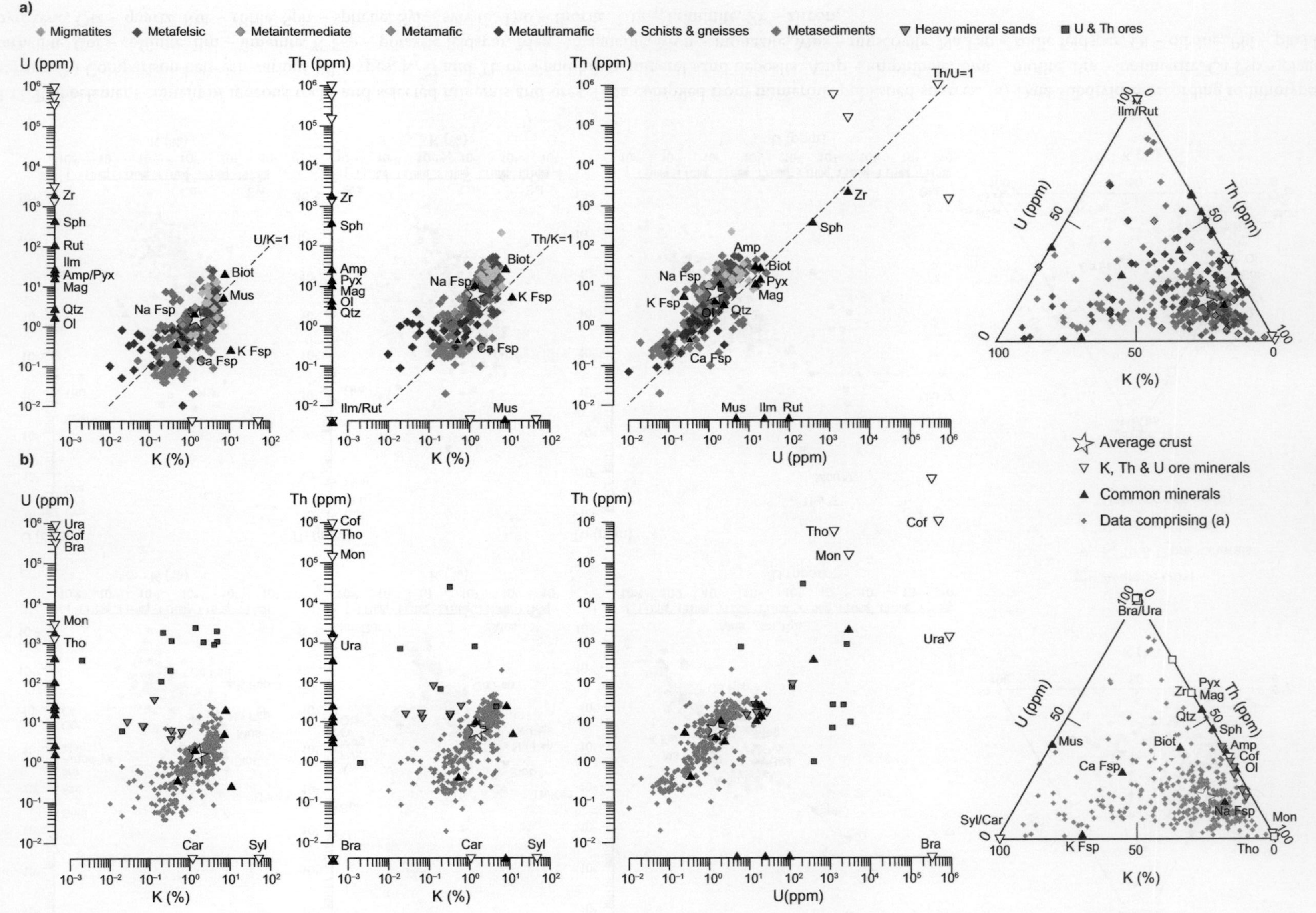

Figure 4.13 Radioelement content of metamorphic rocks and selected minerals and ores. Data compiled from numerous published sources. (a) Data subdivided according to lithotype and geochemistry. (b) Comparison between various lithotypes, K, U and Th ores and heavy mineral sand deposits. See caption to Fig. 4.12 for abbreviations.

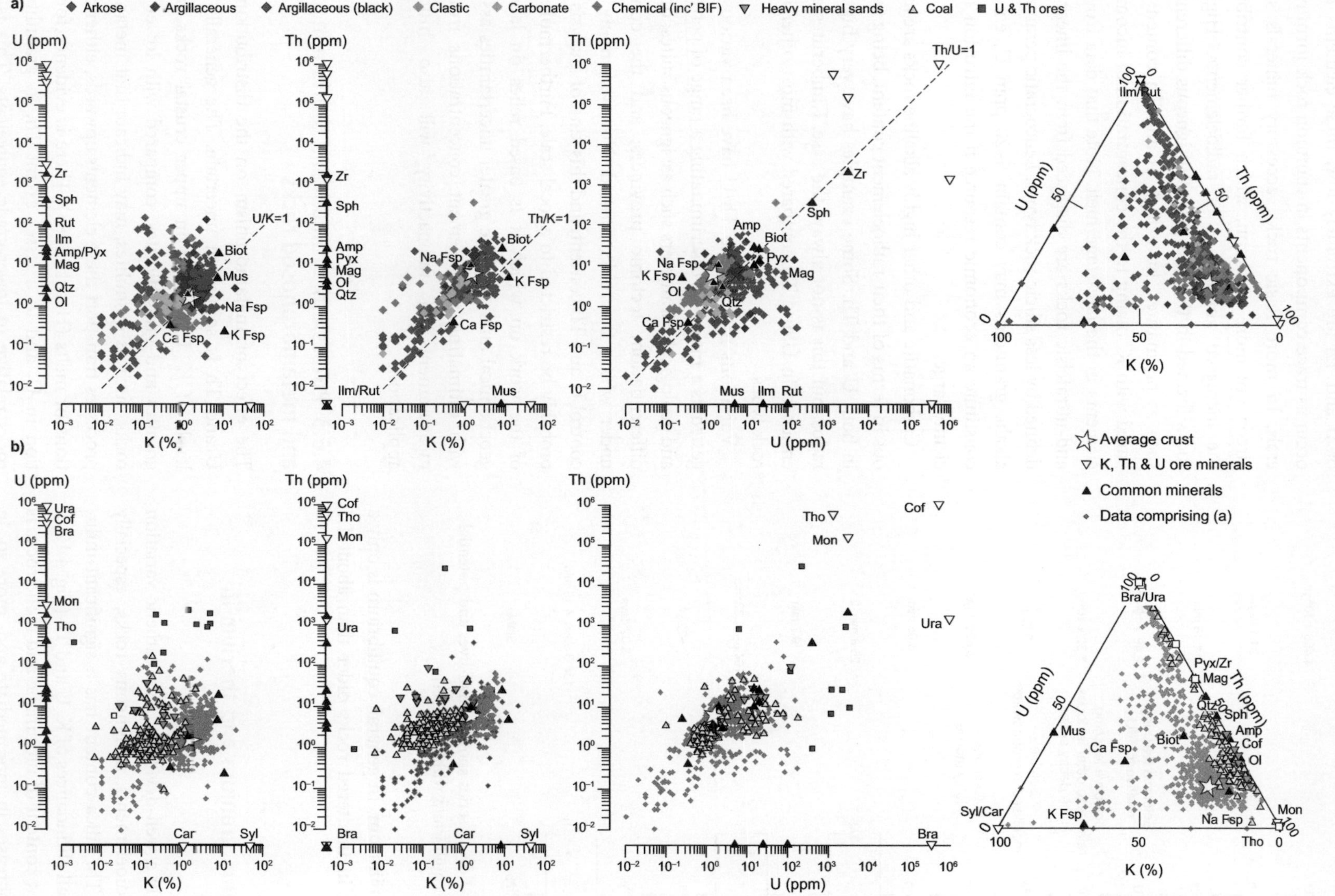

Figure 4.14 Radioelement content of sedimentary rocks and selected minerals and ores. Data compiled from numerous published sources. (a) Data subdivided according to lithotype. (b) Comparison between various lithotypes, K, U and Th ores and heavy mineral sand deposits. See caption to Fig. 4.12 for abbreviations.

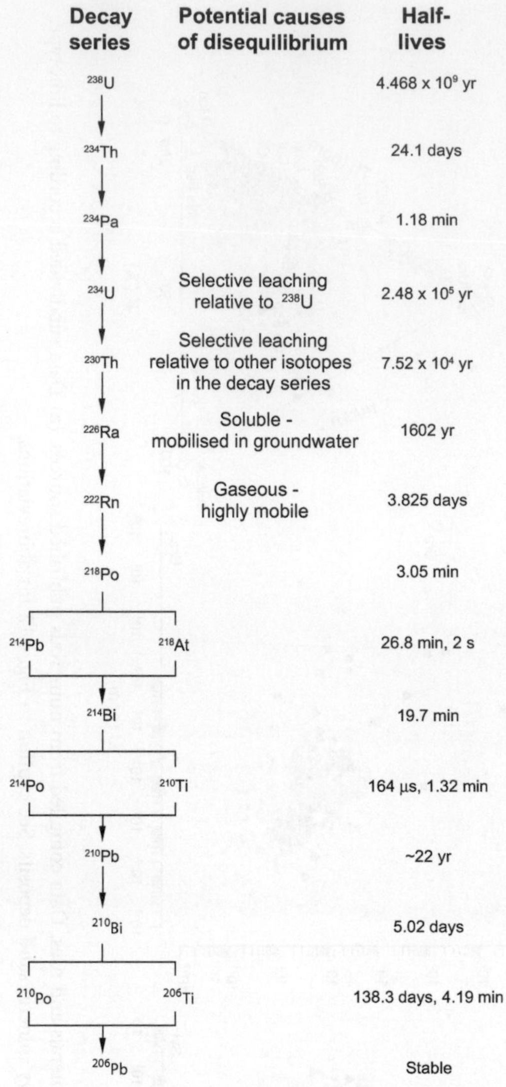

Figure 4.15 The ^{238}U decay series showing half-lives and potential mechanisms for disequilibrium within the series.

likelihood of equilibrium. In general, equilibrium is only a safe assumption in unaltered rocks older than about 1.5 million years.

4.6.2 Potassium, uranium and thorium in igneous rocks

There is generally a well-developed sympathetic variation in the three radioelements in igneous rocks, especially between U and Th. Silica content has a significant influence on the overall abundances of K, U and Th (Fig. 4.12a). The radioelement content of ultramafic and mafic rocks is very low, but increases in intermediate and, more so, in felsic rocks. The increase is due to the greater abundance of feldspars and micas.

In mafic and ultramafic rocks, where the concentrations of U and Th are extremely low, these elements tend to occur as trace components in common rock-forming minerals. In more felsic rocks, accessory minerals such as zircon are more common, which further contributes to the increased abundance of radioelements. High-silica rocks formed in the final stages of igneous differentiation, such as pegmatites, have high U and Th contents compared with K, owing to the concentration of incompatible elements in the late-stage melt. Note that data from felsic and ultrafelsic rocks are displaced from the linear trends defined by less silicic rock types. Leucocratic peraluminous alkalic granitoids may contain 5–20 ppm U, enough to constitute an economic resource if the intrusion is sufficiently large.

Carbonatite and other highly alkaline rocks are anomalous in terms of their radioelement content, being enriched in both U and Th. Some examples have very high Th/U ratios, but this is not always the case. Kimberlites are also enriched in U and Th compared with most other igneous rock types.

Variations in the ratio Th/U have been variously suggested as a means of discriminating a range of petrological and geological parameters such as: igneous suites; degree of differentiation; tectonic provinces; and the conditions under which the rocks formed. However, many factors control U and Th content, and this kind of analysis should probably be restricted to a local scale. Furthermore, much of the work on which it is based relies on laboratory geochemical analyses. The greater uncertainties associated with estimating radioelement concentrations from field measurements of radioactivity will also hinder its application.

4.6.3 Potassium, uranium and thorium in altered and metamorphosed rocks

The effect of metamorphism on the distribution of K, U and Th is somewhat uncertain. The generally higher levels of K, U and Th in upper crustal rocks, such as granites and granodiorites, compared with lower crustal rocks, such as granulites, may indicate that metamorphic processes transport these elements upwards, either as solutions or melts (Haack, 1983). There is evidence for depletion in U, and to a lesser extent Th, in granulite-facies rocks relative to lower-grade equivalents. This is due to fluid loss and increased mobility of these elements at high temperature and pressure, enhanced by U often being loosely bound on grain boundaries and internal surfaces. The data that are available suggest that, in general, metamorphism does not greatly affect radioelement content, so the concentrations in the rocks mainly reflect the protolith. For example, Roser and Nathan (1997) show that there is no significant variation in the concentrations of K, U or Th in a turbidite sequence whose metamorphic grade varies from lower greenschist to amphibolite facies. The same applies to amphibolite-facies gneisses and associated granulites analysed by Fernandes *et al.* (1987).

The radioelement distribution of metamorphic rocks shows a correlation between U and Th (Fig. 4.13), but the relationship between K and the other two elements is somewhat less than in igneous rocks, indicating some element mobility. As would be expected, the data reflect the radioelement content of the precursors. Felsic gneisses are among the most radioactive rocks, as are migmatites. Mafic gneisses and amphibolites are much less radioactive. Metasediments, including schists, pelites, phylites and slates, plot in similar positions to gneisses, reflecting their disparate original compositions. Most metacarbonates are weakly radioactive, but a few may have some U enrichment.

4.6.4 Potassium, uranium and thorium in sedimentary rocks

The distribution of K, U and Th in sedimentary rocks is complex, being influenced by the composition of the parent rock, the processes of the sedimentary cycle and the different geochemical properties of each element. The radiometric response is determined dominantly by the feldspar, mica and clay-mineral contents of the rocks, and supplemented by heavy minerals containing Th and U. The correlation in the radioelement content of sedimentary rocks is less marked than for igneous rocks, but still apparent (Fig. 4.14). The lesser correlation is due to mobilisation of the radioelements in the secondary environment.

For clastic sediments, key factors affecting radioactivity are the nature of the source and maturity. A source rich in radioactive material is, not surprisingly, likely to produce a radioactive sediment. The lower end of the range corresponds with sediments composed primarily of quartz or carbonate grains. As the fine-grained fraction increases so too does radioactivity due to increasing amounts of clay-minerals. Greywackes have radioelement concentrations similar to their source, but as sediment maturity increases quartz becomes increasingly dominant with an associated decrease in radioelement content. Heavy minerals such as monazite, sphene and zircon are resistant to mechanical weathering and, if present, will contribute significantly to the observed radioactivity. In heavy mineral sand deposits these minerals constitute the dominant source of radioactivity, so these deposits are enriched in U and Th relative to K compared with other sediment types. Feldspathic and glauconitic sandstones are clastic sediments that are anomalously radioactive, dominantly owing to K.

Pure carbonates have low radioactivity, especially if dolomitic, but when they contain organic matter they may have relatively high levels of U. The Th content of carbonates is low since it cannot enter the carbonate lattice easily, although studies have linked enhanced Th content in black shales to diagenetic carbonate. Some marine black shales have U concentrations of hundreds, and locally thousands of parts per million (Cobb and Kulp, 1961), but most black shales have radioelement concentrations similar to other types of argillaceous rock. Phosphatic sediments of marine origin may have particularly high U content.

Chemical sediments are generally poorly radioactive. Banded iron formation normally has very low radioelement concentrations, although they appear Th-rich in radiometric data owing to thorium's affinity with iron oxides (see Section 4.6.5). Most evaporites are similarly poorly radioactive, but become more radioactive when potassic minerals such as sylvite and carnallite are present.

Figure 4.14b shows that coal has low radioelement content, especially K content. Occasionally U content can be quite high, but in general coals are amongst the least radioactive sediments.

4.6.5 Surficial processes and K, U and Th in the overburden

For the simplest situation of residual cover, the radiometric response of the cover material is related in a simple way to the underlying bedrock. Of course, it is unlikely that there will be any direct relationship where the cover comprises transported material, so in these cases mapping the bedrock with radiometrics is impossible. Even if the cover is residual, pedogenesis and regolith formation will affect radioelement content.

Some of the many and complex secondary near-surface processes that affect the radiometric response from a single lithotype are summarised in Fig. 4.16. The most important processes are those that remove, transport and deposit K, U and Th. These processes may be sufficient to create

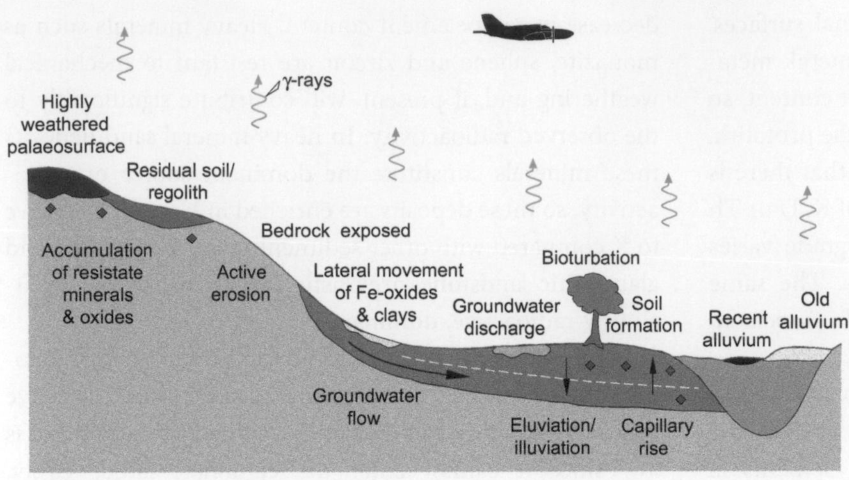

Figure 4.16 Schematic illustration of near-surface processes that affect radiometric responses. Redrawn, with permission, from Wilford *et al.* (1997).

ore-grade U concentrations in surficial materials. Also important are changes in density/porosity that affect the ability of the γ-rays to escape from the source rock, and also facilitate the loss of mobile isotopes, promoting disequilibrium.

Weathering decomposes the main K-hosting minerals in the order biotite, alkali feldspar, muscovite. The K so liberated is taken up in minerals such as illite, and to a lesser extent other clay minerals. Of the main U- and Th-bearing minerals, only zircon and monazite are stable during weathering. Uranium liberated by weathering tends to occur in authigenic iron oxides and clay minerals, or may go into solution as a complex ion. Thorium liberated by weathering may be found in oxides and hydroxides of iron and titanium oxides, and also clays. Similar to U, Th can be transported and absorbed on colloidal clays and iron oxides.

The relationship between the radiometric response of the bedrock and its weathered equivalents depends on factors such as rainfall, temperature, and pH and salinity of water, all of which affect the weathering process and the behaviour of the radioelements. Depending on the local conditions, opposite effects may result. K-feldspar is relatively resistant to weathering, resulting in a relative increase in K content early in the weathering process. However, under lateritic weathering conditions, kaolinisation may cause substantial loss of K. Uranium is not easily liberated by weathering when it is trapped within heavy minerals such as zircon, but when it is present in less resistant minerals, or located along grain boundaries, it may be lost relatively easily. Weathering of shales tends to result in the loss of K, but less so U and Th, especially during lateritic weathering. On the other hand, silicification tends to preserve the K content. Sediments with low initial radioelement concentrations, such as limestones and clean sandstones, generally produce soils with similarly low elemental concentrations. Recent organic sediments may be rich in uranium.

The relative and absolute abundances of the three radioelements for rocks, and the overburdens derived from them, from a deeply weathered Australian terrain and a glaciated Canadian terrain are shown in Fig. 4.17. Although these are two geologically different terrains, similar behaviour is observed in both regions. The relative concentrations of the radioelements are largely unaltered, but in virtually every case the cover has significantly lower concentrations of the radioelements than the bedrock. This reduction may in fact be even greater than those suggested by the data (Dickson and Scott, 1997). In other places the concentrations in the soils may exceed that of the protolith. This is common for soils formed above mafic and ultramafic rocks, and iron formations, with U and Th exceeding those of the source. This is thought to be due to the concentration of U and Th in iron oxides during weathering.

The distribution of the three main radioactive elements is not constant in the soil/regolith profile (Fig. 4.18). The fact that cover materials are more prone to erosion than lithified materials means that the vertical distribution of elements in the sequence determines the radiometric response, because dissection may expose different parts of the soil/regolith profiles. In the case of regolith, the presence of clays or iron oxides at different depths can lead to anomalous concentrations of U and Th. Concentration of Th in iron-rich materials, such as ferricrete and pisoliths, is also a complicating factor.

Other near-surface phenomena that may be of significance are related to the hydrological setting. In areas of

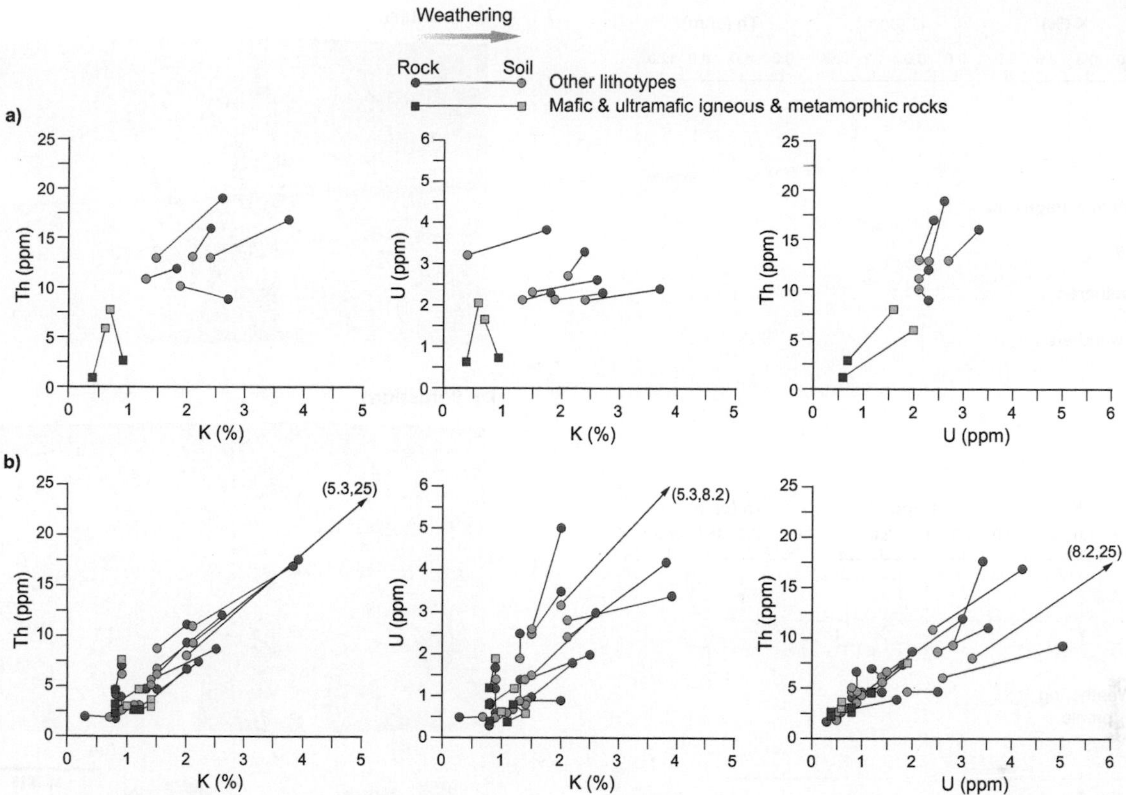

Figure 4.17 Abundances of K, U and Th compared for bedrock and adjacent overburden. (a) Australia; bedrock compared with soils. (b) Canadian shield; bedrock compared with glacial till. Based on data in Charbonneau *et al.* (1976) and Dickson and Scott (1997).

groundwater discharge, U and Th abundance may increase because of the effects of dissolved Ra. Saturated materials are denser than unsaturated materials, which affects the ability of γ-rays to escape into the atmosphere. Permafrost also subdues the radiometric response, as does snow cover. Vegetation may also be important since it increases attenuation. This is greatest for woody material, and normally the vegetation needs to be forest-like to affect the radiometric response significantly.

The surface drainage system has a significant effect on the surface distribution of the three radioelements. The minerals carrying the radioelements can be transported through the system, forming a useful radioactive tracer of the drainage. Similarly, wide-area flooding distributes the radioelements over large areas, both as mineral grains and in solution to be redeposited through evaporation. The effect is to increase the background concentration levels of the radioelements over large areas, often seen as a background 'haze' in K, U and Th imagery.

In summary, the effects of cover on the radioelement distribution are as diverse and as complicated as the process of weathering itself. Consequently, there are no universal relationships between bedrock and weathered equivalents, although some local consistencies are likely. Normally, field studies of the near-surface environment are required in order to establish local conditions and processes, and to properly understand the effects of the cover on the radiometric response.

4.6.6 Potassium, uranium and thorium in mineralised environments

Figures 4.12, 4.13 and 4.14 show that the K, U and Th minerals, and orebodies containing these minerals, have substantially higher radioelement content than any kind of host rock. There are reports in the literature of depletion or enhancement of virtually every combination of the three radiometric elements in hydrothermally altered zones, although U and Th are normally less affected by alteration than K.

The variation in K and Th through the porphyry Cu–Au mineralisation at Goonumbla–North Parkes in the Lachlan Foldbelt of New South Wales, Australia, is shown in Fig. 4.19. Potassium content is higher in both the phyllic and potassium alteration zones. Importantly, weathering greatly decreases the K content. Uranium content is

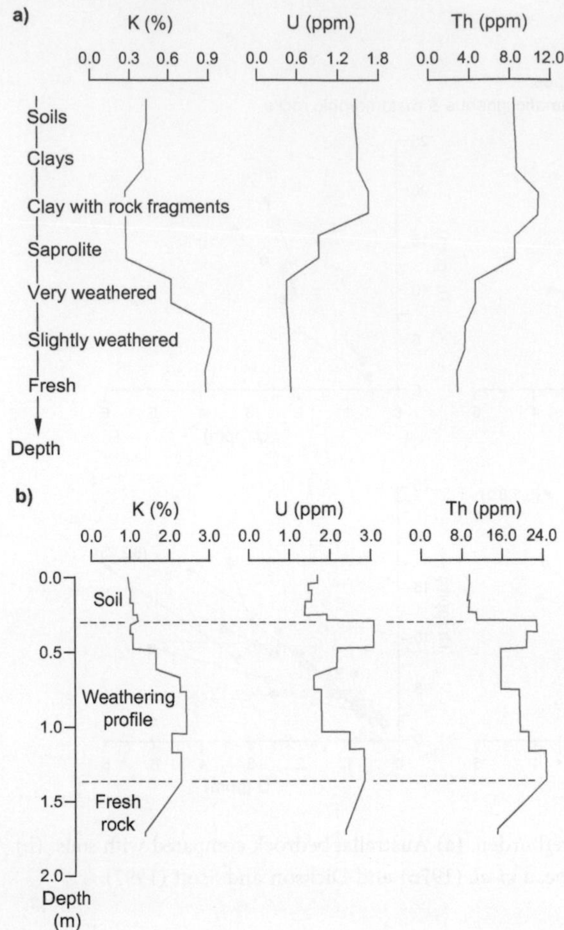

Figure 4.18 Variations in radioelement concentrations in the soil/regolith profile. (a) Radioelement concentrations through the weathering profile of a basalt in Victoria, Australia; (b) in a soil profile overlying an adamellite in New South Wales, Australia. Redrawn, with permission, from Dickson and Scott (1997).

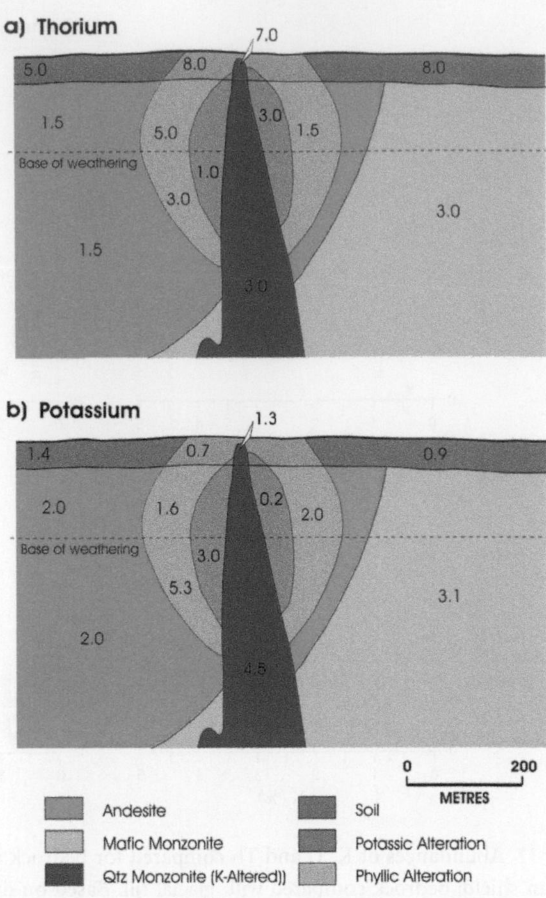

Figure 4.19 Summary of K and Th concentrations in the vicinity of the Goonumbla–North Parkes Cu–Au porphyry deposits. Variation in (a) eTh (in ppm) and (b) K (in %). Redrawn, with permission, from Dickson and Scott (1997).

consistent, but Th is greater in the altered rocks and even more so in the weathered zone and overlying soil. Clearly, the depth of weathering critically affects the radiometric response of the mineralisation, the deposit coinciding with a K-poor zone if weathered and a K-rich zone if unweathered material is exposed. Again, the near-surface environment is shown to be of vital importance to radiometric responses.

A style of mineralisation for which some very strong alteration-related radiometric responses have been reported is epithermal gold–silver deposits. Intense hydrothermal alteration is usually associated with potassium enrichment. The levels of U and Th are generally not affected, or may be slightly depleted. Geochemical data from both altered and unaltered rocks from the Waihi-Waitekauri region in New Zealand are shown in Figure 4.20. Thorium content is unaffected by alteration, U is slightly depleted and K content is significantly increased. The K/Th ratio is shown to be mostly greater in altered rocks than in their unaltered equivalents owing to pervasive adularia and illite alteration at the core of the alteration zones. The degree of K enrichment decreases with distance from the mineralised veins. Examples of the radiometric responses of this area are described in Section 4.7.3.2.

The correlation between potassium content and degree of alteration at the Iron King massive Pb–Zn–Cu–Au–Ag sulphide deposit in Arizona, USA, is demonstrated in Fig. 4.21. The abundance of potassium increases by a factor of about 4, and occurs in micas associated with sericitic alteration.

4.7 Interpretation of radiometric data

The reduction of radiometric survey data aims to produce a spatial dataset that shows the actual ground

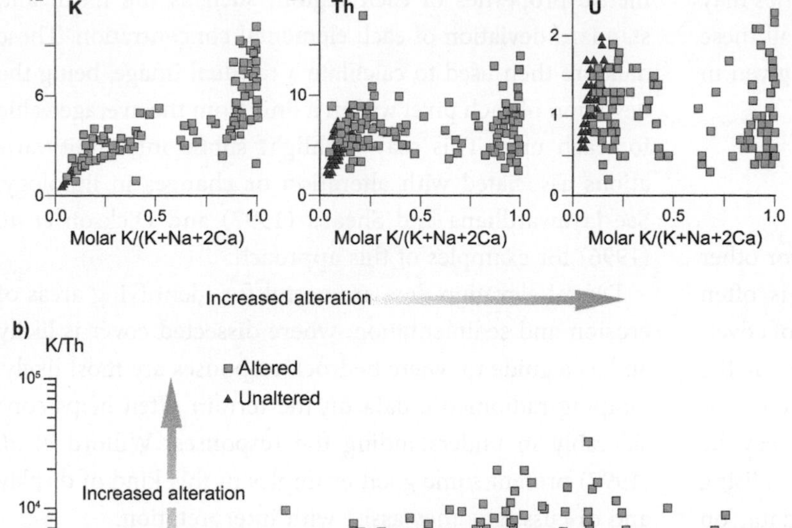

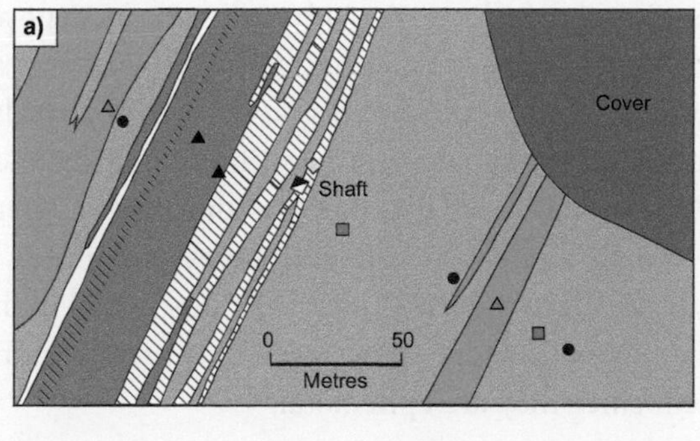

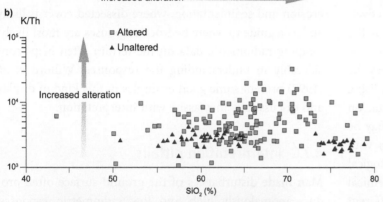

Figure 4.20 Geochemical data from altered and unaltered rocks from the Waitekauri area. (a) Variation in radiometric element content with hydrothermal alteration. (b) Variation in K/eTh ratio with silica content. Redrawn, with permission, from Booden *et al.* (2011).

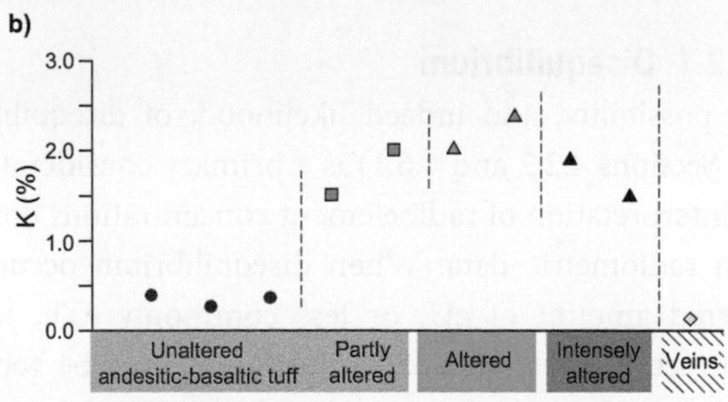

Figure 4.21 The Iron King Pb–Zn–Cu–Au–Ag sulphide deposit. (a) Geological map of the area around the deposit. The symbols designate sample locations and degree of alteration. (b) Correlation between K content and degree of alteration in tuffs. Redrawn, with permission, from Moxham *et al.* (1965).

concentrations of potassium (K), uranium (U) and thorium (Th) in the survey area. In practice this is not entirely achieved because of limitations in the various calibration and correction procedures, so the interpreter must be aware of these limitations and account for them in the analysis of the data. More encouragingly, one limitation of the radiometric response makes it easier to interpret. Being controlled by materials in the top few tens of centimetres, radiometric data are free of the complications of superimposed responses from greater depths, unlike the situation for gravity, magnetic and electrical data. Furthermore, the radiometric response has a very simple relationship to its source, again unlike gravity, magnetic and electrical data, because the transition from background to anomaly is very abrupt and coincides with the edges of the source, facilitating accurate mapping of near-surface contacts.

An understanding of the distribution and behaviour of K, U and Th in the geological environment, and in particular their occurrence in different lithotypes and the possibility for their redistribution by secondary geological processes, is essential for producing a geologically meaningful interpretation of radiometric data. Since the response originates in the near-surface, the radiometric properties of the overburden and the effects of surficial geological processes are particularly important. Unfortunately, these are often complex and not well understood, more than negating the advantages of the favourable response characteristics in most terrains. A further complication is that eU and eTh concentrations are derived from γ-emissions of daughter elements; this allows the possibility of disequilibrium, since the differing chemical and physical properties of elements in the decay series may lead to their separation. A detailed description of all these aspects of the distribution of the radioelements is given in Section 4.6.

4.7.1 Interpretation procedure

Analysis of radiometric data is less complex than for other types of geophysical data, although the process is often hindered by high noise levels caused by the effects of cover, and deficiencies in the acquisition and reduction of the survey data. In principle, areas with anomalous concentrations of one or more of the radioelements may be identified as being potentially indicative of mineralising environments. Geological interpretation of the data, on the other hand, can be complex. Geological maps may be created (radiolithic mapping) by assigning regions with similar radioelement concentrations to pseudo-geological units. However, like virtually every physical/chemical property relevant to geophysical surveying, the different rock types show a range of K, U and Th concentrations, and unequivocal identification of specific lithotypes is not possible. As shown in Figs. 4.12 to 4.14, many rock types are low in all three radioelements; and others, such as granite and shale, may have similar concentrations of K, U and Th. Often there are subtle changes in radioelement content reflecting lithological subdivisions which are not evident in geological inspections. It is quite common, for example, to map from the radiometric data zoning and multiple intrusive phases in large granitic intrusions.

The influence of surficial cover and weathering processes on the radiometric response means it is strongly advisable to make an integrated interpretation using other complementary datasets. An example of this kind of interpretation for mineral exploration is provided by Dickson *et al.* (1996). The obvious datasets to combine with radiometrics are multispectral remote sensing and aerial photographs because they also are only sensitive to surface material. Multispectral data can be used to create images showing the distribution of clays and iron oxides, both of which influence the distribution of K, U and Th. Images showing the distribution of vegetation may indicate where the radiometric response may be subdued.

An integrated interpretation can be part of a more sophisticated strategy designed to identify anomalous regions. All available data, including elemental ratios and other multichannel remote sensing data, are used to map lithotypes and rock units, and then to determine the radiometric properties of each region, such as the mean and standard deviation of each elemental concentration. These data are then used to calculate a residual image, being the deviation of each pixel within a unit from the average value for each unit. This may highlight subtle intra-unit variations associated with alteration or changes in lithology. See Jayawardhana and Sheard (1997) and Dickson *et al.* (1996) for examples of this approach.

Digital elevation data are useful for identifying areas of erosion and sedimentation, where dissected cover is likely and as a guide to where bedrock responses are most likely. Draping radiometric data on the terrain often helps considerably in understanding the responses. Wilford *et al.* (1997) present some good examples of this kind of display and discuss how they assist with interpretation.

4.7.2 Interpretation pitfalls

Man-made disturbances of the ground surface often produce anomalously 'high' and 'low' radiometric responses. Some examples include: the subsurface rocks exposed by quarrying and open-pit mining, tailings from mines and processing plants, transported gravels used for roads and railway lines, phosphate-based fertilisers and atomic fallout. Consulting aerial photographs during the interpretation should allow most of these types of features to be recognised. Also important are variations in soil moisture, surface water and snow, all of which may present more of a problem since they are ephemeral.

4.7.2.1 Disequilibrium

The possibility, and indeed likelihood, of disequilibrium (see Sections 4.2.2 and 4.6.1) is a primary consideration in the interpretation of radioelement concentrations obtained from radiometric data. When disequilibrium occurs, the inferred amount of eU, or less commonly eTh, will be incorrect. The estimated concentration may be too high or too low, depending on whether the location is one where the mobile component of the relevant decay series has been removed or has accumulated. It is possible for there to be only low γ-emissions from a uranium deposit, making it difficult to detect with a γ-ray detector. Also, through the solubility of Ra and mobility of Rn, it is possible to produce ^{214}Bi in locations remote from its parent ^{238}U source to produce a uranium radiometric response in a non-uranium bearing location. Note that the extent of the resulting migration must be larger than the footprint of the radiometric survey to be resolved by the survey. Levinson and Coetzee (1978) describe in detail the implications of disequilibrium for uranium exploration. There is no simple way to identify areas of disequilibrium. Geological settings where it is most likely to occur include areas where groundwater is discharged; reducing environments such as swamps; and situations where porosity is increased, say by physical weathering, to encourage loss of Rn gas.

4.7.2.2 Topographic effects

The reduction of radiometric survey data described previously (see Section 4.4) assumes that the source of the radiation is a flat horizontal surface located below the sensor. The approximately exponential decrease in radiation intensity with increasing height, owing to attenuation (see Section 4.4.6), means that radiometric measurements are particularly sensitive to source geometry.

Ground radiometric measurements are particularly prone to the effects of terrain forms in the immediate vicinity of the measurement. Relief in the terrain causes variations in the distance between the sensor and the γ-ray sources and in the size of the source area (Fig. 4.22). When measurements are made in a gully or valley, the sources of radiation in the rock walls are closer to the detector, causing the readings to be higher than if they were made on flat ground. A measurement made adjacent to a cliff or mine bench will be about 50% higher than a measurement made on flat ground formed of the same material. Conversely, when measurements are made on a ridge the laterally offset sources of radiation are more distant from the detector, causing a lower reading.

As with other airborne geophysical measurements, it is important for the aircraft to maintain constant terrain clearance in order to minimise false radiometric responses. As described in Section 2.6.3.2, this is not always possible,

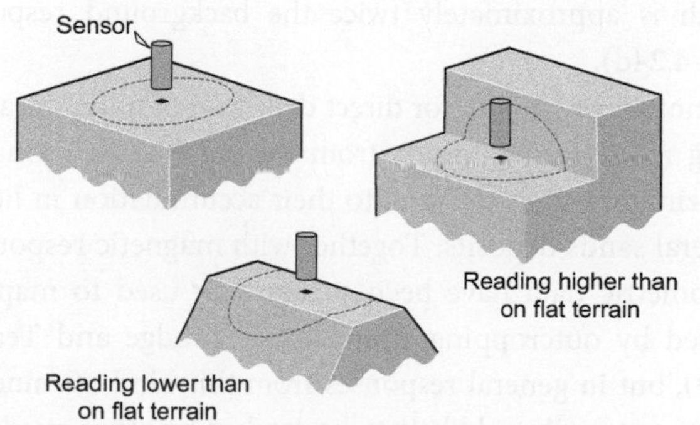

Figure 4.22 Schematic illustration of the radiation source areas of various terrain forms for ground radiometric measurements.

with increases in terrain clearance causing a decrease in the amplitude of the radiometric response and vice versa.

Methods have been developed for upward- and downward-continuation of radiometric data, i.e. to calculate how the data would appear for a different source–detector separation, which partly address topographic effects. However, their routine application is not current practice, probably because of the relatively high noise levels of survey data. Recognising topography-induced artefacts is one of the main reasons that topographic data are required in the interpretation of radiometric data.

4.7.3 Responses of mineralised environments

For many types of deposit, the mineralisation itself, associated alteration zones and lithotypes favourable for mineralisation may all have anomalous radioelement compositions, either in a particular radioelement or a particular ratio of elements. Because of their large size, alteration zones are the most likely candidates for direct detection by radiometric surveys.

Radiometric responses associated with mineralised environments vary from very clear to very subtle. In the latter case, they may be disguised by apparently similar responses of, for example, weathering or changes in overburden type and thickness. Even highly radioactive deposits can be completely concealed by thin cover, but even when the mineralisation does not outcrop there is sometimes anomalously radioactive material at the surface which can be detected. This may be bedrock material brought to the surface and transported and exposed by erosive processes, or may be due to a U-halo produced by the mobilisation of uranium by gaseous diffusion or transportation in groundwater. The anomalous zone may extend over a larger area than the deposit itself, drawing attention to the area.

It is worth investigating all ratios and combinations of K, eU, eTh and TC in an attempt to identify anomalies and anomalous signatures. In all cases, ternary images of the radioelements and composite elemental ternary images (see Section 4.5.3) are effective for targeting anomalous areas for further analysis using individual channel ratios. A description of the radiometric responses from some common types of mineral deposit, and techniques for detecting and mapping them, follows.

4.7.3.1 Mineralisation

Radiometric surveys have led to the discovery of many deposits of uranium of disparate types. Some examples include: the giant alaskite-hosted Rössing deposit in Namibia (Berning, 1986); various unconformity-associated deposits in the Pine Creek Geosyncline in northern Australia, e.g. Ranger (Smith, 1985); deposits in the alkaline igneous complex at Poços de Calderas in Brazil, e.g. Cercado (Forman and Angeiras, 1981); and the calcrete-hosted Yeelirrie deposit in Western Australia (Dentith *et al.*, 1994). In addition to the direct detection of highly radioactive U mineralisation in the U-channel data, radiometric surveys have been widely used to identify terrains with above-average U content. Within these terrains, areas with anomalously high eU/K and eU/eTh ratios may be targeted for follow-up; see Darnley (1972) for examples.

Figure 4.23 shows radiometric data from the Uranium City area, in the Athabasca Basin of Saskatchewan, Canada. Note the similarity between K and TC data, indicating the former to be the dominant source of radiation in the area. There are elevated values of U on this survey line, but the eU/K and eU/eTh profiles resolve

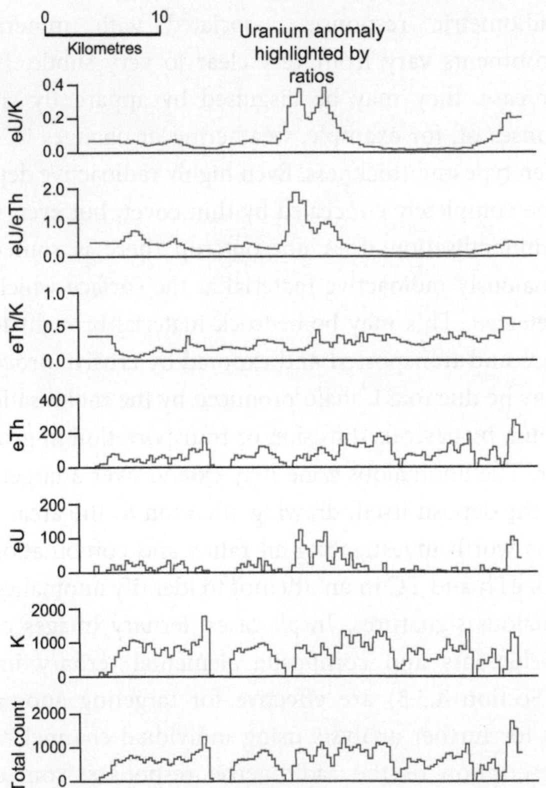

Figure 4.23 Profiles of radiometric data, and various channel ratios from the Uranium City area. The total count data are for an integration period of 1 s, the other data are counts measured over a 5 s integration period. Anomalies in the uranium channel are clearly defined by the various uranium ratios. Redrawn, with permission, from Darnley (1972).

an anomalous response which is possibly due to increased concentrations of U. An alternative thesis is that K and Th have been removed with normal levels of U remaining to produce anomalously high ratios. The lack of TC response is quite commonly observed, normally owing to a reduction in K, and/or U, as Th increases. Data on eU, eU/eTh and eU/K are important for detecting uranium occurrences. However, we emphasise here the point made previously in Section 4.6.1: that the possibility of disequilibrium in the ^{238}U decay series needs to be considered when interpreting U concentrations obtained from γ-ray radiometric data.

The Yeelirrie U deposit, in the northern Yilgarn Craton of Western Australia, comprises carnotite mineralisation in calcrete. The host rocks are Cainozoic sediments within a palaeochannel, with the uranium leached from surrounding radioactive Archaean granitoids. The orebody is a more or less continuous horizontal lenticular zone approximately 9 km in length and 0.5 to 1.5 km wide. It averages about 3 m in thickness and occurs 4 to 8 m below the surface. Sandy and loamy overburden, ranging in thickness from zero to a few metres, occurs above the mineralised calcrete.

The deposit was discovered by a regional airborne magnetic and radiometric survey (1600 m line spacing, 150 m survey height) in 1968. Figure 4.24 shows recent airborne data, although still with reconnaissance specifications (400 m line spacing, 80 m survey height). The data were acquired after the development of a small open pit into the deposit, and so contain responses related to the resultant ground disturbance. The host palaeochannel and the higher-amplitude responses from the mineralised areas within it are clearly seen. Ground radiometric data, acquired soon after discovery, clearly delineate the extent of the orebody (based on a cut-off of 0.1% U_3O_8 over 2 m thickness), corresponding with the 400 counts/minute level which is approximately twice the background response (Fig. 4.24d).

Another possibility for direct detection of mineralisation using radiometrics results from the stability of monazite and zircon, which can lead to their accumulation in heavy mineral sands deposits. Together with magnetic responses, radiometric data have been successfully used to map Th carried by outcropping strand lines (Mudge and Teakle, 2003), but in general responses from this kind of mineralisation are weak and likely to be masked by other overlying sediments. Figures 4.12b to 4.14b show that the radioelement content of heavy mineral sands is higher than that of

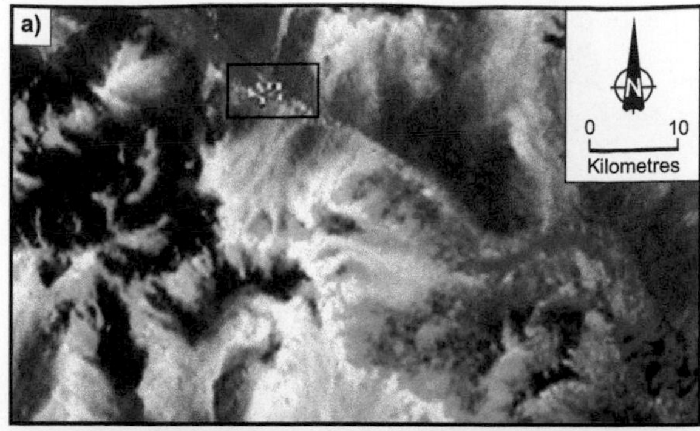

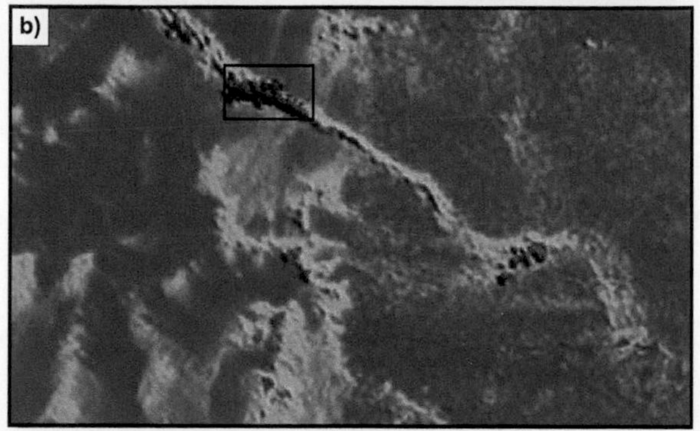

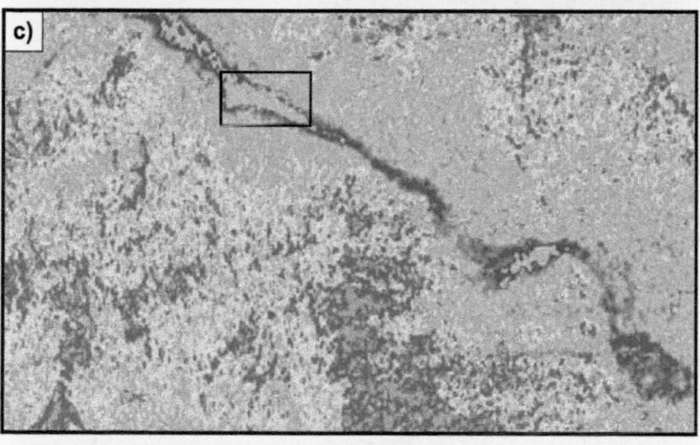

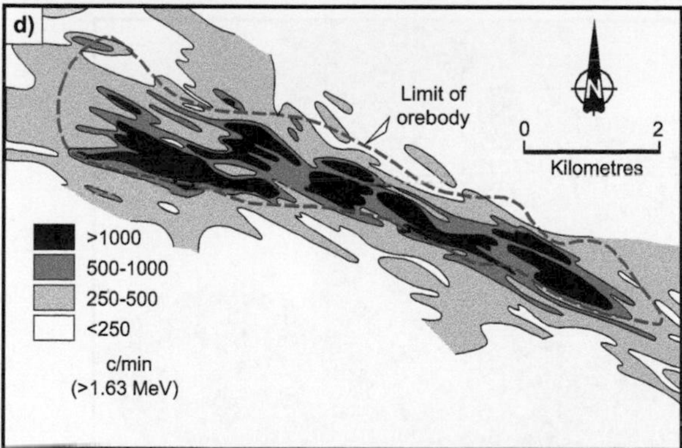

Figure 4.24 Radiometric data from the Yeelirrie calcrete U deposit. (a) Ternary image, (b) eU image, (c) eU/(K+eU+eTh) ratio image, (d) contour map of ground readings, in the vicinity of the orebody, of γ-rays with energies greater than 1.6 MeV. This will include emissions associated with both U and Th. Rectangle in (a–c) shows location of (d). Part (d) redrawn, with permission, from Dentith *et al.* (1994). Airborne data are used with the permission of Geoscience Australia.

most sedimentary rocks, but it is comparable to that of many types of crystalline rocks.

As mentioned previously, coal has low radioelement content, especially K content, but occasionally U content can be quite high (Fig. 4.14b). In general coals are amongst the least radioactive sediments, so composite radiometric ratios may be required to identify subtle anomalous signatures.

4.7.3.2 Alteration zones

Hydrothermal alteration is another potential source of radiometric responses, although for a realistic chance of detection the alteration zone must be large. Dickson and Scott (1997) suggest a width of about 1 km for deposits in Australia. If it is less than this, as for example in structurally controlled hydrothermal gold mineralisation, detection from the surface or above is unlikely, although responses may be detected in γ-logs.

There has been success in detecting the potassic alteration haloes associated with porphyry copper deposits (e.g. Shives *et al.*,1997), epithermal gold–silver deposits (Irvine and Smith, 1990; Feebrey *et al.*, 1998; Morrell *et al.*, 2011) and to a lesser extent volcanogenic massive sulphide deposits (e.g. Moxham *et al.*, 1965 and Shives *et al.*, 2003). Shives *et al.* (1997) recommend the eTh/K ratio as the best means of recognising potassic alteration zones. The depth of erosion is an important control on the radiometric response since the alteration surrounding the mineralisation forms zones, not all of which are radiometric.

Example – Waihi-Waitekauri epithermal Au–Ag mineralisation

The geophysical responses of epithermal Au-Ag mineralisation in the Waihi-Waitekauri region of the North Island of New Zealand are described by Morrell *et al.* (2011). This example demonstrates the use of radiometrics for mapping the surface geology and zones of alteration. These deposits mostly comprise andesite-hosted quartz veins with alteration haloes, up to 15 km^2 in extent, of pervasive clay alteration, potassium metasomatism, magnetite destruction and sulphide mineralisation. The magnetic responses of the alteration zones are discussed in Section 3.9.5. Geochemical data were described in Section 4.6.6. The radiometric responses of the alteration zones are shown in Fig. 4.25. The K radiometric image (Fig. 4.25e) shows high values in the south which are coincident with outcropping ignimbrites and high values to the east which are associated with alluvial deposits. Lower values in the centre and west of the survey area are associated with unaltered

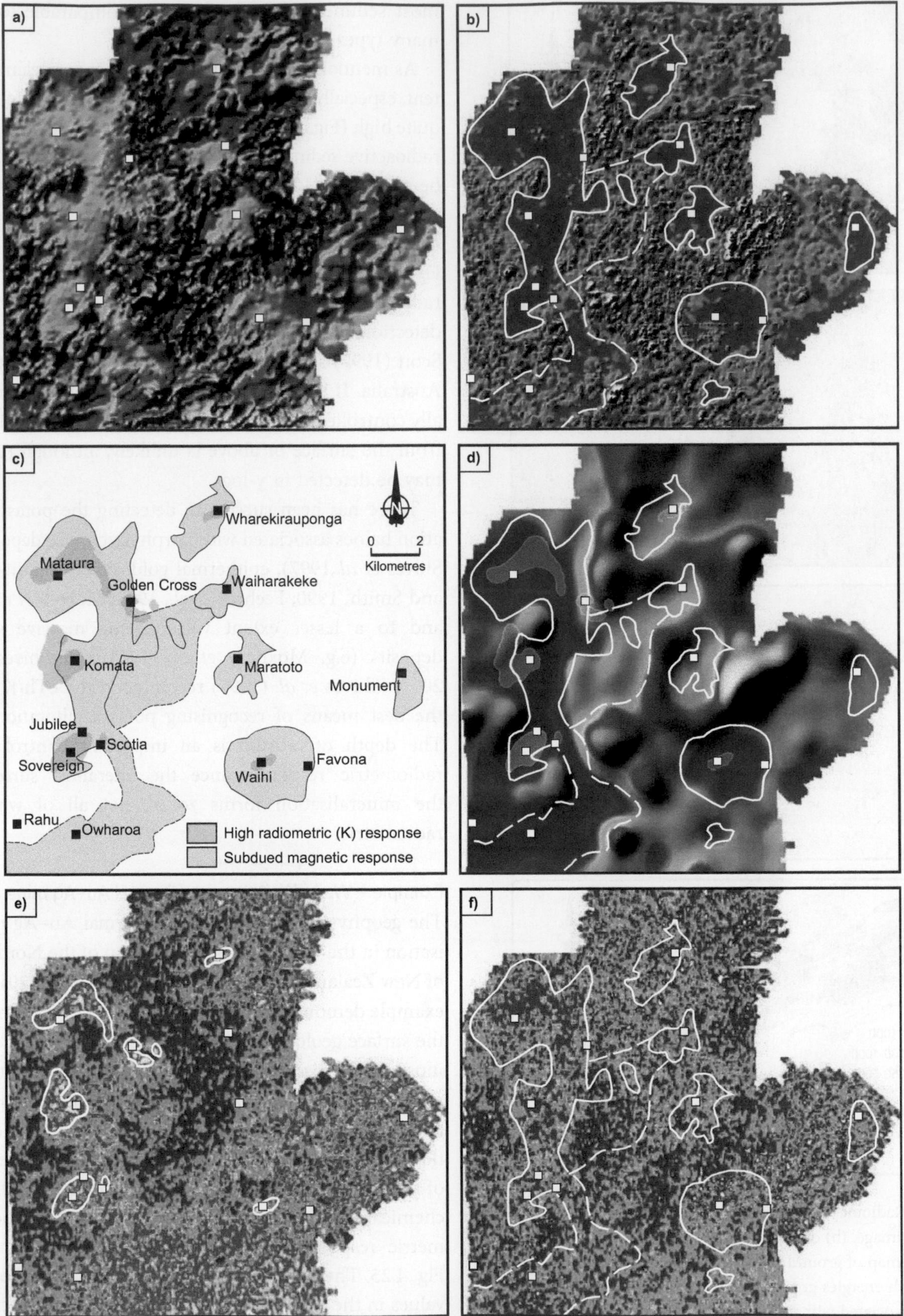

Figure 4.25 Radiometric and magnetic data from the Waihi-Waitekauri epithermal Au–Ag field. (a) Total magnetic intensity (TMI), (b) analytic signal of TMI, (c) summary of geophysical anomalies, (d) low-pass wavelength filtered analytic signal, (e) K-channel radiometrics, (f) K/eTh radiometrics. Redrawn, with permission, from Morrell *et al.* (2011).

andesites. Strong responses are associated with most of the deposits in the region. The K/eTh data (Fig. 4.25f) are less affected by lithological variations and by factors such as vegetation. A broad area of higher ratio coincides with the main zone of subdued magnetic response extending from the Mataura area to the Sovereign area. Elsewhere, the correlation is not as good but there is also a zone of higher values coincident with the anomalous magnetic response surrounding the Waihi deposit. Individual deposits may coincide with very high K/eTh values, for example the Waihi and Wharekirauponga deposits, owing to the intense K metasomatism at the centre of the hydrothermal alteration zones.

4.7.3.3 Host rocks

Mapping K, eU and eTh concentrations and their relative abundances, as assessed from their various ratios, can also be useful for locating favourable host rocks for mineralisation. The anomalous absolute and relative concentrations of U and Th in kimberlites and carbonatites (see Fig. 4.12) have led to the use of radiometrics in exploration for commodities associated with these rock types. In some cases they lack a discernible magnetic response. Ford *et al.* (1988) present an excellent example of the Th channel response from a carbonatite in Ontario, Canada. Also, granitoids associated with tin–tungsten mineralisation in New South Wales, Australia, are shown by Yeates *et al.* (1982) to be enriched in U, and sometimes exhibiting anomalously high eU/eTh ratios. Host sequences for calcrete and sandstone-type uranium deposits may have distinctive radiometric responses (see Fig. 4.24). Banded iron formation normally has very low radioelement concentrations. However, they can appear Th-rich in radiometric data because of Th, liberated by weathering of other materials, becoming concentrated in iron oxides (see Section 4.6.5). Uranium may concentrate in the iron oxides to produce a U response. Iron pisolite deposits may also exhibit similar responses.

4.7.4 Example of geological mapping in a fold and thrust belt: Flinders Ranges

The Flinders Ranges in eastern South Australia are an area with significant topographic relief and generally good exposure. They are a succession of Neoproterozoic clastic and carbonate sediments that form a fold and thrust belt, and contain sediment-hosted base-metal mineralisation. Outcrop is extensive in the west of the study area and this, combined with differences in radioelement content within the succession, makes for an exceptional example of lithological/stratigraphic mapping using radiometrics.

Figure 4.26a shows a ternary image of the radiometric data. The near layer-cake succession creates a pattern of sub-parallel zones in the data which are associated with the different outcropping lithotypes. Resolution of the stratigraphy is so good that it is possible to map the stratigraphic succession accurately and identify lateral changes directly from the radiometric data. The successions in four locations are shown in Fig. 4.26b.

The following features are indicated on Fig. 4.26a. Where diapirs intrude the succession, the resulting cross-cutting relationships are clearly seen (A). These units are potassium-rich and therefore have a distinctive red appearance. The layer-cake stratigraphy combined with radiometric responses which emanate only from surface materials produce 'model' outcrop patterns of folding and faulting. Particularly clear fold closures are labelled (B). Selected faults are labelled (C); the lateral offset of the stratigraphy is the main evidence for their presence. In several places there are subdued radiometric responses from the faults themselves, possibly related to sediments deposited by water courses following the fault plane. Repetition of the stratigraphic succession at (D) is indicative of thrust faulting.

In the east of the study area bedrock is concealed by various types of unconsolidated cover. The radiometric response of the cover is very different from the bedrock. The pattern of responses reflects the local drainage, with a creek bed (E) and various fans (F) evident. In some places the bedrock which sourced the cover can be identified from the similarity in the colour (radioelement content) of their responses. The unconformity between basement and cover can be traced (G), and cross-cutting relationships are evident in places.

The value of terrain models in the interpretation of radiometric data is demonstrated by the images shown in Fig. 4.27. In Fig. 4.27a responses due to sediments in a creek bed and adjacent slope deposits are well defined; the nearby exposed bedrock is clearly the source of detritus. The relationships between topography and the individual units comprising the bedrock are also seen in the radiometric data (Fig. 4.27b). The good correlation between the two data types is the main reason why the radiometric data are able to map the geology of the rugged terrain accurately

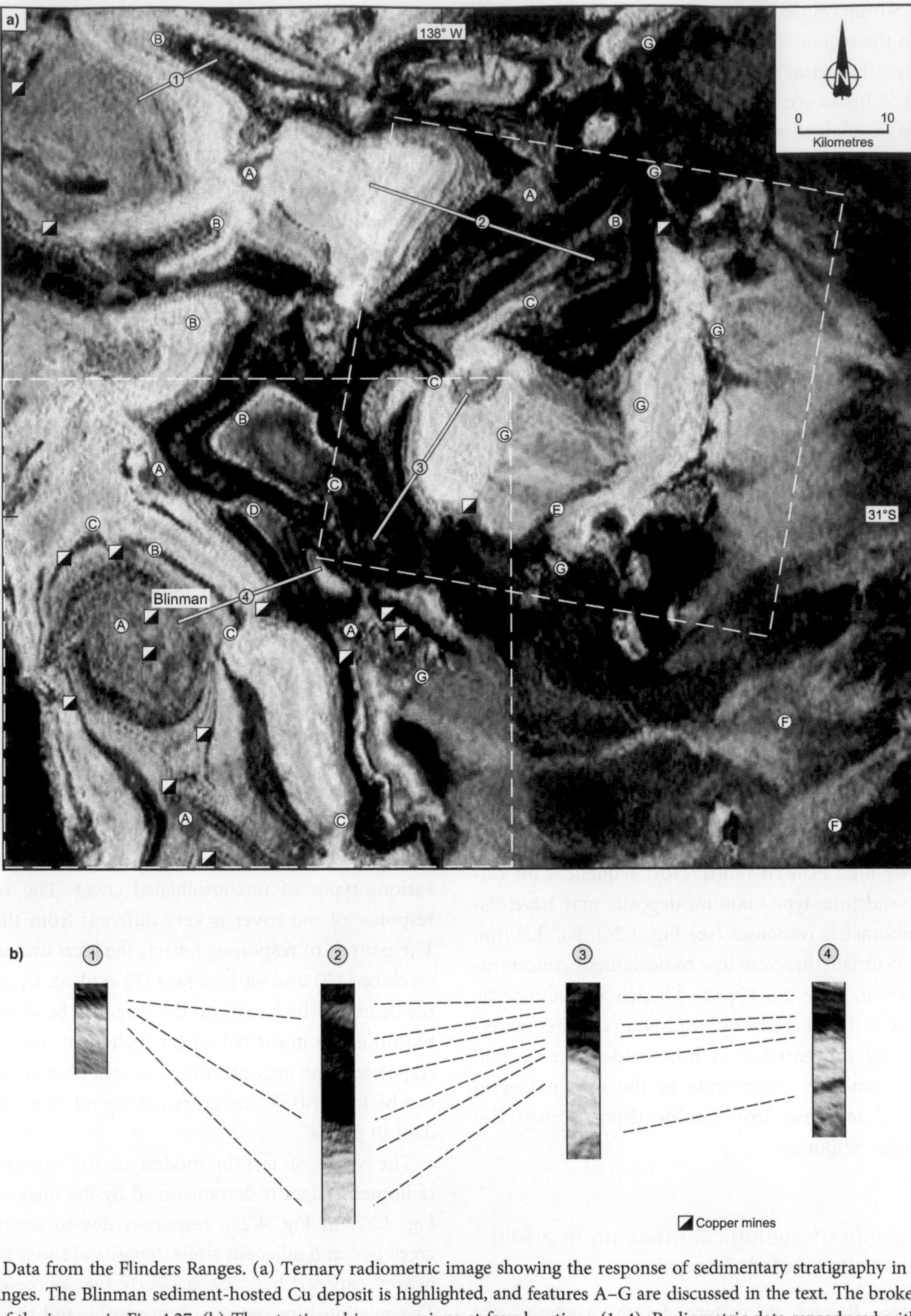

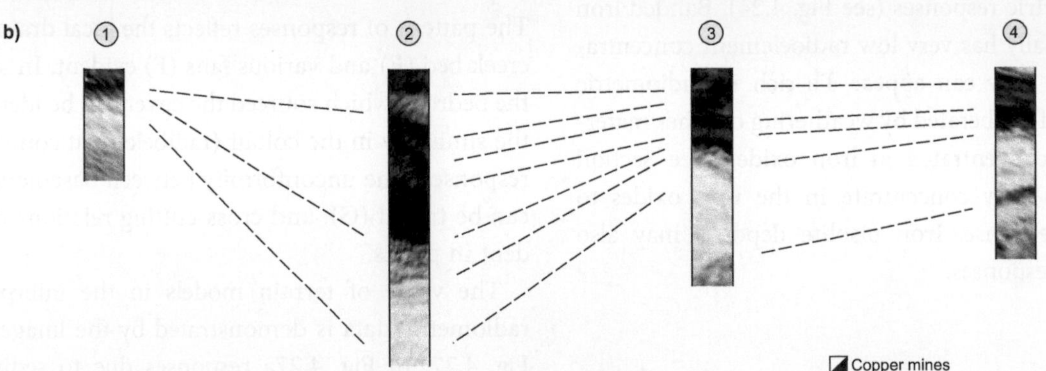

Figure 4.26 Data from the Flinders Ranges. (a) Ternary radiometric image showing the response of sedimentary stratigraphy in part of the Flinders Ranges. The Blinman sediment-hosted Cu deposit is highlighted, and features A–G are discussed in the text. The broken lines mark the extent of the images in Fig. 4.27. (b) The stratigraphic successions at four locations (1–4). Radiometric data reproduced with permission from Geoscience Australia.

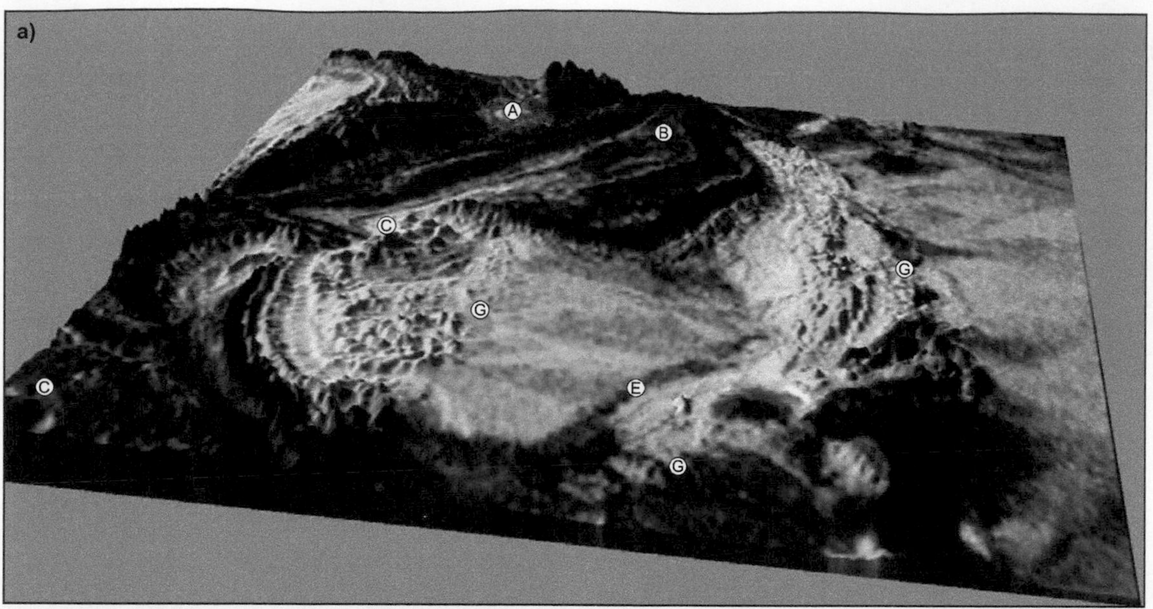

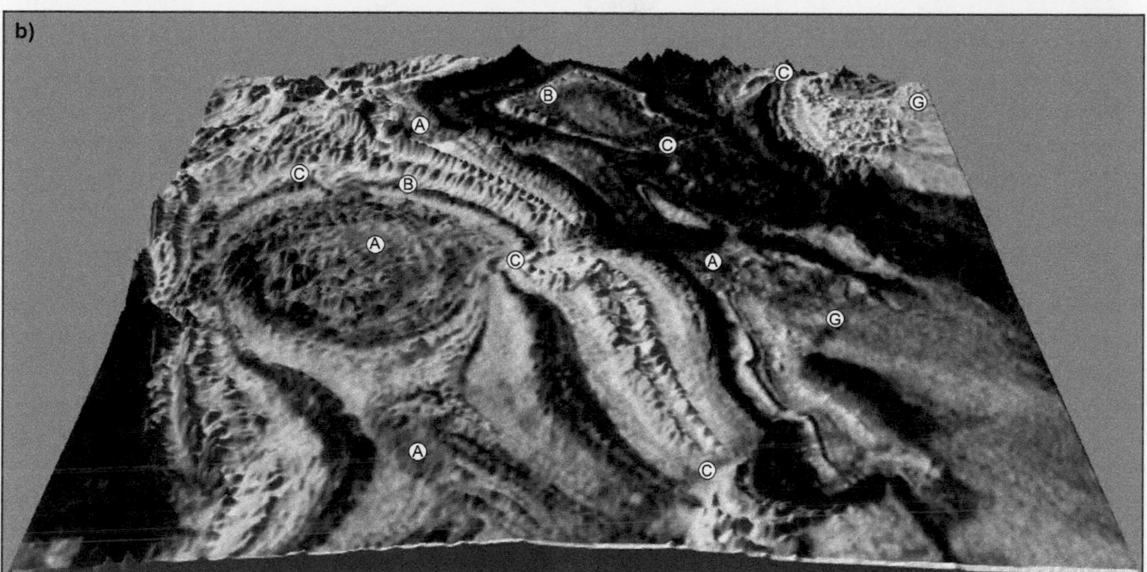

Figure 4.27 Subsets of the radiometric data in Fig. 4.26 draped on terrain models produced from the Shuttle Radar Topography Mission (Farr and Kobrick, 2000). Features A–G are discussed in the text.

over a large area, and it allows topographical features to be related to lithotypes. The slope deposits in the east are again clearly seen.

4.7.5 Interpretation of γ-logs

Logging variations in natural radioactivity on drillcore or continuously downhole is done primarily to detect differences in lithology. Almost every type of geological environment is amenable to γ-logging. This is demonstrated by the disparate geological settings from which the downhole logs shown in Fig. 4.28 are taken, and the generally strong correlation between the logs and the lithological variations.

In sedimentary rocks, it is usual for the γ-log to oscillate between two 'baselines', the higher level known as the *shale baseline*, the lower known as the *sand baseline* (Fig. 4.29). In this respect the γ-log is similar to the self-potential log (SP; see Section 5.5.3.2). The shale baseline represents responses primarily from K-bearing minerals such as mica. The sand baseline reflects the subdued responses from clastic and carbonate sediments which contain few radioelements. Both

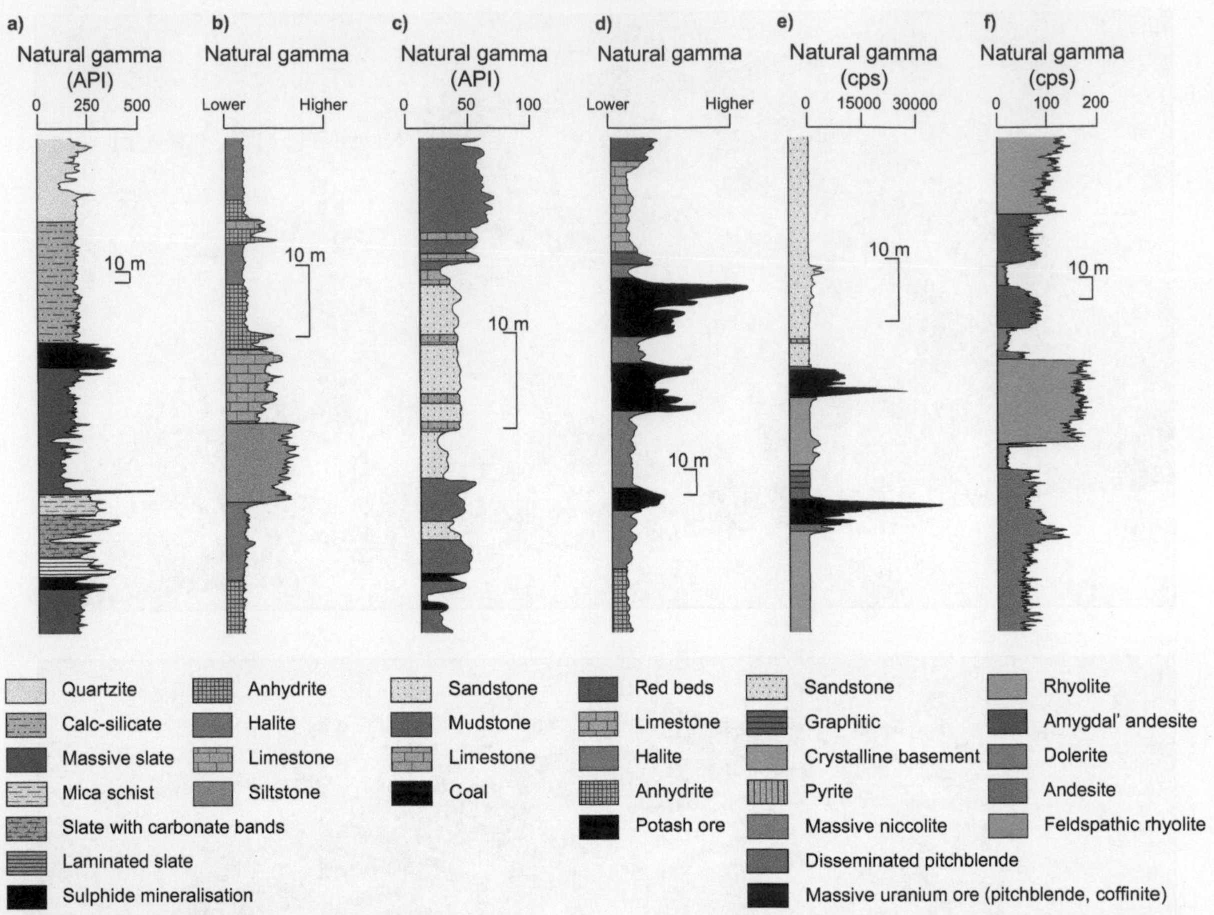

Figure 4.28 Examples of γ-logs and their relationship to lithology from a range of mineralised geological environments. (a) Dugald River Zn–Pb–Ag deposit, Queensland, Australia. The succession comprises various metasediments. Note the different responses from the sulphide mineralisation, caused by variations in its shale content. Based on a diagram in Mutton (1994). (b) Herb River Member, Nova Scotia, Canada. The discrimination between the different types of sedimentary rock is excellent, although the different types of evaporite cannot be distinguished. Based on a diagram in Carter (1989). (c) Coal measure sequence, UK. The coal horizons have a typically low response. Based on a diagram in Emery and Myers (1996). (d) Part of the Elk Point Group, including the Prairie Evaporite Formation, Saskatchewan, Canada. Sylvite is responsible for the high γ-response of the ore horizons. Based on a diagram in Klingspor (1966). (e) Cluff Lake area, Saskatchewan, Canada. As expected, the U mineralisation is associated with extremely high readings. Based on a diagram in Mwenifumbo *et al.* (2004). (f) Buchans Mine area, Newfoundland, Canada. The different igneous rock types are clearly distinguished, with higher readings from more felsic lithologies. Based on a diagram in Killeen *et al.* (1997).

the SP and γ-logs are useful indicators of facies/sedimentary environments, which are inferred from changes in the shapes of the log responses. For example, a fining-upwards sequence is represented by a gradual change from the level of the sand baseline to that of the shale baseline. Facies analysis has been successfully applied in exploration for mineralisation whose setting is controlled by facies variations, notably sandstone-type U mineralisation. This kind of interpretation is a specialised topic beyond our scope; a good summary is provided by Rider (1996).

Gamma-logs are ideal for identifying zones of radioactive alteration and mineralisation, usually U mineralisation (Fig. 4.28e). Details about downhole γ-ray logging for uranium exploration can be found in IAEA (1982). Gamma-logs, carefully calibrated to chemical assays, can produce *in situ* assays of equivalent U_3O_8 (eU_3O_8) grades intersected by the drillhole faster and more economically than assays based on chemical analysis. Calibration accounts for drillhole diameter, drillhole casing type and thickness, the presence or absence of fluids, and the ore-zone angle of incidence with the drillhole (Lehmann, 2010). However, it is a fundamental requirement that the mineralisation be in equilibrium (see Section 4.6.1), and if there are significant emissions from K and Th then spectral logs are required. Løvborg *et al.* (1980) describe an example of resource evaluation using radiometric logging. Problems caused by radioactive disequilibrium can be overcome in downhole logging by directly measuring the concentration of U in the rock formation with a method known as *prompt fission neutron* (PFN) *logging*.

The γ-log is an effective indicator of subtle changes in the composition of apparently similar lithotypes, so these logs may also be extremely useful for stratigraphic correlation. In the Hamersley iron-ore province in northern Western Australia, ferruginous chert horizons, giving rise to distinctive γ-responses within the otherwise lithologically monotonous succession of iron formation (Fig. 4.30), allow reliable correlations to be made at the mine-scale and over hundreds of kilometres.

Gamma-logs are usually an integral component of many types of multiparameter logs. Urbancic and Mwemifumbo (1986) describe the use of radiometric logs as part of a multiparameter log suite to investigate the alteration associated with a gold deposit in the Larder Lake region of Ontario, Canada.

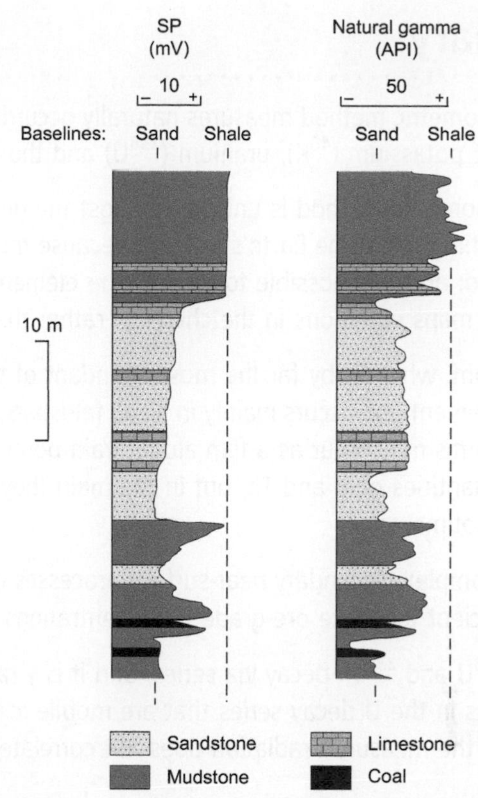

Figure 4.29 Self-potential (SP) and γ-logs through a coal-measure sequence. Note the similarity of the two logs; both vary between sand and shale baselines. Based on diagrams in Emery and Myers (1996).

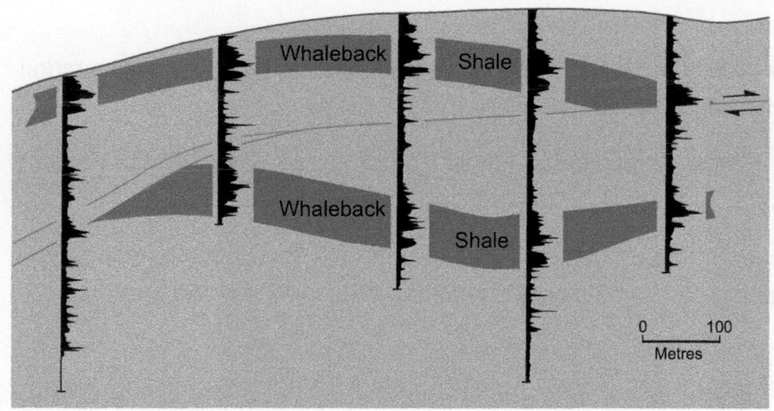

Figure 4.30 Gamma-logs through iron formations at Jimblebar in the Hamersley iron-ore province of Western Australia. The distinctive responses from the Whaleback Shale allow the structure of the area to be defined with confidence. Redrawn, with permission, from Kerr *et al.* (1994).

Summary

- The radiometric method measures naturally occurring radioactivity in the form of γ-rays and specifically from radioactive decay of potassium (^{40}K), uranium (^{238}U) and thorium (^{232}Th).
- The radiometric method is unique amongst the geophysical methods in that the responses originate from only the top few centimetres of the Earth's surface, because the γ-rays can only pass through a few centimetres of rock before being absorbed; and it is possible to identify the elemental sources of the radiation and quantify their concentrations. The method maps variations in the chemical rather than the physical characteristics of the survey area.
- Potassium, which is by far the most abundant of the three radioactive elements in the geological environment, is an alkali element and occurs mainly in alkali feldspars, micas and illite. Uranium can form either an oxide or a silicate and in these forms may occur as a film along grain boundaries. Quartz, feldspar and other rock-forming minerals may contain trace quantities of U and Th, but in the main they occur in accessory minerals such as zircon, monazite, sphene, rutile and xenotime.
- Many complex, secondary near-surface processes can remove, transport and deposit K, U and Th. These processes may be sufficient to create ore-grade U concentrations in surficial materials.
- Both ^{238}U and ^{232}Th decay via series, and it is γ-rays from daughter elements in the series that are detected. Daughter elements in the U decay series that are mobile in the geological environment may cause disequilibrium, whereby the level of the measured radiation does not correlate with the abundance of U present.
- Radioactive decay is a random process, and measurements involve counting the γ-ray photons, using a process known as scintillation, over a finite integration time. The longer the integration time the more accurate is the measurement.
- Gamma-ray spectrometers record the γ-ray count at different energies allowing the source of the radiation to be determined, i.e. K-, U- or Th-rich source.
- Gamma-ray detectors have an acquisition footprint that depends on the source–detector separation and whether the detector is in continuous motion.
- The reduction of radiometric data is based mostly on complex calibration procedures. Radiation originating from radon gas in the atmosphere is particularly difficult to compensate for.
- Concentrations of K, U and Th are often correlated in most rock types, so ratios of the concentrations of each element are powerful tools for detecting anomalous regions.
- Radiometric data are best interpreted in conjunction with multispectral remote sensing data and digital terrain data. These ancillary data are especially useful when mapping cover material and regolith, and when working in rugged terrains.
- Radiometric data are conventionally displayed as ternary images with K displayed in red, Th in green and U in blue.
- Mineralisation itself, associated alteration zones and lithotypes favourable for mineralisation may all have anomalous radiometric compositions, detected either in a particular radioelement or in a particular ratio of the elements.

Review questions

1. What is meant by the terms half-life and Compton scattering, and why are these important for radiometric surveys?
2. Compare and contrast individual channel values and channel ratios. How can they be used in the interpretation of radiometric survey data?

3. Describe how the radiometric element concentration of igneous rocks varies as rock chemistry varies from ultramafic to ultrafelsic.
4. Describe the nature and distribution of K, U and Th in overburden material.
5. Give some examples of mineral deposit types which may be directly detected by radiometric surveys.
6. What is disequilibrium and why is it important?
7. What is the difference between a scintillometer and a spectrometer? Explain stripping ratios and how they are determined.
8. Calculate the statistical measurement error for the following count levels measured over one second: 10, 40, 100, 400 and 1000. How does the error change with count level and how can it be reduced for a given detector?
9. Describe the circle-of-investigation and how this can be used to select the most appropriate survey height.
10. What is the source of atmospheric radon gas and how are its effects removed during reduction of radiometric survey data?

FURTHER READING

Dickson, B.L. and Scott, K.M., 1997. Interpretation of aerial gamma-ray surveys – adding the geochemical factors. *AGSO Journal of Australian Geology & Geophysics*, 17(2), 187–200.

This paper provides an in-depth description of the interpretation of radiometric data. It demonstrates just how difficult it is to predict the radiometric response of mineralisation.

Durrance, E.M., 1986. *Radioactivity in Geology, Principles and Applications*. Ellis Horwood.

A detailed but easily understandable description of radioactivity relevant to the geosciences.

Haack, U., 1982. Radioactive isotopes on rocks. In Angenheister, G. (Ed.), *Physical Properties of Rocks. Landolt-Börnstein: Numerical Data and Functional Relationships in Science and Technology, Group V: Geophysics and Space Research*, 1b, Springer, 433–481.

A comprehensive compilation of data and references on the abundance and behaviour of K, U and Th in the geological environment.

Minty, B.R.S., 1997. Fundamentals of airborne gamma-ray spectrometry. *AGSO Journal of Australian Geology & Geophysics*, 17(2), 39–50.

Minty, B.R.S., Luyendyk, A.P.J. and Brodie, R.C., 1997. Calibration and data processing for airborne gamma-ray spectrometry. *AGSO Journal of Australian Geology & Geophysics*, 17(2), 51–62.

Together these papers provide an in-depth description of the acquisition and processing of radiometric data.

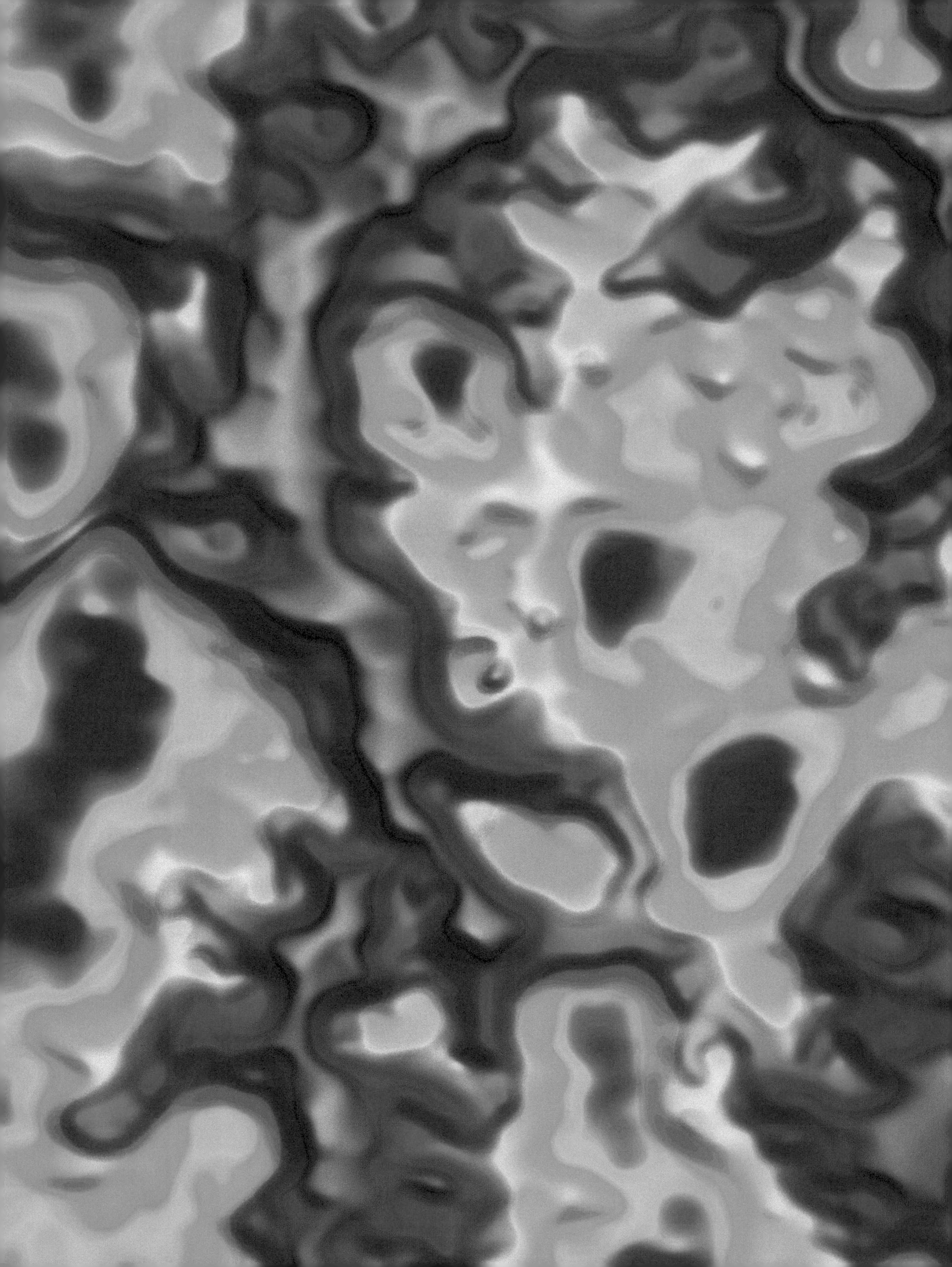

CHAPTER

5 Electrical and electromagnetic methods

5.1 Introduction

There is a wide variety of geophysical methods based on electrical and electromagnetic phenomena that respond to the electrical properties of the subsurface (Fig. 5.1). Some are passive survey methods that make measurements of naturally occurring electrical or electromagnetic fields and use rather simple survey equipment. Others are active methods that transmit a signal into the subsurface and use sophisticated multichannel equipment to derive multiple parameters related to the electrical properties of the subsurface. EM and electrical surveys are routinely undertaken on and below the ground surface, but electrical surveys require contact with the ground so only EM measurements are possible from the air.

Survey methods that involve the measurement of electrical potentials (see Sections 5.5 and 5.6), associated with the flow of subsurface current, by direct electrical contact with the ground, are collectively known as *electrical methods*. These include the *self-potential* (SP), *resistivity*, *induced polarisation* (IP) and *applied potential* (AP) methods. With the exception of the SP method which measures natural potential, all the others depend on the electrical transmission of the current into the ground. *Electromagnetic* (EM) methods use the phenomenon of electromagnetic induction (see Section 5.7) to create the subsurface current flow and measure the magnetic fields associated with it. A variant of the EM method uses high- (radio and radar) frequency electromagnetic waves (see online Appendix 5) with the results resembling the seismic reflection data described in Chapter 6.

Electrical and EM methods both provide information about electrical *conductivity*, a measure of the ease with which electrical currents flow within the subsurface. An associated parameter is *resistivity*, which is the inverse of conductivity and a measure of the difficulty with which electrical currents flow (see Section 5.6.2). In particular, electrical methods depend upon the contrast in electrical properties and respond well to regions where current flow is inhibited, but these methods may be unable to determine the absolute conductivity accurately, especially of small targets. EM methods respond primarily to the absolute conductivity of the ground and the dimensions of that distribution, rather than just the conductivity contrast, and are most sensitive to regions where current flow is least inhibited, i.e. conductive targets.

The rock-forming minerals that form the basis of the geological classification of rocks exert little influence over the electrical properties of rocks. The main control is porosity and the contents of the pore space, so rock electrical properties are extremely variable and hard to predict geologically. Anomalous electrical properties are observed from many types of ore minerals and their occurrences, but responses are widely variable depending upon the geological environment. As with most types of geophysical data, a direct correspondence

◀ Map of the EM time constant (τ) from an airborne survey over a palaeochannel near Balladonia, south-east Western Australia. The area shown is 19 km wide. Data provided by the Geological Survey of Western Australia, Department of Mines and Petroleum. © State of Western Australia 2013.

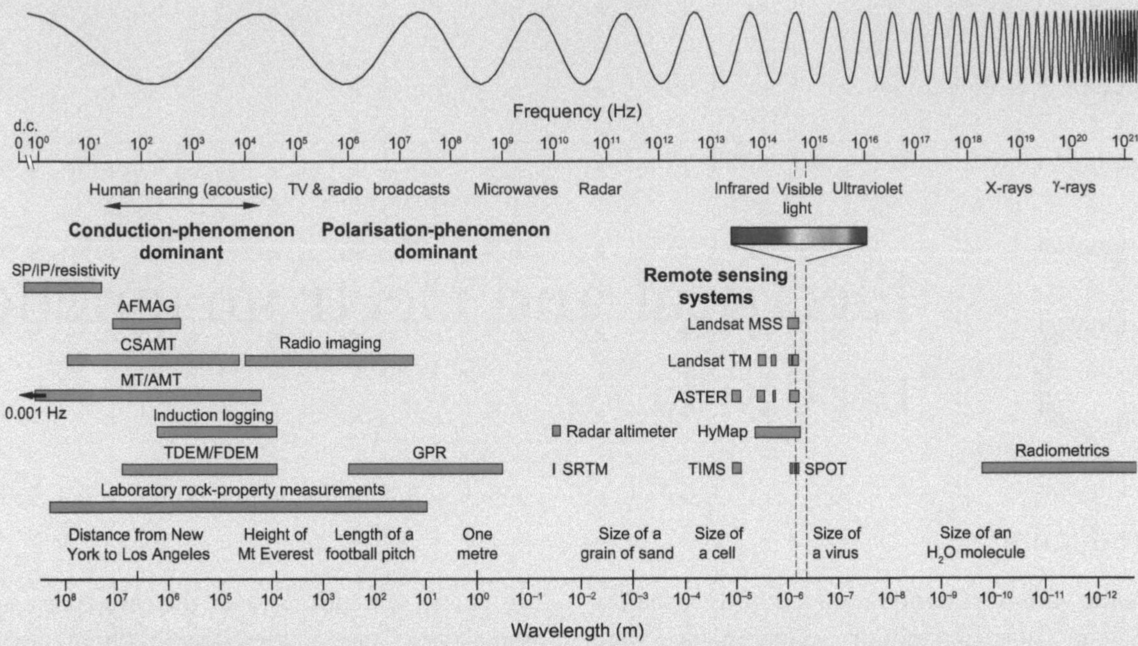

Figure 5.1 The electromagnetic spectrum and the frequency bands used by the various geophysical methods. AFMAG – audio-frequency magnetics, AMT – audio-frequency magnetotellurics, ASTER – Advanced Spaceborne Thermal Emission and Reflection Radiometer, CSAMT – controlled source audio-frequency magnetotellurics, FDEM – frequency domain electromagnetics, GPR – ground penetrating radar, HyMap – Hyperspectral Mapper, IP – induced polarisation, Landsat MSS – Landsat Multispectral Scanner, Landsat TM – Landsat Thematic Mapper, MT – magnetotellurics, SP – self-potential, SPOT - Satellite pour l'Observation de la Terre, SRTM – Shuttle Radar Topography Mission, TDEM – time domain electromagnetics, TIMS – thermal infrared multispectral spectroscopy. Compiled from various sources including Scott *et al.* (1990) and Lowrie (2007).

between lithological differences and the geophysical response is not to be expected.

In mineral exploration, electrical and EM methods are used at regional and prospect scale for direct detection of electrically anomalous targets, in particular metal sulphide and metal oxide mineralisation. Information about the geometry, dimensions and electrical characteristics of the sources of anomalies can be obtained. Another common application is in mapping the internal structure and thickness of the near-surface materials, e.g. regolith or unconsolidated sedimentary cover. Geological mapping using airborne EM techniques is increasing, following the continuing developments in surveying techniques and modelling of data. Since electrical methods cannot be implemented from the air, they have a far lesser role in regional mapping. In common with all other geophysical methods, improved interpretational tools are being developed, mostly based on inverse modelling (see Section 2.11.2.1), that produce images of the subsurface distribution of the various electrical properties. These developments have encouraged more widespread and increasing use of the electrical and EM methods.

There is significant overlap between electrical and EM methods, not only in the physics and geology controlling the Earth's response, but also in terms of the equipment used and how the data are utilised in mineral exploration. This chapter begins with a summary of aspects of the physics of electricity and electromagnetism required to understand and interpret geophysical data in a geological context. This is followed by a summary of the electrical properties of geological materials, the ultimate control on the geophysical responses. Next is a description of some generic aspects of the in-field measurement of electrical and EM phenomena. Following this is a series of descriptions of the different electrical and EM methods commonly used for minerals exploration and geological mapping. Two less used techniques, magnetometrics and frequency-domain magnetotelluric electromagnetics, are described in online Appendices 3 and 4, respectively.

Necessarily, the mathematical aspects of the electrical and EM methods are beyond our scope. For a comprehensive treatment the reader is referred to Zhdanov and Keller (1994). A review of the application of both methods in resource exploration is provided by Meju (2002).

5.2 Electricity and magnetism

An electric current is a flow of electrically charged particles: an analogy can be drawn with the flow of water, i.e. the movement of H_2O molecules. The circuit through which the electric current flows is represented by a 'plumbing' system through which fluid water flows. We use this analogy and the concept of an 'electrical fluid' throughout our description of the properties and behaviour of electricity.

Electromagnetism is the interaction between electricity and magnetism, an in-depth understanding of which requires advanced mathematical skills. Fortunately, the principles of electromagnetism as applied to geophysical surveying can be adequately understood from a purely qualitative description.

5.2.1 Fundamentals of electricity

An electric current is made up of moving charges. In solids it is loosely bound outer-shell electrons that move, a process known as *electronic conduction.* In liquids and gases it is electrons and/or ions (positively and negatively charged) that move; this process is known as *ionic conduction.* A dielectric or displacement current may also be created by distortion of an atomic nucleus and surrounding electrons. Normally the electrons are symmetrically distributed about the nuclei of the atoms forming a substance. If the atoms acquire a consistent asymmetrical form there must be a net movement of charge through the substance, i.e. an electric current. In most geophysical situations the contributions of displacement currents are negligible, but they are important contributors to the geophysical response of radar and radio frequency EM methods (see online Appendix 5).

5.2.1.1 Electric fields

Electrically charged bodies have an *electric field* associated with them and are therefore affected by other electric fields, just as bodies with mass (i.e. everything) create and are affected by gravitational fields. In the same way that a mass has a gravity field everywhere around it, an electric charge is surrounded by an electric field. In both cases the fields act on distant objects; masses always attract, but electrical charges may be positive or negative, so the forces may be either attractive or repulsive: like charges repel each other and unlike charges attract.

Electric fields are defined in terms of electrical *potential.* For an isolated compact electrical charge, known as a *pole,* its electric field is symmetrical about the pole with the *equipotential surfaces*, i.e. surfaces created by joining points having the same electrical potential, forming concentric spheres (Fig. 5.2a). The potential decreases inversely proportionally with distance from the pole. The gravitational potential energy of a body increases when it is moved against the influence of gravity, for example when lifted from the floor to a high shelf. Similarly, a charged particle may increase its electrical potential energy when moved

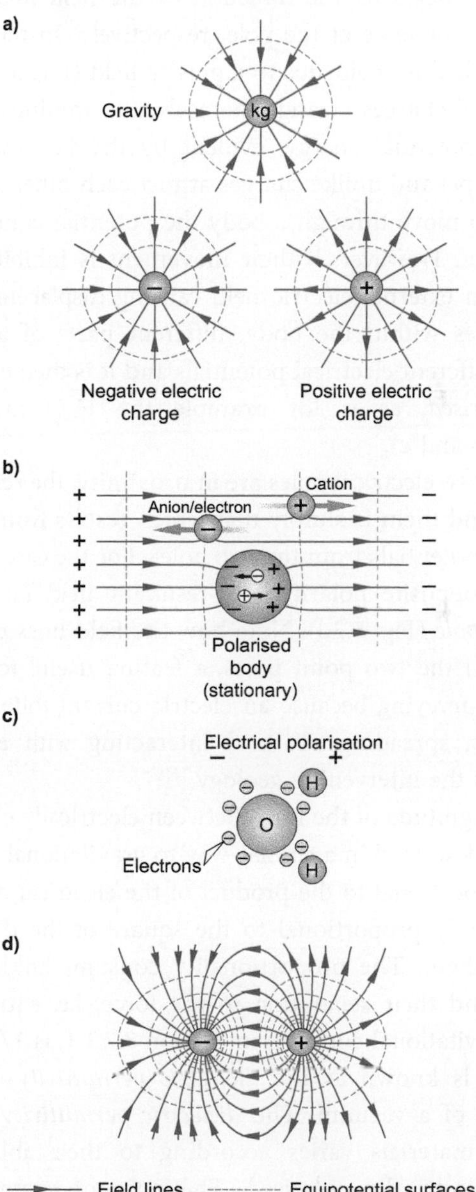

Figure 5.2 Electric charges and their fields. (a) The electric field of an isolated electric pole. Note the similarity to the gravity (potential) field of a mass. (b) Polarisation of a body in an external electric field due to movement of electrical charges within the body. (c) The polarised H_2O molecule. (d) The field of an electric dipole.

against the action of an electric field, for example when a positive charge is moved closer to another positive charge so that the repulsive force between them increases. The direction a charge moves to minimise its energy defines the lines of force or the imaginary *field lines* of the electric field. By convention, the field lines are sketched with arrows showing the direction of the field pointing from positive to negative, i.e. the direction of movement of a positive charge. The field lines radiate from the pole and are everywhere normal to the equipotential surfaces. Since the pole may be positive or negative, the direction of the field lines may diverge or converge at the pole, respectively. In the latter case, this field is analogous to a gravity field (Fig. 5.2a).

Electrical charges arrange themselves to minimise their electrical potential energy, evident by the fact that like charges repel and unlike charges attract each other. If they are free to move through a body the potential is uniform throughout. However, if their movement is inhibited, or there is an external electric field causing displacement of the charges within the body, different parts of a body acquire different electrical potentials and it is then *electrically polarised*, as is for example the H_2O molecule (Figs. 5.2b and c).

When two electrical poles are in proximity, the resulting field around them is simply that which results from summing the potentials from the two poles. For the case of two poles of opposite polarity, the resultant field forms an *electric dipole* (Fig. 5.2d). Note how the field lines diverge away from the two point poles, a feature useful for geophysical surveying because an electric current follows the field lines, spreading out and interacting with a large volume of the intervening geology.

The magnitude of the force between electrically charged bodies is described in a similar way to gravitational forces, being proportional to the product of the electrical charges and inversely proportional to the square of the distance between them. The proportionality constant linking the charges and their separation to the force, i.e. equivalent to the gravitational constant in Section 3.2.1.1, is $1/(4\pi\varepsilon_0)$ where ε_0 is known as the *dielectric permittivity* of free space, i.e. of a vacuum. The *dielectric permittivity* (ε) of different materials varies according to their ability to become electrically polarised. The ratio of a material's permittivity to that of free space is known as the *dielectric constant* (κ) and given by the expression:

$$\kappa = \frac{\varepsilon}{\varepsilon_0} \tag{5.1}$$

Unlike gravity fields, the variation of an electric field is significantly affected by the properties of the materials through which it passes. However, this is only significant for geophysical methods using very high-frequency electromagnetic waves (see online Appendix 5).

5.2.1.2 Potential difference, current and resistance

It is convenient to describe the properties of electric currents using the analogy of a flow of water through the simple plumbing system shown in Fig. 5.3a. Clearly, if the

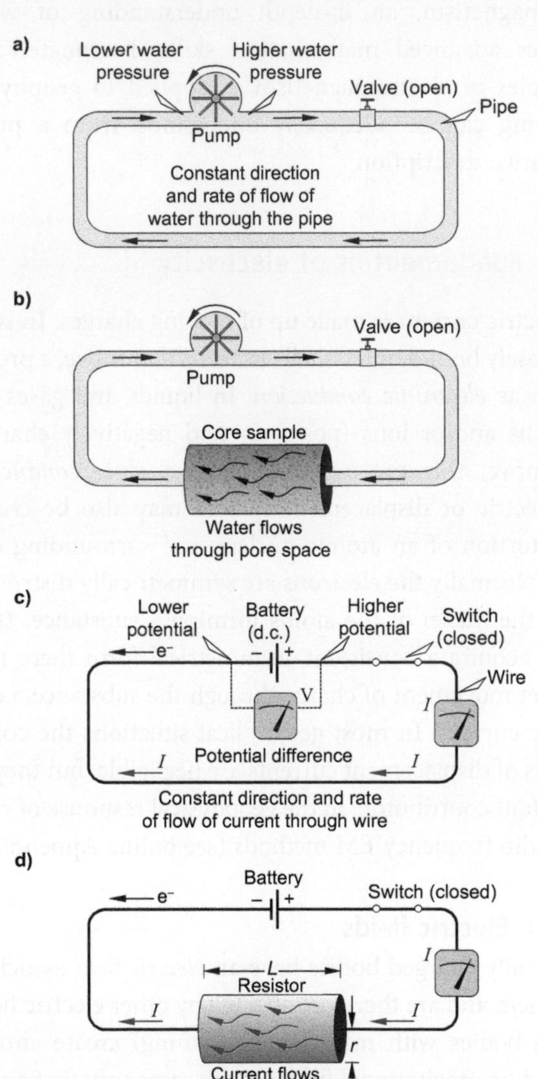

Figure 5.3 Simple plumbing systems (a and b) and d.c. electrical circuits (c and d). (a) Water driven around a closed plumbing system by a pump. (b) Plumbing system in (a) with a porous-rock core sample included through which the water is forced. (c) Electrical circuit comprising a battery and closed loop of wire around which the current flows. (d) Electrical circuit in (c) with the addition of a resistor, hindering the flow of current, and analogous to the plumbing system in (b). V – voltage, and I – current.

water is to flow there must be some form of driving force (the influence of gravity on the pressure and flow of the water will be ignored here). In this case it is the difference in water pressure created by the pump. The water flows away from the area of higher pressure and towards areas of lower pressure. As long as the pump maintains the pressure difference, the flow of water continues. The flow may also be stopped by the valve, which if shut breaks the continuity of fluid flow through the system. Continuity of the system is a second requirement for flow to occur.

Let us now modify the system so that the water is forced to flow through a cylinder of homogeneous rock encased in a water-tight seal (so that the water does not flow out of the sides of the rock specimen) (Fig. 5.3b). The flow of water through the pipes can be thought of as occurring without hindrance, but the ease with which the water flows through the cylindrical rock specimen depends on two factors:

- The porosity and permeability of the rock. If the specimen is a highly porous and permeable sandstone, the flow of the water is hindered less than if it has low porosity and permeability as might occur in, say, a gneiss. In other words, the ease of flow is a function of some inherent characteristics of the rock. Even when highly porous, the tortuosity of the pore system will affect the ease with which flow can occur, with flow paths that are complex and convoluted causing more energy to be lost through processes such as friction.
- The dimensions of the rock specimen. Increasing the length of the cylinder will make it harder to drive the water through the longer specimen. On the other hand, increasing the diameter of the cylinder presents a larger cross-sectional area of pores for the water to flow through, so its passage will be easier.

Electrical storage batteries and generators create electrical potential energy. In a battery this is done through a chemical reaction, in a generator it is done mechanically through electromagnetic induction (see Section 5.2.2.2). The resulting electrical potential energy is available to drive charges through an electrical circuit external to the potential energy source, i.e. to produce a flow of electric current. Consider now the analogous electrical circuits of our plumbing systems (Figs. 5.3c and d). The battery (or it could be a generator) can be thought of as an electrical pump. Just as the water pump creates a difference in water pressure, the battery creates a difference in electrical pressure (electrical potential energy) known as the *electromotive force* (emf). It produces a potential difference measured in volts (V), which is often referred to as simply the voltage. This is what causes movement (flow) of charges through the wires connected to the battery. Recall that it is the movement of charges that constitutes an electric current. Current (I) is measured in amperes (A) and is quantified in terms of the amount of electric charge that flows in one second, i.e. the rate of flow.

The movement direction of the charges depends on their polarity. Negatively charged electrons move from the lowest to the highest potential in a circuit, i.e. electrons move from the negative terminal through the external circuit to the positive terminal of a battery. However, by convention, the direction of current flow is taken as the flow of positive charges from the more positive to the more negative potential. The electric current flows as long as the electrical pressure or force (voltage) is maintained and is sufficient to overcome any hindrance to the flow. This requires that the switch, equivalent to the hydraulic valve, is closed to complete the circuit.

Consider now the amount of charge passing through the cross-sectional area of the wires. It is quantified as the *current density* and, assuming that the current is distributed uniformly across the area, is a measure of the amount of current passing through a unit area and specified in units of A/m^2. Increasing the current increases the number of charges flowing and increases the current density, and similarly, decreasing the cross-sectional area of the wires (thinner wires) increases the current density.

Just as the cylinder of rock resisted the flow of water, a substance will also resist the flow of charges. A *resistor* has been included in the circuit in Fig. 5.3d and its associated *resistance* (*R*) to the flow of current is a form of electrical *impedance*. If it is increased whilst maintaining the same electrical potential, then the current flow is reduced. As with the water analogy, the actual resistance to the flow of electric current is partly a function of the geometry of the substance and partly due to its inherent properties. Recall that the pressure difference created by the pump was constant, with the flow of water through the specimen dependent on its porosity and permeability and its dimensions. As with the flow of water, increasing the cross-sectional area reduces the electrical resistance, whereas increasing the length increases the resistance. To make a meaningful statement about how much a substance opposes the passage of electric current, it is necessary to remove the geometrical effects of the specimen from electrical measurements made on it to obtain the equivalent that would result from a unit cube of the material. This is known as the *resistivity* (ρ) of the material and is obtained by applying a geometric correction factor (k_{geom}) to the measured resistance, as follows:

$$\rho = Rk_{geom} \tag{5.2}$$

For the case of a cylindrical specimen

$$k_{geom} = \frac{\text{Cross-sectional area}}{\text{Length}} = \frac{\pi r^2}{L} \tag{5.3}$$

where r is the radius of the cylinder and L its length with both measured in metres.

The unit of electrical resistance is the ohm (Ω) and the unit of resistivity is the ohm-metre (Ω m). It is also common to quantify the ease with which electric current may flow in terms of the reciprocals of these parameters, respectively *conductance* and *conductivity*. Their units of measurement are, respectively, siemen (S) and siemens per metre (S/m). Conductivity is commonly designated by the Greek letter σ.

It is obvious in the plumbing analogy that increasing the water pressure causes an increase in the flow of water around the plumbing system; the greater the resistance to the flow, the smaller will be the flow, and vice versa. In terms of the equivalent electrical parameters, the current (I) is proportional to the potential difference (V), applied across a resistance (R) and inversely proportional to the resistance and given by:

$$\text{Current} = \frac{\text{Voltage}}{\text{Resistance}} \quad \text{or } I = \frac{V}{R} \tag{5.4}$$

This linear relationship between current and voltage is known as *Ohm's Law* and it quantifies the intrinsic relationship between potential difference, current and resistivity. It is often presented in a rearranged form as:

$$\text{Voltage} = \text{current} \times \text{resistance} \quad \text{or } V = IR \tag{5.5}$$

Stated in terms of the resistivity and geometry of the body:

$$V = I\frac{\rho}{k_{geom}} \tag{5.6}$$

It is clear from Ohm's Law that the difference in potential across parts of a body will be less if it has low resistivity. For a perfect conductor (ρ = zero), all points within it will be at the same potential. In the plumbing analogy, the perfect conductor is a void within which the water pressure is constant throughout.

5.2.1.3 Conductors, semiconductors and insulators

Materials are considered to be insulators if their conductivity is less than 10^{-8} S/m, and conductors if their conductivity is greater than 10^5 S/m. Between these two arbitrary boundaries are semiconductors and electrolytes. Conductors, such as metals, contain many loosely bound electrons and so the electric current can flow comparatively easily when an electric field (a potential difference) is applied. The equivalent materials in the plumbing analogy are porous and permeable formations through which water passes freely. Insulators, also called non-conductors or *dielectrics*, do not have free electrons so the flow of the current is much more difficult, with dielectric conduction the usual mechanism. These materials are equivalent to rocks with minimal porosity and permeability.

Semiconductors, and electrolytes, resemble the conductors when a static electrical potential is applied, in that conductivity involves movement of electrons. However, they have fewer free electrons and more energy is required to move them, so their conductivity is lower. It is also critically dependent on even very small amounts of impurities which cause imperfections in the mineral's crystal lattice where there may be unbalanced charges and potentially mobile electrons. Conductivity is also dependent upon temperature; increasing the temperature increases the energy of the electrons so they can move more easily and increase the conductivity. At low temperatures semiconductors behave more like insulators.

5.2.1.4 Alternating current

When a constant potential difference is applied to a circuit, the current flows in one direction only and is referred to as *direct current* (d.c.). Continuously varying the potential in a sinusoidal manner creates an alternating back-and-forth flow, known as *alternating current* (a.c.). Analogous plumbing and a.c. electric circuits are shown in Fig. 5.4. In this case the water pump creates a back-and-forth water flow. Similarly, in a.c. circuits the varying potential produced by the generator causes the charge carriers to oscillate around a point in the circuit. The rate of sinusoidal oscillation defines the *frequency* of the alternating current (see online Appendix 2).

The alternating current also experiences resistance to its flow, but the situation is more complex than for the case of a direct current. In the context of the description of conductivity, resistivity and voltage given so far, it would be expected that the variations in potential difference and current would occur simultaneously, i.e. they would be exactly *in phase* (Fig. 5.5a) (see online Appendix 2). However, electrical phenomena known as *inductance* and *capacitance* cause the variations in current to occur out-of-step

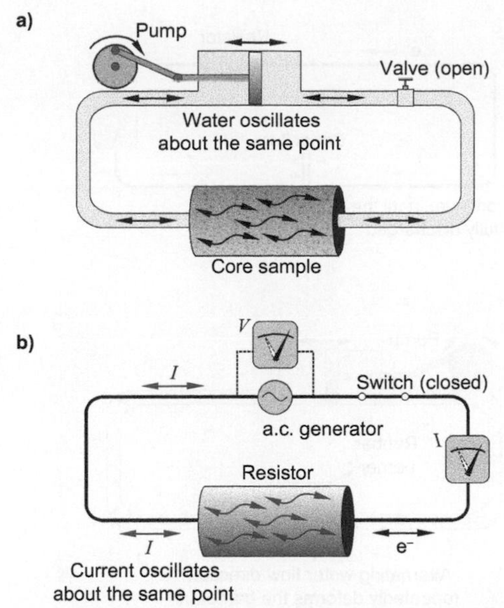

Figure 5.4 Simple plumbing system (a) and analogous a.c. electrical circuit (b). The oscillating piston causes the water flow to oscillate about points in the system. (b) The alternating current generator causes the charge carriers to oscillate in a similar way forming an a.c. current. V – voltage, and I – current.

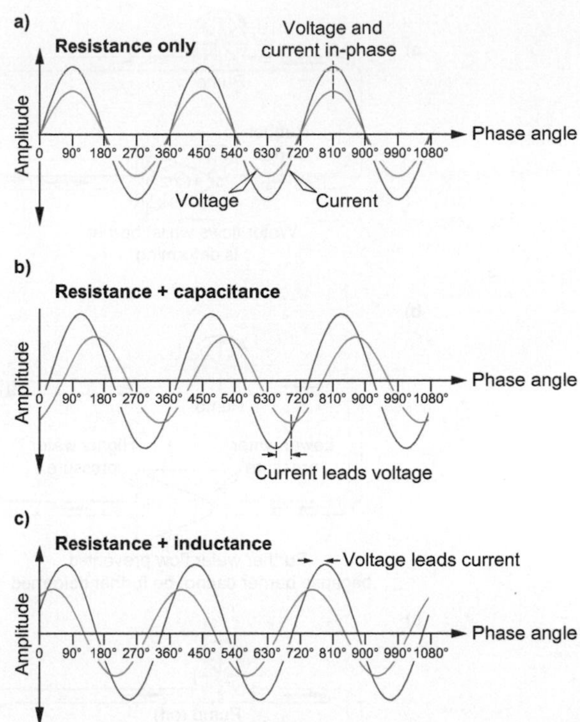

Figure 5.5 The relationship between sinusoidal variations in the voltage across, and current through, (a) a resistance; and (b) a capacitance and (c) an inductance in a resistive circuit.

with the variations in the applied potential, i.e. a *phase shift* is introduced. Capacitance involves the energy of stored charges and is described next. Inductance is described in Section 5.2.2.

5.2.1.5 Capacitance

Consider now the plumbing system in Fig. 5.6a. It contains an elastic rubber barrier that can be deformed, but not penetrated, by the water under pressure. Importantly, this means the continuity of the system is interrupted, albeit not as completely as it would be if a valve were closed. When the pump is turned on the water flow causes the barrier to expand (Fig. 5.6a). This continues until the pump is unable to deform the barrier any further (Fig. 5.6b). At this point the water ceases to flow and there is a pressure difference maintained across the barrier. If the pump is now switched off the water will flow for a short time in the opposite direction as the pressure difference dissipates (Fig. 5.6c).

The electrical analogy of the elastic barrier is a *capacitor*, a circuit element capable of storing electrical charge. It comprises two conductors of large surface area, e.g. plates, with an intervening insulator. Because the conductors are not in electrical contact, charge can be stored on their surfaces. Just as the water next to the barrier is maintained under pressure as long as the pump is running, the capacitor stores electrical charges on its plates, each of opposite polarity and carrying an equal number of charges. The uneven distribution of charges means the capacitor is electrically polarised. The charges flow into the capacitor from the applied potential through a process known as *charging* (Fig. 5.6d). Just as the volume of water confined by the barrier is limited by the pump's ability to force water against the increasing resistance presented by the barrier's elasticity, a capacitor is also limited in the amount of charge it can store. It takes time for the capacitor to fully charge, then the current ceases to flow in the circuit. When the applied potential is disconnected; the capacitor holds this charge as long as the charges have no path along which to flow. Connecting an external circuit across the capacitor provides a current path, the charges being driven by the mutually repulsive forces acting between the like charges on each plate. The current flow *discharges* the capacitor. The time taken for the capacitor to charge, and to discharge, is dependent on the size (capacitance) of the capacitor and resistance of the external circuit connected to it. Note that the discharge current flows in the opposite direction to the charging current (Fig. 5.6e).

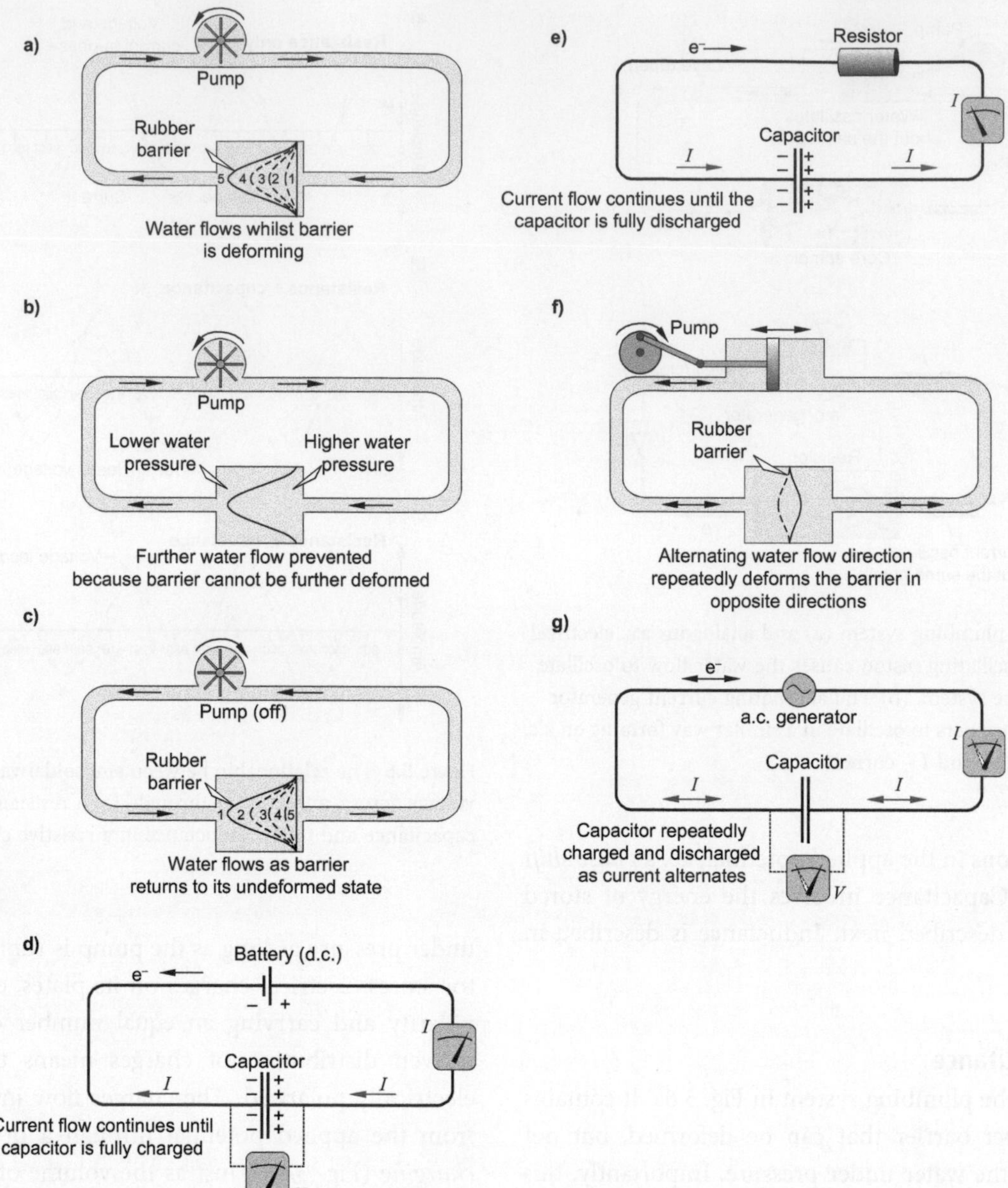

Figure 5.6 Effects of a capacitor in electrical circuits demonstrated using simple plumbing systems. (a–c) Plumbing system analogous to a d.c. circuit during charging of a capacitor. Labels 1 to 5 indicate stages of deformation of the rubber barrier with increasing time: (a) charging, (b) fully charged and (c) discharging. (d) and (e) Charging and discharging of a capacitor in a d.c. circuit. (f) Plumbing analogy for a capacitor in an a.c. circuit. (g) Alternate charging and discharging of a capacitor in an a.c. circuit. V – voltage, and I – current.

Consider now the situation for alternating current. In our plumbing analogy the elastic rubber barrier is now in a system where the direction of water flow is continuously alternating (Fig. 5.6f). Its response to the flow will vary according to the rate or frequency of the alternation. The degree of deformation will be greater for lower-frequency alternations because there is more time for expansion to occur, and the more the barrier is deformed the greater the pressure required for the process to continue. Logically, the amount of deformation will decrease as the frequency of alteration increases, because then there is less time available between pressure pulses for expansion to occur. So for higher frequencies the average resistance, or impedance, to flow presented by the barrier will be less, owing to the lesser effort required to partially deform the barrier. The analogous electrical circuit is shown in Fig. 5.6g.

It is clear then that a capacitor in a d.c. circuit, after a brief charging period, blocks the flow of current. However, the back-and-forth flow of a.c. is not prevented, although it is hindered. This opposition to current flow is known as *capacitive reactance* and is a form of impedance. It results in a phase difference between the voltage applied to the capacitor and the current flowing into it; the current leads the variation in voltage (Fig. 5.5b), opposite to that caused by inductive reactance (see Section 5.2.2.2). The lead in current varies between 0° and 90° (being greatest when the resistance of the external circuit is least).

The above description suggests the possibility of measuring the capacitance of a circuit in three ways: firstly, by monitoring current flow after a direct current is discontinued; secondly, by comparing the resistance of the circuit at different frequencies of alternating current; and thirdly, by measuring the phase difference between an alternating voltage and its associated current flow. Capacitance effects are fundamental in determining the induced polarisation responses of rocks and minerals, as described in Section 5.3.2, and all three methods are used in geophysical surveying to measure it (see Section 5.6.3).

5.2.2 Fundamentals of electromagnetism

When electrical charges move, i.e. when an electric current is flowing, a magnetic field is formed around them and the intensity (strength) of the field is proportional to the magnitude of the current. For current flowing through a straight length of wire, the magnetic field is described by circular field lines concentric to the current (Fig. 5.7a). The direction of the magnetic field is dependent on the direction of current flow and described by the 'right-hand' rule (Fig. 5.7a). The strength of the field decreases with increasing distance from the current flow (the wire).

When a current-carrying wire is formed into a single circular loop (Fig. 5.7b and c) the field is strongest inside and in the plane of the loop, where the field lines point in the same direction and are parallel to the axis of the loop. The direction of the current flow determines the direction of the magnetic field. When a current-carrying wire is formed into a series of electrically connected loops it forms a coil. The magnetic fields of each turn combine to form the field of the coil (Fig. 5.8a). Increasing the number of turns increases the length of the coil and the strength of the field, as does increasing the current. As shown in Fig. 5.8b the magnetic field of a coil is like that of a bar magnet (see Section 3.2.3.1). The north pole of the magnetic field is again found using the right-hand rule (Fig. 5.7a). If the coil is grasped with the right hand so the fingers point in the direction of current flow through the coil, the thumb points to the north pole. The field of an infinitely small bar magnet approaches that produced by a single circular current loop (Fig 5.7c). Creating magnetic fields using current-carrying coils and loops of wire is an integral part of the EM method.

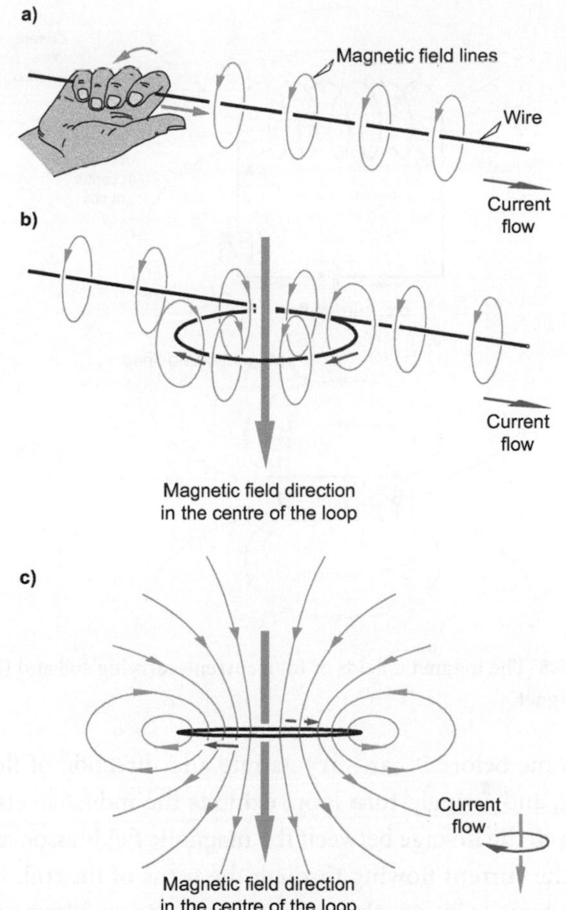

Figure 5.7 Magnetic field of an electric current. (a) Current flowing through a straight segment of wire, and described by the right-hand rule where the thumb points in the direction of current flow and the curled fingers point in the direction of the circular magnetic field lines; (b) flowing through a circular loop of wire. (c) The magnetic dipole field of an isolated current-carrying loop.

5.2.2.1 Inductance

Inductance is the tendency of a circuit to oppose any change in the current flowing through it owing to the effects of energy stored in the magnetic fields associated with the moving charges. Using the plumbing analogy, it is analogous to the inertia of the water that must be

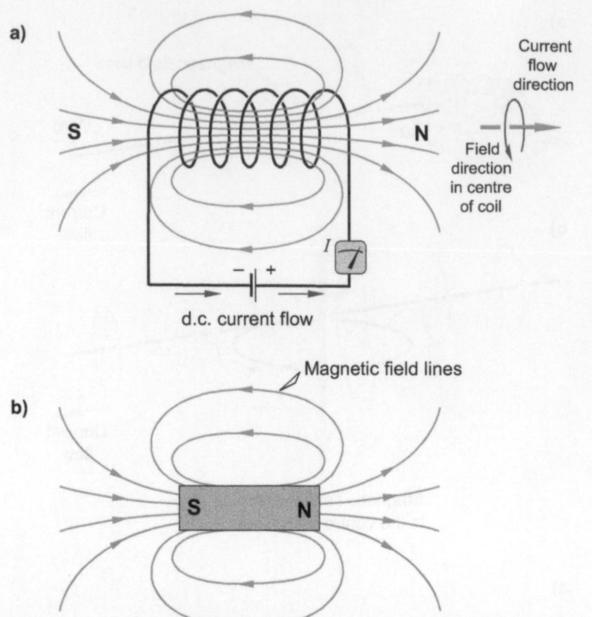

Figure 5.8 The magnetic fields of (a) a current-carrying coil and (b) a bar magnet.

overcome before it can vary its rate and direction of flow. A coil, and a single-turn loop, exhibits the inductive effect owing to the linkage between the magnetic fields associated with the current flowing through the turns of the coil. It is dependent only on the geometry of the coil/loop and increases with increasing number of turns, and is known as its *self-inductance.* Coils and loops are known as *inductances* or *inductors.*

In terms of a steady direct current, there is no change in the current flow through a coil except when it is abruptly turned on or off. These changes create a 'counter' voltage or 'back emf' that opposes the change in applied voltage and causes variations in the current flow to lag behind variations in the voltage (Fig. 5.5c). In contrast, when a.c. is flowing, each cycle of the continuously changing current is opposed by the inductance, which hinders the current flow. The level of this hindrance increases with the rate of change, i.e. with increasing frequency. The effect is known as *inductive reactance* and is a form of impedance. The lag in current varies between 0° and 90° (being greatest when the resistance of the external circuit is least), opposite to that caused by capacitive reactance. The effects of inductance, and the back emf, are important issues in the creation of the magnetic fields used in EM surveying (see Section 5.7.1.2).

5.2.2.2 Electromagnetic induction

Here we investigate the relationships between time-varying magnetic and electric fields. An emf can be induced into a coil by changing the strength of a magnetic field experienced by the coil through a phenomenon known as *electromagnetic induction.* Figure 5.9a shows two coils in close proximity to each other. One coil, which we will refer to as the *transmitter coil,* has a current passing through it. The associated magnetic field will be called the *primary* magnetic field. The emf induced in the second coil is monitored. This coil will be referred to as the *receiver coil.*

The magnitude of the induced emf (ε) in volts (V) is proportional to the rate of change of the magnetic field (ψ) experienced by the coil, i.e. it depends on how fast the magnetic field changes. The induced emf is given by the expression,

$$\varepsilon = -\frac{d\psi}{dt} \tag{5.7}$$

This is known as *Faraday's Law of electromagnetic induction.*

Consider the case of d.c. flowing through the transmitter coil (Fig. 5.9a). Before the current is turned on there is no changing primary magnetic field so there is no induction in the detector coil. When the current is turned on a steady-state magnetic field is rapidly established. Only at the instances of turn-on and, later, at turn-off does the receiver coil experience a change in the magnetic field. Simultaneously and instantaneously, an emf pulse is induced into the receiver coil causing a current to flow. At other times when a steady-state current is established, the magnetic field experienced by the receiver coil is once again constant so there is no emf induced in it. A crucial characteristic of the induced emf is that it causes a current to flow in the receiver coil whose magnetic field attempts to prevent the field around the coil from changing. The negative sign in Eq. (5.7) indicates that the induced emf opposes the change in the primary magnetic field. Induction at the instance of a change-in-state of a magnetic field is the basis of time domain EM measurements (see Section 5.7). Note that the polarity of the induced emf is opposite at turn-on and turn-off. In the description of EM methods in Section 5.7 we describe the induction processes at turn-off, when the induced current flow attempts to recreate the primary magnetic field.

Consider now the case of a.c. flowing through the transmitter coil (Fig. 5.9b). The continually varying current produces a magnetic field that continuously changes in amplitude and direction. The receiver coil continually experiences the changing magnetic field, not just when the current is turned on and off, but for all the time that

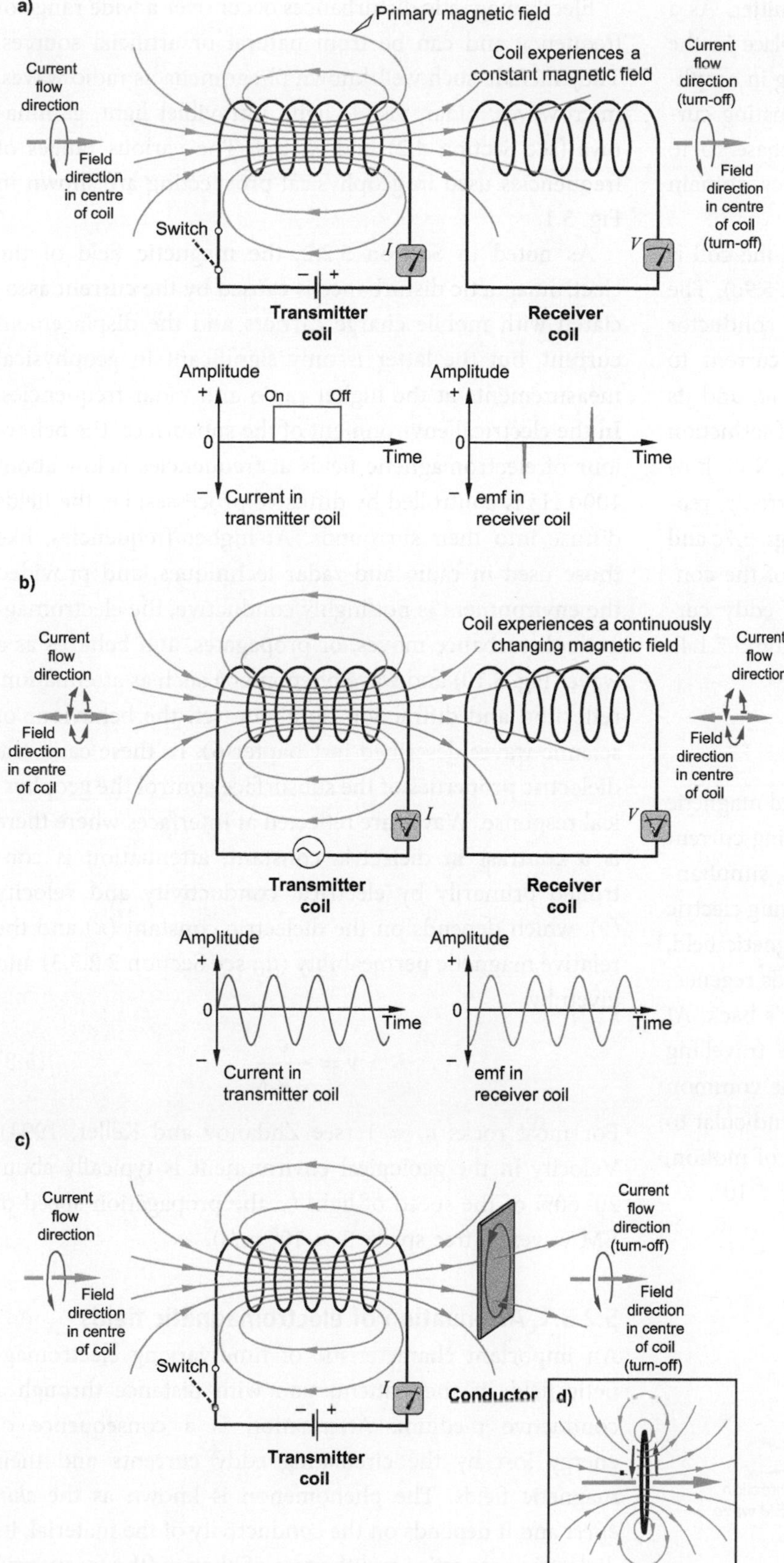

Figure 5.9 Inducing an emf in a coil using magnetic fields created by (a) an intermittent d.c. current and (b) a continuous a.c. current. The relationship between the transmitter current and the receiver voltage is shown. (c) Induction of eddy currents in a conductor. The magnetic field of the eddy currents approximates the primary magnetic field. (d) Magnetic field of current carrying loop of wire.

the alternating current is flowing in the transmitter. As a consequence, induction is continually taking place in the coil so the induced emf is continually changing in amplitude and polarity at the frequency of the alternating current in the transmitter coil, but 90° out-of-phase to it. Continuous induction is the basis of frequency domain EM measurements (see Section 5.7.1).

Electromagnetic induction also occurs when the coil is replaced by a body of conductive material (Fig. 5.9c). The time-varying magnetic field intersecting the conductor induces an emf into the conductor causing current to circulate in it. This is known as an *eddy current*, and its strength and direction of flow at the instant of induction are also governed by Faraday's Law (Eq. (5.7)). Note how the magnetic field associated with the eddy currents, represented by that of a loop of wire as shown in Figs. 5.7c and 5.9d, mimics the primary field in the vicinity of the conductor. We describe further characteristics of eddy currents in terms of geophysical surveying in Section 5.7.1.4.

5.2.3 Electromagnetic waves

Electromagnetic fields comprise both electric and magnetic fields which are inextricably associated. A varying current is surrounded by a varying magnetic field and, simultaneously, the varying magnetic field induces a varying electric field which, in turn, gives rise to a varying magnetic field, and so on. Oscillating electric and magnetic fields regenerate each other, effectively riding on each other's back. At suitably high frequencies they move as waves travelling away from their source, a radio wave being the common example. The two fields oscillate in planes perpendicular to each other and perpendicular to their direction of motion, forming a transverse electromagnetic wave (Fig. 5.10).

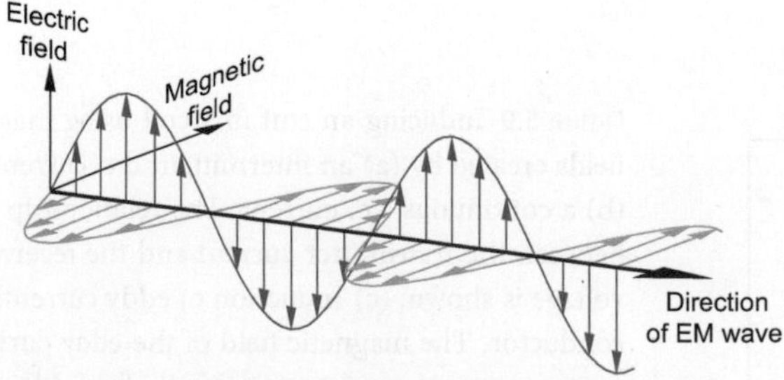

Figure 5.10 Schematic illustration of an electromagnetic wave illustrating its component electric and magnetic fields which fluctuate normal to each other and in the plane normal to the direction of propagation.

Electromagnetic disturbances occur over a wide range of frequency and can be from natural or artificial sources. They include such well-known phenomena as radio waves, microwaves, radar, visible light, ultraviolet light, gamma-rays (see Section 4.2) and X-rays. The various ranges of frequencies used in geophysical prospecting are shown in Fig. 5.1.

As noted in Section 5.2.2, the magnetic field of the electromagnetic disturbance is caused by the current associated with mobile charge carriers and the displacement current, but the latter is only significant in geophysical measurements at the higher radio and radar frequencies. In the electrical environment of the subsurface, the behaviour of electromagnetic fields at frequencies below about 1000 Hz is controlled by diffusive processes; i.e. the fields diffuse into their surrounds. At higher frequencies, like those used in radio and radar techniques, and provided the environment is not highly conductive, the electromagnetic disturbance moves, or propagates, and behaves as a wave (Fig. 5.10) and wave phenomena such as attenuation, reflection and diffraction dominate (cf. the behaviour of seismic waves described in Chapter 6). In these cases the dielectric properties of the subsurface control the geophysical response. Waves are reflected at interfaces where there is a contrast in dielectric constant; attenuation is controlled primarily by electrical conductivity and velocity (v), which depends on the dielectric constant (κ) and the relative magnetic permeability (μ_r; see Section 3.2.3.3) and given by:

$$v = \frac{c}{\sqrt{\mu_r \kappa}} \tag{5.8}$$

For most rocks $\mu_r \approx 1$ (see Zhdanov and Keller, 1994). Velocity in the geological environment is typically about 20–60% of the speed of light (c, the propagation speed of EM waves in free space, 3×10^8 m/s).

5.2.3.1 Attenuation of electromagnetic fields

An important characteristic of time-varying electromagnetic fields is their attenuation with distance through a conductive medium. Attenuation is a consequence of energy lost by the circulating eddy currents and their magnetic fields. The phenomenon is known as the *skin effect* and it depends on the conductivity of the material, its dielectric properties and the rate of change (the frequency) of the field. It determines the penetration of electromagnetic fields into a medium and strongly influences the depth of investigation of EM measurements in geophysics.

For a sinusoidally varying field penetrating an infinite uniform conductive medium, and where the effects of the dielectric properties are very small compared with the effects of conductivity, skin effect is quantified in terms of a parameter known as *skin depth* (δ). It is the distance over which the electromagnetic field's amplitude is attenuated by $1/e$ (i.e. 37%) of its surface value and is given by:

$$\delta = \frac{1}{\sqrt{\pi \mu \sigma f}} \tag{5.9}$$

where μ is the magnetic permeability (see Section 3.2.3.3) of a homogenous medium (henry/m), σ is conductivity in S/m, f is frequency in hertz and the skin depth is in metres. The magnetic permeability of most rocks is nearly the same as that in a vacuum ($\mu \approx \mu_0 = 4\pi \times 10^{-7}$ henry/m) (see Zhdanov and Keller, 1994) allowing the expression to be written as:

$$\delta = \frac{503.8}{\sqrt{\sigma f}} \tag{5.10}$$

Note that this is not the limiting distance/depth of the field's penetration, but a convenient measure of the attenuation in terms of distance or depth. The equations show that EM fields attenuate faster in materials of high conductivity and that attenuation is greater for fields of higher frequency. Figure 5.11 shows skin depths for frequency and conductivity ranges relevant to geophysical measurements, and where the material acts chiefly like a conductor and not a dielectric. At radio and radar frequencies the dielectric properties significantly influence attenuation (see online Appendix 5).

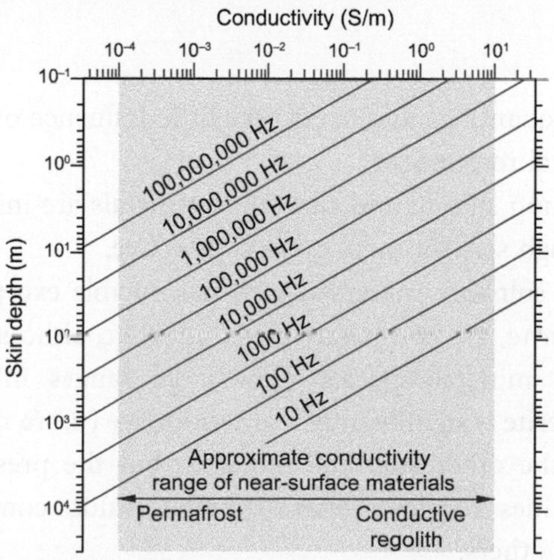

Figure 5.11 Skin depth for the range of frequencies used in EM surveying and for the range of electrical conductivities found in the geological environment.

5.3 Electrical properties of the natural environment

The most important electrical property influencing geoelectrical measurements is electrical conductivity, and its reciprocal resistivity, since it controls the responses measured in the resistivity, electromagnetic and self-potential methods. Also significant are electrical polarisation, a phenomenon related to capacitance and measured by the induced polarisation method, and other dielectric properties which control the responses measured in high- (radar-) frequency EM methods.

5.3.1 Conductivity/resistivity

Making representative measurements of electrical conductivity/resistivity on samples of natural materials is particularly difficult. In porous rocks, electrical properties are often predominantly controlled by the fluids occupying the pore space, which may not be retained during sampling or properly reproduced during the measuring process. Also, electrical properties may be highly heterogeneous, introducing a scale-dependence and possibly unassailable problems of representative sampling. For example, fractures or mineralised veins in the rock may provide the main path for electrical current flow, so measurements on vein/fracture-free specimens are unlikely to be representative of the whole rock. When multiple measurements are available, the largest conductivity is most likely to be representative of the overall conductivity.

The electrical conductivity of natural substances varies greatly, more than any other physical property relevant to geophysical surveying. The conductivity and resistivity ranges of various rocks and selected minerals are shown in Fig. 5.12. Note that the scale is logarithmic and spans no less than 22 orders of magnitude. Even individual minerals show variations in conductivities/resistivities that span several orders of magnitude, often exceeding those exhibited by the various rock types (Fig. 5.13). In fact, there is little agreement amongst published sources as to the limits of the ranges because of factors such as: different methods of measurement; whether individual crystals or aggregates were tested; the controlling effects of cracks and even small quantities of impurities and lattice defects in individual crystals; and the variation in electrical properties related to the direction of current flow through individual crystals or the sample as a whole, i.e. electrical anisotropy (see Section 5.3.1.4). From a practical perspective,

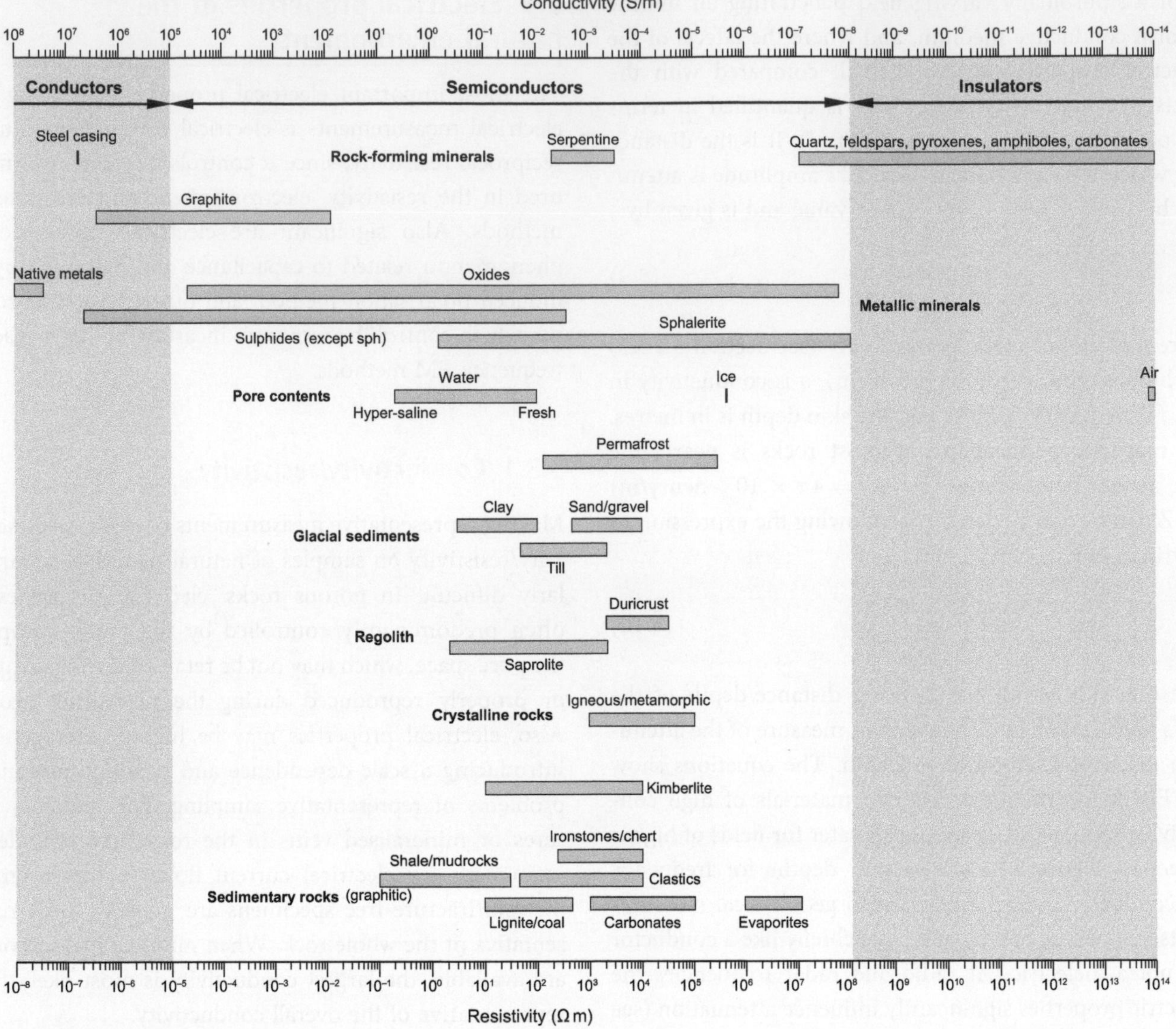

Figure 5.12 Typical ranges in conductivity/resistivity for some common minerals, rock types and near-surface materials. Data compiled from various published sources.

determining the electrical properties of individual minerals is not critically important since this information cannot be easily used to predict the properties of the mineralogical aggregates that are bodies of mineralisation or the common rock types.

5.3.1.1 Properties of rocks and minerals

The following generalisations from Fig. 5.12 are important in terms of geophysical measurements for mapping and exploration targeting:

- Most materials in the geological environment are semiconductors (see Section 5.2.1.3). Exceptions are the highly conductive native metals, but they occur in very small quantities and exert very little influence over geophysical responses.
- Common silicate and carbonate minerals are insulators, although serpentine is more conductive.
- Metal sulphide minerals, with the notable exception of sphalerite, are comparatively conductive, although individual mineral species show large ranges in values. Sphalerite is significantly less conductive (more resistive) than the other sulphide minerals, but the presence of impurities may lower its resistivity to values comparable to the others.
- Metal oxides are normally less conductive than the sulphides, and their range in conductivity overlaps that of the main rock types, so the physical-property contrast

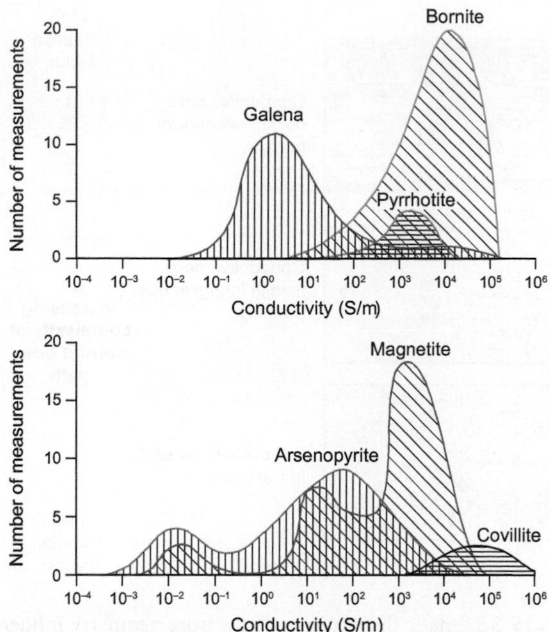

Figure 5.13 Frequency histograms of electrical conductivities of selected ore mineral species. Redrawn, with permission, from Parkhomenko (1967).

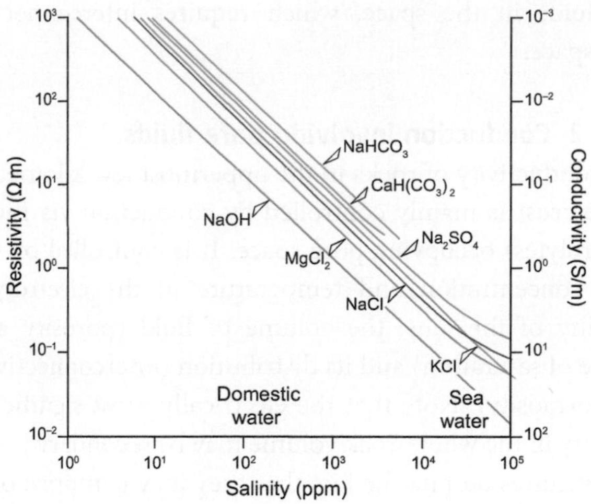

Figure 5.14 Electrical conductivity of selected pure salt solutions as a function of concentration for solutions at 20 °C. Redrawn, with permission, from Keller (1988).

required for geophysical surveying may be absent. Magnetite is generally one of the more conductive species, and haematite is normally a very poor conductor, although impurities can increase its conductivity considerably.

- Of the likely contents of pore space, air and ice are poor conductors, and water is a comparatively good conductor. Pure water has low conductivity, but naturally occurring waters are more conductive because of the impurities they contain, with conductivity increasing as salinity increases (Fig. 5.14). Consequently, the water table is expected to be associated with a conductivity contrast, being more resistive in the underlying unsaturated zone, as are deeply weathered water-filled structures such as faults and shear zones. Although ice is a poor conductor in the natural environment there is usually also water present.
- Electrical resistivity is not diagnostic of rock type. The ranges in conductivity/resistivity of individual rock types tend to lie between those of the rock-forming minerals and groundwater, and are generally closer to the latter, especially if they have significant porosity. This is because electrical properties tend to be dominated by the most conductive components of the rock (see below).
- Sedimentary rocks are generally more conductive than igneous rocks owing to their higher porosity and moisture content. Metamorphic rocks exhibit more variable conductivity/resistivity because this group has a wider range of porosity, which varies from values similar to highly porous sedimentary rocks to the impervious crystalline rocks.
- Materials rich in clay minerals, e.g. saprolite and mudrocks, are amongst the most conductive rock types. Clay minerals become very conductive when wet, and their presence in rock pores increase the rock's conductivity considerably.
- Dry evaporites are amongst the most resistive rock types.
- Graphite is an important source of enhanced conductivity. It can extend the conductivity of rocks in which it occurs, notably shales, towards values comparable to massive sulphide mineralisation. Graphite is a mineral whose electrical properties are highly anisotropic, i.e. they depend on the direction of current flow relative to the crystallographic structure. Conductivity measured parallel to cleavage is far greater than that measured normal to cleavage.
- The electrical properties of massive mineralisation tend towards those of the comparatively conductive constituent minerals, again because bulk electrical properties tend to reflect the most conductive constituents.
- Coals have electrical properties similar to other lithotypes with which they are likely to be associated.

The electrical properties of rocks can be understood in the context of two basic forms of electrical conduction: through the matrix, which requires conductive mineral species to be interconnected; and ionic conduction through the fluids in the space, which requires interconnected pore space.

5.3.1.2 Conduction involving pore fluids

The conductivity of rocks in the uppermost few kilometres of the crust is mainly controlled by conduction via fluids (electrolytes) occupying pore space. It is controlled by the type, concentration and temperature of the electrolyte; mobility of the ions; the volume of fluid (porosity and degree of saturation) and its distribution (interconnectivity and tortuosity). Note that the electrically most significant porosity in the whole-rock volume may be secondary joints and fractures despite the fact that they may comprise only a very small fraction of the volume. They can form excellent current flow-paths and are an important source of electrical conductivity in fractured igneous, metamorphic and sedimentary rocks, demonstrating that electrical conduction through pore fluids is not confined only to porous sedimentary rocks. The electrical properties of sedimentary rocks, as occur in hydrocarbon basins, have been the subject of most study, because of routine downhole logging of electrical properties by the petroleum industry. As outlined below, electrical logs are a means of estimating porosity, an important hydrocarbon reservoir property.

The resistivity of a clean porous medium (i.e. one free of shale and clays) saturated with a saline pore fluid is empirically found to be proportional to the resistivity of the fluid and so defines the *formation resistivity factor* (F), or *formation factor*, given by *Archie's equation* (Archie, 1942) as

$$F = \frac{\rho_{\text{saturated-rock}}}{\rho_{\text{pore-fluid}}} = \frac{a}{\phi^m} \qquad (5.11)$$

where $\rho_{\text{saturated-rock}}$ is the resistivity of the fully saturated rock and $\rho_{\text{pore-fluid}}$ that of the saturating pore fluid (see below for definition of ϕ, a and m).

Formation factor is a dimensionless measure of the passive role played by the framework of matrix material on the overall rock conductivity. It varies between 5 and 500, with porous sandstone having a value of around 10 and poorly permeable limestone usually exhibiting a value of several hundred. The more direct the current path, the lower the formation factor and the closer the bulk resistivity of the rock is to the pore fluid resistivity (F approaches one). Formation factor is fundamentally dependent upon a rock's grain-scale structure (Fig. 5.15) and, as would be expected and illustrated in Fig. 5.16, is a function of the fractional porosity (ϕ). Empirical measurements show a reciprocal relationship between F and ϕ (Eq. (5.11))

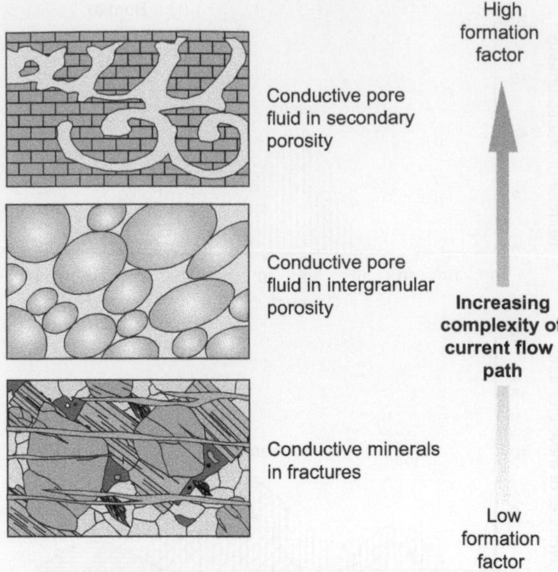

Figure 5.15 Schematic illustration of how pore geometry influences the length and complexity of the path of electrical conduction, and how this is related to formation factor.

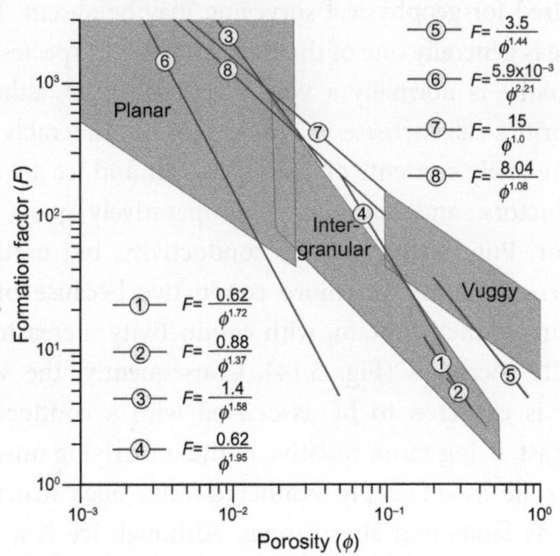

Figure 5.16 Formation factor and porosity fields for different types of porosity. Also shown are a number of empirically derived 'average' relationships (see Table 5.1). Curve (6) is for granitic rocks as quoted by Katsube and Hume (1987). Curves (7) and (8) are for gabbros (Pezard *et al.*, 1991; Ildefonse and Pezard, 2001). Based on a diagram in Keller (1988).

involving two parameters: the constant a and the *cementation exponent* (m). These parameters are controlled by the nature of the pore morphology, so they are related to the lithology and texture of the rock, and ultimately its geological history (diagenesis etc.). Both parameters vary

Table 5.1 Mean values of the Archie parameters for various lithotypes.

Lithology	*a*	*m*
Weakly cemented detrital rocks, such as sand, sandstone and some limestones, with a fractional porosity range from 0.25 to 0.45, usually Tertiary in age	0.88	1.37
Moderately well-cemented sedimentary rocks, including sandstones and limestones, with a fractional porosity range from 0.18 to 0.35, usually Mesozoic in age	0.62	1.72
Well-cemented sedimentary rocks with a fractional porosity range from 0.05 to 0.25, usually Palaeozoic in age	0.62	1.95
Highly porous volcanic rocks, such as tuff, aa and pahoehoe, with fractional porosity in the range 0.20 to 0.80	3.5	1.44
Rocks with fractional porosity less than 0.04, including dense igneous rocks and metamorphosed sedimentary rocks	1.4	1.58

Source: Keller (1988)

significantly across the three rock classes and, although they may be fairly constant for the same lithotypes in a particular area, they are very likely to have different values elsewhere. High tortuosity leads to high values of m, lower values being associated with simpler pore geometries such as fracture-dominant. It varies from 1.3 for packed sandstones to as high as 2.3 in well-cemented clastic rocks.

Numerous empirical and theoretical models have been developed to relate grain structure and bulk conductivity with modifications to the parameters described above, or have included additional parameters, for example the degree of water saturation. However, there is no universal relationship between bulk electrical conductivity and easily measurable rock properties, so Archie's equation continues to be widely used. Table 5.1 lists common values of the parameters a and m for a variety of rock types.

5.3.1.3 Conduction involving the matrix

Archie's equation assumes all conduction is via the pore fluid, with the matrix having only a passive role by controlling the geometry of the conducting pathways. This is reasonable since the majority of minerals are insulators. However, the minerals forming the matrix may also provide a conductive path, this being most significant when porosity is small and/or the pore fluid is weakly conductive. Conduction can occur through the mineral grains and, significantly, via electrochemical interactions between the grains and the pore fluid.

Few rocks contain enough conducting mineral grains for this to be the main conduction mechanism, but those that do often contain significant quantities of potentially economic metal-bearing minerals. Invariably mineral deposits contain numerous mineral species, and it is the electrical properties of the mineralogical aggregate, rather than the type or abundance of an individual species, that control electrical conductivity of the deposit. However, a rough correlation between conductive mineral content and conductivity may be observed. Figure 5.17 shows the relationship between conductivity and the proportion of sulphide minerals present from several deposits. When a conductive mineral occurs as isolated grains this mineral species will contribute little to the overall conductivity of the deposit. As expected, there is little correlation between conductivity and the amount of sulphides when they are disseminated, but when the sulphides occur as veinlets the increased sulphide content causes an increase in conductivity.

The important contributors to electrical conduction in the rock volume are those grains whose distribution creates an electrically continuous network, even if the species in question is not the most conductive or abundant. Clearly, mineralogical texture is a key property in determining conductivity, as it is for water-bearing rocks.

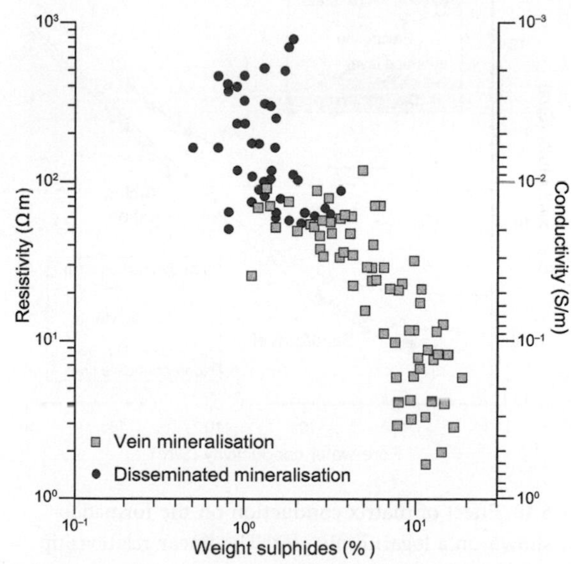

Figure 5.17 Relationship between conductivity/resistivity and sulphide content. Data from various porphyry copper deposits in the southwestern USA. Redrawn, with permission, from Nelson and Van Voorhis (1983).

Clay minerals may also influence conductivity, a fact first recognised by the failure of Archie's equation in rocks that the petroleum industry refers to as *shaly sands*. In these rocks the formation factor, being the slope of the curves shown in Fig. 5.18, changes with the degree of conduction through the matrix minerals. The substitution of ions within the sheet-like structure of clay minerals causes substantial negative charge to accumulate on the grain surfaces. To balance this charge, cations are adsorbed onto the grain surfaces. The surface ions are not bound to the crystal structure and may be exchanged with other ions held in solutions that come in contact with the grain. This alters the type and concentration of ions in the pore fluid. All minerals exhibit this surface conduction effect to some extent, notably zeolites and organic material, but clay minerals are the most important contributors (Fig. 5.18).

The influence of clay minerals depends on their species, grain size, volume and distribution through the rock mass, and the pore surface area. Their large surface area relative to their weight means even quite small amounts of clay minerals can significantly increase bulk conductivity, particularly when their grain size is small. The importance of clay minerals as conduction paths increases as the conductance through the pores diminishes.

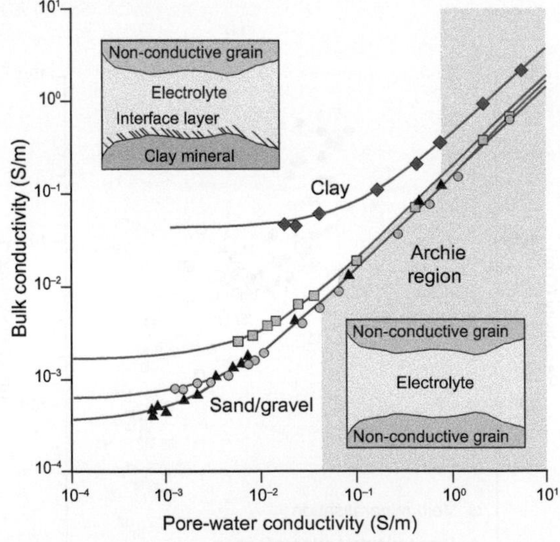

Figure 5.18 Effect of matrix conduction on the formation factor shown on a logarithmic plot. The linear relationship between bulk conductivity and pore fluid conductivity breaks down where matrix conduction is significant. Where clay minerals are present, conduction occurs through an interface layer on the margins of the clay grains. Redrawn, with permission, from Schön (1996).

5.3.1.4 Geological controls on conductivity

The degree of electrical interconnectivity between the conductive constituents is the major influence on electrical conductivity. The effects of common geological processes on the electrical conductivity/resistivity of rocks can be understood in terms of their effect on conduction paths. If porosity and permeability are increased, conductivity is expected to increase. Processes that reduce these parameters, such as compaction, diagenesis and metamorphism, have the opposite effect. In general, weathering increases porosity and permeability, but the creation of clay minerals may have the opposite effect.

Geological processes that affect connectivity through conductive mineral grains include recrystallisation during metamorphism, which may enhance conductivity through changes in the rock's texture which increase the number of connected conductive grains. Also, sulphides deform comparatively easily during metamorphism, allowing them to form interconnecting conductive networks. Silicification, even if not intensive, can reduce the conductivity of massive sulphide mineralisation to effectively zero despite the presence of conductive mineral species: see, for example, Gunn and Chisholm (1984).

Rocks tend to be layered, so most rocks are electrically anisotropic. The effect occurs at a range of scales, from individual grains through to that due to fabrics such as bedding and schistosity, and at the scale of discontinuities such as joints. The largest conductivity is usually parallel to the fabric or discontinuities, since grains are more elongated in this direction and there is more continuous pore space.

Clearly there are no guarantees about the electrical properties of rocks and minerals in the natural environment. Even massive sulphide mineralisation may be highly resistive, particularly if sphalerite is part of the assemblage. The responses of conductive graphite-bearing rock types are likely candidates for being mistaken as the responses of conductive mineralisation. Rocks with significant pore space, especially when filled with saline water, will also appear as conductivity anomalies.

Like most other physical properties, conductivity/resistivity is not diagnostic of lithology, and a geophysical interpretation and a geological interpretation from the same area may be in conflict. The geophysical response is primarily reflecting the response of the pore space or perhaps a minor mineral phase, whereas it is the rock-forming minerals that are the primary basis for the geological description.

5.3.2 Polarisation

When a potential difference is applied to a rock, charges of opposite polarity become concentrated in different parts of the rock causing it to be polarised and exhibit capacitor-like behaviour (see Section 5.2.1.5) (Fig 5.19). When this potential is removed, the capacitor discharges by the movement of the polarising charges through the surrounding electrical pathways. The movement of the charges is hindered by the electrical resistance of the pathways and, in accordance with

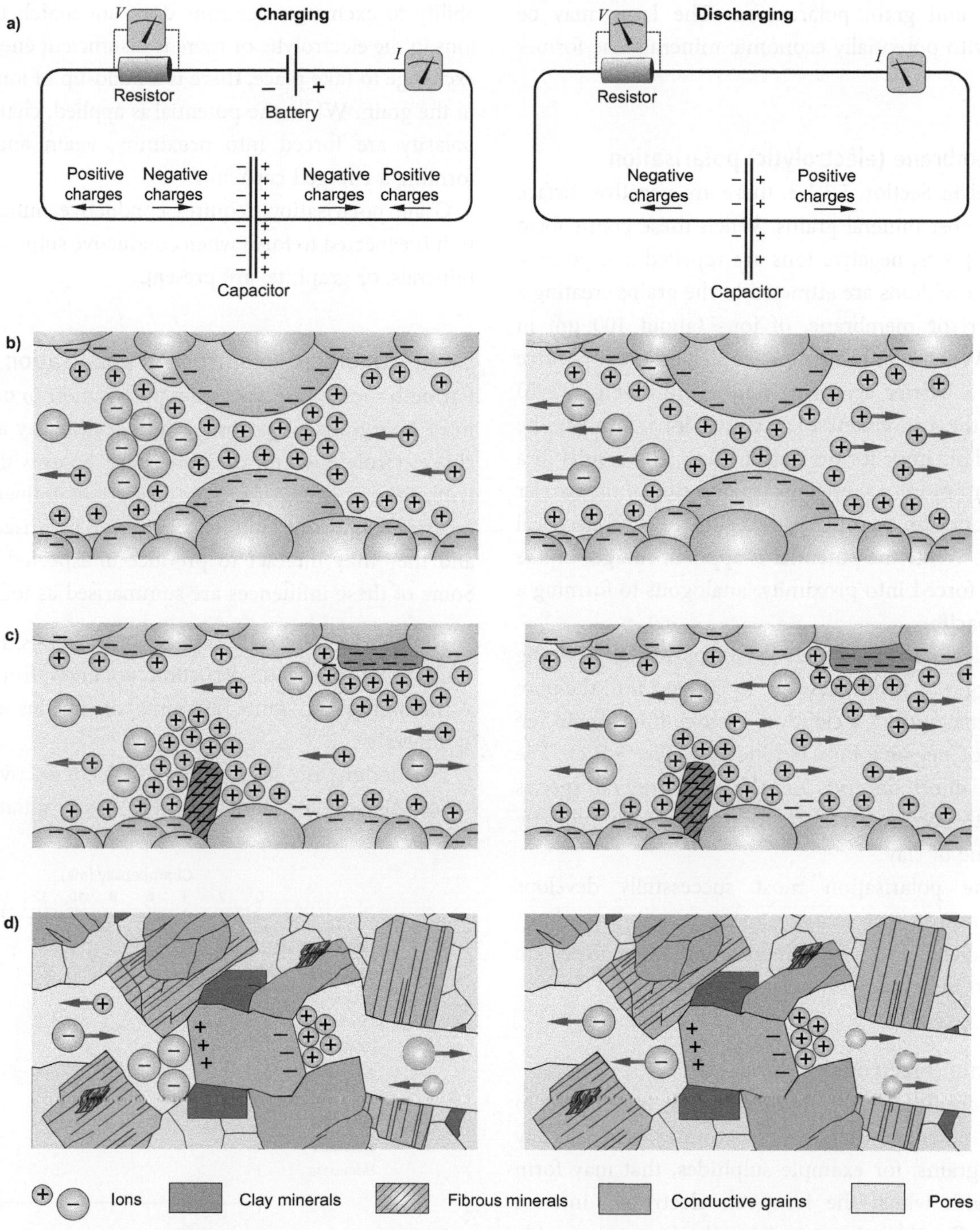

Figure 5.19 Capacitive properties illustrated by the storage (charging) and subsequent discharge of electrical charges. (a) Conductive metal plates in proximity forming a capacitor, (b) membrane polarisation due to the charge barrier formed by ions on grain surfaces at the narrowing of a pore, (c) membrane polarisation due to charge build-up around clay and fibrous mineral grains, and (d) grain polarisation due to charge accumulation on either side of an electrically conductive grain and involving conduction through the grain.

Ohm's Law, potential differences form across segments of the pathways which decay with dispersal of the charges. It is this decaying voltage that is measured by induced polarisation (IP) surveys (see Section 5.6.3).

There are two mechanisms by which the capacitor-like behaviour occurs in the natural environment: membrane polarisation and grain polarisation. The latter may be associated with potentially economic minerals, the former is not.

5.3.2.1 Membrane (electrolytic) polarisation

As described in Section 5.3.1.3, there are negative surface charges on most mineral grains. When these grains form the walls of pores, negative ions are repelled and positive ions in the pore fluids are attracted to the grains creating a surface layer, or membrane, of ions (about 100 μm in thickness). When a pore narrows sufficiently, the positive ions create a barrier across its entire width (Fig. 5.19b) preventing the movement of negative ions through the pore fluid. This impedes the current flow and results in a local concentration of these ions on one side of the barrier and a deficiency on the other side, i.e. there is an electrical polarisation. While the potential is applied, charges of like polarity are forced into proximity, analogous to forming a charged capacitor.

Clay and fibrous minerals have particularly strong surface charges so when these are in contact with the pore space they may attract a cloud of positive ions. Again the movement of negative ions is hindered (Fig. 5.19c). The polarisation effect depends on the clay mineral species present, with maximum effect occurring with typically ~10% volume of clay.

Membrane polarisation most successfully develops where the grain surface charge is greatest, so it is predominantly associated with clay minerals and where pores are small. It also increases with pore fluid salinity.

5.3.2.2 Grain (electrode) polarisation

When ions are flowing through pore fluids in response to an applied potential they may encounter electronically conductive grains, for example sulphides, that may form a barrier with which the ions can electrically interact (Fig. 5.19d). For the current to flow, the ions on one side of the barrier must exchange electrons with the conductive grain, which in turn exchanges electrons with ions on the other side of the barrier. The electrical circuit locally consists of a combination of ionic and electronic conduction and involves a chemical reaction between the mineral and the solution (the pore fluid).

The exchange of electrons results in the conductive grain becoming polarised, i.e. opposite charges accumulate on opposite sides of the grain. Ions of opposite polarity (sign) will tend to concentrate next to the grain boundary. If its ability to exchange electrons does not match that of the ions in the electrolyte, or there is insufficient energy for the exchange to take place, there is a build-up of ions adjacent to the grain. While the potential is applied, charges of like polarity are forced into proximity, again analogous to forming a charged capacitor.

Grain polarisation requires conductive mineral grains, so it is expected to form when conductive sulphide or oxide minerals, or graphite, are present.

5.3.2.3 Geological controls on polarisation

For both membrane and grain polarisation to occur, there must be significant porosity and permeability and a suitable electrolyte within the pore space. So as with conductivity, the rock's texture is an important influence. Many factors are known to influence electrical polarisation effects and they may interact to produce unexpected responses. Some of these influences are summarised as follows:

- The type of electronic conducting minerals. Figure 5.20 illustrates the wide variation obtained from samples containing the same amounts of various conductive minerals.
- The amount and distribution of the conductive material. For disseminated material the polarisation response

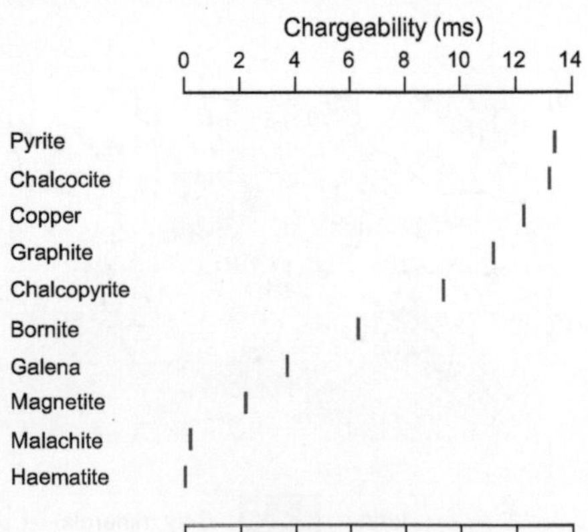

Figure 5.20 Chargeability of 1% volume concentration of a variety of conductive ore minerals measured using a square-wave pulse of 3 s with the decay integrated over a period of 1 s. See Section 5.6.3 for explanation of the measurement of induced polarisation. Based on data in Telford *et al.* (1990).

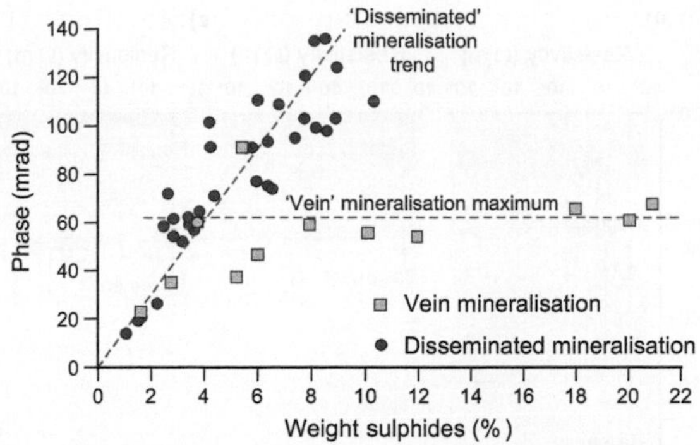

Figure 5.21 Relationship between IP response, measured as a phase shift, and sulphide content for both disseminated and vein style mineralisation. See Section 5.6.3 for explanation of the measurement of induced polarisation. Note the 'saturation' response when veinlet mineralisation exceeds about 5% by weight. Redrawn, with permission, from Nelson and Van Voorhis (1983).

varies roughly proportionally with the amount of conductive material present. For vein material there appears to be a critical amount of conductive material above which the response ceases to increase, i.e. a 'saturation' response is reached (Fig. 5.21).

- The size and shape of conductive grains. Experiments suggest there may be an optimum grain size determined by the texture and electrical properties of the host and the grains themselves. In general, increasing grain size reduces the polarisation response since the surface area available for interaction with ions decreases.
- The size and shape of pores. As well as controlling porosity and permeability, the pore geometry determines whether conductive grains and clay minerals make contact with pore fluids. Increasing permeability decreases the polarisation response because of the larger pore throats and/or associated decreases in clay content.
- The type, distribution and volume fraction of clay minerals. The greatest polarisation responses occur when the clay mineral content is about 5–10% by volume, depending on the mineral species present and cation-exchange capability. This is due to the requirement that both clay minerals and a suitable pore channel for the ions are present. As clay mineral content increases the number of channels decreases.
- The type and salinity of the electrolyte in the pores and the degree of water saturation. Enough ions are required to allow the various mechanisms to occur, but too many allow alternative conduction paths. Also, increasing matrix conductivity diverts current flow away from ionic conduction mechanisms.
- Temperature becomes most significant when it is sufficiently low that the pore waters are entirely or partly frozen. Kay and Duckworth (1983) showed that freezing enhances polarisation, explainable in terms of frozen pore water changing the porosity and tortuosity of the available pore space.

An obvious conclusion from the above is that the proportions of rock-forming minerals in the matrix have negligible control on electrical polarisation and, consequently, there is generally no correlation between it and rock type.

5.3.3 Dielectric properties

The dielectric constant (see Section 5.2.1.1) is a measure of a material's ability to be electrically polarised, and is an important control on the responses of high-frequency EM surveys (see online Appendix 5). Figure 5.22 shows the range in dielectric constant for various minerals and rocks and the materials that occur within pore space. The rock-forming minerals show comparatively little variation. However, metallic sulphide and oxide minerals have significantly higher values, indicating that occurrences of these minerals may be detectable using radar methods. The most significant aspect of the data is the extremely high values for water. The data graphically illustrate that dielectric constant is predominantly a function of the presence, or absence, of water; compare wet and dry sands, for example. Mineralogy (and therefore lithotype) exerts little influence, other than as an indirect control on porosity and potential water content. Even though heavy minerals such as zircon and monazite have higher values than quartz, which is likely to make up their host sediments in beach sand deposits, it is unlikely mineralisation will be detectable in water-saturated sediments because of the dominant influence of the pore water.

5.3.4 Properties of the near-surface

Electrical and EM measurements of the ground are strongly influenced by the electrical properties of the near-surface. The electrical properties of this zone differ markedly from deeper regions, even where bedrock is exposed. Significant lateral and vertical changes in electrical properties can occur.

In geophysical terms, a conductive (low-resistivity) surface layer is known as *conductive overburden*, and

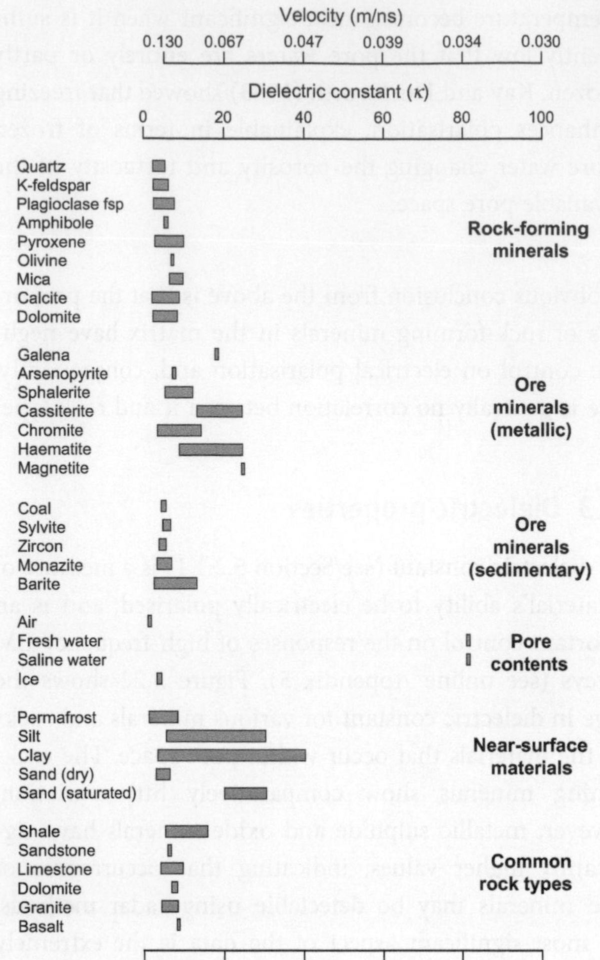

Figure 5.22 The ranges in dielectric constant and electromagnetic wave velocity for some commonly occurring minerals, rocks and near-surface materials measured at radar frequencies. Note the high values for the different forms of water. Data compiled from various published sources and velocities calculated from Eq. (A5.1). Note the non-linear velocity scale.

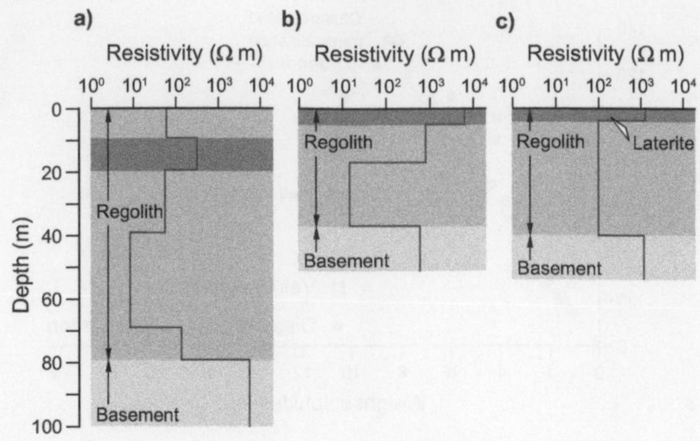

Figure 5.23 The variation of electrical resistivity vertically through regolith. (a) Western Australia (based on a diagram in Emerson *et al.* (2000)), (b) Burundi (based on a diagram in Peric (1981)), (c) Goias, Brazil (based on a diagram in Palacky and Kadekaru (1979)). (a) Based on petrophysical measurements, (b) and (c) based on resistivity soundings.

resistivity may be just a few ohm-metres, and less than 1 Ω m in some places. Conductive overburden is the norm where present-day or past tropical weathering conditions have produced deeply weathered landscapes and created a thick regolith. Also, saline groundwater has a significant influence on the conductivity of the near-surface. Both features are particularly widely spread in Australia and South America. This conductive zone can completely mask the electrical responses of features beneath it. Furthermore, variations in it over small areas produce responses that superimpose geological noise on the responses of target conductors, which can be easily mistaken for deeper conductive bodies.

At the other end of the spectrum, highly resistive near-surface conditions make the acquisition of good-quality electrical measurements difficult owing to difficulties in establishing current flow through the highly resistive ground, but this is not a problem for inductive EM measurements. These conditions occur in areas of permafrost and ferricrete.

5.3.4.1 Regolith

Much progress has been made in terms of understanding, classifying and utilising regolith in the exploration process (Anand and Paine, 2002). A detailed discussion of different types of regolith and their geophysical characteristics is beyond our scope, but profiles of resistivity/conductivity versus depth through regolith in different areas do show some characteristics in common. Figure 5.23 shows the variation in resistivity with depth through several regolith profiles, obtained using a variety of geophysical methods and survey types. The most common feature is the resistive surface soil and/or laterite, which are often indurated and ferruginous (forming ferricrete). Underlying these is very conductive saprolitic material which may be of considerable thickness, beneath which is less weathered, more resistive bedrock that passes, rapidly or gradually, down into comparatively unweathered higher-resistivity protolith. Clay minerals in the saprolite contribute significantly to its higher conductivity. The strong resistivity/conductivity contrasts between neighbouring geological/weathering horizons, which form distinct electrical layers, is a characteristic of the regolith.

The profiles in Fig. 5.23, although typical, are only representative of a particular area. Factors responsible for variations in the thickness and conductivity of the various

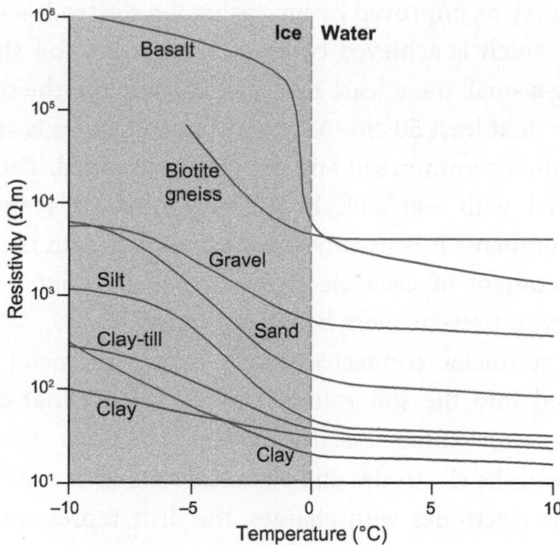

Figure 5.24 Variation of resistivity with temperature close to 0 °C for various rocks types, illustrating the increase in resistivity associated with frozen pore contents. Redrawn, with permission, from Scott *et al.* (1990).

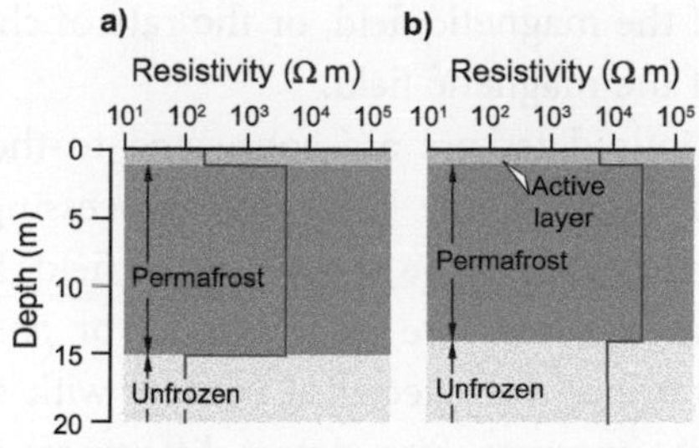

Figure 5.25 The variation of electrical resistivity vertically through permafrost in Quebec, Canada, during summer. (a) For marine clayey silt and (b) for glacial till. Data obtained from resistivity soundings. Based on diagrams in Seguin *et al.* (1989).

layers include multiple weathering fronts, salinity of the local groundwater, depth to the water table, variations in porosity, silicification, lateral variations in the regolith stratigraphy, and the nature of the protolith and transported materials overlying the ferruginous layer (Palacky, 1988; Emerson *et al.*, 2000).

5.3.4.2 Permafrost

Since the main mechanism for the conduction of electricity through geological materials is by ionic conduction through groundwater, freezing the water, as occurs in the formation of permafrost, causes their electrical resistivity to increase (Fig. 5.24). However, the electrical properties of permafrost are rarely consistent owing to such factors as variations in the amount of unfrozen water, its salinity and the presence of massive ground ice. A review of published accounts suggests a continuum from the resistivity of massive ice towards that of fresh water.

Permafrost may vary in thickness from a few metres to several hundred metres. Its gross resistivity structure consists of a shallow zone of higher resistivity overlying a zone of lower resistivity (Fig. 5.25). It is underlain by unfrozen material, and in summer conditions there may also be an overlying 'active zone' of melting. Layers containing more liquid material may also occur within the permafrost. The resistivity contrasts between these different layers depend primarily on the porosity and the salinity of the water occupying the pores. If the unfrozen rock has high porosity and/or saline water in its pore space, a larger contrast occurs at the base of the permafrost than if the rocks have a lesser pore volume.

5.4 Measurement of electrical and electromagnetic phenomena

Active electrical and EM methods produce their own electrical or electromagnetic fields, respectively. This requires a source of electrical power, a control unit and an antenna. Power is supplied from batteries or a portable generator, and the control unit provides time synchronisation and varies the input to the antenna with time to suit the particular kind of measurements being made. The energy is transmitted into the ground via the antenna. In electrical surveys the antenna consists of a pair of *transmitter electrodes* in electrical contact with the ground, allowing the current to flow through the subsurface between the electrodes. They are also referred to as *current electrodes.* In EM surveys the antenna may comprise a long straight wire grounded at one end with the return current path via the ground, or a large, usually rectangular, loop of wire with one or a few turns, or may be circular and comprise multiple turns of wire forming a coil. Current is transmitted through the wire antenna, and the associated magnetic field induces electric currents in the subsurface (see Section 5.7.1). In EM methods it is common practice to refer to the antenna as the transmitter (Tx). In electrical surveys the term transmitter tends to be applied to the actual control unit.

Survey measurements in both active and passive methods are recorded using one or more detectors linked to a recording unit. Different kinds of electrical and EM surveys require detectors that can measure the magnitude and direction of the electrical potential, the magnitude or strength of the magnetic field, or the rate of change of the strength of the magnetic field.

The various detectors are connected to the recording unit within which some basic data processing may take place to enable data to be viewed in the field. In electrical surveys the detectors are the *receiving* or *potential electrodes* and are also in electrical contact with the ground. As the name suggests, they detect differences in electrical potential (it is difficult to measure absolute electrical potentials in the field). Special non-polarising electrodes are required for some types of survey (see Section 5.4.1). Detectors for EM surveys may be magnetometers or wire antennae (sometimes loops, but usually coils) which are normally referred to as the *receiver* (Rx). Wire antennae consist of small portable multi-turn coils of wire wound either on rigid air-cored forms of a metre or two in diameter, as used in airborne EM, or as small compact forms wound on ferrite or metal cores suitable for ground and downhole EM surveying. In electrical surveys, the term receiver tends to be used to describe the actual recording unit.

In the past, different controllers and recorders were used for different kinds of survey, but modern digital survey equipment has the capability to produce and record signals for all the common types of electrical and EM surveys. Accurate time synchronisation between the controller and recording systems is required. This may be achieved by a wire connection between them or synchronisation of their internal clocks, which are automatically synchronised to the time signals broadcast by the constellations of satellites forming global positioning systems.

Recording the orientation of the transmitter and receiver antennae and magnetometers is imperative in EM methods because the parameters measured are vector quantities (see Section 2.2.2). Individual sensors may include spirit-bubble levels to allow manual orientation, or electronic means of recording orientations, as in airborne and downhole equipment. The positional and orientation data are recorded, along with the electrical and electromagnetic signals, by the recording unit, these essential ancillary data being required in the analysis of the survey data.

5.4.1 Electrodes

Transmitting electrodes for electrical surveys may be nothing more than a metal stake driven into the ground, but when the ground is highly resistive or the signal is weak this may not lead to sufficient current flow into the ground. This may be improved by increasing the electrode's surface area, which is achieved by using aluminium foil sheeting lining a small pit at least 1 m^2 in area, dug into the soil to a depth of at least 50 cm. A large volume of water is applied, and often common salt and detergent are added. The foil is covered with wet soil. In sandy porous soils and arid environments it is usually necessary to maintain the moisture content of each electrode during the course of the survey to prevent them becoming resistive.

A voltmeter connected across a pair of metal stakes pushed into the soil will measure a voltage that quickly drifts with time. Electrochemical reactions occurring between the electrodes and chemical salts in the soil *polarise* the electrodes with charges, the drift representing the progression of the reactions. Accurate measurement of difference in potential requires the use of *non-polarising electrodes*. These consist of a metal immersed in a solution of one of its salts in a porous container so that the solution leaks to the ground. The metal is connected to the voltmeter, and the solution maintains the electrical connection between the metal and the ground. The typical combination is copper in saturated copper sulphate solution contained in a porous ceramic pot.

For downhole surveys, the downhole transmitting electrode can be a copper pipe about 20 mm in diameter and 2 m in length. The non-polarising downhole potential electrodes can be pieces of oxidised lead forming a Pb–PbO electrode.

5.4.2 Electrical and electromagnetic noise

All electrical and electromagnetic systems are susceptible to electrical interference from both environmental and methodological noise (see Section 2.4). Sources of methodological noise include variations in the arrangement of the electrodes, EM loops, coils and magnetic sensors during the survey which cause variations in coupling (see Section 5.7.1.3) with the signal being measured. Great efforts are taken to minimise this form of system noise in both ground and airborne systems.

In general, environmental noise levels are high when compared with the geological responses, and much care is needed during data acquisition to reduce the noise level and maximise the signal-to-noise ratio (see Section 2.4). Noise levels are often the limiting parameter in recognising a target response in survey data.

The various electrical and EM methods make measurements at different frequencies (see Fig. 5.1), and noise

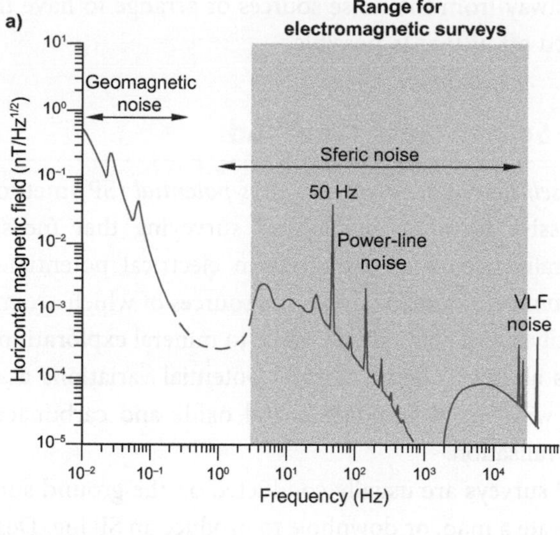

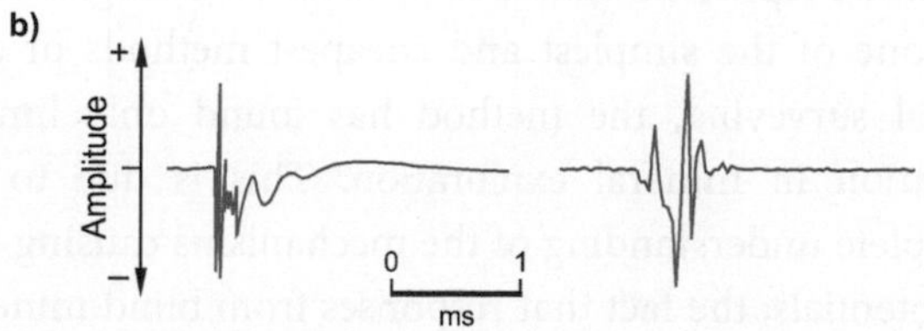

Figure 5.26 Noise in different frequency bands. (a) The electromagnetic noise spectrum. VLF – very low frequency. Redrawn, with permission, from Macnae *et al.* (1984). (b) Two examples of sferics. Redrawn, with permission, from Macnae *et al.* (1984) and Buselli and Cameron (1996).

levels in the different frequency bands vary markedly (Fig. 5.26). Some types of noise are restricted to particular parts of the EM spectrum, so digital filters are included in the survey equipment to attenuate the noisy part of the spectrum (see *Frequency/wavelength* in Section 2.7.4.4).

5.4.2.1 Environmental noise

Geological sources of noise are usually in the form of variable, and especially high, conductivity of the near-surface (see Section 5.3.4). Examples include regolith, permafrost, swamps, lakes, rivers and palaeochannels. These can produce strong responses, particularly when they are close to the system transmitter and/or receiver. When a conductive overburden is present, the ability of electrical and EM surveys to detect electrically anomalous zones in the bedrock can be severely compromised.

Long-wavelength variations, on which target responses are superimposed, are generally less of problem with electrical and EM methods than, for example, gravity and magnetic methods (see Section 2.9.2). These variations may, however, have to be removed from SP data and downhole electromagnetic data. Masking of responses of interest may occur where there are conductive geological features which are not of interest. Examples include conductive faults and shears, rock units such as shales and graphitic zones, and the contacts between rocks with contrasting conductivities across which the background response changes.

Atmospheric noise

Non-geological sources of environmental noise are time-varying natural electromagnetic fields originating from the magnetosphere, the region around the Earth which includes the atmosphere and the ionosphere. The time variations occur over a wide frequency range. Because of their very low frequency they penetrate to great depths (see Section 5.2.3.1), inducing circular current flow systems, several thousand kilometres across, into the Earth's crust and mantle. The currents flow as horizontal layers and are known as *telluric currents*.

Below 1 Hz the fields are due mainly to current systems set up in the ionosphere by solar activity. They remain fixed in position with respect to the Sun and move around the Earth as it rotates. From 1 Hz to 10 kHz the currents are mainly from the atmosphere and mainly due to pulses known as *sferics* (Fig. 5.26b), which are caused by lightning discharges associated with worldwide thunderstorm activity in the lower atmosphere. They are random and vary throughout the day, and they show pronounced seasonal variation with activity decreasing away from the equator since the major sources are in tropical equatorial regions. Storm activity is nearly always occurring somewhere on Earth, and their fields propagate in the Earth-ionosphere cavity to great distances.

Telluric currents are a source of noise in resistivity, IP and most EM surveys, particularly in conductive environments, although they are the signal for magnetotelluric measurements (see online Appendix 4). Noise due to geomagnetic phenomena, i.e. micropulsations occurring below 0.1 Hz (see *Micropulsations* in Section 3.5.1.1), is of sufficiently low frequency to be important at only the very low frequencies used for magnetotelluric measurements. The strength of sferics at a particular site depends on the location, strength, distribution and density of the lightning strikes at the time, and the season. They are a major source of noise in electrical and EM surveys, so surveying in equatorial regions may be more desirable during winter months when noise levels are lower. Also, the horizontal component of the sferic noise is usually stronger than the vertical component. Being pulses, sferics have a broad frequency bandwidth (see online Appendix 2) and so can be difficult to remove from the survey data. Usually some kind of de-spiking algorithm is used (see Section 2.7.1.1). An example of sferics in EM data, and the removal of this noise, is discussed in Section 2.7.4.5.

Another common type of environmental noise is motion of receiver coils and associated wiring in the Earth's magnetic field. The motion induces spurious time-varying responses into the data. In ground surveys the motion is usually caused by the wind and in airborne surveys it is the continuous turbulent movement of the system.

Cultural noise

The main source of man-made, or cultural, noise is the electric power transmission systems operating at a fundamental frequency of 50 Hz (60 Hz in some countries) (Fig. 5.26). Various electrical devices associated with the power transmission system generate currents at multiples of the fundamental frequency. These flow through the powerlines and, along with the 50/60 Hz currents, they electromagnetically induce electric currents in nearby conductors, e.g. wire fences, pipelines, railway lines, metal buildings and borehole casing, and also EM receiving coils. In some areas the return current flow in the power transmission system is via the ground, and these ground currents can be so strong as to render electrical and EM measurements useless.

Radio transmitters are another common source of interference. The naval VLF communication stations, operating in the 20 kHz region, are a significant noise source for many geophysical EM systems (Fig. 5.26).

Cultural features are also a source of spurious anomalies. The responses are strong, narrow and coincident with the highly conductive feature causing them. They appear in survey data either as 'one-line' anomalies associated with spot features such as metal drill casing and metal buildings, or as linear anomalies extending across several survey lines and related to powerlines, pipelines, railway lines etc. They are not normally masked by the response of conductive overburden. The sources can usually be identified from surface maps and aerial photographs.

Cultural noise is a serious problem for all electrical and EM methods when surveying in mine environments and close to built-up areas. It is sometimes necessary to work well away from the noise sources or arrange to have them turned off, if that is possible.

5.5 Self-potential method

The *self-potential* or *spontaneous-potential* (SP) method is a passive form of geophysical surveying that measures naturally occurring variations in electrical potential due to *spontaneous polarisation*, the sources of which are many and diverse. The method's value to mineral exploration lies in its ability to detect natural potential variations associated with metal sulphide, metal oxide and carbonaceous mineralisation.

SP surveys are usually conducted on the ground surface to create a map, or downhole to produce an SP log. Despite being one of the simplest and cheapest methods of geophysical surveying, the method has found only limited application in mineral exploration. This is due to the incomplete understanding of the mechanisms causing natural potentials; the fact that responses from blind mineralisation are easily masked by overburden; and the presence of large potentials produced by water movement, ubiquitous in some environments.

In this section, sources of potential variations in the natural environment, relevant to SP surveying, are described first. This is followed by a description of SP data acquisition and processing, interpretation methods, and finally some examples of SP responses from mineral deposits.

5.5.1 Sources of natural electrical potentials

Natural electrical potentials are caused by a number of mechanisms involving movement of charge carriers, mostly ions in groundwater. The moving charge carriers comprise an electric current so there is an associated potential difference (see Section 5.2.1.2). The physical and chemical processes involved are not well understood. Electrical potentials produced by sources other than mineralisation are a form of environmental noise in mineral exploration SP surveys.

5.5.1.1 Non-mineralised sources

Electrofiltration or streaming potentials are associated with the flow of groundwater through porous media. The potentials are produced by many mechanisms, which include interactions between ions and the pore walls, and the different flow velocities of the different types of ions, and also depend on the electrical properties of the pore fluid and the country rocks. Potentials of tens to hundreds of millivolts are typical, but their amplitude may be affected by rainfall, albeit with sufficient time lag to allow this to translate into changes in groundwater flow (Ernstson and Scherer, 1986). Potential anomalies occur where groundwater flows across the boundaries between materials with different electrical properties. Anomalies may be positive or negative, with variations in the potentials often correlating with topography, higher ground being almost always more negative than lower ground. Several authors have reported measuring negative potentials of around 2000 mV over high ground.

Differences in the concentration and mobility of ions in groundwater are another source of electrical potentials. Potential differences occur when electrolytes having different concentrations of ions come into contact. This is particularly relevant for SP logging when drilling fluid encounters natural groundwaters. The diffusion rates of ions in solution are related to their mobilities; so in electrolytes composed of different types of ions, their different mobilities produce a heterogenous mixture of cations and anions causing *liquid-junction* potentials. These are usually tens of millivolts. *Nernst* or *shale* potentials occur when the two fluids are separated by semipermeable formations. The negative charges on clay minerals in the formations restrict the diffusion of negative ions in the solution, but not the positive ions. This leads to an uneven distribution of ions, and the resulting potential differences are usually a few tens of millivolts.

The adsorption of ions onto the charged surfaces of mineral grains produces positive anomalies of up to about 100 mV. Anomalies of this type occur in association with quartz veins, pegmatites and concentrations of clay minerals. Biogenic causes of SP anomalies are also known, possibly caused by ion selectivity and water pumping by plant roots. These cause decreases in potential of up to about 100 mV.

5.5.1.2 Mineralised sources

Potential anomalies ranging from hundreds to more than a thousand millivolts occur in association with bodies of disseminated and massive sulphide, graphite, magnetite, anthracite coal and manganese mineralisation. These are known as *mineral* or *sulphide potentials*, and they form the signal in mineral SP surveys. The causes of the potential variations associated with coal are poorly understood, and even those due to metallic mineralisation are not fully explained.

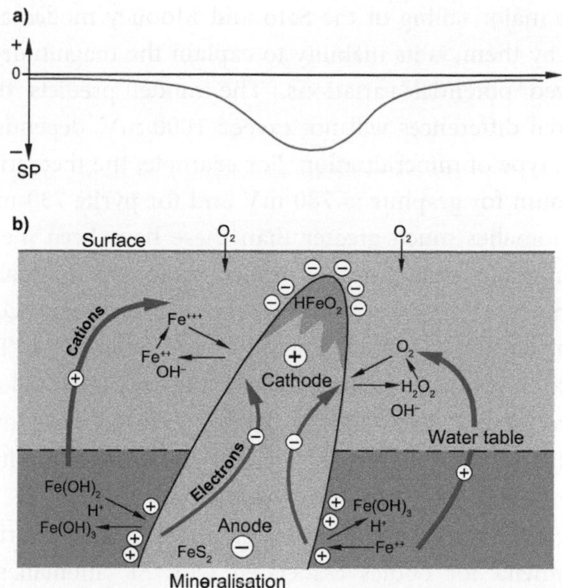

Figure 5.27 Schematic illustration of an electrochemical mechanism for the self-potential produced by an iron sulphide body. Based on a diagram in Sato and Mooney (1960).

The most comprehensive explanation of mineral potentials is that of Sato and Mooney (1960) and involves electrochemical interaction between the mineralisation and the groundwater, as shown schematically in Fig. 5.27. In the Sato and Mooney model, the mineralisation straddles regions with differing oxidation potential, most likely regions above and below the water table, and this is assumed in the following description. Potential differences are formed by the different, but complementary, electrochemical reactions occurring above and below the water table. Below the water table there is an anodic half-cell reaction with ions in the solution in contact with the mineralisation being oxidised to release electrons. Above the water table, ions in solution are reduced by a cathodic half-cell reaction, so a supply of electrons is required. This causes the shallower outer-part of the mineralisation to have a negative potential relative to the deeper outer-part. The free electrons created by the reactions below the water table are electronically conducted upwards through the (electrically conductive) mineralisation to take part in the reduction reactions above the water table. The mineralisation has a passive role acting as a conduit from anode to cathode for the electrons involved in electronic conduction. In the surrounding electrolytes, current is passed by ionic means. Weathering of the near-surface part of the mineralised body removes it from the self-polarisation process.

The major failing of the Sato and Mooney model, and noted by them, is its inability to explain the magnitude of observed potential variations. The model predicts that potential differences will not exceed 1000 mV, depending on the type of mineralisation. For example, the theoretical maximum for graphite is 780 mV and for pyrite 730 mV, but anomalies much greater than these have been measured in many areas. Another problem is that SP anomalies are known to be associated with apparently poorly conducting bodies, including disseminated mineralisation. The shortcomings of the Sato and Mooney model have long been recognised, and although developments and improvements have been suggested, no profoundly different alternative models have been proposed.

The failure to identify a consistent set of geo-electrical phenomena for bodies associated with SP anomalies – considering the almost infinite variety of geometrical and electrical properties of mineralisation, and chemical and electrical properties of the surrounding geology and groundwater – suggests that there are various mechanisms responsible for the observed potential variations. The source of each anomaly may be unique to the body of mineralisation, as suggested by the observation that where there are adjacent and apparently identical bodies of mineralisation one may produce an SP anomaly whilst the other does not.

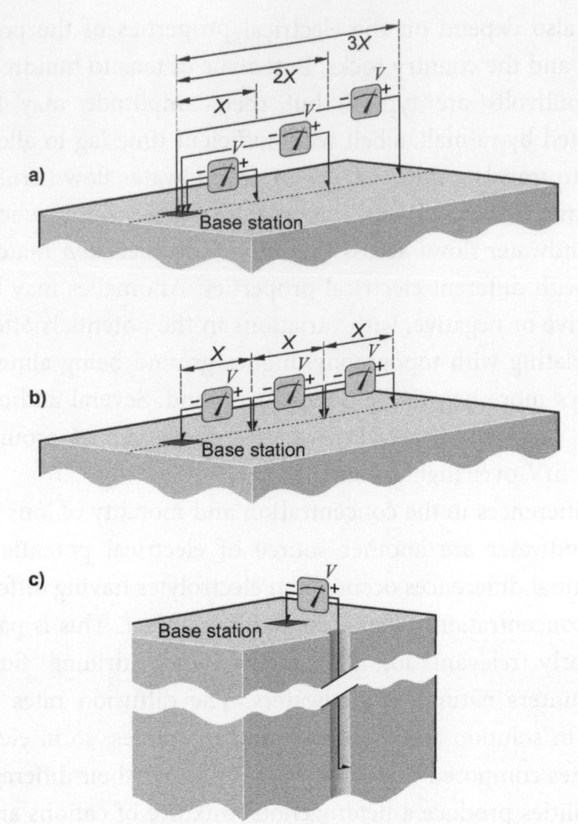

Figure 5.28 Electrode configurations for SP surveying. Surface surveying using the (a) fixed-base and (b) gradient modes. (c) Downhole survey mode. Potential differences are measured relative to a base station.

5.5.2 Measurement of self-potential

The SP method requires only the simplest of field apparatus: a pair of non-polarising electrodes (see Section 5.4.1), a reel of insulated wire and a high-impedance voltmeter with a resolution of at least 1 mV. Alternatively, measurements may be made using a modern resistivity/IP receiver, which offers the convenience of digital storage.

5.5.2.1 Surface surveys

Surface SP surveys usually comprise a series of parallel lines oriented perpendicular to strike and spaced to suit the resolution required, but generally 50 to 100 m apart for compact 3D targets. Station spacing varies from a few metres to a few tens of metres.

Surveys are conducted in one of two ways (Fig. 5.28a and b). In the fixed-base procedure, one electrode is placed at the base station and connected to the negative terminal of the voltmeter. The second (roving) electrode, connected to the positive terminal of the voltmeter through the reel of wire, is moved along the survey line and the potential differences between the electrodes are recorded. The procedure is repeated for each survey line, and a tie line, intersecting the survey lines at their 'local' base stations, is surveyed in the same manner. All the measurements are then referenced to a 'master' base station located in an area of low potential gradient, and usually assigned a potential of zero volts.

The second survey procedure measures potential differences with a constant electrode separation, usually about 10 m. Both electrodes are moved along the survey line and the potential differences recorded. The total potential variation can be obtained by summing the measured differences between stations, although measurement errors are cumulative.

5.5.2.2 Downhole surveys

For downhole SP surveys (Fig. 5.28c) the local base station is some arbitrary point on the surface and measurements are made at various depths downhole, commonly with measurement spacing of a few centimetres to a few metres.

SP is one of the quickest, simplest and cheapest forms of downhole logging, although measurements can only be made in uncased, water-filled, drillholes. If necessary, the local base can be tied to other surface or downhole surveys allowing the creation of potential-variation cross-sections or even volumes.

5.5.3 Display and interpretation of SP data

Self-potential maps can often be very complicated and difficult to interpret in terms of the geology, owing to the poor understanding of the origins of the potential variations. Responses of interest are often superimposed on regional trends, for example caused by streaming potentials or telluric currents (see *Atmospheric noise* in Section 5.4.2.1). Removal of the regional response is done using the same methods as applied to other types of geophysical data (see Section 2.9.2).

5.5.3.1 Surface data

SP responses of economic significance are usually large in amplitude compared with other responses, so they are easy to identify in a contour map or basic image of the measured potential.

It is helpful to interpret surface SP data in conjunction with other types of data in order to identify spurious responses. Topographic data help in identifying streaming potentials; geological maps, magnetic data and satellite spectral measurements help in recognising effects related to changes in geology or vegetation. Since faults may act as conduits for groundwater flow, these may be associated with SP responses. Graphite is an excellent target for SP surveys because of its conductivity. For the same reason, black shales and other conductive lithologies may produce strong responses, but these may be of no economic significance. Conductive cover can attenuate or even completely mask bedrock responses, and changes in the thickness or type of cover may control groundwater flow and produce variations in electrical potential.

The subsurface electrochemical mechanisms that give rise to economically significant anomalies are dipolar (i.e. the sources have both positive and negative electrical poles), so both positive and negative responses can be expected in SP anomalies. A general characteristic of these anomalies, regardless of source shape, is that vertically dipping bodies have a single negative anomaly located over their negatively polarised shallow portions (Fig. 5.29a and d), and dipping bodies also exhibit an additional lower-amplitude positive peak in the dip direction related to the deeper pole of positive polarity: see for example Fig. 5.29c and f.

The overall anomaly shape on the map reflects the overall shape of the source. Assuming a particular shape for the source, e.g. a polarised sphere, rod or thin sheet (the latter two polarised in the direction of dip) information about the source, such as its depth and dip, can be determined. Responses for these simple source models can be computed to model survey data. El-Araby (2004) describes a method where the most appropriate source-shape is estimated, and also provides a comprehensive reference list covering methods of inverse modelling SP responses.

Figure 5.29 demonstrates the general characteristics of SP anomalies with computed responses and field data from various mineral deposits for the widely applicable thin-sheet source geometry. An actual example of this kind of source is the Joma pyrite deposit in Trøndelag, Norway (Logn and Bølviken, 1974). The mineralised body comprises predominantly pyrite, with some chalcopyrite, pyrrhotite and sphalerite within a carbonate-bearing greenstone unit. A potential cross-section across it is shown in Fig. 5.29g. The downhole data, which have been smoothed to emphasis the effects of the electrochemical processes (see Fig. 2.37), clearly define a negative pole near the surface and a positive pole at depth.

Sometimes the observed SP anomaly reflects polarisation of only that part of a larger mass exposed to the water table and the zone of oxidation. Anomaly analysis often shows that the centre of the polarised source coincides with the base of the regolith or weathered horizon, consistent with electrochemical theories for SP mechanisms. Although anomaly width increases with increasing depth of the source, a wide anomaly in an area with the limited depth of oxidation may indicate a wider rather than a deeper source. In areas with severe topography the SP anomaly may be laterally offset, downslope, from its source.

5.5.3.2 Downhole data

The downhole SP log is one of the standard logs used by the petroleum industry, its main use being the qualitative discrimination of permeable and impermeable strata and the calculation of the electrical resistivity of the pore waters in the rocks surrounding the drillhole. An excellent introduction to SP logging as used by the petroleum sector is given by Rider (1996). Examples of SP logs from various geological environments and types of mineralisation are shown in Fig. 5.30.

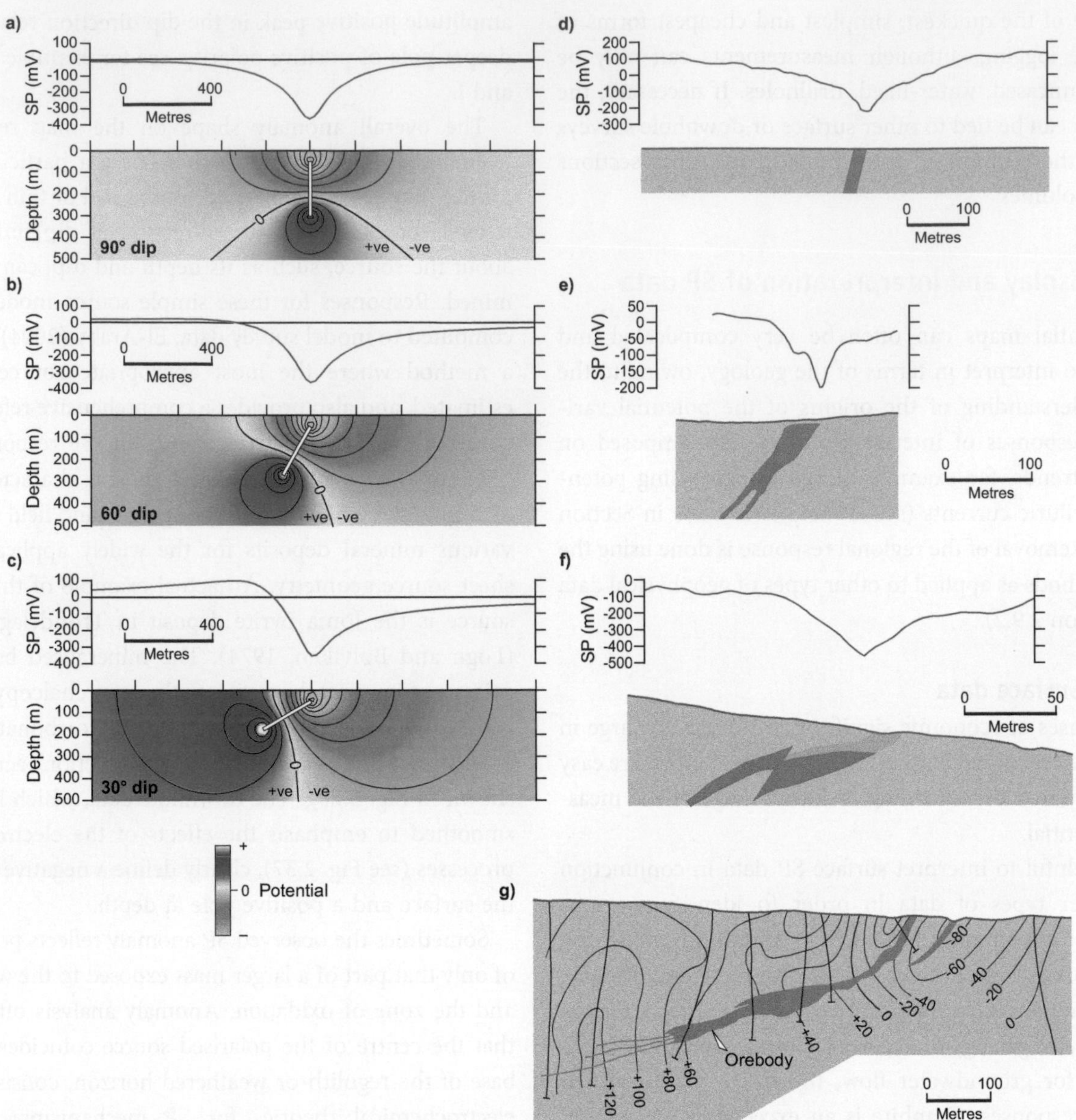

Figure 5.29 SP anomalies associated with sheet-like bodies. (a–c) Computed potential on a cross-section of a polarised sheet for a range of dip. (d) Sargipalli graphite mineralisation in Sambalpur, Orissa, India. Based on a diagram in Madhusudan *et al.* (1990). (e) New Insco copper-bearing massive sulphide deposit, Quebec, Canada. Based on a diagram in Telford and Becker (1979). (f) A copper-bearing massive sulphide body in the Mirdita Zone of Albania. Based on a diagram in Frasheri *et al.* (1995). The dark and light grey areas highlight the massive and disseminated mineralisation, respectively. (g) Contours of downhole and surface potentials (mV) in a cross-section through the Joma pyrite deposit, Norway; see text for details. Redrawn, with permission, from Logn and Bølviken (1974).

Potential variations associated with changes in lithology are common in drillholes, being caused by liquid-junction and shale potentials (see Section 5.5.1.1). As a consequence, the measured SP tends to vary between two 'baseline' values, the sand baseline and shale baseline in a similar way to γ-logs (see Section 4.7.5), which correspond with permeable and impermeable formations, respectively. If there is no difference between the ionic concentrations in the water in the drillhole and the water in the adjacent formations, then SP variations of this type will not be detected.

In soft-rock terrains, the SP log may be used for hole-to-hole correlation and facies analysis. Galloway and Hobday (1983) describe the use of SP logs to map facies in sediments hosting various types of mineralisation including coal and sandstone-type uranium deposits. In hard-rock terrains, the SP log can be useful for identifying changes in lithology (see for example Urbancic and Mwenifumbo

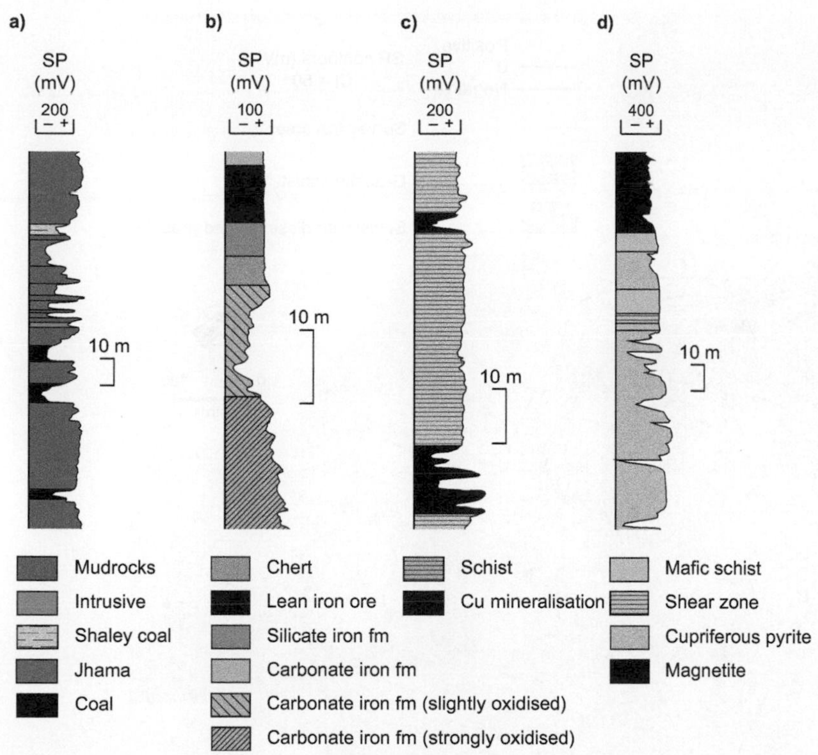

Figure 5.30 SP logs through various types of geology and mineralisation. (a) Jharia Coalfield, Jharkhand, India. Jhama is burnt coal resulting from intrusion of igneous rock. Note how the log varies between two baseline values (see Section 5.5.3.2). Based on a diagram in Kayal (1979). (b) Oxidised iron formation, Cuyuna Iron Range, Minnesota, USA. Oxidation of the formations results in variations in SP response. Based on a diagram in Hansen (1970). (c) Kimheden Cu deposit, Lapland, Sweden. Note the erratic variations within the relatively conductive oxide and sulphide mineralisation. Based on a diagram in Parasnis (1970). (d) Copper mineralisation in the Singhblum copper belt, Bihar, India. SP responses are subdued in the host rock and erratic in the mineralisation. Based on a diagram in Kayal *et al.* (1982).

(1986)), but is generally less useful than most other kinds of logs. Conductive mineralisation generally coincides with significant variations in SP (see for example Becker and Telford (1965)), which allows intersections to be readily identified from the logs. Responses from less conductive lithotypes are more subdued and harder to predict.

As with surface SP data, it may be necessary to differentiate between localised fluctuations and smoother variations, i.e. a form of regional response removal (see Section 2.9.2). In mineralised sections, the short-wavelength variation in the logs is a response from potential differences caused by electric current transfer across the interface between the mineralisation and host rock. The longer-wavelength variation reflects potential variations within the country rock, cf. the Sato and Mooney electrochemical model.

5.5.4 Examples of SP data from mineral deposits

The following example SP surveys demonstrate the use of the SP method for exploring for and investigating two quite different types of mineral deposit.

5.5.4.1 Almora graphite deposits

This example demonstrates application of the SP method as a quick, cheap and effective means of exploring for graphite in the Almora district in the Himalayan Region of Uttarakhand, India (Srivastava and Mohan, 1979). The SP method has been extensively used in the region, both alone and in association with other electrical methods. Figure 5.31 shows the results from a survey designed to detect high-grade graphite mineralisation within graphitic schist horizons. The local geology consists of a folded, faulted and metamorphosed sequence of schists and quartzites. The SP data have been effective in mapping the graphitic horizon between outcrops, the associated anomalies having amplitudes of –200 to –600 mV. Selected profiles allow the dip of the source to be determined (cf. Fig. 5.29). Drilling of the main SP anomalies intersected significant occurrences of graphite.

5.5.4.2 Safford porphyry copper deposit

This example demonstrates the application of SP to the mineralogically complex environment of a large porphyry copper deposit; illustrating the different responses obtained from the sulphide-rich ore and the various sulphide-rich alteration zones, and structure. Figure 5.32 shows SP data from the Safford porphyry Cu deposit in the Lone Star district of Arizona, USA. The geology of the deposit is described by Robinson and Cook (1966), who also show a contour map of the SP responses. This allows the SP data to be compared with such features as alteration, structure and the sulphide mineralisation itself.

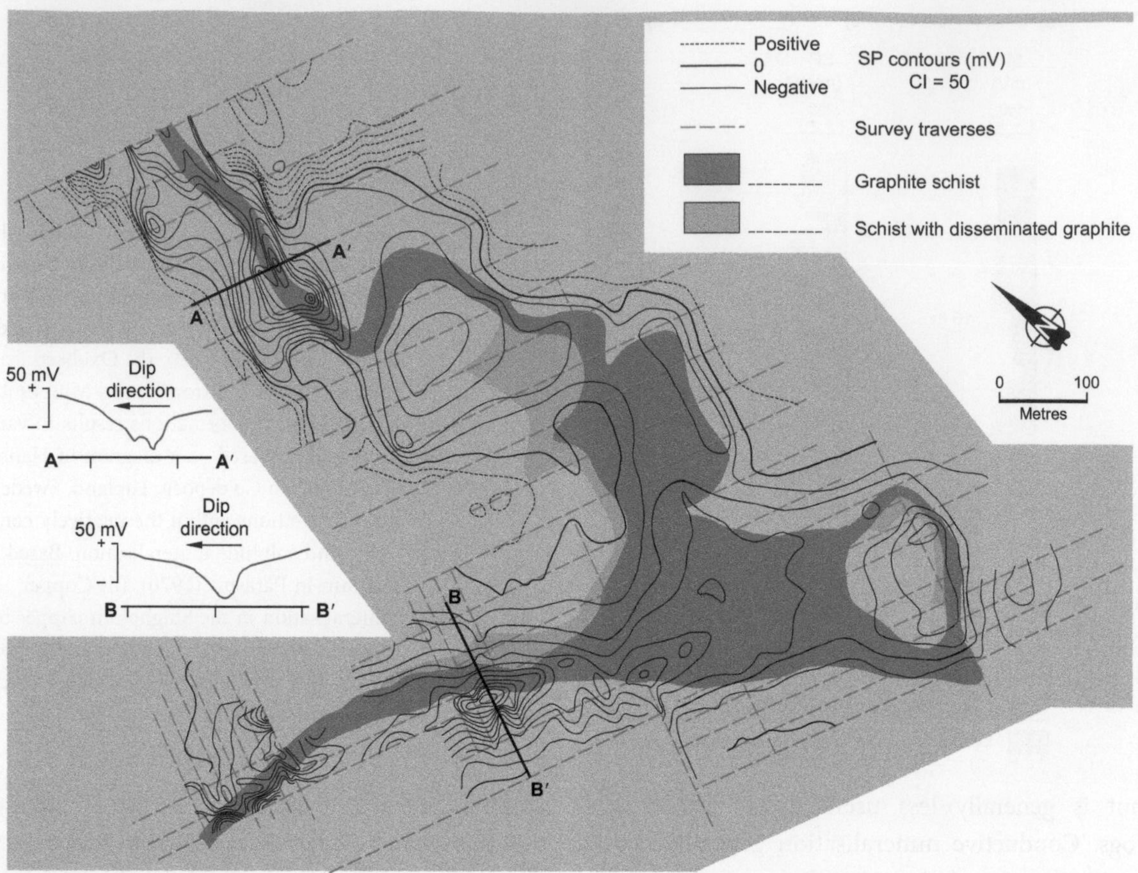

Figure 5.31 Contours of the surface SP over the Almora graphite deposits. The form of the local geology is revealed by the negative anomalies associated with graphitic horizons. CI = 50 mV. Redrawn from Srivastava and Mohan (1979), with permission of the Director General, Geological Survey of India.

The Safford deposit occurs in an igneous complex. There is extensive alteration in the area around the deposit. Mineralisation occurs in a central zone of sericite and biotite alteration surrounded by concentric zones of chloritic and propylitic alteration. Sulphide mineralisation also forms a roughly concentric pattern, but is only loosely correlated with the alteration zones. A core of dominantly copper sulphides is surrounded by a zone dominated by pyrite, these extending to the approximate limit of the chloritic alteration zone. The main primary sulphide minerals are pyrite and chalcopyrite in the form of disseminations and veins, with the former the most abundant. In the ore zone, pyrite content is 0.2 to 1.0%, but in the surrounding alteration haloes it ranges from 4 to 8%. The primary ore contains about 0.7% chalcopyrite, decreasing to about 0.4% in the pyrite-dominated zones. Northeast-trending faults and shear zones are the dominant structures in the deposit, these having acted as conduits for hydrothermal fluids. The peak SP response of 500 mV is a very strong anomaly and coincides with a sulphide-rich zone where there is much faulting. The orebody lies within this anomaly, but the anomaly maximum is to the south of it, possibly owing to the distribution of sulphides at depth. Other alteration zones in the vicinity are also associated with SP anomalies.

The SP data from Safford successfully map the sulphides/alteration around the orebody, but cannot distinguish the economically significant from the barren. The authors comment that SP responses are significantly diminished by cover. These observations are typical of SP results in general. Corry (1985) describes various factors that affect the SP responses from porphyry style mineralisation.

5.6 Resistivity and induced polarisation methods

Resistivity and induced polarisation (IP) surveys have much in common in terms of data acquisition, display and interpretation. Resistivity-only surveys may be undertaken, as is widely practised in groundwater and

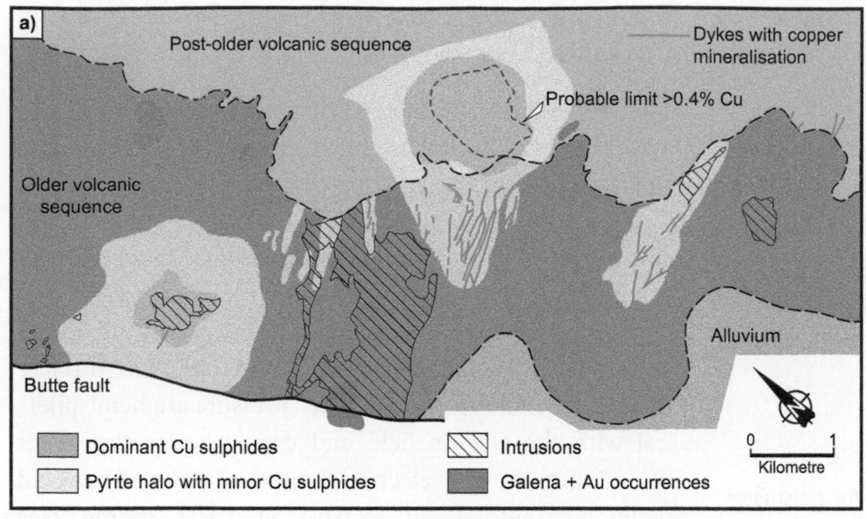

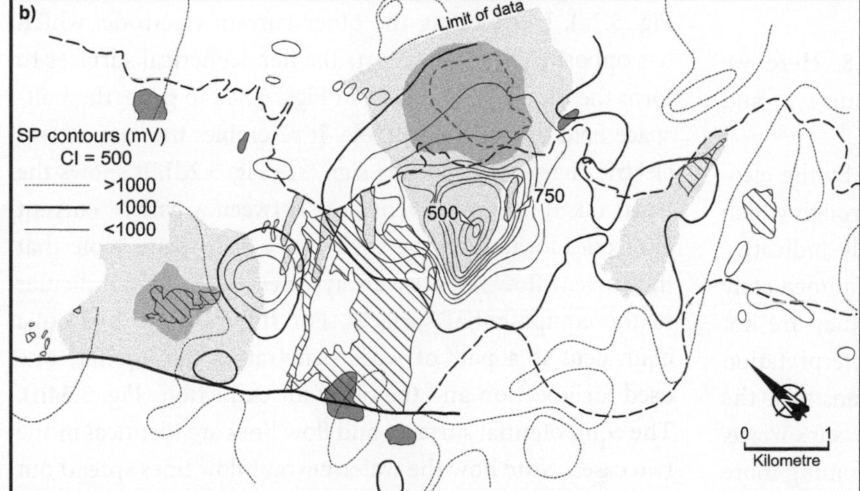

Figure 5.32 Safford porphyry Cu deposit. (a) Summary geological map showing the distribution of sulphide mineralisation. (b) Contours of surface SP. Note the correspondence between the main SP anomaly and the mineralisation within dykes. Redrawn, with permission, from Robinson and Cook (1966).

environmental studies. In mineral exploration, measurements are usually made of both parameters, resistivity data being acquired as part of the IP surveying methodology, with the analysis of each complementing the other.

In resistivity/IP surveying (Fig. 5.33), an emf produced by a battery or a portable generator is applied to the ground via a pair of electrodes known as the *current electrodes*, which form the *current* or *transmitter dipole*. An electric field is formed in the ground, resulting in current flow through the subsurface, which is measured. Differences in electrical potential between selected locations, owing to the subsurface current flow, are measured using a second pair of electrodes known as the *potential electrodes*, which make up the *potential* or *receiver dipole*. The layout or configuration of the four electrodes is known as an *electrode array*. At each survey station, the transmitted current and the measured potential difference across the receiver dipole are recorded along with the location of the four electrodes. From these data the resistivity of the subsurface can be calculated, as described in Section 5.6.2. The IP response is also recorded and the various IP parameters calculated as described in Section 5.6.3.

Lateral and vertical variations in resistivity/IP properties of the subsurface can be mapped by moving the array around the survey area and by changing the relative positions of the electrodes making up the array. Data may be acquired and presented in the form of profiles (1D), maps or cross-sections (2D), or volumes (3D). Like many types of geophysical data, assigning a depth to a particular measurement can be problematic. Usually the potential and the current electrodes are located on the surface of the ground, but one or more electrodes may be located in one or more drillholes which can lead to significant improvements in target detection and resolution. A common form of downhole resistivity surveying is the applied potential method described in Section 5.6.9. Also common is downhole

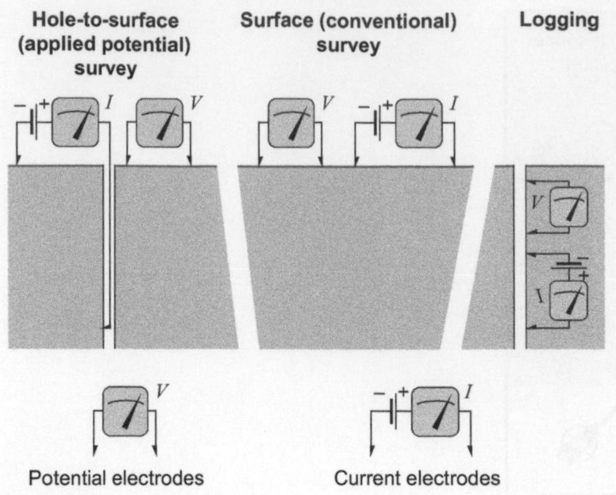

Figure 5.33 Electrode configurations commonly used for resistivity/IP surveying.

resistivity logging, described in Section 5.6.8. Here we primarily discuss surface (conventional) surveys and logging.

Resistivity/IP measurements are influenced by the electrical properties of the large volume of rock through which the current passes, so they are not necessarily indicative of the electrical properties of the material immediately below the measurement point. Furthermore, they are not displayed in geology-like form, so complex interpretation and modelling techniques are required to transform the measurements into electrical models of the subsurface. Inversion modelling of 3D data volumes is becoming more common with the increasing mass-acquisition of detailed 3D resistivity/IP data volumes. These techniques produce results in a geology-like form, following a similar trend to the potential field and EM methods.

Instead of measuring potential differences associated with the subsurface current flow, the magnetic fields associated with these currents can be measured by a class of surveying techniques referred to here as *magnetometric methods*. These include *magnetometric resistivity* (MMR) and *magnetic induced polarisation* (MIP), and a set of related 'total-field' techniques known as *sub-audio magnetics* (SAM). There are some advantages to measuring magnetic fields: for example, they are less affected by conductive overburden. Also, magnetometric methods are responsive to both highly conductive targets and to weakly conductive, electrically connected mineralisation, such as disseminated mineralisation and sphalerite-rich ores, which can be poor targets for conventional electrical and EM measurements. These methods are described in online Appendix 3.

5.6.1 Electric fields and currents in the subsurface

To assist in visualising the subsurface electric field and associated current flow we make use here of the plumbing analogy of electricity introduced in Section 5.2.1. Figures 5.34a to c show the situation for one current electrode in isolation located on the surface of an electrically homogeneous subsurface, known as a *half-space*. The hydraulic equivalent is a shallow well injecting water into a homogenous flat-topped aquifer (Fig. 5.34d). The surfaces of equal electrical-potential/water-pressure are hemispherical with the electric field and current/water flow lines diverging from the electrode/injection point. The field resembles that of an isolated electrical charge (see Fig. 5.2a). Introducing the other current electrode, which has opposite polarity, distorts the hemispherical surfaces to form the dipole field shown in Figs. 5.34e to g, i.e. the half-space field of a current dipole. It resembles the case of two electric charges of opposite sign (see Fig. 5.2d). It shows the paths taken by the current flow between a pair of current electrodes located on the surface of a half-space. Note that the current flow lines are always everywhere perpendicular to the equipotential surfaces. For this case the hydraulic equivalent is a pair of wells penetrating the aquifer, one used for injection and the other for extraction (Fig. 5.34h). The equipotential surfaces and flow lines are identical in the two cases. Note how the water/current flow lines spread out widely between the wells/electrodes to occupy a large volume of material. Clearly then, electrical properties inferred from the measurements are an 'average' of a large volume of subsurface material.

Variations in the subsurface resistivity/conductivity alter the shape of the equipotential surfaces which in turn alter the path of the current flow. Referring to Fig. 5.35, the equipotential surfaces conform to the shape of a conductive body, causing the current to follow the line of least resistance, i.e. the flow lines are deflected toward conductive bodies. This is an illustration of *current gathering* or *current channelling* (see Section 5.7.2.4). Contrastingly, current flows 'around' resistive bodies. Groundwater behaves in an equivalent way, with flow being concentrated in porous and permeable formations. The diagrams in Fig. 5.35 are equally valid whether treated as maps or cross-sections. Clearly then, measuring potential variations on the surface or downhole will allow zones of anomalously high or low electrical conductivity/resistivity to be detected.

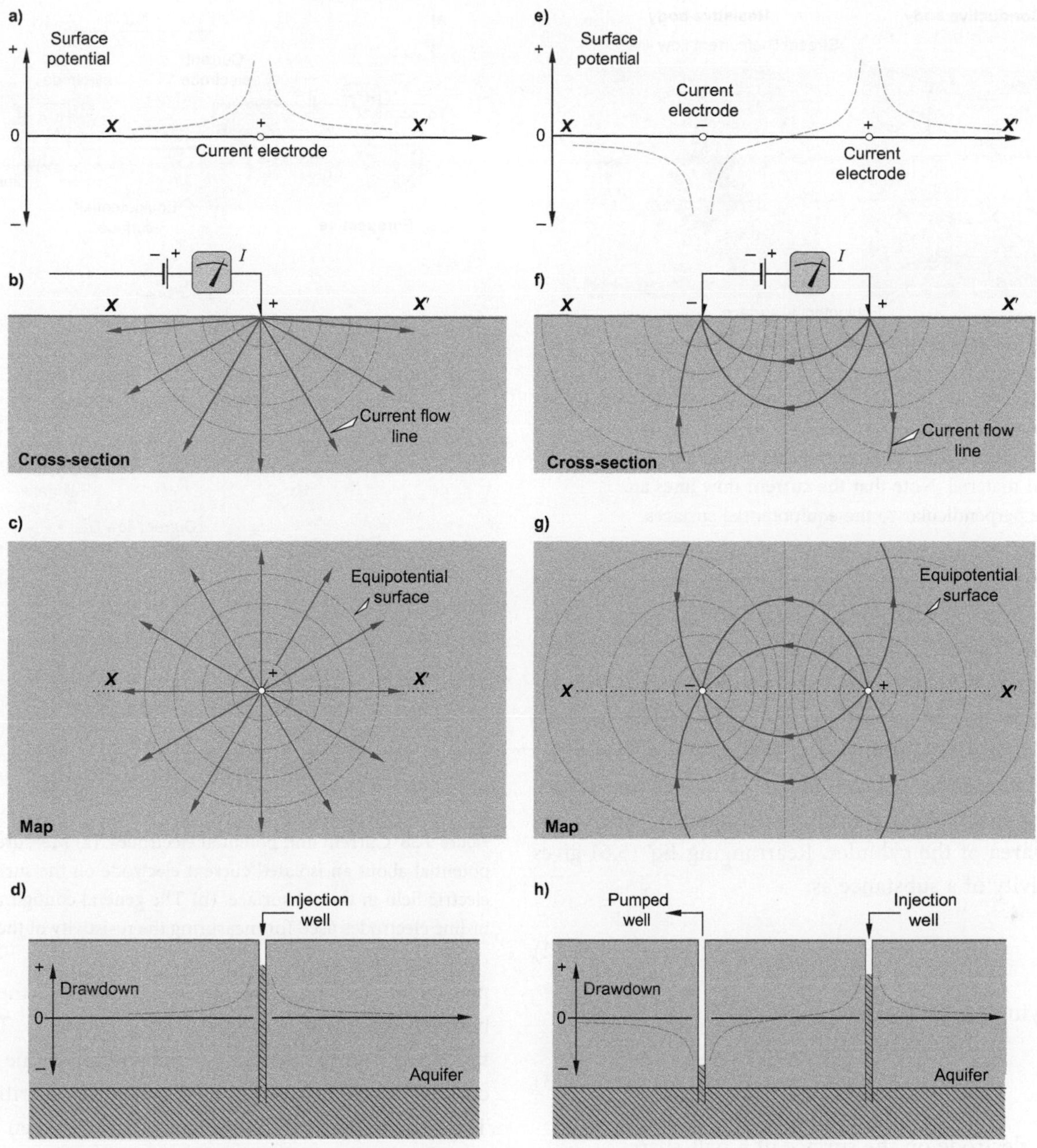

Figure 5.34 Electric field in a half-space and the associated surface potential formed by: (a–c) an isolated pole and (e–g) a dipole. The hydraulic equivalents for both are shown in (d) and (h), respectively. Potential is measured with respect to zero potential, located at infinity.

5.6.2 Resistivity

In resistivity/IP surveying the electrical properties of the subsurface are determined by measuring the current passing through the ground via the current (transmitter) electrodes, and measuring the resultant potential difference produced between the potential (receiver) electrodes. The results depend on both the electrode configuration and the actual subsurface distribution of the electrical properties with respect to the electrode locations. Resistivity surveys map resistivity/conductivity in the subsurface.

In Section 5.2.1.2 we described how the resistivity (ρ) of a body could be determined by accounting for its geometry using a geometric correction factor (k_{geom}) (Eq. (5.2)). We also saw how measurements of the current (I) flowing through a resistor and the resultant potential difference (ΔV) across it could be used to calculate its resistance (R) using Ohm's Law (Eq. (5.4)).

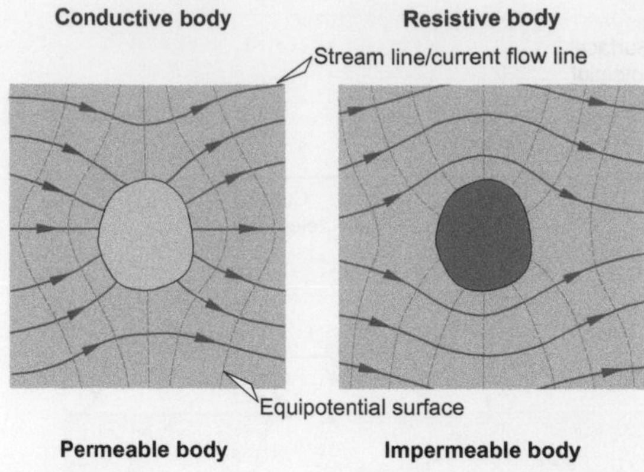

Figure 5.35 Distortions of the equipotential surfaces and associated hydraulic/electric current flows due to zones that are more permeable/conductive and less permeable/conductive than the background material. Note that the current flow lines are everywhere perpendicular to the equipotential surfaces.

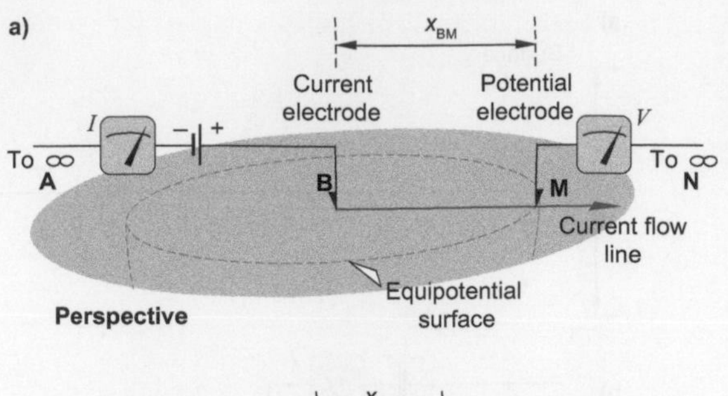

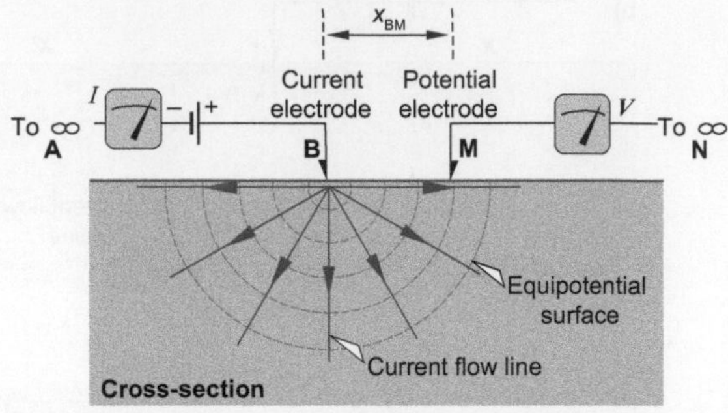

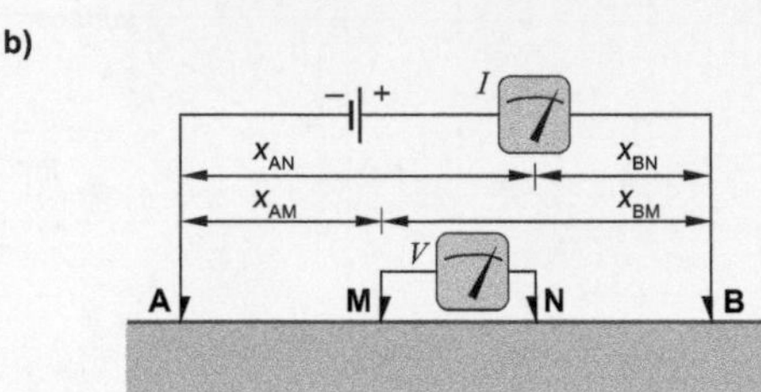

Figure 5.36 Current and potential electrodes. (a) Measurement of the potential about an isolated current electrode on the surface and its electric field in the subsurface. (b) The general configuration of in-line electrodes used for measuring the resistivity of the subsurface.

Consider the theoretical situation of an isolated current electrode on the surface of an electrical half-space producing a hemispherical field in the subsurface (Fig. 5.36a). The potential (V) is measured at a point located a distance X from the current electrode, and with respect to zero potential located at infinity. The distance X is equivalent to the length of the cylinder in Eq. (5.3) and the surface area of the hemisphere ($2\pi X^2$) is equivalent to the cross-sectional area of the cylinder. Rearranging Eq. (5.6) gives the resistivity of a substance as:

$$\rho = \frac{V}{I} k_{\text{geom}} \tag{5.12}$$

and applying Eq. (5.3) gives:

$$\rho = \frac{V}{I}\left(\frac{2\pi X^2}{X}\right) = \frac{V}{I}(2\pi X) \tag{5.13}$$

so for an electrode on the surface of a half-space

$$k_{\text{geom}} = 2\pi X \tag{5.14}$$

Rearranging Eq. (5.13) gives the potential at any point a distance X from the electrode as:

$$V = \frac{I\rho}{2\pi X} = \frac{I\rho}{2\pi}\left(\frac{1}{X}\right) \tag{5.15}$$

Now consider the simplest situation for the resistivity method: a half-space with both current electrodes and both potential electrodes located anywhere on the surface (Fig. 5.36b). When describing electrode arrays, convention has it that the current electrodes are labelled as 'A' and 'B' and the potential electrodes as 'M' and 'N'. Survey parameters can then be defined in terms of the distances between electrodes, designated X_{AB} and X_{MN}. The potential at any point is the sum of the potentials due to the two current electrodes, which have opposite polarities (signs). The resultant potential difference (ΔV) between the potential electrodes due to each current electrode can be obtained by applying Eq. (5.15) for each electrode and is given by:

$$\Delta V = \frac{I\rho}{2\pi}\left(\frac{1}{X_{\text{AM}}} - \frac{1}{X_{\text{BM}}} - \frac{1}{X_{\text{AN}}} + \frac{1}{X_{\text{BN}}}\right) \tag{5.16}$$

Resistivity is obtained by rearranging Eq. (5.16) as follows:

$$\rho = \frac{2\pi\Delta V}{I}\left(\frac{1}{X_{\text{AM}}} - \frac{1}{X_{\text{BM}}} - \frac{1}{X_{\text{AN}}} + \frac{1}{X_{\text{BN}}}\right)^{-1} \tag{5.17}$$

Equation (5.17) gives the true resistivity of an electrically homogenous subsurface. That part of it representing the effects of the electrode separations is the geometric factor given by:

Table 5.2 Geometric factors for the common surface arrays and several downhole arrays.

Surface arrays	Geometric factor (k_{geom})
Pole–pole	$2\pi X_{BM}$
Pole–dipole	$2\pi n(n+1)X_{MN}$
Dipole–dipole	$\pi n(n+1)(n+2)X_{MN}$
Wenner	$2\pi X_{MN}$
Schlumberger	$\pi \frac{(X_{AB}/2)^2}{X_{MN}}, X_{MN} \leq X_{AB}/20$
Gradient	$2\pi \frac{L^2}{X_{MN}G}$ $G = \frac{(1-U)}{[V^2+(1-U)^2]^{3/2}} + \frac{(1+U)}{[V^2+(1+U)^2]^{3/2}}$ $U = x/L,\ V = y/L,\ L = X_{AB}/2$
Downhole arrays	
Normal log	$4\pi X_{BM}$
Lateral log	$4\pi \frac{X_{BM}X_{BN}}{X_{MN}}$
Applied potential	$4\pi X_{BM}$

$$k_{geom} = 2\pi\left(\frac{1}{X_{AM}} - \frac{1}{X_{BM}} - \frac{1}{X_{AN}} + \frac{1}{X_{BN}}\right)^{-1} \quad (5.18)$$

Table 5.2 gives k_{geom} for a number of commonly used electrode arrays, i.e. reduced forms of Eq. (5.18).

5.6.2.1 Apparent resistivity

Equation (5.17) gives the true resistivity (ρ) of an electrically homogenous subsurface. As described in Section 5.6.1, any resistivity inhomogeneity in the subsurface will distort the electric field and cause the measured potential difference to differ from that due to a homogeneous subsurface. In this case the resistivity is known as the *apparent resistivity* (ρ_a) of the subsurface, because it assumes that the measurements are made on an electrically homogeneous subsurface, even though this is very unlikely to be the case. The apparent resistivity depends on the true resistivity distribution of the subsurface and the electrode configuration used for the measurement. Transforming a set of apparent resistivity data into the true resistivity distribution of the subsurface is the fundamental challenge for interpretation techniques, which we discuss in detail in Section 5.6.6.

5.6.3 Induced polarisation

Resistivity is measured during the application of a constant d.c. current. Induced polarisation is measured with a varying current, either d.c. or a.c. As described in Section 5.2.1.5, capacitance can be measured in three ways: by measuring the decay of the potential after switching off a d.c. current, by comparing apparent resistivity obtained with a.c. currents of two different frequencies, and finally by comparing the phase between an applied a.c. current and a measured a.c. potential difference. The first is a form of time domain measurement; the other two are forms of frequency domain measurement. Although the frequency domain and time domain polarisation parameters are different, they both produce the same anomaly shapes which are comparable. Induced-polarisation surveys measure a range of parameters to quantify the electrical polarisability of the subsurface.

5.6.3.1 Time domain measurements

In the time domain, the transmitter alternately turns on, producing a steady current, and then off. A graph of the signal from the transmitter is a square wave (Fig. 5.37a; see online Appendix 2). Note that the polarity of the current is reversed between successive 'on' cycles to cancel the effects of residual polarisation and to reduce the effects of ground and telluric currents (see *Atmospheric noise* in Section 5.4.2.1). The duration of the on and off periods is selected after field tests prior to commencing the survey. As described in Section 5.2.1.5, it takes time for the electrical polarisation to occur, so the on-time must be long enough to achieve sufficient polarisation but no longer, since this will increase the time required to complete the survey. The on-time and off-time is typically in the range 1 s to 4 s, usually 2 s.

Figure 5.37a shows the variation in potential measured when the subsurface is electrically polarisable. When the current is turned on, the potential immediately increases sharply, then more slowly before reaching a steady value. This is known as the *primary voltage* (V_p), and is that used for the calculation of apparent resistivity. The gradual increase is associated with 'charging' of the subsurface capacitor. When the transmitted current is turned off the reverse occurs, i.e. there is an initial sharp drop in potential and then a gradual decay. This *secondary voltage* (V_s) is dependent on the polarisation properties of the ground and is associated with the 'discharging' of the capacitor. It is the measurement of this decay that is the basis of the time domain IP method.

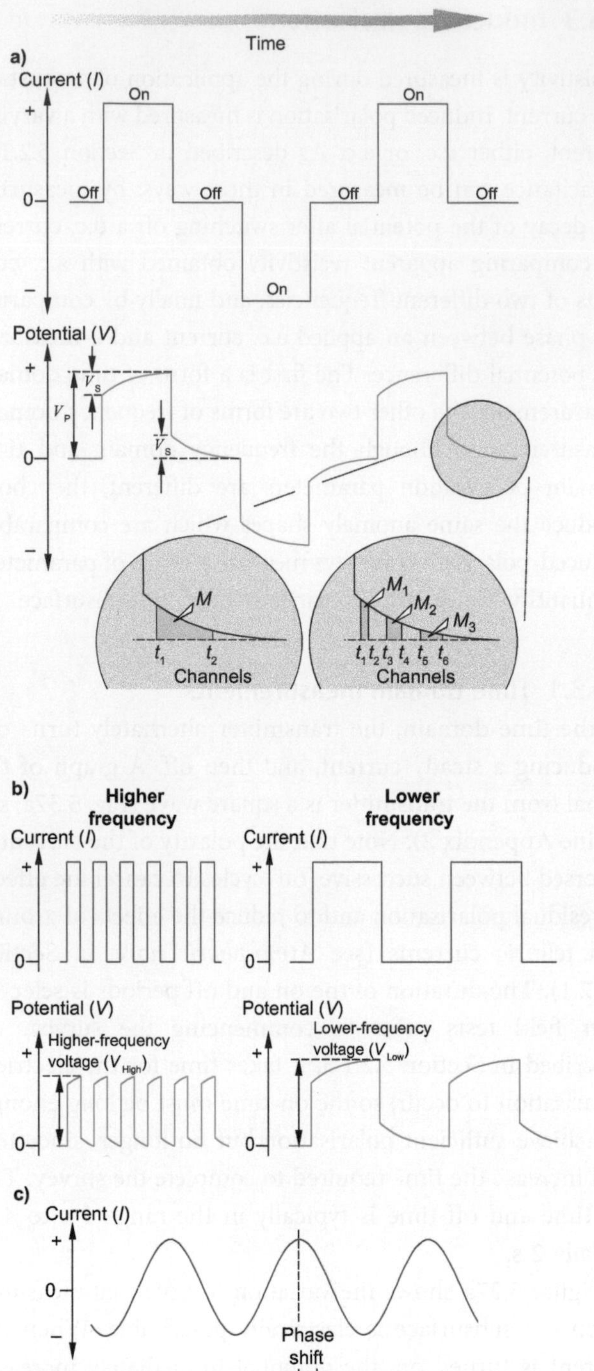

Figure 5.37 Signal waveforms used in resistivity and IP measurements. (a) The time domain bipolar square wave signal and the distorted wave measured by the receiver. Polarisation effects produce the slow rise and decay in the received signal. The time intervals used to measure apparent chargeability (M) are shown shaded. (b) The frequency domain dual-frequency square wave signals and the distorted waves measured by the receiver. (c) The frequency domain sine wave signal and the phase-shifted wave measured by the receiver.

Measurement of the decaying secondary voltage consists of a number of discrete voltage measurements at selected decay times. Usually the measurements are an average taken over very small time-intervals, or channels, so they approximately represent the area under the decay curve for each measurement channel. Amplitude of the secondary voltage is dependent upon the amplitude of the primary voltage which depends on, apart from the subsurface resistivity, the transmitted current. To account for this, the decay voltage for each channel is normalised (divided) by the primary voltage. This is known as the *chargeability* (M); it has dimensions of time and is usually quoted in milliseconds. The value of M depends on the measurement period and also on the width of the transmitted pulse, increasing as either parameter increases. Values of M for all the decay channels represent the polarisation decay. The entire polarisation–decay cycle is repeated a number of times so that the primary and secondary voltages can be stacked, or averaged (see Section 2.7.4.1), by the receiver, after correcting for the changing polarity of the current, to improve the signal-to-noise ratio of the measurements. Modern instruments can measure the secondary decay over a large number of channels to provide an accurate definition of it. Also, the time-width of the channels, M_1 etc. in Fig. 5.37, can be adjusted which can make it difficult to directly compare actual measurements from different surveys.

As for resistivity measurements (see Section 5.6.2.1), the measured quantity corresponds to the *true chargeability* of the ground only when the subsurface is electrically homogeneous; otherwise it is the *apparent chargeability*. A slow or long decay is indicative of highly polarisable material, so anomalously large chargeability at late decay times is considered significant.

5.6.3.2 Frequency domain measurements

In the frequency domain, a.c. currents of different frequencies are transmitted and the change in apparent resistivity between each frequency is used as a measure of electrical polarisation. In its most sophisticated form, frequencies extending across the spectrum from 0.01 Hz to 1000 Hz are used. This is known as *spectral IP* (SIP) (Wynn and Zonge, 1975; Pelton *et al.*, 1978), also referred to as *complex resistivity*. The method has been successful in discriminating various sulphide and oxide minerals in a number of deposit types. It takes considerable time to make spectral measurements and as a consequence the method remains chiefly in the realms of research and is not commonly used in geophysical prospecting.

In 'conventional' frequency domain surveys, current is transmitted as a very low-frequency alternating square wave, in the range 0.1 to 3 Hz, but usually 0.1 Hz. Measurements are made at two frequencies, the higher usually three times the lower frequency, and the apparent resistivity of the ground is calculated for both (Fig. 5.37b). The resistivity at the lower frequency is taken as the apparent resistivity of the ground. The difference in resistivity at the two frequencies, relative to the resistivity at the higher frequency, is used to calculate the induced polarisation parameter known as the *percentage frequency effect* (PFE) given by:

$$\mathrm{PFE}(\%) = \frac{\rho_{\mathrm{low}} - \rho_{\mathrm{high}}}{\rho_{\mathrm{high}}} 100 \tag{5.19}$$

where ρ_{low} and ρ_{high} are the resistivities at the lower and the higher frequencies, respectively.

A related IP parameter is the *metal factor* (MF). This is the PFE normalised (divided) by the resistivity measured at the lower frequency. It is intended to remove variations in PFE related to the host rock resistivity and to highlight zones of anomalous 'metal content'. In conductive terrains, however, the parameter is dominated by the effects of the low resistivity of the host rock, resulting in spurious MF anomalies in regions where there is no significant increase in polarisation. For this reason MF is not normally used.

There is also a *delay* or *phase shift* (see online Appendix 2) between the transmitted sine wave current and the measured voltage at each frequency that can also be used as an indication of electrical polarisation (Fig. 5.37c). The phase shifts at two frequencies can be combined into a single parameter known as the *relative phase shift* (RPS), which is relatively immune to EM-coupling effects (see Section 5.6.7.2), and given by:

$$\mathrm{RPS} = \phi_{\mathrm{low}} - \phi_{\mathrm{high}} \tag{5.20}$$

where ϕ_{low} and ϕ_{high} are the phase shifts at the lower and the higher frequencies, respectively, and have units of degrees.

5.6.4 Measurement of resistivity/IP

Fundamental to the acquisition of electrical data are two basic variables: array geometry, both in terms of electrode spacing and their relative positions, and array location relative to the target. As would be expected, that part of the subsurface closest to the array exerts the greatest influence on the measured parameters. It follows that by positioning the array in different locations information about different parts of the survey area is obtained.

5.6.4.1 Depth penetration

The distance between the current electrodes controls the influence that features at depth have on the measurement. Figure 5.38a is a cross-section showing the distribution of current flowing through a half-space; cf. Fig. 5.34f. The numbers on the lines show the percentage of the total current flowing above the line. The 50% line has a maximum depth equal to half the current dipole length ($X_{AB}/2$) and almost 90% flows above a depth equal to three times the dipole length.

When presenting and interpreting the data, a *pseudo-depth* for a given reading can be assigned, which is a function of the array geometry and the electrode separations (see Section 5.6.5.1). By 'expanding' an array (increasing the electrode separations about the same

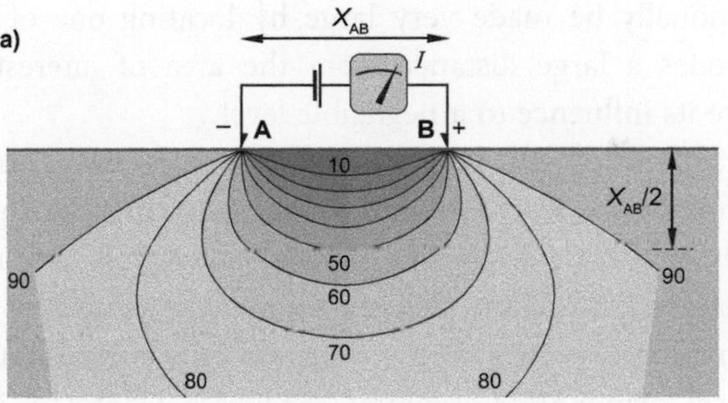

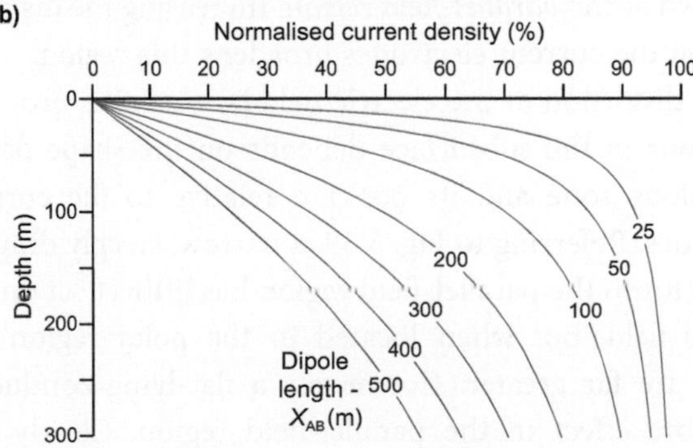

Figure 5.38 The half-space current distribution in the axial section of a surface dipole. (a) A current dipole of length X_{AB}. The numbers on the lines show the percentage of current flowing above the line. (b) Plot of percentage current that flows above a given depth. Based on a diagram in Robinson and Coruh (1988).

location) a *vertical electrical sounding* may be created, which is a representation of the vertical variation in electrical properties. This is achieved by increasing the separation of the current and potential dipoles, or increasing the length of both of them, or most usually, increasing the length of just the current dipole.

It follows that a traverse of measurements made with the separations and relative positions of the electrodes maintained will comprise a profile of readings pertaining to a constant pseudo-depth. A set of such traverses can be used to create a map. By surveying along the same traverse with different electrode spacings the data from individual traverses can be combined to create a *pseudosection*, or for a series of traverses a pseudovolume.

5.6.4.2 Target detection

The electric field created by the electrodes of a current dipole comprises several distinct regions with different and important characteristics (see Figs. 5.34f and g). The electric field close to each current electrode is little affected by the field of the other more distant electrode and the situation resembles that due to a single current electrode. This is known as the *polar region*. The polar region can intentionally be made very large by locating one of the electrodes a large distance from the area of interest to reduce its influence to a negligible level.

That part of the field distant from the two current electrodes, by at least several times their spacing (the dipole length), is known as the *dipole-field region*. Midway between the two electrodes, the equipotential surfaces are approximately vertical and the current flow approximately uniform, planar and horizontal (see Figs. 5.34f and g). This is known as the *parallel-field region*. Increasing the distance between the current electrodes broadens this region.

The distortion of the electric field by electrical property variations in the subsurface depends on the shape of the anomalous zone and its position relative to the current electrodes. Referring to Fig. 5.39, a narrow, steeply dipping conductor in the parallel-field region has little effect on the electric field, but when located in the polar region the effects are far greater. Conversely a flat-lying conductor has more effect in the parallel-field region. Clearly, for optimal detection of targets of a given geometry, the array needs to be positioned so that the targets are within the appropriate part of the electric field, e.g. the polar, parallel-field or dipole-field regions. This leads to the use of arrays with different relative locations of the current and potential electrodes, some configurations creating a large parallel-field regions etc.

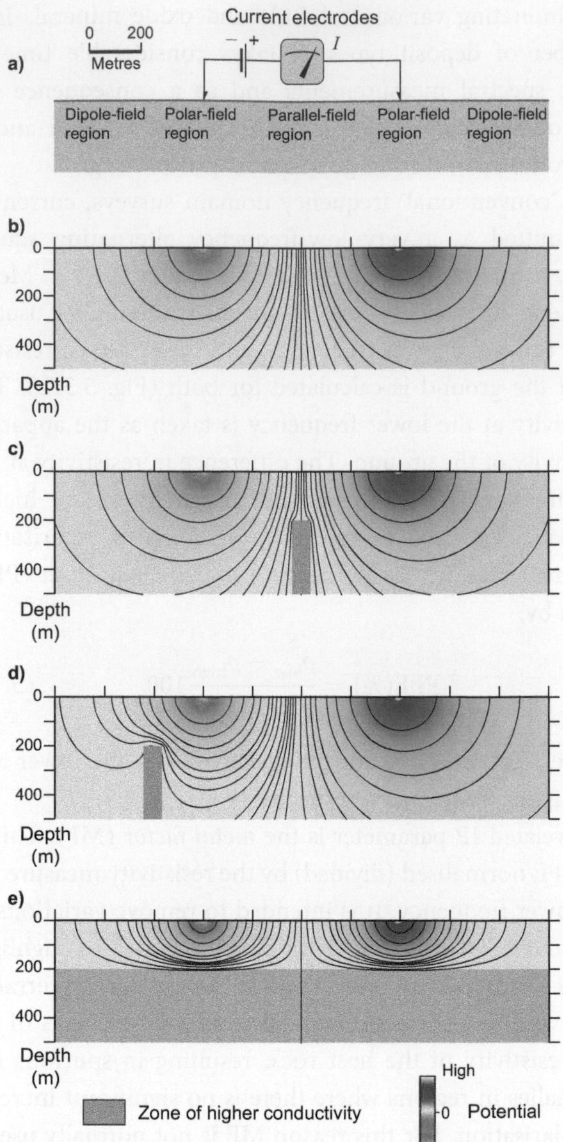

Figure 5.39 The electric field in the axial section of a surface current dipole. (a) The various field regions of the dipole. (b) The electrical field for a half-space, for a conductive vertical dyke in (c) the parallel field and (d) the dipole region, and (e) for a conductive layer at depth. Contour interval is variable and is the same for all models.

5.6.4.3 Resolution

The relative spacing of the potential electrodes largely controls the lateral resolution of the data; put simply, the electrical properties of the material lying between the two electrodes are averaged so the edges of anomalous zones whose widths are less than this spacing will not be properly defined. Vertical resolution is controlled by sensitivity to layers with different electrical properties.

5.6.5 Resistivity/IP survey practice

The equipment required for field resistivity measurements is not complex. IP measurements, on the other hand, require a transmitter capable of generating time-varying potentials, typically several thousand volts to ensure that the potential differences produced over the survey area are stronger than the ambient electrical noise level, and a receiver that can digitally process a large volume of time-varying voltage measurements. Modern instruments can measure several electrical parameters simultaneously and, monitor and attenuate electrical noise.

In order to obtain reliable measurements, particular attention must be paid to the preparation of both the current and potential electrodes. For IP surveys, minimising contact resistance in order to maximise the current flow and maximise the amplitude of the received signal is particularly important for measuring the weaker secondary time domain IP decay potentials (see Section 5.6.3.1). This is achieved by using foil in pit electrodes (see Section 5.4.1). Potential electrodes must be non-polarising (see Section 5.4.1). For resistivity surveys, electrochemically inactive stainless-steel stake electrodes are sufficient.

5.6.5.1 Electrode arrays

In principle, resistivity/IP measurements can be made with any configuration of the four electrodes, and with either all or some of the electrodes on the surface or below the surface. With the deployment of a large number of surface electrodes in a grid network over the target area, and the use of multichannel receivers, a large number of measurements can be made efficiently. Electrodes do not have to be repeatedly moved; instead changes to the position of the array, and the spacing and relative positions of the electrodes, are achieved by linking the various electrodes to the various recorder channels.

The electrodes are usually arranged in-line, i.e. along the survey traverse, this being logistically convenient. The commonly used electrode configurations for surface surveying are shown in Fig. 5.40. Note that to create the effect of an isolated electrode (a pole) the other electrode in the dipole is positioned at sufficient distance so as to not influence the measurement. This is typically greater than ten times the spacing of the other electrodes. For the *pole–dipole* and *dipole–dipole arrays*, the dipole spacing is usually set as an integer multiple (n) of the dipole length, and for the dipole–dipole array the current and potential dipoles are usually the same length. In the *Schlumberger array* the gradient of the potential at the midpoint of the current dipole is required in order to maximise resolution of horizontal layers, so the potential dipole needs to be much smaller than the current dipole. The *gradient array*, an extension of the Schlumberger array, allows surveying on lines parallel to the current dipole. Figure 5.40 shows for each array the nominal location for assigning the measurement, relative to the electrodes, and the pseudo-depth (Z) for plotting the data.

The gradient array is commonly used for mapping. All the other arrays can be used for profiling and soundings, and maps can be created by combining adjacent traverses.

As described in Section 5.6.2, resistivity is calculated from the current and the measured potential, and a geometric factor is applied to compensate for the arrangement of the electrodes. Geometric factors obtained from Eq. (5.18) for the arrays described here are shown in Table 5.2.

5.6.5.2 Selecting an array

The key characteristics of an electrode array are its lateral and vertical resolutions, depth of investigation, ease with which data can be acquired, ease with which its responses can be interpreted, signal-to-noise ratio (see Section 2.4) and EM-coupling characteristics (see Section 5.6.7.2).

Geometrically symmetrical arrays, i.e. the symmetry of the electrode layout about the midpoint, produce a symmetrical response along the profile over a source with symmetrical geometry and are easier to interpret. Maximising what may be small signal amplitude at the potential dipole is also an important issue. All of these characteristics can vary widely between the different arrays, and for each array they vary according to the electrical structure of the subsurface.

Increasing the current proportionally increases the amplitude of the measured potential so currents as high as a few tens of amperes may be used in IP surveys. The use of computer modelling during survey design can assist in evaluating the nature of the responses from the various arrays and help in selecting the most appropriate array for the targeted subsurface structures.

Surveying with the pole–pole and pole–dipole arrays is fast, the latter requiring slightly more effort to move the third electrode. Both arrays find application in reconnaissance work. The dipole–dipole array also requires more complex logistics, but it produces good lateral resolution of steeply dipping features, although signal amplitude may be low. The array is commonly used for detailed work with the dipole spacing parameter (n) ranging typically from 1 to about 8, although larger spacing is sometimes used

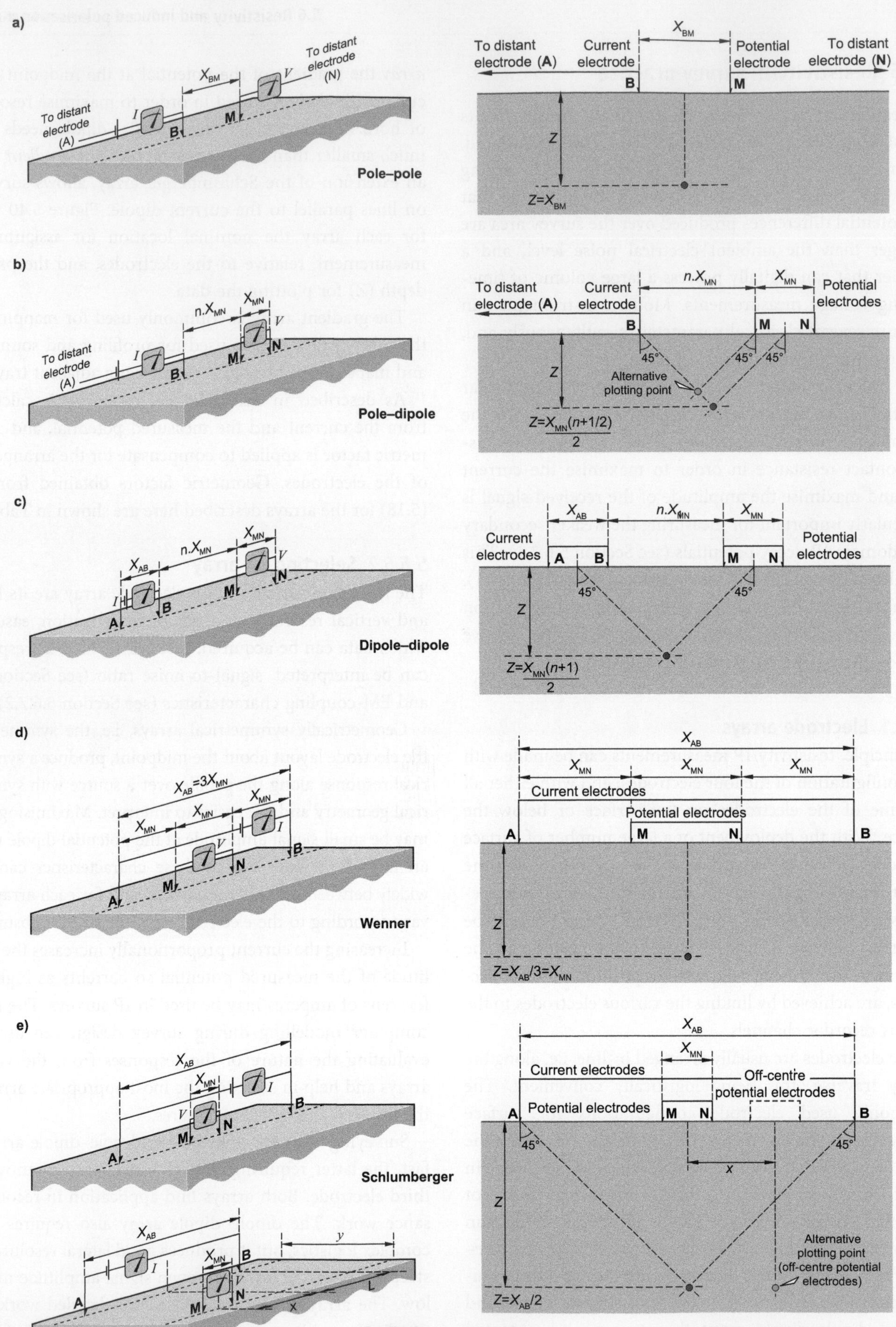

Figure 5.40 The commonly used electrode arrays and plotting conventions for soundings and pseudosections: (a) pole–pole, (b) pole–dipole, (c) dipole–dipole, (d) Wenner, (e) Schlumberger and gradient arrays.

despite the diminishing increase in depth penetration. EM coupling (see Section 5.6.7.2) is small for the two laterally offset dipoles.

For the gradient array, the large current dipole requires some time and effort to lay out; but once set up, a large volume of the ground is energised and the large parallel-field region below the central part of the current dipole allows a large area to be surveyed rapidly. The uniform horizontal subsurface current flow in the measurement region produces simple anomaly shapes. The response of a steeply dipping target much thinner than the potential dipole length is generally poor. The current dipole can be relocated along strike and further surveys conducted to provide continuous coverage of a large area. The gradient array is highly prone to EM coupling, a problem particularly in highly conductive areas (see Section 5.6.7.2).

Potential measurements in the Wenner and Schlumberger arrays are made in the large parallel-field region of the current dipole, which allows good resolution of horizontally layered electrical structures. Consequently, both arrays are mainly usually used for vertical electrical soundings (see Section 5.6.6.1). The Wenner array requires more complex logistics as all of the equispaced electrodes are moved.

In general, electrode arrays that allow rapid surveying require less manpower and are therefore favoured for reconnaissance work, but they provide limited depth information. Conversely, those arrays providing more resolution and depth information, where measurements are made with multiple electrode spacing, are slower and require more manpower, adding to survey cost. The benefits of the additional information cannot be overemphasised as this is essential for reliable interpretations.

5.6.5.3 Downhole surveying

It is necessary to distinguish between downhole electrical surveys, where any of the electrode arrays described for surface surveying are deployed within drillholes, and electrical logging where the electrodes are within an instrument, a sonde, which makes continuous readings as it is raised through the drillhole. An essential requirement for downhole electrical work is that the drillholes contain water to ensure electrical connection between the electrodes and the wall rocks. Electrodes for downhole surveying are described in Section 5.4.1.

Downhole surveying investigates the region beyond the drillhole environment. Some of the electrodes may be located downhole (hole-to-surface array) or all of them located beneath the surface, and either in the same drillhole (in-hole array) or distributed between several drillholes (hole-to-hole array). This places them closer to the target zone, increasing the response of the target considerably and increasing the resolution of close-spaced targets. This is particularly useful in areas with conductive overburden (see Section 5.6.7.1). It also allows greater flexibility in focusing the survey to a particular depth and in a particular direction. A common downhole surveying technique is the applied potential method described in Section 5.6.9. Descriptions of other forms of drillhole resistivity/IP surveys are provided by Mudge (2004).

Downhole logs of both resistivity and IP parameters are often acquired in metalliferous environments (see Section 5.6.8). Resistivity-only measurements are routinely made in drillholes, and the most common electrode configurations are the '*normal*' and '*lateral*' arrays shown in Figs. 5.41a and b. A third form of downhole surveying measures *single-point resistance.* Here one current electrode is located at the

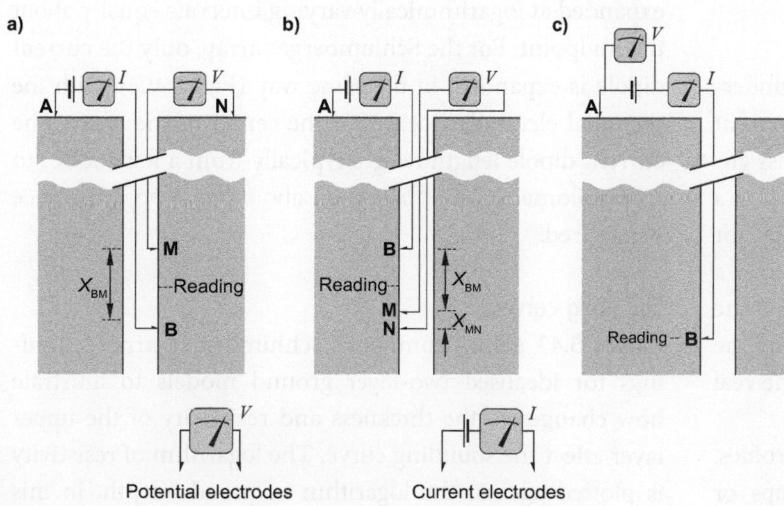

Figure 5.41 Some common downhole logging arrays. (a) Normal array, (b) lateral array and (c) single-point resistance array. See Table 5.2 for geometric factors.

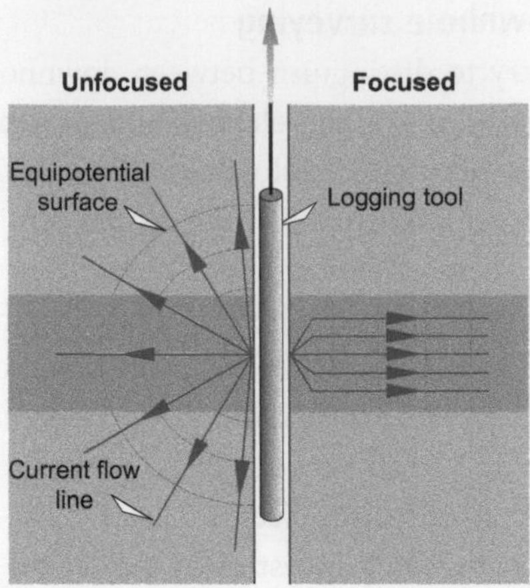

Figure 5.42 Schematic illustration of the 'focused' potential field and current flow lines of a focused logging array compared with that of an unfocused array.

surface, the other within the drillhole (Fig. 5.41c). The overall resistance of the circuit, comprising the electrical equipment, wires and the current paths from drillhole wall to surface, is measured. Since this is to a large extent a function of the conditions in the drillhole, the data can only be interpreted qualitatively. A more sophisticated form of electrical logging uses potential electrodes to 'focus' the current flow into a thin disc centred on the drillhole (Fig. 5.42). This increases the along-hole resolution of the data, but requires more sophisticated logging equipment. There are numerous logging tools of this type, with varying resolutions and depths of penetration, developed by the petroleum industry for well-logging in sedimentary environments. Electrical logging of coal seams is a major application in the mining industry.

5.6.6 Display, interpretation and examples of resistivity/IP data

When interpreting electrical data it is essential to understand that individual resistivity/IP measurements represent an average of a large volume of material and that assignment of a measurement to a point in the subsurface is a matter of convenience rather than physical reality. Only for the case of an electrically homogeneous subsurface, i.e. a half-space, is the measurement a true representation of the actual subsurface properties. For all other situations the measurement represents a gross simplification of the real situation.

One-dimensional data are presented as profiles. Two-dimensional data comprise pseudo-depth maps or pseudosections which may be displayed as images or contours. The true form of the subsurface electrical structure is usually not easily recognisable in the various displays. As a consequence, modelling, and in particular inverse modelling (see Section 2.11.2.1), is an integral part of resistivity/IP interpretation.

Since depth control of the electrical measurements is of profound importance in their interpretation, we first describe the interpretation of electrical soundings before describing the analysis of profile and pseudosection data, the latter having elements of both profiling and soundings. Finally we describe analysis of map data. Our description focuses on resistivity data but the same principles are applicable to IP data. There are a number of pitfalls commonly encountered in the interpretation of all types of electrical data which are discussed in Section 5.6.7.

Increasingly resistivity/IP data are being acquired using a large number of electrodes deployed in a grid to map a volume of the subsurface. For example, White *et al.* (2001) describe a survey over a sulphide-rich copper–gold porphyry body, which allowed a detailed 3D model of the deposit to be created.

5.6.6.1 Vertical electrical soundings

Vertical electrical soundings are surveys that specifically seek to define the vertical variation in electrical properties. As noted in Section 5.6.4.1, soundings can be made with any array by focusing the measurement to deeper regions. The measured parameters are plotted against some function of the electrode separation, i.e. the pseudo-depth (Z) in Fig. 5.40.

The Wenner and Schlumberger arrays are commonly used for soundings (see Section 5.6.5). In the Wenner array (Fig. 5.40d), the spacing (X_{MN}) between all the electrodes is expanded at logarithmically varying intervals equally about the midpoint. For the Schlumberger array, only the current dipole is expanded, in the same way (Fig. 5.40e), with the potential electrodes located at the centre of the array. The current dipole length ranges typically from a few metres to a few kilometres, if information about the deep subsurface is required.

Sounding curves

Figure 5.43 shows computed Schlumberger array soundings for idealised two-layer ground models to illustrate how changes in the thickness and resistivity of the upper layer affect the sounding curve. The logarithm of resistivity is plotted against the logarithm of pseudo-depth, in this

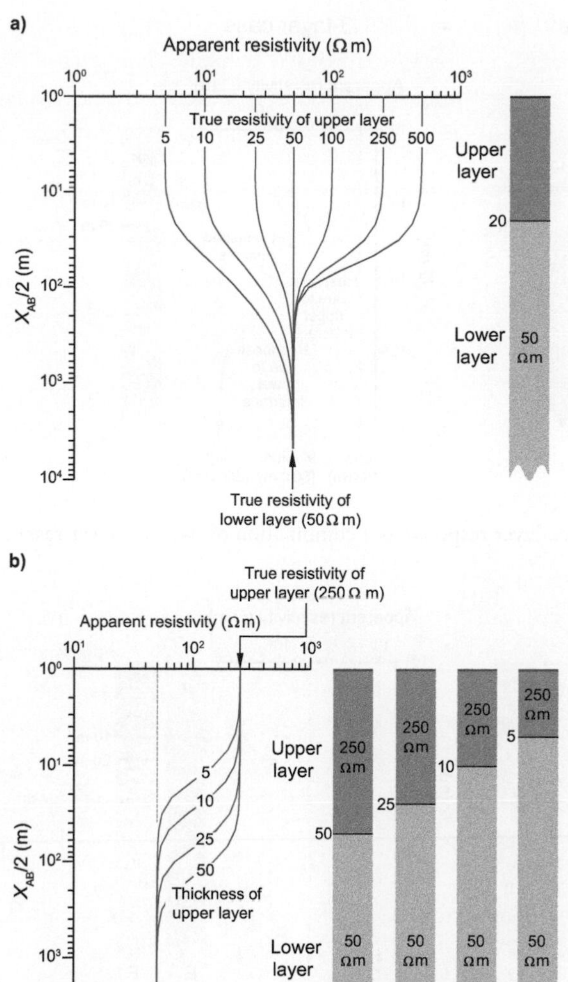

Figure 5.43 Schlumberger array VES resistivity curves. The curves are computed for a two-layer ground showing the change in response for (a) a range of upper layer resistivity with constant thickness, and (b) a range of upper layer thickness with constant resistivity.

case half the current dipole length ($X_{AB}/2$). The measured resistivity at small electrode separations approaches that of the upper layer, because then the current is chiefly confined to the top layer. As the array expands the influence of underlying layers on the measurements increases, so that at larger separations the measured resistivity approaches that of the lower layer. Clearly then, it is easy to identify the relative resistivities of the two layers in the sounding curve, but the depth to the interface between them is another matter. Notice how the change in the curve occurs at increasingly greater pseudo-depths as the upper layer resistivity falls below that of the lower layer (Fig. 5.43a). This is simply due to more of the current flow taking the path-of-least-resistance, preferring to flow in the lower-resistivity upper layer. When the resistivities are kept constant and the thickness of the upper layer varied, the curve behaves as would be expected and the change in apparent resistivity occurs at increasingly larger $X_{AB}/2$ as depth increases (Fig. 5.43b).

Responses from three (or more) layers can be thought of as a series of superimposed two-layer responses (Fig. 5.44). The component two-layer responses, affected by the relative resistivities and interface depths, combine to produce a sounding curve that less resembles the resistivities, and, in particular, the thicknesses of the various layers. Figure 5.45 shows sounding curves for the six (A–F) possible combinations of three-layer relative resistivities. The true resistivities of both the upper and lower layers can be fairly accurately determined from the soundings but information about the middle layer is elusive. In particular, when the resistivity of the middle layer is intermediate between those of the upper and lower layers (models E and F) the response is effectively identical to the two-layer case, so the middle layer can be transparent unless it is particularly thick. This is an illustration of a limitation of electrical soundings; they are ambiguous (see Section 2.11.4). Put simply, the same set of observations may be created by significantly different electrical layering, with obvious consequences for interpretations.

There is an extensive literature on the interpretation of electrical soundings, and forward and inverse modelling software are widely available. These are usually based on a 1D model (see *One-dimensional model* in Section 2.11.1.3) which assumes that the electrical structure of the ground is a series of horizontal layers of constant resistivity and chargeability and constant thickness. Many algorithms attempt to account for ambiguity in the interpretation by producing a range of possible solutions.

Departure from a 1D electrical structure is a serious problem for electrical soundings. This is likely to occur at larger array dimensions because the assumption of homogeneous horizontal layers often breaks down in the larger volume of ground energised by the array. Lateral variations in electrical properties and topography cause spurious inflections in the sounding curve, which can be mistaken for layering (Pous *et al.*, 1996).

A common application of electrical soundings in mineral exploration is investigation of the overburden, especially its thickness and internal variation in electrical properties, often to facilitate an understanding or, if

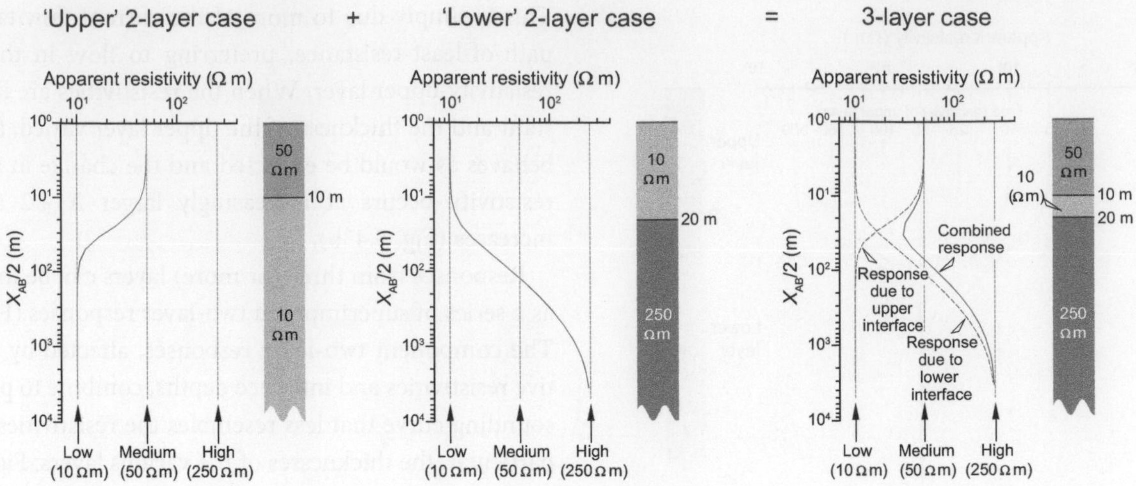

Figure 5.44 Computed Schlumberger array VES resistivity curves showing a three-layer response as a combination of two two-layer responses.

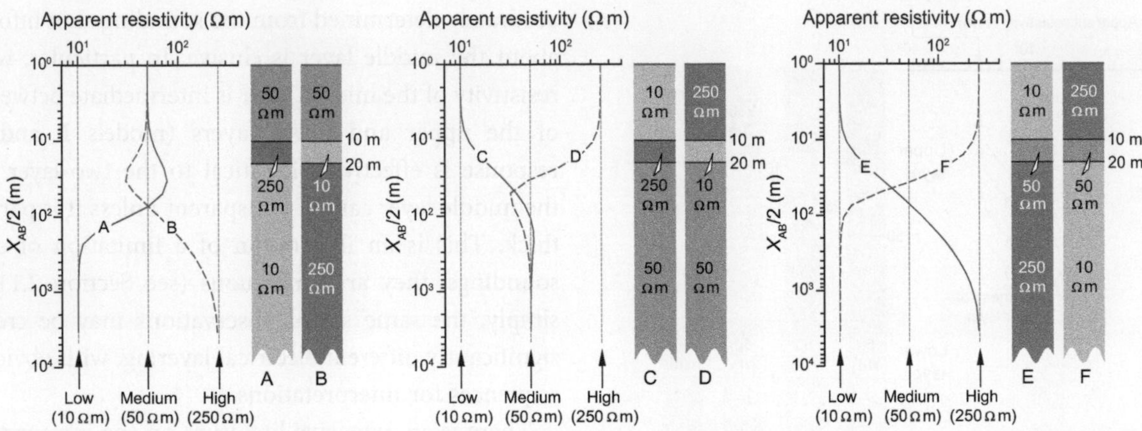

Figure 5.45 Schlumberger array VES resistivity curves computed for a three-layer ground illustrating the various forms of the responses for the six (A–F) possible combinations of relative resistivity.

possible, removal of its effects from other forms of geophysical data.

Example – Palmietfontein kimberlite

Resistivity surveys in and around the Palmietfontein kimberlite pipe, North West Province, South Africa, illustrate their application to mapping the extent and internal structure of kimberlites (da Costa, 1989). The Palmietfontein kimberlite is large for a kimberlite, covering an area of about 12 ha. There has been little erosion of the pipe and the yellow ground, blue ground and hardebank are all preserved. These layers have distinctly different physical properties. The yellow ground, being hydrated and oxidised kimberlite, is mainly composed of clay minerals and has the lowest resistivity of the three layers. The fresh rock comprising the hardebank is the most resistive, and the partially weathered and serpentinised blue ground is of intermediate resistivity. Clearly, the depth of erosion is expected to be a fundamental influence on the electrical responses of kimberlites. In addition to responses from the kimberlite itself, preferential erosion often causes the pipes to be overlain by a layer of young sedimentary materials, which may also produce geophysical responses useful for locating kimberlites.

Vertical electric soundings were made within and outside the pipe to investigate changes in electrical properties with depth (Figs. 5.46c and d). The Schlumberger array was used with a maximum current electrode separation (X_{AB}) of 280 m for the sounding centred within the pipe (VES1), with the sounding of the country rock outside the pipe (VES32) made to a separation of 1200 m. Both sounding curves show a basic three-layer medium–low–high response (cf. Fig. 5.45, curve B) and were interpreted using 1D models. The country rock sounding was

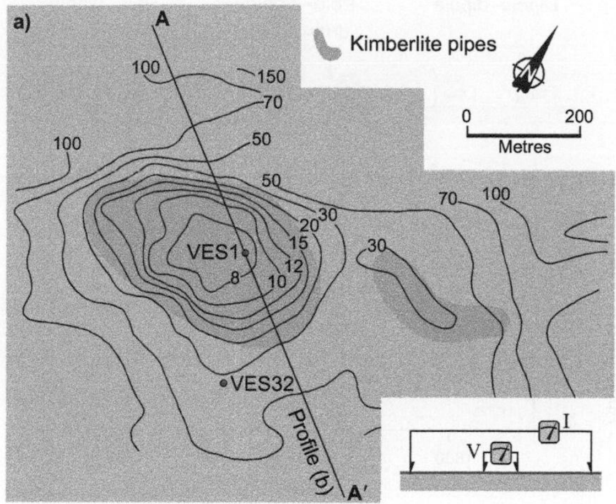

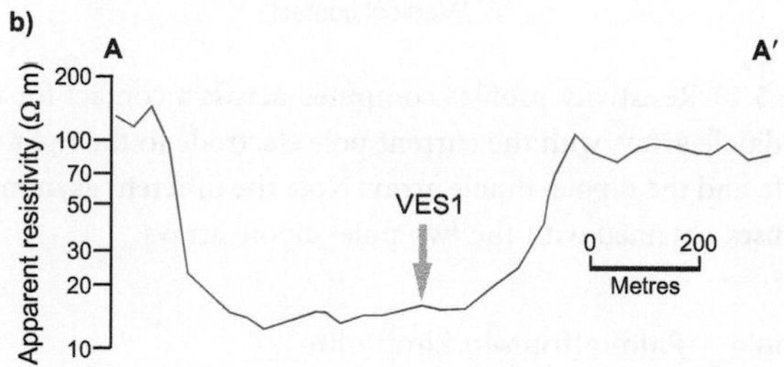

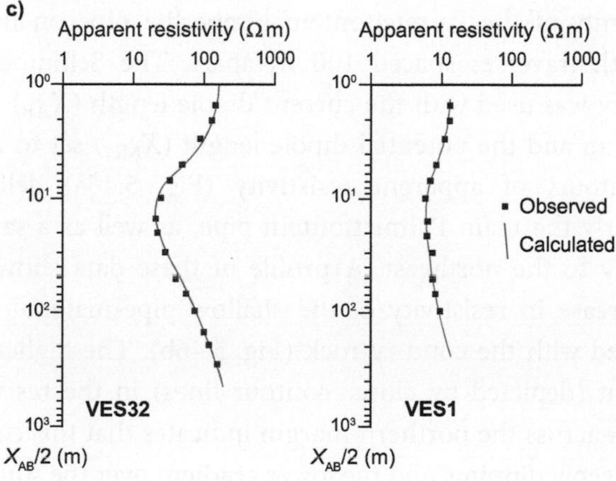

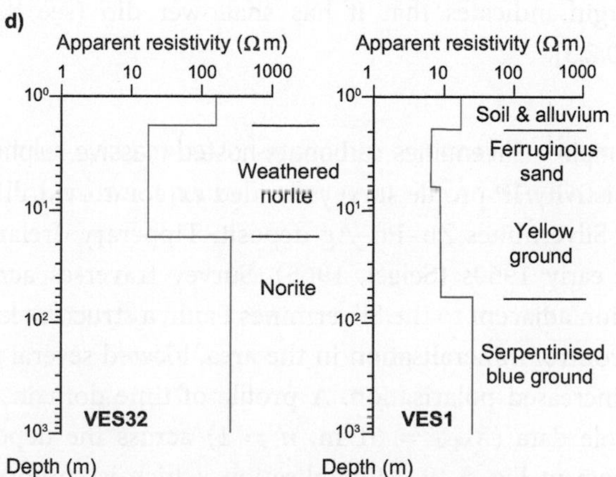

Figure 5.46 Schlumberger array resistivity data from the Palmietfontein kimberlite pipe. (a) Contours of apparent resistivity (Ω m) from profile surveys. Locations of profile A–A′ and vertical electrical soundings VES1 and VES32 are shown. (b) Resistivity profile A–A′, (c) sounding curves VES1 and VES32 and (d) their interpretations. See text for details. Based on diagrams in da Costa (1989).

interpreted to be due to a thin layer of resistive surficial material (thickness = 1.8 m, ρ = 160 Ω m) overlying weather bedrock (thickness = 16 m, ρ = 17 Ω m) and underlain by fresh bedrock (ρ = 260 Ω m). The sounding from within the pipe was modelled using four layers, although the middle two are similar in resistivity, which gives the appearance of a three-layer ground. The upper layer was interpreted as surficial material (thickness = 1.9 m, ρ = 13.5 Ω m). The next layer was also interpreted as surficial material, specifically as ferruginous sand known to occur in the area (thickness = 4.3 m, ρ = 5 Ω m). This is underlain by the yellow ground (thickness = 52 m, ρ = 7 Ω m), which is underlain by the more resistive blue ground (ρ = 20 Ω m). The outer points of the sounding were not modelled since the current electrodes were located outside the pipe, violating the 1D assumption inherent in the model. The higher resistivity of the country rock causes an increase in apparent resistivity in this section of the sounding curve.

The interpreted depth to blue ground is 58 m and is within about 20% of the actual depth of these rocks, intersected at 44 m and 45 m in nearby drillholes. The gradational nature of the contact is probably the cause of this large discrepancy. The contrasting resistivities of the various components of the kimberlite are typical and are the reason for the routine use of electrical and EM methods in kimberlite exploration (Macnae, 1979).

5.6.6.2 Profiles

In profiling mode, the survey line and the array are usually oriented across the geological strike and the array moved systematically along the line to produce a profile of the subsurface resistivity and polarisation parameters. Profiling surveys can be conducted with any of the electrode arrays described in Section 5.6.5. The spacing or the length of the dipoles can be varied in order to change the investigation depth.

Examples of profiles for the commonly used pole–dipole and dipole–dipole arrays are shown in Fig. 5.47. The data are presented as profiles of electrical parameter

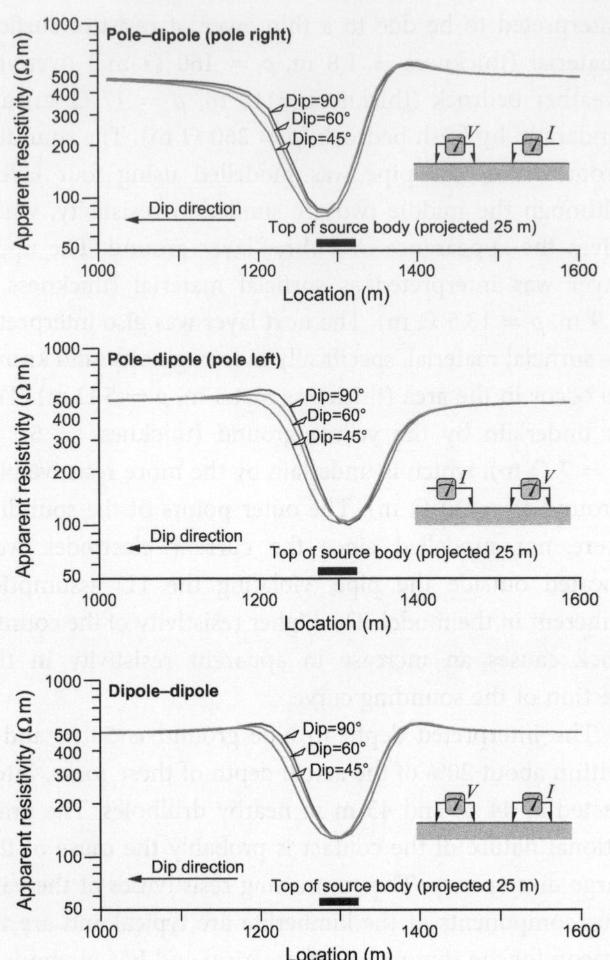

Figure 5.47 Resistivity profiles computed across a conductive body, for a range of dip, for the asymmetrical pole–dipole array with the current pole-electrode located to the right and to the left of the potential dipole, and for the symmetrical dipole–dipole array.

versus array location (see Fig. 5.40). Qualitative interpretation is quite simple and concentrates on identification of anomalous responses. Difficulties occur owing to the asymmetrical responses produced by geometrically asymmetrical arrays, as shown by the computed responses in Fig. 5.47 for the different current pole–electrode locations in the pole–dipole array. It can be difficult to interpret dips reliably from these. Note from the figure that for the symmetrical dipole–dipole array the asymmetric response is due only to the dip of the body. Other difficulties include complex responses at contacts between units with different electrical properties (Fig. 5.48) and the fact that the depth of penetration varies, being less in conductive ground. In the contact example, the depth of investigation is much less in the more conductive ground than in the more resistive terrain on the other side of the contact.

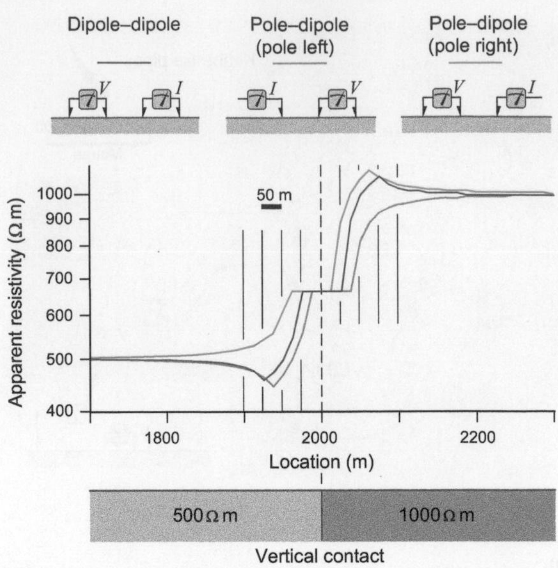

Figure 5.48 Resistivity profiles computed across a contact for the pole–dipole array, with the current pole electrode to the right and to the left, and the dipole–dipole array. Note the different asymmetrical responses obtained with the two pole–dipole arrays.

Example – Palmietfontein kimberlite

Resistivity profiling surveys were also conducted in the vicinity of the Palmietfontein kimberlite pipe on north–south traverses spaced 100 m apart. The Schlumberger array was used with the current dipole length (X_{AB}) set to 150 m and the potential dipole length (X_{MN}) set to 20 m. Contours of apparent resistivity (Fig. 5.46a) delineate clearly the main Palmietfontein pipe, as well as a satellite body to the northeast. A profile of these data shows the decrease in resistivity of the shallow pipe-material compared with the country rock (Fig. 5.46b). The higher gradient (depicted by closer contour lines) in the resistivity data across the northern margin indicates that this contact is steeply dipping, and the lower gradient over the southern margin indicates that it has shallower dip (see Section 2.10.2.3).

Example – Silvermines carbonate-hosted massive sulphide

Resistivity/IP profile surveys guided exploratory drilling of the Silvermines Zn–Pb–Ag deposit, Tipperary, Ireland, in the early 1960s (Seigel, 1965). Survey traverses across a region adjacent to the Silvermines Fault, a structure known to control mineralisation in the area, located several zones of increased polarisation. A profile of time domain pole–dipole data (X_{MN} = 61 m, n = 1) across the deposit is shown in Fig. 5.49. Mineralisation, which is conformable with dip, occurs in gently dipping dolomitic limestone adjacent to the Silvermines Fault. Both massive and disseminated sulphides are present, the latter containing a high percentage of pyrite. Anomalously high apparent resistivity and chargeability occur over the deposit, with the maximum chargeability and minimum resistivity occurring over the shallowest part of the mineralisation. A subsidiary chargeability response further to the north is probably due to pyrite in an overlying chert horizon. Apparent chargeability profiles obtained with the electrode spacing set to 30 m and 122 m (changing both the dipole length and spacing) are shown in Fig. 5.49b. Note the change in lateral resolution and response amplitude for the different electrode separations.

5.6.6.3 Pseudosections

Measurements from multiple electrode separations on the same traverse can be displayed and analysed as individual profiles, as shown in Fig. 5.49. More conveniently, the data can be displayed and analysed as a single entity in the form of a pseudosection (see Section 2.8.1). The lateral position and pseudo-depth of the measured values in the pseudosection conform to the data plotting convention for each array (see Fig. 5.40).

We describe here the pseudosections produced only by the dipole–dipole array; similar procedures and types of characteristics apply to the other in-line arrays. Plotting of the pseudosection is illustrated in Fig. 5.50. The measurement from each dipole spacing (n) is assigned to the point midway between the two active dipoles and located at the intersection of two lines subtending an angle of 45° from the surface midpoint of the current and potential dipoles. Larger values of n correspond to greater depth of investigation and are therefore plotted in lower regions of the pseudosection. Measurements for the same value of n plot on the same horizontal line. Their lateral position is determined by moving the two dipoles laterally whilst maintaining their separation (Fig. 5.50b), so the horizontal spacing of the plotted data is equal to the dipole length (X_{MN}).

Figure 5.51 shows the responses of several simple models of subsurface resistivity variations for the dipole–dipole array displayed as pseudosections. It is obvious that the source geometries are not replicated by the pseudosections. The pseudosection response, in general, takes the form of an inverted 'V' shape. This is referred to as a '*pants-legs*' response. The reason for the pants-legs response is shown in Fig. 5.52. The zone with anomalous electrical properties below the current (or potential) electrodes affects readings regardless of the location of the other electrodes. The plotting of the responses at locations extending diagonally downwards results in the two pants legs.

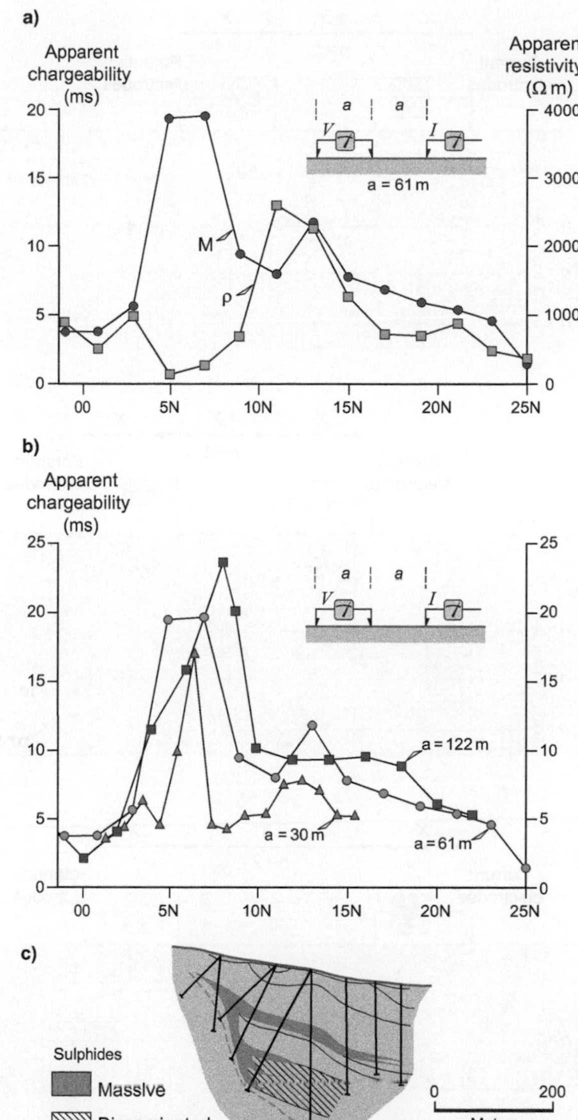

Figure 5.49 Silvermines Zn–Pb–Ag deposit. (a) Pole–dipole array apparent resistivity and chargeability profiles from the Silvermines Zn–Pb–Ag deposit. (b) Chargeability profiles from the same survey line for a range of dipole length and spacing. Note the change in amplitude and lateral resolution between the profiles. Redrawn, with permission, from Seigel (1965).

Referring to Fig. 5.51, it is possible to determine some characteristics of the source from the pseudosection. For example, a vertical source produces a broader and smoother pants-legs response as its depth increases. Dip produces an asymmetric response, although determining the dip direction can be problematic since the controls on the nature of the asymmetry are complicated. Responses

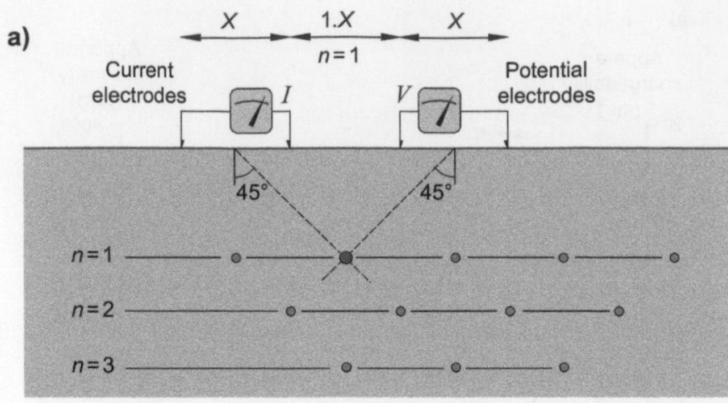

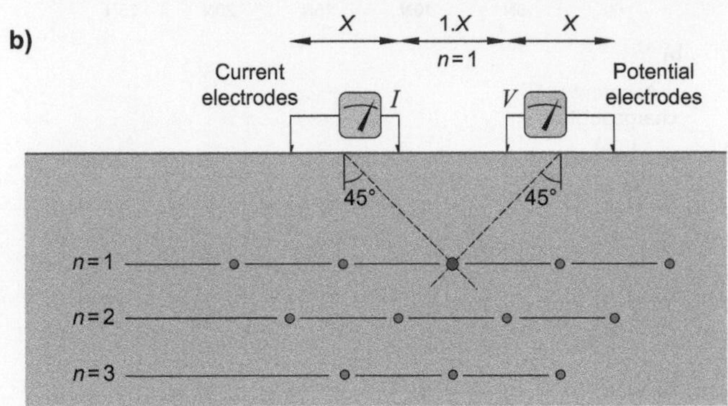

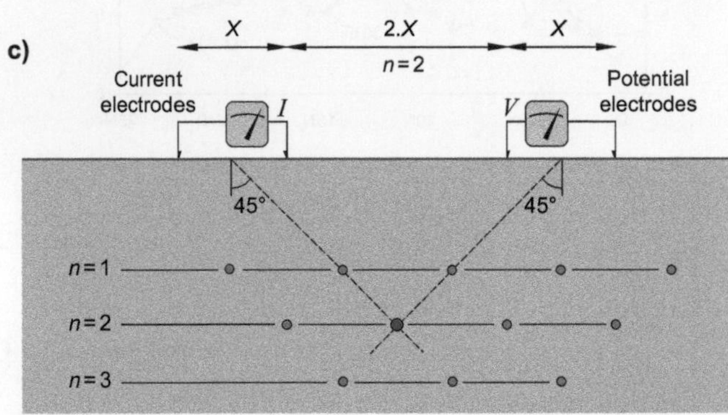

Figure 5.50 Creation of a pseudosection for the dipole–dipole array. A grid of data points is created by varying the positions and spacings (n) of the current and potential dipoles. See text for detailed description.

from two adjacent sources overlap and may result in the maximum response occurring at a location which coincides with neither body. Two closely spaced sources may appear like a single broader source. This is an important observation since it may lead to erroneous positioning of drillholes. A resistivity contact, representing a geological contact also produces a pants-legs response. One 'leg' is of 'high' values extending across the zone of higher resistivity, and the other 'leg' is of 'low' values extending across the lower resistivity side of the contact. The surface location of the contact is somewhere between the two sloping 'legs'.

In addition to the above complications, the location of the electrodes with respect to the anomalous zones has a significant effect on the measured response, particularly the shallower zones in the upper region of the pseudosection. Also, data from a survey line in rugged terrain will be further distorted when projected onto the straight line pseudosection plot. It cannot be over-emphasised that a pseudosection is not a true 'picture' of the electrical structure of the subsurface, but a 'mathematical abstraction' that requires interpretation and modelling in order to transform it into a (hopefully) true depth section of the subsurface. Interpreting pseudosections without the aid of model responses can produce unreliable results.

Inverse modelling (see Section 2.11.2.1) is routinely used in the interpretation of resistivity/IP pseudosections (Oldenburg and Li, 1994). Despite the electrical complexities of the ground, and the various inherent resolution limits of the various survey arrays and the interpretation techniques, many complex subsurface electrical distributions can be reduced to simple but plausible models allowing identification of target zones. Limitations imposed by mathematical complexities also affect the accuracy of models, but non-uniqueness predominates (see Section 2.11.4).

In the following examples demonstrating the application and interpretation of resistivity/IP pseudosections, the original pseudosection data have been modelled using the inversion algorithm of Oldenburg and Li (1994). Terrain has been accounted for where necessary but otherwise the inversions were unconstrained.

Example – Estrades volcanogenic massive sulphide

The Estrades Cu–Zn–Au deposit, Quebec, Canada, is described by Bate *et al.* (1989). The deposit occurs in an Archaean greenstone belt within the Abitibi Subprovince and consists of a series of sub-cropping precious metal-bearing massive sulphide lenses within a felsic pyroclastic unit. The host sequence comprises a sub-vertical volcanic sequence comprising mainly mafic to intermediate lavas. The massive sulphide body, which is conformable with the stratigraphy, reaches 5 m in width and contains pyrite, chalcopyrite and sphalerite. Disseminated sulphides, mainly pyrite, occur in the footwall. The deposit occurs beneath 12 to 40 m of glacial overburden. This example clearly demonstrates how multiple lenses of polarisable mineralisation (the sulphides) can produce a fairly simple IP response. The rather accurate multi-body model can only be developed now because of the large amount of

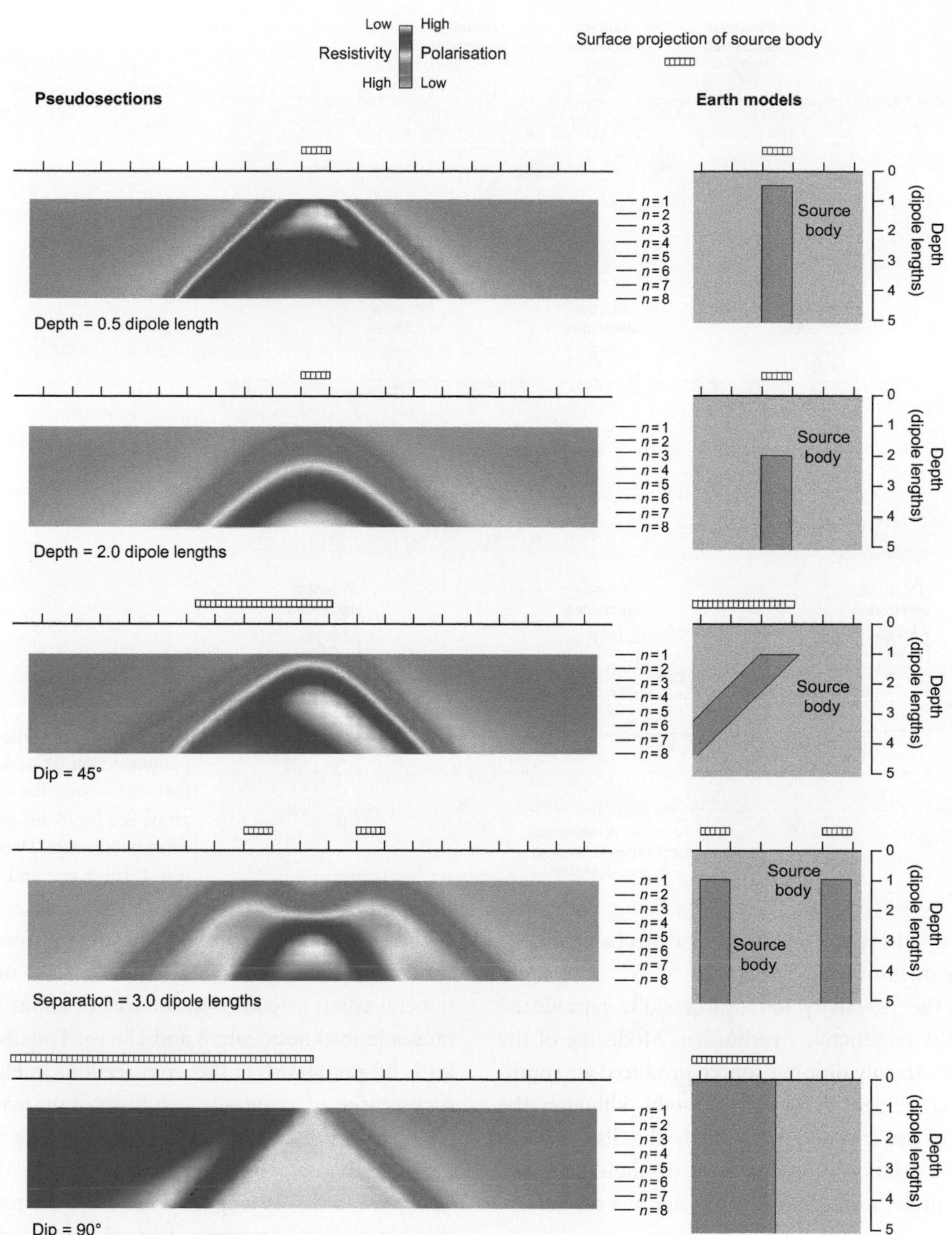

Figure 5.51 Computed dipole–dipole array resistivity responses of some simple 2D conductivity models. Horizontal scale is one dipole length per division.

drillhole geology available. In addition it is a good example of conductive overburden response in resistivity data.

The deposit was discovered using EM methods and investigated further with a time domain resistivity/IP survey using a pole–dipole array (Fig. 5.53). The apparent resistivity pseudosection shows high conductivity at low *n* values. This is the typical response of conductive overburden (see below), with the lateral variations usually reflecting changes in overburden thickness and/or resistivity. However, in this area they may also be due to the conductive mineralisation being in contact with the overburden (see Section 5.3.4). In contrast, the apparent chargeability pseudosection has no obvious relationship to the overburden. It shows a clear positive 'pants-legs' anomaly

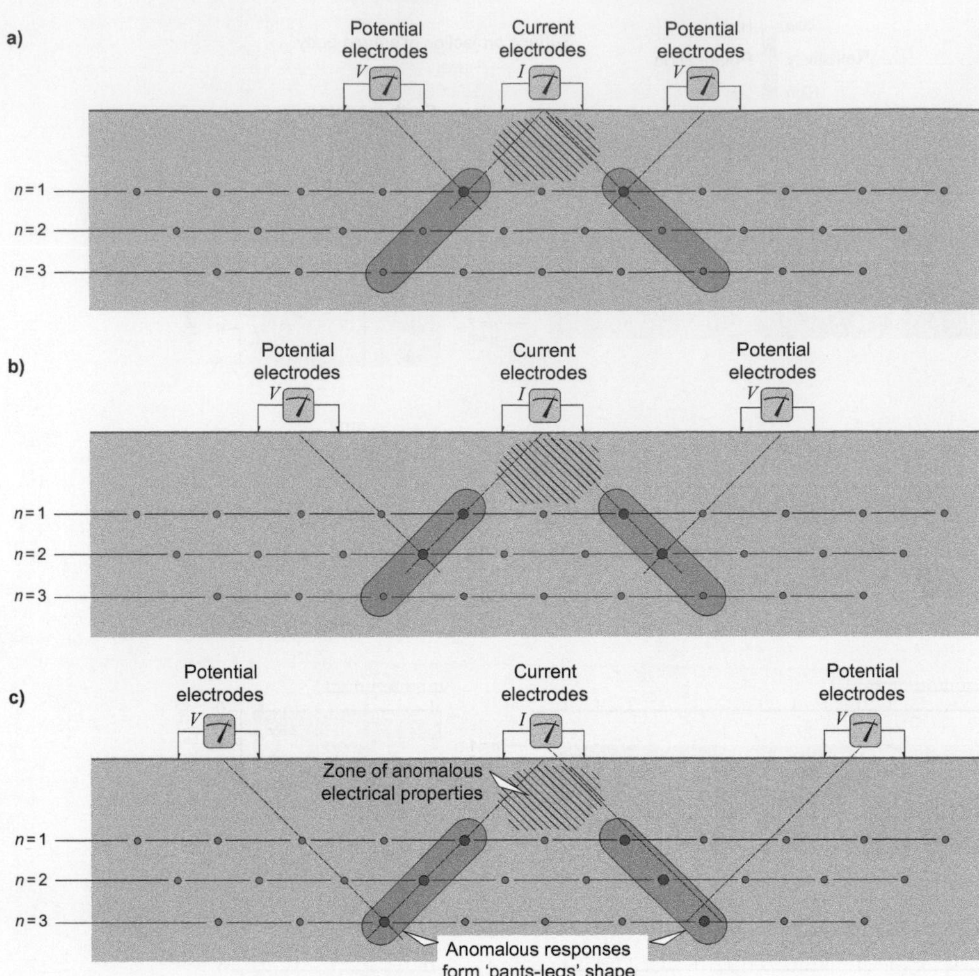

Figure 5.52 Explanation of the pseudosection pants-legs response illustrated using the dipole–dipole array. See Fig. 5.40c for details of the array parameters. Data shown for (a) $n = 1$, (b) $n = 2$ and (c) $n = 3$.

located directly above the ore zone and probably chiefly caused by the disseminated sulphides.

Modelling the resistivity data successfully reproduced the form of the conductive overburden. Modelling of the IP data using a steeply dipping source produced an anomaly roughly coincident with the orebody, although the main source of the anomaly is probably the disseminated footwall mineralisation. Nevertheless, drillholes targeted using the IP model would have intersected the ore zone.

Example – Red Dog black-shale hosted massive sulphide

Located in Alaska, USA, the Red Dog deposit is a shale-hosted polymetallic massive sulphide containing reserves of lead, zinc and silver (Moore *et al.*, 1986). Geophysical surveys across the deposit included resistivity/IP and are described by Van Blaricom and O'Connor (1989). This example illustrates the response of a shallow, flat-lying, target in the presence of significant topographic effects.

The principal sulphides present are sphalerite, pyrite, marcasite and galena, with silica and barite the main gangue. In addition to shale, the host sequence contains clastic sedimentary rocks and chert. The main zone of mineralisation extends over an area of about 1.5 km^2, and ranges in thickness from 8 and 158 m. The distributions of lead and zinc shown in the cross-sections in Fig. 5.54 reflect occurrences of economic sulphides, but pyrite will also contribute to the geophysical response. The time domain resistivity/IP data were acquired with the dipole–dipole array with a dipole length of 122 m. Both pseudosections show 'pants-legs' responses centred between stations 4000 and 5200. The strong responses in the lower n values indicate that the source is shallow. The width of the uppermost part of the anomaly indicated a comparatively wide source. The mineralisation occurs in a topographic low which can be expected to affect the data (see Section 5.6.7.3). However, modelling confirms the presence of a conductive and polarisable body having similar depth extent to that of the deposit. The lack of response to the west probably reflects the lower grades of mineralisation here, the best mineralisation occurring at station 4300.

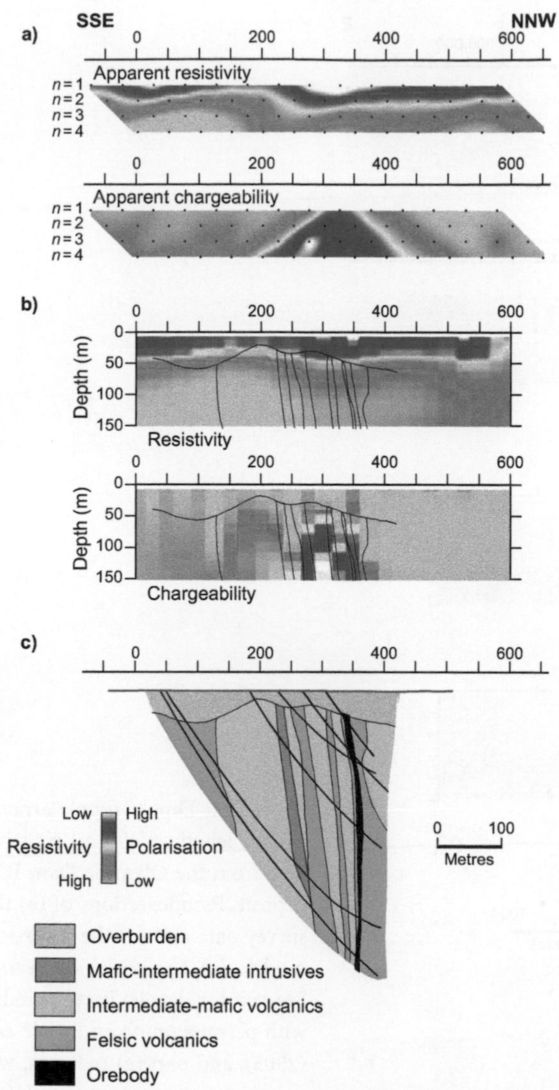

Figure 5.53 Pole–dipole array (dipole length = 50 m) resistivity/IP data from the Estrades massive sulphide deposit. Pseudosections of (a) the survey data and (b) the inversion models. (c) The geological section. See text for details. Based on diagrams in Bate *et al.* (1989).

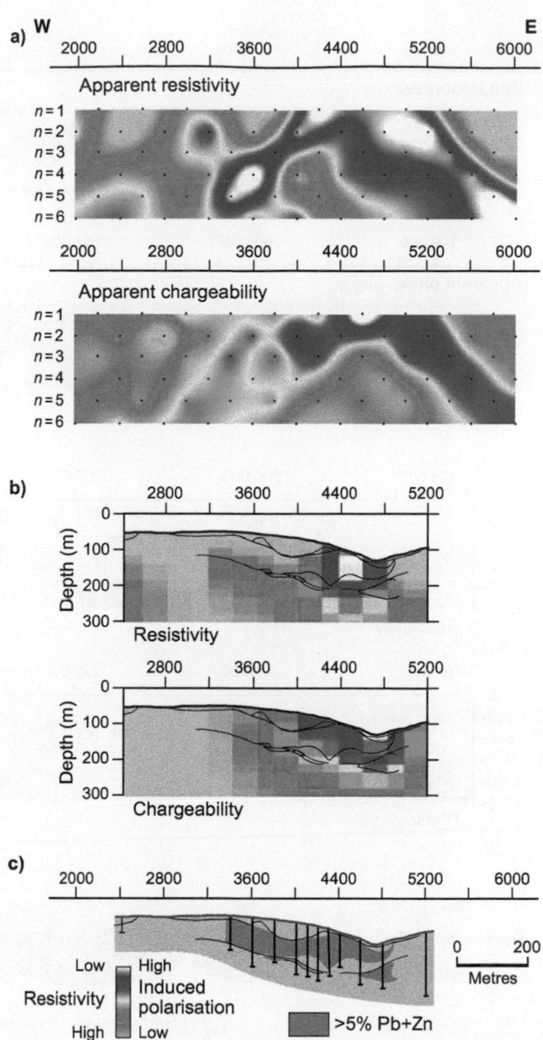

Figure 5.54 Dipole–dipole array (dipole length = 122 m) resistivity/IP data from the Red Dog massive sulphide deposit. Pseudosections of (a) the survey data and (b) the inversion models. (c) The geological section. See text for details. Based on diagrams in Van Blaricom and O'Connor (1989).

Example – Olympic Dam iron oxide copper gold deposit

The giant Olympic Dam Cu–U–Au–Ag and rare-earth-elements deposit, in South Australia, is located in a hydrothermal breccia complex beneath several hundred metres of unmineralised Proterozoic sedimentary rocks (Reeve *et al.*, 1990). Gravity and magnetic data, interpreted in the context of a conceptual geological model, led to its discovery. Resistivity/IP surveys were trialled after discovery; selected data are presented here as examples illustrating responses from a very deep target under overburden. The most comprehensive description of geophysical surveys at Olympic Dam is provided by Esdale *et al.* (2003).

Figure 5.55 shows frequency domain data acquired using the dipole–dipole array. Measurements of the phase polarisation parameter were made at multiple frequencies to allow removal of EM-coupling effects (see Section 5.6.7.2). The surveys were hindered by conductive drill casing but the data presented here were not overly affected. The phase data show a deep, broad, positive response impinging the lower n values, indicating that depth to the top of the source is about one dipole length, i.e. 400 m. The apparent resistivity data show a more focused anomaly, but again appearing in the shallowest part of the pseudosection. Inversion of these data defined a broad conductive body closely coincident with the Olympic Dam Breccia Complex. Depth to the top is accurately predicted by the

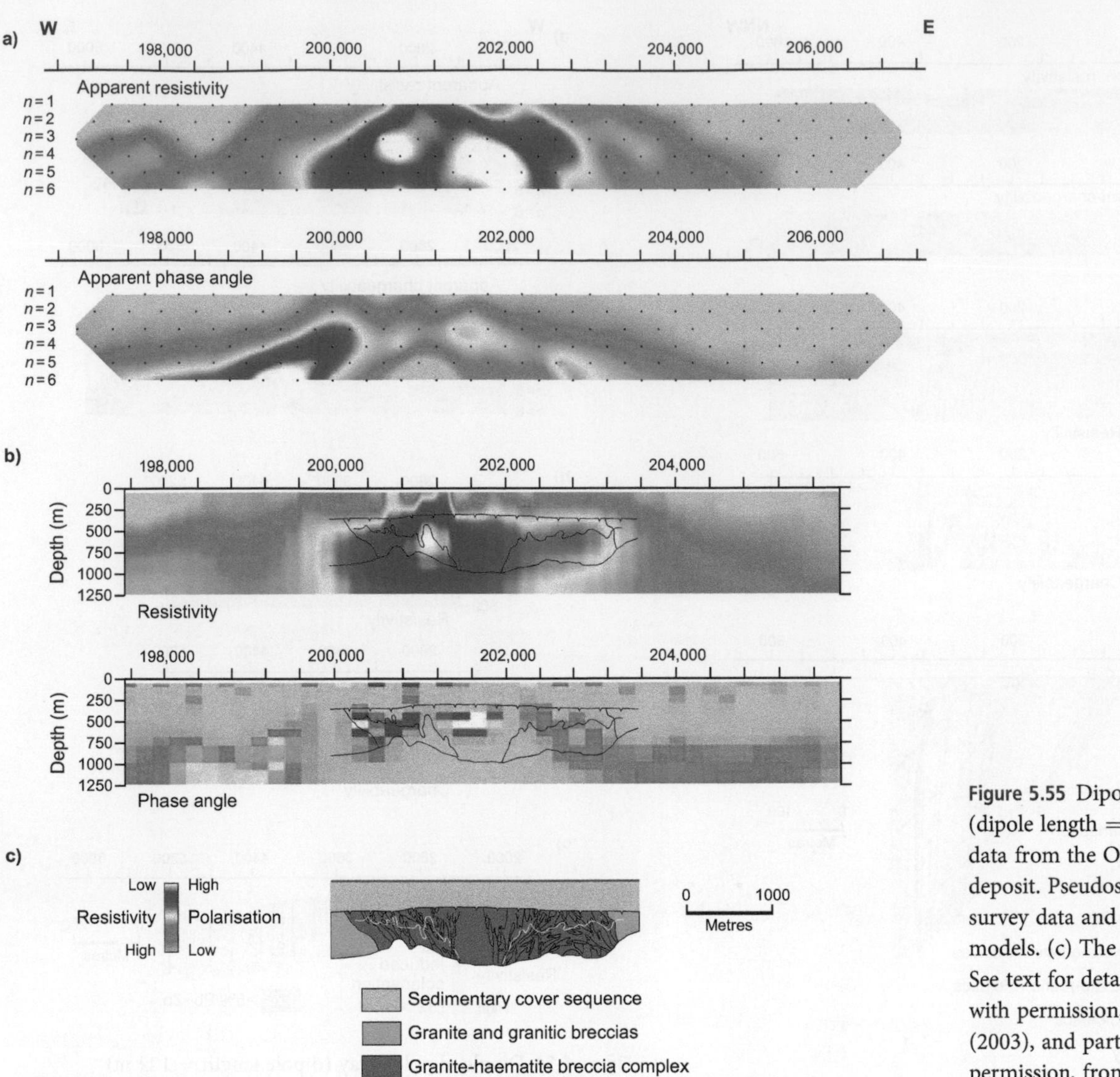

Figure 5.55 Dipole–dipole array (dipole length = 400 m) resistivity/IP data from the Olympic Dam IOCG deposit. Pseudosections of (a) the survey data and (b) the inversion models. (c) The geological section. See text for details. Part (a) redrawn, with permission, from Esdale *et al.* (2003), and part (c) redrawn, with permission, from Reeve *et al.* (1990).

model. Petrophysical measurements showed that zones of low resistivity are due to saline groundwater within highly porous haematitic breccias. The strongest IP sources occur within the conductive body. Their origin is uncertain but it may be haematite within the breccia complex. Clearly, resistivity/IP surveys are capable of defining responses from deep targets, but there is no response from the economic component of the Olympic Dam deposit.

5.6.6.4 Maps

The interpretation of electrical parameter maps mainly involves identification of anomalous responses. The gradient array is commonly used for mapping an area and anomalies are chiefly coincident with their source, with strike of the source corresponding with the orientation of the response, if elongated. Dip may be inferred from the nature of the gradients, steeper gradients being on the up-dip side of the source body (see Section 2.10.2.3). The following two examples demonstrate electrical mapping in mineralised terrains using the gradient array.

Example – Pine Point carbonate-hosted massive sulphide

The first example demonstrates the detection of sulphide ores in contoured chargeability data. The data are from the Pine Point region in the Northwest Territories, Canada, where there are numerous small carbonate-hosted zinc-lead deposits (Rhodes *et al.*, 1984). The host sequence is a Devonian reef complex, and there is glacial overburden with an average thickness of about 15 m. The orebodies consist of sphalerite, marcasite, galena, pyrite and sometimes pyrrhotite. Gangue is calcite and dolomite. The ores are not electrically conductive and do not produce

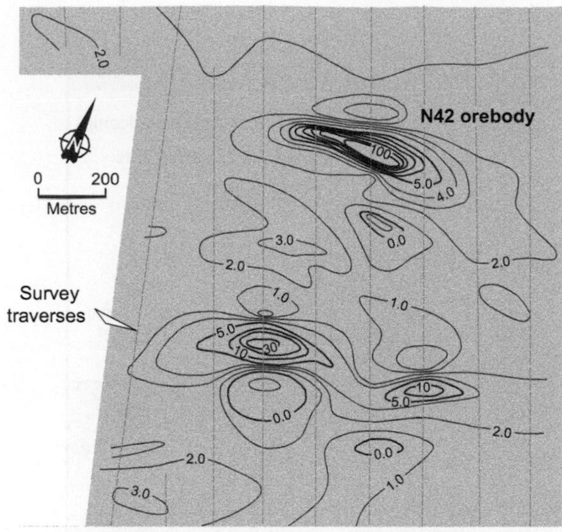

Figure 5.56 Contours of chargeability (ms) from gradient array resistivity/IP survey of the Pine Point MVT deposits (potential dipole length = 61 m). Redrawn, with permission, from Lajoie and Klein (1979).

self-potential or electromagnetic responses (Lajoie and Klein, 1979). However, a reconnaissance gradient-array resistivity/IP survey defined chargeability anomalies of about 7 times the background response associated with the N42 orebody, although the resistivity responses were variable. Figure 5.56 shows the chargeability map produced from the results of a subsequent gradient-array survey with a potential dipole length of 61 m. In addition to the response from the N42 orebody, distinct responses from three previously unknown orebodies to the south were defined.

Example – Pajingo epithermal gold deposits

Hoschke and Sexton (2005) describe various electrical and EM surveys during exploration of the Pajingo epithermal system, near Charters Towers, Queensland, Australia. Low sulphidation epithermal alteration and veining occur in intermediate volcanic rocks over an area of about 150 km^2. Conductive cover sediments cover about 80% of the area. Gold occurs within quartz veins 0.5 to 3 m wide. Alteration is variable but tends to be silicic near the mineralisation and dominated by clay minerals further away.

Gradient array resistivity surveys were undertaken to map the basement geology. Data were acquired as a series of square blocks with side of 600 or 800 m and current electrodes 1200 m or 1600 m apart. Line spacing was 40 m and the potential dipole was 20 m in length. Figure 5.57 shows a subset of the data. The zones of silicification which contained the mineralised veins were mapped as linear resistive zones (100s to 1000s of ohm-m) against a background with a resistivity of about 50 Ω m. The resistivity data, along with aeromagnetics and geological data, were used to site the drillhole that discovered the Cindy deposit under 5–15 m of cover, and subsequently the Nancy and Vera deposits.

5.6.7 Interpretation pitfalls

All the electrical parameters measured in resistivity/IP surveys are subject to a variety of spurious responses from a number of sources. The most common include conductive overburden, electromagnetic fields associated with the wires connecting the electrodes, undulating topography, electrical anisotropy of the subsurface material itself and cultural interference. Their responses distort genuine target anomalies and can masquerade as target anomalies. They can be serious interpretation pitfalls unless procedures are adopted to identify and account for them in the interpretation of survey data. Man-made interference and cultural anomalies are common to all electrical and EM measurements and are discussed in Section 5.4.2.

5.6.7.1 Overburden effects

The electrical properties of the near-surface (see Section 5.3.4) have a disproportionate effect on electrical property measurements made with electrodes on the surface. Electric current 'prefers' to flow via the path of least resistance, so where the overburden is highly conductive the transmitted current will concentrate in the more conductive layer with less entering the underlying higher-resistivity material. Increasing the thickness of the surface layer and increasing its conductivity causes a greater proportion of the current to flow in it, with even less penetrating the underlying material.

Telluric currents (see *Atmospheric noise* in Section 5.4.2.1) are induced in flat-lying conductive overburden and are a common source of noise that often obliterates the weaker polarisation currents. Larger current dipoles are required to penetrate the overburden; and larger current and longer polarisation times in the time domain, to say 4 seconds, help increase IP signal levels. Conductive overburden also significantly increases EM-coupling effects (see Section 5.6.7.2).

Dipole–dipole pseudosections of modelled responses of conductive overburden are shown in Fig. 5.58. As demonstrated by the figure, the limited penetration of current

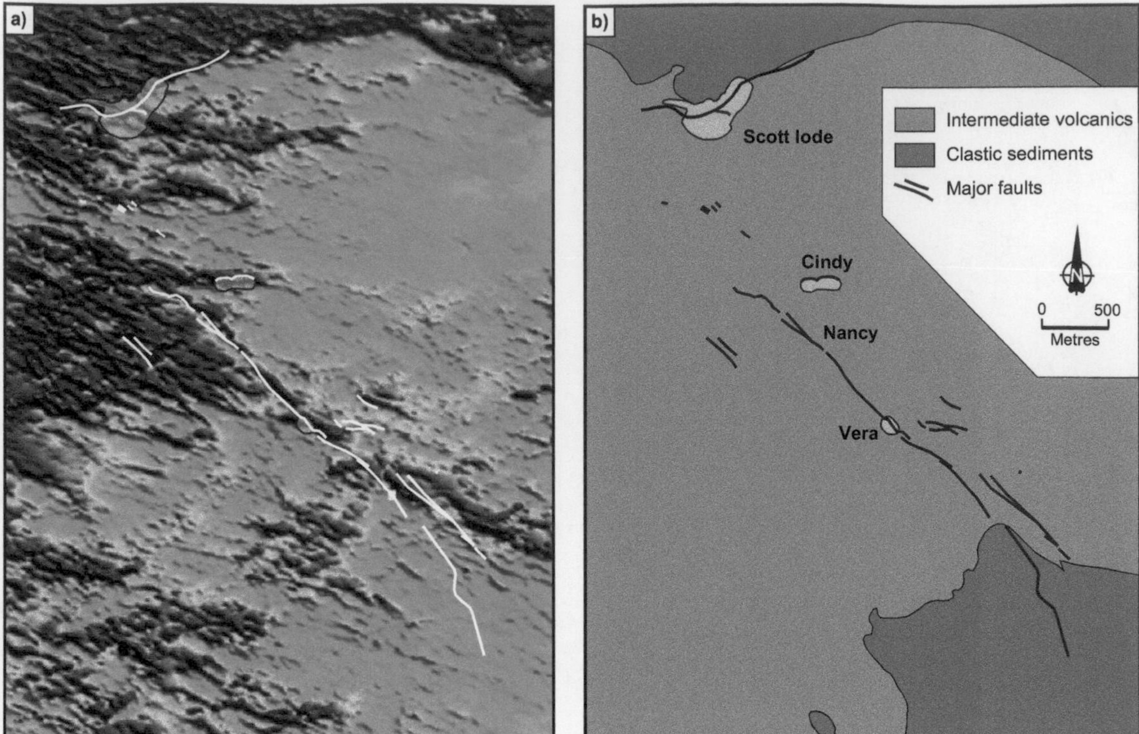

Figure 5.57 Pajingo epithermal system. (a) Image of gradient array resistivity data. Data courtesy of Evolution Mining. (b) Geological map. Redrawn, with permission, from Hoschke and Sexton (2005).

through the conductive layer means that targets in the underlying region produce weak responses, which are often obliterated by the background noise. Where the conductive body is in contact with the overburden, the current may be channelled into it and a stronger response produced. Changes in overburden conductivity and thickness create spurious responses that can resemble those of potentially economic significance. Overburden conductivity and thickness can be obtained from electrical measurements made on drill samples or determined from vertical electrical soundings, and are useful in the analysis of the survey data. Ideally, this information should be available for computer modelling during survey design to assist in optimising the survey parameters.

Where the overburden is comparatively resistive, such as the tills in the glaciated region of the North Hemisphere, there is less of a problem, but it may still mask economically significant responses (Reynolds, 1980).

5.6.7.2 Electromagnetic coupling

Measurement of IP parameters is subject to interference related to changes in current flow in the wires connecting the current electrodes. Eddy currents are induced in the conductive ground and into the wires connecting to the receiver (see Section 5.2.2.2). The effect is known as *electromagnetic coupling*, EM-coupling or inductive coupling and causes spurious IP-like effects, which may overwhelm the responses due to electrically polarisable materials in the subsurface (Wynn and Zonge, 1975). Grounded cultural features such as wire fences and pipelines can also contribute to coupling. The effect is exacerbated in highly conductive terrains and in areas of highly conductive overburden. Resistivity measurements are unaffected.

The removal of EM-coupling, or *decoupling* the IP parameters, has been a major area of IP research. A number of schemes have been devised to calculate the effect and correct the IP measurements, all with various limitations. For the case of time domain measurements, it is often assumed that EM-coupling has little effect in the late decay times, so the late-time secondary voltages are preferred for mapping the IP parameters. Techniques that attempt to correct the full decay-series voltages are described by Fullagar *et al.* (2000). In the frequency domain, correction techniques depend on the changes in the measured parameters with frequency. Coupling-related responses vary more rapidly with frequency than normal IP effects so the coupling component can be estimated and removed to resolve the IP effect.

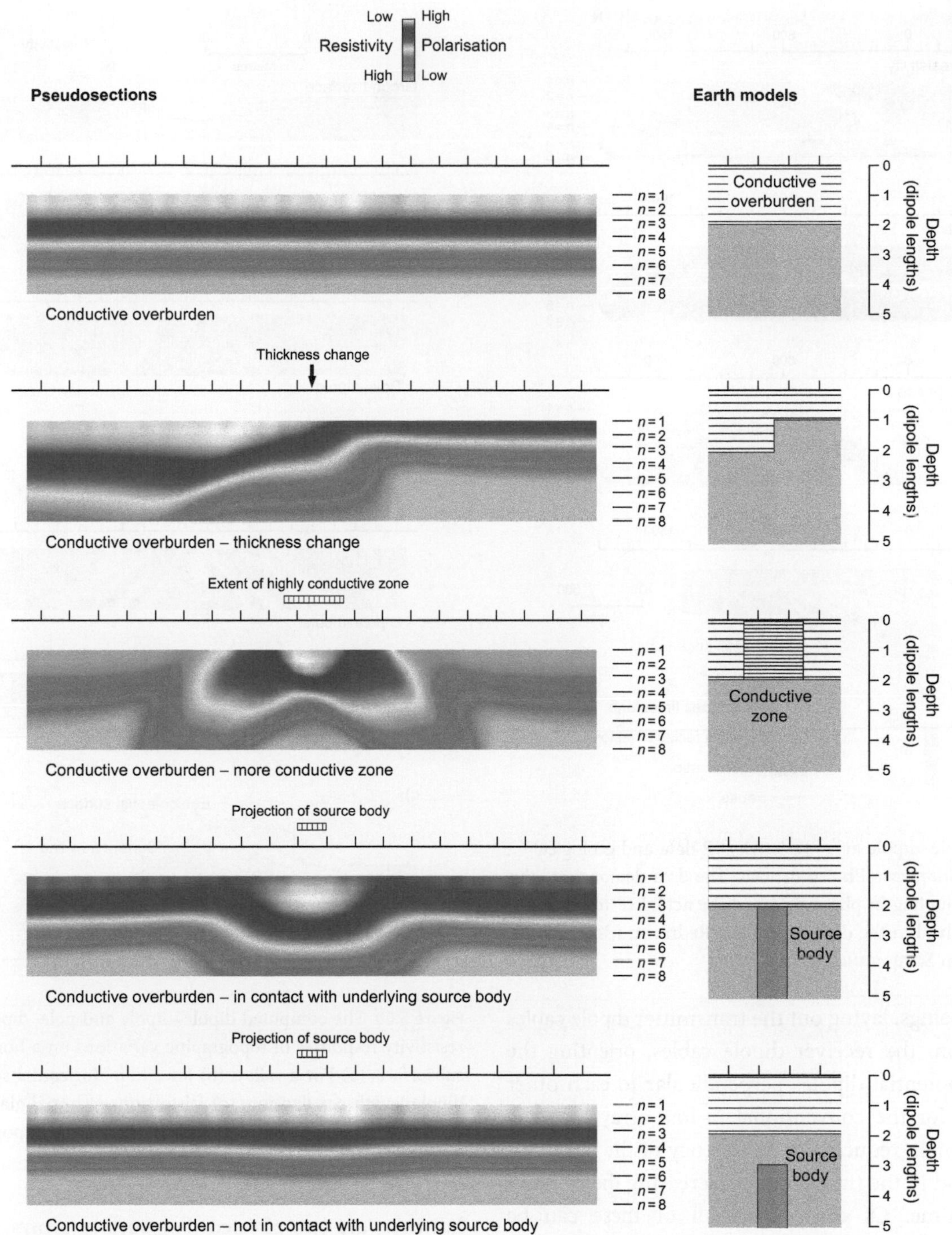

Figure 5.58 The computed dipole–dipole array resistivity responses of some simple 2D conductivity models with conductive overburden. Horizontal scale is one dipole length per division.

Figure 5.59 shows how EM-coupling appears in IP phase pseudosection data. These data are from the Goongewa (formerly Twelve Mile Bore) MVT-type Pb–Zn deposit, Canning Basin, Western Australia (Scott *et al.*, 1994). Since the effects are greater for larger dipole separations, the raw phase shows a consistent increase with dipole separation (n) and, therefore, an apparent increase in IP effect with depth. The decoupled data resolve the 'pants-legs' anomaly due to the polarisable mineralisation.

All electrode arrays are prone to EM-coupling, to some degree. The Schlumberger and gradient arrays, with their long cable runs to the distant current electrodes, are highly susceptible to EM-coupling. Strategies for minimising the effect include: minimising the dipole lengths, minimising

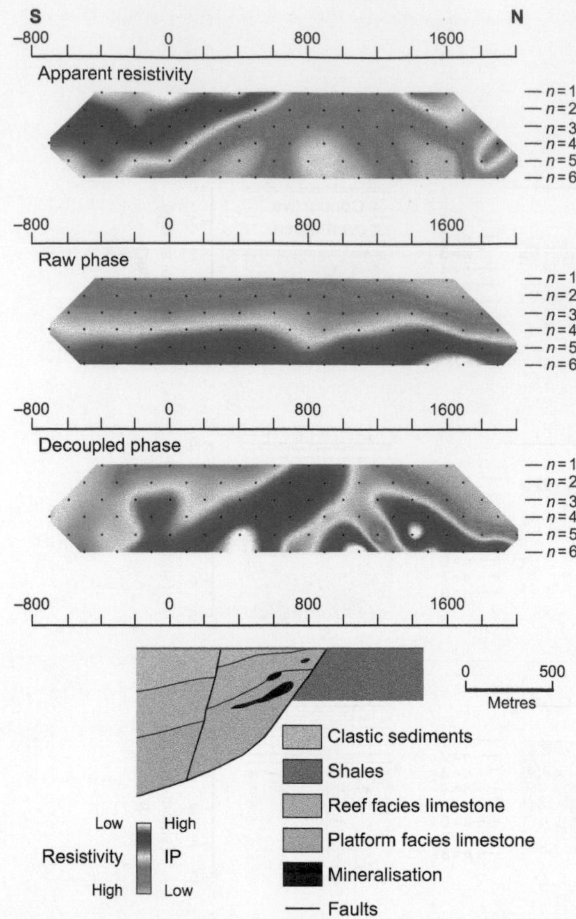

Figure 5.59 Dipole–dipole array resistivity/IP data and geological data from the Goongewa Pb–Zn deposit. The data demonstrate the effects of EM-coupling in phase data and the actual IP response of the ground resolved in the decoupled data. Redrawn, with permission, from Scott *et al.* (1994).

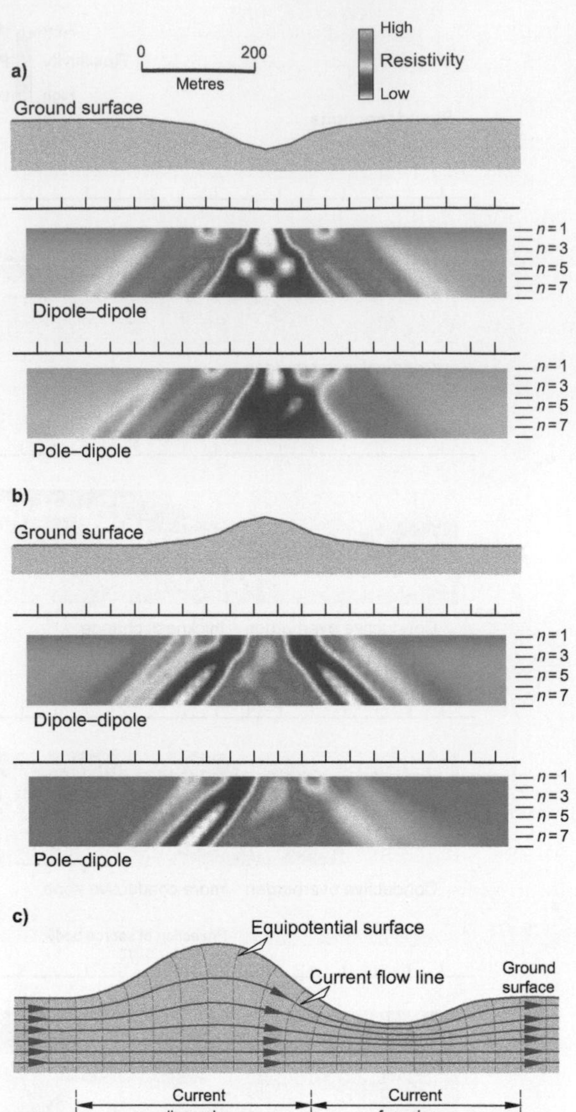

Figure 5.60 The computed dipole–dipole and pole–dipole array resistivity responses of topographic variations on a homogeneous subsurface. (a) For a valley; (b) for a hill. Horizontal scale is one dipole length per division. (c) Distortions of horizontal current flow and associated equipotential surfaces due to topographic variations.

the dipole spacings, laying out the transmitter dipole cables well away from the receiver dipole cables, orienting the current and potential dipoles perpendicular to each other (not possible for the conventional in-line arrays); in the frequency domain reducing the frequency of the transmitted current and in the time domain increasing the primary polarisation time. Of course, not all of these can be achieved without jeopardising other survey parameters, such as depth of investigation and lateral resolution.

5.6.7.3 Topographic effects

Apparent resistivity calculations based on Eq. (5.17) assume that the electrode array is located on a flat survey surface. Departures from this assumption, when electrodes are located on undulating topography and man-made disturbances to the terrain, create spurious variations in resistivity (Fox *et al.*, 1980). Induced polarisation parameters, however, are less affected. For surface arrays such as the gradient and Schlumberger arrays, potential measurements are made inside the current dipole where the equipotential surfaces are generally vertical, so the current flow is generally horizontal and parallel to the surface. Hills cause the equipotential surfaces, and the current flow, to diverge, and valleys cause them both to converge (Fig. 5.60c). The distortion in the equipotential surfaces produces variations in the measured potentials unrelated to variations in the subsurface electrical properties. Valleys produce high apparent resistivity and hills have the opposite effect. The effects of topography on dipole–dipole and pole–dipole pseudosections are shown in Fig. 5.60, the responses resembling those from shallow bodies.

There are no means of accurately removing terrain effects from resistivity/IP data, the practical approach to interpretation being to include topographic variations in the computed model. Of course, it is perfectly plausible for an economic target to be responsible for a change in topography, owing to greater or decreased resistance to weathering, the anomalous electrical responses drawing attention to the area.

5.6.7.4 Anisotropy effects

Electrical anisotropy affects all electrical and EM measurements and is discussed in Section 5.3.1.4. In resistivity/IP surveying for mineral targets, the electrode array is usually orientated in a fixed direction for the full survey area, usually perpendicular to strike. It is only anomalous variations in the measured parameters along the survey direction that are of interest for detecting targets, so anisotropy is usually of little concern in field surveys. However, its effect will be seen where, say, the dip of a rock formation is different in part of the survey area causing its foliation to change orientation relative to the electric field of the survey array. Changes in resistivity and polarisation due to anisotropy would then be expected.

Anisotropy is also likely to be noticed when comparing the absolute values of measured parameters from different arrays where the principal direction of current flow in each could be different: for example chiefly horizontal flow for measurements using the Wenner, Schlumberger and gradient arrays, and significantly vertical flow for the dipole-dipole array. Comparison of the measured electrical properties is difficult, as it also is when comparing EM conductivity data with electrical resistivity data owing to the principally different directions of current flow in each method.

5.6.8 Resistivity/IP logging

Resistivity/IP and resistance logging in hard-rock terrains is mainly undertaken to define the electrical properties of a known succession and to aid in the interpretation of electrical and EM surveys. In soft-rock terrains the aim is more often to establish stratigraphic correlations and in some cases infer subtle changes in physical characteristics indicative of facies changes. Resistivity/IP and single-point resistance logging for these purposes are demonstrated in the following two examples.

5.6.8.1 Example of resistivity/IP logging – Uley graphite deposit

At Uley in South Australia, mineralisation occurs in a graphite schist within a succession of schist and gneiss, beneath a cover of laterite and calcrete. Graphite ore zones are up to 12 m thick and are stratiform and strata-bound. The host rocks are strongly weathered. Numerous electrical and EM surveys have been made in the area (Barrett and Dentith, 2003).

Figure 5.61 shows the downhole logs of various electrical properties from a drillhole through the graphite mineralisation. A correlation between IP anomalies, SP highs and carbon assays is evident. The data indicate that a chargeable source has been intersected at a depth of about 40 m. This is consistent with the results from modelling of surface IP data. The logs also show that there is little relationship between mineralisation and resistivity; this is inconsistent with IP and subsequent EM surveys and is probably due to the differing sample volumes measured in surface surveys compared with the downhole measurements.

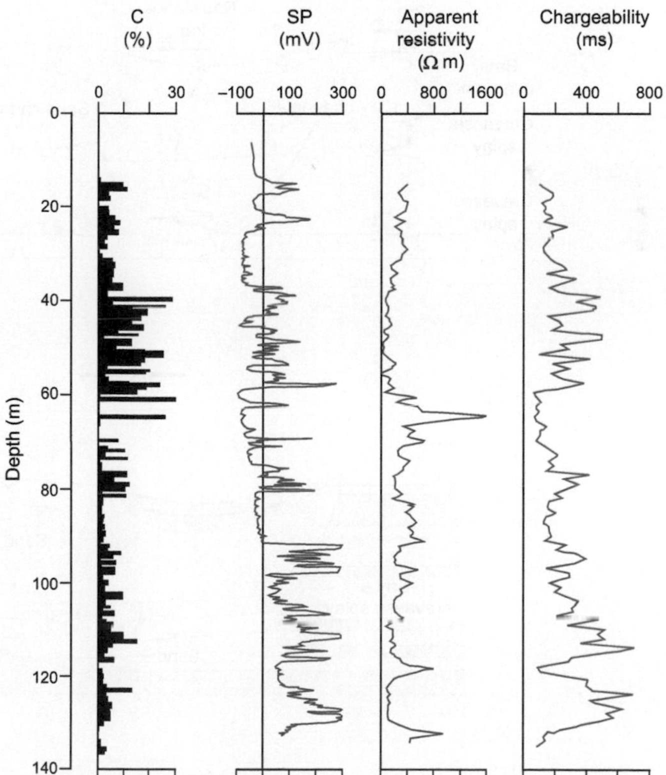

Figure 5.61 Assay data and downhole logs from the Uley graphite deposit. The resistivity/IP data were acquired using the dipole–dipole array (dipole length = 0.75 m). Redrawn, with permission, from Barrett and Dentith (2003).

5.6.8.2 Example of resistance logging – Franklin roll-front uranium

An example demonstrating the use of single-point resistance logs for facies mapping in a soft-rock uranium environment is shown in Fig. 5.62. The drill section crosses a deltaic point bar near the Franklin uranium mine in Texas, USA (Cossey and Frank, 1983). The distribution of mineralisation in roll-front type uranium deposits is critically dependent on redox conditions, which in turn are related to sedimentary facies. At the Franklin Mine, uranium mineralisation occurs in deltaic point-bar sediments. The character of the resistance logs can be used to infer positions within subsets of the point-bar sequences, allowing directions to uranium mineralisation to be inferred.

5.6.9 Applied potential/mise-à-la-masse method

The applied potential (AP) or mise-à-la-masse (MALM) method is a commonly used downhole form of electrical surveying (see Section 5.6.5.3). The technique involves applying a potential to an electrically conductive body and then mapping variations in the potential in the surrounding rock mass to determine the gross geometry of the conductor. Mise-à-la-masse, a French term literally translated as excitation of the mass or body, is widely used, but we prefer to use applied potential, since it is more descriptive of the method.

The target conductor may either outcrop or have been intersected in the subsurface through drilling or mining. In a mining context, the conductive body is usually massive sulphide mineralisation, but other applications of the method include mapping conductive faults and zones of contaminated groundwater. Potential measurements may be made at the surface, in drillholes or in underground workings. The method can also be used to establish whether conductive bodies identified at different locations, usually drillhole intersections, are electrically connected, i.e. part of the same electrical mass. The method is

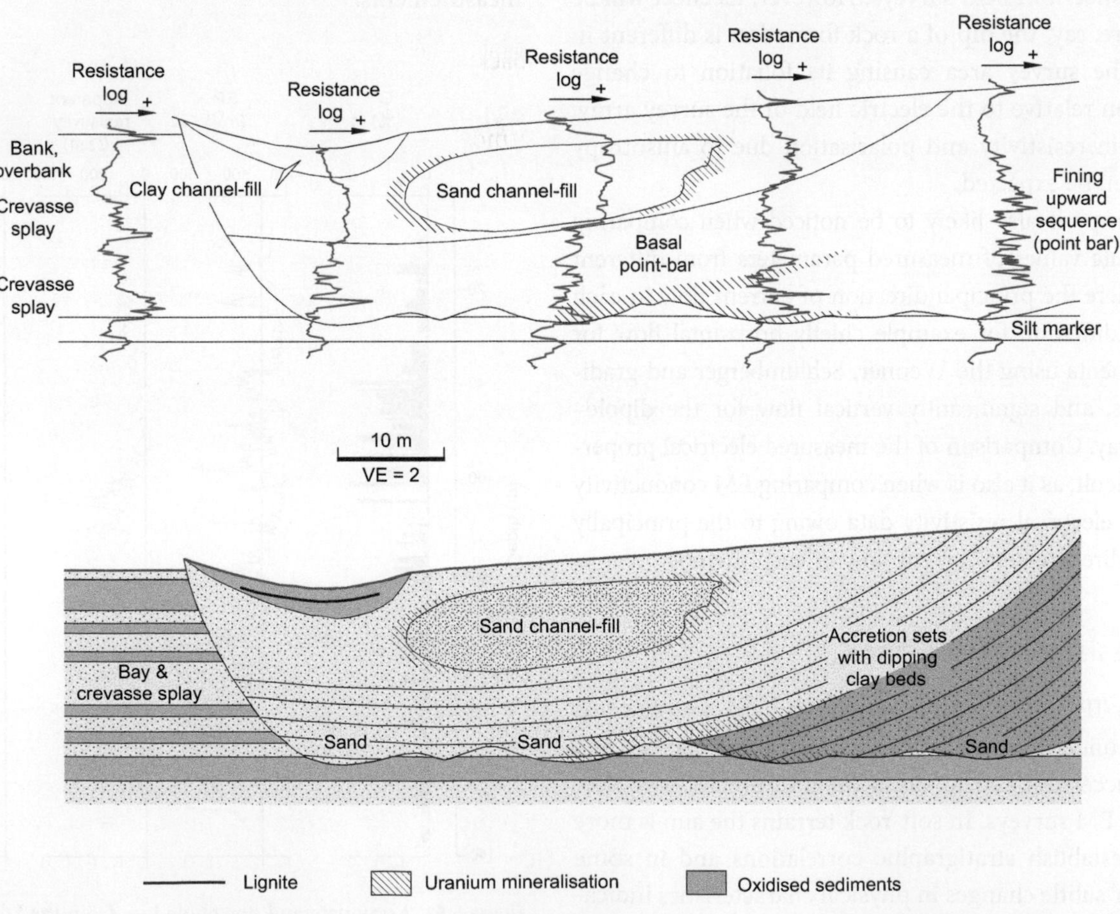

Figure 5.62 Single-point downhole resistance logs from a drill section of the Franklin roll-front type U deposit. See text for details. VE – vertical exaggeration. Redrawn from Cossey and Frank (1983), *AAPG Bulletin* (AAPG ©1983), and reprinted by permission of the AAPG whose permission is required for further use.

comparatively cheap and in appropriate situations can provide useful results.

5.6.9.1 AP survey practice

The survey equipment required is the same as that for resistivity/IP surveying (see Section 5.6.4). Details of the electrodes are described in Section 5.4.1. To conduct an AP survey, an electrical potential from the transmitter used for resistivity/IP work is applied to a pair of current electrodes. The electrode configuration is the pole–pole array (see Section 5.6.5) with the current pole-electrode located (buried) in the conductive target body (Fig. 5.63). It is the means by which the potential is applied to the target conductor, so that the conductor acts as a giant buried current electrode. The other current electrode is located remote from the survey area so that it is effectively at infinity where it has minimum influence on the electric field (the shape of the equipotential surfaces) in the vicinity of the conductor.

Mapping the electric field about the conductive body will reveal its presence and its shape (see Fig. 5.35). This is done by making potential measurements with a pair of non-polarising electrodes, one of which is established as a 'fixed' reference electrode and located at an 'infinite' distance, from the body. Measurements are made with the other 'roving' potential electrode moved systematically over the survey area (including down drillholes), which is usually centred on the energised conductor. The potential difference is measured relative to the distant electrode and the measurements normalised for variations in the current transmitted into the body (the voltage is divided by the current transmitted during the measurement), so the results are expressed in units of V/A. Sometimes the data are transformed to apparent resistivity using the geometric correction factor given in Table 5.2 (X_{BM} is the distance between current pole-electrode (B) and the 'roving' potential electrode (M)). IP parameters can also be measured.

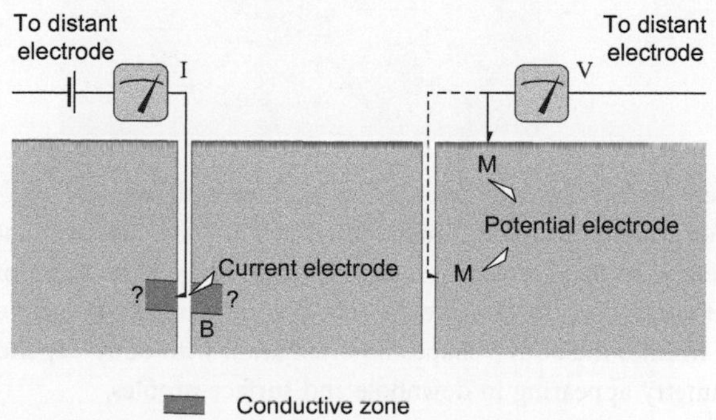

Figure 5.63 Schematic illustration of the applied potential method for surface and downhole measurements.

In order to confirm anomalous responses related to the conductive body, the whole survey procedure can be repeated with the buried current electrode located outside the conductor. For example, for the case of an electrode in a drillhole, it may be raised above the body by say, 10 m. The potential distribution about the pole-electrode is mapped and used to confirm that anomalous features observed with the electrode embedded in the conductor are related to the conductive body and not due to the background geology. This background response can be removed from the original results to enhance the resolution of anomalous features, a form of regional–residual separation (see Section 2.9.2).

Measurements are normally made on an equidimensional grid, or along survey traverses perpendicular to the expected strike of the conductor, and a map of the surface potential produced. Station spacing is typically 5 to 50 m with survey lines spaced of 25 to 100 m apart. Potential measurements can also be made down adjacent, uncased and water-filled, drillholes; typically at intervals 5 to 10 m, or closer to increase resolution in the vicinity of rapid changes in conductor geometry.

5.6.9.2 Display and interpretation of AP data

Maps and cross-sections of the measured potential are normally presented as contours instead of images, because contours generally provide better resolution of subtle gradients than images. The amplitudes and shapes of the contours, and the gradients between them (the spacing of the contours), are interpreted in terms of conductor geometry (see Section 2.10.2). AP data can be interpreted using computer models and can be quite reliably interpreted using qualitative methods. As mentioned previously, the main application of the AP method is mapping of conductive bodies at prospect scale, based on observed variations in electrical potential. An increase in potential indicates the presence of an electrically connected conductor, and gradients in the field reflect conductor geometry and electrical contacts.

A common application of the AP method is to assess the significance of a conductor intersected by a drillhole. If the intersection is part of a small body at great depth, the equipotential surfaces will be spherical and centred about the current electrode. Contours of the potential field on any plane, such as the ground surface, will be circular and centred about the projection of the current electrode's location (Fig. 5.64a). The same will be true for any cross-sections. However, if the intersection is part of a large body, the potential field will reflect the shape and electrical continuity of the body. Contours of equal potential mapped on a planar surface will approximately correspond with the projected shape of the body (Fig. 5.64b). It is generally not possible to determine accurately which particular contour corresponds with its edges. Also, the position of the current electrode projected onto the survey plane will not necessarily correspond with the maximum observed potential.

If the plane in which measurements are made is a horizontal surface, the strike direction of the body can be inferred from the direction of elongation of the contours. The dip and/or plunge of the body can also be estimated from the gradients of the potential field, which can be assessed from the spacing of the potential contours (cf. Fig. 2.43). They will be more widely spaced in the direction of dip, which is also perpendicular to their elongation. Similarly, the contour spacing in the direction of elongation indicates the plunge direction. Equivalent information about the attitude of the body can be inferred from measurements made on a cross-section.

Information about the attitude of a conductive body is easily obtained from AP surveys provided the measurements are comparatively close to the body. As the distance between the measurement surface and the body increases, the correspondence between the equipotential contours and the body's geometry diminishes. When the distance is large compared with its lateral dimensions, the contours become circular and symmetrical about the body, i.e. at great distance (depth) the body looks more like a point.

A common form of AP survey involves measuring the electrical potential at locations adjacent to the central drillhole containing the current electrode, which may be within surface or underground drillholes, or even mine galleries. There is no fundamental difference between this approach and that described above for surface measurements. From the contour plots in Fig. 5.64 it can be seen that the highest potentials occur in the conductive body containing the current electrode. If a second drillhole were to intersect the same body the maximum potential measured in this hole would coincide with the intersection(s), indicating that the intersections are electrically connected. On the other hand, and very importantly, the absence of potential maxima coincident with any other intersections of potentially conductive material in the drillhole indicates that these intersections are not electrically connected to the

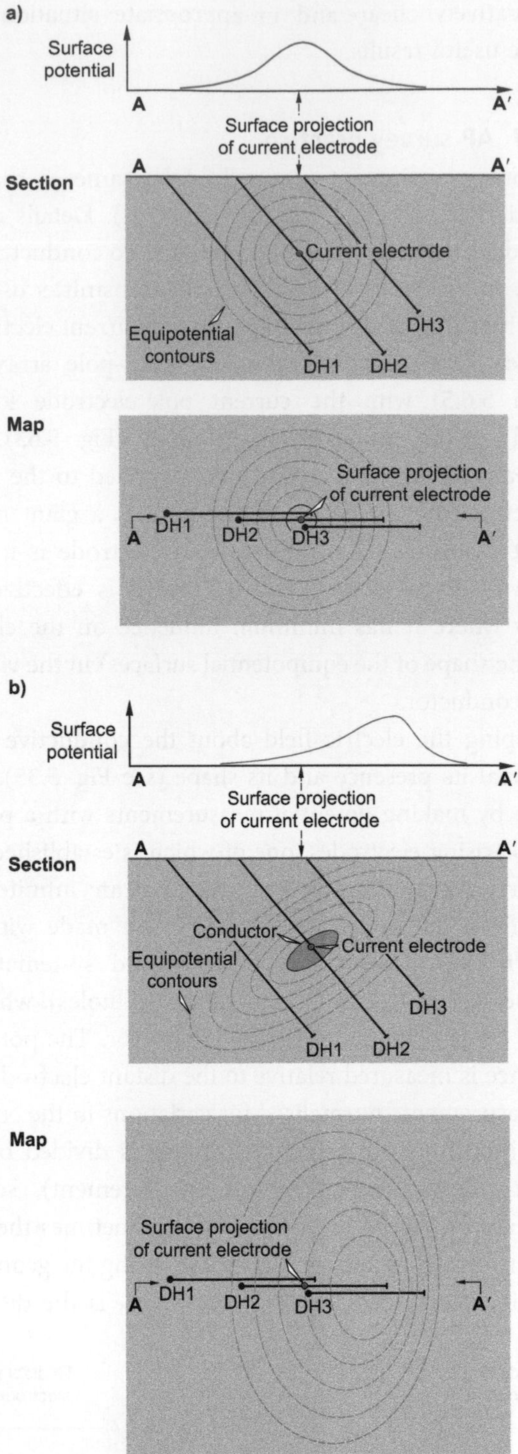

Figure 5.64 Electric field about a buried current electrode. (a) For a point electrode, equipotential surfaces are spherical and concentric to the electrode, producing symmetrical potential profiles downhole and on the surface. (b) The field of a large conductive body mimics the body's shape and orientation with body-dependent asymmetry appearing in downhole and surface profiles.

body in which the current electrode is embedded. This ability to determine electrical connectivity between two conductive zones can be a very powerful tool for evaluating the extent and dimensions of a conductive body, and a useful aid to a drilling programme. By systematically placing the current electrode in different intersections in different drillholes, and measuring potential variations in other drillholes, the correct correlations between intersections may be established. Some possible situations of this type are illustrated in Fig. 5.65.

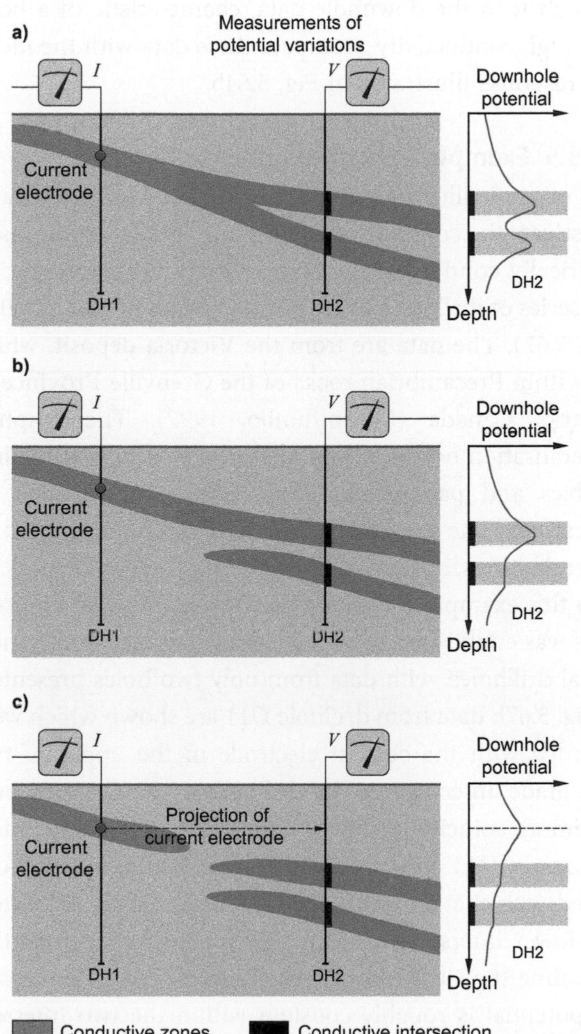

Figure 5.65 Schematic illustration of downhole applied potential logs used to establish electrical continuity of conductive stratigraphic units between drillhole intersections for three different cases. (a) Both intersections electrically connected to the energised intersection; (b) connectivity only to the upper intersection; and (c) both intersections electrically disconnected from the energised intersection. Electrical connectivity is indicated where the potential is strongest.

We demonstrate the application of the AP method and the interpretation of both surface and downhole data with examples from two different types of mineral deposit.

5.6.9.3 Interpretation pitfalls

Although the interpretation of AP data is, in principle, straightforward, it suffers all the same pitfalls experienced by other electrical arrays (see Section 5.6.7) and these should be taken into account. For example, anisotropy is common in obviously layered rocks such as well-bedded sedimentary successions and metamorphic rocks such as schists and gneisses. When there is no significant conductor present, the equipotential contours will be elongated in the direction of highest conductivity and their centre may be displaced up-dip, and no longer centred above the current electrode. The current electrode can be deliberately located away from conductors to allow these affects to be assessed. Conductive overburden and variations within it also complicate the interpretation, and undulating topography distorts the electric field, but when necessary these effects can be accounted for in computer models (Oppliger, 1984).

When conductors are small or disrupted by faulting, or when measurements are taken close to their margins, the patterns of the potential anomalies are far more complex than for the simple body shape shown in Fig. 5.64, and as illustrated in Greenhalgh and Cao (1998). When the intention of the survey is to establish continuity between conductors, it is quite possible that the electrical continuity between two massive sulphide intersections is via a water-saturated fault zone rather than a continuous body of mineralisation. Similarly, an apparent lack of electrical continuity between two intersections in what is the same body of mineralisation may be caused by a zone of poorly conductive minerals, whether gangue or sulphides such as sphalerite (an electrical insulator). Quite minor faulting can also break the electrical continuity.

5.6.9.4 Example – Woodlawn polymetallic massive sulphide

This example illustrates the response from a moderately dipping conductive massive sulphide body with complex geometry and high contrast with its host rocks. Figure 5.66 shows the results of an AP survey at the Woodlawn Cu–Pb–Zn deposit in New South Wales, Australia (Templeton *et al.*, 1981). The deposit is of volcanic-hosted massive-sulphide type, and the ore forms a west-dipping lens in a succession of pelitic and pyroclastic rocks and

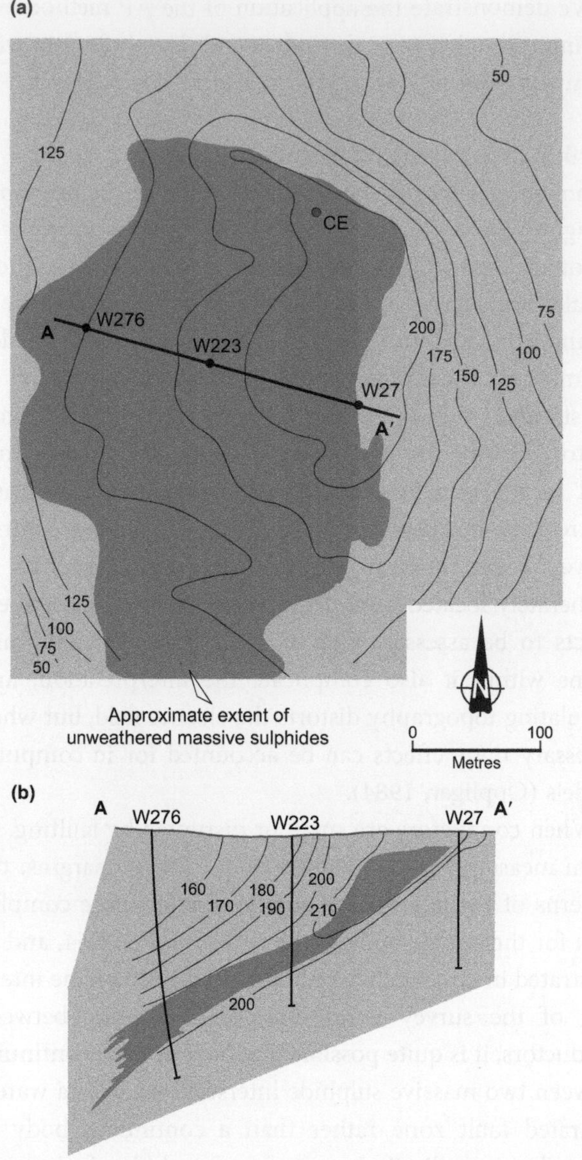

Figure 5.66 Applied potential data from the Woodlawn Cu–Pb–Zn deposit. (a) Surface potential contours (mV/A) with dip direction of the conductive body indicated by the potential gradient along section A–A′. (b) Downhole potential contours (mV/A) on section A–A′. CE is the projected location of the buried current electrode. See text for details. Redrawn, with permission, from Templeton *et al.* (1981).

basic intrusives that have been metamorphosed to greenschist facies. The major ore minerals are pyrite, sphalerite, galena and chalcopyrite. The exploration history of the Woodlawn deposit is described by McKay and Davies (1990). The current electrode was located in a drillhole intersecting the orebody near its northern margin. It was positioned in the centre of a 36 m intersection of massive sulphides at a vertical depth of 46 m. Potential measurements were made on the surface and down three drillholes.

Equipotential contours of the surface measurements define the extent of the orebody extremely well, with the highest potentials coinciding with the shallowest part of the body (Fig. 5.66a). Its westerly dip is clearly indicated by the wider spacing of the contours to the west, and the northwesterly plunge of the body is also suggested. The downhole potential measurements very effectively outline the dipping orebody. The maximum potentials occur in the vicinity of the intersections, demonstrating the electrical continuity between them and the intersection containing the current electrode (Fig. 5.65b). Note the low gradient in the surface potentials over the massive mineralisation, and through it in the downhole data, characteristic of a body with high conductivity. Compare these data with the idealised response illustrated in Fig. 5.64b.

5.6.9.5 Example – Victoria graphite deposit

This example illustrates the use of hole-to-hole AP data to investigate the continuity between multiple intersections of electrically conductive high-grade graphite mineralisation in a series of drillholes along a strike length of about 200 m (Fig. 5.67). The data are from the Victoria deposit, which lies within Precambrian rocks of the Grenville Province in Ontario, Canada (Mwenifumbo, 1997). The graphite mineralisation occurs within a sequence of highly silicified marbles and paragneisses that have been intruded by pegmatites. The graphite occurs as massive lenses and as disseminations.

In this example continuity between high-grade intersections was established from potential variations within individual drillholes, with data from only two holes presented. In Fig. 5.67b data from drillhole D11 are shown which were recorded with the current electrode in the upper of two high-grade intersections in drillhole D45. The potential maximum coinciding with the upper of two graphite intersections in D11 shows this horizon to be electrically connected with that in which the current electrode is located. The lower intersection has a significantly lower potential, indicating that it is not electrically connected. Notice that the potential is roughly constant within the two intersections because of their high conductivity. Compare these data with Fig. 5.65b. The second dataset (Fig. 5.67c) shows the potential measured in drillhole D47, which contains four intersections of high-grade mineralisation. The current electrode is in the upper of two intersections made by drillhole D48. Potential maxima coincide with two of the four intersections, indicating electrical continuity between these and the intersection in D48. Potential is lower in the

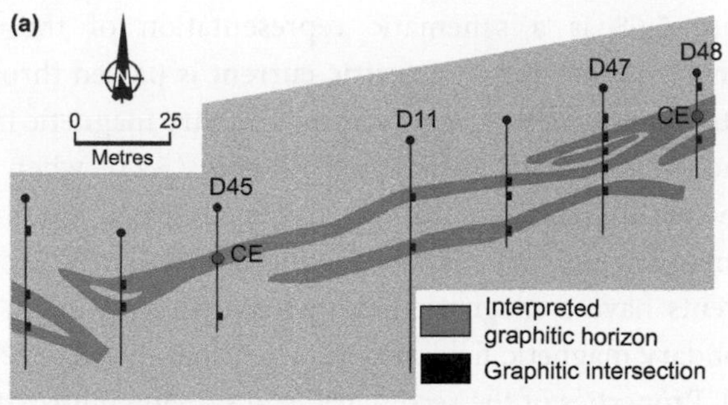

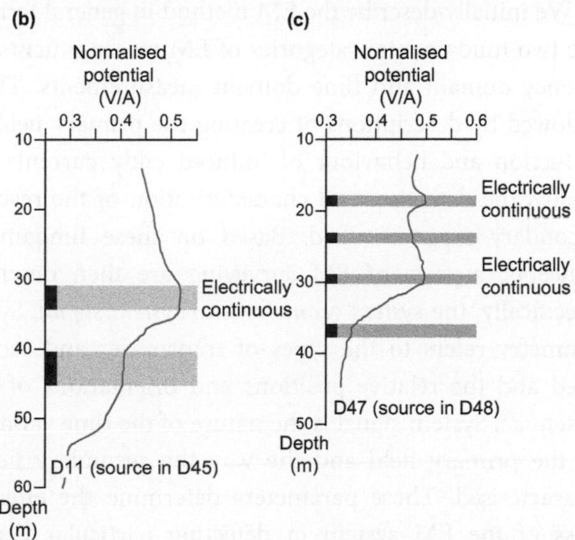

Figure 5.67 Downhole applied potential logs from the Victoria graphite deposit. (a) Drill section and stratigraphy as interpreted from the drillhole intersections. Locations of the buried current electrode labelled as CE. (b) Profile of potential in drillhole D11 with the current electrode in the upper of two high-grade intersections in drillhole D45. (c) Profile of potential in drillhole D47 with the current electrode in the upper of two intersections in drillhole D48. See text for details. Based on diagrams in Mwenifumbo (1997).

other two intersections, indicating they are electrically isolated. Compare these data with Fig. 5.65a and b.

Continuity between the various intersections was established by the AP survey, leading to the interpretation shown in Fig. 5.67a. The resulting interpretation was different, in important aspects, from the contemporaneous geological model and was crucial to the decision to begin mining in the area between drillholes D11 and D48.

5.7 Electromagnetic methods

Electromagnetic (EM) surveys as used by the minerals industry are chiefly a type of active geophysical method, i.e. they use artificially created electromagnetic fields. The measurement of naturally occurring EM fields is less common (see online Appendix 4 for a description of these methods). Surveys are conducted from the air (airborne electromagnetics; AEM), on the ground surface, and in drillholes (downhole electromagnetics; DHEM). Although primarily used as an exploration tool, there is increasing use of EM in-mine, the intention being to map an orebody accurately prior to, and during, mining to reduce the amount of delineation drilling required. The basic aim of EM surveying is to map spatial variations in electrical conductivity with the data presented in the form of pseudo-maps, cross-sections or volumes showing the conductivity variations in a continuous form. The data may also be used to infer the location of a 'target', i.e. a discrete zone of conductivity approximated by some simple shape.

New survey and interpretation systems are continually evolving, and EM theory related to geophysical surveying is continually being developed. The result is a complex and diverse science. Making and understanding EM measurements in areas of conductive overburden (see Section 5.3.4) is especially challenging. Future developments are aimed at improving signal detection and resolution of more subtle conductivity contrasts, better depth penetration, especially beneath conductive overburden, and reducing the cost of EM surveying. Considerable work is ongoing in the mathematically complex area of improving data interpretation tools, which includes the development of associated software for building more accurate and complex 3D conductivity models of the ground from ever-increasing data volumes. These developments allow better integration of geological information and a greater role for the geologist in the analysis of the data than is the case at present.

There exists a plethora of EM systems, instrumentation and interpretation techniques, past and present. Confusingly, survey results are dependent upon many variables, in particular the specific EM system used. We have attempted to elucidate the principles and practice of this complex area of exploration geophysics, with emphasis on currently available systems and survey practice. The reader is referred to Klein and Lajoie (1992), van Zijl and Köstlin (1986) and Nabighian (1991) for descriptions of some older EM methods, including contemporaneous survey procedures and interpretation techniques.

5.7.1 Principles of electromagnetic surveying

The EM method is based on the principle of electromagnetic induction described in Section 5.2.2.

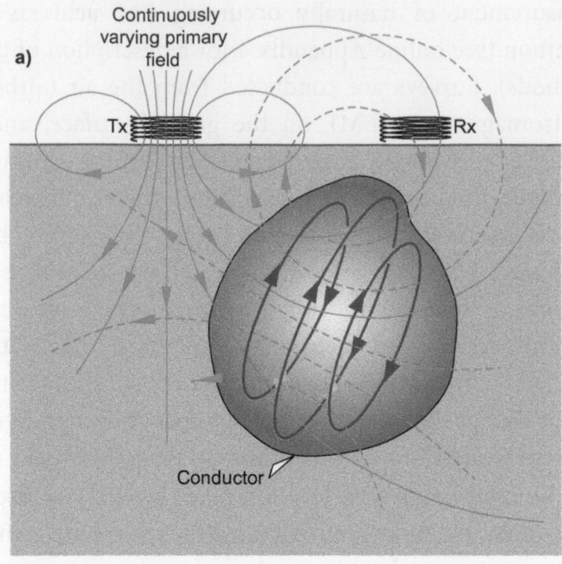

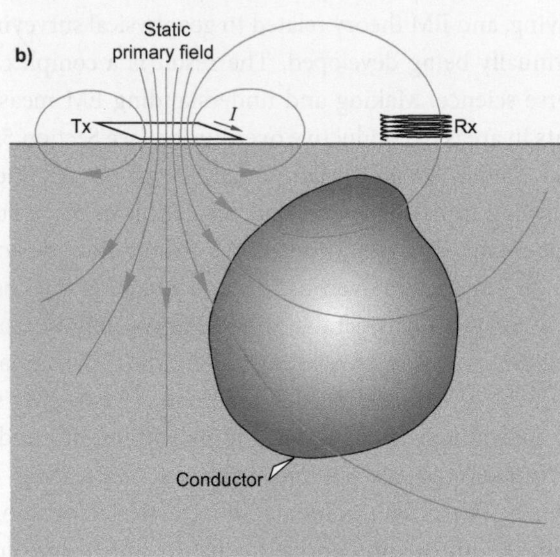

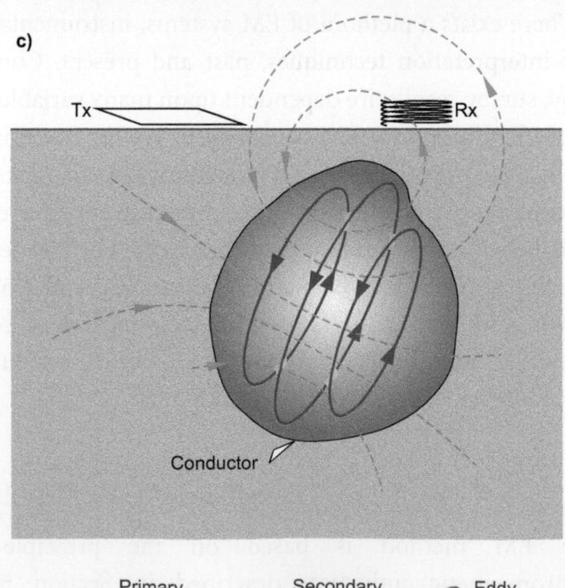

Figure 5.68 is a schematic representation of the EM method. A time-varying electric current is passed through the transmitter (Tx) to produce the primary magnetic field. In accordance with Faraday's Law (Eq. (5.7)), when the time-varying magnetic field intersects electrically conductive material eddy currents are induced within it. The eddy currents have a magnetic field associated with them, the secondary magnetic field, which is detected by the receiver (Rx). Properties of the secondary field provide information about the conductor.

We initially describe the EM method in general terms of the two fundamental categories of EM measurements: frequency domain and time domain measurements. This is followed by descriptions of creating the primary field, the induction and behaviour of induced eddy currents and, finally, the detection and characterisation of the resulting secondary magnetic field. Based on these fundamental concepts, aspects of EM surveying are then described: specifically, the *system geometry* and *system signal*. System geometry refers to the types of transmitter and receiver used and the relative positions and orientations of their antennae. System signal is the nature of the time variations in the primary field and the way the secondary field is characterised. These parameters determine the effectiveness of the EM system in detecting particular discrete conductors, mapping variations in a wide range of conductivity and resolving shallow and deep targets.

5.7.1.1 Time domain and frequency domain EM

Electromagnetic systems vary the primary magnetic field with time in one of two ways, as described in Section 5.2.2.2, which leads to two classes of EM systems: time domain (TDEM) and frequency domain (FDEM) systems.

In the time domain the change in the primary magnetic field is produced by either abruptly turning off or turning on a steady (d.c.) current. A pulse of current is induced in a conductor. The eddy currents circulate in the conductor for a short period and quickly decay as they lose energy.

Figure 5.68 Schematic illustration of an EM system used for geophysical investigations. (a) The frequency domain case of a continuously varying primary field. The diagram shows the situation for a primary field increasing in strength, so the secondary field has the opposite direction to oppose the change. (b) The time domain case with a steady-state primary field shown before turn-off. (c) Induced eddy currents and their secondary field following turn-off. To oppose the change in the primary field, the secondary field (broken lines) has the same direction. The receiver (Rx) is either a coil, shown here, or a magnetic sensor. Adapted from Grant and West (1965).

Their strength and duration depends on the electrical properties and geometry of the conductor, so monitoring their decay provides information about the subsurface conductivity. In our description of TDEM methods we describe the induction processes at turn-off.

In the frequency domain a continuous sinusoidal (a.c.) current is used to produce a continuous sinusoidally varying primary magnetic field, and usually at several frequencies, inducing sinusoidally varying eddy currents into a conductor. They circulate in the conductor and are out of phase with respect to the primary field. Their strength and phase depend on the properties of the conductor, so by measuring these properties information can be obtained about the subsurface conductivity.

FDEM systems measure the weak secondary field in the presence of the stronger primary field. Features of the resultant of the primary and secondary fields combined are measured, which reduces the system's sensitivity to small variations in the secondary field. Furthermore, variations in the orientation and position of the receiver with respect to the transmitter cause variations in the strength of the primary field at the receiver and are a significant source of noise. In contrast, TDEM measurements are, in principle, made when the primary field is turned off (variations to this arrangement are described in Sections 5.7.1.6 and 5.7.1.7), so the receiver sensitivity is maximised to detect the weaker secondary signals. In addition, variations in the orientation and location of the receiver with respect to the transmitter are not a source of noise. TDEM systems offer superior performance over FDEM systems.

FDEM systems were developed before TDEM systems, but because of inferior performance in resolving conductors in host rocks with variable conductivity and beneath conductive cover, they have been almost completely replaced in mineral exploration by TDEM systems. For this reason, our description of EM methods is almost entirely focused on time domain methods. Frequency domain FDEM systems still find application in shallow engineering, archaeological and groundwater investigations. Their use in mineral exploration is limited to airborne surveys to map shallow conductive targets, e.g. kimberlites (Reed and Witherly, 2007) and magnetotelluric methods, as described in online Appendix 4.

5.7.1.2 Creating the primary magnetic field

The primary magnetic field in EM methods is usually created by using a large loop of wire (see Fig. 5.7 and Section 5.2.2.1). The strength of the magnetic field is quantified by the magnetic dipole moment (m) given by:

$$m = nIA \tag{5.21}$$

where I is the current in amperes, n the number of turns in the coil or loop and A the area of the coil/loop in square metres. The dipole moment has units of A m^2.

Increasing the number of turns, and/or increasing the current in a loop, increases its dipole moment. The larger the transmitter's dipole moment the stronger is the primary field and the stronger the eddy currents it induces. This improves the signal-to-noise ratio of the secondary field measurement. There are several advantages in using large loops as transmitters. Equation (5.21) shows that a large loop has a larger dipole moment than a smaller loop, and increasing the size of the loop spreads the field over a larger volume of the ground. The strength of the field of small coils and loops decreases as 1/distance^3 from the loop, but as approximately 1/distance^2 for a large loop, so deeper targets can be detected.

Transmitter currents of up to several hundred amps are used and are limited by the level of the back emf produced at the instant of turn-off (see Section 5.2.2.2). The back emf is determined by the rate of turn-off of the current and the inductance of the loop, which is proportional to the square of its dimensions (L^2) and the number of turns. A large multi-turn loop produces a large back emf, preventing the instantaneous step turn-off of the magnetic field. The current must be turned on, or off, at a slower rate over a period known as the *ramp time*. The magnetic field changes at a less desirable rate and has a detrimental effect on the EM response.

5.7.1.3 Primary-field to conductor coupling

The concept of coupling is illustrated in Fig. 5.69 where the large current-carrying loop represents the transmitter, producing the primary field, and the conductor is represented by a small loop of wire. The only possible flow path for the induced current is around the small loop, and the strength of the current is dependent on the strength of the component of the primary field parallel to the axis of the loop; so the current is greatest when the plane of the loop is perpendicular to the primary field (see Section 5.2.2.2), a relationship labelled as 'very well coupled' in the figure. There is no current flow in the small loop where the magnetic field is parallel to the plane of the loop, i.e. where the field does not cut the loop. The loop is then said to be *null coupled* with the field. Note how the coupling changes around the transmitter loop for various locations and orientations of the small loop.

The same principle applies to subsurface conductors, where the location, attitude and geometry of a buried conductor with respect to the primary magnetic field determine the coupling between them. Figure 5.70 illustrates this with a large transmitter loop on the ground surface and sheet-like conductors in the subsurface. Depending on its location and orientation relative to the loop, the conductor may be *well coupled*, *poorly coupled* or *null coupled* with the primary field produced by the loop. At any location, maximum coupling (i.e. maximum eddy current flow) and the strongest secondary magnetic field occur when the primary field is perpendicular to the plane of the conductive body. Establishing a geometry whereby the target is intersected by the primary magnetic field at as high an angle as possible is fundamental to creating a strong eddy current system. Optimum coupling between transmitter and conductor is fundamental for the detection of the conductor.

Once detected, poor coupling and null coupling are used to advantage to provide diagnostic information about the location and attitude of a conductor with respect to the primary field. This is achieved by relocating the transmitter loop relative to the conductor and observing changes in its response. For example, in Fig. 5.70a the vertical-dipping conductor is well coupled to the loop when the loop is located to the side of the conductor, and null coupled when the centre of the loop is located directly above it (Fig. 5.70b). A flat-lying conductor in the same location is very well coupled (Fig. 5.70c), and poorly coupled when the loop is to the side of the conductor (Fig. 5.70d).

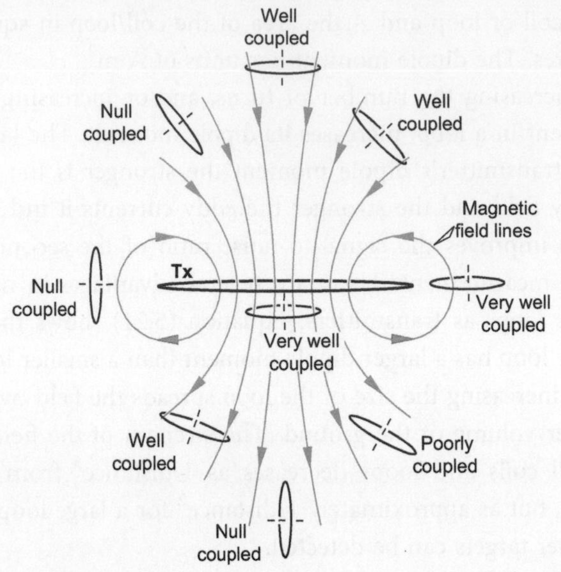

Figure 5.69 Variation in coupling between conductors, represented by small loops, and the magnetic field of a large current-carrying loop (Tx) at various locations around the Tx. Maximum coupling occurs where the magnetic field intersects the plane of the conductor loop perpendicularly.

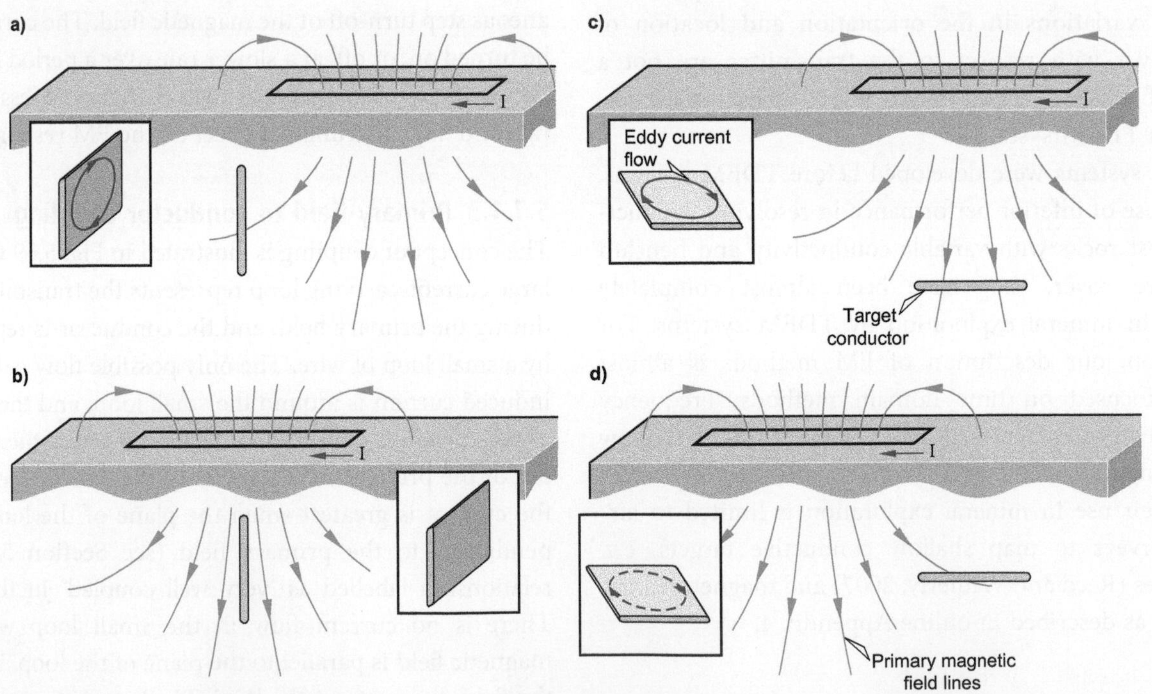

Figure 5.70 Variations in coupling to vertical and flat-lying sheet-like conductors in the subsurface for various locations of a horizontal loop on the surface. The conductor is (a) well coupled, (b) null coupled, (c) very well coupled and (d) poorly coupled.

5.7.1.4 Eddy currents

Induction of eddy currents (see Section 5.2.2.2) occurs instantaneously and simultaneously with the variations in the primary field and in all of those conductors in the subsurface where a component of the primary field intersects them perpendicularly (see Section 5.7.1.3). In accordance with Faraday's Law (Eq. (5.7)), the faster the variation in the primary field the stronger the eddy current induced. For the reasons described in Section 5.2.2.2, in a homogeneous equidimensional conductor eddy currents flow in closed circular paths in the plane perpendicular to the field causing them. The extent to which this actually occurs depends on the size, shape and electrical homogeneity of conductive zones in the subsurface.

In contrast to the sheet-like bodies described above which have only the dip plane in which the eddy currents can flow, irrespective of the direction of the primary field intersecting the body (Fig. 5.71a), in sphere-like bodies with homogeneous conductivity the eddy current circulation depends only on the direction of the primary field (Fig. 5.71b). Changing the field direction by moving the transmitter loop changes their orientation, i.e. a different current system is set up. Even for a single loop position,

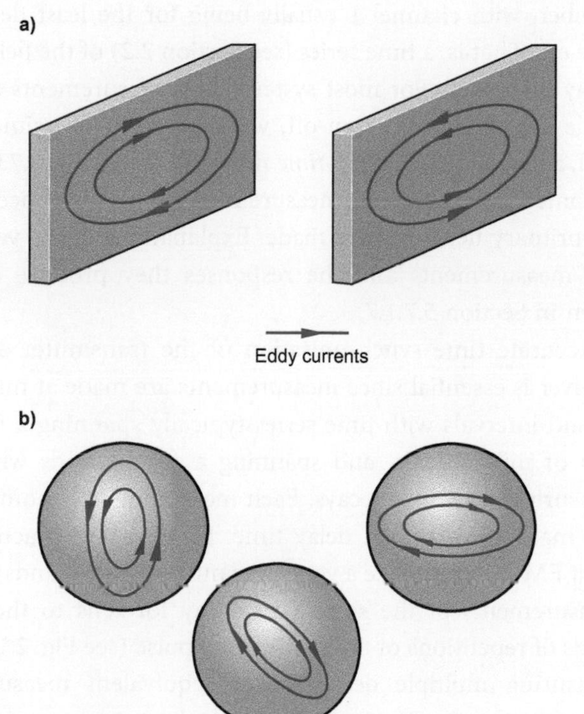

Figure 5.71 Orientation of eddy currents in homogeneous conductors. The current system is oriented in the plane (a) of sheet-like conductors, and (b) perpendicular to the primary field direction in spherical conductors.

thick bodies with 3D geometry may have several eddy current systems created within them (Fig. 5.71b), which tend to merge and reorientate in the body with time.

In reality, conductors in the geological environment are often electrically heterogeneous. Variable mineralogy, textures and pore contents produce electrical anisotropy and inhomogeneity (see Section 5.3). Currents can be channelled through the more conductive parts of a heterogeneous conductor, and anisotropy will influence orientation of flow paths. Also, multiple zones of higher conductivity in a conductor have their own local eddy currents which may merge at later times into a single current system, and variations in body thickness and shape will affect the current flow. For the case where the primary field is stronger over a small part of a conductor, the eddy currents will flow in that part of the body. The current system may then be controlled by the geometry of the primary field rather than the geometry of the whole conductor. The system of eddy currents in a conductor is also influenced by the changing eddy current flows in overburden, conductive host rocks and neighbouring conductors (see Section 5.7.2).

Initially, after the primary field turn-off (or turn-on), the eddy currents essentially flow around the surface of the conductor. With increasing time the eddy currents lose energy mostly as heat due to the resistance of the conductor and become weaker. This causes their magnetic field to change, causing adjacent regions of the conductor to experience a changing magnetic field. This creates a mechanism for a diffusive inward migration of the circulating current, the migration and decay being slower with increasing conductivity. Figure 5.72 illustrates this effect for an electrically homogeneous conductor.

Since the decay of the eddy currents depends on the body's electrical properties and its shape, and because they are flowing in different parts of the target with increasing time, analysing the associated changes in the secondary magnetic field can provide information about the distribution of the conductivity. Soon after the primary field is turned off, the strength and flow path of the currents are dependent mainly on the shape of the conductor and not its conductivity and, therefore, give good diagnostic information about conductor geometry. At later time, the eddy currents flow deeper within the body and so depend on the conductivity of its interior, although for bodies with very high conductivity skin effects (see Section 5.2.3.1) restrict the current flow to chiefly the surface region. The diffusion of eddy currents in conductivity structures common in the geological environment is described in Section 5.7.2.

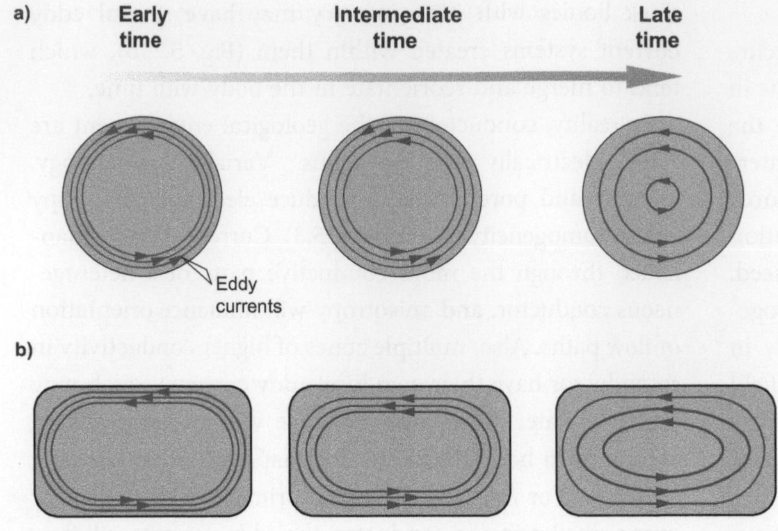

Figure 5.72 Eddy current distribution with time. (a) In spherical and (b) in sheet-like conductors. Redrawn from McNeil (1980) with permission of Geonics Ltd, Mississauga.

5.7.1.5 Receiver sensors and field components

The decay of the eddy currents is monitored by measuring the secondary magnetic field. The EM receiver may measure the strength of the secondary magnetic (**B**) field, in which case the magnetic sensor is often referred to as a *B*-field sensor. These are rapid-sampling low-noise magnetometers. Alternatively, the EM receiver may be a coil, or a larger loop, in which case the voltage induced in it is proportional to the time rate of change of the strength of the decaying magnetic field (see Section 5.2.2.2), i.e. its time derivative d**B**/d*t*. A coil is often referred to as a d*B*/d*t* sensor. Techniques for obtaining *B*-field data from d*B*/d*t* measurements are used in airborne survey systems and are described in Section 5.9.1.2.

Both types of sensors sense the secondary magnetic field in a particular direction, i.e. a particular component of the vector field (see online Appendix 1). The component in the vertical is known as the *Z component*; the horizontal component in the direction of the survey line, i.e. the along-line component, is known as the *X component*; and the across-line horizontal component is the *Y component*. In DHEM a set of three components specific to the drillhole is used (see Section 5.8.1.3).

The orientation of the secondary magnetic field varies with location around the conductor, and at any location it may vary with time owing to the migrating flow paths of the eddy currents. Coupling between the sensor and the secondary field will, in general, be different for each component. The three-component measurements fully describe the orientation of the secondary field at the measurement location and the relationships between the component responses provide diagnostic information about conductor location, geometry and attitude.

5.7.1.6 On-time and off-time measurements

Recall from Section 5.7.1.1 that time domain EM is based on measuring the response of the ground to an abrupt change in the primary magnetic field. The strength of the secondary magnetic field is measured as a function of time since the primary field turn-off (or turn-on), known as the *delay time*. Measurements for a given delay time are referred to as *channels*, and are often referred to by their number, with channel 1 usually being for the least delay time etc. That is, a time series (see Section 2.2) of the field's decay is created. For most systems the measurements are made after the pulse turn-off, when there is no primary field, and are known as *off-time* measurements (Fig. 5.73a). In some systems *on-time* measurements, in the presence of the primary field, are also made. Explanations of the various measurements and the responses they produce are given in Section 5.7.1.7.

Accurate time synchronisation of the transmitter and receiver is essential since measurements are made at millisecond intervals with time series typically spanning a few tens of milliseconds, and spanning a few seconds when measuring very slow decays. Each measurement is nominally made at a defined delay time. However, in practice, most EM systems make a very large number (thousands) of measurements of the secondary decay for tens to thousands of repetitions of the transmitted pulse (see Fig. 2.19). Measuring multiple decays allows equivalent measurements in the time series to be stacked to suppress noise (see Section 2.7.4.1). This is particularly important at late delay times when the secondary signal is very weak and usually obscured by noise. The stacked measurements are merged into a smaller number of receiver 'windows', also

a)

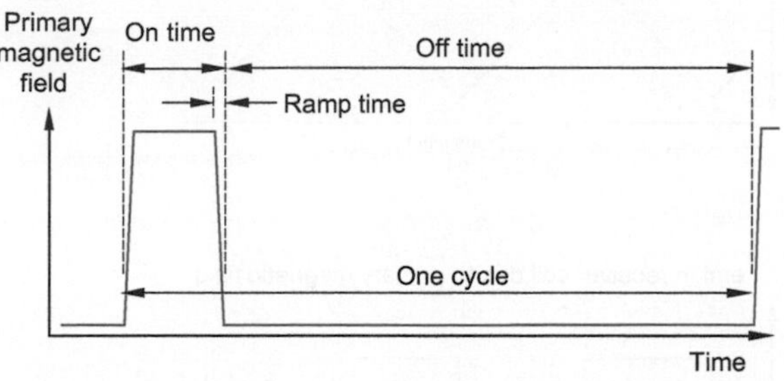

b)

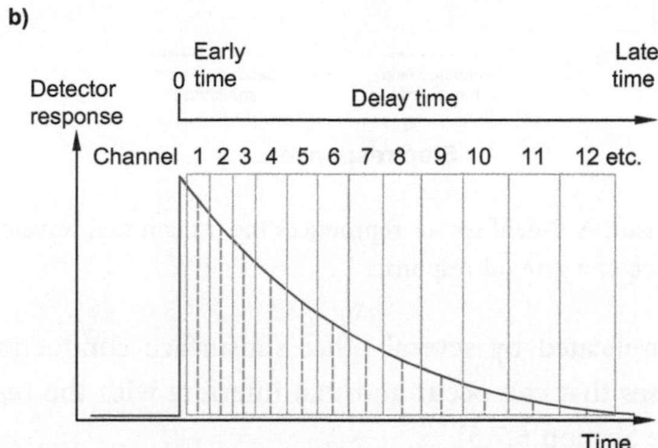

Figure 5.73 Schematic illustration of the basic time domain EM transmit–receive cycle showing (a) the on-time of the primary pulse, the ramp-time for its turn-off and the off-time period. (b) The receiver channels for measuring the decaying secondary field at various delay times during the off-time period.

referred to as channels, which have a finite time-width that increases with delay time. The channel amplitudes form the final representation of the decaying secondary field.

The channels are approximately logarithmically spaced in time. More closely timed channels are required at early times to accurately record this rapidly changing part of the decay than are needed for the more slowly changing late-time decay (Fig. 5.73b). The delay time for each channel and the period over which the decay is recorded are set by the system base frequency (see Section 5.7.3.1).

On-time measurements are critically affected by variations in the location of the receiver with respect to the transmitter (as for FDEM measurements; see Section 5.7.1.1) which cause variations in the measurements. Accurate monitoring of system geometry is essential in order to minimise this source of system noise. Off-time measurements are less affected.

5.7.1.7 Measuring step and impulse responses

Measuring the decaying secondary magnetic field with a *B*-field sensor after a step change in the primary field produces the *step response* of the ground. In other words, the measured decay is how the ground responded following a step change in the primary field. In the same way, using a *B*-field sensor to measure the ground's response to an impulse change in the primary field, i.e. an infinitely narrow time pulse, produces the *impulse response* of the ground.

Recall from Section 5.7.1.5 that a coil measures the rate of change of a magnetic field (**B**), i.e. its time derivative (d**B**/d*t*). Replacing the *B*-field sensor with a coil produces the derivative or slope of the step response; and given that the time-derivative of a step function is an impulse, the measured response is the impulse response of the ground. A coil can also be used to obtain the step response by varying the primary field as an infinitely long triangular pulse instead of as a step, because the triangular pulse is the integral (reversal of the time-derivative) of the step function. In this case it is the total field that is measured, i.e. the primary field plus the secondary field, during the variation of the primary field. The responses obtained with the various combinations of waveforms and sensors are summarised in Table 5.3. Note that for a variety of reasons it is not possible to transmit these ideal, or theoretical, primary field waveforms. In practice, a series of finite-width pulses are transmitted and the measured responses, depending on the nature of the pulses (the primary field variation), have many of the attributes of the step and impulse responses. It is also possible to calculate *B*-field data from d*B*/d*t* measurements made during both the on- and off-times of the primary signal pulse, a procedure used in AEM systems (see Section 5.9.1.2).

Figure 5.74 shows waveforms comprising series of square and triangular pulses and the responses measured with a coil in the absence of the ground response (a *B*-field sensor response is the same as the transmitted waveform). In this case series of 'near-perfect' steps and impulses are recorded by the sensors. The ground response distorts the measured waveforms, which then contain the information of interest about the electrical properties of the subsurface.

Table 5.3 **The measured responses for various combinations of receivers and perfect primary waveforms.**

Primary field	Receiver	
	B-field (magnetometer)	d*B*/d*t* (coil)
Step	Step	Impulse
Impulse	Impulse	
Triangle		Step

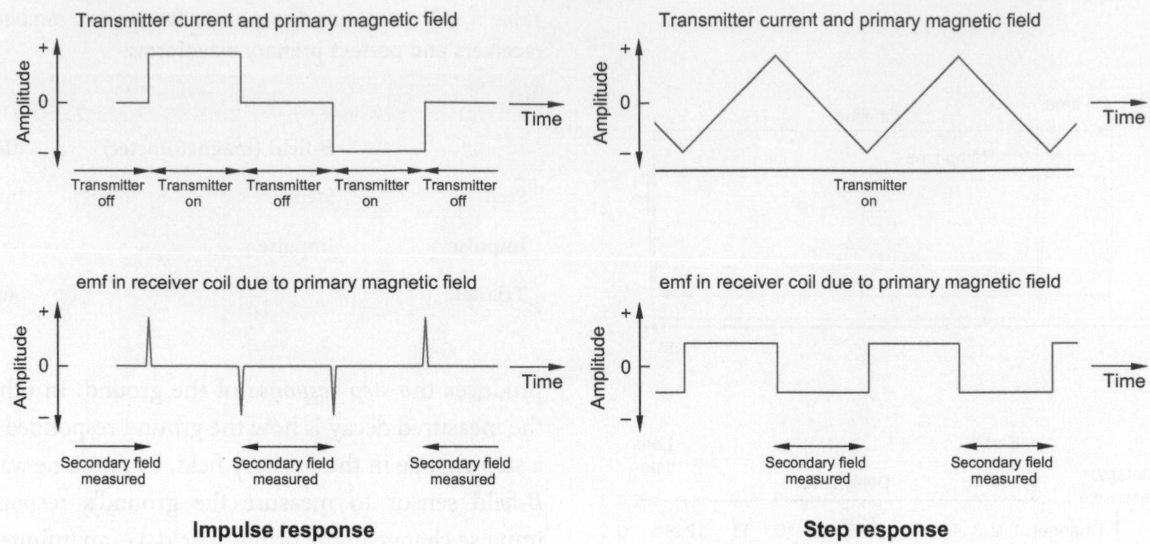

Figure 5.74 Transmitter waveforms and the responses obtained with a coil sensor. A *B*-field sensor reproduces the transmitted waveform. The (approximate) impulse and step response waveforms are shown in the absence of a ground response.

Until comparatively recently, *B*-field sensors capable of making low-noise measurements of the weak time-varying secondary field were unavailable. Instead, coils were used to obtain the impulse response from a square-like pulse of the primary field, and the step response was obtained by the transmission of a triangular pulse. Most of the modern TEM systems produce both *B*-field and d*B*/d*t* data. The nature of the responses obtained critically depends on the actual waveform transmitted, but the respective responses contain many of the attributes of the ideal step and impulse responses. Without further qualification, we refer to these as the step and impulse responses throughout our description of TDEM.

There are significant differences between the step and impulse responses and the information they contain about a conductor, which are described in Section 5.7.2. An example of d*B*/d*t* and *B*-field data from the same location is presented in Section 5.7.6.

5.7.2 Subsurface conductivity and EM responses

Time domain EM responses of the subsurface can be complex, but they can be understood in terms of the responses of an electrically homogeneous background, a conductive overburden (if present) and localised 'target' regions of contrasting conductivity. Key to understanding TDEM responses is that at the instant of current turn-off, eddy currents are created at all interfaces across which there are contrasts in electrical conductivity, and that couple to the primary field. The observed decay is due to a combination of their individual decays. In practice the situation is further complicated by several other subsurface conduction processes that can occur and that interfere with the responses (see Section 5.7.6).

5.7.2.1 Homogeneous subsurface

It is convenient to begin the description of TDEM responses using the simplest case of an electrically homogeneous ground where the only contrast in electrical properties occurs at the ground surface, i.e. the ground–air interface. This is known as a *half-space* (see *Half-space model* in Section 2.11.1.3). It is a useful model to introduce the key concepts of diffusion, smoke ring, diffusion depth and decay rate which describe the electromagnetic diffusion process and are used for the analysis of the measured secondary decay.

Recall from Section 5.2.2.2 that initially eddy currents flow so as to try to oppose any change in the primary field, i.e. they attempt to maintain the primary field everywhere as it was prior to turn-off. For the case of the surface transmitter loop, the eddy current is an image of the loop (Fig. 5.75a). When the loop is elevated above the ground, as in AEM surveying, the eddy current at the instant of turn-off will be laterally more expansive than the loop.

Figures 5.76a to c show a cross-section through a horizontal transmitter loop on the surface of a half-space and the induced migrating eddy current at progressively greater delay times for a step turn-off of the primary field. As described in Section 5.7.1.4, immediately upon its creation the eddy current begins to expand and migrate outward and downward, losing energy rapidly and causing the region to experience a changing magnetic field.

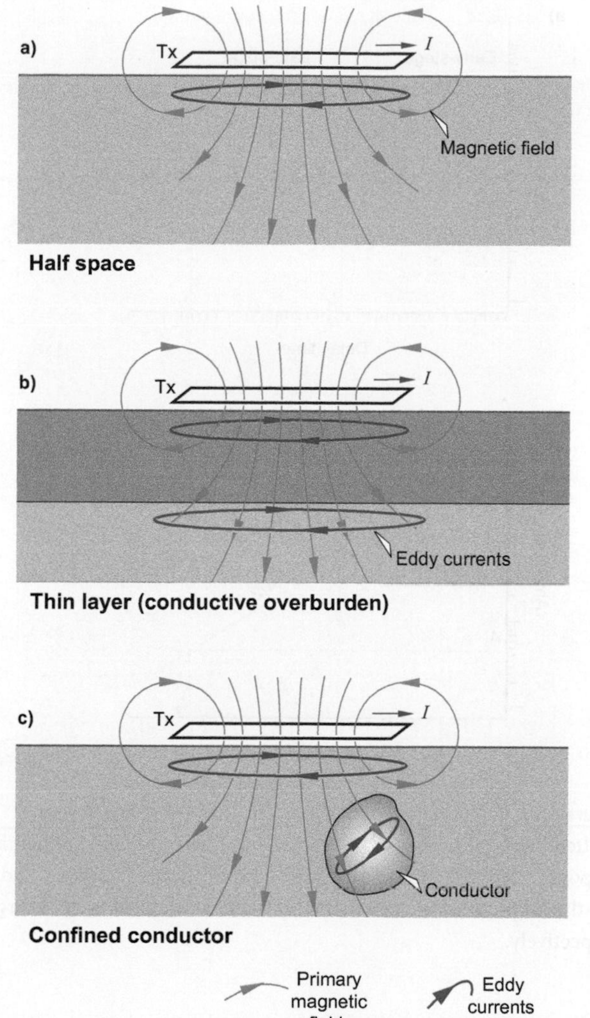

Figure 5.75 Induction of eddy current systems. (a) Into conductive half-space, (b) into conductive half-space with conductive overburden and (c) into conductive half-space hosting a discrete conductor.

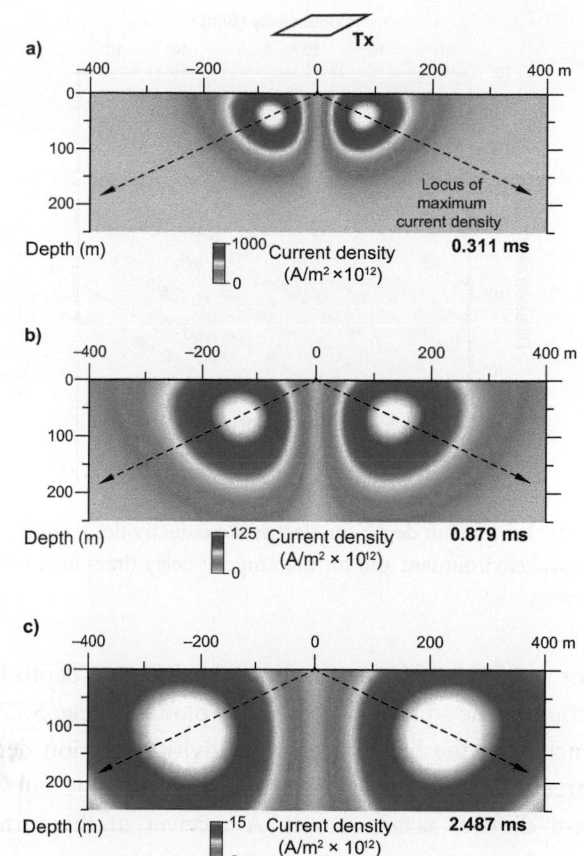

Figure 5.76 Schematic illustration of the diffusion of the eddy current system in a conductive half-space at different delay times. (a) Early, (b) mid and (c) late delay times. The data shown are computed for a half-space conductivity of 0.1 S/m. Based on a diagram in Reid and Macnae (1998). The current density decreases rapidly with time so different colour scales are used in each part of the diagram.

The result is a doughnut-shaped zone of current flow below the loop referred to as a '*smoke ring*'. It is a deformed image of the loop that becomes more blurred with time. As the system continues *diffusing* into the subsurface, the current and the speed with which it moves decrease. After a very short initial period, the smoke ring expands further and moves downwards at an angle of approximately 30°. The amplitude of the response depends strongly on the conductivity and the velocity with which the current system moves away from the transmitter loop.

An important aspect of *diffusion* in a half-space is that the only boundary confining the expanding current system is the ground–air interface; the system is otherwise unconfined and free to expand in all other directions. The half-space is referred to as an *unconfined conductor*.

Diffusion depth

The depth to the maximum current density (see Section 5.2.1.2) at a particular delay time (t) is known as the *diffusion depth* (d); it is a measure of skin depth (see Section 5.2.3.1) in the time domain and depends on the conductivity (σ) and magnetic properties of the ground, described by the magnetic permeability (μ; see Section 3.2.3.3). It is given by the expression:

$$d = \sqrt{\frac{2t}{\mu\sigma}} \tag{5.22}$$

where t is in seconds, d in metres and σ in siemens/metre. For most rocks μ is nearly the same as that of a vacuum ($\mu = \mu_0 = 4\pi \times 10^{-7}$ henry/m) (see Zhdanov and Keller, 1994) so the expression can be rewritten as:

$$d = 1261.6\sqrt{\frac{t}{\sigma}} \tag{5.23}$$

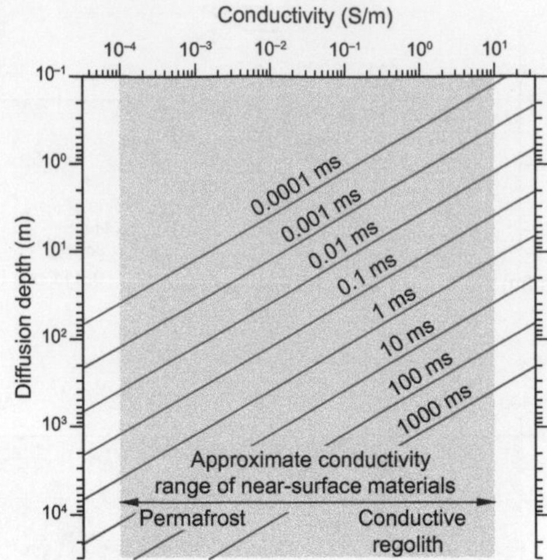

Figure 5.77 Diffusion depth for electrical conductivities found in the geological environment and for the range of delay times used in EM surveying.

For a given half-space conductivity, diffusion depth is a function of the square-root of t. It is plotted in Fig. 5.77 as a function of the half-space conductivity. Diffusion depth is larger for more resistive ground because the current system diffuses faster into it. A receiver at the surface senses the smoke ring attenuating quickly, i.e. a rapid decay in signal amplitude as the distance to the moving smoke ring rapidly increases. In resistive ground it is necessary to make measurements at early delay times in order to detect the rapidly expanding smoke ring. Its velocity decreases with increasing conductivity; in other words, diffusion is a slow process in conductive ground. As a consequence, measurements made at early delay times pertain to shallower depths in conductive environments than in resistive ones. Note that without the conductivity being known, delay time is an unreliable indicator of the depth to which particular measurements pertain.

Late-stage response

As described in Diffusion depth above, the current system initially expands rapidly. At late times when it has diffused to a very large distance from the transmitter loop, the current system limits its own expansion and its position changes very slowly, i.e. its velocity decreases to a low value. This creates a large region where the secondary field is vertical and of fairly constant amplitude everywhere. This is known as the *late-stage response* of the half-space, and the amplitude of the secondary field decays with a power law, i.e. the signal varies with delay time at the rate of t^{-k} where k is the *power-law constant.*

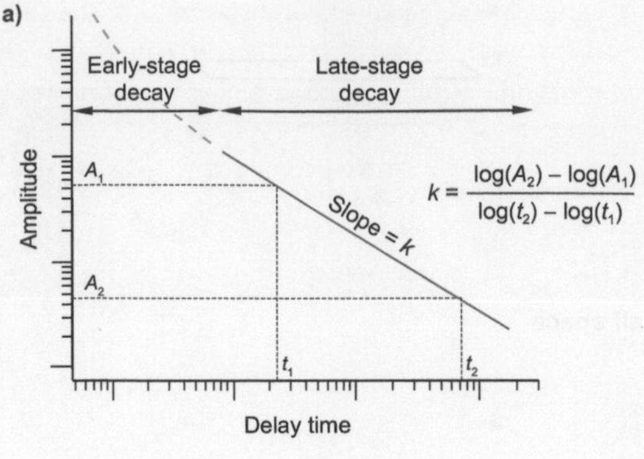

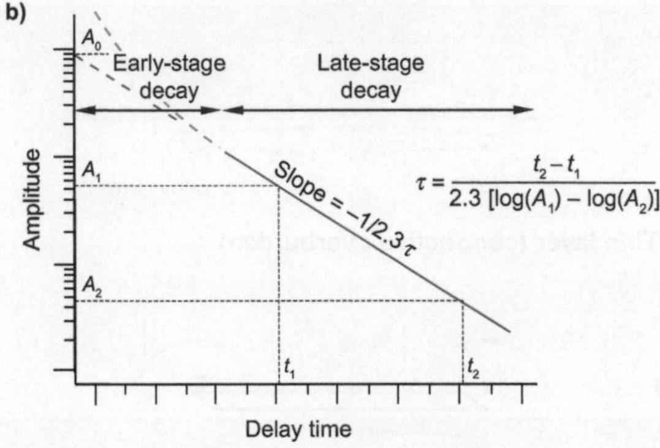

Figure 5.78 TDEM secondary decay for (a) power-law response plotted on log–log axes with decay constant k and (b) exponential response plotted on log–linear axes with time constant τ. A_1 and A_2 are the amplitudes of the secondary field at delay times t_1 and t_2 respectively.

A graph of the logarithm of the signal amplitude on the vertical axis versus the logarithm of the delay time on the horizontal axis shows the power-law decay as a straight line with slope equal to the decay constant ($-k$) (Fig. 5.78a). The value of k depends upon whether an impulse or step response is being measured and on the component of the secondary magnetic field measured, and is independent of the conductivity. Values of k for a half-space response are shown in Table 5.4. Note that the late-stage response of a half-space is the same everywhere, so it does not matter where the receiver is in relation to the transmitter loop.

5.7.2.2 Thin layer

A thin flat-lying layer with higher conductivity than the surrounding rocks is a commonly occurring conductivity structure in the geological environment that couples well to a horizontal loop. It commonly takes the form of a surface layer, referred to as *conductive overburden* (see Section 5.3.4), and produces a very strong EM response because the loop is close to it.

Table 5.4 **Values of the power-law constant (*k*) for the thin layer and half-space responses, and for current (*I*) -channelling within them, for horizontal (*X* and *Y*) and vertical (*Z*) component measurements. Note that the decay in the horizontal components is faster than that of the vertical component.**

	Step		Impulse	
	X and *Y*	*Z*	*X* and *Y*	*Z*
Thin layer	−4	−3	−5	−4
I-channelling		−4		−5
Half-space	−2	−1.5	−3	−2.5
I-channelling		−2.5		−3.5

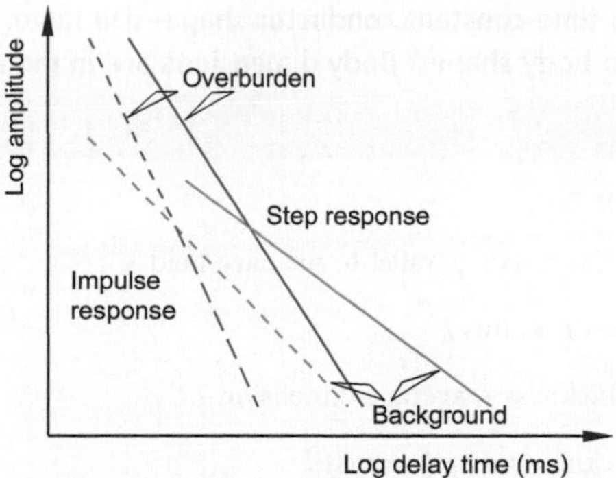

Figure 5.79 The power-law decays of conductive overburden (a thin surface layer) and conductive background (a half-space) for the step and impulse responses. Actual decay constants (slope values (*k*) in Fig. 5.78a) are given in Table 5.4.

In this case eddy currents are induced into the conductive layer at the conductivity interface of its upper surface (Fig. 5.75b). Diffusion is confined to within the layer and, for a homogeneous layer of infinite lateral extent, the current system is free to quickly expand laterally. It is an unconfined conductor with response decaying as a power law (Fig. 5.78a) with decay constant (*k*) larger than that of a half-space, i.e. the decay is faster than that of a half-space. The decay in a thin layer is controlled by the product of its conductivity and thickness, but not by these parameters separately, i.e. the *conductivity–thickness product*, also known as the *conductance* (see Section 5.2.1.2). Values of *k* for a thin layer are shown in Table 5.4.

As the thickness of the layer increases, the diffusion begins to experience vertical migration downward, as well as outward, from the upper interface (closest to the source loop), and the response approaches that of a homogeneous half-space. Where the subsurface is multi-layered, induction occurs simultaneously in each conductive layer and their decaying responses interact. Their conductance and depth, and the spacing between the layers, determine the amplitude of their responses and the resolution of each in the measured decay. Thin resistive layers have low conductance and tend to be invisible.

Figure 5.79 shows the responses measured in the presence of conductive overburden. The relative decay rates of the thin layer and half-space responses means that at early delay times the strong response is chiefly due to the overburden, so early-time measurements are necessary for resolving the rapidly decaying overburden response. At later times when the overburden response has diminished, the weaker, slower-decaying response of the deeper eddy current system in the underlying basement, i.e. the conductive background, becomes apparent. Further aspects of near-surface conductive material and its responses are described in Section 5.7.6.1.

5.7.2.3 Confined conductor

When a homogenous subsurface contains a discrete zone of contrasting conductivity, two eddy current systems are created when the primary field is turned off (Fig. 5.75c). One current system is induced in the background material, as described in Section 5.7.2.1, and the other induced in the discrete conductor. Those in the conductor try to reproduce the primary field in the vicinity of the conductor, then immediately begin to diffuse over and through the body with their expansion being confined by its boundaries, which have a significant influence on the nature of the transient decay. It is known as a *confined conductor*, also referred to as a *discrete conductor*.

When the conductor is in a high-resistivity environment and there is no interaction with its surroundings, the initial surface eddy current flow is dependent in a complex way on the geometry of the conductor. With time the current system migrates through the body and, for an electrically homogeneous body, the late-stage decay is exponential. The amplitude (A) as a function of delay time t for the step response is then given by:

$$A(t) = A_0 e^{-t/_\tau} \quad (5.24)$$

and for the impulse response by:

$$A(t) = -\frac{1}{\tau} A_0 e^{-t/_\tau} \quad (5.25)$$

where A_0 is the apparent initial amplitude of the exponential decay which is dependent upon the conductor's shape,

Table 5.5 The time-constant conductor shape–size factor *S* for some common body shapes. Body dimensions are in metres.

Body shape	*S*
Sphere, radius *r*	r^2
Cylinder, radius *r*, axis parallel to primary field	$1.71r^2$
Disc, thickness *t*, radius *r*	$1.79tr$
Thin plate, thickness *t*, average dimension *L*	tL
2D plate, thickness *t*, depth extent *l*	$2tl$

size and depth, and τ is the *time constant*, the time taken for the signal to decay to $1/e$ or 36.8% of its initial value and has the same units as *t*, usually milliseconds. The time constant τ has the same value for both the step and impulse responses and depends on the conductivity and the effective cross-section of the conductor and is given by:

$$\tau = \frac{\mu\sigma S}{\pi^2} \qquad (5.26)$$

where *S* is the shape-dependent size of the body (m^2), for which formulae for some common body shapes are given in Table 5.5. Other variables are defined in *Diffusion depth* in Section 5.7.2.1. Note from Eq. (5.26) that conductivity–thickness product (see Section 5.7.2.2) is fundamental in determining the response of plate-like conductors.

A graph of the logarithm of the signal amplitude on the vertical axis versus the delay time on the linear horizontal axis shows the exponential decay of Eqs. (5.24) and (5.25) as a straight line with slope proportional to the inverse of τ (Fig. 5.78b). The value of τ for mineralisation ranges typically from about 200 μs to hundreds of milliseconds, and to several seconds for very high-quality conductors; see for example Fig. 5.81. An indication of the shape of the conductor can be obtained from the spatial variation of the response, and its conductivity, for the appropriate model shape, can be estimated using Eq. (5.27) (by rearranging Eq. (5.26)):

$$\sigma = \frac{\tau}{1.27 \times 10^{-7} S} \qquad (5.27)$$

Conductor quality

The time constant τ quantifies the 'quality' of the conductor; a low value is indicative of a poor conductor having low conductivity and/or small size, a high value indicative of a good conductor having high conductivity and/or large size.

The time constant τ and the conductor geometry control the amplitude of the secondary field (Fig. 5.80). Good

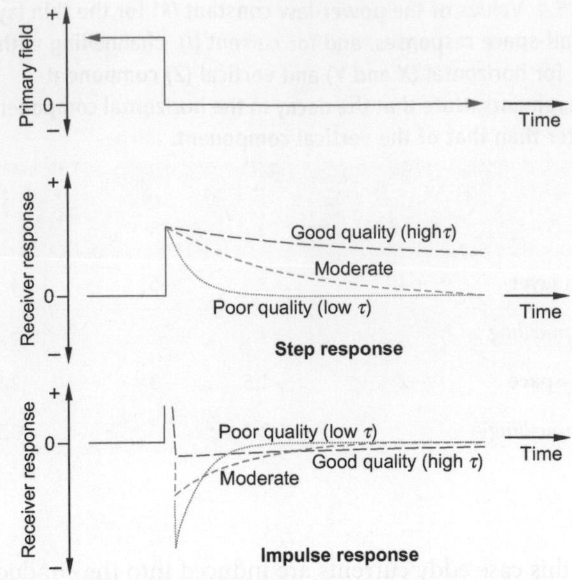

Figure 5.80 The exponential decays of a confined conductor of poor, moderate and good quality schematically illustrated for the step and impulse responses due to a perfect step turn-off of the primary field. The reverse polarity of the impulse response is a consequence of the negative sign in Eq. (5.25). Redrawn, with permission, from West and Macnae (1991).

conductors (large τ) maintain the current system for a long time and are referred to as *late-time conductors.* Poor conductors (small τ) lose the energy faster because of their higher resistivity and are referred to as *early-time conductors.* Note from Eqs. (5.24) and (5.25), and as shown in Fig. 5.80, that for a given body geometry the initial ($t = 0$) amplitude (A_0) of the step response is the same for all values of τ, i.e. it is independent of conductivity and depends on the shape, size and depth of the body. For the impulse response it is also inversely proportional to τ, i.e. it is mainly inversely dependent on the quality of the conductor. So for the case of an impulse of the primary field, poor conductors produce a high-amplitude eddy current flow at first, but the energy is quickly lost. Good conductors produce a weaker eddy current at first but maintain the current system for a longer time. For both responses, measurements to later times allow discrimination between good and poor-quality conductors.

Late-time measurements

When a conductive overburden layer is present and/or the host/country rocks are conductive, their strong and fast decaying power-law responses dominate at early to mid-times (Fig. 5.79) and obliterate the weaker exponential decays of confined conductors. In these situations the confined conductor response will only be detectable at

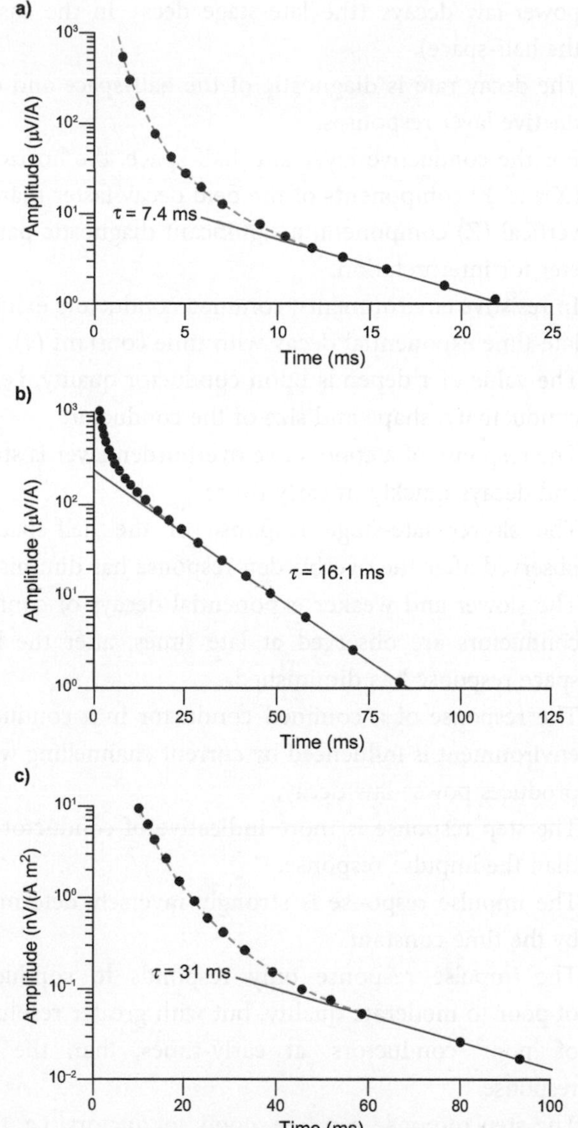

Figure 5.81 Log–linear decay plots showing late-time exponential decay due to target conductors. (a) Ernest Henry deposit, redrawn, with permission, from Webb and Rowston (1995), (b) Trilogy deposit, redrawn, with permission, from Sampson and Bourne (2001), and (c) Eloise deposit, redrawn, with permission, from Brescianini *et al.* (1992). Note the different time scales and decays (see Section 5.7.7.1 and Fig. 5.89 for additional information).

mid- to late times after the overburden and half-space responses have diminished to reveal the slower-decaying target response. This is evident in Fig. 5.81 where the exponential decay due to the conductive mineralisation is only apparent at later times. Late-time measurements are essential for exploring in conductive environments, but poor conductors can remain undetected since their decay may be too rapid to be evident after the overburden and host rock responses have disappeared. Resolving individual responses from the observed composite response, assuming it is due only to the decaying secondary magnetic fields, is described in *Decay analysis* in Section 5.7.5.3.

5.7.2.4 Current channelling

In conductive environments the target conductor is often electrically connected to the conductive host rocks, and possibly also to conductive overburden if present, allowing current to flow between them and their individual eddy current systems to interact. Also, and as described in *Diffusion depth* in Section 5.7.2.1, diffusion in a highly conductive background is slow. In this case, the conductor is energised in a complex manner by three mechanisms: the collapsing primary field of the loop, the varying magnetic field of the slow-moving half-space eddy current system, and the half-space currents flowing through the conductor in an effect known as *current gathering* or *current channelling*. The fields are inextricably coupled.

Recall that current flows via the path of least resistance (see Sections 5.3.1 and 5.6.1), so current channelling only occurs when the conductor is electrically connected to its host rocks, and when the half-space current system is passing through the region of the conductor. It is a function of the background conductivity and distance from the loop to the conductor. In resistive environments it is likely to be seen at early times and for a short time only as the fast-moving background current system passes through, but in conductive environments it persists for a longer period. With increasing distance from the loop, induction in the conductor becomes weaker, but the half-space eddy currents can be stronger and dominate current flow through the electrically connected conductor. The effect is weaker when the conductor is located within the transmitter loop (c.f. Fig. 5.76).

Current channelling increases the amplitude of the response for a surface receiver and broadens the anomaly of the conductor, and produces a faster-decaying power-law response (see Table 5.4). It can usefully increase the response of poor conductors, but the strong responses it produces makes it difficult to distinguish between poor and good conductors. In highly conductive environments, current channelling can completely obliterate the exponential decay with the loss of diagnostic information about the conductor, apart from identifying its location. In a similar way, the expanding current system of a conductive overburden layer can be the source of the channelling current.

A resistive body in conductive host rocks will deflect the half-space currents around itself and can produce a 'conductor response' because of the increased current density in its immediate surrounds.

5.7.2.5 Step and impulse responses

The step and impulse responses (see Section 5.7.1.7) have different characteristics (Fig. 5.80). For confined conductors, the amplitude of the step response (Eq. (5.24)) decreases with decreasing time constant (τ) for all delay times. It is fairly constant for a wide range of conductance but it is strongly determined by conductor geometry and size, i.e. large bodies produce strong responses and small bodies produce weak responses. So step response data are generally quite intuitive to interpret, as the response amplitude for targets of a given size and depth can be predicted. They provide diagnostic information about a conductor for a wide range of conductivity.

In contrast, the amplitude of the impulse response (Eq. (5.25)) at early delay times is also strongly determined by the inverse of τ, the amplitude being stronger for bodies with low τ and decreasing quickly as τ increases. The reverse occurs at late times where the response is higher for higher τ. The impulse response is complexly dependent upon both conductivity and geometry, making the analysis of impulse response data more complex. It responds only to conductors of poor to moderate quality and, significantly, conductors of very high quality are 'invisible' to the impulse response. The signal-to-noise ratio for good conductors is greater in the step response than in the impulse response, and for poor conductors it is greater in the impulse response.

Step response data are required for detecting and characterising high-quality conductors, e.g. large highly conductive zones like massive sulphides, in particular nickel sulphides. Step response systems can see good conductors at greater depth, and at greater distance in DHEM, than impulse response systems. Also, the stronger step response of good conductors at early times allows greater discrimination of targets located under weakly conductive overburden. The impulse response complements the step response by increasing the resolution of conductors of low conductivity and/or small size.

Note that, amongst other parameters, the ability to differentiate bodies of different conductivity is also strongly determined by the shape of the transmitted pulse and the system base frequency (see Section 5.7.3.1).

5.7.2.6 Summary of subsurface EM responses

The key characteristics of TDEM responses are summarised as follows:

- The conductive layer and the homogeneous half-space are unconfined conductors whose responses exhibit power-law decays (the late-stage decay in the case of the half-space).
- The decay rate is diagnostic of the half-space and conductive layer responses.
- For the conductive layer and half-space, the horizontal (*X* and *Y*) components of the field decay faster than the vertical (*Z*) component, a significant diagnostic parameter for interpretation.
- In resistive environments, confined conductors exhibit a late-time exponential decay with time constant (τ).
- The value of τ depends upon conductor quality, i.e. the conductivity, shape and size of the conductor.
- The response of a conductive overburden layer is strong and decays quickly in early times.
- The slower late-stage response of the half-space is observed after the overburden response has diminished.
- The slower and weaker exponential decays of confined conductors are observed at late times, after the half-space response has diminished.
- The response of a confined conductor in a conductive environment is influenced by current channelling which produces power-law decay.
- The step response is more indicative of conductor size than the impulse response.
- The impulse response is strongly inversely determined by the time constant.
- The impulse response only responds to conductors of poor to moderate quality, but with greater resolution of 'poor' conductors, at early-times, than the step response.
- The step response can 'see' good conductors, i.e. those with very large time constants (τ), which are 'invisible' to the impulse response.
- Measurements over a wide range of delay time provide greater ability to discriminate between the various classes of conductors, and between poor and good conductors.

5.7.3 Acquisition of EM data

Conductors with different electrical properties, different shapes and orientations, at different depths, occurring in isolation or near other conductors, and in host rocks with different conductivities, with or without a conductive overburden, are optimally detected in different ways by EM surveys. The ultimate aim of the survey (i.e. whether it is detection of a discrete conductor, such as an orebody, or mapping spatial variations in conductivity, such as a regolith profile) determines the way in which the data are acquired. It is impossible to design a single EM survey that optimally fulfils all possible requirements – one reason why there are many types of EM survey systems available.

As described in Section 5.7.2.5, there are advantages to the step and impulse responses in different circumstances, and different targets will require the decay of secondary fields to be measured over different time intervals. Additional survey parameters that need to be considered are the base frequency, the system geometry and how the data are normalised. Together these parameters influence the system's ability to detect and resolve the geometry and attitude of a conductor. They also strongly determine the system's depth of investigation in a particular environment.

Sometimes EM data are acquired primarily to determine vertical variations in conductivity, i.e. EM soundings, the equivalent of the electrical soundings described in Section 5.6.6.1. The data are treated in the same way as data collected along traverses, albeit with each dataset treated in isolation.

5.7.3.1 Transmitter waveform and base frequency

The primary field waveform is a fundamental control on the response recorded. The waveform can be described by its frequency spectrum (see online Appendix 2), with differently shaped waveforms having different frequency content.

It is impossible to create the ideal step and square pulse variations in the primary field because these would require instantaneous changes in the transmitter loop current, which is prevented by the inductance of the loop (see Section 5.7.1.2). Instead a less-desirable slower rise and fall is produced (Fig. 5.73a). The rise may be linear with time or of some other mathematical relationship; the turn-off is usually a relatively fast linear ramp. There is a corresponding variation in the responses measured. Systems based on the transmission of a triangular waveform are less affected by loop inductance. Figure 5.97 shows the actual system waveforms for a variety of AEM systems. Compare these with the ideal step waveform of Fig. 5.80.

In general, high-powered systems have slower turn off ramp times and lower-powered systems have faster turn-off. The ramp time of the pulse turn-off determines the delay time of the earliest measurement and significantly influences the resolution of fast-decaying conductors, whilst the width of the pulse affects the resolution of slower-decaying conductors.

A key acquisition variable is the repetition rate of the primary field pulse, called the *system base frequency*. Reducing the base frequency creates a primary field with more low-frequency energy. It also increases the transmitter off-time so the secondary field can be measured to later delay times, important for resolving good-quality conductors (see *Conductor quality* in Section 5.7.2.3). Increasing the base frequency has the opposite effect of increasing the high-frequency energy. It also produces a faster pulse turn-off allowing the secondary field to be measured closer in time to the pulse turn-off, important for resolving poor-quality conductors and near-surface features. For mobile, i.e. airborne, systems, the shorter transmit-receive cycle means there is greater lateral resolution because the system travels less during each cycle.

The base frequency of most EM systems can be adjusted to allow the system to be 'tuned' for a particular survey objective, and to minimise sensitivity to powerline interference. Setting the system base frequency to that providing the best resolution of the target sought is fundamental to the success of the survey. A series of repeat line-surveys over known target areas using different base frequencies can help in this regard. Otherwise, modelling results can be used (if sufficient information is available about the actual electrical parameters of the target and the environment).

5.7.3.2 System geometry

The system geometry is the arrangement, spacing, sizes and orientations of the transmitter and receiver. Whether the survey is for reconnaissance work or for prospect-scale surveying will determine the transmitter–receiver configuration used, which also determines the systems lateral resolution.

A coil or loop which lies in the horizontal plane, such that its axis is vertical, is described as *horizontal*. For large loops placed on the ground and for AEM surveys this is the only practical possibility. A horizontal loop couples well with host rocks, horizontal layers and conductors with a wide range of dips. Transmitter loops for ground surveys are rectangular and generally consist of a single turn of insulated wire laid on the surface. The situation for airborne surveys is discussed in Section 5.9.2. Loop size is discussed in *Transmitter loop size* in Section 5.7.3.2.

Most EM systems simultaneously measure both the vertical (Z) and the along-line (X) components of the secondary field (see Section 5.7.1.5). Sometimes the across-line (Y) component is also measured. The perpendicularly orientated components allow better characterisation of conductors than a single component. When the transmitter loop and receiver coil lie in the same plane the configuration has *co-planar* geometry, and it has *co-axial* geometry when their axes are aligned.

Two survey modes are in common use: *moving loop* and *fixed loop*.

Moving-loop mode

For reconnaissance surveying and when the dip direction of a conductor is unknown, the moving-loop mode is very effective and provides good resolution. Both the transmitter and receiver are moved along the survey line with their relative positions fixed. The most common configuration has the receiver located at the centre of the transmitter loop and is known as the *in-loop* configuration (Fig. 5.82a). The data are located to the loop centre. The array is insensitive to the horizontal (*X* and *Y*) components of the secondary fields of horizontal conductive layers and conductive half-space (see Fig. 5.84).

The asymmetric *separated-loop* configuration (Fig. 5.82b) is used in towed-bird AEM systems (see Section 5.9.2.1) and for ground surveying. It produces the best resolution of steeply dipping thin conductors. Separation between the centre of the loop and the receiver for ground surveying is typically twice the loop size. The data are located to the midpoint between the loop centre and the receiver. The laterally separated geometry produces responses dependent on the relative positions of the transmitter and receiver with respect to the conductor. This means that the responses of a dipping conductor obtained from surveying adjacent parallel survey lines in opposite directions (as often occurs with airborne data) will be different and laterally offset (see *Profile analysis* in Section 5.7.5.3).

The loop size in moving-loop mode is typically 50 m to 500 m and the whole array usually moved at an interval of 50% of the loop size. The transmitter is at a different location for each measurement, so coupling to any

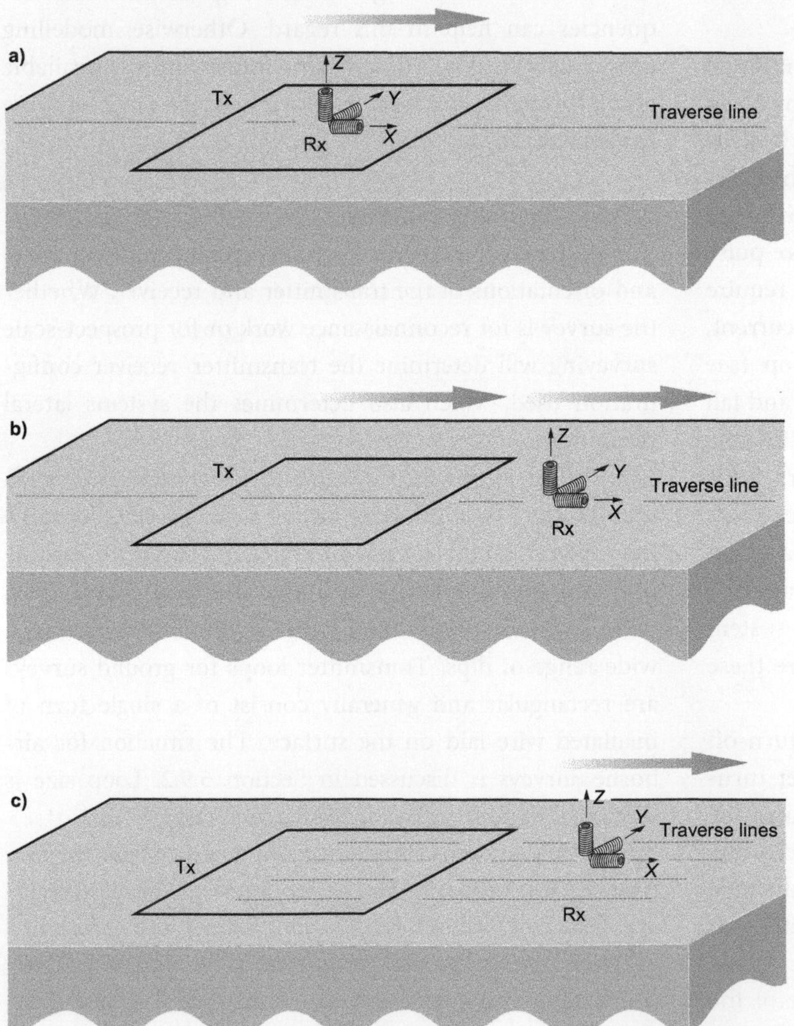

Figure 5.82 The transmitter–receiver configurations for various survey modes. (a) The in-loop and (b) separated-loop configurations of the moving-loop mode, and (c) the fixed-loop mode. Orientations of the *X*, *Y* and *Z* components of the receiver are shown. The *Z* component and the transmitter loop are co-planar in all cases and also co-axial in (a). Tx and Rx are the transmitter and receiver, respectively.

subsurface conductors present (see Section 5.7.1.3) varies along the survey traverse and new eddy current systems are induced at each station (see Section 5.7.1.4 and Fig. 5.70). This is very useful for detecting conductors whose dip and orientations are unknown. A portable generator is used to supply current to the transmitter loop, typically up to 30 A. For a 200 × 200 m single-turn loop the dipole moment (Eq. (5.21)) would be as much as 1,200,000 A m^2. Moving-loop mode provides high resolution of conductors, but anomalies have complex shapes.

Moving-loop mode requires across-line access for laying the loops, which may be logistically difficult in rugged and densely vegetated terrains. The in-loop configuration also requires access to the centre of the transmitter loop.

Fixed-loop mode

For investigating the geometry of a known conductor, or where the dip of the target conductor zone is known, and for deep penetration, the large fixed-loop mode can be used. The transmitter loop is fixed, or stationary, for the duration of the survey. The loop may have dimensions as large as 2000 m by 1000 m, with its longest side oriented parallel to the expected strike of sheet-like conductors and located so as to couple well with the target zone (see Section 5.7.1.3). Prior knowledge about the dip and strike of the conductor is required, possibly obtained from a moving-loop survey, to optimise the location of the loop.

The receiver is systematically moved relative to the transmitter loop, in surface surveys along traverses perpendicular to the loop's longest side and passing through and away from the loop (Fig. 5.82c). In DHEM surveying the receiver is placed at different depths in one or more drill-holes (see Section 5.8.1). The data are located to the receiver location. Survey distance from the side of the loop is limited by the decreasing amplitude of the primary field away from the loop. Since the loop is stationary, it could, if access permits, be energised by a high-power truck-mounted generator producing currents up to about 100 A. For an 800 × 200 m single-turn loop the dipole moment (Eq. (5.21)) would as much as 16,000,000 A m^2.

Fixed-loop configuration provides lower resolution than moving-loop configurations, so three-component measurements (see Section 5.7.1.5) are important for obtaining diagnostic information about a conductor. The survey may be repeated with the fixed loop at different locations relative to the conductor zone to vary the coupling of the primary field with the target. This is useful for detecting conductors with opposite dip direction, for investigating the orientations of the various conductivity planes (current systems) in thick bodies and for resolving close-spaced conductors.

Anomalies measured with a fixed loop have a fairly simple form because the coupling between the loop and the conductor is constant along the survey traverse. Fixed-loop surveying has simpler operational logistics than moving-loop mode as only the receiver is moved along the survey traverse, a distinct advantage when working in rugged and densely vegetated terrains. However, considerable effort may be required in these terrains to lay out the large loop, and access restrictions may limit the survey to just one loop location. Careful planning is required to optimise loop location relative to the target zone.

Transmitter loop size

As described in Section 5.7.1.2, larger transmitter loops produce a stronger EM field penetrating a larger volume of the ground. The increased signal-to-noise ratio at the later delay times increases the depth of investigation. Maximum signal-to-noise ratio and vertical resolution are obtained when the size of the loop is of the same order as the depth being investigated. However, large loops couple well with conductive background rocks and conductive near-surface layers, causing these to produce strong responses, the effects of which can obliterate the target response (see *Late-time measurements* in Section 5.7.2.3). In conductive environments a compromise has to be made between depth of penetration and overburden and background responses. In moving-loop mode, larger loops reduce spatial resolution so, overall, there is a trade-off between depth of investigation and spatial resolution when specifying loop size. Note that a larger loop produces a longer turn-off ramp time (see Section 5.7.1.2) which, and depending on the shape of the primary pulse, determines the delay time of the earliest off-time measurement (Fig. 5.73).

In general, a loop-size to investigation-depth ratio of about 1:3 is suitable in resistive environments, with larger loops required in conductive areas and where thick conductive overburden is present, a ratio of say 1:2. However, the response of conductive overburden and the background increase, but this may not present a problem for a deep probing DHEM receiver distant from the overburden. A surface survey may benefit from using a smaller loop. A sounding may require data to be collected with more than one loop size. The optimum size of the loop for detecting the expected target at a particular depth in the host environment can be determined with computer modelling.

5.7.3.3 EM survey design

Detectability and resolution of a discrete buried conductor depend on the geological environment and the configuration of the EM system used. Parameters that need to be considered include whether the target is a massive conductor or disseminated, the possible range of conductivity, its geometry and orientation, depth of burial and whether it is likely to be in close association with other local or regional conductors. For example, in a particular host environment, a horizontal plate-like conductor may be detectable at greater depths than a similar vertically dipping conductor depending upon the orientation of the primary field at the conductor (Fig. 5.70) and the orientation and location of the receiver etc.; and conductivity of the environment can significantly affect detectability (see Section 5.7.6.1). Despite the fact that the explorer generally has only limited knowledge about the electrical properties of the area to be surveyed, computer modelling techniques can be applied during survey design to estimate the system's depth of investigation and lateral resolution, and the detectability of a particular target in a particular environment for a range of system parameters.

5.7.4 Processing and display of EM data

Since the decaying secondary field is measured at multiple delay times at each survey station (Fig 5.73), the results from an EM survey comprise a multichannel dataset. Ground and downhole EM surveys are usually of comparatively limited extent so the multichannel data volumes obtained are not large. However, AEM datasets can involve significantly large volumes of data.

Reduction of EM data involves normalising the measurements for variations in the strength of the primary field. The data are displayed in two basic ways: as profile plots of variations in channel amplitudes and as displays of conductivity variations in the subsurface.

5.7.4.1 Amplitude normalisation

Among other factors, the strength of the secondary field also depends upon the strength of the primary field, so it is essential that the measurements be corrected for variations in the amplitude of the primary field. The amplitude of each channel is divided, or normalised, by the transmitter current measured at the time of the pulse. Sometimes the measurements are further normalised for the moment of the transmitter loop (see Section 5.7.1.2), the usual practice in AEM systems. For *B*-field sensors (including the recovered *B*-field of AEM systems) the measurement is a magnetic field strength and the normalised measurement has units T/A (where T is teslas and A is amps). For coil sensors, the induced emf is also dependent on the sensitivity of the coil, so it is necessary to normalise the measured voltage for the moment of the coil (see Section 5.7.1.2). The fully normalised measurement has units V/(A N m^2) (where V is volts, A is amps, N is number of turns, m^2 is square metres). Normalisation allows the measurements to be compared with those obtained with other sensors of different sensitivities.

The decrease in primary field strength with distance from the loop in a fixed-loop survey causes the strength of the secondary field to decrease progressively away from the loop, distorting the shape of anomalies. The effect can be reduced by normalising the decay channel amplitudes by the strength of the primary field measured, either at a particular reference station over the target anomaly/conductor or at each survey station. The former is known as *point normalisation* and useful for analysing anomaly shape, and the latter is known as *continuous normalisation* and useful for analysing anomaly amplitude. The actual procedure used varies between the different EM systems, making comparisons of their data difficult.

5.7.4.2 Decay channel amplitudes

Amplitudes of the channel values can be displayed as 1D multichannel profiles showing their variations as a function of location along the survey lines (Fig. 5.83a). For in-loop configuration this is the centre of the loop and for the fixed-loop configuration it is the receiver location. For the separated-loop array it is the midpoint between the loop and receiver. For downhole surveys it is distance measured along the drillhole trajectory. This kind of display is becoming less common for ground and especially AEM data but remains the only way to display DHEM data (see Section 5.8.2).

Decay channel amplitude data typically have large variations (particularly impulse response data) and are usually displayed using a combined logarithmic and linear amplitude scale. Logarithmic scaling suppresses large-scale variations and is used for amplitudes whose absolute value is greater than 1. Linear scaling helps resolve small amplitude variations, useful for the lower-amplitude late-time channels, and is normally used to display amplitudes between +1 and −1.

Profile plots have the advantage of being a true representation of the measurement and they can be rapidly scanned for anomalous responses. Their disadvantage is that they are hard to interpret 'geologically' since the

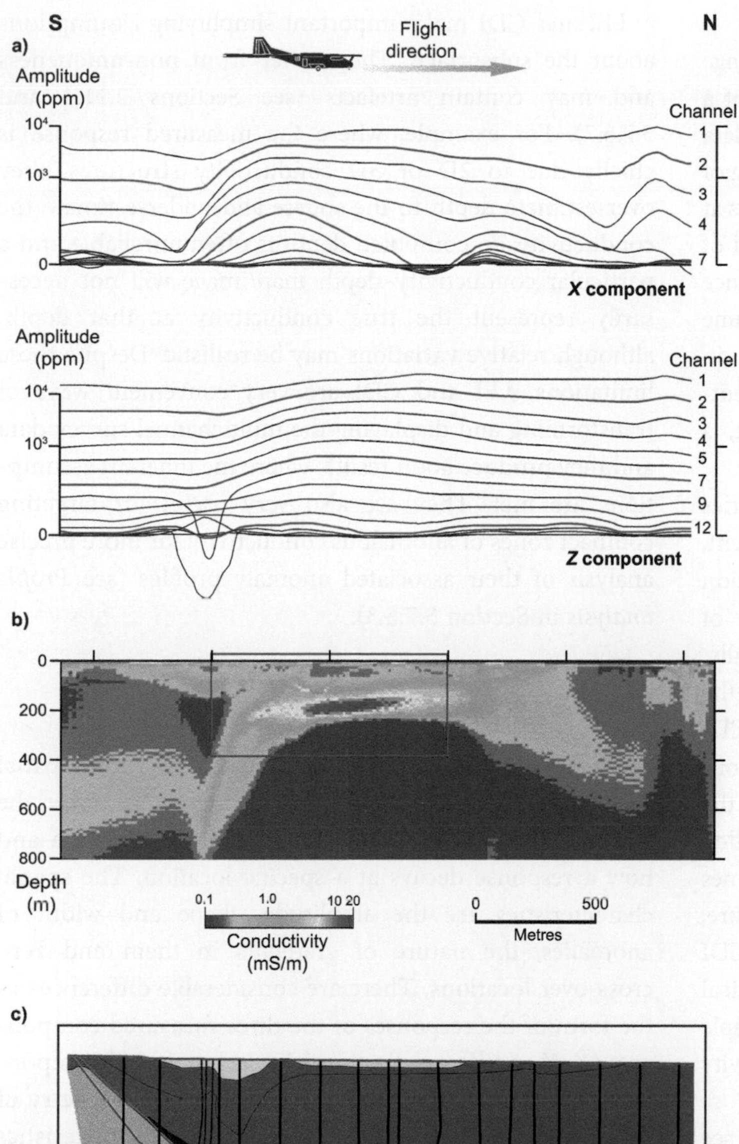

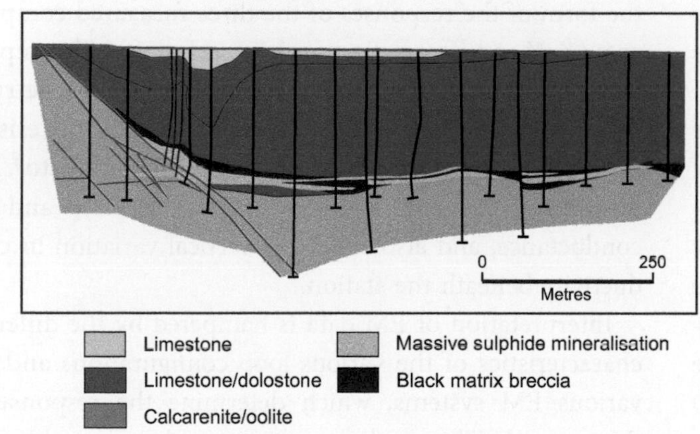

Figure 5.83 GEOTEM (75 Hz) airborne EM data from across the Lisheen carbonate-hosted base-metal deposit. (a) Profile plots, (b) conductivity parasection (the red rectangle shows the extent of part (c)) and (c) geology. Redrawn, with permission, from Nabighian and Asten (2002).

anomalous amplitudes do not resemble the sources of the observed variations. The interpretation of data in this form is described in *Profile analysis* in Section 5.7.5.3.

A second form of amplitude plot is the secondary decay measured at specific stations, which may be analysed to determine power-law decay and exponential time constants to identify specific conductors (see Figs. 5.78 and 5.81).

Another common presentation of amplitude data, when data from multiple traverses are available, is a raster display of the gridded amplitudes of a selected channel (for examples, see Section 5.9.5.1). Computed decay and time constants can also be displayed in this way. Images are interpreted in a qualitative manner as described in Section 2.10.

5.7.4.3 Inferred conductivity

The second form of data display is a coloured-pixel image showing inferred conductivity variations in the form of a map or cross-section (Fig. 5.83b) and occasionally as a data volume. This form of display relies on inverse modelling of the data to compute the conductivity at a range of depths at each measurement location. The advantage of this kind of display is that the results can be readily interpreted since they resemble a cross-section, a geological map or a volume of the subsurface. Two computational methods are in common use: layered Earth inversion (LEI) and current-depth imaging (CDI). Both are computationally fast, a requirement for working with the large AEM datasets.

LEI assumes that the ground is composed of a series of discrete horizontal conductivity layers of infinite extent, i.e. a 1D model (see *One-dimensional model* in Section 2.11.1.3). It produces the thickness and conductivity of the layers, the number of layers being chosen manually. A two-layered model can be very useful for mapping the thickness and conductivity of conductive overburden. CDI determines subsurface conductivity by analysing variations in the measured secondary field decay in terms of the depth–time relationship of the downward moving eddy current 'smoke ring' (see Section 5.7.2.1). It also assumes that the subsurface has a flat-lying conductivity structure, but without the restriction of a fixed number of layers. CDI methods can be more effective in areas where the vertical conductivity distribution is more complex than the simple layered-Earth model, for example where the conductivity changes gradually with depth.

Note that for the case of the in-loop configuration (see *Moving-loop mode* in Section 5.7.3.2), which includes all the helicopter AEM systems (see Section 5.9.2.2), the horizontal (X and Y) components are zero over a horizontally layered Earth; so only the vertical (Z) component can be used to produce the 1D models. The horizontal components contain additional information where the subsurface conductivity distribution is 2- or 3D, in which case 2D/3D modelling is required in order to make use of these data.

For both methods, the individual inversions (models) are merged and smoothed to form a continuous conductivity–depth distribution and displayed as parasections, maps for selected 'depths' and paravolumes (see Section 5.9.5.1). The choice between using LEI or CDI should be strongly guided by the nature of the geological environment. They can be usefully applied where the conductivity structures are sufficiently wide so that they appear locally as 1D in nature; otherwise artefacts are produced.

LEI and CDI make important simplifying assumptions about the subsurface. They suffer from non-uniqueness and may contain artefacts (see Sections 2.11.4 and 5.7.6.7). For example, where the measured response is chiefly due to 2D or 3D conductivity structures, they over-estimate depth to the source and under-estimate the conductivity, so computed depth is often unreliable and a particular conductivity–depth map/image will not necessarily represent the true conductivity at that depth, although relative variations may be realistic. Despite these limitations, LEI and CDI are very convenient ways of transforming and displaying the multichannel survey data and they produce good results where the inherent assumptions are met. They are also very useful for targeting compact zones of anomalous conductivity for more precise analysis of their associated anomaly profiles (see *Profile analysis* in Section 5.7.5.3).

5.7.5 Interpretation of EM data

Multichannel EM data are interpreted using both spatial and temporal variations in the responses, i.e. how the amplitudes of the decay channels vary with position and how a response decays at a specific location. The spatial characteristics are the amplitude, shape and width of anomalies, the nature of gradients in them and zero cross-over locations. There are considerable differences in the form of the responses of the three measured components (X, Y and Z) (see Section 5.7.1.5) but, for all components, their spatial characteristics relate to the geometry of anomalous zones of conductivity. Temporal characteristics provide information about the class of conductor, i.e. unconfined or confined (see Section 5.7.2), and its conductance, and also about the vertical variation in conductivity beneath the station.

Interpretation of EM data is hampered by the different characteristics of the various loop configurations and the various EM systems, which determine the response of the ground. This makes comparison between different EM systems and configurations difficult. Also, coincidence between the step and impulse responses cannot always be expected; each varies differently with the quality of the conductor (see *Conductor quality* in Section 5.7.2.3). For example, a strong impulse response may be obtained from the lower-conductivity disseminated part of a mineral deposit and not at all from the much higher conductivity of its massive zone, the latter producing only a stronger step response.

New methods for displaying and interpreting EM data are aimed at transforming the measured response into a 3D conductivity map of the subsurface, which can be interpreted directly in terms of the geology. This is also a convenient way of comparing data from the various EM systems. However, there are limitations in deriving 3D physical property information from EM data due to the complexity of the mathematics describing the dispersion of EM fields in the ground and the necessary simplifying assumptions about the conductivity structure of the subsurface. Moreover, computing the 3D conductivity distribution of the full survey area can take many hours and sometimes days.

5.7.5.1 Interpretation procedure

Like most types of geophysical data, the interpretation of EM survey data can be confined to outlining conductivity variations related to the various rock formations to assist geological mapping, or identifying anomalous features (targeting), or extended to detailed quantitative analysis of target anomalies to obtain information about their location, geometry and conductivity.

The most common interpretation approach involves identifying various classes of conductors, possibly in the following order: cultural conductors, topographic effects, surficial conductors, formational or regional conductors and then the bedrock conductors. For most cases in mineral exploration, all except the last category are forms of environmental noise. Surficial conductors, most commonly conductive overburden, are usually large in area, producing broad and strong responses (see Section 5.7.6.1). Formational or regional conductors are associated with rock units and rock formations. They typically have a large strike extent and a wide range of conductance and produce a wide range of anomaly amplitude. They include conductive faults and shears, rock units such as shales and graphitic zones, and the contacts between rocks with contrasting conductivities across which the background response changes. Bedrock conductors are localised and possibly also steeply dipping with a wide range of conductivity producing a wide range in anomaly amplitude. They may have a large strike extent and may be associated with formational conductors, whose stronger responses can mask that of the bedrock conductor. The different types of conductor can be resolved using a variety of data processing techniques. A summary of the response characteristics of some forms of conductors is given in Section 5.7.2.6.

The first step in interpretation is to produce conductivity–depth parasections and depth-slice images, and possibly 3D conductivity block models, using LEI or CDI (see Section 5.7.4.3). Ideally LEI and CDI need to be applied to both step and impulse response data (when available) in order to resolve a wide range of conductivity variations. Throughout the interpretation process the interpreter needs to be alert to a number of common interpretation pitfalls, as described in Section 5.7.6.

5.7.5.2 Qualitative interpretation

Maps of decay channel amplitudes and the computed conductivity data can be qualitatively interpreted like any other geophysical dataset: seeking patterns in the data defined by changes in amplitude and/or texture that can be interpreted in terms of the geology. From LEI and CDI products (see Section 5.7.4.3), obvious target conductors can be identified, variations in the thickness and conductivity of overburden can be resolved and areas of thick conductive overburden identified (where penetration into the bedrock will be limited). Features with large horizontal extent, such as formational conductors, major structures and contacts in the host rocks, should be reasonably well resolved in the parasections and depth-slice images. Profile displays of the multichannel data provide the highest resolution of responses from compact sources.

The interpreter needs to be aware always that resolution of the data is strongly dependent on the parameters of the EM system. If delay times are small enough then it should be possible to resolve the surface overburden response (see *Late-time measurements* in Section 5.7.2.3 and Section 5.7.6.1), and possibly variations within it. Resolution of smaller features is dependent on the number of channels recorded and their delay times.

In the case of EM systems where the receiver is separated from the transmitter (see *Moving-loop mode* in Section 5.7.3.2, and see Section 5.9.2.1), the response of a conductor is not only asymmetric, but also depends on which side of the transmitter the receiver is located, relative to the conductor. Furthermore, for a dipping conductor the response will be different for each survey direction. Ideally, adjacent survey lines should be surveyed in the same direction, but this is not always practical. Where adjacent lines have opposite directions, anomalies of the same conductor will be laterally offset across the lines, and will be affected by the dip direction, creating the *herringbone effect* in image and contour displays. It is most obvious with laterally continuous conductors. Sykes and Das (2000) present some examples.

The vast majority of the responses of geological origin will be associated with graphitic and sulphide-rich shales and schists, water-filled structures (faults, porous formations) and zones of increased weathering. Other forms of geophysical data and geological maps may allow the probable source of the responses to be determined, e.g. a fault mapped by magnetic data.

For targeting-related analysis, the exponential time constant (τ) and conductance of confined conductors (see Section 5.7.2.3) can be computed for the full dataset, using Eqs. (5.24) and (5.25), for a conductor of specific geometry and size, usually as a dipping plate model. A map of the mid- to late-time decay is a convenient means of targeting discrete conductors, although inhomogeneities in the conductors and the interfering effects of the overburden and half-space responses will affect accuracy. Stacked channel-profiles of the EM data are required for targeting subtle responses.

Anomalies are then parameterised and classified in terms of their amplitude, width and time constant, and their association with other datasets such as magnetics and gravity. Quantitative interpretation can then concentrate on the features identified as of interest.

5.7.5.3 Quantitative interpretation

Quantitative analysis requires information about the characteristics of the EM system, i.e. the transmitted waveform, system geometry, delay times; and the type of measurement, i.e. impulse or step response. It is imperative that all available geological and topographic information be integrated with the EM data. When modelling EM data, an exact match between the observed and the computed responses cannot be expected, owing to the complexity of variations in electrical properties in the geological environment and the complex interactions between their electromagnetic responses. For example, multiple close-spaced conductors produce complex responses due to their electromagnetic interactions, i.e. their secondary field couples with the neighbouring conductor, making resolution of the individual conductors difficult. This is unlike, say, the gravity method (see Section 3.2) where individual responses are additive. A common example is the complex interaction between a discrete conductor and conductive overburden.

In resistive terrains, and where there is no conductive overburden, quantitative interpretation can be straightforward, producing suitable estimates of the target conductor's parameters (see Section 5.7.2.3). In conductive terrains the target anomaly will include the effects of the conductive environment, making quantitative interpretation difficult (see Section 5.7.6.1). Sometimes it is possible to identify and remove the responses of the overburden and the background rocks, but the effects of current channelling and adjacent conductors can be difficult to quantify.

Profile analysis

Spatial information is obtained from the spatial characteristics of the channel amplitudes, such as shape and width of anomalies, zero cross-over locations and the nature of gradients. From these the shape, depth, width, dip and strike extent of a conductor can be determined. The changing spatial distribution, if any, of the response over the measured delay time, as the eddy current system moves through the body, is related to the geometrical properties of the conductivity distribution in the body (see Section 5.7.1.4). Having determined the shape of the conductor, its conductivity and thickness, i.e. its conductance, can be determined from analysis of its decay (see Section 5.7.2.3).

The spatial response is determined by the shape and orientation of the conductor, the loop configuration used (see Section 5.7.3.2), and whether the target conductor lies beneath a conductive overburden, its depth from it and their relative conductivities. The series of computed model responses shown in Figs. 5.84 to 5.87 illustrate the key characteristics of the responses obtained with three commonly used loop configurations from a finite thin plate conductor in three orientations. *B*-field data, approximating the step response, have been computed for the target conductor set in a high-resistivity background, with and without the presence of conductive overburden. Eddy current flow is restricted to just the dip plane of the plate (see Section 5.7.1.4). The overburden is simulated with a conductive thin plate 20 m below the surface.

Responses for the moving symmetrical in-loop configuration (Fig. 5.84) show that both the X and Z components are zero over the top of a steeply dipping conductor. The responses are symmetrical across a vertical conductor with polarity reversal (change of field direction) in the X component producing a cross-over. Asymmetry is related to dip with the strongest responses in the down-dip direction. The geometrical symmetry of a horizontal plate also produces symmetrical responses, but in this case the Z component peaks over the centre of the plate. As described in Section 5.7.2.2, an infinite conductive layer, e.g. a conductive overburden, produces power-law decay uniformly across the profile. Its strong response obliterates the early-time Z-component response of the target conductor,

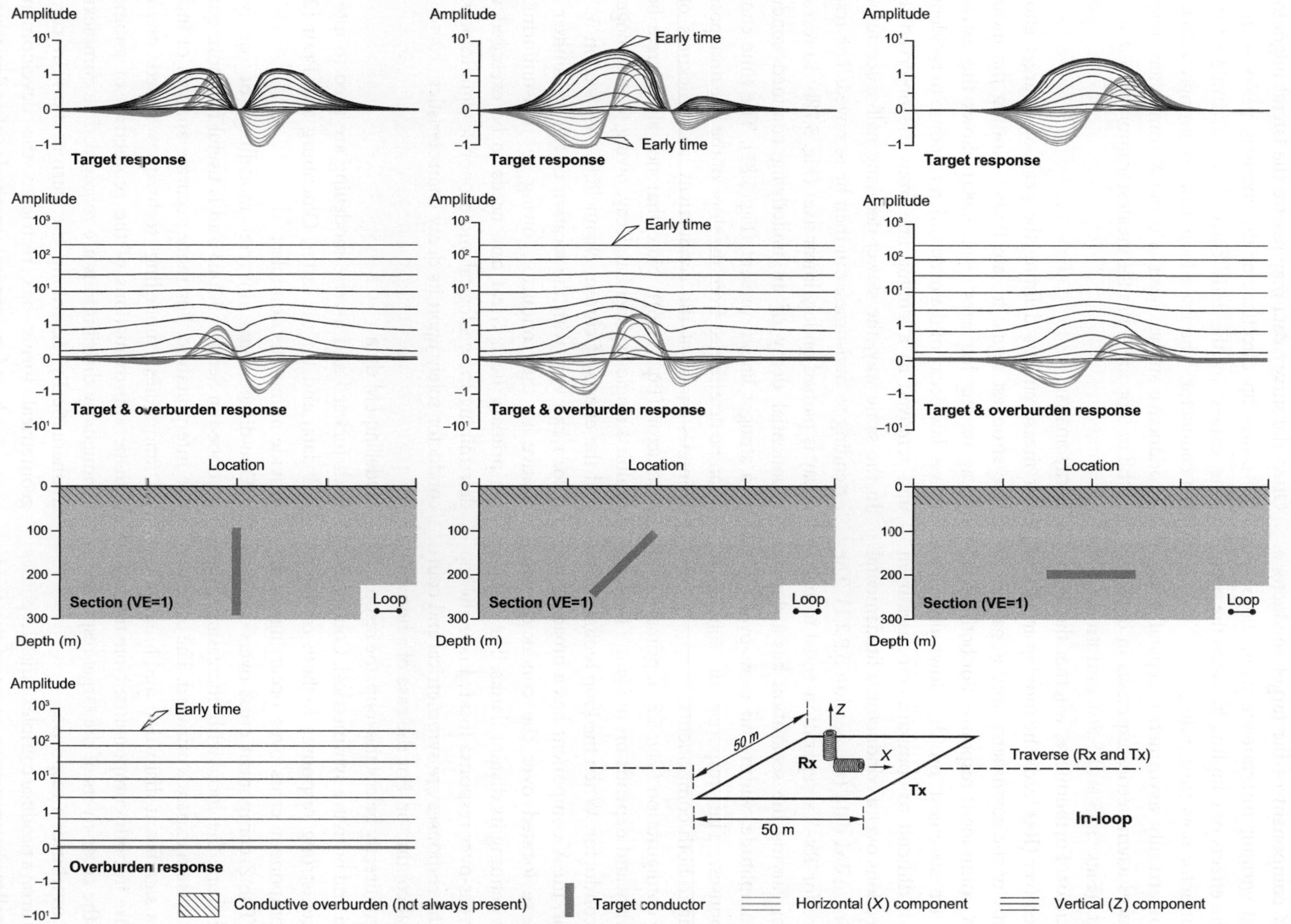

Figure 5.84 Computed model responses for the in-loop configuration: for a finite thin plate conductor in three orientations, a conductive overburden layer, and their combined responses. VE – vertical exaggeration.

for all conductor orientations. Note that for the in-loop configuration that the *X*- (and *Y*-) component response is zero for a horizontal conductive layer and a conductive half-space, so the *X* component of the target conductor's response is resolved without interference from the overburden response (the effects of coupling between the overburden and the conductor notwithstanding).

The moving geometrically asymmetric separated-loop configuration produces asymmetric responses in both the *X* and Z components (Figs. 5.85 and 5.86), and strongly so for a dipping conductor. Furthermore, whether the transmitter (Tx) or the receiver (Rx) leads the moving array also determines the nature of the asymmetry, and in particular the polarity of the *X*-component response. For both orientations, responses are strongest on the down-dip side. Asymmetry makes it difficult to compare responses from opposite survey directions over a conductor, a fundamental issue with towed-bird AEM data (see Section 5.9.2.1). The model responses are for Rx–Tx separation equal to twice the loop dimension; varying the separation has a significant effect on the amplitude, width and cross-over locations of the responses. The response of conductive overburden is strong in both components.

In the fixed-loop configuration (Fig. 5.87), primary field coupling does not vary and depends on the loop's location with respect to the conductor. When the loop is offset from a dipping conductor, the *X* component has a broad asymmetric peak response located over the conductor with polarity cross-overs defining its distant flanks. The *Z* component is a single cross-over response located near the top of the conductor. The responses are symmetrical and coincident over a vertical conductor. For the case of a horizontal conductor located directly below the loop, the responses exchange their form and both are symmetrical. Conductive overburden produces strong responses in both components with the *X*-component cross-over occurring at the centre of the loop. The *Z*-component cross-overs occur at increasing distance from the loop with increasing delay time, as the current system expands outward. The overburden response causes significant distortion and broadening of the target response in both components. Note the difficulty in identifying the cross-overs of the target response in the presence of the overburden response.

The moving-loop arrays produce complex target responses over a relatively smaller distance than those of the fixed-loop configuration, but the broader responses of the latter are simpler to analyse in terms of target parameters. For all arrays, the relationships between the shape and polarity of the *X*- and *Z*-component responses determine whether a conductor is horizontal or is steeply dipping and the amount and direction of its dip, and they locate the conductor. Often the survey data may resolve the lateral migration, in the down-dip direction, of the anomaly peaks with delay time, caused by the eddy current's inward migration into the conductor. The fixed-loop array is strongly affected by a conductive surface layer and the *X*-component response of the in-loop array has the greatest immunity to it.

Decay analysis

For measurements defining the peak of a target anomaly (the strongest signal to late times), plotting the measured decay on log–log axes (Fig. 5.78a) allows the early-time power-law decay of the overburden response to be identified and removed as a sloping line from the observed decay. In the same way, the slower-decaying half-space response, extending to mid-times, can then be removed. The residual decay is plotted on log–linear axes (Fig. 5.78b) to reveal the exponential decay of an underlying confined conductor as a straight line, if present (Fig. 5.81). The time constant τ can be determined from the slope of the line and, possibly, estimates made of the conductivity and thickness of the conductor (Eq. (5.26)). Note that not all discrete bodies exhibit a single exponential decay, owing to inhomogeneity and the effects of current channelling (see Section 5.7.2.4). Also, τ can vary with the location of the transmitter loop, relative to the conductor, owing to non-uniformity of the primary field. Great care needs to be exercised when determining the geological and economic significances of a conductor solely upon its decay characteristics.

Modelling EM data

Both forward and inverse modelling are used to interpret EM data, and for the latter, Oldenburg and Pratt (2007) provide numerous examples.

One-dimensional inverse modelling of decays is described in Section 5.7.4.3 and is useful for more qualitative interpretation. For more accurate analysis of individual anomalies, modelling techniques based on more realistic approximations of the geometries of anomalous conductivity distributions are required. The complexity of mathematically describing the current flow in complex geometrical forms has restricted the development of computer-based numerical methods of calculating the response of confined bodies and complex 3D conductivity distributions. An example of a typical graphical display for EM modelling software is shown in Fig. 5.93.

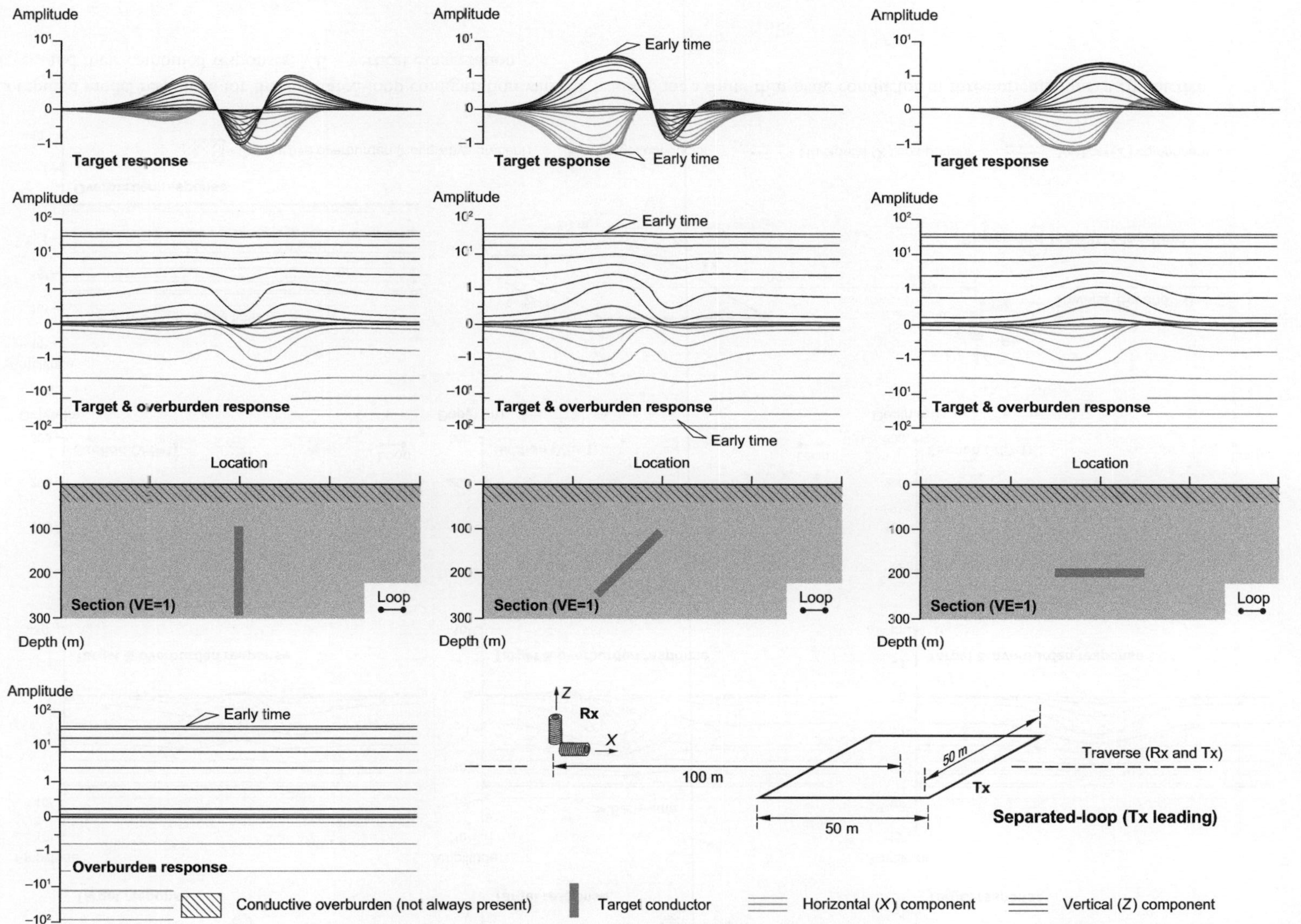

Figure 5.85 Computed model responses for the separated-loop configuration with Tx leading: for a finite thin plate conductor in three orientations, a conductive overburden layer, and their combined responses. VE – vertical exaggeration.

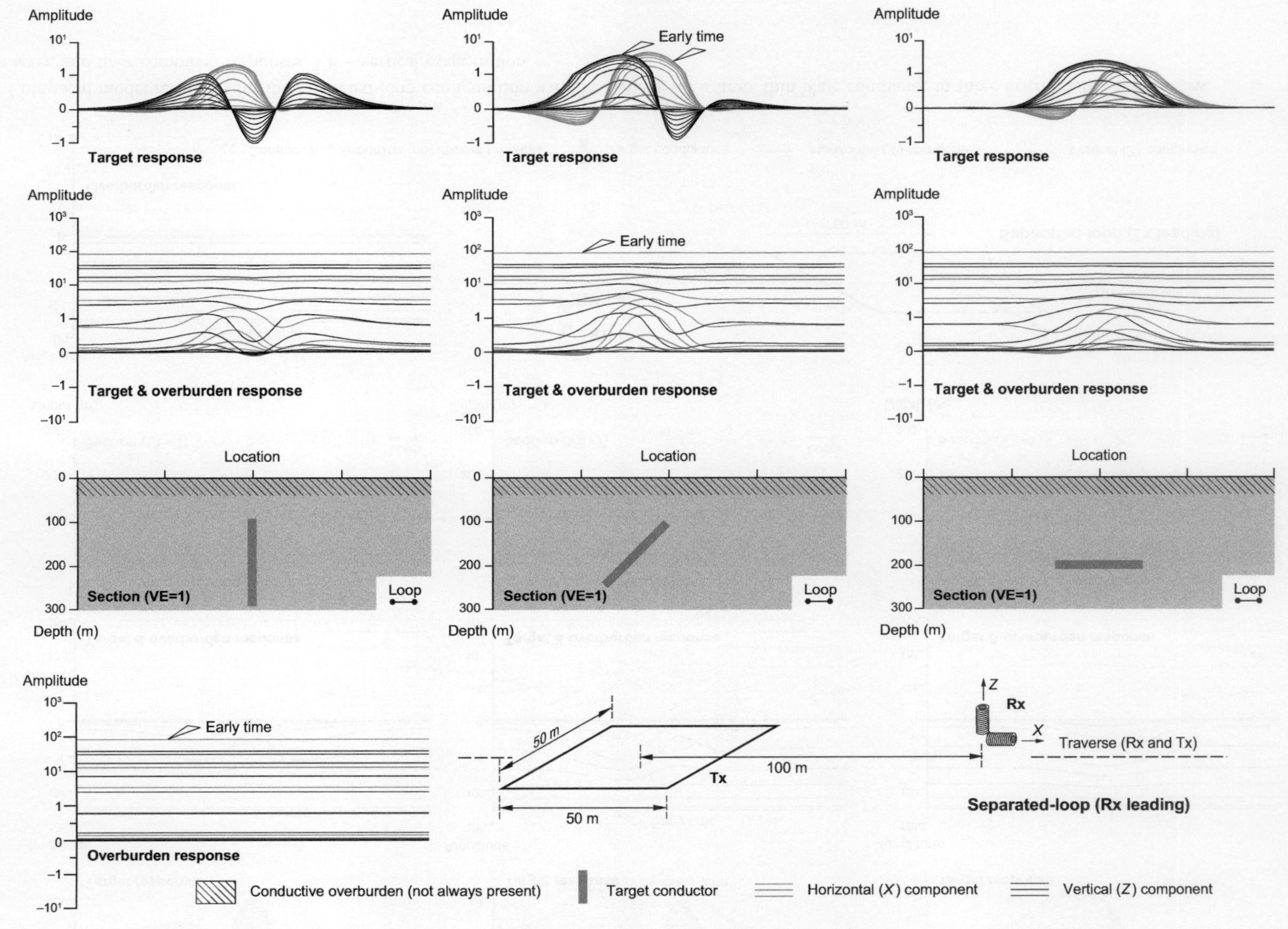

Figure 5.86 Computed model responses for the separated-loop configuration with Rx leading: for a finite thin plate conductor in three orientations, a conductive overburden layer, and their combined responses. VE – vertical exaggeration.

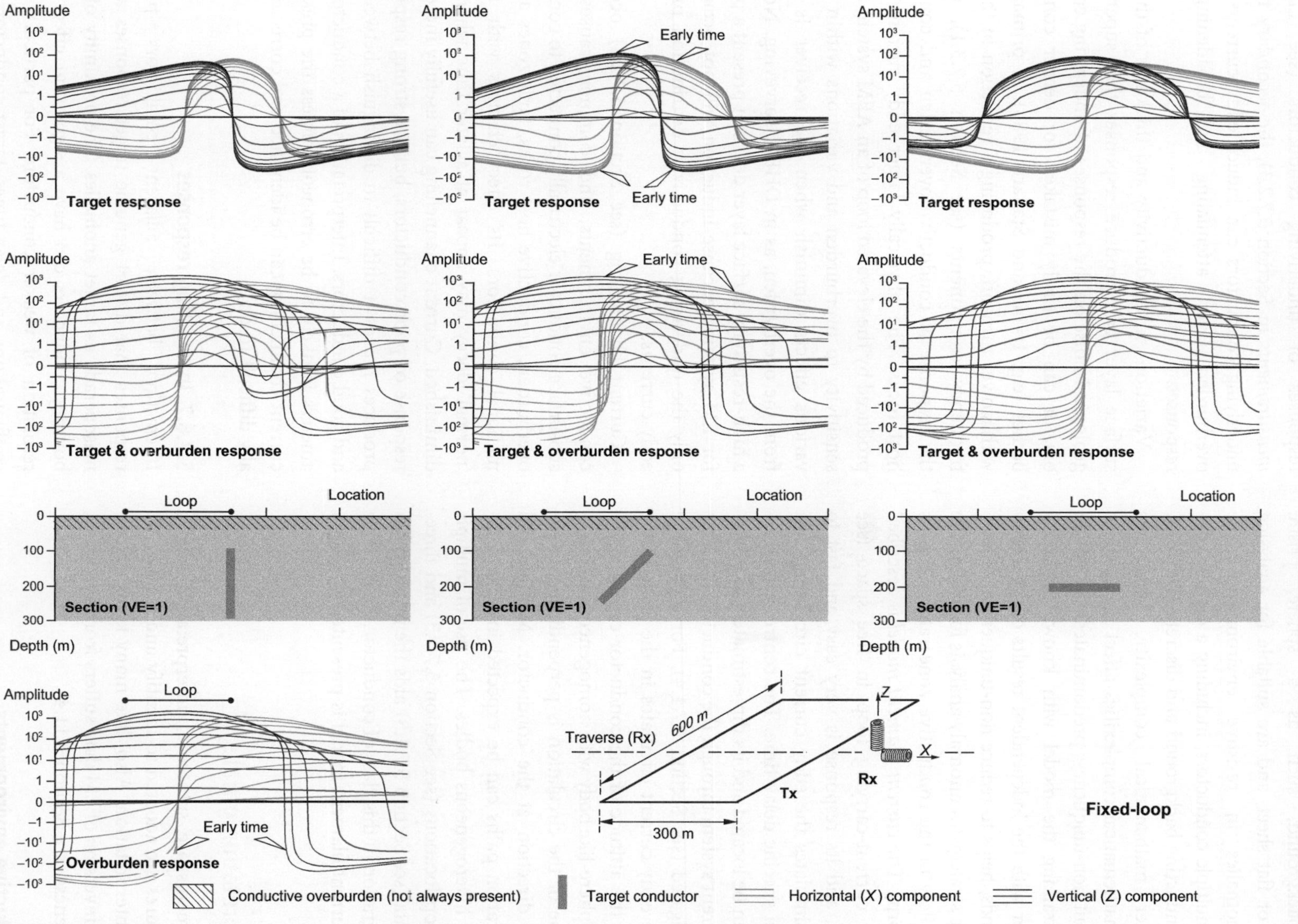

Figure 5.87 Computed model responses for the fixed-loop configuration: for a finite thin plate conductor in three orientations, a conductive overburden layer, and their combined responses. VE – vertical exaggeration.

The EM responses can be rapidly computed for parametric models (see Section 2.11.1.1) set in a non-conducting background, such as a sphere, prism, dipping plate or flat sheet, and are suitable for analysing individual anomalies in resistive environments. The responses of multiple conductors, including a discrete conductor in a conducting background and discrete 3D bodies, involve greater mathematical complexity. Limitations imposed by mathematical complexities affect the accuracy of the result, but non-uniqueness predominates (see Section 2.11.4). Constraining the model with known geological information, or with the independent results of other geophysical methods, helps to reduce non-uniqueness.

An alternate strategy for anomaly analysis focuses on the eddy current flow in the conductive zone rather than its geometrical form. The *current filament model* is described by a circular current-carrying loop in free space (see Section 5.2.2), and its response is very easy and fast to compute. It simulates the eddy current circulating in a conductor at a specific delay time. The centre of the current system can be located and its dip estimated; so migration of the current system through the conductor with time can be investigated (see Section 5.7.1.4). For a plate-like conductor the eddy current circulates in the plane of the conductor, so the attitude of the conductor can be determined. For a sphere-like body with homogeneous conductivity, the plane of the circulation is perpendicular to the primary field direction at the conductor. Multiple and complex migration paths can be expected in large complexly shaped heterogeneous bodies. The use of multiple transmitter loop locations (see Section 5.7.1.3) and three-component data (see Section 5.7.1.5) aids the investigation of current migration in this class of conductor. An example of using the current filament model is presented in Section 5.8.3.1.

5.7.6 Interpretation pitfalls

EM measurements are subject to interpretation pitfalls unless procedures are adopted to identify and account for them in the interpretation. Also, like many kinds of geophysical data, inversion of EM data suffers from ambiguity, or non-uniqueness (see Section 2.11.4).

5.7.6.1 Conductive environments

The response of conductive overburden and examples of its effect on the target response are demonstrated by the model responses in Figs. 5.84 to 5.87. In addition to the overburden response obscuring the weaker early-time responses of underlying conductors (see *Late-time measurements* in Section 5.7.2.3), the secondary fields of underlying conductors can induce eddy currents in the overburden layer, attenuating and broadening their responses.

Variations in conductivity and thickness of the near-surface layer can produce responses that superimpose geological noise on the responses of underlying conductors, or can be easily mistaken for deeper conductive bodies (e.g. Irvine and Staltari, 1984). A primary field with slower turn-off, producing a reduction in the high-frequency components (see Section 5.7.3.1), reduces the response of conductive overburden and conductive host rocks. The physically larger eddy current system produced by the elevated loop of an AEM system reduces sensitivity to overburden and variations within it. The various effects diminish when the receiver is distant from the overburden, as in DHEM surveying. Note that a high-resistivity surface layer does not present a problem for EM systems because induction will not occur in it; only the underlying conductivity distribution produces eddy currents.

Current channelling (see Section 5.7.2.4) occurs in conductive environments, and common causes are a subsurface conductor electrically connected to conductive overburden, conductive host rocks, fault zones and formational conductors. Its effect changes with time; a 'normal' decay may appear at a later time when it has diminished. Current channelling can usefully increase the response of poor conductors, but the strong responses it produces makes it difficult to distinguish between poor and good conductors. Interpretation of a conductor's parameters is likely to be erroneous unless the presence of current channelling can be identified and corrected (usually difficult to do).

5.7.6.2 Interacting responses

Interaction between adjacent conductors produces responses that distort genuine target responses and can masquerade as target anomalies. The proximity of neighbouring conductors can have a significant effect on the resolution of weaker conductors. For particular system configurations and delay times, target conductors can be electrically shielded by larger neighbouring conductors with larger time constants, an example being a discrete target conductor hosted in a large formational conductor. It is also possible for a large body having a large time constant to shield a smaller body with smaller time constant, a common problem when a target conductor is hosted by a large formational conductor.

5.7.6.3 Cultural and topographic effects

As described in Sections 2.4.1 and 5.4.2.1, man-made features such as fences and powerlines will produce EM responses. These can be identified using aerial photographs and topographic maps and should be noted at the time of data acquisition.

Where survey lines pass over hills and valleys, typical in AEM surveys, hills usually appear as conductive features. Responses are likely in early delay times. The response from hills is stronger when the survey aircraft passes closely over the peak of the topography, and is usually stronger in low-level AEM data. Care should also be taken where responses coincide with a 'step-like' change in ground clearance, such as a scarp, that can cause artefacts in 1D inversion models (see Section 5.7.6.7). Topographic effects can be identified by observing correlations between apparent conductivity images and topographic data. There is no practical way of removing topographic effects other than to include the topography in computer models and account for its effect in the interpretation of the survey data.

5.7.6.4 Induced polarisation

It is generally assumed in EM that the conductivity of the ground is frequency independent, i.e. its conductivity does not change with frequency or delay time. If this is not the case, electrically polarisable conductors and polarisable ground become charged (see Section 5.3.2) at early delay times by the circulating eddy currents (Flis *et al.*, 1989). At later times, when the eddy currents are weaker, the polarisation discharges through the conductive ground, and the magnetic field associated with the decaying discharge current is measured by the EM system. The discharge has opposite polarity to the induced eddy currents and reduces the amplitude of the measured secondary response. At late delay times, IP effects can produce anomalous decay rates or even reverse the polarity of the measured secondary decay (Fig. 5.88). It is usually difficult to distinguish between IP and EM effects in EM data, and procedures for recognising, interpreting and correcting IP effects are currently the subject of ongoing research.

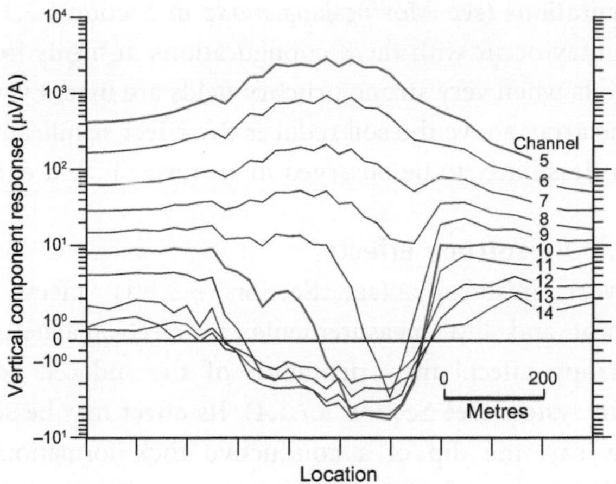

Figure 5.88 Negative amplitudes in TDEM data indicative of IP effects. Data were collected with 100 m square moving loop and a near-coincident vertical component coil sensor. Channel delay times range from 0.25 to 1.45 ms. Redrawn, with permission, from Flis *et al.* (1989).

5.7.6.5 Superparamagnetism

An external magnetic field causes magnetic domains in ferromagnetic material to align with it (see Section 3.2.3). There is a time lag between removal of the field and the magnetic dipoles returning or relaxing to their original state. The effect is known as *magnetic viscosity* or *superparamagnetism* (SPM). The strong magnetic field of a transmitter loop causes very fine ferromagnetic grains in the soil to behave in this way. The changing magnetic field associated with the relaxation can be observed during the measurement period of most TDEM systems. It is a logarithmic process in the step response; so in the impulse response it exhibits power-law decay (Fig. 5.78a) with decay constant (k) equal to −1 (actual measurements range from −0.8 to −1.2). Barsukov and Fainberg (2001) observed anomalous variations in SPM in weathered materials above a placer gold deposit and a nickel sulphide deposit.

SPM effect, if present, is evident after the faster decaying thin-layer and half-space responses have disappeared (see Section 5.7.2). It persists to later delay times and causes the ground to appear to be more conductive than it actually is, i.e. apparent resistivity is lower than expected. SPM effect is stronger with increasing primary field strength and can be measured within several metres of a ground transmitter loop (Buselli, 1982), a possibility in fixed-loop mode (see *Fixed-loop mode* in Section 5.7.3.2). It is usually not observed with the commonly used in-loop (with the receiver at the centre of the loop) and separated-loop configurations (see *Moving-loop mode* in Section 5.7.3.2), but it may occur with these configurations in highly iron-rich soils when very strong primary fields are used. Elevating the array above the soil reduces the effect significantly, so it is less likely to be observed in airborne TDEM data.

5.7.6.6 Anisotropy effects

Electrical anisotropy (see Section 5.3.1.4) affects all electrical and EM measurements. In electromagnetism, anisotropy affects the orientation of the induced eddy current system (see Section 5.7.1.4). Its effect may be seen where, say, the dip of a conductive rock formation is different in part of the survey area causing its foliation to change orientation and orientation of the induced eddy current system. A change in conductivity due to anisotropy would then be expected.

5.7.6.7 Inversion artefacts

The 1D assumption underlying LEI and CDI (see Section 5.7.4.3) can lead to seriously inaccurate models of the geoelectric section when the electrical properties of the ground are significantly heterogeneous and 3D in nature. The inversion may model the data very well even though the conductivity distribution in the subsurface is not 1D (Ellis, 1998).

Inversion results are reliable where the assumption of one-dimensionality holds, i.e. where the ground is approximated by a flat-lying layered conductivity structure. Otherwise the parasection may contain conductivity zones and apparent targets which are artefacts of the inversion process. Where there are lateral changes in conductivity, non-existent steeply dipping conductive zones appear (see Fig. 5.83b). These may be the edges of flat-lying conductors, vertical features such as dyke-like conductors or steeply dipping conductivity contrasts such as occur near faults. When steeply dipping conductors are present they are usually detected but not properly defined in terms of shape or depth (Wolfgram *et al.*, 2003).

5.7.7 Examples of EM data from mineral deposits

As would be expected, EM surveys are primarily used to directly target mineralisation which would be expected to be conductive: most commonly massive base metal sulphides (SEDEX, VMS; e.g. Bishop and Lewis, 1992), magmatic nickel deposits, e.g. King (2007), and also graphitic shear zones associated with unconformity-style uranium mineralisation (Powell *et al.*, 2007). Other targets include graphite deposits and supergene manganese mineralisation (see Section 5.9.5.1). EM surveys may be used to identify potentially mineralised environments such as kimberlite pipes. Examples of downhole and airborne EM responses are described in Sections 5.8.3 and 5.9.5, respectively. Here we describe some examples of ground surveys.

5.7.7.1 Massive sulphide deposits

Figure 5.89 shows ground TDEM responses from four massive sulphide deposits. These show anomalous responses due to the respective target conductors, as well as a variety of other secondary responses.

Data from the Ernest Henry IOCG deposit located near Cloncurry, Queensland, Australia, are shown in (Fig. 5.89a). The geophysical characteristics of the deposit are described by Webb and Rowston (1995) and Asten (2000). Copper–gold mineralisation occurs in the matrix of brecciated volcanics with an associated zone of supergene mineralisation occurring beneath about 45 m of conductive sediments. A moving-loop survey was conducted using coincident square loops (analogous to the in-loop configuration described in *Moving-loop mode* in Section 5.7.3.2) 100 m in size, and d*B*/d*t* data were acquired. It shows a weak and localised response (A) due to the supergene mineralisation. The anomaly is characteristic of a shallow dipping plate (see Fig. 5.84b). The main body of mineralisation is too poorly conductive to be detected beneath the conductive overburden, the response of which (B) extends to at least channel 11 along the whole profile. The subtle peak in the measured response to the left of the main anomaly (C) may be due to the zone of deeper weathering. The geological section shown is located 350 m from the geophysical profile shown.

Trilogy (Fig. 5.89b) is a polymetallic deposit comprising semi-massive sulphide and stringer mineralisation in silicified carbonaceous phyllite/slate (Sampson and Bourne, 2001) in the Proterozoic Mount Barren Basin near Ravensthorpe, Western Australia. The d*B*/d*t* data from a moving in-loop survey using an 80 m square transmitter loop show a broad response (A) to late times which diminishes slowly to the right of the section indicating the dip direction of the conductive mineralisation. The weak late-time responses to the left, (B) and (C), are probably related to shallow mineralisation. The response (D), increasing in width with delay time and disappearing by the latest times, may be related to the depression in the weathered zone. The responses of the conductive weathered zone, and possibly including that of the host phyllite (the half-space

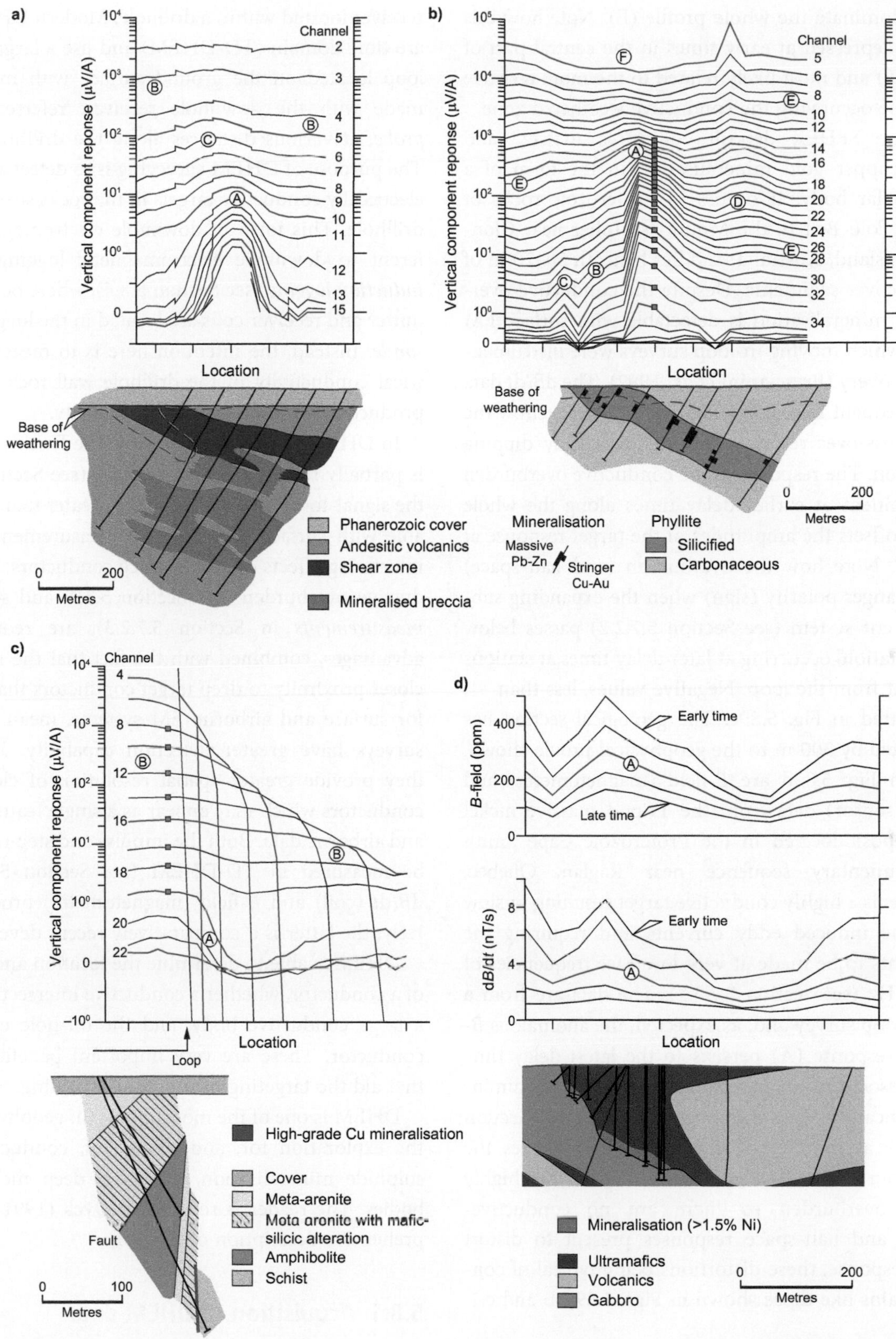

Figure 5.89 Ground EM data from various mineral deposits. (a) Ernest Henry IOCG deposit, redrawn, with permission, from Webb and Rowston (1995) and Asten (2000), (b) Trilogy VMS deposit, redrawn, with permission, from Sampson and Bourne (2001), (c) Eloise SEDEX deposit, redrawn, with permission, from Brescianini *et al.* (1992) and (d) Tripod nickel sulphide deposit, redrawn, with permission, from Osmond *et al.* (2002). The blue squares in (b) and (c) are the points defining the decays shown in Fig. 5.81. Note the different vertical scales.

response), dominate the whole profile (E). Note how this response is depressed at early times in the central part of the profile (F) and most likely related to the more resistive silicified zone occupying the conductive weathered zone.

The Eloise SEDEX deposit comprises massive and stockwork copper–gold mineralisation in the form of a dipping tabular body. It occurs in Proterozoic rocks of the Eastern Fold Belt in the Mount Isa Inlier near Cloncurry, Queensland, Australia, and lies beneath 50–70 m of conductive cover sediments. Despite the conductive overburden, the mineralisation is detectable with surface EM surveys, of which moving in-loop surveys were instrumental in its discovery (Brescianini *et al.*, 1992). The d*B*/d*t* data from a subsequent fixed-loop survey (Fig. 5.89c) show the late-time cross-over response (A) of the steeply dipping mineralisation. The response of the conductive overburden (B) is prominent at earlier delay times along the whole profile and offsets the amplitudes of the target response at those times. Note how the overburden (and half-space) response changes polarity (sign) when the expanding subsurface current system (see Section 5.7.2.2) passes below the survey station, occurring at later delay times at stations more distant from the loop. Negative values, less than −1, are not plotted in Fig. 5.89c. The geological section has been projected by 300 m to the geophysical profile shown.

Shown in Fig. 5.89d are *B*-field (magnetometer) and d*B*/d*t* (coil sensor) data from the Tripod massive nickel sulphide deposit located in the Proterozoic Cape Smith volcano-sedimentary sequence near Raglan, Quebec, Canada. This is a highly conductive target resulting in slow decay of the induced eddy currents and requiring the measurements to be made at very low base frequencies of 5 and 1.67 Hz (see Section 5.7.3.1). The data are from a moving in-loop survey and, as expected, the anomalous *B*-field (step) response (A) persists to the latest delay time whilst the associated d*B*/d*t* (impulse) response has diminished significantly (see *Conductor quality* in Section 5.7.2.3). The asymmetry of the anomalies indicates the dip of the source. This is a resistive terrain without highly conductive overburden so there are no conductive-overburden and half-space responses present to distort the target response, these distortions being typical of conductive terrains like those shown in Figs. 5.89a, b and c.

5.8 Downhole electromagnetic surveying

The term downhole (or drillhole) electromagnetics (DHEM) is used here to describe EM systems with a receiver located within a drillhole. Modern DHEM systems are time domain (TD-DHEM) and use a large transmitter loop located on the ground surface with measurements made with the downhole receiver, referred to as the *probe*, at various distances along the drillhole trajectory. The purpose of DHEM surveying is to detect and delineate electrically conductive targets in the rocks surrounding the drillhole. This form of downhole electromagnetics is different to downhole electromagnetic logging, known as *induction logging* (see Section 5.8.4), where both the transmitter and receiver coils are located in the logging tool, the *sonde*. Instead, the intention here is to measure the electrical conductivity of the drillhole wall rocks *in situ* and produce a downhole log of conductivity.

In DHEM the receiver is below the Earth's surface so it is partially shielded from EM noise (see Section 5.4.2), so the signal-to-noise ratio is usually greater than that obtainable with surface and airborne measurements. Also, the interfering effects of near-surface conductors, such as conductive overburden (see Section 5.3.4, and see *Late-time measurements* in Section 5.7.2.3), are reduced. These advantages, combined with the fact that the receiver is in closer proximity to deep target conductors than is possible for surface and airborne EM systems, mean that DHEM surveys have greater detection capability. Furthermore, they provide greater spatial resolution of closely spaced conductors which may appear as a single feature in surface and airborne data. Both the impulse and step responses can be measured in TD-DHEM (see Section 5.7.1.7), with d*B*/d*t* (coil) and *B*-field (magnetometer) probes, respectively; the latter is a comparatively recent development.

DHEM is able to determine the location and orientation of a conductor, whether a conductive intersection is part of a large conductive body, and the off-hole extent of the conductor. These are very important benefits of DHEM that aid the targeting of subsequent drilling.

DHEM is one of the most important geophysical tools in the exploration for, and mining of, conductive massive sulphide mineralisation, especially deep nickel sulphide bodies. The reader is referred to Dyck (1991) for a comprehensive description of DHEM.

5.8.1 Acquisition of DHEM data

In most DHEM systems the transmitter loop is located horizontally on the ground surface (Fig. 5.90). In order to obtain a more favourable orientation for coupling to a particular target (see Section 5.7.1.3), the loop can be located on the slopes

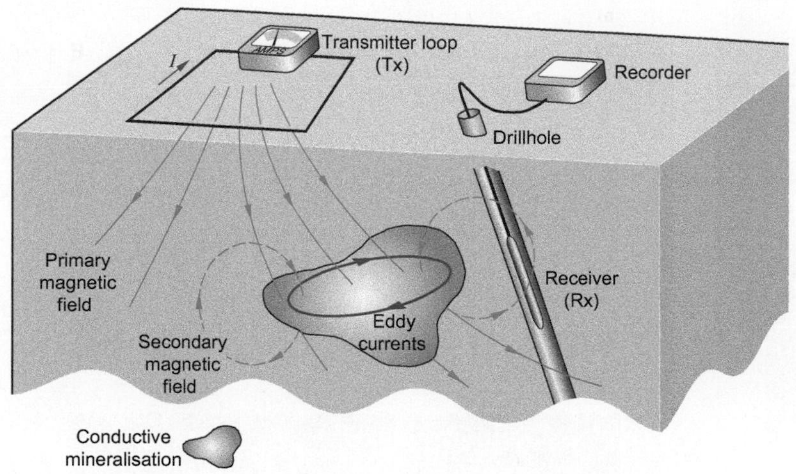

Figure 5.90 Typical time domain DHEM survey system using a large transmitter loop (Tx) located on the surface. Directions of the eddy current flow and its associated secondary magnetic field are shown for the period immediately after turn-off of the primary field shown. Redrawn, with permission, from Killeen (1997a).

of rugged terrain (Mudge, 1996). Where there are adjacent mine workings, it may be possible to locate the loop within the workings (Doe *et al.*, 1990). The receiver is usually lowered down the drillhole, although for in-mine surveys it may be moved along drillholes of almost any orientation, including upwardly orientated drillholes. These arrangements are very useful because, in principle, there are no spatial restrictions on the location of conductors relative to the DHEM system, so optimal coupling with a target conductor can often be achieved (Figs. 5.69 and 5.70). This is unlike surface and airborne surveying where both the transmitter loop and receiver are usually co-planar and parallel to the ground surface, and the conducting regions lie below the EM system.

DHEM probes are usually capable of operating to depths exceeding 1 km, with the most robust probes capable of up to 3 km depth. Transmission of the primary field signal and data recording are achieved using the same equipment as used for surface surveys (see Section 5.7.3), with the additional probe orientation data also recorded. Downhole EM surveying can be conducted in dry holes and even in those cased with plastic piping. Magnetic and highly conductive steel casing and drilling rods in the drillhole cause strong interference to EM systems and prevent the use of DHEM.

The radius of investigation around the drillhole is determined by the minimum detectable signal level, which is determined by the strength of the primary field, the loop location (coupling) with respect to the target conductor, receiver sensitivity, background noise and the type of response measured. Step response systems have greater radius of investigation for good conductors than impulse response systems (see Section 5.7.2.5). A conductor in the vicinity of the drillhole, i.e. a 'near-miss', can be detected, and in favourable conditions good conductors more than 1 km from the drillhole can be detected.

Compared with surface and airborne EM surveys, downhole data are collected over a relatively small area with surveying often confined to just a single drillhole. The emphasis then is on optimising survey procedures to ensure the detection and determination of the geometry, position and orientation of any conductor in the area from the single downhole survey traverse. Optimally achieving these objectives requires the acquisition of three-component measurements of the secondary magnetic field from within several drillholes and with different transmitter loop positions.

5.8.1.1 Variable field coupling

Surveying with transmitter loops in different locations around the drillhole creates primary magnetic fields with different orientations and changes the coupling with any conductors present (see Section 5.7.1.3). Typically five loop positions are used (Fig. 5.91a) with the downhole survey repeated for each, and differences in the shape, amplitude and sign of the secondary field are sought. The intention is to ensure that all potential conductors in the vicinity of the drillhole are energised. Data may be recorded in multiple drillholes, and possibly by using just a single transmitter loop (Figs. 5.91b and c). When surveying multiple drillholes using a single transmitter loop, differences in coupling of the secondary magnetic fields and the downhole receiver, between the different drillholes, constrain the location and orientation of an energised conductor. The different responses obtained from conductors using these survey procedures, including when poor or null coupling occurs, provide

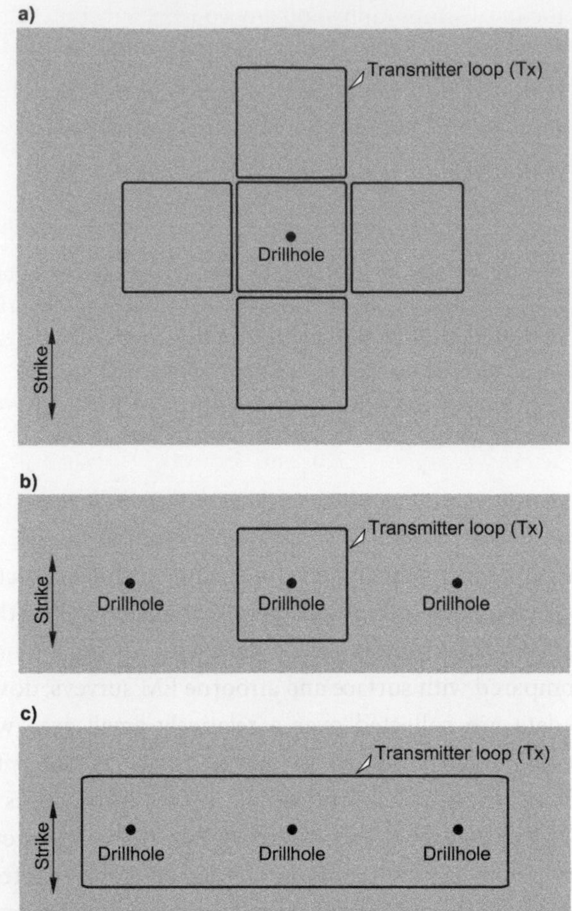

Figure 5.91 Typical layout of transmitter loops for DHEM surveying. (a) Multiple loops for surveying a single surface drillhole; (b) and (c) single loop with surveys conducted in multiple drillholes.

information critical for determining the target's location and orientation. This method also assists in laterally discriminating multiple close-spaced conductors.

Multiple loops are an integral part of surveys using axial component probes (see Section 5.8.1.2), but three-component probes (see Section 5.8.1.3) can reduce the need for multiple loops and multi-hole surveys.

5.8.1.2 Axial-component measurement

Historically, the probe used in TD-DHEM was a single multi-turn coil oriented along the drillhole axis, i.e. an along-hole or axial component sensor. It measures the strength of the field in the direction of the drillhole at the point of the measurement, component *A* in Fig. 5.92. The orientation of the probe (the drillhole) at each measurement point is required for analysis of the data, and is obtained by other means. The measured response does not provide unique directional information about any conductors that might be detected. The results have a

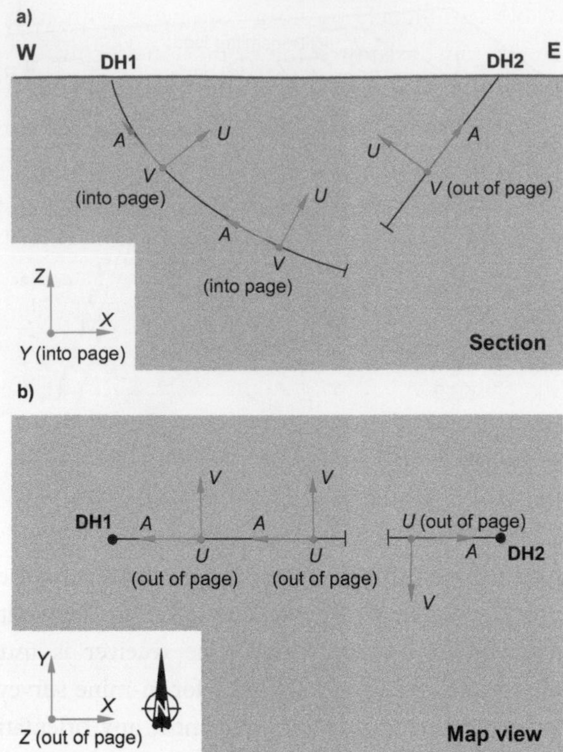

Figure 5.92 The drillhole coordinate system (*A*, *U*, *V*) reference frame for the vector components measured in DHEM. *A* is the axial component, and *U* and *V* the perpendicular cross-hole components. (a) Drill section showing the reference frames specific to each drillhole and how the absolute direction of the components varies when the drillhole trajectory is curved (DH1) or changes direction (DH2). (b) Map view of the reference frames for the drillholes shown in (a). *X*, *Y* and *Z* are coordinates of the absolute (geographic) reference frame. See text for definition of the orientations of the components.

rotational ambiguity, meaning that the distance to a conductor and its orientation can be determined, but not its azimuth relative to the axis of the drillhole. This requires the survey to be repeated with the transmitter loop in different locations so as to vary the coupling with the conductors (see Section 5.8.1.1). Few, if any, DHEM surveys are conducted now with just a single axial component probe.

5.8.1.3 Three-component measurements

A DHEM probe containing three mutually perpendicular coils or magnetic (*B*-field) sensors to measure two perpendicular cross-hole or radial components in addition to the axial component is known as a *three-component probe.* Three-component measurements provide a complete determination of the orientation of the primary and the secondary magnetic fields at the measurement location. It is possible to uniquely locate a conductor with just one transmitter loop, assuming the primary field couples with the conductor. Not having to survey the drillhole with several transmitter loop positions significantly reduces the time and cost of DHEM surveying. Nevertheless, it is good practice to survey the drillhole with several transmitter loops, and to survey neighbouring drillholes with the same transmitter loop, to confirm the interpretation.

The orientations of the three measured field components are fundamental to the analysis of the data. In order that the data can be analysed in a consistent reference frame, changes in the orientation of the drillhole and the probe must be monitored and their effects corrected for. Magnetic sensors and accelerometers in the probe provide the necessary information. The axis of the drillhole provides the reference for the analysis of the three components of the EM field, which are termed *A*, *U* and *V*. The axial component (*A*) is always orientated along the drillhole axis, reckoned positive upwards (Fig. 5.92). The radial (*U*) component lies in the plane of the drillhole and perpendicular to the drillhole axis, and is reckoned positive upward. It is vertical in a horizontal drillhole, and in a vertical drillhole it is horizontal and can point in a direction to suit the interpreter. The other radial component (*V*) is perpendicular to the *U* component and horizontal, with positive polarity as shown in (Fig. 5.92). The measured EM field is visualised and interpreted in the *A*, *U*, *V* coordinate space. Some workers refer to the *A*, *U* and *V* axes as *Z*, *X* and *Y*, respectively, and may change the polarity (sign) of the *Y* axis.

Clearly the orientation of the *A*, *U* and *V* reference frame varies in absolute direction depending upon the azimuth and inclination of the drillhole. The EM data can be further rotated through the inclination and azimuth of the drillhole to resolve the measured EM field in an absolute reference frame, such as relative level (RL), easting and northing (*X*, *Y*, *Z* in Fig. 5.92). This can be useful when working with data from several drillholes in the vicinity. At prospect scale, drillholes are often drilled in the same direction (azimuth) and the transmitter loops are usually laid out along this direction, so the *A*, *U* and *V* coordinate system is usually adequate for a small survey area.

5.8.1.4 DHEM survey practice

In DHEM surveying two sides of the rectangular transmitter loops are usually orientated parallel to the drillhole azimuth (Fig. 5.91). Given that most drilling is directed across strike, it means that the transmitter loops will be aligned with the local stratigraphy and any concordant conductors, although coupling and access considerations may dictate otherwise. Loop size is determined using the same criteria described for ground surveys (see *Transmitter loop size* in Section 5.7.3.2).

Drillhole locations, orientations and depths are usually known when designing DHEM surveys, as are the most likely locations and attitudes of targets. Forward-modelling techniques can therefore be very effective in optimising the transmitter loop size, the loop layouts and the number of loops required. The primary field from the loops can be computed and plotted to show their coupling with known conductors (see Fig. 5.70) and expected conductors at other locations around the drillhole; for example, see Bishop (1996). Anomaly shapes and polarities can be predicted for each component of the secondary field, and survey procedures optimised to record the expected responses. Where multiple drillholes are available, those offering the optimum response from the target conductors can be identified and selected for surveying.

In the event that the direction to the target is unknown, a transmitter loop centred over the drillhole (Fig. 5.91b) to energise the ground equally around it is often the best starting point. Additional loops can be deployed based on results obtained from the starting loop.

5.8.2 Display and interpretation of DHEM data

DHEM surveys are characterised by large numbers of multichannel profiles of three-component data acquired from a multiplicity of drillholes and transmitter loops potentially located anywhere within the rock volume of interest. This contrasts with data from ground and airborne EM surveys which are in the form of an ordered series of parallel survey lines with measured responses from targets located below the survey surface, and which can be displayed as 2D conductivity maps and sections. In DHEM it is generally not possible to make composite conductivity images of multiple drillhole data, so specialised presentation, visualisation and modelling software is required. The data are still presented in profile form (see *Profile analysis* in Section 5.7.5.3) but within a 3D volume display which allows the relative positions of the relevant drillholes, conductors and transmitter loops to be displayed (Fig. 5.93).

In DHEM surveying, conductors can be located anywhere around the receiver producing a wide range of responses. The data often have a large range in amplitude,

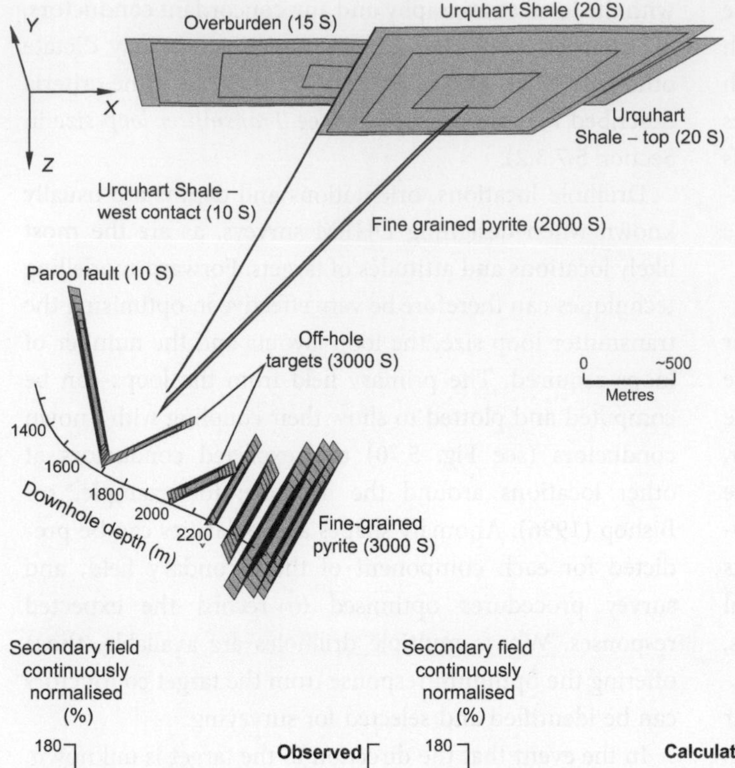

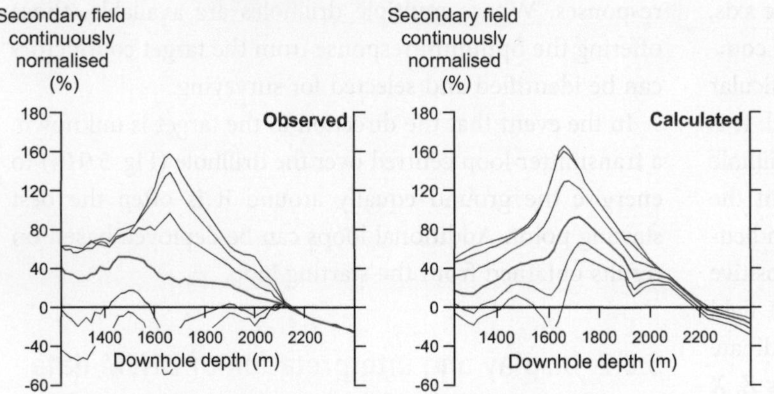

Figure 5.93 Presentation of DHEM data in 3D as used for data modelling. The locations of conductors relative to the drillhole trajectory are shown along with the observed and calculated data. Data are from the Mount Isa Cu mine. Redrawn, with permission, from Jackson *et al.* (1996).

with well-coupled conductors close to the drillhole producing particularly strong responses. The measurements may also include complex responses due to closely spaced and interacting conductors, and artefacts due to deviations in the drillhole trajectory (probe orientation changing). Interpretation proceeds by identifying the responses of in-hole and off-hole conductors (see Section 5.8.2.1). A series of computed downhole profiles for all three components from one or more drillholes located at various locations around and intersecting a model conductor is an essential interpretation aid.

5.8.2.1 In-hole and off-hole responses

Conductors detected by DHEM are classified as either in-hole conductors, those intersected by the drillhole, or off-hole conductors, those located away from the drillhole. The three-component response is diagnostic of these two classes and their identification is fundamental in the interpretation of DHEM data (Fig. 5.94). Profiles of each component show symmetry and change in polarity around the conductor. The distribution of the current flow in the conductor depends on the size, shape and orientation of the induced current system at any particular delay time (see Section 5.7.1.4). For in-hole conductors the responses vary according to whether the drillhole has intersected the central region of the conductor or its edge, because this means the hole is in a different location relative to the migrating eddy currents.

Based on the relationships between component responses and their polarities, the location, orientation and size of the source conductor with respect to the drillhole can be interpreted. The computed model responses shown in Fig. 5.94 illustrate the key characteristics of the axial (A) and radial (U) component responses (see Section 5.8.1.3), in four vertical drillholes, obtained with a single surface transmitter loop located directly above a horizontal finite thin plate conductor. The direction of U is ambiguous in a vertical drillhole (Fig. 5.92); it is taken as positive

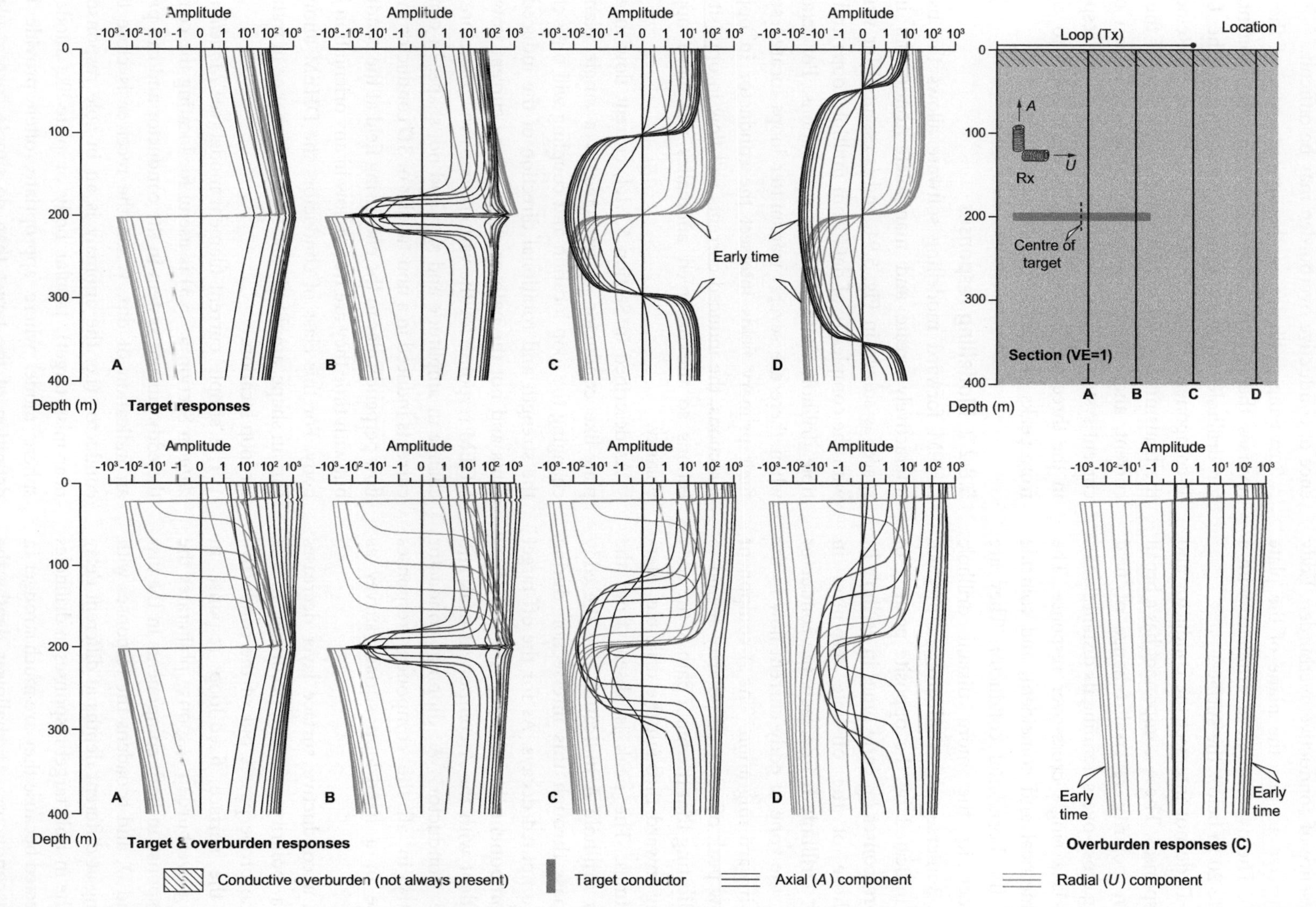

Figure 5.94 Computed DHEM model responses for the axial and horizontal components in four drillholes (A–D) for a horizontal finite thin plate conductor located directly below the surface transmitter loop, conductive overburden layer, and their combined responses. Overburden response shown for drillhole C. VE – vertical exaggeration.

to the right in Fig. 5.94. *B*-field data, approximating the step response, have been computed for the target conductor set in a high-resistivity background, and with and without the presence of conductive overburden. Eddy current flow is restricted to just the plane of the plate (see Section 5.7.1.4). The overburden is simulated with a conductive thin plate 20 m below the surface.

Drillholes C and D do not intersect the conductor and produce off-hole responses. The *A*-component has a broad peak response with polarity reversal (change of field direction) producing cross-overs defining its distant flanks. The *U*-component is a single cross-over response. The responses are symmetrical and coincident, and coincide with the plane of the horizontal conductor. They are weaker and broader in the more distant drillhole D. Drillholes A and B intersect the conductor and produce in-hole responses, indicated by the opposite polarity of their *A*-component responses (at early times in drillhole B) compared with those of the off-hole responses in drillholes C and D. Drillhole B intersects the conductor near its outer edge, the zone of eddy current flow in the earlier stage of its inward migration. The *A*-component also shows a narrow peak of reversed polarity at mid to late delay times indicating that the main path of the eddy current is migrating inward through the drillhole intersection at these times. The peak is absent from the profile of the inner drillhole A indicating that the eddy current does not pass through this intersection during the period of the computed decays. As for the off-target drillholes, the *U*-component in drillholes A and B show a cross-over coincident with the *A*-component peak and the plane of the conductor. A dipping conductor produces asymmetry in all the component responses analogous to those of a fixed-loop surface survey as shown in Fig. 5.87.

The response of a conductive surface layer decreases with depth with a polarity reversal occurring in the *U*-component as the probe passes out of the conductive layer. Similar to the surface fixed-loop response in Fig. 5.87, the strong overburden response obliterates the early-time target response in both components in the distant drillholes C and D, and broadens the responses with cross-overs occurring at different depths at different delay times. In contrast, the in-hole target responses in drillholes A and B are less affected because they are much stronger in the vicinity of the conductor. At shallower depths the overburden responses overwhelm the target responses, causing a reversal in polarity of the *U*-component response. The diminishing overburden response with depth means it produces less distortion of the signal from deeper conductors. Identifying shallow targets in the presence of conductive overburden may be difficult.

Surveying drillholes on the opposite edge of the body reverses the polarity of the horizontal component, and for drillholes located on the adjacent edges the *U* and *V*-components interchange their responses. Relocating the transmitter loop can change the direction of the eddy current, and in thick conductors the orientation of the current system. This changes the strength of the responses in the three components and can cause them to change from peaks to cross-overs, and vice versa.

5.8.2.2 Modelling responses

DHEM forward modelling software allows the user to interactively create and manipulate conductors in 3D space, as shown in Fig. 5.94, and is essential for working with the complexity of data from multiple loops and multiple drillholes with different orientations. For example, when there are several transmitter loops located so that their primary fields intersect the conductor in opposite directions, the induced currents will flow in opposite directions so the observed anomalies will have opposite polarity.

As described in Section 5.7.1.4, current flow in a thin plate-like conductor is restricted to a single plane, so changing the loop location and coupling will only change the strength and rotational direction of the induced currents and not the path or shape of the current flow. The DHEM response for the various loops will, therefore, only change in amplitude and sign and not shape. In contrast, currents induced in a non-tabular or 3D conductor initially flow perpendicular to the energising field at the conductor, but with time they are free to flow in any orientation in the body. For this class of conductors the DHEM anomalies vary in shape, amplitude and sign with change in transmitter loop location.

The 'simple' current-filament model (see *Modelling EM data* in Section 5.7.5.3) is useful for locating the centre of the eddy current system in the conductor and can provide an indication of dip. When the receiver is close to the conductor (i.e. the anomaly is an in-hole response or a near-miss target), tabular body or plate-like models, or sphere model where appropriate, often provide better definition of the target than do simple current-filament approximations. Inverse modelling methods (see Section 2.11.2.1) can be used to refine the model.

5.8.2.3 Interpretation pitfalls

A number of responses need to be accounted for when interpreting DHEM data. They relate to conductive overburden, conductive host rocks, current channelling and induced polarisation. These all apply to EM data in general and are described in Section 5.7.6.

In DHEM the response of an overburden layer resembles a long-wavelength regional response decreasing in amplitude downhole, and upon which are superimposed the shorter-wavelength target responses (Fig. 5.94). These can be separated using filtering techniques (see Section 2.9.2). The effects of variations in the near-surface conductivity depend on their coupling with the primary field. They will vary with the different loop locations, and from drillhole to drillhole.

The response of current channelling (see Section 5.7.2.4) can be strong in DHEM as the downhole probe can be in close proximity to the current channelling paths.

5.8.3 Examples of DHEM responses from mineral deposits

DHEM is widely used in mine- and prospect-scale exploration for massive sulphide mineralisation, especially nickel sulphides. Good examples include those presented by Jackson *et al.* (1996) and King (1996).

5.8.3.1 Balcooma volcanogenic massive sulphide

The use of TD-DHEM during the exploration of a massive sulphide body is demonstrated in this example. It illustrates the nature of the axial and three-component DHEM impulse response of multiple conductors, shows the benefits of using multiple transmitter loops and measuring three-component data, and demonstrates the analysis required to interpret the data in terms of the location and geometry of the conductors.

The Balcooma volcanogenic massive sulphide is located in northern Queensland, Australia (Huston and Taylor, 1990). It is a polymetallic deposit and is most significant for its copper content. An outcropping gossan led to its discovery. Host rocks are metasedimentary and metavolcanic rocks (amphibolite facies), including schistose metagreywackes, pelites and tuffs. Mineralisation occurs as four separate lenses, referred to as zones 1 to 4, containing massive, disseminated and stringer sulphides (Fig. 5.95a). The zones are elongate ellipses striking to the northeast, parallel to the local strike, and plunging at 15 to 25 degrees towards the southeast.

Hughes and Ravenhurst (1996) describe the DHEM surveys. Initially, only axial component data were recorded in drillhole SH20 using five transmitter loops (Fig. 5.95b). The dB/dt data are shown for the later delay times, channels 10 to 15 (3.5 ms after pulse turn-off), and show off-hole responses from two conductors. Note from Fig. 5.95c how the responses are very weak from the northern loop, indicating that this loop is poorly coupled to the conductors. The anomaly at about 75 m depth coincides with the known 'zone 1' disseminated mineralisation; the anomaly is probably due to the conductivity of continuous stringer or massive mineralisation further south in the lens. The consistent polarity of the southern, central and western loop responses, and the reverse polarity from the eastern loop, suggests that the conductor lies to the west of the eastern transmitter loop. This interpretation is consistent with the known geology.

The deeper response, at about 225 m depth, did not coincide with known mineralisation, but is possibly due to the along-strike extension of 'zone 4' mineralisation intersected more than 100 m to the south of SH20. The largest responses are from the central and southern loops, suggesting that the conductor is located south of the drillhole. The response from the eastern loop is weaker than that from the western and central loops, suggesting that the conductor is nearer these latter two loops. However, the opposite polarity of the response from the western loop suggests that the conductor is located between the western and central loops.

The response from the deeper conductor was modelled using a current filament model (see *Modelling EM data* in Section 5.7.5.3), but the data allowed the filament to be located either north or south, or above or below the drillhole 'for most loops'. This indicates that, for each loop, the orientation of the current system is being controlled by the direction of the primary field through the conductor rather than always being constrained by the plane of a plate-like conductor. This suggests that the conductive mineralisation in this zone is a thick body rather than a thin plate.

To better constrain the location and orientation of the conductor, cross-component data were recorded to supplement the existing axial-component data. Drillhole SH20 was surveyed using only the central transmitter loop. With the aid of computed model responses similar to those shown in Fig. 5.94, the three-component data (Fig. 5.95d) suggest that the conductor is above and mostly to the south of the drillhole. The peak response of the A-component,

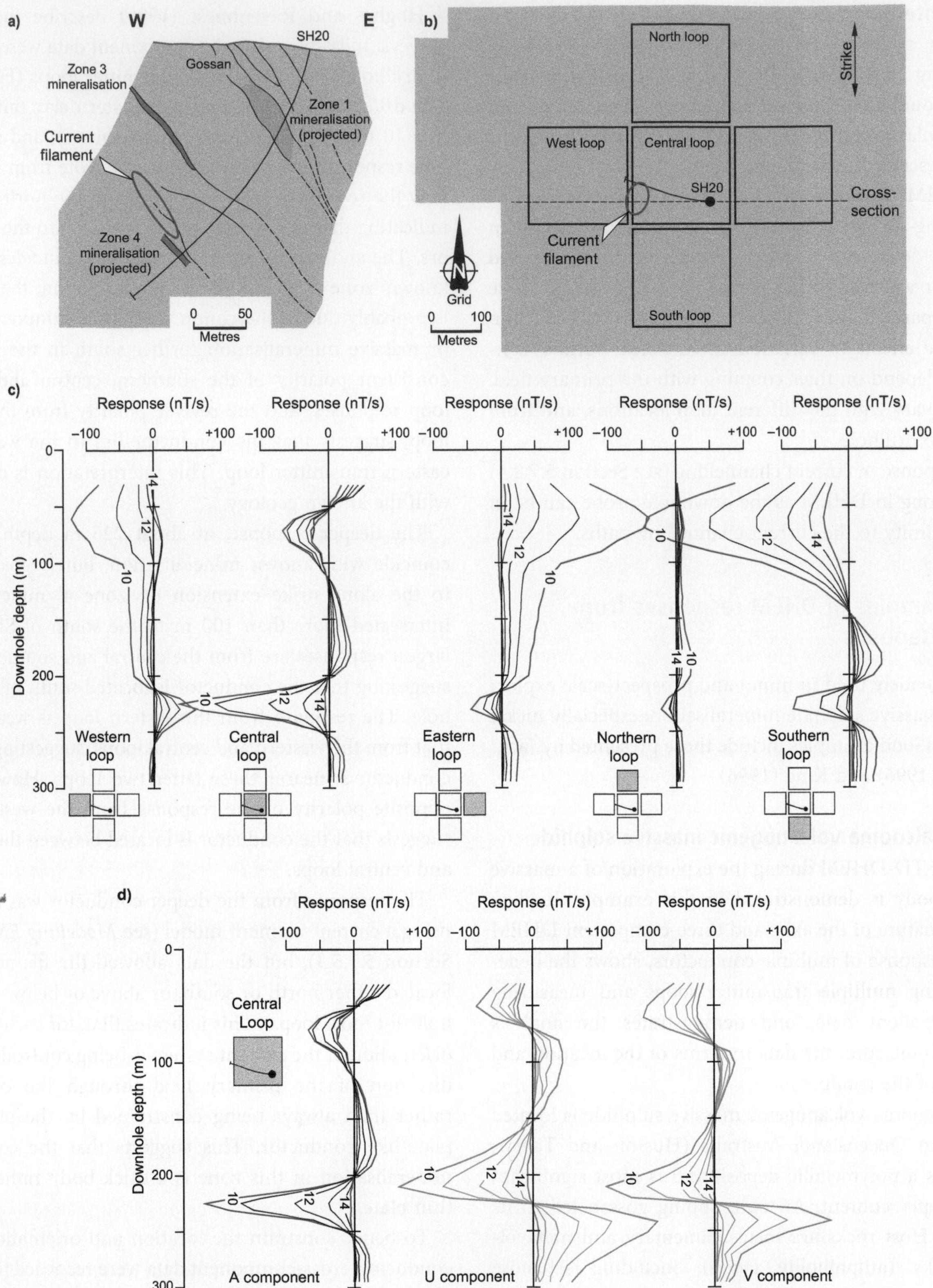

Figure 5.95 DHEM data from Balcooma massive sulphide. (a) Geological cross-section, (b) location of the five transmitter loops around drillhole SH20, (c) axial component late-time data from the five transmitter loops, (d) three-component late-time data from the central transmitter loop. All loops are 200 m × 200 m in size. Based on diagrams in Hughes and Ravenhurst (1996).

coincident with cross-overs in both the U- and V-components, gives the downhole position of the plane of the conductor, and the symmetry of all three components indicates that it is oriented perpendicular to the local direction of the drillhole trajectory. The combination of the polarities of the three components, defined in Fig. 5.92a, uniquely locates the eddy current system relative to the drillhole. Modelling of the data with a current filament model showed that an east-dipping, north-northeast trending conductor fitted the survey data (Fig. 5.95a). Drilling based on this interpretation intersected 'zone 4' mineralisation. Both the multi-loop axial and single-loop cross-component DHEM surveys demonstrate that this zone of mineralisation does not extend north of drillhole SH20.

5.8.4 Induction logging

Induction logging is a form of downhole frequency domain EM surveying (see Section 5.7.1.1) which measures the electrical conductivity of the rocks forming the walls of the drillhole. Like all EM methods, it is most effective for logging conductive zones rather than resistivity contrasts within resistive zones, although the logs generally correlate closely with electrical resistivity logs (see Section 5.6.8). Inductive methods can be used in dry and plastic-cased drillholes; this is a distinct advantage over electrical methods which require water in the drillhole for electrical connection to the wall rocks.

The logging tool contains transmitter and receiver coils oriented co-axially with the tool's long axis, which is coincident with the drillhole axis (Fig. 5.96a). In its basic form the transmitter coil, which carries alternating current at typically 200 Hz to 40 kHz, induces eddy currents in the surrounding rock that, for homogeneous rocks, flow concentrically around the drillhole. The circulating current flow is restricted in its along-hole extent, i.e. the current distribution approximates a disc centred on the probe. This determines the along-hole resolution of the data, typically about 10 cm. The magnetic fields associated with the eddy currents induce an alternating secondary voltage in the receiver coil and its phase-lag (see Section 5.7.1.1), with respect to the primary signal, is used to infer conductivity of the drillhole environment. Penetration into the wall rocks increases with increasing transmitter–receiver coil separation, and is also strongly influenced by the frequency of the transmitted current and the conductivity of the wall rocks, i.e. skin-effects (see Section 5.2.3.1). Data typically relate to the electrical properties of the rock volume within about a 1 m radius of the drillhole.

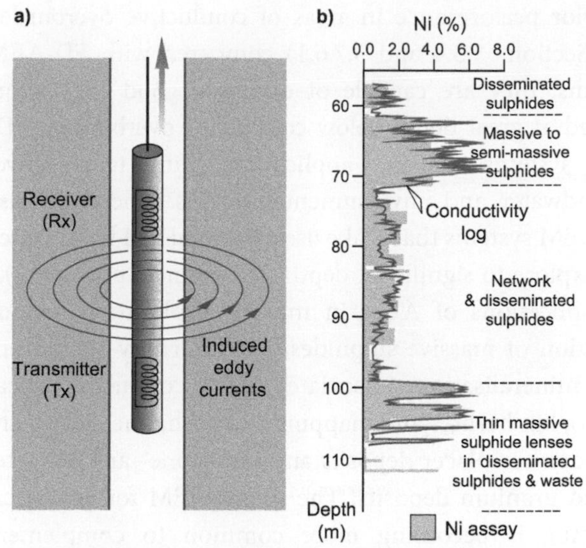

Figure 5.96 Downhole induction logging. (a) Schematic illustration of a downhole induction logging tool. (b) Correlation between an inductive conductivity log and Ni assays of drillcore. Data are from the Enonkoski Ni sulphide mine in Eastern Finland. Host rocks are mafic and ultramafic lithotypes. Redrawn, with permission, from Hattula and Rekola (1997).

Inductive conductivity logs are interpreted in much the same way as electrical resistivity logs (see Section 5.6.8). One particular application is assisting in grade control measurements of massive nickel sulphide deposits. As illustrated in Fig. 5.96b, there can be strong correlations between nickel content and electrical conductivity allowing ore-grade to be rapidly and economically assessed with induction logging.

5.9 Airborne electromagnetic surveying

Like all airborne geophysical methods, airborne electromagnetic (AEM) surveying offers the ability to survey large areas quickly and economically, but with lower resolution and less penetration than ground-based EM surveys. AEM systems were originally developed in the frequency domain (FD-AEM) to detect conductive massive sulphide bodies within the resistive rocks of the Precambrian shield of Canada. The subsequent need to explore for other types of targets and explore other kinds of geological environments combined with developments in EM systems has led to higher-sensitivity time domain (TD-AEM) systems now being used almost exclusively for mineral exploration and geological mapping. In particular, TD-AEM systems have superior performance in areas of conductive overburden (see Sections 5.3.4 and 5.7.6.1) compared with FD-AEM systems; they are capable of detecting good conductors located at great depth below conductive overburden. FD-AEM systems now find application mainly to shallower groundwater and environmental studies. There are also FD-AEM systems that make use of natural EM fields which can explore to significant depth (see online Appendix 4).

Applications of AEM in mineral exploration include detection of massive sulphides, unconformity-style uranium mineralisation (associated with conductive shear zones), kimberlites and mapping palaeochannels as potential hosts for placer deposits and sandstone- and calcrete-hosted uranium deposits. The use of AEM for geological mapping is becoming more common to complement magnetic and gravity surveys, although AEM remains significantly more costly than aeromagnetics (see Section 1.2.3). AEM has a significant role in mapping groundwater and soil salinity. These are important applications in mine development; moreover, soil moisture and salinity are usually the source of the EM responses measured when mapping palaeochannels and the weathered upper portions of some types of mineral targets, such as kimberlite pipes.

We describe the principles of TD-AEM systems and their applications in mineral exploration with an emphasis on the types of systems in current use. The reader is referred to Klein and Lajoie (1992) for descriptions of some older AEM systems, including frequency domain systems, survey procedures and interpretation techniques.

5.9.1 Acquisition of AEM data

Fixed-wing and helicopter AEM systems are in use with both configurations comprising a large horizontal transmitter loop and small receiver coils (dB/dt sensors) measuring both the along-line horizontal (X) component and the vertical (Z) component of the field (see Section 5.7.1.5). A few systems also measure the across-line horizontal (Y) component. The horizontal orientation of the transmitter loop is dictated by the practicalities of fitting it to an aircraft. A consequence is that coupling (Fig. 5.70) is optimum for conductors with large horizontal dimensions located immediately below the loop, and ideal for measuring the background conductivity (see Section 5.7.2.1). Many potential mineral targets do not fulfil this criterion, but nevertheless, good coupling can be achieved to steeply dipping targets located in the side regions of the loop.

AEM systems operate in the same way as ground EM systems, but with some obvious distinctions. The transmitter loop is located at a height above the ground so the eddy current induced at the instant of the pulse turn-off (see Section 5.7.2.1) is laterally more expansive and locally weaker than for the same loop on the ground surface. In addition, the loop is necessarily smaller than that typically used in ground surveys, the transmitter and receiver are further from the target and the measurements are made from a continuously moving platform. Consequently, the measurements are inherently noisier so more attention is required to maximise the signal-to-noise ratio in both the instrumentation and the data reduction. AEM systems generally have a more powerful transmitter than ground systems and some also have a multi-turn loop, which increases its inductance (see Section 5.7.1.2). The lower-flying helicopter systems place the transmitter loop closer to the ground counteracting geometrical attenuation of the loop's field; so some helicopter systems produce a stronger primary signal at the target than is achievable with some of the higher-powered, higher-flying, fixed-wing systems. The continuously moving AEM system 'averages' an elongate zone of the subsurface parallel to the flight line to produce a single reading, which is exacerbated by longer measurement periods, i.e. lower base frequency. There is the added complication of continuously changing coupling between the transmitter and any conductors present, and between the secondary magnetic field and the receiver. These effects reduce lateral resolution in the survey line direction. The final along-line data interval is set by the system timing and the speed of the aircraft.

Applications of AEM in mineral exploration range from conductor detection to conductivity mapping at a range of depths, and in both conductive and resistive environments. It is difficult to build an AEM system that is fully multi-purpose because of engineering considerations. For this reason there exist a variety of TD-AEM systems with individual systems tending to be optimised for a selected range of applications. Characteristics that vary between systems include the transmitter waveform, signal frequency, receiver sampling rate, type of response measured by the receiver, primary field strength, and signal enhancement and noise suppression applied to the measured signal.

5.9.1.1 System waveform

A wide variety of transmitted pulses and system waveforms are used by the various TD-AEM systems (Fig. 5.97). Like ground TDEM systems, the actual width of the pulse and

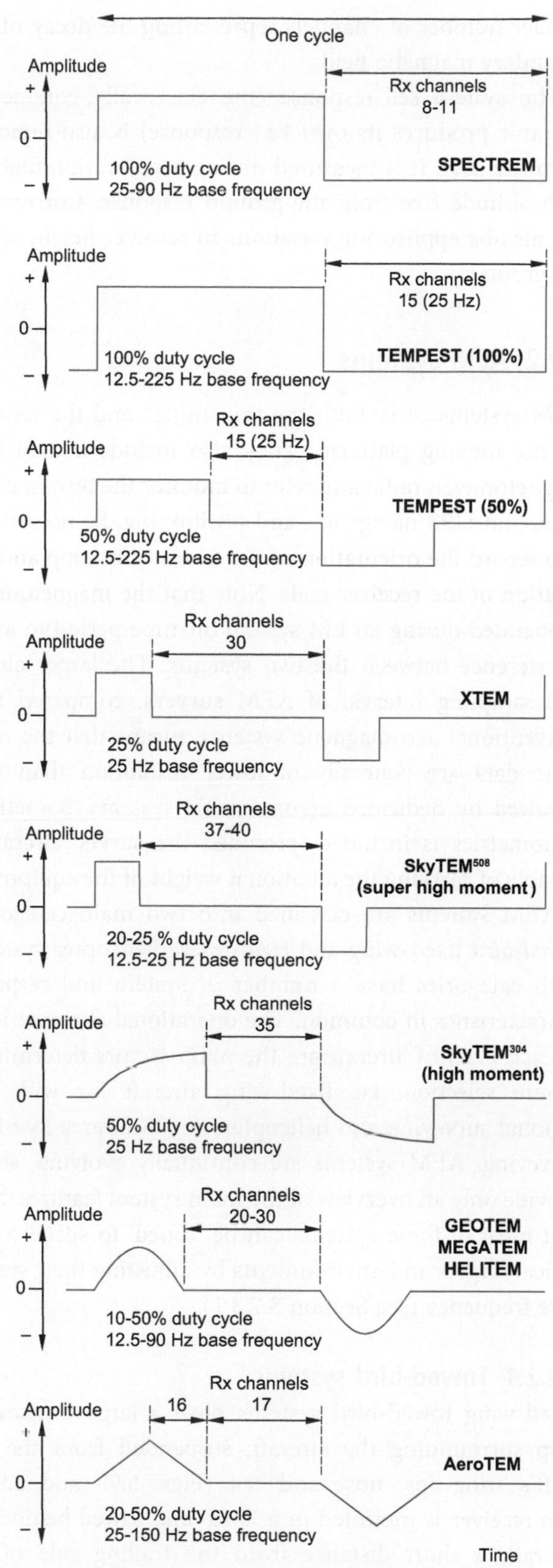

Figure 5.97 Schematic illustrations of one cycle of the system waveforms, including the response measurement window, of various TD-AEM systems. TEMPEST (100%) is obtained from the actual transmitted waveform shown as TEMPEST (50%). Note that the actual amplitude of the transmitted pulse varies widely between the systems and is not depicted in the figure. Details of each are as available at the time of our writing and the reader is referred to the various system providers for updated information about individual systems. SPECTREM waveform reproduced with the permission of Spectrem Air Ltd; GEOTEM, MEGATEM, HELITEM and TEMPEST waveforms reproduced with the permission of CGG; XTEM waveform reproduced with the permission of GPX Surveys; SkyTEM waveforms reproduced with the permission of SkyTEM Surveys; and details of the AeroTEM waveform from Balch *et al.* (2003) and Huang and Rudd (2008).

the decay period over which measurements are made are set by the system base frequency (see Section 5.7.3.1). The actual periods of the AEM waveforms shown in the figure can be determined for their respective base frequencies (see online Appendix 2). For most AEM systems the base frequency can be changed and the system response optimised for different geological targets (for example see Fig. 5.100).

Increasing the amplitude of the primary pulse increases the strength of the secondary decay and reduces noise, particularly at late times, allowing detection of deeper conductors and slow-decaying high-conductivity conductors (see *Conductor quality* in Section 5.7.2.3). There are practical limitations to the dipole moment that can be created in AEM system, owing to restrictions on both the physical size and weight of the transmitter loop and the equipment needed to generate the current. As explained previously (Section 5.7.1.2), the inductance of the loop has a significant and detrimental effect on the ramp turn-off. Consequently, most high-power systems have a slow pulse turn-off with the decay measurements commencing at a relatively long time after the start of the turn-off. This reduces resolution of the early-time response of the near-surface and the rapidly decaying responses from conductors with lower conductivity (see Section 5.7.2). In contrast, low-power systems have a faster turn-off allowing the secondary decay measurements to commence closer to the start of the pulse turn-off. They produce better resolution of near-surface conductors and fast-decaying low-conductivity conductors, but at the expense of reduced penetration (because of reduced signal-to-noise ratio).

5.9.1.2 Measuring the step and impulse responses

Both *B*-field and coil sensors are used for making step and impulse response measurements in ground and downhole EM systems (see Section 5.7.1.7). To avoid the complexities of deploying a separate B-field sensor, AEM systems obtain both responses solely from dB/dt measurements made with a coil sensor in one of the following three ways.

Firstly, recall from Section 5.7.1.7 that for systems transmitting a step-change in primary field, the off-time dB/dt measurement is the impulse response of the ground. As an alternative to using a separate B-field sensor, the B-field measurement can be obtained by mathematically integrating the dB/dt measurements and adding the initial value of the magnetic field at the beginning of the decay period. Known as the *recovered B-field*, it requires that dB/dt measurements be also made during the on-time of the transmitted pulse. The measurement obtained with this quasi B-field sensor has (depending on the shape of the transmitted pulse) many attributes of the step response.

Secondly, and recalling Section 5.7.1.7, for systems transmitting a triangular pulse the dB/dt measurements made during the pulse on-time are the step response. The off-time measurements approximate the impulse response.

The third method involves making a very large number (usually thousands) of equi-timed dB/dt measurements continuously during the full period of the system waveform, known as *full-waveform sampling*. The measurements include the strength of the primary field and the decaying secondary field. The precisely known primary field is mathematically removed from the measurements and then any response of the ground can be computed from the highly sampled time series, usually the step and impulse responses (see Section 5.7.1.7).

Similarly to ground EM systems, AEM receiver measurements are normalised for the transmitter loop moment, the transmitter current at the time of the measurement (see Section 5.7.4.1) and the receiver coil moment.

5.9.1.3 Noise suppression

The receivers in TD-AEM systems make a large number, usually thousands, of measurements during the measurement period. A variety of noise suppression algorithms are applied to these data post-survey to attenuate system noise, powerline noise and sferics. Stacking is applied to improve the signal-to-noise ratio (see Section 2.7.4.1). However, given that the survey platform is continually in motion during the measurement period, a compromise is required between achieving an acceptable signal-to-noise ratio and the inevitable reduction in resolution and sensitivity due to the motion. Too many stacks distort the data, reducing both the temporal and spatial resolutions. Throughout the process the large volume of measurements are reduced to a smaller number of channels representing the decay of the secondary magnetic field.

The system self-response (the electrically conducting airframe produces its own EM response) is also removed from the data. It is measured during a calibration flight at high altitude free from the ground response. Corrections can also be applied for variations in receiver height above the ground.

5.9.2 AEM systems

AEM systems carry both the transmitter and the receiver on the moving platform. They also include a total field magnetometer, radio altimeter to monitor the terrain clearance, and GPS navigation and positioning. Some systems also record the orientation of the transmitter loop and the location of the receiver coils. Note that the magnetometer is operated during an EM system off-time period to avoid interference between the two systems. The larger along-line sampling interval of AEM surveys, compared with conventional aeromagnetic systems, means that the magnetic data are generally of lower resolution than data acquired by dedicated aeromagnetic systems. Sometimes radiometrics is included, provided the survey aircraft is capable of carrying the additional weight of the equipment.

AEM systems are classified into two main categories: *towed-bird* fixed-wing and *rigid-frame* helicopter systems. Both categories have a number of system and response characteristics in common. The operational characteristics of each class of aircraft are the main factors determining system selection: i.e. fixed-wing aircraft for wide-area regional surveying and helicopters for local-area low-level surveying. AEM systems are continually evolving, so we provide only an overview of the main system features. Note that most of these systems can be 'tuned' to specific geological targets and environments by adjusting their system base frequency (see Section 5.7.3.1).

5.9.2.1 Towed-bird systems

Fixed-wing towed-bird systems have a large transmitter loop surrounding the aircraft, suspended from the aircraft's wing-tips, nose and tail (Figs. 5.98 and 5.99a). The receiver is mounted in a 'bird' and towed behind the aircraft, a short distance from the trailing side of the transmitter loop, in the separated-loop configuration (see *Moving-loop mode* in Section 5.7.3.2). Oscillatory motions of the towed receiver bird, i.e. yaw, pitch and roll, produce changes in the measured strength of the secondary field

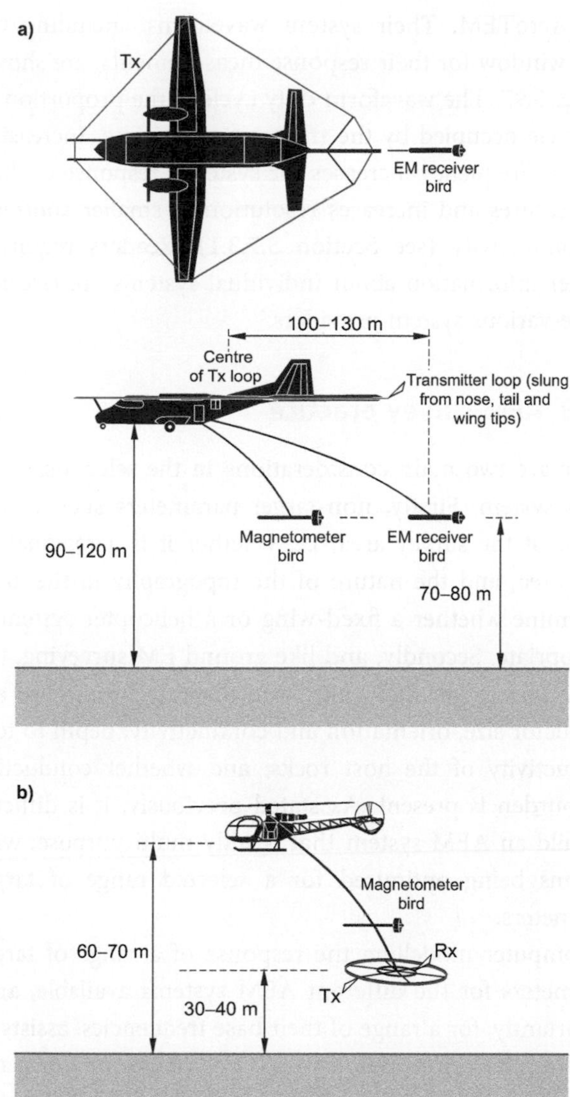

Figure 5.98 Schematic illustrations of (a) fixed wing and (b) helicopter AEM systems.

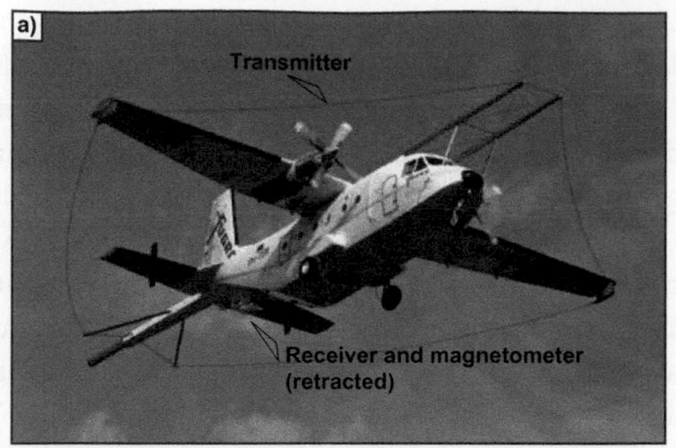

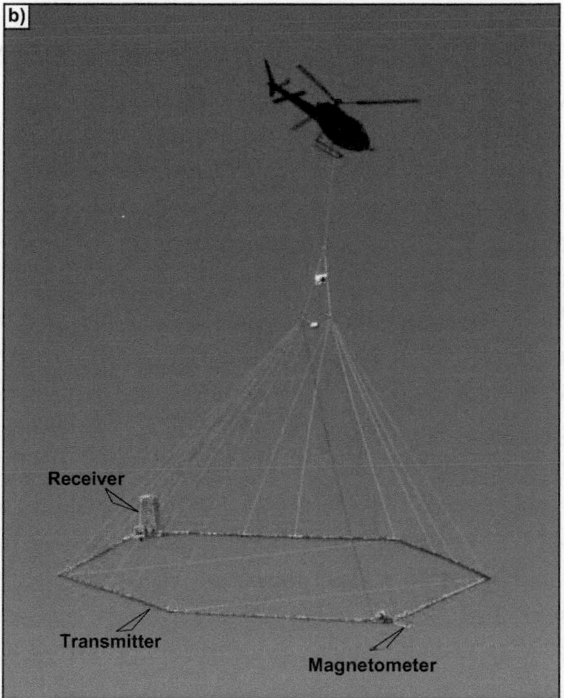

Figure 5.99 Examples of AEM systems. (a) TEMPEST fixed-wing towed-bird system. Courtesy of CGG. (b) SkyTEM helicopter system. Courtesy of SkyTEM Surveys.

which appear as noise in the measured response. This, along with the noise from all other sources, reduces system sensitivity at late decay times. Furthermore, variations in transmitter to receiver separation cause variations in the on-time measurements (see Section 5.9.1.2). The monitoring of bird location and motion with respect to the transmitter loop allow a correction to be applied to the data.

The separated-loop configuration has asymmetric geometry. The array changes orientation on reciprocal survey headings so the response is asymmetric and there is an offset in its location between survey lines flown in the opposite direction (as demonstrated by the differences in the model responses in Figs. 5.85 and 5.86). Also, the polarity of the *X*-component data is reversed. These are not major problems when analysing individual line-profiles, but are a significant disadvantage when compiling 2D images of the measured parameters, as processing artefacts are caused by the asymmetry. When displaying *X*-component data, their polarity for alternate headings may be reversed to facilitate inter-line comparison of responses.

Fixed-wing towed-bird AEM systems fly at a survey speed of 180–250 km/h (50–70 m/s) at 90–120 m above the terrain with the towed receiver 70–80 m above the ground (Fig. 5.98a). The along-line data interval is typically about 15 m. The area of the transmitter loop and the number of turns in it vary from system to system, with dipole moments ranging from about 500,000 to 2.2 million A m^2.

5.9.2.2 Rigid-frame systems

Rigid-frame systems have the horizontal transmitter loop and the receiver coils attached to a single rigid structure made of lightweight non-conducting material. This maintains the separation and orientation of the transmitter loop and receiver coils ensuring constant coupling between them, which is essential when making measurements during the transmitter on-time (see Section 5.9.1.2). Some systems make use of a collapsible lightweight suspension system. We also classify these as rigid-frame systems. The whole structure is suspended below the aircraft (Figs. 5.98b and 5.99b). The receiver coils are located either on the vertical axis of the loop or in the near-null position of the primary field near the edge of the transmitter loop (Fig. 5.99b). Any offset between the receiver coils and the centre of the loop is small compared with the survey height so they can be considered coincident. The responses are the same as those from the in-loop configuration used for ground surveys (see *Moving-loop mode* in Section 5.7.3.2).

The in-loop configuration offers the distinct advantage, over the separated-loop configuration of towed-bird systems, of producing sharper anomalies with simpler shapes that are independent of survey line direction, i.e. there is no directional asymmetry (see Section 5.9.2.1). It also has less sensitivity to conductive overburden than the towed-bird configuration. These features are demonstrated by the model responses in Figs. 5.84 when compared with those in Figs. 5.85 and 5.86.

As with all helicopter systems, surveying is possible in rugged terrains and at lower survey heights and speeds than is achievable with fixed-wing aircraft, which leads to increased signal strength and higher spatial resolution. Helicopter systems fly more slowly at 75–120 km/h (20–33 m/s) and lower at 60–70 m above the terrain with the EM system suspended 30 m below the helicopter, i.e. at about 30–40 m above the ground (Fig. 5.98b). The along-line data interval is typically 2–7 m. The area of the transmitter loop and the number of turns in it vary from system to system, with dipole moments ranging typically from about 100,000 to 600,000 A m^2, and sometimes as high as 2 million A m^2.

5.9.2.3 Examples of AEM systems

A variety of TD-AEM systems are in use at the time of writing. These include fixed-wing systems known as SPECTREM, TEMPEST, GEOTEM and MEGATEM, and helicopter systems known as XTEM, SkyTEM, HELITEM and AeroTEM. Their system waveforms, including the time-window for their response measurements, are shown in Fig. 5.97. The waveform duty cycle is the proportion of the cycle occupied by the transmitter on-time. Increasing the base frequency increases the system's response to shallow features and increases resolution of smaller contrasts in conductivity (see Section 5.7.3.1). Readers requiring further information about individual systems are referred to the various system providers.

5.9.3 AEM survey practice

There are two main considerations in the selection of an AEM system. Firstly, non-target parameters such as the extent of the survey area, i.e. whether it is a regional or local area, and the nature of the topography in the area, determine whether a fixed-wing or a helicopter system is appropriate. Secondly, and like ground EM surveying, the main parameters that determine the target response are conductor size, orientation and conductivity; depth to top; conductivity of the host rocks; and whether conductive overburden is present. As stated previously, it is difficult to build an AEM system that is fully multi-purpose, with systems being optimised for a selected range of target parameters.

Computer modelling the response of a range of target parameters for the different AEM systems available, and, importantly, for a range of their base frequencies, assists in selecting the appropriate system. The ability to adjust the system base frequency (see Section 5.7.3.1) allows the system to be 'tuned' to a particular geological application; for example see Fig. 5.100. Furthermore, measurements made over a longer decay period provide more information for discriminating between poor and good conductors (see *Conductor quality* in Section 5.7.2.3). Modelling can set the maximum depth of detection of a body of given size, orientation and conductivity, and quantify the resolution possible for the available AEM systems. It also helps survey planning to establish the nature of the overburden and the background, possibly by initially undertaking petrophysical studies and/or ground surveys in the proposed AEM survey area.

Survey parameters are usually set to suit the requirements of the AEM system; magnetics and radiometrics are secondary considerations. Survey lines should be orientated perpendicular to the regional strike as this provides better coupling to target conductors, which usually have the same strike. Line spacing is determined by the strike length of

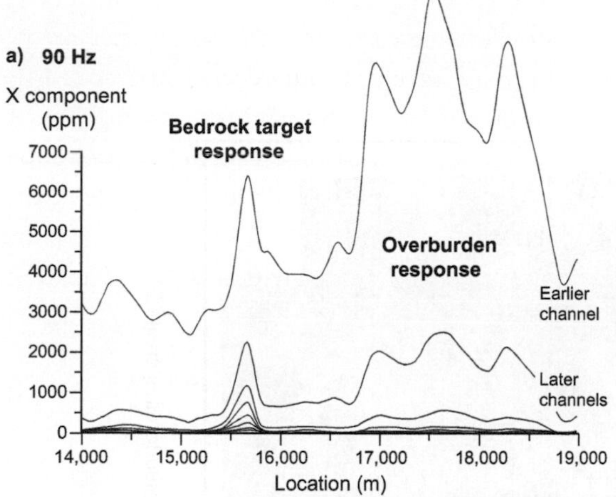

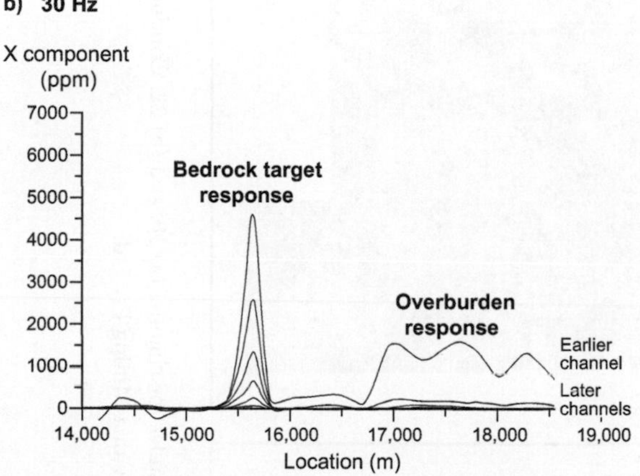

Figure 5.100 GEOTEM responses for two different system base frequencies. (a) 90 Hz and (b) 30 Hz. Note the enhanced target response and reduced overburden response at the lower frequency. Redrawn, with permission, from Smith and Annan (1997).

the target, and the spacing should be set so that at least two, and preferably three, survey lines pass over the target. Survey line spacing is usually chosen as a compromise between survey cost and the expected strike length of target conductors with terrain clearance kept as low as possible whilst maintaining safe operations.

5.9.4 Display and interpretation of AEM data

Display and interpretation of AEM data is essentially identical to that for ground data (see Sections 5.7.4 and 5.7.5), although the characteristically large data volumes have led to AEM data being routinely displayed as imaged conductivity distributions obtained by inversion of the data (see Sections 2.11.2.1 and 5.7.4.3). The analyses otherwise proceed in an identical way to that for ground data (see Section 5.7.5). Anomaly shapes for the in-loop and separated-loop configurations described previously (see *Profile analysis* in Section 5.7.5.3) apply also to AEM data.

5.9.4.1 Interpretation pitfalls

AEM data have the same interpretational problems as EM data in general and are described in Section 5.7.6. In addition, variations in survey height above the ground, common in rugged terrains, cause variations in the size and strength of the primary field in the ground with subsequent effect on lateral resolution and strength of the secondary field. Integrating survey height data and terrain data with the EM data allows these effects to be recognised for what they are. The survey-direction dependent asymmetry of the towed-bird separated-loop response needs to be accounted for when working with profile data from adjacent survey lines.

5.9.5 Examples of AEM data from mineralised terrains

Witherly (2000) reviews 50 years of AEM surveying and credits AEM with directly aiding in the discovery of more than 80 mineral deposits. The most common targets are massive metal sulphides. Examples of AEM responses from nickel sulphide deposits are described by Wolfgram and Golden (2001).

5.9.5.1 Butcherbird Supergene manganese deposit

This example illustrates the use of AEM for mapping shallow conductivity structure and for targeting conductive manganese mineralisation in the Mesoproterozoic Collier Basin of Western Australia. d*B*/d*t* (impulse response) and magnetic data were acquired with the XTEM helicopter system along survey lines spaced 200 m apart. Manganese oxide mineralisation occurs as sub-horizontal sheets several hundred metres in length usually within a few tens of metres of the surface. Figure 5.101 shows the magnetic data and gridded TDEM channel amplitude data. These data are chosen to illustrate shallow, intermediate and deeper responses. CDIs were created from the profile data (Fig. 5.102) and maps were created by combining the results for selected pseudo-depths (Fig. 5.101). Known mineralisation at Yanneri was shown to coincide with a conductive response, and other similar target responses were identified in the data. Subsequent drilling discovered additional manganese mineralisation. Responses related to non-economic targets were also mapped, for example a palaeochannel (a source of water) and a pyritic shale unit.

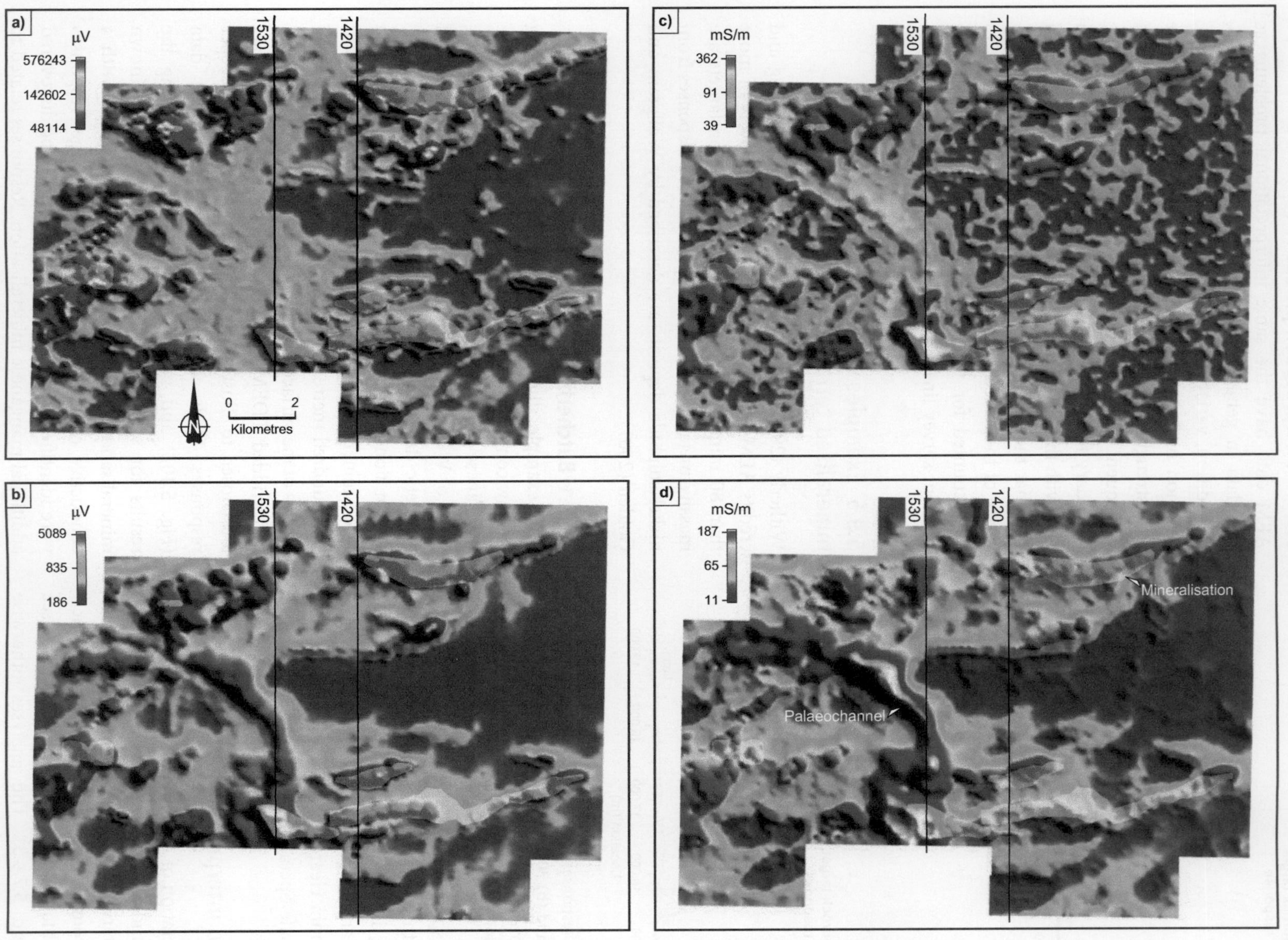

Figure 5.101 XTEM AEM data from the Butcherbird Mn deposit. Images of amplitudes for channel (a) 10, and (b) 22. Conductivity pseudomaps created from CDIs at computed depths of (c) 20 m and (d) 40 m. Data courtesy of Montezuma Mining Pty Ltd.

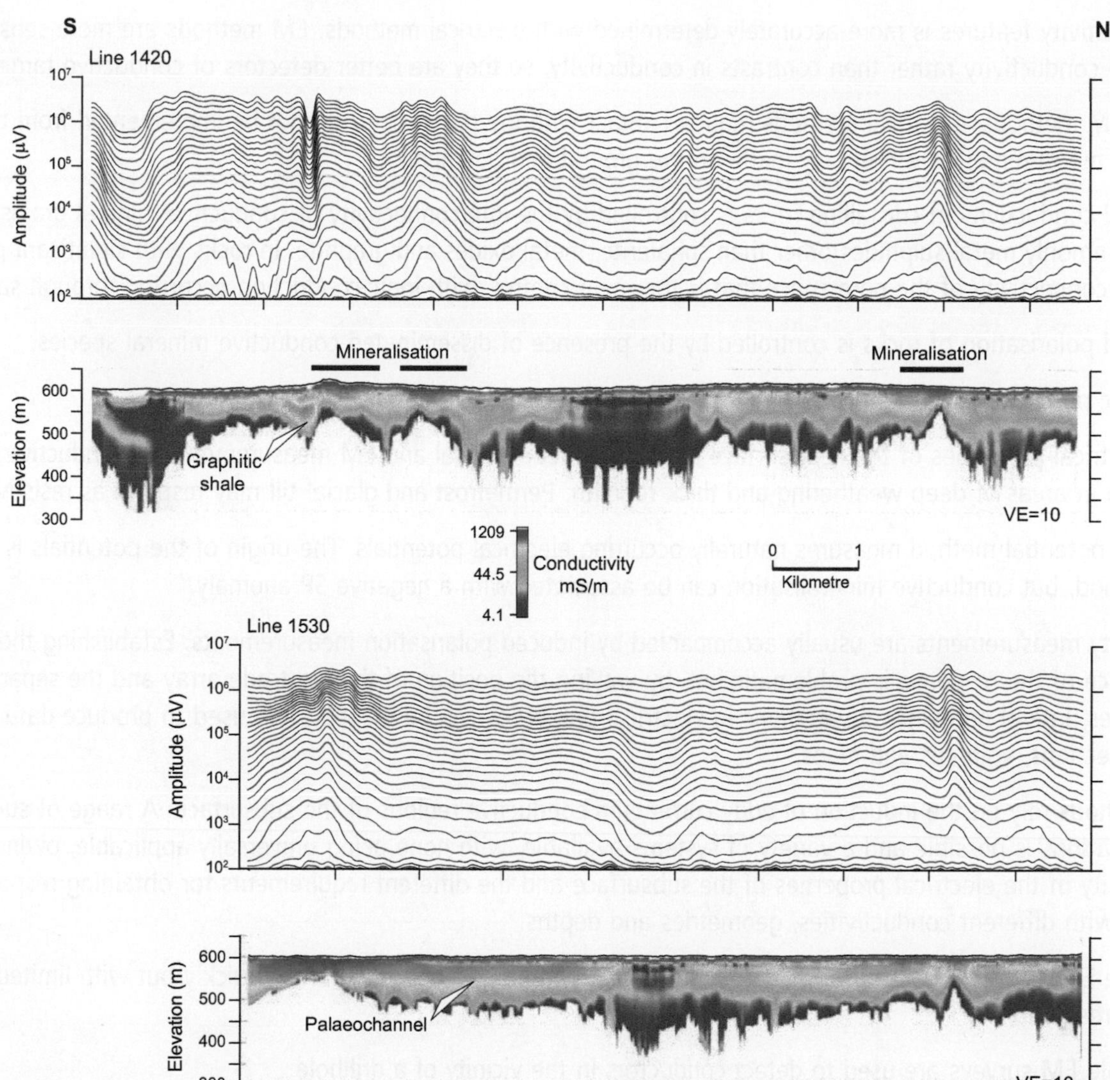

Figure 5.102 XTEM profiles and CDIs from two survey lines over the Butcherbird Mn deposit. See Fig. 5.101 for locations. VE – vertical exaggeration. Data courtesy of Montezuma Mining Pty Ltd.

Summary

- Electrical and electromagnetic (EM) methods respond to the electrical properties of the subsurface.
- The most commonly measured property is electrical conductivity or its reciprocal, resistivity. This controls the response measured by the resistivity, magnetometric, self-potential and applied potential methods, which are electrical methods, and also the EM method.
- The induced polarisation method is an electrical method that responds to the ability of the ground to become electrically polarised, i.e. to have capacitor-like properties.
- Radar-frequency EM methods respond to the dielectric properties of the ground.
- Electrical methods are sensitive to contrasts in resistivity, irrespective of the absolute resistivities involved, so they have higher resolution than EM methods and are appropriate for mapping changes in resistivity/conductivity. The resistivity of

high-resistivity features is more accurately determined with electrical methods. EM methods are more sensitive to the absolute conductivity rather than contrasts in conductivity, so they are better detectors of conductive targets.

- Unlike EM methods, electrical methods require contact with the ground and cannot be implemented from the air. Both types of measurement are routinely made on the ground and downhole.
- Electrical conductivity in crystalline rocks is controlled by the interconnectivity of conductive mineral grains, which are most commonly metal sulphides other than sphalerite, metal oxides and graphite. In rocks with significant porosity it is the interconnectivity of the pore space that is the main control, with ionic conduction occurring through saline fluids.
- Electrical polarisation of rocks is controlled by the presence of disseminated conductive mineral species.
- Dielectric properties of rocks are strongly controlled by their water content.
- The electrical properties of the near-surface strongly affect electrical and EM measurements. A conductive layer is common in areas of deep weathering and thick regolith. Permafrost and glacial till may respond as resistive layers.
- The self-potential method measures naturally occurring electrical potentials. The origin of the potentials is not well understood, but conductive mineralisation can be associated with a negative SP anomaly.
- Resistivity measurements are usually accompanied by induced polarisation measurements. Establishing the 'depth' of the source of the response is problematic but, by varying the position of the electrode array and the separation of the electrodes, lateral and vertical variations in electrical properties can be mapped and used to produce data pseudosections, volumes and maps.
- EM methods rely on the induction of eddy currents in conductive regions of the subsurface. A range of survey configurations is possible and a variety of systems available, with none being universally applicable, owing to the complexity of the electrical properties of the subsurface and the different requirements for obtaining responses from targets with different conductivities, geometries and depths.
- Airborne EM surveys are used for reconnaissance exploration, covering large areas quickly but with limited lateral and vertical resolution.
- Downhole EM surveys are used to detect conductors in the vicinity of a drillhole.
- Quantitative interpretation of electrical and electromagnetic data differs for the various survey configurations and systems, and is hindered by non-uniqueness. Sophisticated forward and inverse modelling methods are used and must take into account survey and system parameters, and the effects of terrain. An exact match between the observed and the computed responses cannot be expected, owing to the complexity of variations in electrical properties in the geological environment and the complex interactions between their electromagnetic responses.

Review questions

1. What are the two main ways in which electric current flows through rocks?
2. Why is the presence of a conductive overburden so significant for electrical and EM methods?
3. What are thought to be the main causes of self-potential anomalies?
4. What factors should be considered when selecting an electrode array for a resistivity/IP survey?
5. Explain how a pseudosection is created for one of the common electrode arrays. Why does this lead to 'pants-legs' anomalies?
6. Explain how the depth of investigation can be varied in electrical and EM methods.

7. Describe the relationship between the primary and secondary magnetic fields.
8. What is meant by the terms 'step response' and 'impulse response' in EM methods?
9. How would you recognise the response from a confined conductor in time domain electromagnetic data?
10. How would you differentiate between in-hole and off-hole responses in downhole electromagnetic data?
11. What factors control the depth penetration of an electromagnetic field?
12. What kind of electrical and EM surveys would be most suitable for detecting the following kinds of nickel mineralisation? (a) Disseminated sulphide; (b) massive sulphide; (c) lateritic.

FURTHER READING

Bishop, J. R. and Lewis, R. J. G., 1996. Introduction to the special volume on DHEM. *Exploration Geophysics*, 27, 37–39.

Eadie, T. and Staltari, G., 1987. Introduction to downhole electromagnetic methods. *Exploration Geophysics*, 18, 247–351.

The journal Exploration Geophysics has produced these two thematic volumes on DHEM, volume 27 (2/3) in 1996 and volume 18 (3) in 1987. These contain both theory and case study papers with numerous examples.

Goldie, M., 2002. Self-potentials associated with the Yanacocha high-sulfidation gold deposit in Peru. *Geophysics*, 67, 684–689.

A description of extreme SP variations in a gold camp in Peru, interesting both as a case study and as a demonstration of the incomplete understanding of the causes of variations in self-potentials.

Harris, N.C., Hemmerling, E.M. and Mallmann, A.J., 1990. *Physics: Principles and Applications.* McGraw-Hill.

This is a first year university physics text. Its description of electricity and electromagnetism is straightforward. The structure is that of a traditional physics book, so it includes some basic mathematics. There is no mention of geophysical applications.

Macnae, J., 2007. Developments in broadband airborne electromagnetic in the past decade. In Milkereit, B. (Ed.), *Proceedings of Exploration 07: Fifth Decennial International Conference on Mineral Exploration*. Decennial Mineral Exploration Conferences, 387–398.

A useful summary of the challenges facing AEM.

Nabighian, M. N., 1991. *Electromagnetic Methods in Applied Geophysics*, Volume 2, *Application: Parts A and B*. Society of Exploration Geophysicists, Investigations in Geophysics 3.

This large technical volume of electromagnetic geophysics gives a comprehensive description of the TDEM method. There have been significant developments in TDEM techniques since its publication.

Pelton, W.H. and Smith, P.K., 1976. Mapping porphyry copper deposits in the Philippines with IP. *Geophysics*, 41, 106–122.

The age of the paper is reflected in the way the data are displayed and interpreted. However, it is still a classic illustration of resistivity/IP responses from a very important style of mineralisation.

Sumner, J. S. 1976. *Principles of Induced Polarization for Geophysical Exploration*. Elsevier Scientific.

This is a classic text describing the principles of resistivity/IP techniques. It is intended for geophysicists but is not overly mathematical.

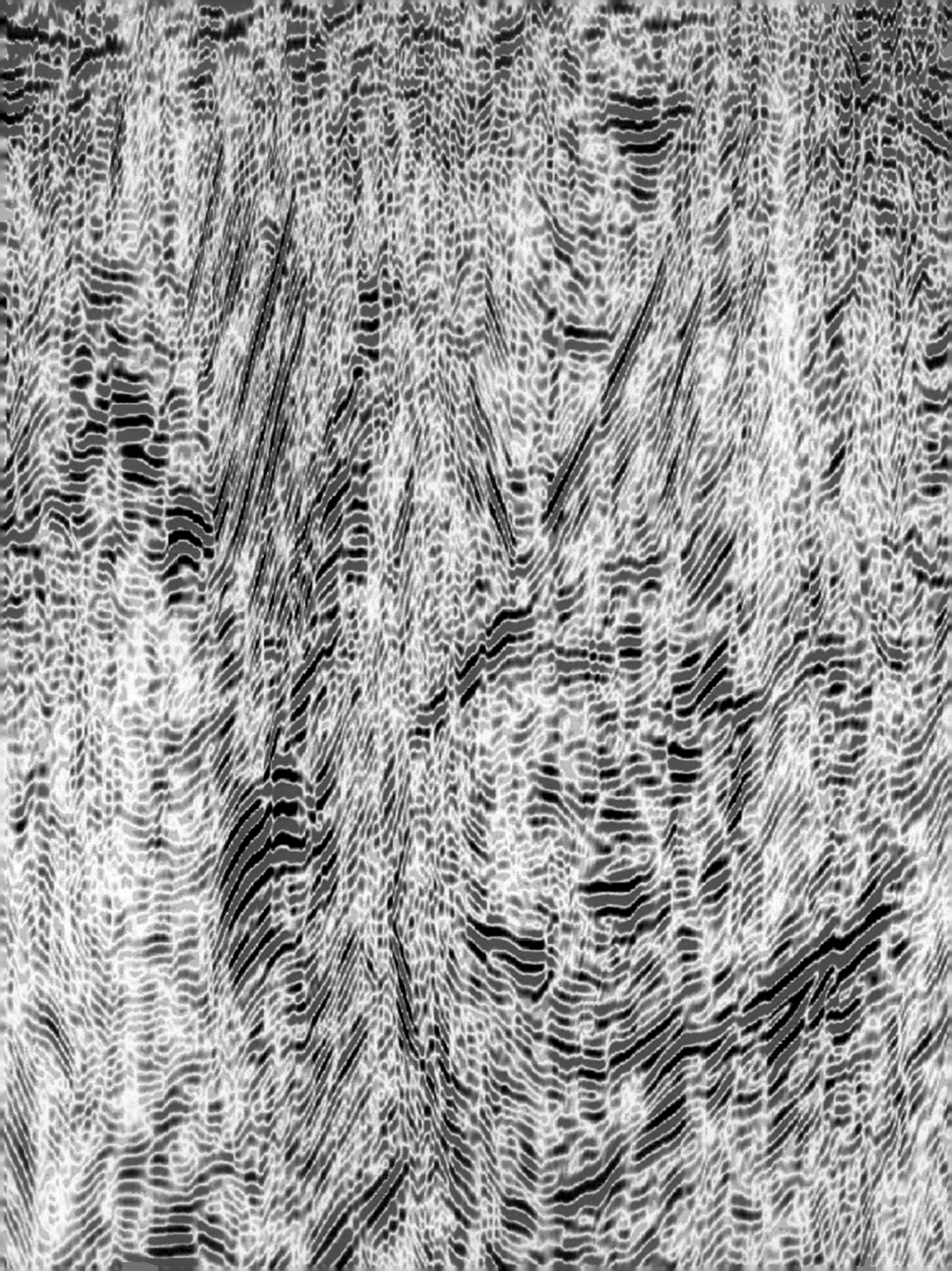

CHAPTER

6 Seismic method

6.1 Introduction

The seismic method is an active form of geophysical surveying that uses elastic waves to investigate the subsurface. The waves are created by a source and propagate through the subsurface before being recorded by detectors that measure deformation of the ground (Fig. 6.1). The deformation of the ground as a function of time since the waves were created comprises a time series, which is called a *seismic trace*. The passing of a seismic wave appears as a deflection of the trace, referred to as an *arrival* or an *event*. The path of the waves from source to detector is controlled by the elastic properties of the material through which they travel. Discontinuities in the elastic properties deflect and divide the seismic waves so that the detectors record a series of waves that have taken different paths through the subsurface. It is by identifying these different arrivals and analysing their travel times and amplitudes that the nature of the subsurface can be inferred.

Most seismic surveys use sources and detectors located on, or near, the Earth's surface. Surface seismic surveys are of two types. Surveys that particularly exploit waves reflected at elastic discontinuities are known as *seismic reflection* surveys, whilst those based on waves deflected so as to travel parallel to the discontinuities are known as *seismic refraction surveys*. Of all the geophysical methods, seismic reflection surveys provide the most detailed information about the subsurface, albeit at the greatest cost (see Fig. 1.2). Reflection survey data are displayed in a pseudosection/pseudovolume form where the travel times of the seismic waves provide an approximate indication of depth; but accurate time-to-depth conversion is difficult. Refraction surveys provide less detailed information, detecting only major changes in elastic properties, although the positions of these boundaries can usually be accurately determined.

The seismic reflection method is most effective where geological boundaries are laterally continuous and sub-horizontal, as occurs in sedimentary basins. As a consequence, it is the mainstay of the petroleum industry where it is used to obtain information about subsurface structure and stratigraphy to depths of several kilometres. It is used during both exploration and exploitation of petroleum resources and is the most important geophysical method in terms of the number of surveys conducted, personnel involved and expenditure. Within the mining industry it is the coal sector that makes significant use of the seismic reflection method, as do potash miners (the potash occurs in evaporite horizons). In both cases the method is primarily used during resource exploitation. The metalliferous sector makes much less use of the seismic method, although its use is becoming more common. The limited use is because the method is less suited to the geologically complex environments in which many kinds of metallic mineral deposits occur, and because the method is very expensive.

Compared with reflection surveys, seismic refraction surveys are considerably cheaper. Major users of the seismic refraction method are the engineering and

◄ Seismic reflection data from the Broken Hill area, Australia. The section is approximately 2.5 km in length. Data reproduced with the permission of Geoscience Australia.

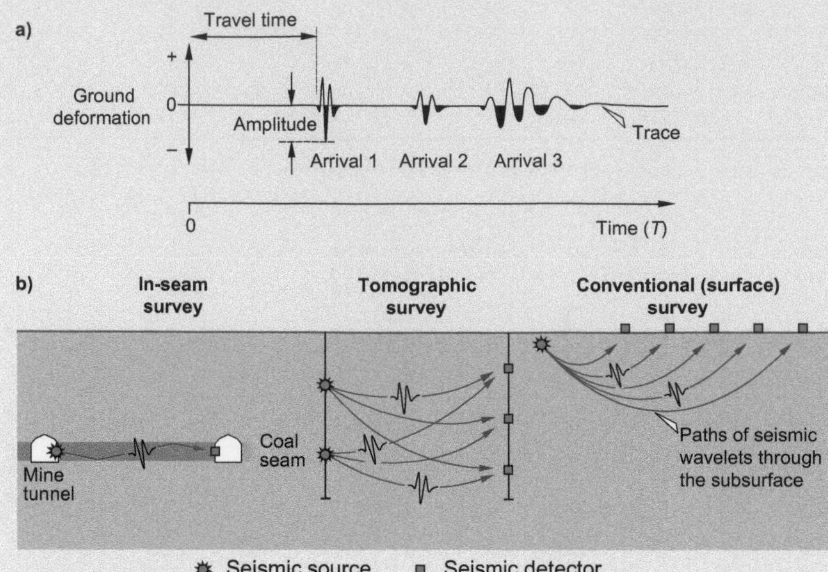

Figure 6.1 Seismic method. (a) A simple seismic trace showing three seismic wave arrivals and the travel time and amplitude of the first of the arrivals. (b) Schematic illustration of three forms of seismic survey.

environmental industries, who use it to investigate the structure of the shallow subsurface. The method's primary use by the mining industry is during exploration for mineralisation in unconsolidated near-surface materials, e.g. placer deposits.

Alternative forms of seismic surveying used by the mining industry involve detectors and/or seismic sources located in the subsurface, with access provided by drillholes or underground workings (Fig. 6.1). *Tomographic* seismic surveys can map mineralisation between drillhole intersections and are used for exploration at a prospect scale and during mining. *In-seam* seismic surveys use seismic waves that are deliberately 'guided' through a coal seam to determine its characteristics prior to mining.

Compared with other geophysical methods, the seismic method, in any of its forms, is currently not widely used in mineral exploration and production, so a very detailed description of all aspects is not justified here. We begin with a description of seismic waves and how they are affected by variations in the physical properties of the subsurface. This is followed by descriptions of some generic aspects of seismic data acquisition and display, and then the processing of seismic reflection data is described. The geological controls on seismic responses are described next, followed by the interpretation of seismic reflection data. Seismic refraction surveying has distinctly different processing and interpretation methods and is described in online Appendix 6. Finally, in-seam and tomographic surveying are briefly described. In our description of seismic methods we rely heavily on terminology describing waves. It is recommended that readers unfamiliar with these terms consult online Appendix 2 for details.

6.2 Seismic waves

The waves used in the seismic method are of short duration; they are actually wavelets rather than continuous waves. As described in online Appendix 2, a wavelet can be defined in terms of the interference of a series of waves of different amplitude, frequency and phase. Therefore, a wavelet, unlike a wave, does not have a single frequency etc. The range of frequencies making up a wavelet define its bandwidth, although normally the wavelet will have a dominant frequency/period. The wavelet's *dominant period* (P_{Dom}), or time taken for one cycle, is the reciprocal of its *dominant frequency* (f_{Dom}). Its *dominant wavelength* (λ_{Dom}), the length of one cycle, is dependent on the wave's velocity (V) (the speed at which it is travelling) and is obtained from the expression

$$\lambda_{\mathrm{Dom}} = \frac{V}{f_{\mathrm{Dom}}} = VP_{\mathrm{Dom}} \tag{6.1}$$

There are several different types of seismic wave. The two main categories are *body waves* and *surface waves*, the former constituting the signal in seismic surveying and the latter being a form of methodological noise (see Section 2.4.2). The most important difference between them is that surface waves propagate in the vicinity of a boundary of a volume of material, e.g. the surface of the Earth in a seismic survey. In contrast, body waves can travel through the interior of the volume as well as along its surfaces, i.e. they can penetrate into the subsurface and can provide information about its characteristics.

6.2.1 Elasticity and seismic velocity

As a seismic wave propagates through the subsurface, the rock is temporarily deformed, or strained. The amount of strain is small and of very short duration (except in the immediate vicinity of the seismic source) and under these conditions the strain is proportional to the applied stress. After the passage of the wave, the rock regains its original shape, i.e. it displays elastic behaviour. It can be useful to think of wavelets as localised packets of elastic strain energy travelling through the rock.

The response of a material to different kinds of strain is described using *elastic constants*. Put simply, high values of these constants indicate a greater resistance to deformation, i.e. the material is more rigid, whilst ductile materials have lower values. A set of elastic moduli quantify a material's response to different kinds of strains. These include the bulk modulus (κ), a measure of the material's ability to resist compression; the shear modulus (μ), related to shear strains; and the axial modulus (Ψ), describing the response to a uniaxial stress when there is no strain in other directions (Fig. 6.2a).

Different kinds of seismic wave cause different types of strains as they propagate. The speed with which a body wave travels through a material depends on the density (ρ) and the relevant elastic modulus of the material.

$$\text{Speed}_{\text{bodywave}} = \sqrt{\frac{\text{elastic modulus}}{\rho}} \qquad (6.2)$$

Seismic speed is often directionally dependent (seismic anisotropy) and is usually greatest parallel to any planar fabric within the rock. Also, noting that velocity is speed in a specified direction, it is correct to define a material's seismic velocity in terms of both speed and direction. In practice, speed and velocity are used interchangeably, and to comply with common practice we will use the term velocity throughout our descriptions of the seismic method.

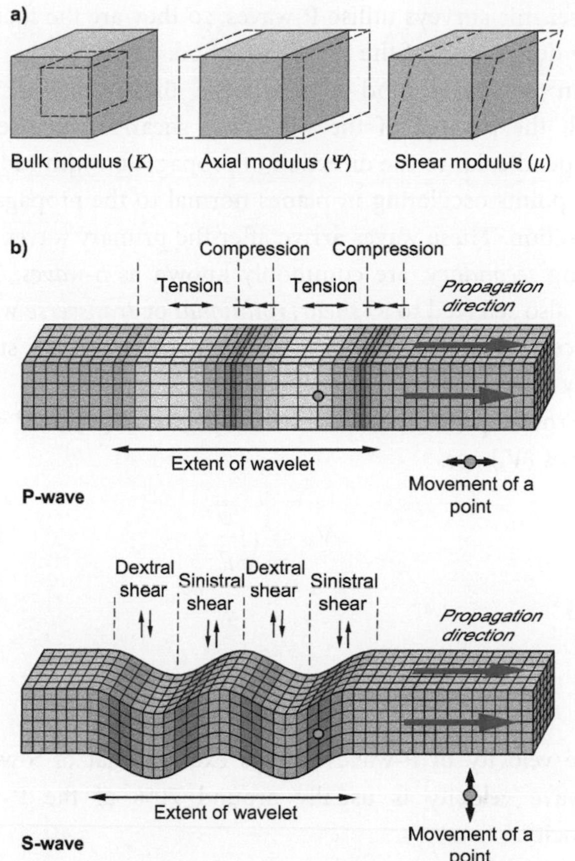

Figure 6.2 Elastic deformation of a material. (a) The three elastic moduli quantifying the response of a cube of material to compression, uniaxial strain and shear strain. (b) Deformation of the propagating medium associated with the passage of a P-wave (uniaxial strain) and an S-wave (shear strain); redrawn, with permission, from Bolt (1976).

For a more detailed description of elastic properties and the mechanisms by which seismic waves travel, see Dobrin and Savit (1988) and Lowrie (2007).

6.2.2 Body waves

There are two types of body wave (Fig. 6.2b). The first type is like sound waves. As they propagate, the rocks undergo a series of uniaxial compressions and tensions causing a point in the subsurface to oscillate along the direction of propagation. These waves travel more quickly than other kinds of seismic wave, so they are the first to be detected and are known as *primary* (or P-) waves. They are also referred to as *compressional, irrotational* or *longitudinal* waves. P-waves can travel through solid material such as rock and also liquids such as pore water. The vast majority of seismic surveys utilise P-waves, so they are the focus of our description of the seismic method.

In the second kind of body wave, the strain associated with the passage of the wave is a shear strain oriented perpendicular to the direction of propagation with individual points oscillating in planes normal to the propagation direction. These waves arrive after the primary waves and, being *secondary*, are commonly known as *S-waves*. They are also referred to as *shear*, *rotational* or *transverse waves*. Since the propagation of S-waves involves a shear strain, they cannot pass through fluids (liquids and gases).

From Eq. (6.2), the velocities of P-waves (V_P) and shear waves (V_s) are:

$$V_P = \sqrt{\frac{\psi}{\rho}} \tag{6.3}$$

and

$$V_S = \sqrt{\frac{\mu}{\rho}} \tag{6.4}$$

The velocity of P-waves always exceeds that of S-waves. S-wave velocity is usually around 70% of the P-wave velocity.

6.2.3 Surface waves

The most important surface wave in seismic surveying is the *Rayleigh wave*, informally called *ground roll*. During the passage of a Rayleigh wave, the motion of a point is elliptical in the plane containing the vertical and the direction of wave propagation (Fig. 6.3). The associated

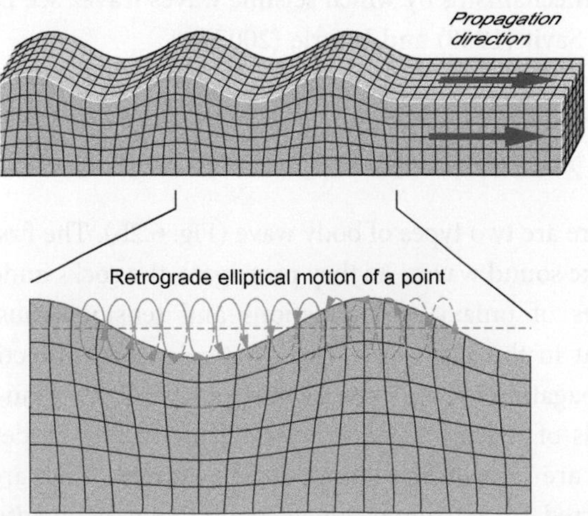

Figure 6.3 The deformation associated with the passage of a Rayleigh wave. Redrawn, with permission, from Bolt (1976).

deformation involves shear strain, so these waves cannot exist in a fluid. The amplitude of Rayleigh waves decreases exponentially with depth below the surface. In this respect they are similar to water waves; a swimmer can easily dive under the surf at the beach.

Rayleigh waves have complex, long-lived, multicycle and continually changing forms, owing to the interference of different frequency components travelling at different velocities, referred to as *dispersion*. They generally have larger amplitudes, more complex waveforms and longer wavelengths than body waves. Considerable effort is made during seismic surveying and data processing to reduce this major source of noise. Although the velocities of Rayleigh waves are frequency dependent, they are always less than the velocity of S-waves, commonly around 90%.

6.3 Propagation of body waves through the subsurface

As the seismic wavelet travels through the subsurface, it is changed by variations in elastic properties of the subsurface, which are related to changes in the geology. To understand how it is changed we describe firstly the fundamental concepts of tracing the propagation path using wavefronts and rays. This allows us to describe why the amplitude of the wavelet changes with time, how the direction of travel is changed and, most importantly, how the wavelet can be split into different parts which travel along different paths, allowing a single source to produce many body wave arrivals at each detector.

6.3.1 Wavefronts and rays

Seismic waves propagate away from their source, in a way approximately analogous to the expanding circle of ripples created when a stone is thrown into a pond. The ripples are wavefronts, defined as the envelope of the locations that a particular point on the seismic wavelet has reached at a given time. The progress of the seismic wavelet through the subsurface can be monitored by determining the shape and location of the wavefronts at different times.

In the simple situation where the seismic source is located at depth in a homogeneous material, i.e. one through which seismic waves travel with the same velocity in all directions, a body-wave wavefront will be spherical and centred on the source. When the source is at the surface a body-wave wavefront forms a hemisphere (Figs. 6.4a and b). The radius of the sphere/hemisphere

a)

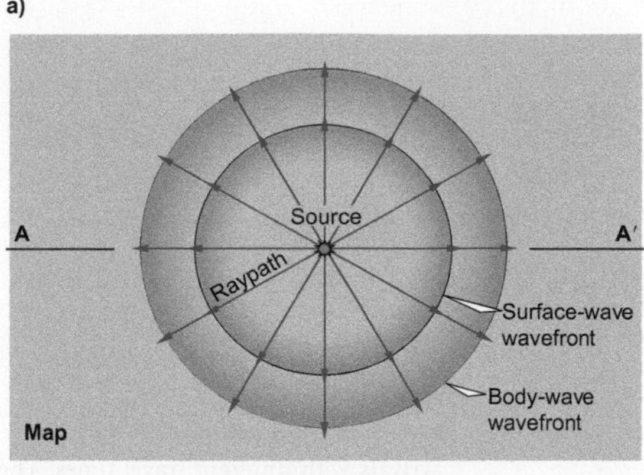

b)

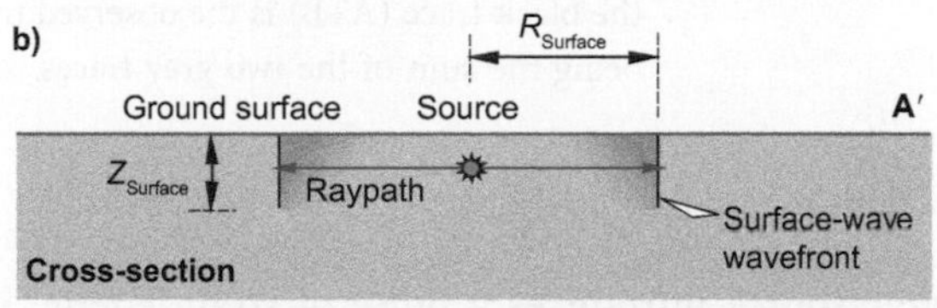

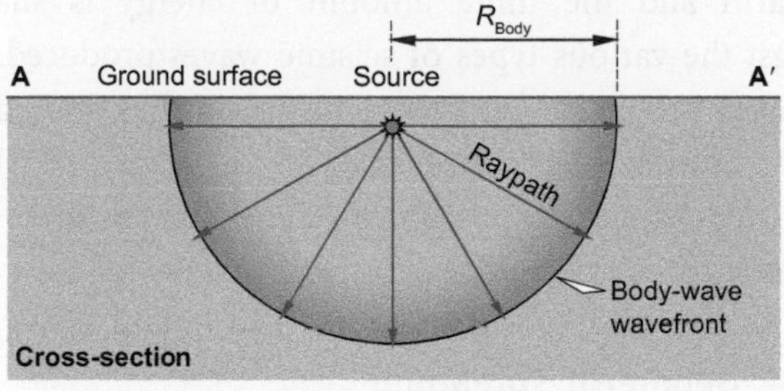

c)

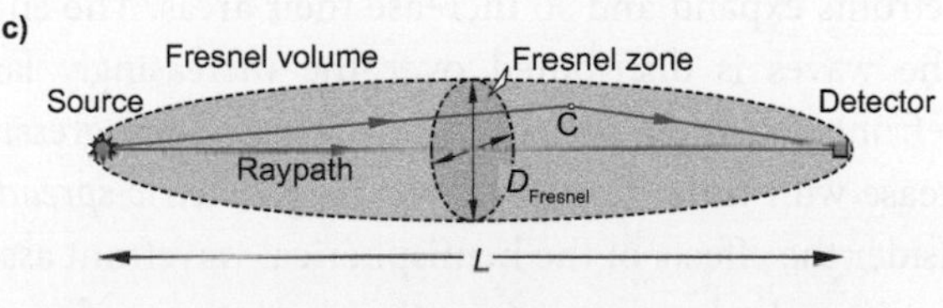

d)

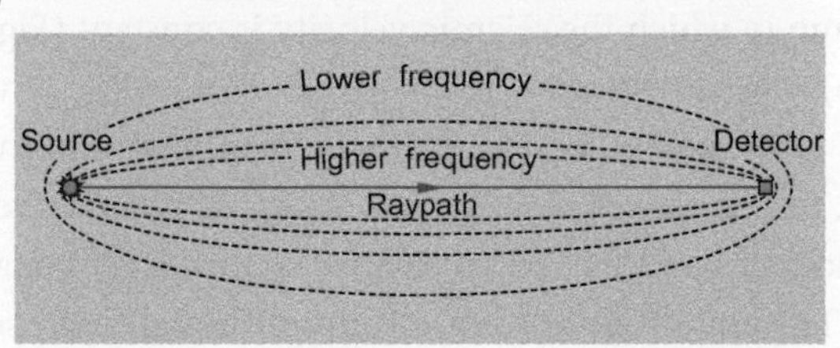

Figure 6.4 Rays and wavefronts for surface and body waves produced by a source located at the ground surface. (a) Map and (b) cross-sections, (c) first-order Fresnel volume (ellipsoid) and Fresnel zone (circle) for a direct raypath through a homogeneous isotopic medium and (d) geometry of the Fresnel volume at different frequencies. (d) Redrawn, with permission, from Kvasnicka and Cerveny (1996; Fig. 1, p. 139).

increases with propagation time and is equal to the product of the wave's velocity and the travel time. A surface-wave wavefront, being confined to the vicinity of the surface, is approximated by a vertical cylinder, again centred on the source. Surface waves have lower velocity than body waves so the two wavefronts rapidly separate, with the surface wavefront inside that of the body wave.

Wavefronts are an actual physical phenomenon but, although they are physically and mathematically precise, using them to qualitatively describe the paths of seismic waves through the subsurface can become cumbersome when any realistic degree of complexity is introduced. More useful in these circumstances are *raypaths* (often just called *rays*). Raypaths are a geometrical construct depicting the path and direction of a seismic wave as arrowed lines (Fig. 6.4a and b) perpendicular to the wavefront. In homogenous material all the raypaths are straight lines, but they will be curved when the elastic properties change along the travel path. Unlike wavefronts, seismic rays do not actually exist, but they are a powerful tool for illustrating the propagation of waves and their interaction with the subsurface.

6.3.2 Fresnel volume

There can be many similar raypaths that direct the seismic waves to the same detector. For two distinct arrivals to be recognisable within a seismic trace, their travel times should differ by more than half the dominant period of the arrivals (Fig. 6.5). If the difference is less than this, a single arrival, albeit with an increased period/wavelength, is observed.

Referring to Fig. 6.4c, consider the direct (shortest distance) raypath from a source to a detector that is travelled in time $T_{\text{source–detector}}$. Another ray passing through an intermediate point C, will have a travel time that is $T_{\text{source–C}} + T_{\text{C–detector}}$. When this travel time differs from that of the direct raypath by less than half the dominant period of the seismic signal (so the difference in distance travelled is less than half the wavelength), the arrivals will constructively interfere (see online Appendix 2) and will not be individually identified within the seismic trace. The volume of material within which this occurs is known as the *Fresnel* (pronounced Fre-nel) *volume* and a cross-section of the Fresnel volume, perpendicular to the direct raypath, is known as a *Fresnel zone*. The Fresnel volume is an important concept in understanding how the seismic waves interact with the subsurface.

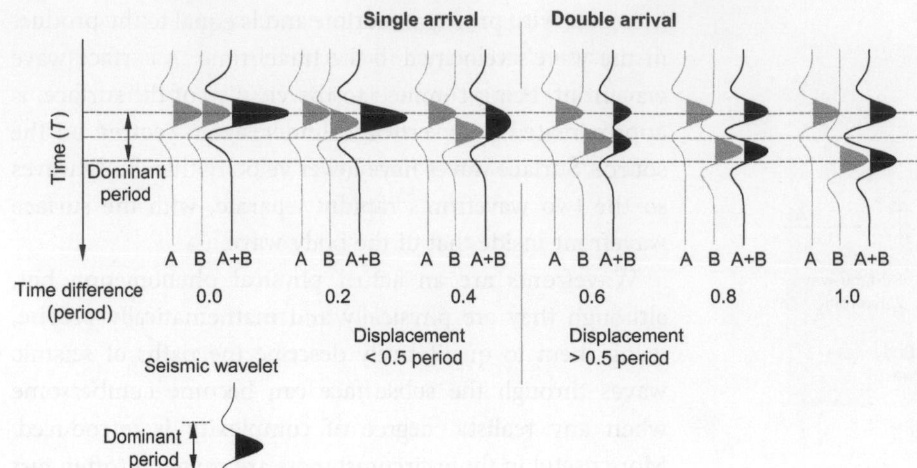

Figure 6.5 Summation of two identical arrivals with different travel times. The grey traces (A and B) show the two arrivals; the black trace (A+B) is the observed trace, being the sum of the two grey traces.

In homogeneous material, where all raypaths are straight lines, the Fresnel volume is an ellipsoid of revolution and Fresnel zones are circles (Fig. 6.4c). The maximum diameter of the Fresnel zone (D_{Fresnel}), assuming the distance between source and detector (L) is much greater than the dominant wavelength of the seismic wavelet, is given by the approximate expression:

$$D_{\text{Fresnel}} \approx \sqrt{L\lambda_{\text{Dom}}} = \sqrt{LV/f_{\text{Dom}}} = \sqrt{LVP_{\text{Dom}}} \qquad (6.5)$$

where f_{Dom} is the dominant frequency and P_{Dom} the dominant period of the signal, V the velocity at which it travels and λ_{Dom} the dominant wavelength. From this expression it is clear that the ellipse is broader for longer paths and longer wavelengths. Reducing the wavelength, i.e. increasing the frequency, reduces the area of the Fresnel zones (Fig. 6.4d). A ray (having no actual width) can be thought of as representing a signal of infinite frequency. In reality the size of the Fresnel volume varies for each component frequency of the seismic wavelet; representing the Fresnel volume by the dominant frequency of the wavelet is an approximation.

A useful working description of the Fresnel volume is that it is the volume of material that is sampled by the seismic wave on its journey through the subsurface to a particular detector, and the characteristics of the wave, when detected, are determined by the elastic properties of the entire volume. Variations in the elastic properties within the Fresnel volume are 'averaged' and cannot be separated from the volume as a whole. The dimensions of the Fresnel volume and Fresnel zones are a measure of the smallest resolvable feature in the subsurface (see Section 6.7.1.1).

6.3.3 Seismic attenuation

A seismic source introduces a pulse of strain energy into the Earth and the finite amount of energy is shared amongst the various types of seismic waves produced. As the waves propagate into the subsurface they lose their energy, causing their amplitudes to decrease and their shapes to change. Two basic mechanisms are involved: geometric spreading and absorption.

6.3.3.1 Geometric spreading

As seismic waves move away from their source, the wavefronts expand and so increase their areas. The energy in the waves is distributed over the increasingly larger wavefronts, so the amplitudes of the waves progressively decrease with time. This is known as *geometric spreading*. Consider the effects of the hemispherical wavefront associated with a body wave originating near the surface of a medium in which the seismic velocity is constant (Fig. 6.4a and b). The surface area of a hemisphere is $2\pi R^2_{\text{body}}$, where R_{body} is its radius, i.e. the distance from source to wavefront. In this case the energy per unit area of the wavefront is linked to its radius by an inverse square ($1/R^2$) relationship. The amplitude of a wave is proportional to the square root of its energy, so geometric spreading causes body-wave amplitude to vary inversely proportional to R_{body}, i.e. a $1/R$ relationship. Geometric spreading changes the amplitude, but not the shape, of a seismic wavelet as it propagates through the medium (Fig. 6.6a). Note the very rapid initial decrease in amplitude followed by a slower decrease (the $1/R$ relationship).

The wavefront of a surface wave is circular on the ground surface and extends to depth Z_{surface}, i.e. it forms

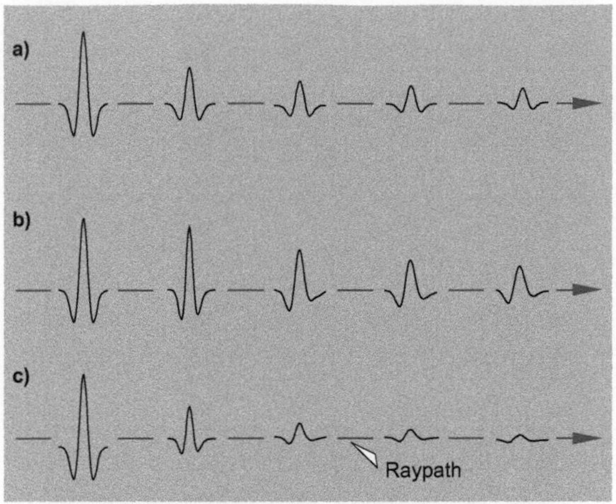

Figure 6.6 Distortion of a wavelet propagating through a homogenous medium. (a) By geometric spreading, (b) by absorption; (c) the combined effects of geometric spreading and absorption.

a vertical cylinder (Fig. 6.4a and b). The radius of the wavefront is $R_{surface}$, so its surface area is $2\pi R_{surface}Z_{surface}$. The energy of the wave at a point on the wavefront is then inversely proportional to $R_{surface}$, i.e. a $1/R$ relationship, and the wave's amplitude is inversely proportional to $1/\sqrt{R}$. Clearly then, attenuation due to geometric spreading is greater for body waves than for surface waves, which is one reason that surface waves are some of the highest-amplitude arrivals seen on seismic recordings (for examples, see Figs. 6.13 and 6.19).

6.3.3.2 Absorption

Rocks are not 'perfect' elastic materials, so when a seismic wavelet travels through them some of its energy is lost through a phenomenon known as *absorption*. The principal cause is friction at grain boundaries and cracks, although the exact mechanism is not well understood. Absorption may be described in terms of several parameters, but the most common are the *quality factor* (Q) and the *absorption coefficient* (α) for a particular wavelength (λ). These parameters are related via the expression:

$$Q = \frac{\pi}{\alpha\lambda} \qquad (6.6)$$

The absorption coefficient represents the proportion of energy lost during transmission through a distance equivalent to one wavelength. The amplitude of each cycle of the wave is reduced so that its amplitude is equal to some percentage of the preceding cycle; so absorption causes amplitude to decrease exponentially.

Higher-frequency (shorter-wavelength) waves are increasingly attenuated because these travel through a greater number of cycles in a given distance than lower-frequency (longer-wavelength) waves. Consequently, the different frequency components making up the seismic wavelet experience different levels of absorption. A practical demonstration of absorption is the sound of loud music being played at the other end of the street. The bass and drums generate lower-frequency sounds whilst the guitars and vocals tend to produce higher frequencies which undergo greater attenuation. The result is that at a distance the sound of the music is dominated by the beat and the melody is lost. Recall that a seismic wavelet is the result of interference of waves of different frequencies. The greater attenuation of higher frequencies changes its frequency content with distance travelled, causing the amplitude and shape of the wavelet to change (Fig. 6.6b). The greater loss of higher frequencies causes the wavelet to broaden, i.e. its dominant period increases. This is an important phenomenon since as the dominant period increases, Fresnel volumes become larger (Fig. 6.4c) and resolution decreases (see Section 6.7.1).

Figure 6.6c shows the combined effects of geometric spreading and absorption on a seismic wavelet. The effects of both increase with the distance travelled through the subsurface by the wavelet. This must be accounted for when planning a seismic survey, as it is necessary to ensure that the source produces sufficient energy across a frequency range that can penetrate to the required depth. Figure 6.7 shows typical frequency ranges for various types of seismic survey. Note the lower frequencies used for deep penetrating crustal-scale surveys compared with those of shallower probing surveys. Tomographic surveys, which involve waves travelling much shorter distances, can use much higher frequencies and, therefore, have higher resolution.

6.3.4 Effects of elastic property discontinuities

It is the interaction of seismic waves with changes in elastic properties and density of the subsurface that ultimately provides information about the structure of the subsurface. For conventional (surface) seismic surveys, it is essential that the downward travelling waves created by the source encounter changes in the subsurface that 'turn them around' by some means so that they return to the surface where they will be recorded by the detectors. This occurs through any or all of three fundamental processes:

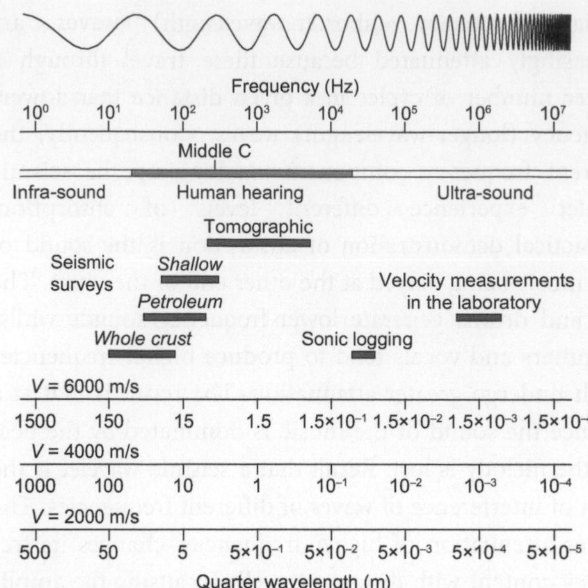

Figure 6.7 The range of frequencies used for different types of seismic survey and for *in situ* and laboratory velocity measurements. The quarter wavelength is a measure of the resolution of a seismic dataset (see Section 6.7.1). For comparison, the human ear responds to sound over a frequency range extending from about 20 to 20,000 Hz.

diffraction, reflection and refraction. Note that it is customary to refer to seismic wave arrivals that have been diffracted as simply 'diffractions', reflected arrivals as 'reflections', etc., even though they are all the results of the processes rather than the processes themselves.

For refraction to occur, only a change in seismic velocity is required. For diffraction and reflection to occur there must be a change in a physical property called *acoustic impedance* (ζ) which depends on the P-wave velocity (V_P) and density (ρ):

$$\zeta = \rho V_P \tag{6.7}$$

Clearly, a difference in acoustic impedance can be caused by changes in either or both of velocity and density. Of course, it is also possible for an increase in velocity and a decrease in density, or vice versa, to combine so that acoustic impedance remains the same, but usually these two properties vary sympathetically (see Section 6.6).

We illustrate diffraction, reflection and refraction in the subsurface with a series of consecutive 'snapshot' images of the P-wave wavefronts as time increases after activation of the seismic source (Figs. 6.8 and 6.10). The shade of grey in the figures represents the deformation caused by the wavelet, with areas of compression and dilation represented by lighter and darker shades of grey, respectively. In the final

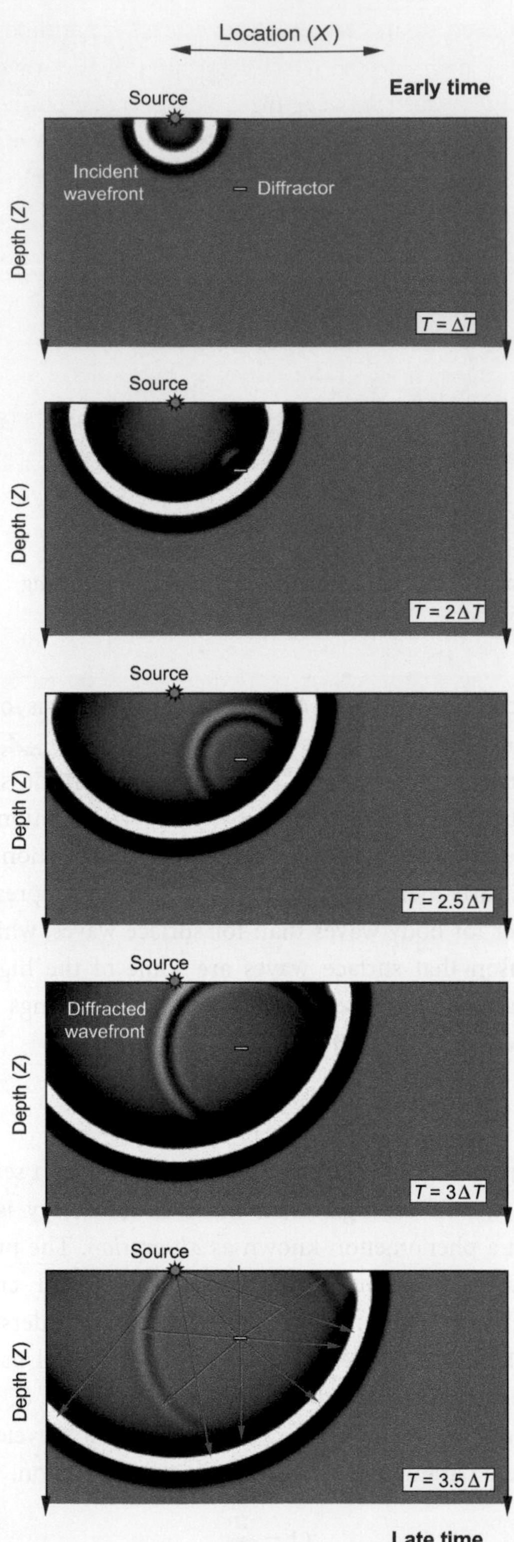

Figure 6.8 Time (T) 'snapshot' images showing the progression of a source-generated incident wavefront encountering a diffractor in the subsurface. Summary raypaths are superimposed on the image at $T = 3.5\Delta T$.

part of each figure, raypaths have been superimposed to show the paths taken by selected parts of the wavefront. The actual wavefronts are three-dimensional, but a single vertical cross-section through the subsurface containing the source is sufficient to show what is occurring.

6.3.4.1 Local discontinuities

Diffraction is the scattering of seismic waves when a wavefront encounters a diffractor, which is a localised discontinuity in the subsurface's acoustic impedance whose boundary curvature is large and/or whose dimensions are smaller than a Fresnel zone (see Section 6.3.2). This might be a localised change (a point diffractor) or the sharp edge of a larger discontinuity (typically the faulted edge of a surface).

Figure 6.8 shows wavefronts and rays within an Earth model comprising a point diffractor within an otherwise homogeneous subsurface. As expected from Fig. 6.4, the wavefront created by the source, the *incident* wavefront, is hemispherical (semicircular in the plane of the cross-section). It expands uniformly from the source, both laterally and downwards, until it encounters the diffractor. Here it is scattered, the scattered waves forming a second wavefront which is spherical (circular in the plane of the cross-section) and centred on the diffractor. The raypaths of the incident wavefront are radial and centred on the source. The diffracted rays form a similar pattern centred on the diffractor.

The diffracted wavefront expands in all directions, and part of it encounters the ground surface and may be recorded by the seismic detectors: that is, the requirement for the incident wave to be turned around has been met. A diffracted wave is of low amplitude, compared with the incident wave, since it derives its energy from only a small fraction of the latter. The maximum amplitude occurs in the area immediately above the diffractor since this is the shortest travel path and so less attenuation has occurred. In fact, in order to see the diffracted wavefront in Fig. 6.8, amplitude scaling was applied to attenuate the stronger incident wavefront.

6.3.4.2 Continuous discontinuities

The effects are more complicated when elastic properties and density change across a laterally continuous interface. The key phenomenon is the partitioning (splitting) of the incident wave into two separate waves: the reflected and transmitted waves. The directions and the relative amounts of energy in the two new waves depend on the nature of the change in acoustic impedance and velocity at the interface, and vary across the incident wavefront owing to variation in the direction from which it intersects the interface. The different effects are described in turn using rays; but we emphasise that they are inter-related components of the same overall process.

Figure 6.9 shows the changes in directions of the raypath at the interface, which depends on the change in velocity

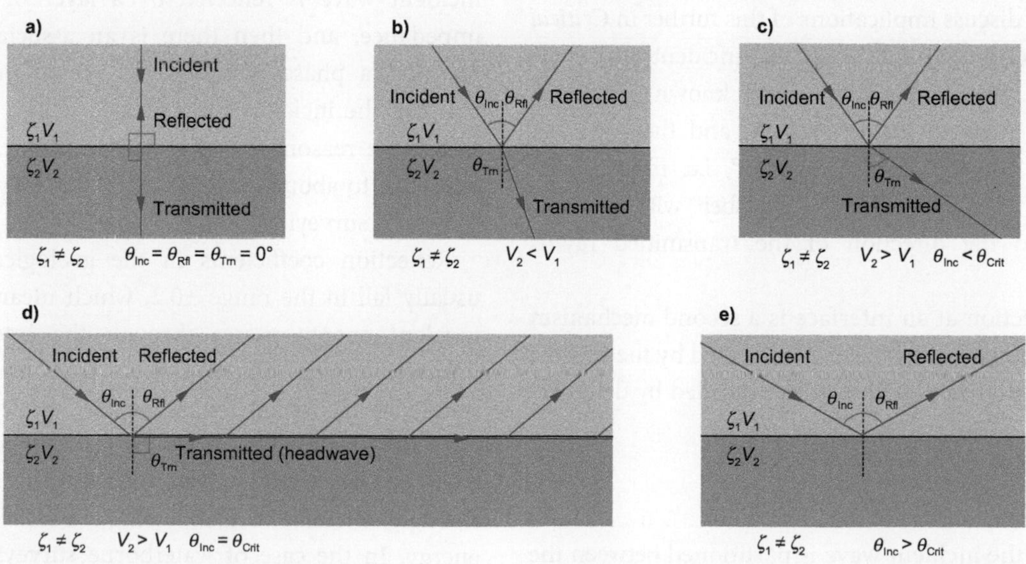

Figure 6.9 Ray diagrams showing the effects of an interface across which there is a change in acoustic impedance. (a) Normal incidence, (b) oblique incidence with velocity decreasing across the interface, (c) oblique incidence with velocity increasing across the interface, (d) critical refraction and (e) total internal reflection.

between the regions above and below the interface. The directions of the incident, reflected and transmitted (refracted) rays are measured with respect to the normal (perpendicular) to the interface (the vertical for the case of a horizontal interface), and these angles are known as the *angles of incidence* (θ_{Inc}), *reflection* (θ_{Rfl}) and *transmission* (θ_{Trn}) (or *refraction*), respectively. They are defined by Snell's Law:

$$\frac{\sin\theta_{Inc}}{V_1} = \frac{\sin\theta_{Rfl}}{V_1} = \frac{\sin\theta_{Trn}}{V_2} \quad (6.8)$$

From this, the relationships between the directions of the incident, reflected and transmitted rays are given by:

$$\frac{\sin\theta_{Inc}}{\sin\theta_{Trn}} = \frac{V_1}{V_2} \quad (6.9)$$

and

$$\frac{\sin\theta_{Inc}}{\sin\theta_{Rfl}} = 1 \quad (6.10)$$

For the reflected rays, the wave remains within the same layer and travels at velocity V_1; and in accordance with Snell's Law, the angle of incidence is equal to the angle of reflection. The transmitted wave travels through the second layer with a different velocity (V_2), and in a different direction, i.e. it is refracted. When V_2 is less than V_1 the transmitted raypath is deflected (refracted) towards the normal (Fig. 6.9b), and when V_2 is greater than V_1 the transmitted raypath is refracted away from the normal, i.e. refraction brings the ray closer to being parallel with the interface (Fig. 6.9c). We discuss implications of this further in *Critical refraction* in Section 6.3.4.2. When the incident ray is perpendicular to the interface (Fig. 6.9a), known as *normal incidence*, the angle of incidence is 0° and the angles of reflection and transmission are also 0°, i.e. the reflected ray coincides with the incident ray, albeit with opposite directions, and the direction of the transmitted ray is unchanged.

Clearly reflection at an interface is a second mechanism by which the down-going wavefront created by the seismic source is turned around so it can be recorded by detectors at the surface.

Energy partitioning

The energy of the incident wave is partitioned between the transmitted wave that passes through and into the underlying layer, and the wave that is reflected and remains within the upper layer. This means that the down-going wavefront generated by the source loses energy. This is another form of energy loss associated with wave propagation in addition to geometric spreading and absorption (see Section 6.3.3). The relative amount of energy, and therefore amplitude, of the transmitted and reflected waves depends on the angle of incidence and the change in physical properties across the interface. Basically, the greater the contrast in elastic properties, the greater is the proportion of incident energy that is reflected. Conversely, the smaller the acoustic impedance contrast, the more energy that is propagated across the boundary.

For P-waves at normal incidence, the ratios of the amplitudes of the reflected (A_{Rfl}) and the transmitted (A_{Trn}) waves, with respect to that of the incident waves (A_{Inc}), are known as the *reflection coefficient* (RC) and *transmission coefficient* (TC), respectively. They depend only on the acoustic impedances each side of the interface and are given approximately by:

$$\mathrm{RC} = \frac{A_{Rfl}}{A_{Inc}} = \frac{\zeta_2 - \zeta_1}{\zeta_2 + \zeta_1} \quad (6.11)$$

$$\mathrm{TC} = \frac{A_{Trn}}{A_{Inc}} = \frac{2\zeta_1}{\zeta_2 + \zeta_1} \quad (6.12)$$

Note that

$$\mathrm{RC} + \mathrm{TC} = 1 \quad (6.13)$$

The reflection coefficient can only vary between −1 and +1. An absolute value of 1 corresponds to all the energy being reflected by the interface. A negative RC occurs when the incident wave is reflected by a layer of lower acoustic impedance, and then there is an associated reversal of polarity (a phase-shift of 180°; see online Appendix 2) between the incident and reflected waves. These relationships give reasonably accurate results for angles of incidence up to about 15°, which is the usual case for seismic reflection surveying.

Reflection coefficients in the geological environment usually fall in the range ±0.2, which means nearly all the incident energy passes through the interface into the underlying layer. Under 'normal circumstances', a reflection can be recognised when the contrast in acoustic impedance produces a reflection coefficient of roughly 0.05. It can be difficult to image the geology below a strongly reflective horizon, owing to the lack of transmitted energy. In the case of waterborne surveys, the water–air and water–water bottom interfaces are important physical property contrasts. The reflection coefficient at the former is very close to −1, the latter between 0.33 and 0.67

depending on the nature of the bottom. The obvious implication from this is that seismic waves generated in the water will tend to stay in the water, reverberating in the water body rather than entering the underlying sediments or rocks (see *Multiples* in Section 6.5.1.1).

Critical refraction

Referring to Fig. 6.9c, increasing the angle of incidence (θ_{Inc}) increases the angle of transmission (θ_{Trn}) and, eventually, a *critical angle* of incidence (θ_{Crit}) is reached where θ_{Trn} is 90° (Fig. 6.9d). The transmitted raypath is then along the interface. This is a phenomenon known as *critical refraction* and can only happen if the velocity below the interface is greater than that above, i.e. $V_2 > V_1$. From Snell's Law (Eq. (6.8)), and since the sine of 90° is 1, then

$$\sin \theta_{\mathrm{Crit}} = \frac{V_1}{V_2} \qquad (6.14)$$

Critical refraction is important because it represents a third way that the down-going waves created by the seismic source can be returned to the surface. Recall that a seismic wave is a packet of strain energy which deforms the rocks as it passes through them. The critically refracted wave is travelling at the higher velocity of material below the interface. Although it is within the underlying higher-velocity region (V_2), it still affects the lower-velocity region (V_1) above the interface. The two regions are physically continuous, so deformation associated with the wave in the area immediately below the interface must affect the adjacent area immediately above. An important consequence of this is that the wave, and its associated deformation, is travelling 'too fast' for the lower-velocity upper region. This causes the disturbance at the boundary to act as a mobile source of seismic waves as it travels along the interface at the (higher) velocity of the underlying region. The resulting upward travelling wavefront in the lower-velocity layer is planar and the associated raypaths are at the critical angle. Seismic waves created in this way are known as *headwaves.*

The concept of critical refraction and headwave arrivals is a simplification. In reality, and especially in the near-surface, velocities are rarely constant nor are interfaces planar. Instead velocity varies continuously and increases with depth. In these circumstances, transmitted waves may penetrate a short distance into the deeper higher-velocity region following curved raypaths roughly parallel to the 'refracting' interface. These are referred to as *diving waves.* It is common, although something of a generalisation, to refer to these and headwave arrivals simply as 'refractions'.

Total reflection

When the angle of incidence is greater than θ_{Crit}, the wave does not cross the interface and is totally reflected (Fig. 6.9e), i.e. there is no transmitted wave. These are known as *post-critical reflections* and, since no energy is 'lost' to a transmitted wavefront, they can have relatively high energy and hence large amplitude. Total reflection is an important phenomenon for the in-seam surveys described in Section 6.8.1. For these surveys, the seismic source is actually located within a coal seam. The seam forms a waveguide because the energy cannot pass through its upper and lower boundaries, being totally reflected by the boundaries, so the waves are trapped within the seam. The wave is guided by the seam and is known as a *guided wave.*

Wavefronts at an interface

Following on from the description of reflection and transmission, given mostly in terms of rays, we now illustrate the phenomenon using wavefronts. Consider first an interface where the acoustic impedance changes owing to the velocity below the interface being lower than that of the layer above. The wavefront snapshot images in Fig. 6.10 show how, on encountering the interface, the incident wavefront is partitioned into two separate wavefronts. Note that because the velocity below the interface (V_2) is lower than that above (V_1), the wavelength of the transmitted wave in the lower layer is less than that of the incident wave (see online Appendix 2 for the relationship between frequency, velocity and wavelength). The upward travelling reflected wave travels at the same speed as the incident wave so its wavelength is unchanged, but its polarity is reversed as shown by the relative locations of the darker and lighter regions in the images. From a physical perspective, areas of compression and tension (see Fig. 6.2b) are interchanged.

We now change our model of the subsurface so that the velocity and acoustic impedance increase below the interface, i.e. V_2 is greater than V_1. This model allows the downward moving wavefront to return to the surface via critical refraction. Referring to the wavefronts in Fig. 6.10b, again the incident wavefront is partitioned into reflected and transmitted wavefronts, but with some important differences from the example shown in Fig. 6.10a. Firstly, the increase in acoustic impedance at the interface produces a positive reflection coefficient (see *Energy partitioning* in

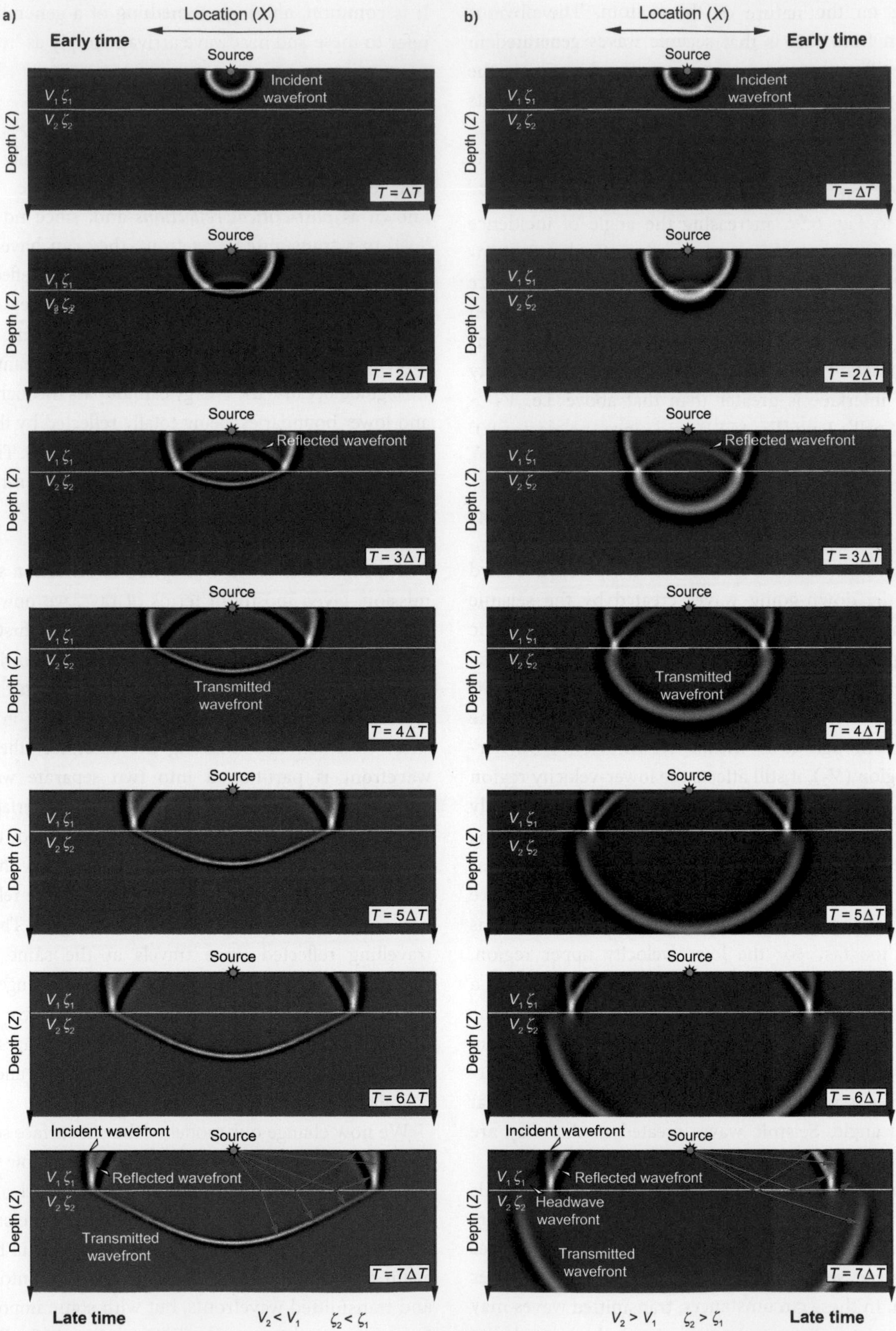

Figure 6.10 Time (T) 'snapshot' images showing the progression of a source-generated incident wavefront encountering an interface in the subsurface. Summary raypaths are superimposed on the image at $T = 7\Delta T$. (a) Acoustic impedance and velocity below the interface are lower than those above; (b) both parameters are higher, allowing the creation of headwaves.

Section 6.3.4.2), so the polarity of the reflected wavefront is now the same as the incident wavefront. Secondly, the transmitted wave now travels through a medium of higher velocity, so its wavelength increases and its greater rate of expansion causes the transmitted wavefront to separate from the reflected wavefront. Its higher velocity of propagation leads to the creation of headwaves at the interface (see *Critical refraction* in Section 6.3.4.2).

6.3.4.3 Relationship between diffractors and reflectors

An important relationship between diffractors and reflectors is illustrated in Fig. 6.11. Here the subsurface contains a series of horizontally aligned, but variably spaced, point diffractors. The figure shows snapshots, all at the same time since the source was 'fired', of a wavefront that has encountered the line of diffractors. When the diffractor spacing is wide, individual diffracted wavefronts are seen, which overlap and interfere. As the diffractor spacing decreases, it is clear that the interference is increasingly destructive (see online Appendix 2) except at the margins of the area containing the wavefronts. The resulting wavefield approaches that produced by a continuous interface, as shown in the lowermost part of the figure. This result is of importance since the concept of a reflector being equivalent to many closely spaced diffractors is useful for explaining a number of crucial aspects of seismic reflection responses (see *Post-stack time migration* in Section 6.5.2.5, and also see Section 6.7.1.1).

6.4 Acquisition and display of seismic data

Seismic surveys commonly use large numbers of detectors, sometimes many hundreds in the case of reflection surveys, to record the response of the subsurface to waves produced from the seismic source. Data acquisition will comprise recordings with the source and detectors at many different locations. The configuration of source and detectors is known as the *spread*.

6.4.1 Seismic sources

The ideal seismic source creates a wavelet that has short duration and high energy. Short duration improves survey resolution (see Section 6.7.1), and higher energy produces a wavelet with higher amplitude making it easier to detect the seismic signal within background noise. The greater the distance the waves must travel from the source to the detectors (equating to greater depth penetration for

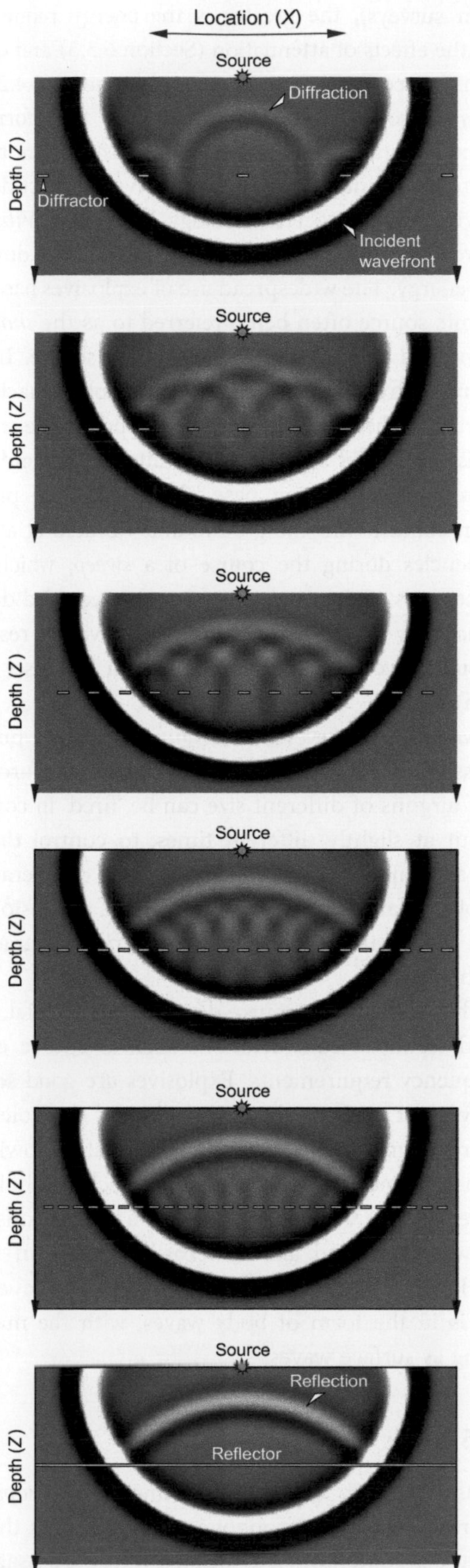

Figure 6.11 Time 'snapshot' images showing the interaction of an incident wavefront with an increasing number of closer-spaced diffractors. The travel time is the same in every case.

reflection surveys), the greater is the energy required to counter the effects of attenuation (Section 6.3.3) and energy partitioning (see *Energy partitioning* in Section 6.3.4.2).

A seismic source's ultimate function is to deform the Earth and generate elastic waves. The most common seismic sources for land surveys are explosives, impact devices and a vibrating impact technology known as *Vibroseis.* Explosives and impact devices produce a short duration pulse of energy. The widespread use of explosives has led to the seismic source often being referred to as the *shot*, and the term *fired* meaning activation of the source. Impact devices range from a sledge hammer to mobile machinery incorporating a heavy mass which is dropped to the ground. Vibroseis introduces seismic energy into the ground via a vehicle-mounted vibrating base plate. The base plate is placed in contact with the ground and vibrated at a range of frequencies during the course of a *sweep*, which lasts several seconds. This necessitates that the recorded data be mathematically transformed into the equivalent response that would have been recorded from an impulse (short duration) source.

For waterborne surveys, an *airgun* releasing a pulse of compressed air into the water is a common source. An *array* of airguns of different size can be 'fired' in combination, but at slightly different times, to control the frequency and energy of the wavelet and to counteract the effects of gas (air) bubbles, whose periodic expansion and collapse create the equivalent of a series of gradually diminishing pulses.

The choice of seismic source depends on financial, logistical and technical considerations such as source energy and frequency requirements. Explosives are good sources of P-waves but require expensive drilling of shotholes, and the source signature is not always consistent, owing to different geological conditions local to each shothole. On the other hand, Vibroseis produces a more consistent source waveform and also can easily operate in urban areas. However, only a fraction of the seismic energy created is in the form of body waves, with the majority occurring as surface waves.

6.4.2 Seismic detectors

Land-based seismic detectors are known as *geophones.* These are a form of microphone which converts the velocity of the ground motion associated with the passage of a seismic wave into a voltage response. Large-scale land surveys require hundreds to thousands of detectors, so logistics and survey economics require that these are cheap, small, light and easy to deploy. Normally only the vertical component of ground velocity is measured since this corresponds with the direction of motion of the ground for P-waves travelling steeply upwards (see Section 6.2.2), but also means that Rayleigh waves produce strong responses (see Section 6.2.3). Waterborne surveys use *hydrophones* in which the sensor is a pressure-sensitive piezoelectric element that produces a voltage in response to pressure variations associated with the propagation of P-waves through the water.

A seismic detector rarely consists of only a single geophone or hydrophone. The recording is the combined response of a number of geophones/hydrophones known as an *array* or *group.* The position of the detector is taken as the centre of the group and their spacing is known as the *group interval.* For the case of hydrophones, they are mounted within a flexible waterproof tube, usually filled with oil of a specific density to create neutral buoyancy, and towed in the water behind a moving vessel. This is known as a *streamer.*

Appropriate arranging of the individual detectors allows the group to have a direction-dependent response, designed to favour seismic waves travelling upwards from below rather than those which have travelled laterally and, therefore, have not sampled the subsurface. The combining of the outputs from multiple detectors is a form of stacking which suppresses random noise (see Section 2.7.4.1).

6.4.3 Displaying seismic data

As noted previously, the deformation associated with the passage of seismic waves at a detector location is displayed in the form of a seismic trace. This is a graph representing the velocity of the movement of the ground (or pressure variation in water) at the location of the detector as a function of time. The resulting time series begins at the instant ($T = 0$) that the seismic waves were generated by the source (see Fig. 6.1). The detector output is digitised using a sampling interval chosen to avoid aliasing (Section 2.6.1).

6.4.3.1 Traces

A simple line graph of ground velocity (or pressure variations) is referred to as a *wiggle trace* (Fig. 6.12a,i). An alternative display is the variable-area display, where the wave is shaded when it exceeds a specified threshold value (Fig. 6.12a,ii). Sometimes the trace itself is omitted leaving just the shaded areas (Fig. 6.12a,iii). Also commonly used

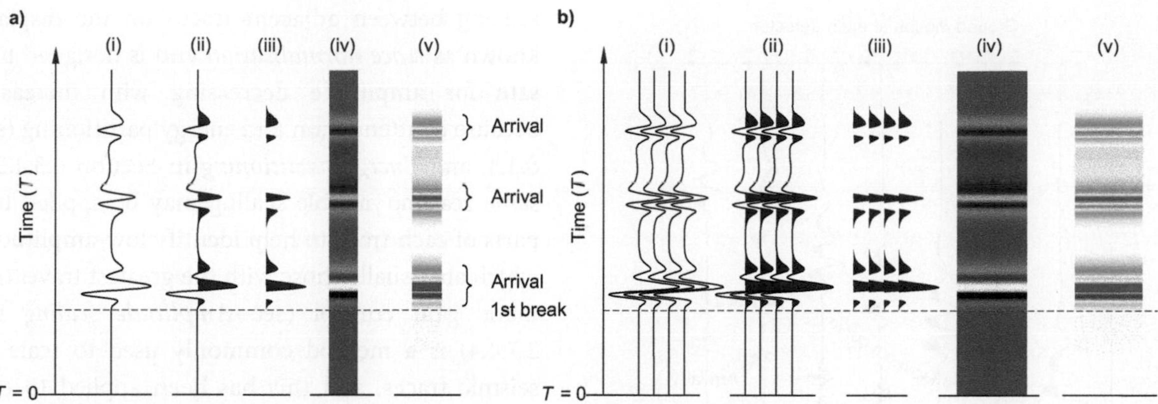

Figure 6.12 Common methods of displaying seismic data. (a) Single traces and (b) multiple traces illustrating the continuity of arrivals between adjacent traces in different forms of display. See text for details.

are pixel-based displays which represent variations in amplitude as changes in colour. Particularly common are grey-scale images (Fig. 6.12a,iv) and colour schemes used to highlight the polarity of the trace, i.e. whether the response is positive or negative (Fig. 6.12a,v). These commonly use a look-up table (see Section 2.8.2.2) comprising mostly two colours, red and blue in this case, with a thin region of a third colour (white) close to the zero value. Different types of arrivals are recognised by correlating equivalent arrivals on traces from adjacent detectors; shading greatly assists with the inter-trace correlation (Fig. 6.12b).

The choice of display is really a matter of personal preference. Pixel displays are most effective when the geology is comparatively simple and when features in the data have a consistent appearance and extend over large distances compared with the detector spacing, such as occurs in a sedimentary basin. This form of display is used in Figs. 6.40 and 6.41. The variable-area display is better at revealing subtle variations in amplitude and is generally preferred when the local geology is more variable and complicated. This form of display is used in Figs. 6.46 and 6.48.

6.4.3.2 Gathers

Various multi-trace displays, referred to as *gathers*, are used to display seismic data; but the most fundamental is the *shot gather*, a collection of traces from detectors at different locations recording the same source. The traces comprising a gather are plotted on a common time scale, with time increasing either vertically upwards or downwards. Each trace has its own local horizontal axis that quantifies the velocity or pressure variations at the detector. Individual traces are laterally positioned according to the detector's location.

In the case of a shot gather, the locational information used to position each trace is the horizontal distance between the source and the detector that gave rise to the trace (Fig. 6.13). This distance is called the *offset*. This form of display is favoured since it is a *time versus distance* (T–X) graph, allowing arrivals of the same type to be correlated between traces and their *moveout* to be determined. Moveout is the increase in travel time with offset and is a fundamental property used to determine whether the arrivals have been reflected, diffracted etc. in the subsurface (see Section 6.3.4). In the schematic example shown in Fig. 6.13a there are three types of arrival and they all have linear moveout, i.e. travel time is proportional to offset. The larger the slope of the line approximating the moveout, the slower is the velocity of the seismic waves. Velocity can be obtained from the slope of a line representing the moveout ($V = 1/\text{slope}$). In the example, the three moveout curves pass through the origin. This combined with linear moveout is characteristic of *direct arrivals*, seismic waves that have travelled along the 'direct' path, parallel to the surface between source and detector (Fig. 6.13b).

Figure 6.13c shows an actual shot gather. Various kinds of arrivals are labeled. Both kinds of body wave are seen, the greater velocity of the P-waves (see Section 6.2.2) causing their moveout to be less. The moveout of the surface waves is greater because these waves travel more slowly than body waves. Note their large amplitude and long duration compared with the body-wave arrivals (see Section 6.2.3). The arrival with the slowest velocity is the *air wave*. This is a P-wave that has travelled not through the ground but through the air. As noted in Section 6.6.1.2, the velocity of air is significantly less than that of geological materials, so the moveout is large.

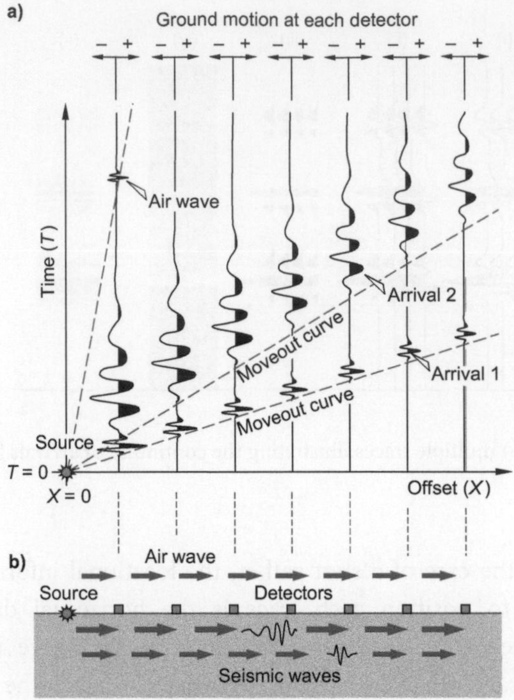

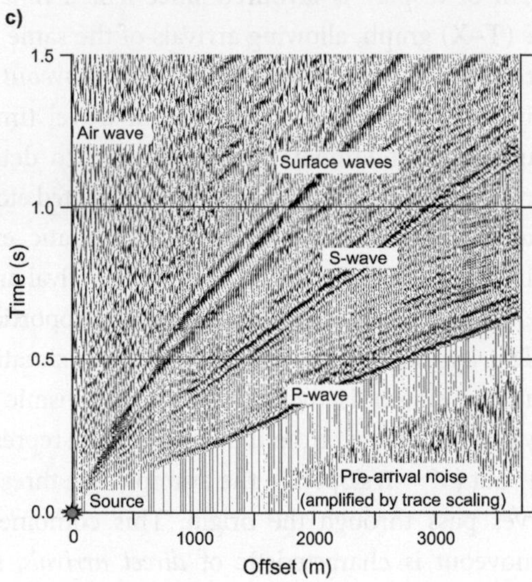

Figure 6.13 Seismic recording. (a) Schematic shot gather; a set of traces recording a single seismic source. The seismic traces have a common time scale but their relative horizontal locations represent the offset (distance) between source and detector. (b) Schematic cross-section showing the seismic waves travelling along direct paths through the subsurface from source to detectors. (c) Example shot gather showing different kinds of seismic arrival; redrawn, with permission, from Roberts *et al.* (2003).

6.4.3.3 Trace scaling

The amplitude of each seismic trace may be scaled prior to display so that its maximum amplitude spans a certain width on the presentation, usually some fraction of the spacing between adjacent traces on the display. This is known as *trace normalisation* and is designed to compensate for amplitude decreasing with increasing offset because of attenuation and energy partitioning (see Section 6.3.3, and *Energy partitioning* in Section 6.3.4.2). For the same reason, variable scaling may be applied to different parts of each trace to help identify low-amplitude features, which are usually those with the greatest travel time. Automatic gain control (see *Amplitude scaling* in Section 2.7.4.4) is a method commonly used to scale individual seismic traces, and this has been applied to the data in Fig. 6.13c.

The loss of absolute amplitude information caused by trace scaling is acceptable since, in most cases, interpretation depends on the recognition of the type of arrival and its travel time, and not absolute amplitudes. If amplitude information is to be used, then the scaling applied must be taken into account.

6.5 Seismic reflection method

The seismic reflection method is based on seismic waves that have been reflected, and to a lesser extent diffracted, in the subsurface. The basic aim of seismic reflection surveying is to map changes in acoustic impedance in the subsurface, i.e. to resolve the depth and thickness of layers having different seismic velocity and/or density.

Reflected arrivals are usually of low amplitude. Complex data acquisition and processing methodologies are needed to detect and enhance these weak signals whilst suppressing all other kinds of arrivals and other forms of noise. Those interpreting seismic data only need to be aware of the basic principles of data acquisition and data processing and, in particular, how they affect the interpretation of the data. The resolution of particular geological features may critically depend on the design of the survey, and accurate interpretation of the data may require knowledge of the processing applied.

Here we summarise the fundamentals of the acquisition and processing of reflection data recorded on land, focusing on the creation of a dataset comprising equivalent zero-offset recordings. The description is purely qualitative, omitting details of the mathematics underlying the various procedures. In so doing, significant simplifications have been made. Readers requiring a comprehensive description of these subjects are referred to Evans (1997) and Yilmaz (2001).

6.5.1 Data acquisition

A key aspect of the processing of seismic reflection data transforms the data, collected as a series of shot gathers (see Section 6.4.3.2) with finite offsets between the sources and detectors, into zero-offset form. This is equivalent to having the source and a single detector at the same location; and in this case only reflections with raypaths perpendicular to the reflector can be recorded. Consequently, these data are sometimes called *normal-incidence* data. This process, known as *stacking*, results in processed data with a greatly increased signal-to-noise ratio. The need to stack the data controls the way the data are acquired since it is necessary to record reflections from the same places in the subsurface, but with different source and detector positions. Before describing data acquisition we briefly discuss some aspects of the raypaths and travel times of reflected seismic waves.

6.5.1.1 Travel times of reflected arrivals

The travel times of reflections from a series of horizontal planar contrasts in acoustic impedance (ζ) are shown in Fig. 6.14. The spread is linear, i.e. the detectors are deployed in a straight line, and the source is at the end of the line (an *end-on* spread). The moveout of reflected arrivals is non-linear in shot gathers and the travel time curves do not pass through the origin; instead they intercept the time axis at T_0, the zero-offset ($X = 0$) reflection time. This corresponds to the vertical down-and-return raypath having normal incidence to the reflector (the angle of incidence (θ_{Inc}) from Eq. (6.8) is 0°). Referring to Fig. 6.14a, where there is a single reflecting interface, T_0 is given by:

$$T_0 = \frac{2Z}{V_1} \tag{6.15}$$

The '2' in the equation occurs because the seismic waves travel down to the reflector and back to the surface, i.e. they travel a distance equal to twice the depth (Z). For this reason, zero-offset travel times are often called *two-way time* (TWT). The depth Z can be obtained by rearranging Eq. (6.15):

$$Z = \frac{T_0 V_1}{2} \tag{6.16}$$

Using the path ABC in Fig. 6.14a as an example, the travel time at non-zero offset ($X > 0$) is:

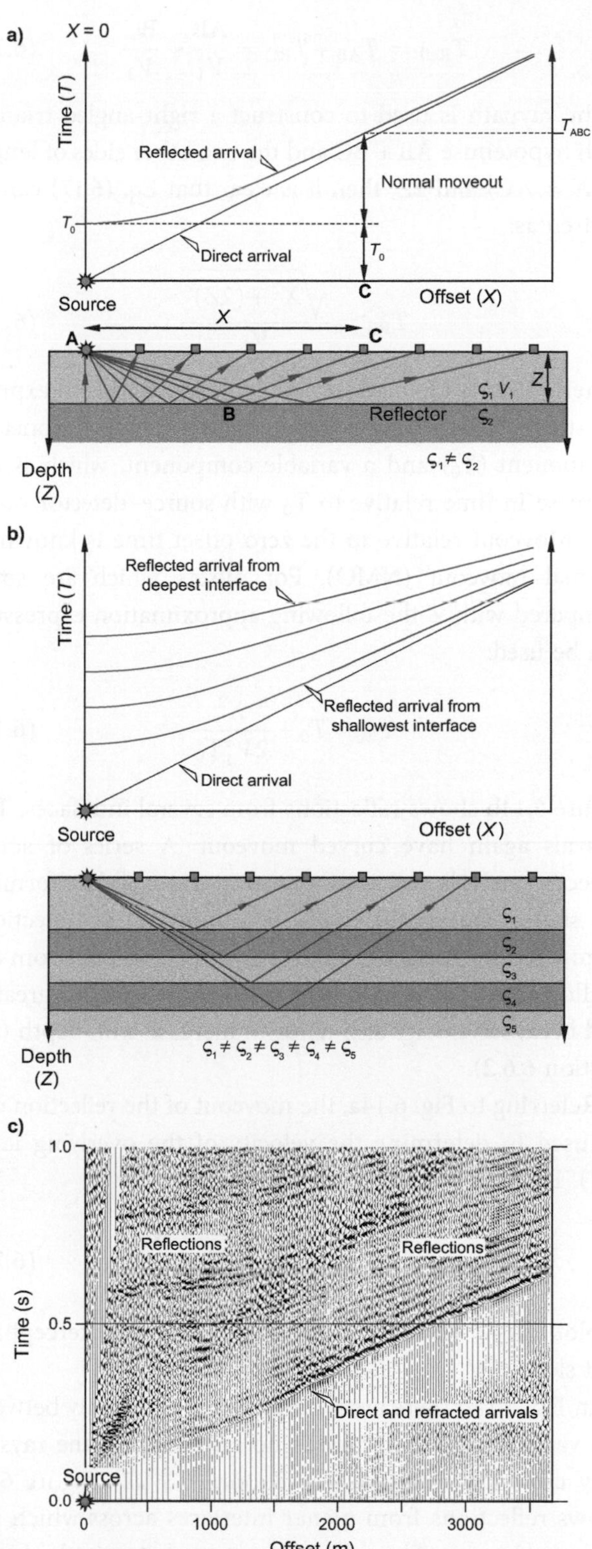

Figure 6.14 Time–distance plots and raypath diagrams for reflections from horizontal interfaces. (a) Single interface and b) multiple interfaces with the same layer velocities. See text for details. (c) Shot gather showing reflected arrivals. The data are the processed form of those in Fig. 6.13. (c) Redrawn, with permission, from Roberts *et al.* (2003).

$$T_{\text{Refl}} = T_{\text{AB}} + T_{\text{BC}} = \frac{\text{AB}}{V_1} + \frac{\text{BC}}{V_1} \quad (6.17)$$

If the raypath is used to construct a right-angled triangle with hypotenuse AB + BC and the two other sides of length X (X = AC) and $2Z$, then it is clear that Eq. (6.17) can be written as:

$$T_{\text{Rfl}} = \frac{\sqrt{X^2 + (2Z)^2}}{V_1} \quad (6.18)$$

When discussing stacking it is convenient to express the travel times of reflections in terms of a constant component (T_0) and a variable component, which is the increase in time relative to T_0 with source–detector offset (X). Moveout relative to the zero-offset time is known as normal moveout (NMO). For offsets which are small compared with Z the following approximation expression can be used:

$$T_{\text{Rfl}} \approx T_0 + \frac{X^2}{2V_1^2 T_0} \quad (6.19)$$

Figure 6.14b shows reflections from several interfaces. The arrivals again have curved moveout. A series of actual reflected arrivals can also be seen in the real data forming the shot gather in Fig. 6.14c. The moveout of reflections from deeper interfaces is markedly less than that from the shallow interface. This is because T_0 (Eq. (6.19)) is greater, and because velocity also tends to increase with depth (see Section 6.6.2).

Referring to Fig. 6.14a, the moveout of the reflection can be used to determine the velocity of the overlying layer (V_1). From Eq. (6.18):

$$T_{\text{Rfl}}^2 = \frac{4Z^2}{V_1^2} + \frac{X^2}{V_1^2} = T_0^2 + \frac{X^2}{V_1^2} \quad (6.20)$$

A plot of ${T_{\text{Rfl}}}^2$ versus X^2 is a straight line with intercept ${T_0}^2$ and slope $1/{V_1}^2$.

In Fig. 6.14b there are no differences in velocity between the various layers so there is no refraction of the rays as they cross the interface (see Section 6.3.4.2). Figure 6.15 shows reflections from planar interfaces across which the velocity changes. The velocity contrasts between the layers refract the raypaths of the reflections from the deeper interface, changing their lengths and travel times. The effect is greater at longer offsets and for larger velocity contrasts. These are the minimum travel time paths. Also shown in the figure are the equivalent non-refracted (minimum-distance) raypaths. In the presence of velocity contrasts, the travel time for a reflection from a deeper interface is represented by modified forms of Eqs. (6.18) to (6.20). The (constant) velocity of the material overlying the reflector (V_1) is replaced by the *root-mean-square velocity* (V_{rms}) of the layers overlying the reflector (the root-mean-square is a form of average, literally the square root of the mean of the squares of the velocities). The velocity of an individual layer is known as the *interval velocity* (V_{int}). This can be determined for each layer once V_{rms} for reflections from the top and base of the layer are known, using the Dix (1955) equation:

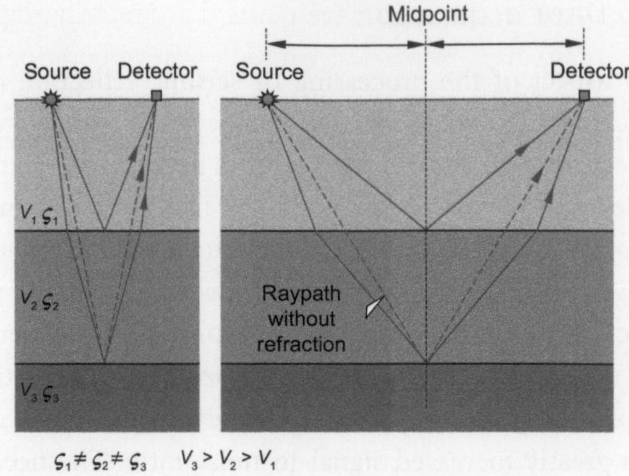

Figure 6.15 Refracted (minimum time) and non-refracted (minimum length) raypaths of reflected arrivals for a three-layer Earth model.

$$V_{\text{int}} = \sqrt{\frac{V_{\text{rms}_n}^2 T_n - V_{\text{rms}_{n-1}}^2 T_{n-1}}{T_n - T_{n-1}}} \quad (6.21)$$

where the subscript n refers to the velocity and time for the reflection from the lower interface of the nth layer, and $n - 1$ the same from the interface above it. A comprehensive description of velocity information derived from reflected arrivals is provided by Al-Chalabi (1974).

Multiples

Arrivals reflected only once during their journey through the subsurface are known as *primaries*. Arrivals that have been reflected more than once during their journey are known as *multiples*. They are categorised as either long- or short-period. The former have paths sufficiently different from their primary that distinct arrivals are observed in the survey data. Long-period multiples from a shallow reflector can easily be mistaken for primaries from a deeper reflector. They are a significant source of methodological noise (see Section 2.4.2), especially in water-borne surveys.

Short-period multiples have paths that are only slightly longer than their primaries, caused by reverberations in thin layers or by a *ghost* reflection of the source at the ground surface. Their travel times are only slightly greater than that of the primaries to the extent that only one (longer-wavelength) arrival is recognisable (see Section 6.3.2), which reduces the resolution of the survey data (see Section 6.7.1).

6.5.1.2 Survey design

A key objective of reflection data acquisition is to create a uniform coverage of subsurface reflection points across the area of interest. It is normal to assume that the interfaces reflecting the seismic waves are horizontal, in which case the surface projection of the point of reflection is halfway (the midpoint) between source and detector (as can be seen in Fig. 6.15). Recall that to enable stacking, data acquisition is designed so that reflections from the same point in the subsurface are repeatedly recorded.

Two-dimensional surveys

For 2D surveys, recordings are made using a linear spread along one or more traverses, ideally forming a regular dip- and strike-line (grid) network of perpendicular lines. An approximate grid arrangement may be all that is possible in areas of limited access, often dictated by the network of roads and tracks in the survey area. For a straight line traverse and an end-on spread source, the reflection points form a straight line and are spaced at half the distance between the individual detectors. Figure 6.16 illustrates 2D seismic profiling using an end source and six detectors (A–F). In reality, the number of detectors would be much larger. For simplicity, it is assumed that the subsurface comprises two layers, both having constant velocity but different acoustic impedance (ζ), and separated by a planar horizontal interface. Only reflected arrivals, with their characteristic curvilinear moveout (see Section 6.5.1.1), are shown in the shot gathers, which are from eight spread locations. Note how reflections from the three coloured reference points on the interface are repeatedly recorded as the spread is moved, i.e. reflections at the red location are recorded by receiver F from source 1, by receiver E from source 2 and so on. The number of occasions when reflections are recorded from the same point on the interface is known as the *fold* of the stack. There are six reflections from the red location so the fold of the stack is six in this case, also described as 600%.

Three-dimensional surveys

When a 3D survey is conducted on land, ideally the spread comprises parallel detector-lines and sources located along a perpendicular line, as shown in Fig. 6.17a. Coverage comprises a grid of reflection points. As shown in Fig. 6.17b and c, recording using various combinations of detector and source positions allows repeated recording of reflections from the same points in the subsurface.

6.5.2 Data processing

The three operations fundamental to the processing of seismic reflection data are stacking, deconvolution and migration. All other processing operations are designed to improve the effectiveness of these three operations, either by attenuating noise or by making the data better conform to the assumptions inherent in the operations.

Deconvolution is a means of manipulating the characteristics of the signal wavelet, in particular reversing some of the detrimental consequences of its passage through the subsurface. *Stacking* combines traces recorded at finite source–detector offsets into a much smaller number of equivalent zero-offset recordings and in so doing improves the signal-to-noise ratios. This means that *post-stack* processing operations are applied to the greatly reduced number of traces, offering considerable savings on computer resources compared with *pre-stack* operations. However, stacking makes important assumptions about the nature of the subsurface and if these are significantly violated the data may be degraded. In these cases pre-stack processing operations are preferable. *Migration* is a process that moves features in the seismic section so that they are in their correct relative positions, and the section better resembles the subsurface geology.

6.5.2.1 Common processing operations

Before describing the three fundamental processing operations we briefly summarise some of the other commonly used processing methods. As with any type of data enhancement, noise suppression involves identifying some property of the noise that differs from the signal. In seismic processing, properties that are exploited include randomness/coherence, frequency, moveout and periodicity. For example, frequency filtering is routinely applied to remove low-frequency surface waves, as is shown in Fig. 2.26. The different velocities of the different types of waves, causes

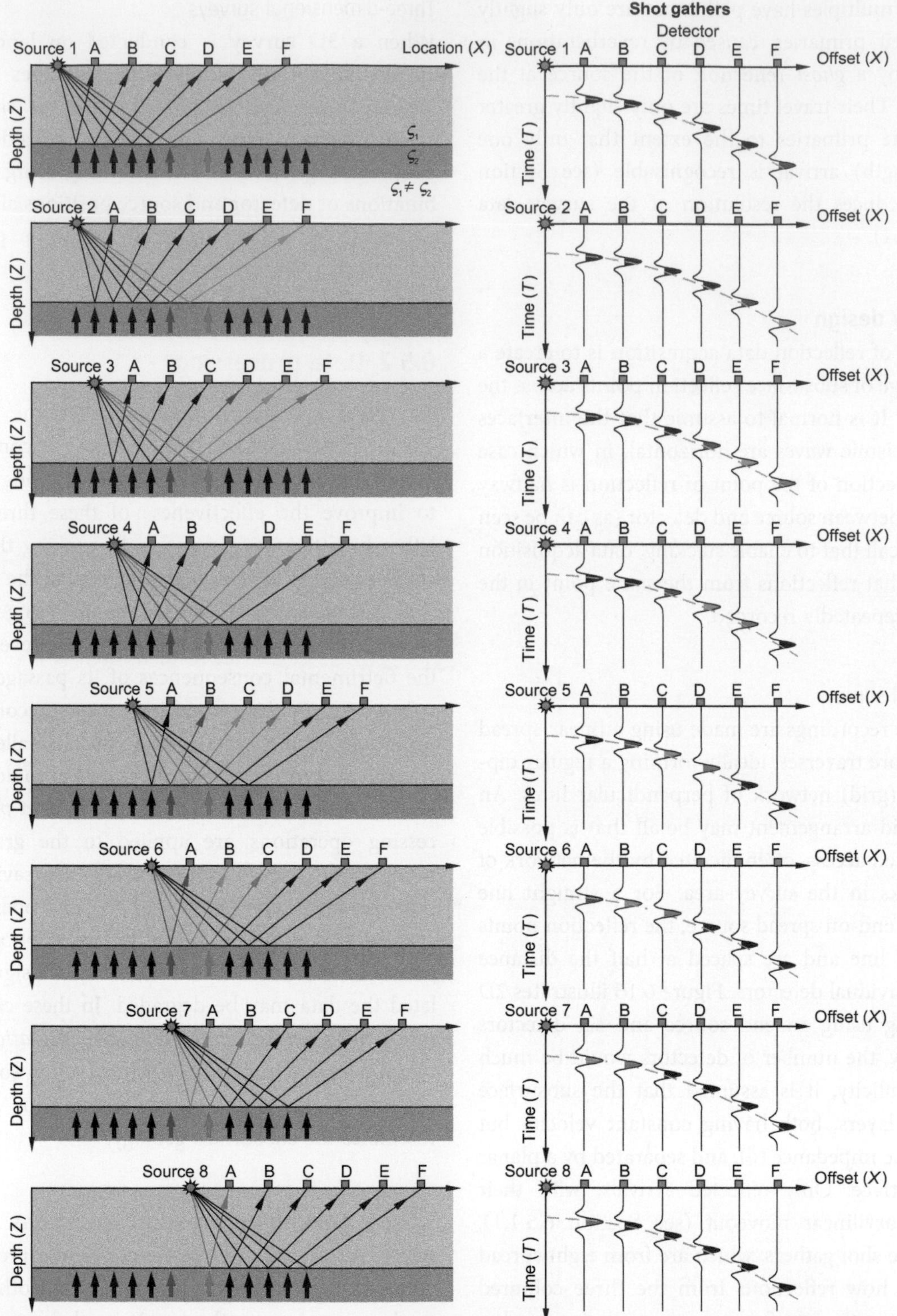

Figure 6.16 Schematic illustration of 2D continuous seismic profiling. The coloured raypaths and traces are associated with three reference points on the reflecting interface. Note the repeated recording of reflections from these points for each shot location. Only reflected arrivals are shown in the shot gathers.

them to have different 'apparent dips' (moveouts) on a shot gather (Fig. 6.13). Trend-based filtering, called *f–k* filtering, may be used to remove those arrivals which are noise (see *Trend* in Section 2.7.4.4).

Pre-processing

Pre-processing is the term used to describe operations carried out to prepare datasets for processing. It involves basic operations such as converting the data to standard

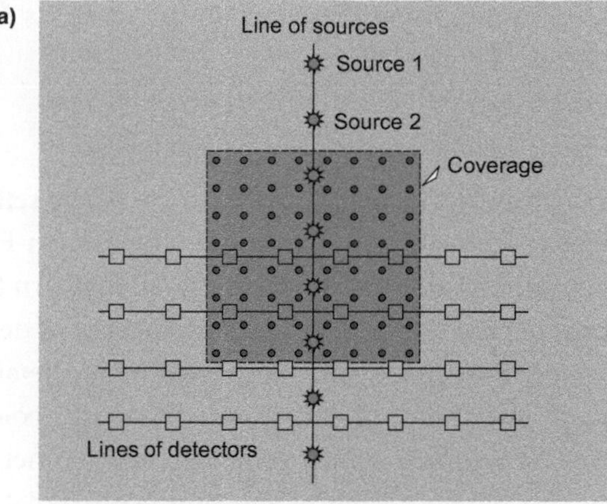

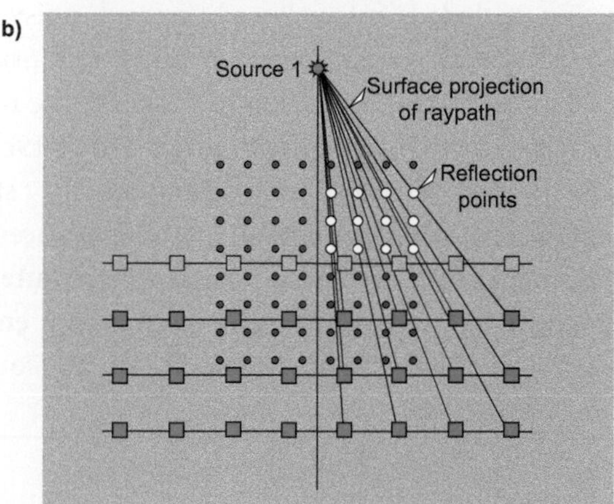

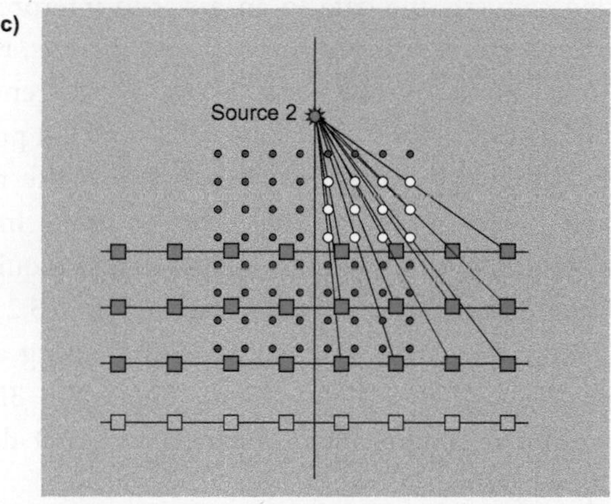

Figure 6.17 Schematic illustration of the acquisition of a 3D seismic dataset. (a) Map of the source locations, detector positions and the reflection points, (b) spread for source position 1 with selected raypaths and the surface projections of reflection points shown, and (c) as for (b) but for source position 2.

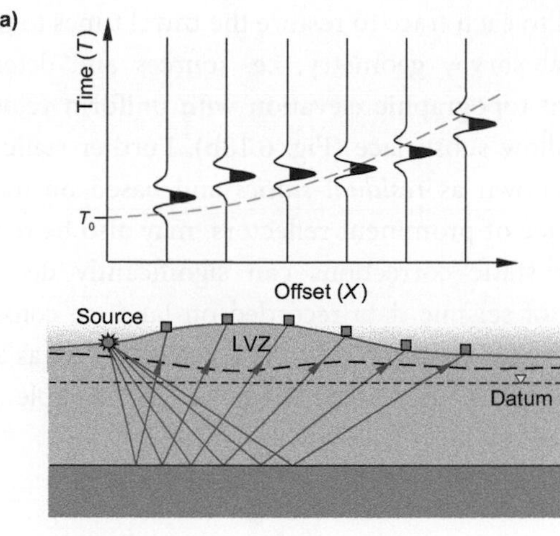

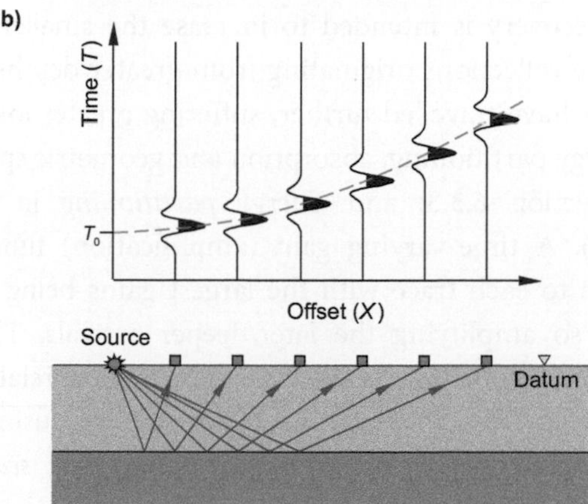

Figure 6.18 Statics. (a) Static effects due to surface topography and the near-surface low-velocity zone (LVZ) cause distortion of the ideal smooth curvilinear moveout (dashed blue line) of a reflection. (b) Static-corrected traces show the equivalent travel times for a horizontal datum surface beneath the LVZ. T_0 – zero-offset reflection travel time.

digital formats, manual inspection to identify obviously unsuitable recordings, and collation and integration of positional and survey geometry information. The direct and critically refracted arrivals (first arrivals) may be manually edited out (muted) since they occur in different parts of the shot gather from all but the longest offset reflections. The airwave arrivals may also be muted.

Static correction

Static effects are variations in arrival times caused by variations in source and/or detector elevations and the effects of the low-velocity zone (LVZ) (Fig. 6.18a). The LVZ occurs at, or near, the surface and is primarily the result of variations in weathering and water saturation (see Section 6.6.5). Velocities tend to be lower than at depth, and lateral changes in the velocity and thickness of the near-surface material may cause local variations in arrival time. Details of near-surface velocity/thickness variations are often derived from analysis of critically refracted arrivals using the methods outlined in online Appendix 6. A time correction, known as the *static correction*, is applied to each trace to restore the travel times to those for an ideal survey geometry, i.e. sources and detectors at constant topographic elevation with uniform velocities in the shallow subsurface (Fig. 6.18b). Further static corrections, known as *residual statics* and based on the lateral coherence of prominent reflectors, may also be required.

Poor static corrections can significantly degrade the quality of seismic data recorded on land, so considerable time is spent ensuring that the corrections are as accurate as possible. Gibson (2011) shows an example of how important static corrections can be.

Scaling

Gain recovery is intended to increase the smaller amplitudes of reflections originating from greater depths. These arrivals have travelled further, suffering greater losses due to energy partitioning, absorption and geometric spreading (see Section 6.3.3, and *Energy partitioning* in Section 6.3.4.2). A time-varying gain (amplification) function is applied to each trace with the largest gains being at later times, so amplifying the later/deeper arrivals. The gain function may be based on some mathematical relationship or derived from the trace itself, for example using automatic gain control methods (see *Amplitude scaling* in Section 2.7.4.4).

6.5.2.2 Deconvolution

Deconvolution is a form of filtering operation and, as described in Section 2.7.4.4, is also called *inverse filtering*. It may be applied several times during a seismic processing sequence, both pre- and post-stack. Deconvolution is used to 'undo' undesirable characteristics of the seismic wavelet resulting from its passage through the subsurface – the effects of the Earth filter. These may involve changing its shape into one that makes interpretation of the data easier, for example changing it to zero phase (see online Appendix 2) or shortening it to improve resolution (see Section 6.7.1).

There are several different types of deconvolution, differing in how the deconvolution operator is determined, which itself is related to how well the source wavelet and the nature of the Earth filter are known. *Predictive deconvolution* is commonly used to suppress multiples and works on the principle that arrivals that are multiples, being periodic, can be predicted from earlier parts of the trace whereas primary reflections cannot.

It is not possible to convert the seismic wavelet perfectly to the desired form, but deconvolution is very effective and an integral part of any seismic processing sequence. A detailed description of deconvolution in all its forms is provided by Yilmaz (2001).

6.5.2.3 Example of pre-stack processing

The improvement in signal-to-noise ratio of the reflections in a shot gather can be very marked. The data in Fig. 6.19 are from the Kristineberg mining area in northern Sweden (Ehsan *et al.*, 2012). The local geology consists of deformed low- to medium-grade metasedimentary and metaigneous rocks and granitoids. Note the curvilinear moveout characteristic of reflections in a processed shot gather. These arrivals are barely visible in the unprocessed data. Data processing included: manually editing out noisy traces, muting of the airwave, frequency filtering to suppress surface waves and deconvolution to compress the reflected arrivals (Fig. 6.19b); static corrections (Fig. 6.19c); then muting of first arrivals in preparation for stacking (Fig. 6.19d). Another example of pre-stack processing is the two shot gathers in Figs. 6.13 and 6.14, the latter being the processed data. Again there is considerable enhancement of reflected arrivals in the processed data compared with the unprocessed data.

6.5.2.4 Stacking

Stacking converts the data to an easier-to-interpret zero-offset form and suppresses non-reflected arrivals, random and transient noise, and most kinds of coherent noise (usually arrivals other than P-wave reflections). It produces a significant improvement in the signal-to-noise ratio of the data. Furthermore, stacking provides useful information about seismic velocities in the subsurface required for migration and depth conversion (see Section 6.5.2.5). For simplicity, we first describe stacking using a 2D dataset although it is also routinely applied to 3D data; the additional complications incurred with 3D data are considered later.

The first step in stacking involves *sorting* the shot gathers from a continuous profile survey (Fig. 6.16) into *common-midpoint* (CMP) *gathers* (Fig. 6.20a and b). These consist of sets of traces whose midpoints (halfway between the source and detector) are at the same location; the number of traces in each gather is equal to the fold of the stack, six in this case. The colour-coded raypaths in Fig. 6.16 are shown, after gathering, in Fig. 6.20a. The different source–receiver offsets means that the raypath lengths and the travel times of the reflections vary, again with a curvilinear moveout (Fig. 6.20b).

The next stage of stacking involves analysis of CMP gathers. The aim here is to combine their constituent traces

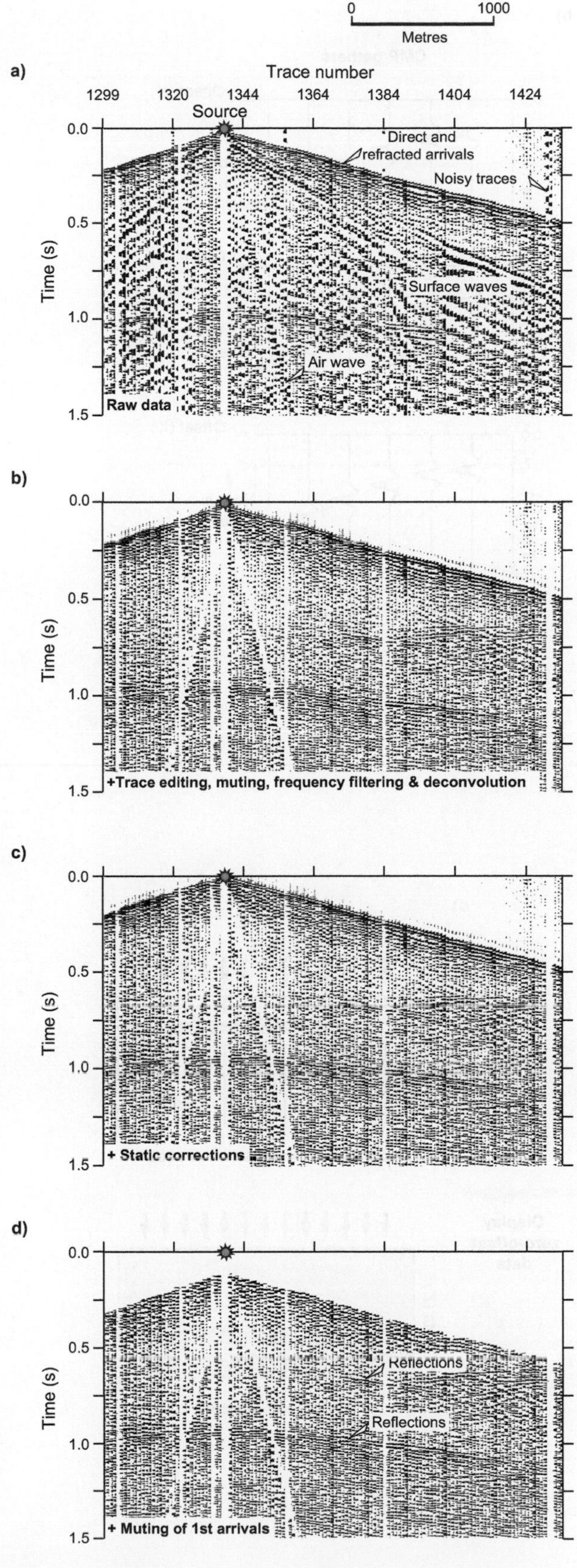

Figure 6.19 Pre-stack processing of a shot gather from the Kristineberg area. (a) Raw data with labels indicating various types of arrival constituting noise; (b–d) data after successive processing operations. Redrawn, with permission, from Ehsan *et al.* (2012).

into a single *stacked* trace which approximates a zero-offset recording made at the gather's midpoint (Fig. 6.20c). A time shift is applied to each trace so that the reflected arrival is shifted to the equivalent zero-offset travel time (T_0). Since the time shift is equal to the NMO (see Section 6.5.1.1) this is called the *normal moveout correction*. After the appropriate corrections have been made to each trace, they are summed together; this is the 'stacking' process itself (Fig. 6.20d). Any random noise present in the individual traces comprising the CMP gather will tend to cancel (see Section 2.7.4.1) and the reflected arrival, common to all traces, will be enhanced. Also, arrivals with moveouts different to the primary reflections (noise) will be suppressed.

Correcting for normal moveout

Key to stacking is the accurate determination of the NMO correction. Recall that the travel times of reflections can be defined in terms of a constant component (T_0) and a normal moveout (NMO) component (see Section 6.5.1.1). The NMO component depends on T_0, the depth to the reflector (Z); the root-mean-square velocity (V_{rms}), which is a function of the overlying interval velocities; and the source–detector offset (X). The problem when seeking to make the NMO correction is that only the offset is known (if the other parameters were known then the seismic survey would be unnecessary, an example of the geophysical paradox mentioned in Section 1.3). This apparently insurmountable problem is addressed by using a combination of two well-established scientific techniques: 'trial and error', and a preconception about the 'right' answer – a strategy that can be described as 'fitting a model to the data'.

Figure 6.21a and b shows a hypothetical CMP gather for a subsurface comprising three (horizontal and planar) reflecting interfaces. The interfaces separate layers that have constant velocity, increasing with each deeper layer, i.e. $V_3 > V_2 > V_1$, and different acoustic impedance (ζ). As shown in Section 6.5.1.1, the moveout of the reflections is less when the interface is deeper. In order to effectively stack the traces making up the CMP gather, it is necessary to flatten each of the three arrivals, but a different time shift needs to be applied to different parts of each trace because the various reflections have different moveout.

There is a variety of ways to determine the NMO correction. The method we describe here has the advantage of being conceptually simple and so is a useful means of demonstrating the concepts involved. The method is based on the creation of *constant velocity scans* of the CMP gathers. These are created by determining the NMO

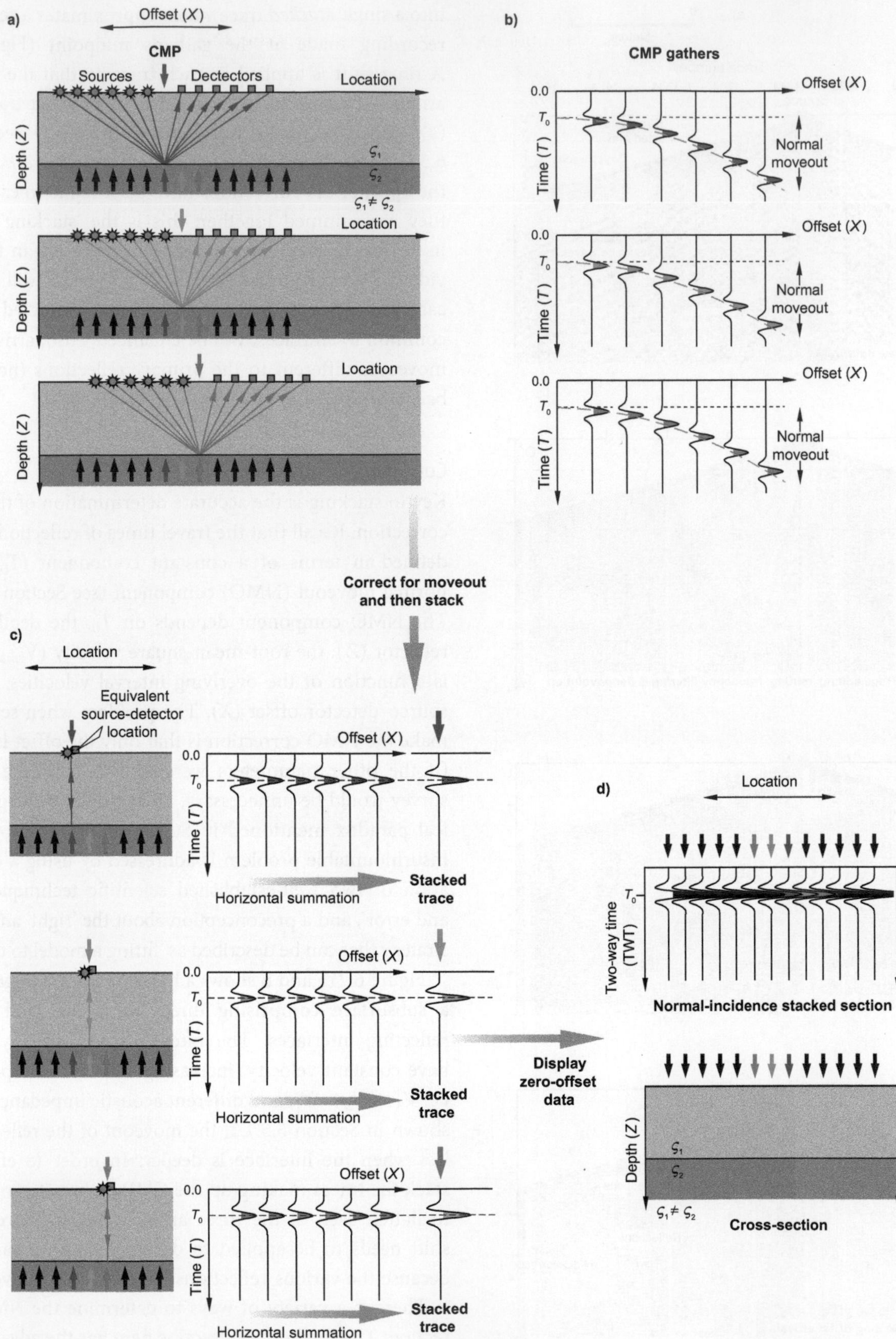

Figure 6.20 Schematic illustration of the procedure for creating a zero-offset stacked seismic section. The colour-coded traces, raypaths and reflection points correspond with those in Fig. 6.16. See text for explanation.

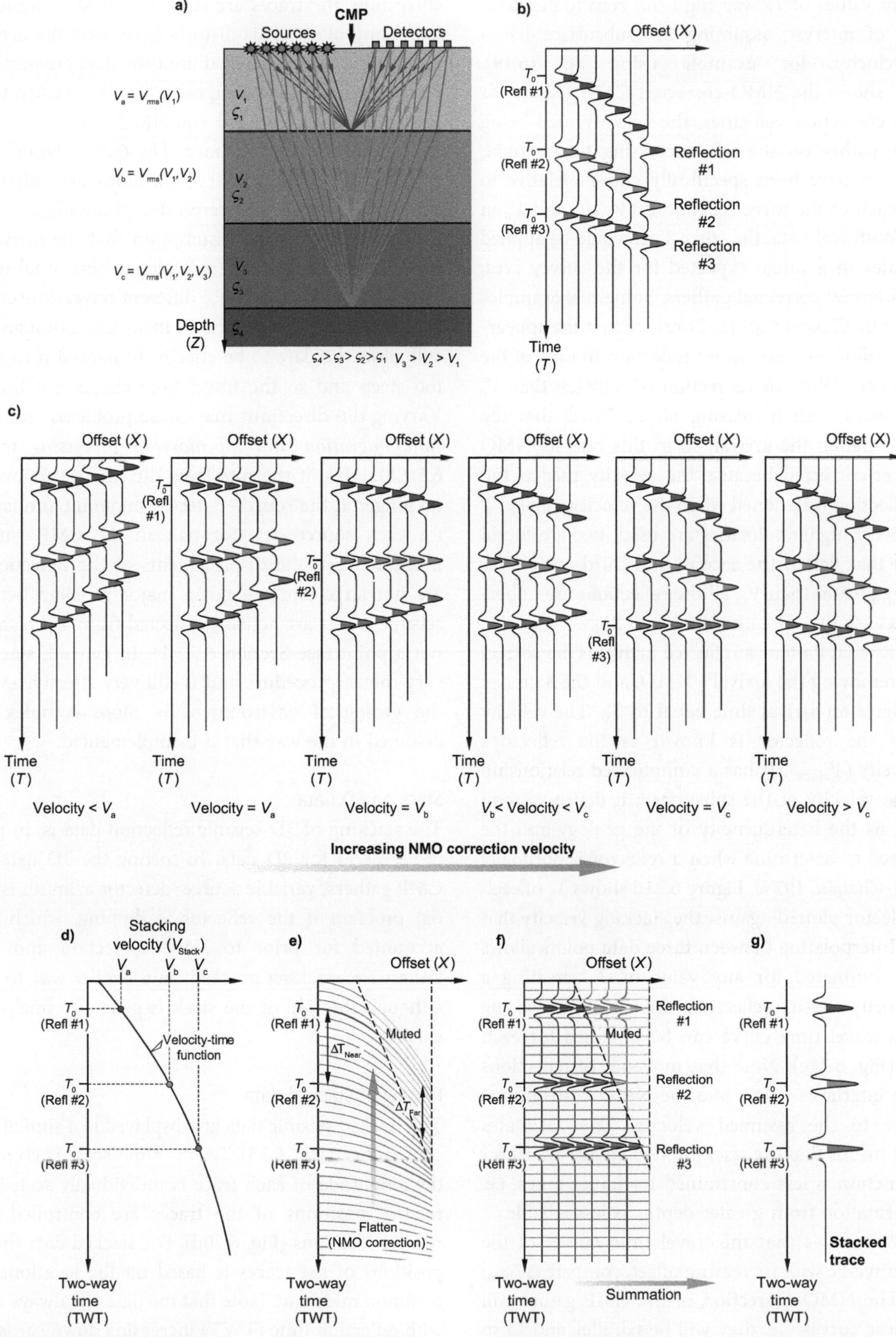

Figure 6.21 Schematic illustration of the procedure for determining the NMO correction. See text for explanation.

correction for values of T_0, varying from zero to the maximum time of interest, assuming the subsurface has a constant velocity, for example using Eq. (6.19). Figure 6.21c shows the NMO-corrected CMP gathers for six different correction velocities, the velocity used being least for the gather on the left, increasing to the right. These velocities have been specifically chosen relative to the V_{rms} of each of the three reflections (V_a, V_b and V_c in Fig. 6.21a). With real data, the correction would be applied using velocities in a range expected for the survey area, creating perhaps 20 corrected gathers. Some real examples are presented by Gibson (2011). Consider first the appearance of the shallowest (least time) reflection in each of the corrected gathers. With the correction velocity less than V_a it curves upward with increasing offset. Recall that the purpose is to flatten the arrival, so in this case its NMO has been over-corrected because the velocity used is too low. This reflection is flattened when the velocity equals V_a. As progressively higher velocities are used, two are found (V_b and V_c) that flatten the second and third reflections. For a velocity greater than V_c, all the reflections are under-corrected because the velocity used is too high.

The velocity that flattens a reflected arrival is the correct velocity for removing the arrival's NMO, and the flattened arrival will have an arrival time equal to T_0. The velocity that flattens the reflector is known as the reflector's *stacking velocity* (V_{stack}). It has a complicated relationship with the true velocity of the subsurface; it depends upon factors such as the heterogeneity of the geology and the algorithm used to determine when a reflector is optimally flattened (Al-Chalabi, 1974). Figure 6.21d shows T_0 of each flattened reflector plotted against the stacking velocity that flattened it. Interpolating between three data points allows V_{stack} to be estimated for any value of T_0, creating a stacking velocity versus reflection time function. Using Eq. (6.19), a travel time curve can be obtained for each value of T_0 (Fig. 6.21e). Note that moveout of reflections from deeper interfaces is less and the NMO correction is less sensitive to the assumed velocity. This facilitates stacking, but means that the stacking velocity versus reflection time function is less constrained for larger times, i.e. velocity information from greater depths is less reliable.

Figure 6.21e shows that the travel time curves of the reflections converge with increasing offset; compare ΔT_{Near} and ΔT_{Far}. The NMO correction of the CMP gather will flatten all these curves, i.e. they will be parallel, and in so doing the primary reflections are flattened in preparation for summation, as shown in Fig. 6.21f. During NMO correction, the traces are stretched in time, more so with increasing offset. This distorts (stretches) the arrivals and when it exceeds a specified limit the data are muted (set to zero) to avoid introducing noise into the stacked trace. The summation of the muted and NMO-corrected traces produces the final stacked trace (Fig. 6.21g). Notice how the amplitudes of the primary reflections are enhanced and that they occur at their zero-offset times (T_0).

Stacking makes the assumption that the arrivals being analysed are reflections from planar horizontal interfaces. Dipping interfaces produce different travel time curves and the estimated velocities will be incorrect, although they will still allow the data to be effectively stacked if dips are not too steep and so the travel time curves not too altered. Varying dip directions may cause problems (see *Pre-stack time migration and dip-moveout processing* in Section 6.5.2.5). Also, if the interfaces dip, the reflection point is no longer at the source–detector midpoint and is different for each source–detector pair in the CMP gather. For moderate dips the displacements of the reflection points are not large and are not a major problem because the seismic waves are actually reflected from a (Fresnel) zone, not a point (see Section 6.7.1.1). In general, stacking is a very robust procedure and is still very effective even when the geological environment is more complex than is assumed in the way that it is implemented.

Stacking 3D data

The stacking of 3D seismic reflection data is, in principle, the same as for 2D data. In sorting the 3D data to form CMP gathers, variable source–detector azimuth is a potential problem if the reflector is dipping, which must be accounted for prior to NMO-correction and stacking. Otherwise, stacking proceeds in a similar way to 2D data, although the fold of the stack is generally smaller for 3D surveys.

Display of stacked data

The stacked seismic data are displayed in a similar way to a shot gather (Fig. 6.13), i.e. the time scale is common but the amplitude of each trace is individually scaled and the relative positions of the traces are controlled by their relative locations (Fig. 6.20d). For stacked data the relative positions of the traces is based on the locations of their common midpoint. Note that the data are always displayed with reflection time (TWT) increasing downwards, so as to mimic a depth section. The coloured traces in the figure identify the normal-incidence stacked traces produced from each gather and the arrows indicate the locations of their common midpoints.

For 2D surveys, the result is a seismic pseudosection, the term being used because the vertical scale is in TWT, not depth. For a 3D survey, a pseudovolume is produced. Compare the cross-section and the pseudosection in Fig. 6.20d. In this simple situation the two closely resemble each other, although the vertical scale is in different units. As is described next, this is not the case in real-world situations.

6.5.2.5 Migration and depth conversion

As described above, the conversion from finite- to zero-offset data transforms the seismic section into a form that begins to resemble the subsurface geology; for example, compare the cross-section and the stacked section in Fig. 6.20d. In fact, normal-incidence data correctly represent the geometry of the subsurface only when the responses are reflections from a horizontal interface and the velocity is constant throughout the entire subsurface (Fig. 6.22a).

Figures 6.22b to g illustrate how the subsurface is misrepresented in the stacked data when these limiting conditions are not satisfied. For example:

- Lateral changes in seismic velocity create apparent vertical displacements in reflectors. Higher velocity decreases the reflection travel times producing an apparently shallower reflector (Fig. 6.22b). The effect is known as *velocity pull-up*. Note the phantom diffractor where the velocity changes.
- Diffractions appear as concave-downwards events, but they originate from 'point' sources and from abrupt terminations or changes in the dip of a reflecting horizon (Figs. 6.22c and d), note the diffractors at the edges of the steps.
- Dipping reflectors are in the wrong place. Their dip is reduced, they are lengthened and they are laterally displaced down-dip (Fig. 6.22e). These effects are exacerbated with increasing dip.
- For the case of folded reflectors (Fig. 6.22f), the diffraction and dipping reflector responses combine to create the most obvious discrepancy between actual reflector geometry and the seismic response. Anticlines appear too open with geologically impossible cross-cutting events on their flanks if folding is tight. Worse still are synclines, where a 'bow-tie' pattern is formed (Fig. 6.22g).

Clearly, realistically complex geological structures will produce seismic responses that show little resemblance to their actual form. The distortions arise because the ray-paths are non-vertical (Fig. 6.22), whereas each trace in the stacked section is plotted vertically. Consequently, the display wrongly implies that the reflector or diffractor responsible for the arrival is located directly below the source–detector midpoint. The various components of the stacked data need to be moved to their correct relative positions to produce a more geologically realistic section. This is achieved using a process known as *migration*.

Depending on the type of migration used, the vertical display axis may remain as two-way time (*time migration*) or it may be converted to depth (*depth migration*). If the data remain in two-way time they may subsequently be *depth converted*. In both cases, information about velocity variations in the subsurface is required, and the variations in velocity must be completely known for the conversion to be totally accurate. Again we are faced with the geophysical paradox (see Section 1.3), i.e. that the information required to fully understand the geophysical response would render the geophysical survey unnecessary. In practice, velocities are estimated from whatever sources of information are available, usually stacking velocities (see *Correcting for normal moveout* in Section 6.5.2.4) and sonic logs (see Section 6.6.9), and collated to create a *velocity model*. Fortunately, some types of migration are not too sensitive to the choice of velocities. In contrast, depth conversion is highly sensitive to this choice. So rather than degrade the data through an incorrect choice of velocities, seismic reflection data are usually left as a TWT pseudosection. An example of depth conversion is presented by Zhou and Hatherly (2004), who worked with data from near the Oaky Creek coal mine, located in Queensland, Australia.

Migration may be applied pre- or post-stack and applied to both 2D and 3D data. Here we describe only *post-stack migration* in any detail. Pre-stack migration is more complex than post-stack migration, but is based on the same concepts. Also, there are many ways to implement migration, varying from the conceptually straightforward to those of considerable mathematical complexity. Again we describe an implementation that is easily understood at a conceptual level. For a comprehensive discussion of the subject of migration the reader is referred to Yilmaz (2001). Lillie (1999) provides a useful, less technical, description.

Post-stack time migration

It was previously demonstrated (see Section 6.3.4.3) that a reflecting interface is equivalent to a series of closely spaced

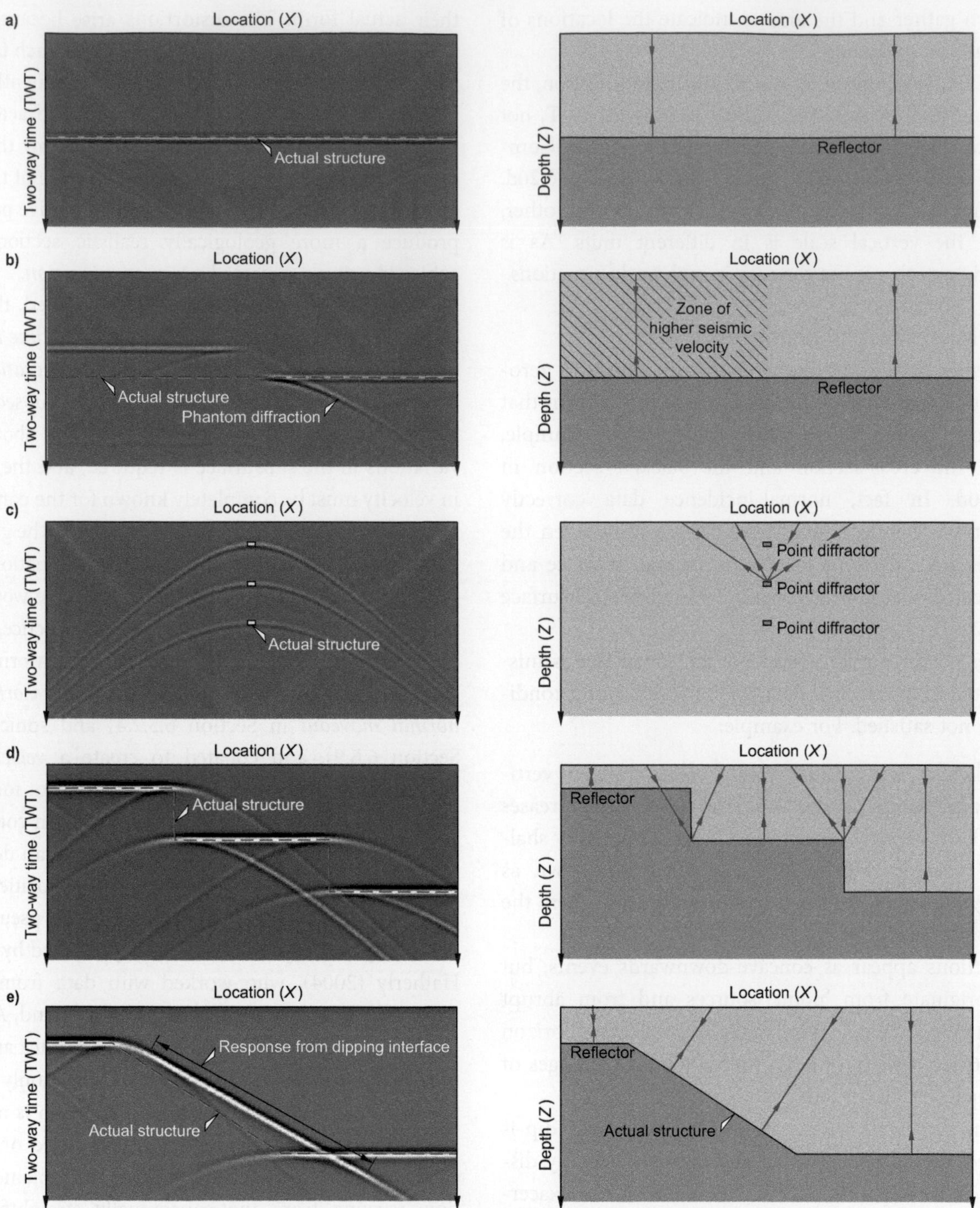

Figure 6.22 Some common subsurface structures and their representation in unmigrated normal-incidence seismic reflection data. Selected raypaths are shown. (a) Horizontal reflector, (b) horizontal reflector with lateral change in velocity, (c) diffractions, (d) horizontal stepped reflector, (e) dipping step reflector, (f) antiform and (g) synform. Velocity is constant in all figures except (b). The seismic velocity is 2 km/s, so that the TWT and depth scales are equivalent, allowing 'actual structure' to be overlaid on the seismic data.

point diffractors. The concept is useful for explaining migration. Figure 6.23a shows zero-offset data produced by a series of equally spaced diffractors located at the same depth. The data are the zero-offset equivalents of the finite-offset wavefields shown in Fig. 6.11. As shown in Fig. 6.22c, each diffractor produces a characteristic hyperbolic response with its maximum amplitude at its apex. Where the hyperbolae intersect they interfere, with the

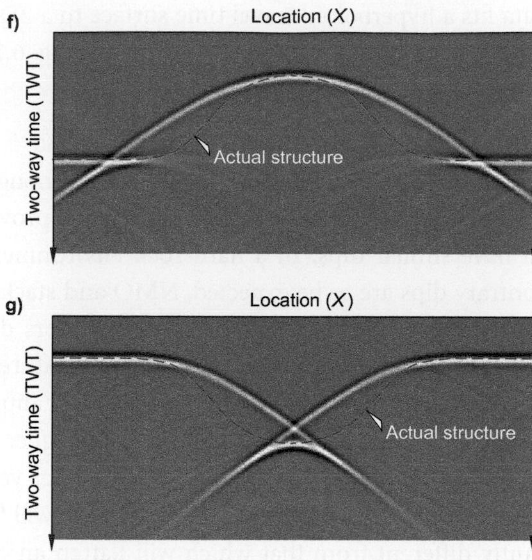

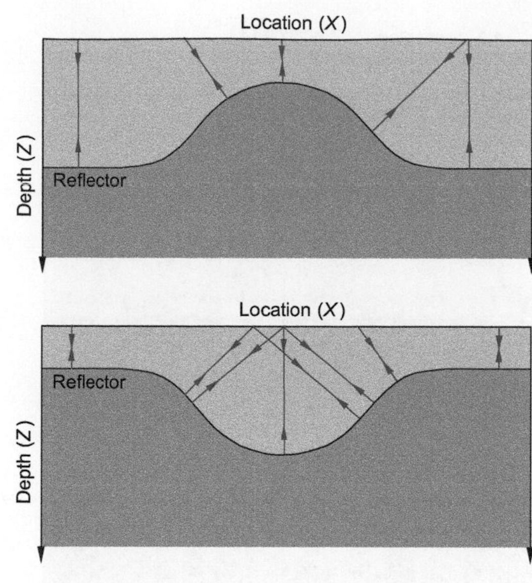

Figure 6.22 *(cont.)*

interference ever more destructive as the spacing between the diffractors decreases. When the spacing is sufficiently small the seismic response takes the form of a horizontally continuous feature coincident with the apices of the individual diffraction hyperbolae. This is the appearance of a horizontal reflector in zero-offset data (Fig. 6.22a).

Figure 6.23b shows the result when the closely spaced diffractors form a dipping line. The interfering hyperbolae produce a continuous 'apparent reflector' whose dip, length and down-dip displacement is displaced from the line of diffractors. This corresponds with the geometrical distortions illustrated for the dipping reflectors in Fig. 6.22e. Figure 6.23b shows that one way to correct the distortions in stacked data, i.e. to migrate the data, is to treat the subsurface as comprising a series of hypothetical diffractors, and therefore the data as consisting of a series of interfering hypothetical diffractions. Each diffraction hyperbola is transformed, or collapsed, to a 'point' located at its apex. Ideally, the data would then correspond with the actual structure of the subsurface.

Figure 6.24a illustrates the migration of a simple zero-offset dataset containing only a single diffraction. The seismic velocity in the subsurface is constant. The process is only shown for the central trace at location *X*. Recall that each seismic trace is a digital times series. For each sample in the central trace, the form of a hypothetical diffraction travel time curve (hyperbola) having its apex at the sample is calculated. This requires information about the velocity of the subsurface, which in this case is known. The amplitudes of the adjacent traces where they are intersected by the travel time curve are summed, and the result is the amplitude of the sample in a new version of the central trace, the *migrated* trace (Fig. 6.24b). The migrated dataset is formed by repeating this process for every sample in every trace in the unmigrated dataset. In this idealised example, the diffraction is perfectly collapsed so that all of its amplitude (energy) is concentrated at its apex. The migrated data resemble the actual form of the subsurface, in this case a single point diffractor, so a key objective of migration has been achieved.

When the process is applied to real data, where the hypothetical diffraction coincides with an actual diffraction the summation process will encounter samples of the same polarity on adjacent traces. The resultant absolute amplitude on the migrated trace will be large and so an arrival will 'be created'. Where a hypothetical diffraction is not coincident with an actual diffraction, the sample values along the hyperbola will be random and will tend to sum to zero. This part of the migrated trace will not contain an arrival.

The process described above is known as *diffraction-summation migration* and, although mathematically simple, can be very effective. A more sophisticated form of this type of migration, known as *Kirchhoff summation*, accounts for changes in both phase and amplitude along the diffraction hyperbolae. The same processes can be

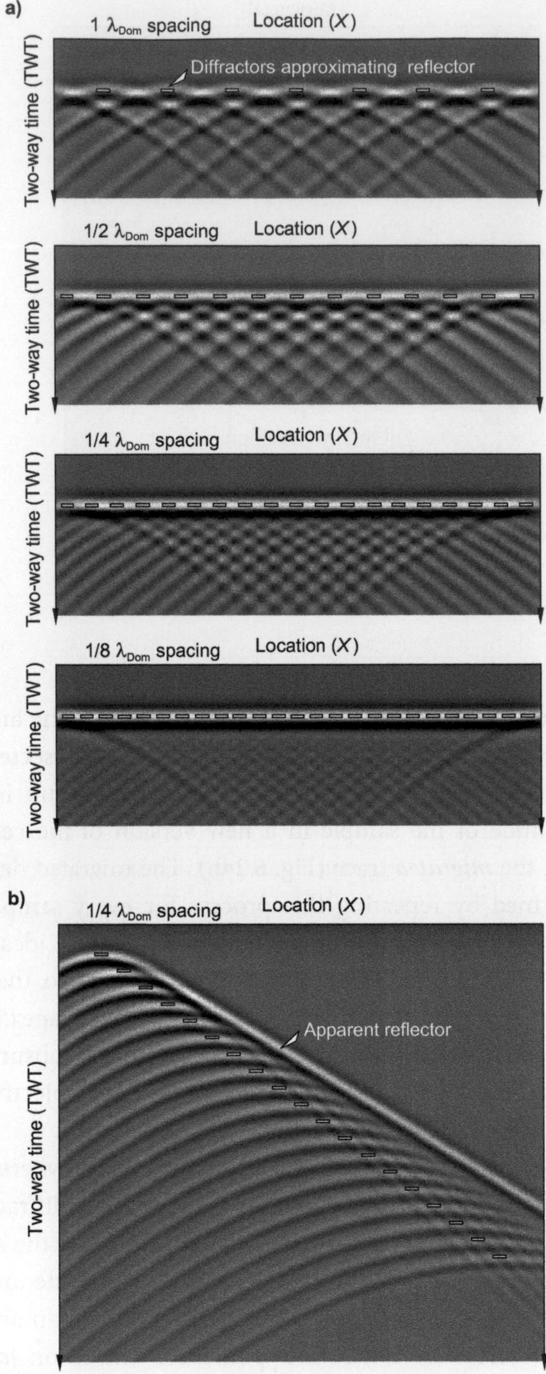

Figure 6.23 A reflector represented as a series of closely spaced diffractors with the spacing specified in terms of the dominant wavelength (λ_{Dom}) of the seismic signal. (a) Horizontal reflector with diffractor spacing decreasing, and (b) dipping reflector. The seismic velocity in the model used to create the seismic data is 2 km/s, so that the TWT and depth scales are equivalent, allowing diffractors to be overlaid on the seismic data.

applied to 3D data. Instead of identifying samples in a line (1D) of traces, constituting a section, that are intersected by a hyperbolic travel time curve, migration of 3D seismic data fits a hyperboloid travel time surface to a 2D distribution of trace locations, i.e. a data volume (Fig. 6.25).

Pre-stack migration and dip-moveout processing

Stacking assumes horizontal reflectors, although it will work reasonably well with dipping horizons provided they all have similar dips. In a hard-rock environment, where contrary dips are to be expected, NMO and stacking act as a dip-filter, preserving arrivals from reflectors dipping in one direction and suppressing those from features dipping in the opposite direction. When the reflector dips towards the source the stacking velocity will be higher than the actual velocity, and vice versa, so the correct velocity for flattening an event dipping in one direction will be significantly different from that which will flatten an event dipping in the opposite direction. Events with similar dips stack fairly well, but at the expense of suppressing those with different dips.

As a general rule, if velocities and structure are sufficiently complex that reflections do not have their ‘usual’ curvilinear moveout, then pre-stack migration is required. Pre-stack is conceptually the same as post-stack migration but must account for the fact that the data are in finite-offset form, so the travel time curves of the arrivals are different than for the equivalent zero-offset data. A major disadvantage of pre-stack migration is that the data volumes are much larger because it precedes the combining of multiple traces into one equivalent zero offset trace. That is, the data volume is larger by a factor equal to the fold of the stack. Migration is a computationally intense process so this is a significant disadvantage. An alternative approach is called *dip-moveout* (DMO), which is a partial migration that is applied after correcting for NMO so that subsequent stacking preserves all arrivals regardless of dip. The preservation of the dipping events means that post-stack migration will produce results that are more realistic. A description of DMO and pre-stack migration is beyond our scope. A straightforward demonstration of the equivalence of pre-stack migration and DMO+post-stack migration (in 2D) is presented by Russell (1998).

Time and depth migration

Whether implemented pre- or post-stack, two basic classes of migration are recognised based on their assumptions about velocity variations in the subsurface. In time migration, as implemented in diffraction-summation migration, only variations in velocity directly above the hypothetical

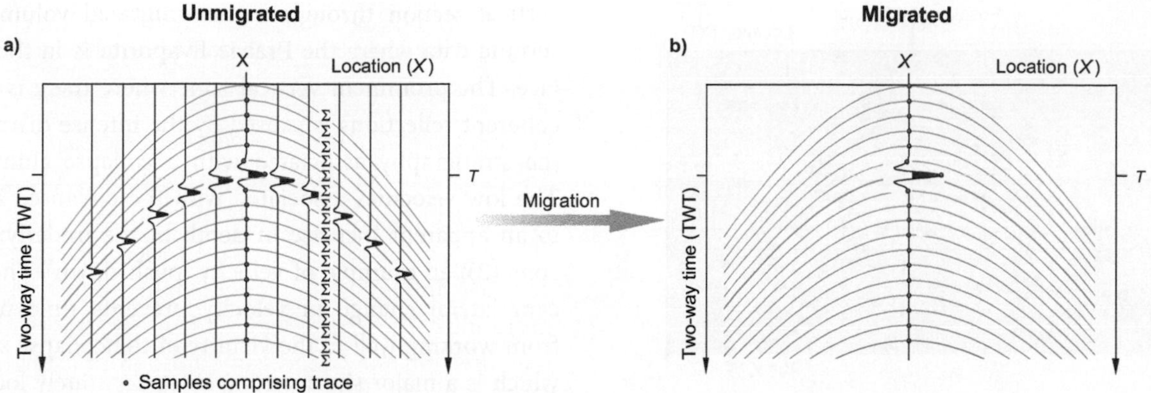

Figure 6.24 Schematic illustration of 2D diffraction-summation time migration. The samples making up the migrated trace are the summation of samples in the unmigrated traces where they intersect the hypothetical diffraction hyperbolae (blue lines).

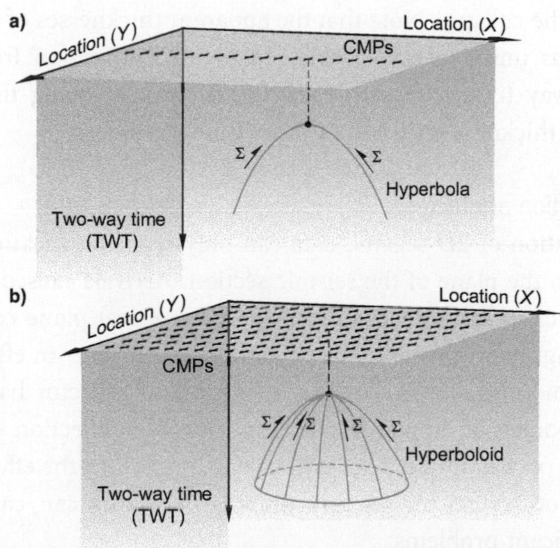

Figure 6.25 Schematic illustration of diffraction-summation migration. (a) In 2D, and (b) 3D. CMP – common midpoint.

diffractor are considered when computing the travel time curves from the 'diffractors'. In so doing, velocity may vary vertically above the diffractor but the variation is assumed to be 1D, i.e. there is no lateral variation. Time migration does partially allow for lateral as well as vertical velocity changes since a different velocity–depth variation can be used to migrate different traces. However, this approach is inadequate where lateral changes in velocity are significant.

The effects of lateral changes in velocity on zero-offset data are illustrated in Fig. 6.22b and for a slightly different situation in Fig. 6.26c. A discontinuous high-velocity layer located above a continuous reflector produces a gap in the reflector, a phantom diffraction and a 'pull-up' of the reflector below the high-velocity layer. Time migration will not perform well in these situations. Figure 6.26a shows a diffraction hyperbola when there is no lateral variation in velocity (V). The event is symmetrical with its apex laterally coincident with the diffractor. In Fig. 6.26b, lateral changes in velocity associated with a dipping interface (there is no change in acoustic impedance ζ) produce an asymmetrical arrival with its apex laterally displaced from the diffractor. Put simply, diffraction-summation time migration applied to these data will use summations along incorrect diffraction travel time curves.

Migration where lateral velocity variations are explicitly considered (e.g. diffraction-summation migration in which velocity variations are taken into account when determining the diffraction travel time curves) is known as *depth migration*. The resulting section has depth rather than time as its vertical axis, so no subsequent depth conversion is required, unlike time migration. Depth migration requires the velocities to be accurately known, and usually considerable time and effort is spent making the velocity model as accurate as possible; otherwise artefacts appear in the data (Herron, 2000). Seismic velocity normally tends to increase with depth and may vary vertical and laterally owing to changes in lithology (see Section 6.6). To create a reliable velocity model, the seismic data need to be initially interpreted in order to create the depth-migration velocity field. The velocity model is then iteratively modified using successive pre- and post-stack migrations until a satisfactory solution is obtained based on the geometry of features revealed by each iteration. This is a time-consuming and expensive process, especially when working in 3D.

The different results obtained from time- and depth-migration are illustrated in Figure 6.27. These data are part

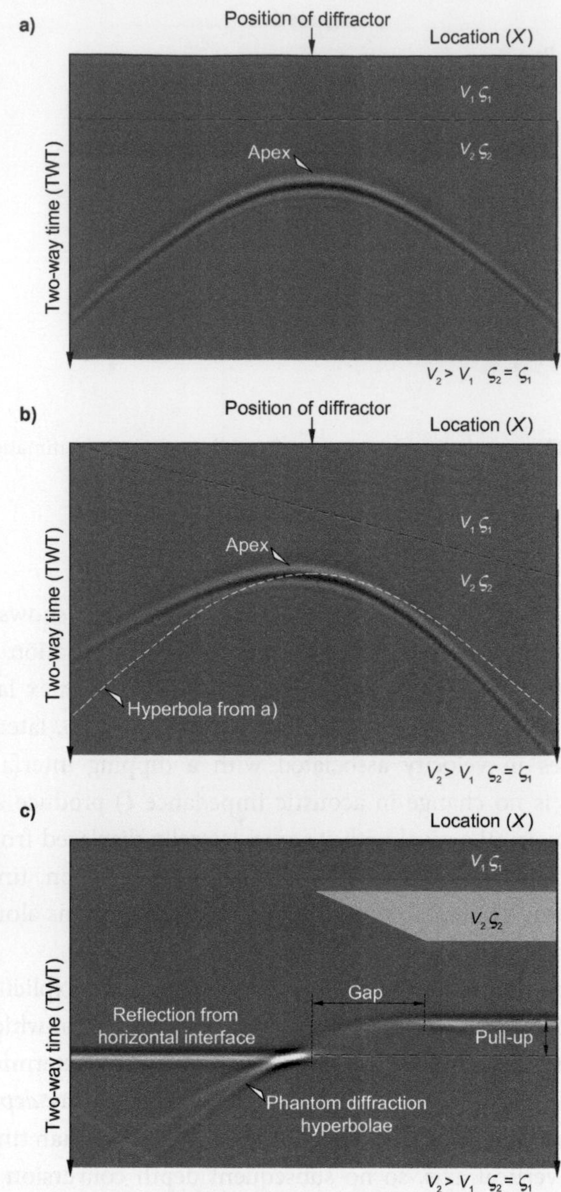

Figure 6.26 The effects of lateral variations in velocity on zero-offset seismic data. (a) Diffraction with no lateral velocity variation, (b) diffraction with an overlying lateral velocity change and (c) distortion of a continuous reflector in the presence of an overlying lateral change in velocity.

of a seismic survey designed to image the Prairie Evaporite Formation, located in Saskatchewan, Canada (Nemeth *et al.*, 2002). The Prairie Evaporite is a major source of potash and is mined using longwall methods. Dissolution of the evaporite horizon causes voids into which the overlying strata falls, creating salt collapse chimneys. These are a major hazard to mining, but can be detected using seismic reflection surveys. The geological environment comprises sub-horizontal sedimentary units and is, therefore, well suited to seismic surveying. Figure 6.27a shows a vertical section through a time-migrated volume of 3D seismic data where the Prairie Evaporite is in the subsurface. The prominent vertical zone where there is a loss of coherent reflections is caused by the intense disruption of the stratigraphy associated with a collapse chimney (1). The low velocities associated with the collapsed area lead to an apparent increase in depth of the underlying horizons (2), an example of velocity 'push-down'. The significant lateral change in velocity prevents time migration from working well in the vicinity of the collapse structure, which is a major shortcoming since accurately locating its margins is essential for mine safety. The depth-migrated data (Fig. 6.27b) provide greater clarity of the subsurface with the extent of the collapse chimney well defined. Furthermore, since the section is now presented in terms of depth, there is no spurious structure in the stratigraphy near the collapse. Note that the apparent thicknesses of the various units have also changed in the conversion from two-way time to depth, their thickness now being their 'true' thickness rather than their 'time' thickness.

Migration artefacts

Migration of 2D seismic data can only reposition features within the plane of the seismic section. Arrivals caused by features in the subsurface outside the vertical plane containing the source and line of detectors comprise an effect known as *sideswipe*. For example, when a reflector has a component of dip across the survey profile, reflection will occur on the up-dip side. When dips are shallow the effects are not severe, but steeply dipping reflectors can cause significant problems.

The width of the diffraction hyperbola in diffraction-summation migration, i.e. how many traces are involved in the summation process, is known as the *migration aperture*. If it is too narrow, the full benefit of summing the hyperbola is not achieved. For example, the steeper the dip of an event the more displaced is its response in the unmigrated data, so it is more likely to lie outside the aperture and not be correctly repositioned. Also, the edges of the seismic data present a problem, since the full extent of the hyperbola will not have been recorded. This causes the quality of the migrated data to deteriorate rapidly towards the edges where it may appear that reflectors have been truncated by a fault.

Seismic data are usually sampled more intensively in time than spatially, as represented by the wider trace interval along the section compared with the closer time sampling interval of the traces. Sometimes steeply dipping

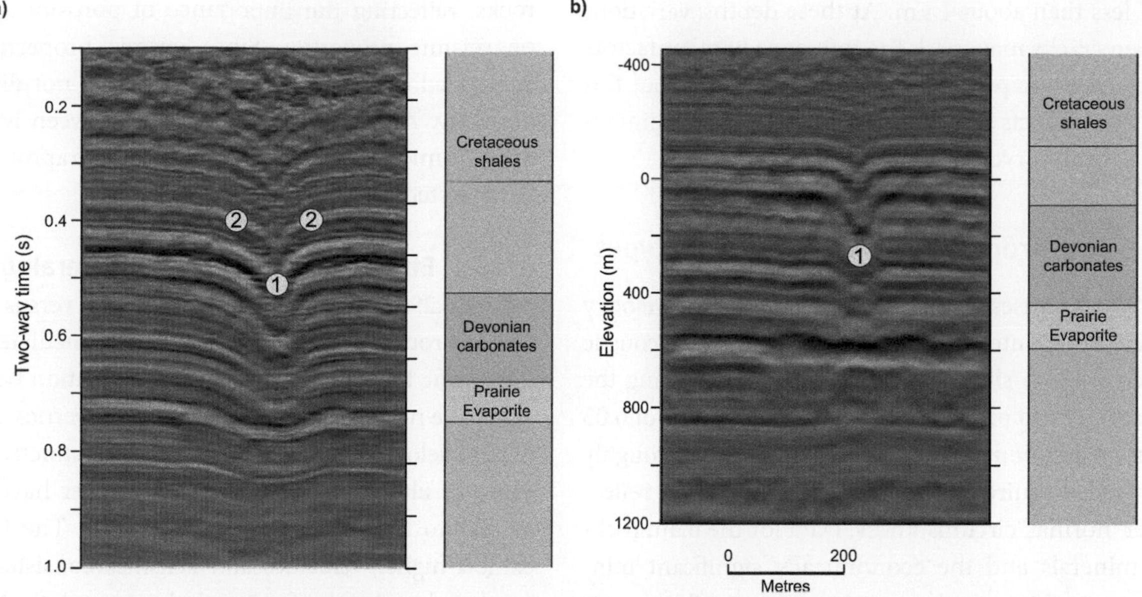

Figure 6.27 Vertical slices through a post-stack migrated 3D seismic volume across a collapse chimney above the Prairie Evaporite Formation. (a) Time migrated, and (b) depth migrated. See text for details. Based diagrams in Nemeth *et al.* (2002).

high-frequency arrivals are not adequately sampled spatially, presenting another problem for migration. If the response is aliased (see Section 2.6.1) dips are misrepresented in the data and so are incorrectly repositioned by migration.

If the migration velocities are incorrect, the shape of the diffraction hyperbola used for the summation will also be incorrect, and obviously so too will be the result. When the migration velocity is too small, it may cause concave-downward arcs to appear, known as *migration frowns*, and the section is *under-migrated*. If the velocities are too high, concave-upward *smiles* result and the section is *over-migrated*. Noise in the data may also cause smiles, and these are commonly seen in the deeper parts of the section where the signal-to-noise ratio tends to be small. A tell-tale sign of incorrect migration is reflectors cross-cutting each other to create geologically implausible structures.

Choice of migration type

In many relatively undeformed soft-rock terrains, post-stack time migration gives acceptable results, whilst in more geologically complex terrains DMO processing is required prior to post-stack migration. Where the geology is most complex, expensive pre-stack depth migration is required. A review of recent publications on seismic reflection surveying in hard-rock mineral provinces revealed that the most common processing strategy involved the use of DMO and post-stack time migration.

6.6 Variations in seismic properties in the geological environment

The seismic response of the subsurface is controlled by variations in seismic wave velocity and attenuation, and changes in acoustic impedance. These are collectively referred to here as *seismic properties*. An understanding of the geological controls on seismic properties is fundamental to making geologically realistic interpretations of seismic data, so we describe the subject here in some detail. Note that only P-wave seismic properties are considered. Since there are far more measurements available for density and velocity than for attenuation, and because the latter is rarely used in interpretations, it is discussed briefly and separately. Also, although geological controls on density affect acoustic impedance, they are only alluded to here as they are described in detail in Section 3.8.

Various geological factors affect seismic velocity and acoustic impedance, and there are numerous laboratory studies focusing on the sedimentary rocks of Mesozoic and Cainozoic sedimentary basins. Many laboratory measurements have been made on igneous and metamorphic rocks under the extreme temperature and pressure conditions expected deep within the Earth's interior. Neither area of investigation is especially relevant to mineral exploration. The discussion below is based on a database of published velocity and density data comprising laboratory measurements made under conditions expected at depths of less than about 1 km. At these depths, variations due to open cracks may mask the influence of other factors that control seismic properties (see Section 6.6.2), but this complication reflects the actual situation with mining-oriented seismic surveys.

6.6.1 Seismic properties of common rock types

Figure 6.28 comprises a series of plots of seismic velocity versus density. Contours showing variations in acoustic impedance are also shown, their interval representing the contrast required to produce a reflection coefficient of 0.05 (see *Energy partitioning* in Section 6.3.4.2). This is roughly the magnitude required to produce a recognisable reflection under 'normal' circumstances. Data for the main rock-forming minerals and the economically significant minerals (average values are shown, given the possible ranges in mineral compositions and the fact that velocity varies with direction in most crystals) and common pore contents are also included.

From Fig. 6.28a it is immediately clear that P-wave velocity is approximately proportional to density, but there is considerable scatter. For example, materials with the 'average' crustal density of 2.67 g/cm^3 (see Section 3.8) may have velocities between roughly 3000 and 6000 m/s, which is nearly half the overall range.

Referring to Fig. 6.28b to d, the velocity versus density data for the main rock classes define a continuous distribution with an overall trend from unconsolidated materials, exhibiting the lowest velocities and densities, through sedimentary rocks to crystalline rocks, which exhibit the highest values. The lower end of the continuum approaches the properties of water, the most common pore fluid. At the high end of the continuum, the data plot in the vicinity of the values for individual rock-forming minerals. As demonstrated below, this occurs because when porosity is small mineralogy is the dominant influence on velocity and density, and as porosity increases the pore fluid becomes an important influence. Consequently, data from sedimentary rocks, which normally have significant porosity, form a continuum from the properties of water to the properties of their most common matrix minerals, namely quartz, feldspars and calcite. The data from igneous and metamorphic rocks, which normally have much lower porosity, occupy the region bounded by the properties of their most common mineral constituents, such as quartz, feldspars, pyroxenes, amphiboles etc. The range in velocity and density is much lower than is seen in sedimentary rocks, reflecting the importance of porosity as a control on seismic properties. Since seismic properties are not controlled only by mineralogy, they are not diagnostic of lithology. A one-to-one correlation between lithologically and seismically defined structure/stratigraphy should not be expected, as discussed in Section 1.1.

6.6.1.1 Effects of chemistry and mineralogy

Figure 6.29a shows variations in velocity versus density for igneous rocks, differentiating felsic, intermediate, mafic and ultramafic types. There is a rough correlation between how silicic the rocks are and their seismic properties. Felsic rocks exhibit velocities and densities that are distinctly lower than those of ultramafic rocks etc. The former have properties similar to those of quartz and feldspar. The latter group exhibit higher velocities and densities consistent with the greater abundance of minerals having relatively high velocity and density in their assemblages, such as olivine, pyroxene and amphibole. The rocks whose chemistry lies between these two extremes have intermediate velocities and densities. This compositional dependency of seismic properties allows igneous or metamorphic layering to produce seismic responses, for example in large mafic intrusions (Adams *et al.*, 1997). Note that the mafic and ultramafic rocks show greater scatter than the other types, with the data extending into the region beyond the lower limits of the velocities and densities of their major constituent minerals. Serpentinisation is most likely to be the cause for this (see Section 6.6.3.1).

Metamorphic rocks behave much like igneous rocks. Those of mafic and ultramafic composition have the highest seismic velocities and densities, whilst metasedimentary rocks, such as marble and quartzite, exhibit the lower velocities of their constituent quartz/dolomite/calcite crystals.

Figure 6.29b shows velocity versus density for sedimentary rocks and unconsolidated sediments. The distinction between lithotypes is less obvious than for igneous rocks, although in general carbonates have higher velocities than siliceous rocks, followed by mudrocks and then the unconsolidated materials.

6.6.1.2 Effects of porosity and pore contents

In Fig. 6.29c the sediments and sedimentary rocks in the database are grouped according to their fractional porosity (ϕ). Comparison with Fig. 6.29b clearly shows that for rocks with significant porosity the main control on seismic properties is their porosity and not their mineralogy.

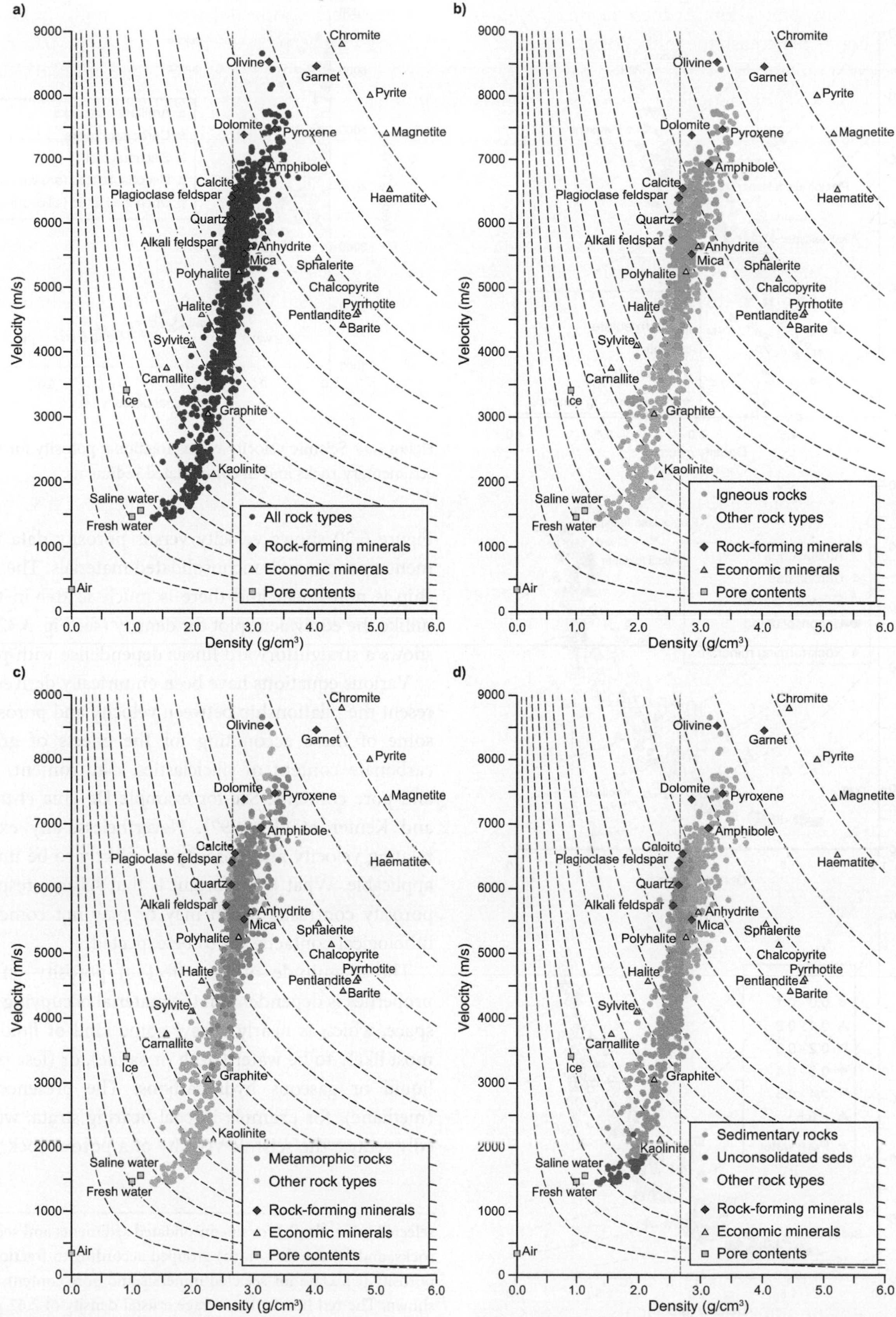

Figure 6.28 Compilation of published seismic velocity and density measurements of rocks for pressure–temperature conditions occurring in the upper 1 km of the Earth's crust. Data for selected rock-forming minerals, economic minerals and pore contents are also included. The red line is the average crustal density of 2.67 g/cm^3. The broken lines are contours of acoustic impedance with their separation representing the contrast required to produce a reflection coefficient of 0.05.

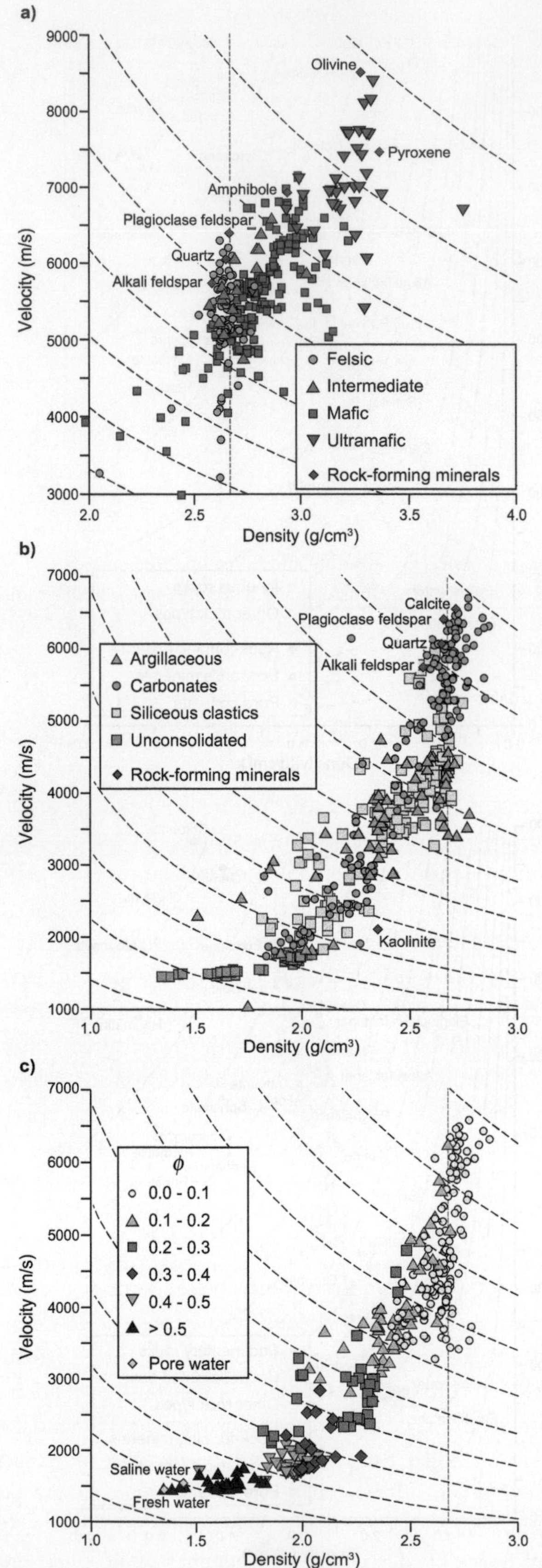

Figure 6.29 Subsets showing detail of the seismic velocity versus density data in Fig. 6.28. (a) For various types of igneous rocks and selected minerals, (b) for unconsolidated sediments and sedimentary rocks, and (c) the data in (b) grouped according to fractional porosity (ϕ). Data for selected minerals and pore contents are also shown. The red line is the average crustal density of 2.67 g/cm^3. The broken lines are contours of acoustic impedance with their separation representing the contrast required to produce a reflection coefficient of 0.05.

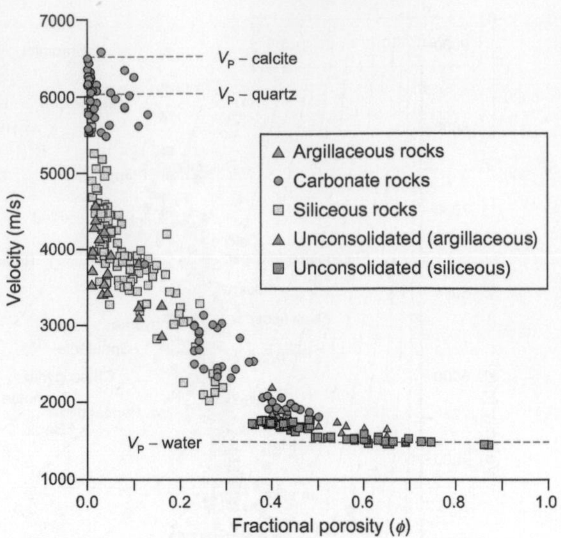

Figure 6.30 Seismic velocity versus fractional porosity for various sedimentary rocks and unconsolidated sediments.

Figure 6.30 shows velocity versus porosity data for sedimentary rocks and unconsolidated materials. The relationship is non-linear and there is much scatter in the data, unlike the equivalent plot for density (see Fig. 3.32), which shows a straightforward linear dependence with porosity.

Various equations have been empirically derived to represent the relationship between velocity and porosity, with some of these accounting for the effects of grain size, carbonate content of siliciclastics, clay content, pressure and pore contents. See for example De Hua *et al.* (1986) and Kenter *et al.* (1997). Nevertheless, any expression relating velocity and porosity is unlikely to be universally applicable. What is important is that seismic responses at porosity contrasts, which may or may not coincide with lithological contacts, are to be expected.

The magnitude of the effect of porosity on seismic properties is dependent on the material occupying the pore space, which is nearly always some kind of fluid. This is most likely to be water (fresh or saline), or (less probably) liquid or gaseous hydrocarbons. The presence of gas (methane), for example in coal-bearing strata, will markedly reduce the seismic velocity of a porous rock, creating

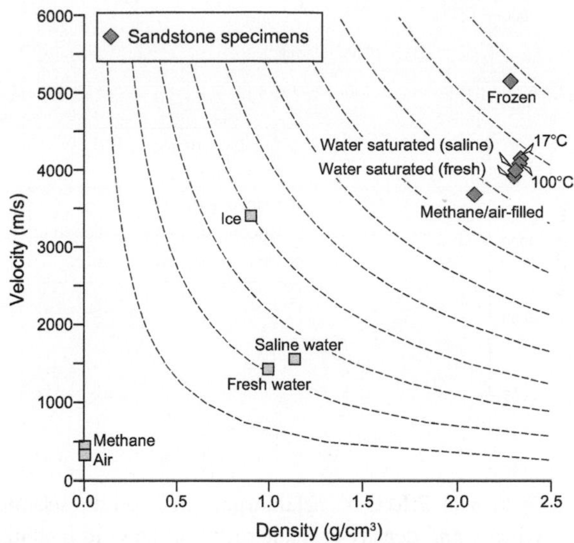

Figure 6.31 Seismic velocity versus density for various pore contents and a clean sandstone containing the different pore contents. Fractional porosity of the samples is about 0.2. Data from Timur (1968, 1977) and Wyllie *et al.* (1962). The broken lines are contours of acoustic impedance with their separation representing the contrast required to produce a reflection coefficient of 0.05.

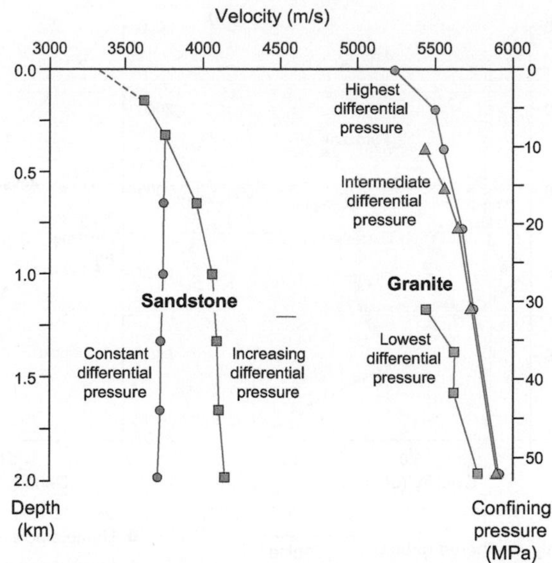

Figure 6.32 Effects of absolute and differential pressure on the seismic velocity of a sandstone and a granite. Based on diagrams in King (1966) and Todd and Simmons (1972).

a significant acoustic impedance contrast with adjacent non-gas-bearing units. Above the water table the pore fluid is most likely to be air, although in cold climates ice, associated with permafrost, is an alternate non-fluid possibility. Velocity versus density data for common pore contents, and for a porous sandstone (the Berea Sandstone from Ohio, USA) in the presence of the various pore contents, are shown in Fig. 6.31. Note the very low velocities and densities of air and methane. Clearly, there are significant contrasts in seismic properties between water-filled (saturated) and gas-filled (dry) rock, but very little difference between specimens containing saline or fresh water. This demonstrates that a significant subsurface seismic property boundary may be associated with the water table.

6.6.2 Effects of temperature and pressure

Figure 6.31 shows that the effects of temperature on a saturated rock are small until the pore fluid freezes, and then velocity increases significantly. This has important implications for seismic surveys conducted in permafrost areas, as changes in velocity of geological origin may be exceeded by those related to the nature of the pore contents.

Figure 6.32 shows how the seismic velocities of a granite and a porous sandstone change with increasing confining pressure. Equivalent depth, assuming lithostatic loading, is also shown. Both rock types exhibit significant increase in velocity with depth, at least to the depths of interest here, although the rate of increase tends to decline significantly at depths of a few kilometres. The rapid change at shallow depth is due to the closure of crack-related porosity. Equivalent variations in density are shown in Fig. 3.33. The primary cause for the increases in velocity and density is the reduction in porosity as confining pressure increases. For crystalline rocks, their generally low porosity tends to be in the form of cracks, which easily collapse under pressure. In granular sedimentary rocks, porosity is not only much greater but also tends to be more stable. All rock types show this general behaviour so their seismic property contrasts tend to be preserved as the rocks are buried.

In rocks with significant porosity, the dominant control on velocity is the *differential pressure*, i.e. the difference between the confining pressure and the pore pressure. When the differential pressure is constant, velocity also tends to be fairly constant as is illustrated for the sandstone in Fig. 6.32. In this case, a minor decrease in velocity occurs. When differential pressure decreases, i.e. the pore pressure increases relative to the confining pressure, velocity normally decreases. Consequently, fluid-filled areas with anomalous pressure, such as over-pressured zones, can give rise to seismic responses.

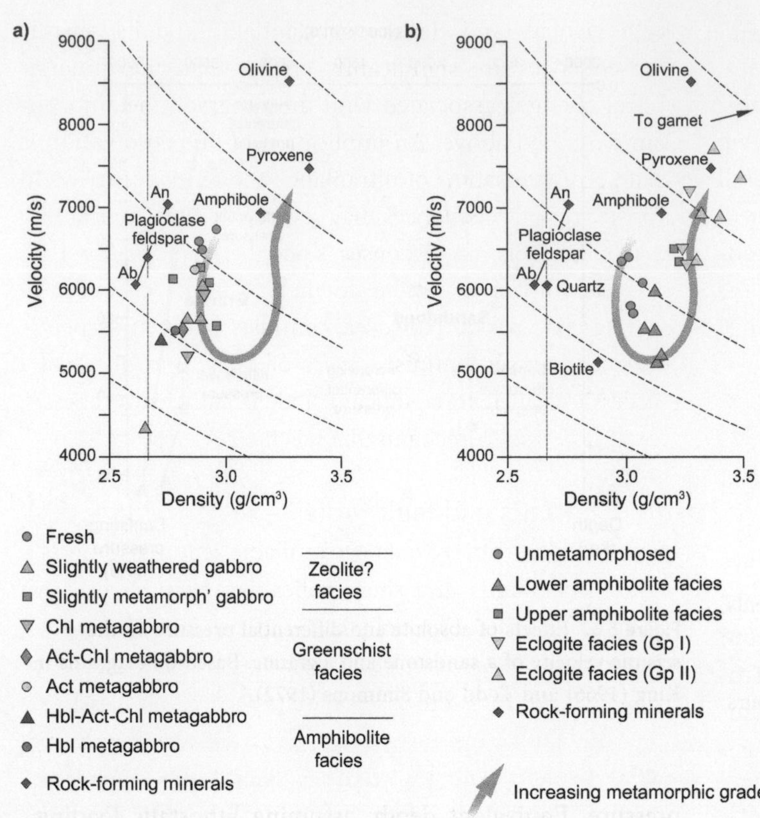

Figure 6.33 Effects of metamorphic grade on the seismic velocity and density of mafic rocks. (a) Low- to medium-grade metamorphism. Based on data in Fox *et al.* (1973). (b) Medium- to high-grade metamorphism. Based on data and diagrams in Hurich *et al.* (2001). Ab – albite, Act – actinolite, An – anorthite, Chl – chlorite, Hbl – hornblende. The red dashed line is the average crustal density of 2.67 g/cm^3. The broken lines are contours of acoustic impedance with their separation representing the contrast required to produce a reflection coefficient of 0.05.

6.6.3 Effects of metamorphism, alteration and deformation

The relationship between metamorphic grade and seismic properties is not well documented and is illustrated here using a mafic protolith. The low porosities of crystalline rocks cause mineralogy to be the dominant control. The effects of initial alteration and low-grade metamorphism are demonstrated with samples of gabbroic rocks from the Atlantic Ocean (Fox *et al.*, 1973), and shown in Fig. 6.33a. Metamorphic grade ranges from zero to amphibolite facies. Development of metamorphic minerals such as albite, actinolite and chlorite at the expense of the original assemblage (plagioclase, pyroxene, olivine) leads to a markedly reduced velocity and, to a lesser extent, density.

The effects of higher grades of metamorphism on mafic rocks are illustrated in Fig. 6.33b which is based on petrophysical measurements made on compositionally similar gabbros and metagabbros from the Grenville Province in eastern Québec, Canada (Hurich *et al.*, 2001). Metamorphic grade ranges from sub-greenschist to eclogite facies, but granulite facies data are not available. As with the oceanic samples, the igneous assemblage consists of plagioclase, olivine and pyroxene. The lower amphibolite facies assemblage comprises plagioclase (more sodic), amphibole, biotite, garnet and pyroxene. There is an associated decrease in velocity and a slight increase in density, the latter in contrast with the data from the oceanic samples. Velocity decreases owing to the change in plagioclase composition (note the lower density and velocity of albite) and the addition of biotite. The transformation to upper amphibolite facies produces a mineral assemblage comprising plagioclase, garnet, pyroxene, plus or minus amphibole. The disappearance of biotite and the increasing volume of garnet cause both velocity and density to increase.

Two groups of rocks are recognised at eclogite facies (Fig. 6.33b) and are thought to be the result of different protolith compositions. In Group I, the change to eclogite facies creates a mineral assemblage comprising plagioclase, garnet, amphibole and pyroxene, with the latter three minerals occurring as a greater proportion of the rock. The large range in physical properties reflects the variable plagioclase content. Two samples have similar properties to amphibolite facies rocks, but one shows a substantial increase in velocity, density and acoustic impedance. Group II is more mafic. It does not contain plagioclase but shows increases in the amounts of garnet, pyroxene and amphibole, which produce significant increases in velocity, density and acoustic impedance compared with the amphibolite facies samples.

These examples demonstrate that where metamorphic reactions replace existing phases with minerals having different properties, there will be a change in the overall seismic properties of the rock. However, the changes in properties are generally quite small and likely to be gradual. A seismic response coinciding with a metamorphic isograd or alteration front may be possible if the change in mineralogy is sufficiently abrupt.

6.6.3.1 Serpentinisation

A metamorphic process that produces substantial changes in seismic properties is serpentinisation. This is the hydrous alteration of olivine and pyroxene to produce the serpentine minerals antigorite, chrysotile and lizardite, and is common in ophiolite complexes, layered igneous complexes and greenstone belts. Most serpentinite is produced by alteration of ultramafic igneous rocks and, commonly, both the relict and alteration mineral assemblages are present. Figure 6.34 shows a compilation of published velocity versus density data from various serpentinised rocks, where the original authors have estimated the degree of serpentinisation. It is clear that serpentinisation reduces

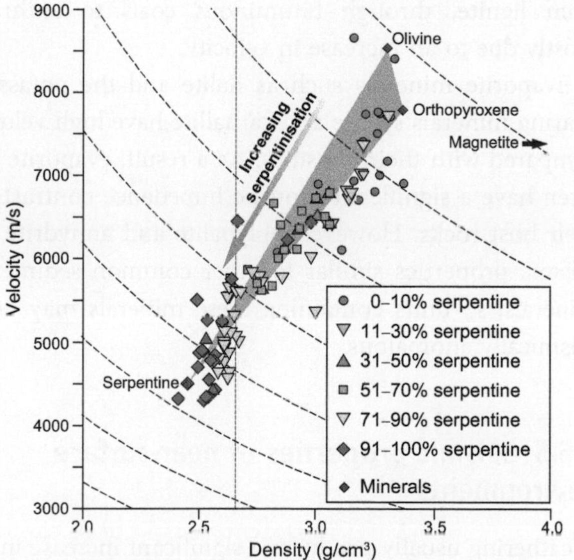

Figure 6.34 Effects of serpentinisation on seismic velocity and density. Also shown are the fields for the main minerals involved in serpentinisation reactions. The shaded area represents the field for aggregates of olivine, orthopyroxene and serpentine. The red line is the average crustal density of 2.67 g/cm^3. Data from numerous published sources. The broken lines are contours of acoustic impedance with their separation representing the contrast required to produce a reflection coefficient of 0.05.

both velocity and density substantially, and its seismic consequences are significantly greater than the mineralogical changes associated with the prograde metamorphism described above. An implication of this observation is that serpentinisation of ultramafic lithologies occurring in non-ultramafic sequences may severely reduce their ability to generate seismic responses, since the seismic properties of serpentinites is similar to that of felsic igneous and sedimentary rocks. However, contrasts between serpentinised and unserpentinised areas of mafic and ultramafic successions will be significant and, if sufficiently abrupt, could give rise to recognisable seismic responses.

6.6.3.2 Faults and fault rocks

Few seismic property data are available, but it is generally agreed that faults and shear zones can be the source of seismic responses, especially in highly deformed hard-rock terrains. Responses can be caused by the juxtaposition of rock types with different seismic properties, but it is likely that the fault structures themselves can be the source owing to their porosity, which is likely to be water-filled. Ductile faulting may form mylonite zones that may be reflective in their own right, owing to their layered structure combined with lithologic diversity and/or greater fabric development (see Section 6.6.6). Arguments have been presented for enhancement of their reflectivity by reduction in velocity related to retrograde mineral assemblages, and by increased velocity related to loss of silica.

Alteration, caused by hydrothermal fluid-flow through the fault zone, is another possible cause of acoustic-impedance contrast. It has been postulated, but not yet categorically demonstrated, that enhanced reflectivity may be characteristic of structures that were the main conduits for hydrothermal mineralising fluids in a mineralised terrain.

6.6.4 Seismic properties of mineralisation

Both metallic mineralisation and non-metallic materials of economic interest, such as coal and evaporites, plot outside the velocity–density field of the common rock types (Fig. 6.35). Sulphide and oxide minerals may have greater or lesser seismic velocities than the rock-forming minerals, but all are significantly denser (Fig. 6.35a). Salisbury *et al.* (1996) show that velocities and densities of massive sulphide ores can be estimated on the basis of mixing lines joining the physical properties of their main constituents (ore minerals and gangue). Figure 6.35 supports this, with

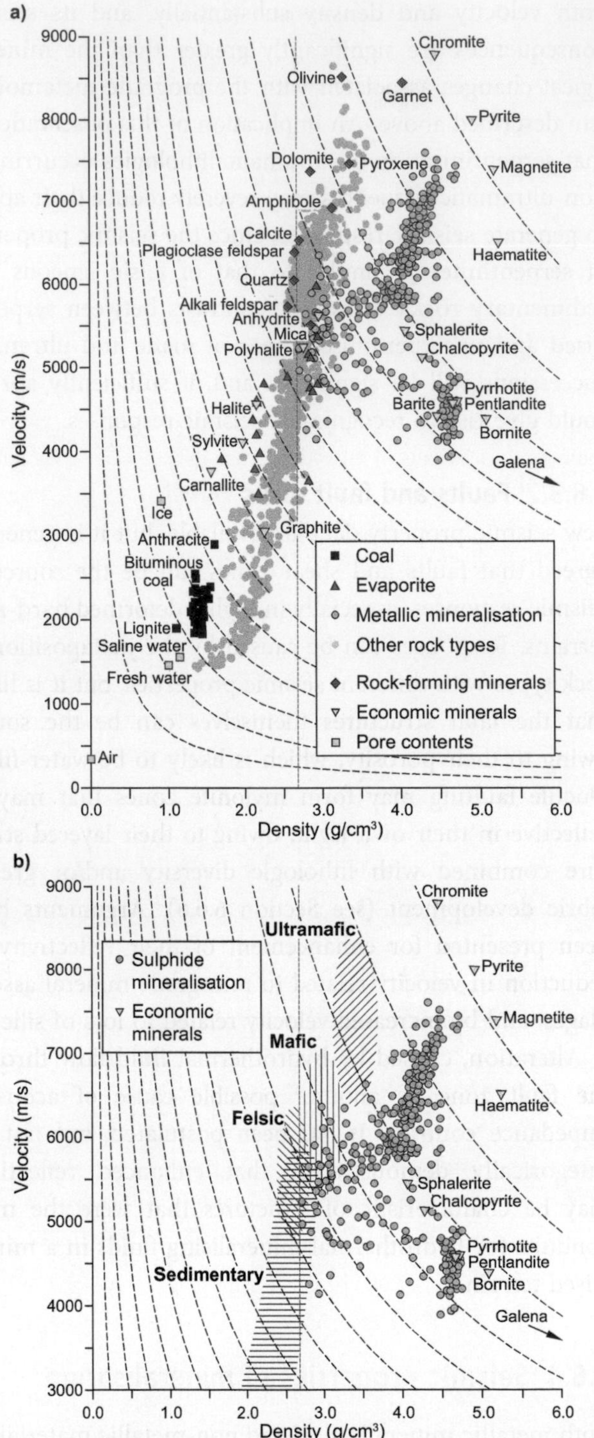

Figure 6.35 Seismic velocity versus density for a variety of economic minerals and types of mineralisation. (a) Integrated with the data from Fig. 6.28 to show the variation in the parameters for mineralisation compared with those of common rocks, and (b) the data for sulphide mineralisation integrated with data for various rock types. Sulphide data from Adam *et al.* (1997), Duff *et al.* (2012), Fullagar *et al.* (1996), Greenhalgh and Mason (1997) and Heinonen *et al.* (2012) and unpublished data supplied to the authors. Coal data from Greenhalgh and Emerson (1986). Evaporite data from Wheildon *et al.* (1974) and Kern (1982).

data for sulphide mineralisation occurring in the area between the common ore and gangue minerals. In most cases, massive sulphide mineralisation has higher velocities and higher densities than its host rocks; but velocities may be lower if sphalerite, chalcopyrite or pyrrhotite constitute a significant proportion of the mineralisation. The contours of acoustic impedance shown in Fig. 6.35b indicate that in some cases there may be no significant difference in acoustic impedance between mineralisation and host rocks. For example, most types of sulphide mineralisation can be detected within a felsic host, but not necessarily all of those hosted in mafic and ultramafic rocks, although serpentinisation would greatly enhance the physical-property contrast between mineralisation and host rocks (see Section 6.6.3.1). Importantly, the relative pyrite content of a sulphide-bearing mineral deposit is a key factor controlling the physical-property contrast between the deposit and its host rocks, owing to the high velocity of pyrite. For the same reason, layers rich in chromite within large mafic intrusions have been targeted using seismic surveys, e.g. Gibson (2011).

Coal has very low velocity and low density compared with the sedimentary rocks within which it occurs, producing a large contrast in acoustic impedance with the surrounding succession. There is a continuum of variation from lignite, through bituminous coals to anthracite, mostly due to an increase in velocity.

Evaporite minerals such as halite and the potassium-bearing minerals sylvite and carnallite have high velocities compared with their densities. As a result, evaporite units often have a significant acoustic impedance contrast with their host rocks. However, polyhalite and anhydrite have seismic properties similar to other common sedimentary minerals, so units containing these minerals may not be seismically anomalous.

6.6.5 Seismic properties of near-surface environments

Weathering usually produces a significant increase in porosity, so seismic velocity and density are progressively reduced in more weathered rocks. This is illustrated in Fig. 6.36 with data from Ishikawa *et al.* (1981). Here

The red line is the average crustal density of 2.67 g/cm³. The broken lines are contours of acoustic impedance with their separation representing the contrast required to produce a reflection coefficient of 0.05.

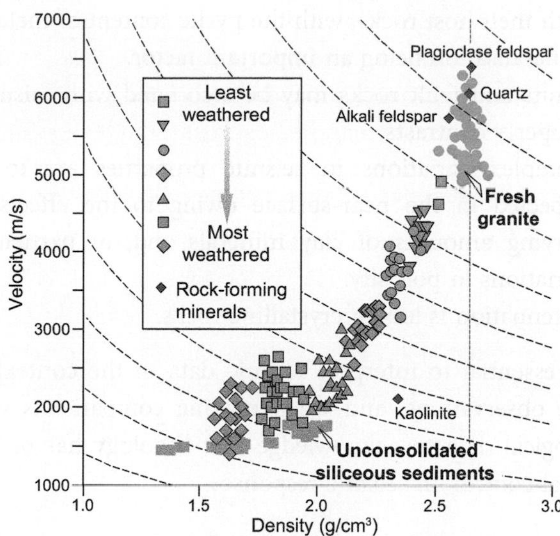

Figure 6.36 Seismic velocity versus density from progressively more weathered granite. Weathered materials data from Ishikawa *et al.* (1981). The red line is the average crustal density of 2.67 g/cm^3. The broken lines are contours of acoustic impedance with their separation representing the contrast required to produce a reflection coefficient of 0.05.

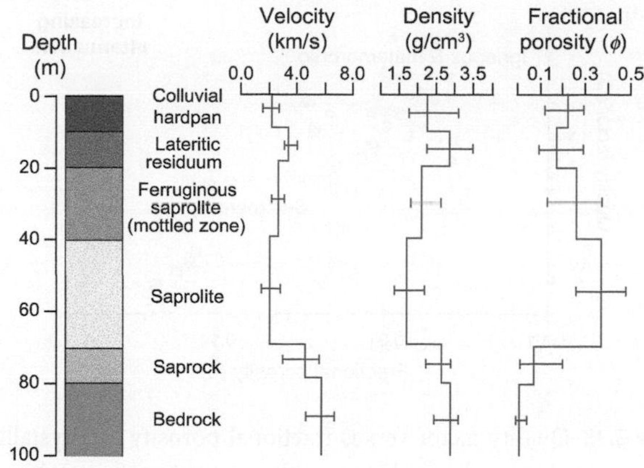

Figure 6.37 Variations in seismic velocity, density and fractional porosity through the regolith in a greenstone terrain in Western Australia. The horizontal bars represent the range in values and the lines are the mean values. Based on diagrams in Emerson *et al.* (2000).

weathered granite is assigned to one of eight categories representing the degree of weathering, based upon several factors including hardness, cohesion and condition of constituent minerals. The data form a continuum from fresh granite through to siliceous unconsolidated materials. The degree of weathering correlates very well with position within the continuum, defining a progression from values close to those of the relevant rock-forming minerals towards those of common pore contents.

The high porosity of weathered materials means that the degree of saturation is an important control on seismic properties. The formation of low-velocity clay minerals may also be important.

The laterites and thick clay-rich regolith that form under tropical weathering conditions are another weathering-related phenomenon of importance. Emerson *et al.* (2000) provide a detailed description of the physical characteristics of the regolith in the Archaean greenstone terrain near Lawlers, in the Yilgarn Craton of Western Australia. Figure 6.37 shows the variation in velocity, density and porosity through the regolith and into underlying bedrock. A surprising result is the high porosities, even in clay-rich saprolite materials. As expected, velocity and density vary approximately in sympathy, both being inversely correlated with porosity. Again, the high porosity suggests that the degree of water saturation will be a key factor, and Emerson *et al.* (2000) estimate that when there is a substantial reduction in water saturation, the velocity of saprolitic layers may decrease by as much as 40 to 50%. The depth profiles demonstrate that the seismic properties of the regolith are complex, with both increases and decreases in properties to be expected. This will be further complicated by lateral changes in the bedrock, variations in the thickness of the various intra-regolith layers, perched water tables and multiple weathering fronts. Importantly, there is likely to be a significant increase in seismic velocity at, or near, the base of the regolith.

6.6.6 Anisotropy

Most rock formations are seismically anisotropic, exhibiting the lowest seismic velocity in the direction perpendicular to any planar fabric and the maximum velocity usually parallel to any linear fabric. Velocity anisotropy may arise from small-scale interbedding of different lithotypes, mineralogical layering or preferred orientation/alignment of mineral grains with intrinsic single-crystal seismic anisotropy. There is evidence that variations in the intensity and orientation of planar fabrics can create sufficient changes in seismic properties to affect the seismic response.

6.6.7 Absorption

Figure 6.38 shows absorption as a function of fractional porosity (ϕ) for crystalline and sedimentary rocks. The data are for a range of frequencies, and the scatter is partly

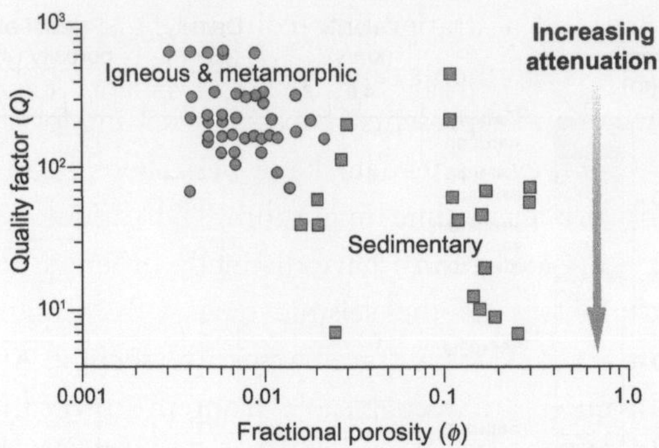

Figure 6.38 Quality factor versus fractional porosity for crystalline and sedimentary rocks. Redrawn, with permission, from Johnson *et al.* (1979).

due to different methods of measurement and state of the samples, e.g. saturated versus unsaturated. As is to be expected, given the role of friction (see Section 6.3.3.2), absorption is greatest in materials where adjacent surfaces can mechanically move in relation to each other. Consequently, absorption is greatest in porous, especially fractured, and poorly cemented rocks and near-surface materials where confining pressure has not closed voids. Cracks eventually close as pressure increases with depth (see Section 6.6.2), so they are then unable to contribute to frictional loss, leading to a reduction in attenuation. The data also show that absorption is much less in crystalline rocks. This is due to their lower porosity and also probably due to the greater rigidity of their interlocked mineral structure.

6.6.8 Summary of geological controls on seismic properties

The important conclusions to be drawn from the above descriptions of the geological controls on seismic velocity and density are:

- When porosity is small, seismic properties are mainly controlled by mineralogy/lithology.
- When porosity exceeds about 10% it becomes the dominant control on seismic properties.
- In porous rocks, the type and state of the pore contents are important controls on seismic properties. Important factors include differential pressure and whether the pore contents are liquid, gaseous or solid.
- Serpentinisation significantly reduces the velocity and density of mafic and ultramafic rocks.
- Most kinds of metallic and sedimentary mineralisation (coal, evaporites) have seismic properties that contrast with their host rocks, with the pyrite content of metallic mineralisation being an important factor.
- Faults and fault rocks may be associated with seismic-property contrasts.
- Complex variations in seismic properties are to be expected in the near-surface owing to the effects of varying amounts of clay minerals and, in particular, variations in porosity.
- Attenuation is less in crystalline rocks.

It is essential to interpret seismic data in the context of these observations, and when making comparisons with geological data to acknowledge that lithology has only a partial control on seismic responses.

6.6.9 Measuring seismic properties

The seismic velocity of a rock sample can be accurately measured in the laboratory. However, the results may not be representative of the rock *in situ* because factors such as pore fluid and its pressure strongly influence velocity; and these factors are hard to reproduce accurately in the laboratory. Also, measurements on small samples require the use of much higher-frequency waves than those used for seismic surveying (see Fig. 6.7), so features in the samples that are too small to affect actual seismic survey data can significantly influence the laboratory measurements.

Downhole measurement of velocity *in situ* is known as *sonic logging* and produces a continuous record of velocity as a function of depth. Sonic logs provide more reliable information than do laboratory measurements on samples, although the frequencies used are still significantly higher than used for seismic surveying (Fig. 6.7). There is also the possibility that the physical properties of the drillhole wall rocks may have been altered by the drilling, so the rock volume influencing the measurements may no longer be entirely representative of the rock formation.

The simpler form of sonic logging involves measuring the travel time of P-waves from one or more sources to pairs of detectors, all mounted on the logging tool (Fig. 6.39). The distance between the detectors is known as the *span* and represents the limit of vertical resolution of the data. The sources 'fire' alternately as the tool is raised up the drillhole and the differences in arrival times of the waves at the detectors are determined. Data may be displayed as either a velocity or slowness (1/velocity) with the result assigned to the depth of the centre of the span. Velocities are displayed as metres per second or kilometres

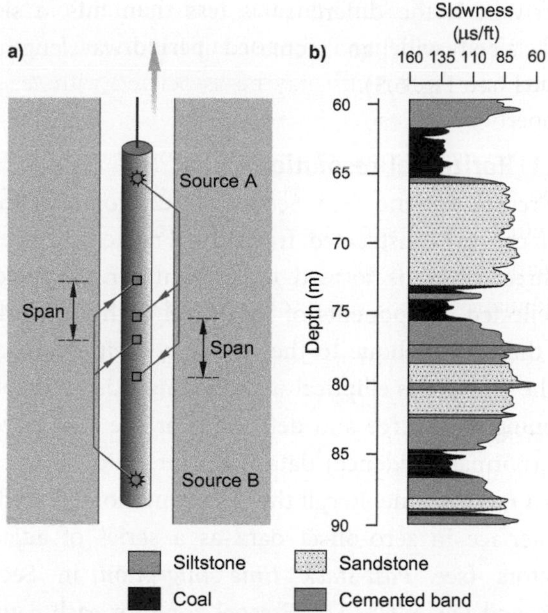

Figure 6.39 Downhole sonic logging. (a) Schematic illustration of a logging tool. (b) Sonic log through a coal-bearing sedimentary sequence. The horizontal scale is slowness (1/velocity) measured in imperial units. (b) Based on a diagram in Davies and McManus (1990).

per second, but slowness is often displayed in microseconds per foot (μs/ft), a practice resulting from the continued use of imperial units by the petroleum industry. The use of multiple sources and detectors reduces effects related to changing drillhole conditions.

More sophisticated is *full waveform* sonic logging which, in addition to measuring arrival times, records what is equivalent to a seismic trace. This allows different types of waves to be recognised and additional seismic properties to be determined, e.g. attenuation. Good descriptions of both types of sonic logging are provided by Rider (1996), and in a hard-rock context, by Schmitt *et al.* (2003).

6.7 Interpretation of seismic reflection data

The interpretation of seismic reflection data is mostly a qualitative process based on the recognition and mapping of reflective horizons. Single 2D dip profiles may be recorded across features of interest, but to determine the structure of a region properly, a network of intersecting survey lines is required. Features that can be reliably recognised are correlated across the seismic sections to build up a 3D map of their form. These may be stratigraphic horizons, stratigraphic intervals or structures such as faults.

Figures 6.40 and 6.41 show seismic data from the Witwatersrand Basin in South Africa. Figure 6.40a shows some data and a stratigraphic column. This well-bedded sequence of sediments and volcanics is comparatively undeformed and represents an excellent setting for seismic reflection surveys. The data have been depth converted following pre-stack time migration. The different stratigraphic units are seen to have distinctly different appearance (character) on the seismic data with stratigraphic reflectors, e.g. (1), easily traced across the section. Angular unconformities are recognisable from the truncation of underlying dipping reflectors (2). In Fig. 6.40b faults are recognised from offsets in events of stratigraphic origin.

The most common form of display shows trace amplitude, but it is also possible to calculate enhanced forms of the data (see Section 2.7.4). In the case of seismic data, the results are referred to as *attributes*, e.g. Manzi *et al.* (2012a). These may comprise a representation of the frequency or phase information in the trace (see online Appendix 2). Alternatively, for 3D data, horizons can be analysed in terms of characteristics such as dip and dip direction.

Although a migrated zero-offset seismic reflection section resembles a geological cross-section, it may be very difficult to relate observed responses to the known geology, notably depth data from drillhole intersections and, if available, the mine geology. Every opportunity should be taken to establish a correlation between the known geology and the seismic response by calibrating, or 'tying', the seismic data to drillhole data, preferably using synthetic seismograms (see Section 6.7.2.1).

6.7.1 Resolution

Seismic reflection sections and volumes constitute the most detailed representations of the subsurface that can be created using geophysical methods. Moreover, the display of the data resembles a picture of the subsurface. When interpreting these data it is particularly important to have an appreciation of the resolution of the data so as not to over-interpret them. As will be demonstrated below, the size of resolvable features in the subsurface is orders of magnitude larger than those that might be recognised in exposures or drillcore.

Seismic resolution is normally quantified in terms of lateral and vertical resolution. In both cases the dominant wavelength of the data is the controlling parameter (see Section 6.3.2). Recall that for two distinct arrivals to be recognisable within a seismic trace, their travel times should differ by more than half the dominant period of

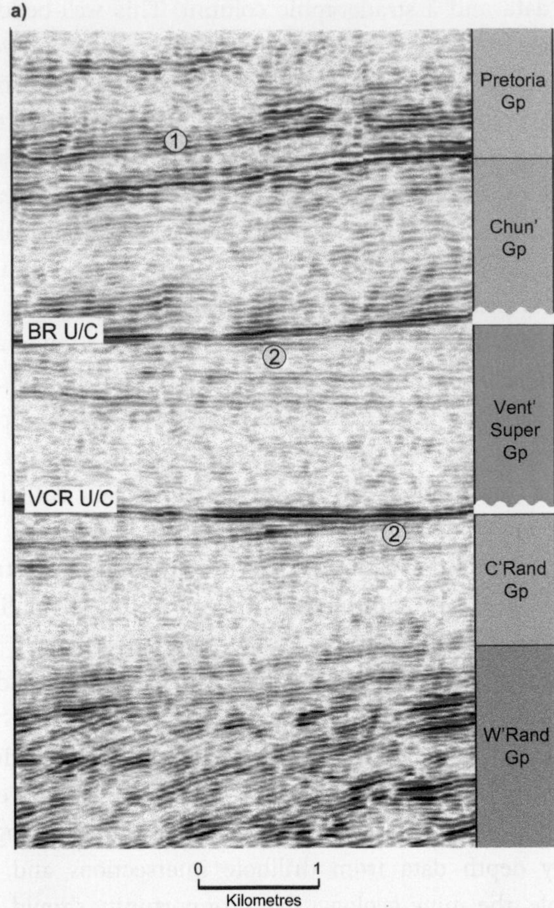

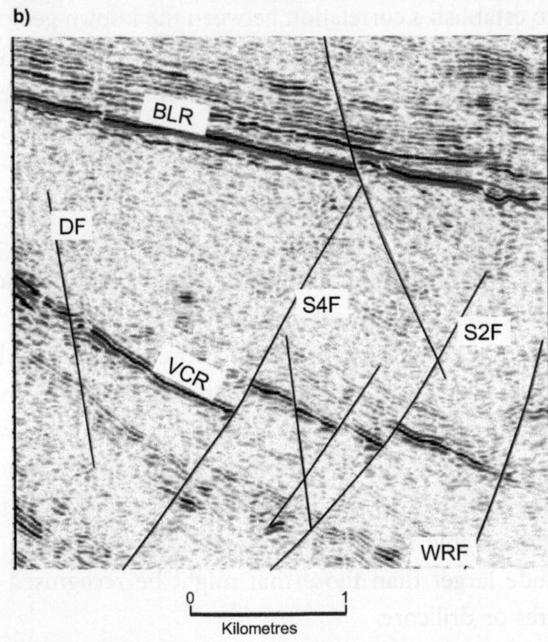

Figure 6.40 Seismic reflection sections data from the Witwatersrand Basin. (a) Correlation between seismic data and stratigraphy, and (b) examples of faults. The Pretoria Group comprises mudstone-sandstone, andesitic lavas, conglomerates, diamictite and carbonate units. The Chunleepoort Group (Chun' Gp) comprises dolomite, banded iron formation and lacustrine deposits. The Ventersdorp Supergroup (Vent' S' Group) comprises bimodal volcanics including tuffs and lavas. The Central Rand Group (C' Rand Gp) comprises quartz arenite and quartz wacke, conglomerate and shale. The West Rand Group (W' Rand Gp) comprises predominantly shale and arenite. BLR – Black Reef Formation, the basal unit in the Chunleepoort Group, DF – Danies Fault, S2F – Shaft 2 Fault, S4F – Shaft 4 Fault, VCR – Ventersdorp Contact Reef, WRF – West Rand Fault. Redrawn, with permission, from Manzi *et al.* (2012a).

the arrivals. If the difference is less than this, a single arrival, albeit with an increased period/wavelength, is observed (see Fig. 6.5).

6.7.1.1 Horizontal resolution

The Fresnel volume (see Section 6.3.2) for a reflected arrival can be constructed from the Fresnel volumes of two direct raypaths located to account for the incident and reflected components of the travel path (Fig. 6.42a). Since these are oblique to the reflector, their intersection with the reflector is elliptical with its long axis in the plane containing the source and detector. For the case of zero-offset (normal incidence) data the intersection is circular and is a Fresnel zone. Recall the representation of a reflecting interface in zero-offset data as a series of adjacent diffractors (see *Post-stack time migration* in Section 6.5.2.5, and Fig. 6.23). The Fresnel zone for each source-detector location comprises all the 'diffractors' that return seismic waves whose travel times differ from the normal-incidence travel time by less than half the dominant period of the seismic signal.

A clear description of calculating Fresnel-zone dimensions of reflectors is given by Lindsey (1989). Based on the geometry in Fig. 6.42b, the normal-incidence raypath represents the shortest path with which the longer lengths of the oblique paths are compared. The diameter of the normal-incidence Fresnel zone (D_{Fresnel}) and its area (A_{Fresnel}) at a specific wavelength (λ) for a planar reflector at depth (Z) are given by:

$$D_{\text{Fresnel}} = 2\sqrt{(Z+\lambda/4)^2 - Z^2} \tag{6.22}$$

$$A_{\text{Fresnel}} = \pi[(Z+\lambda/4)^2 - Z^2] \tag{6.23}$$

D_{Fresnel} is simply twice the length of one side of a right-angled triangle formed by the normal-incidence and oblique raypaths. Note that the term $\lambda/4$ is used because the seismic waves travels the path twice, i.e. down from

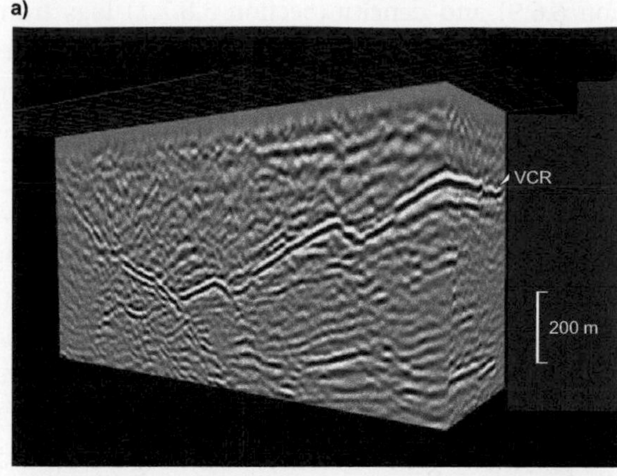

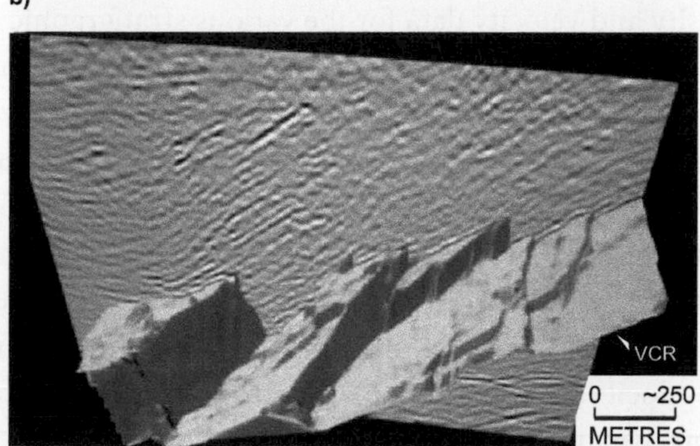

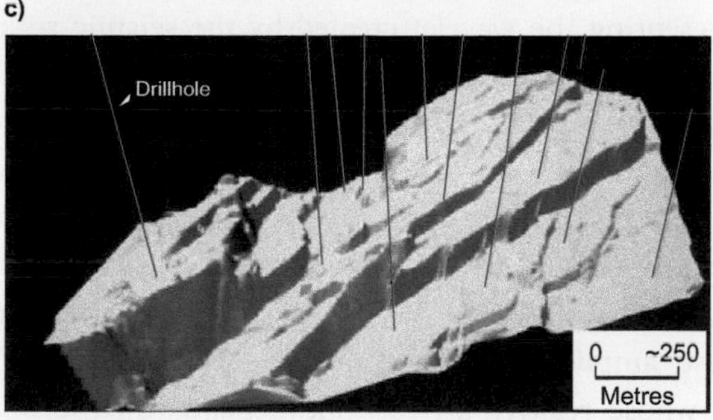

Figure 6.41 Three-dimensional seismic datasets from the Klerksdorp Goldfield. (a) 3D data volume, (b) a vertical section and the horizon map of the Ventersdorp Contact Reef (VCR), and (c) the VCR horizon map. (a) and (c), Redrawn, with permission, from Pretorius (2004) and (b) redrawn, with permission, from Pretorius *et al.* (2003).

source then up after reflection, so the overall path difference is $\lambda/2$. Calculating Fresnel-zone diameters and areas for various combinations of frequency, velocity and depth relevant to mineral exploration shows that diameters are typically a few hundred metres and their areas around 5 to 10 ha. These can be thought of as the minimum area of a discontinuity if it is to produce a reflection in unmigrated data; otherwise it acts as a diffractor. Variations in the

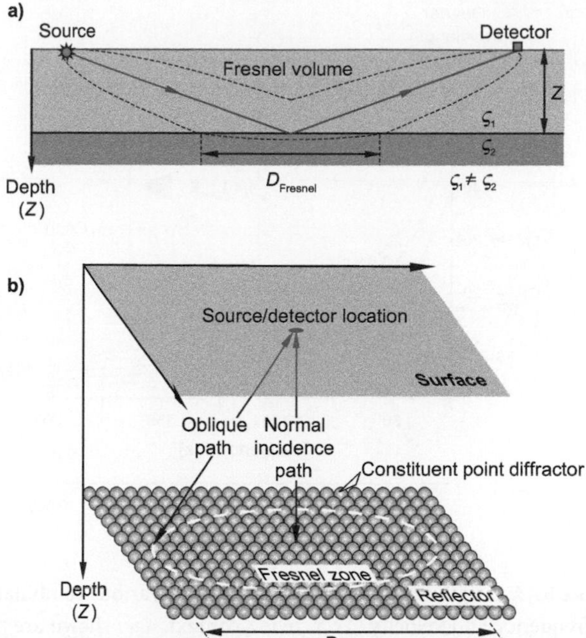

Figure 6.42 Schematic illustration of the reflection Fresnel volume. (a) Cross-section and (b) the Fresnel zone on the reflector surface. The reflector is represented as a series of juxtaposed point diffractors.

properties of a reflector which occur over an area smaller than the Fresnel zone are 'averaged', although in fact the contribution from the central part of the zone is greater than that from marginal areas.

Fortunately lateral resolution is significantly improved by migration which, in addition to repositioning reflections, reduces the Fresnel zone diameter. Ideally, this is to one-quarter of the dominant wavelength ($\lambda_{Dom}/4$), although half a wavelength is a more realistic estimate of what is achieved in practice. Figure 6.43 shows how large this is for parameters typical of mining-orientated seismic data.

6.7.1.2 Vertical resolution

Vertical resolution is defined in terms of the minimum vertical separation required between two reflectors for them to create two separate arrivals on the seismic trace (see Fig. 6.5). For the case of normal incidence, this requires the two reflectors to be separated by a vertical distance that is travelled in a time interval equal to a quarter of the signal's dominant period. This distance is travelled twice during a reflected path and so the overall difference in travel time is half the dominant period. In distance terms, the reflector separation is a quarter of the dominant wavelength ($\lambda_{Dom}/4$), i.e. the same as the ideal post-migration horizontal resolution. Figure 6.43 shows

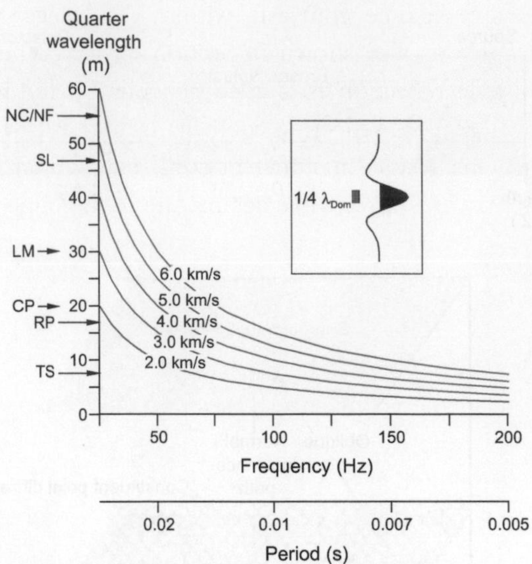

Figure 6.43 Length of a quarter wavelength for various combinations of frequency and velocity. To provide context, also shown are the lengths/heights of various well-known objects. CP – length of a cricket pitch, LM – height of the Lincoln Memorial (Washington DC), NC – height of Nelson's Column (London), NF – height of Niagara Falls, RP – height of the posts at the Twickenham rugby ground (London), SL – height of the Statue of Liberty (New York), TS – height of a two-storey house.

vertical resolution is a few tens of metres in most cases for typical survey parameters.

6.7.2 Quantitative interpretation

A variety of computer software for 2D and 3D forward and inverse modelling of seismic responses is available, mostly from the petroleum sector. Varying degrees of subsurface complexity can be accommodated, from constant-velocity planar layers through to velocity and density distributions of almost any degree of complexity described by values specified on a network of nodes.

The seismic waves may be approximated as rays and, usually, particular raypaths can be specified. Raypaths are fast to compute but a major shortcoming, especially in geologically complex terrains, is that diffraction cannot be easily accounted for in the models. Wave-based algorithms can compute more physically realistic models, but they are much more computationally intensive.

6.7.2.1 Synthetic seismograms

A particularly common and simple form of 1D forward modelling compares seismic observations in the vicinity of a drillhole with the response calculated from sonic (see Section 6.6.9) and density (Section 3.8.7.1) logs from the drillhole. The calculated seismic response is known as a *synthetic seismogram.* It is principally used to correlate reflections with geological interfaces or sequences intersected by the drillhole.

We showed in Section 6.3.4.2 that when seismic waves encounter an interface across which there is a change in acoustic impedance, both reflection and transmission of the waves occur. The relative amplitudes of the reflected and transmitted waves are defined in terms of the reflection coefficient (RC) and transmission coefficient (TC) (see *Energy partitioning* in Section 6.3.4.2). Figure 6.44 shows how a 1D synthetic seismic trace can be calculated. Firstly, density and velocity data for the various stratigraphic units are obtained from density and sonic logs. In the example shown, velocity and density are constant in each layer so changes in acoustic impedance only occur at their boundaries, so these are the only places where the reflection coefficients are non-zero (Fig. 6.44a). The variations in reflection coefficient with depth are then converted to variations as a function of two-way reflection time, using the velocity data (Fig. 6.44b). The plot of reflectivity coefficients versus two-way time is known as the *reflectivity series.* It is convolved (see Section 2.7.4.3), with a function representing the wavelet created by the seismic source to produce the synthetic seismic trace. Various idealised source wavelets are routinely used for calculating synthetic seismograms, in this case it is a type of zero-phase wavelet (see online Appendix 2) known as a *Ricker wavelet.* Figure 6.44b demonstrates how seismic traces consist of repetitions (reflections or echoes) of the source wavelet whose amplitude is scaled by the reflection coefficient and whose relative arrival time depends on the position of the non-zero reflection coefficients in the reflectivity series. When reflection coefficients are negative the polarity of the source-wavelet 'echo' within the trace is reversed.

In reality, there will be many closely spaced non-zero reflection coefficients within the reflectivity series. This means it will be hard to identify the individual occurrences of the source wavelet due to the interference of numerous echoes. This situation is mimicked in Fig. 6.44c by increasing the length of the source wavelet relative to the reflectivity function. The resulting trace becomes increasingly complicated, and is often difficult to relate to specific acoustic impedance contrasts. This is a significant observation since it demonstrates that a 'reflection' in a seismic dataset, such as shown in Fig. 6.40, is not from a single level or bed. Instead, it is a composite response of many

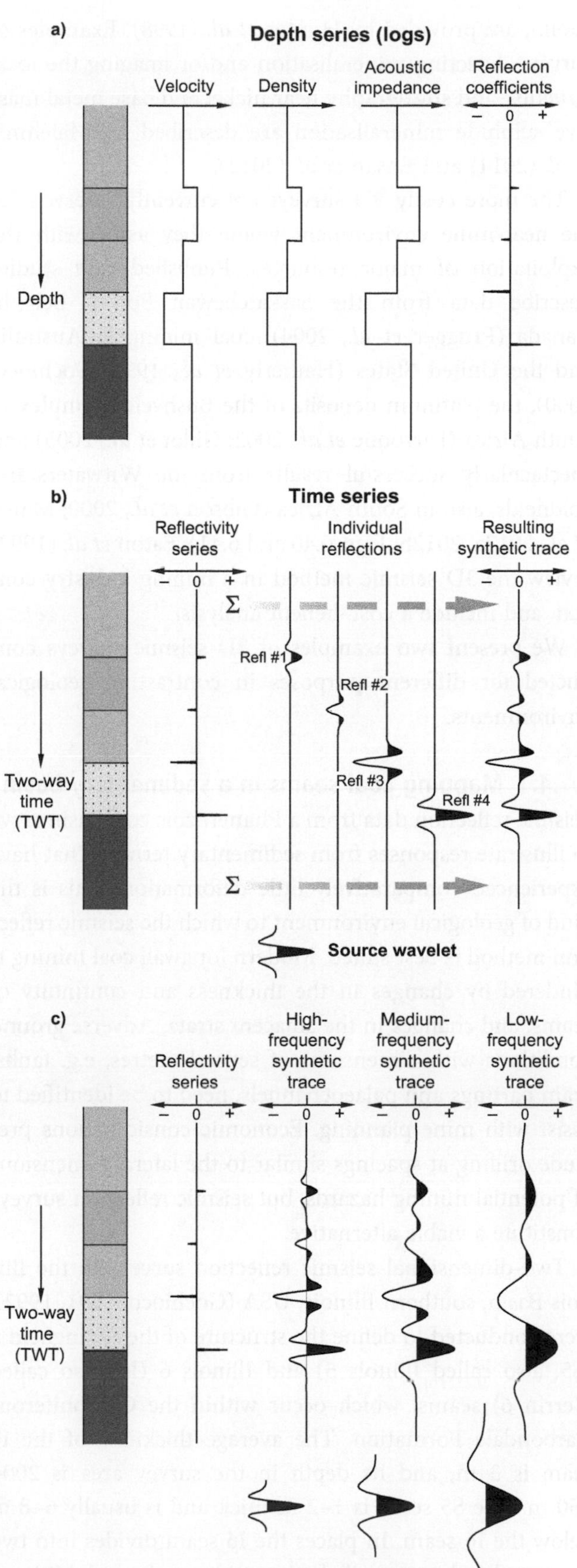

acoustic impedance contrasts within a 'package' of the stratigraphy, as also shown in Section 6.7.4.1. Whether a recognisable reflection is created depends on the magnitude of individual reflection coefficients, and also whether their spacing results in sufficient constructive interference that an arrival is discernible against the background noise.

6.7.3 Interpretation pitfalls

The close resemblance between a seismic section and a geological cross-section makes these data, in principle, one of the geologically easiest to interpret. Nevertheless, the interpreter should always have an appreciation for the resolution of the data and be aware of the possibility of artefacts in the data. Stacking and migration rely on simplifying assumptions about the distribution of seismic properties in the subsurface and, as such, may not work 'perfectly'.

6.7.3.1 Artefacts

Ideally, data processing techniques, notably deconvolution, will have removed multiples (see *Multiples* in Section 6.5.1.1) from the data. Regardless, the interpreter should attempt to identify periodic features in the data that may be due to multiples. At any given location, the travel time of a long-period multiple is an integer multiple of the travel time of its equivalent primary. So if a strong reflection is seen at, say, 0.5 s TWT at a given location, then the interpreter should be suspicious of reflections that arrive at 1.0 s, 1.5 s etc. In the presence of dip, the apparent dip of the multiple increases relative to the primary and may cause it to cross-cut primary reflections with similar TWT. This geologically unlikely situation (the response could theoretically be from a sill) is one way to identify multiples.

Since multiples have to travel further than primaries recorded at the same detector, they lose more energy through attenuation (see Section 6.3.3) and energy partitioning (see *Energy partitioning* in Section 6.3.4.2). This means they are of low amplitude unless associated with reflections from large contrasts in acoustic impedance. For marine surveys the main source of multiples is the water

Figure 6.44 Computation of a 1D synthetic seismic trace. (a) Velocity and density logs and, the acoustic-impedance and reflection-coefficient logs calculated from them. (b) Reflectivity series and the resultant synthetic seismic trace. The synthetic trace comprises four occurrences of the source wavelet. (c) Synthetic seismic traces computed using source wavelets of different dominant wavelength/frequency.

bottom. For land surveys, potential causes of large contrasts, and hence causes of multiples, are massive sulphide/oxide mineralisation, and coal and evaporite horizons (see Section 6.6.4).

The other common types of artefact in seismic data are related to migration. As described in *Migration artefacts* in Section 6.5.2.5, a 2D survey may record responses (sideswipe) from features that are not directly beneath the survey traverse. In geologically complex environments this is likely to be the norm rather than the exception. For example, a dyke or fault trending parallel to the traverse could reflect seismic waves and create a response resembling that from a layer below the traverse. The interpreter needs to be aware of off-traverse features that might cause responses in the data so as to not misinterpret these types of responses. It is common to see 'smiles' in the lower part of the section, owing to the decreasing signal-to-noise ratio. Loss of information means that features at the margin of the data should be treated with extreme caution. Cross-cutting events may be indicative of incorrect migration velocities.

6.7.3.2 Velocity-related distortions

Data presented with the vertical axis in TWT must not be treated as depth sections. Variations in the velocity of the subsurface will produce complex relationships between reflection time and depth. Even in the comparatively simple geological environment of an undeformed sedimentary basin, the complex interaction of lithological- and depth- (mainly porosity-) related changes in velocity will distort the appearance of the geology, as was demonstrated in Fig. 6.27. For data from hard-rock environments the absence of porosity means that depth-related effects are less important, but these are more than compensated for by the seismic consequences of complex lithological and alteration-related effects, for example due to serpentinisation (Fig. 6.34).

The interpreter needs to be aware of the likely velocities of the rock types in the survey area and should be careful when interpreting features that occur below lateral changes in the shallower geology.

6.7.4 Examples of seismic reflection data from mineralised terrains

The use of seismic reflection surveying in mineral exploration and mining is increasing. Examples of shallow surveys, including one associated with exploration for kaolin, are provided by Hunter *et al.* (1998). Examples of surveys detecting mineralisation and/or imaging the local structure and stratigraphy near nickel and base metal massive sulphide mineralisation are described by Malehmir *et al.* (2014) and Ehsan *et al.* (2012).

The more costly 3D surveys are currently reserved for the near-mine environment where they assist with the exploitation of major resources. Published case studies describe data from the Saskatchewan Potash Belt in Canada (Prugger *et al.*, 2004), coal mining in Australia and the United States (Hatherly *et al.*, 1998; Gochioco, 2000), the platinum deposits of the Bushveld Complex in South Africa (Larroque *et al.*, 2002; Gillot *et al.*, 2005) and spectacularly successful results from the Witwatersrand goldfields, also in South Africa (Gibson *et al.*, 2000; Manzi *et al.*, 2012a, 2012b; Figs. 6.40 and 6.41). Eaton *et al.* (1997) review the 3D seismic method in a mining industry context, and include a cost–benefit analysis.

We present two examples of 2D seismic surveys conducted for different purposes in contrasting geological environments.

6.7.4.1 Mapping coal seams in a sedimentary basin

Seismic reflection data from a Phanerozoic coal basin serve to illustrate responses from sedimentary terrains that have experienced comparatively little deformation. This is the kind of geological environment to which the seismic reflection method is best suited. Modern longwall coal mining is hindered by changes in the thickness and continuity of seams, and changes in the adjacent strata. Adverse ground conditions with dimensions of several metres, e.g. faults, seam partings and palaeochannels, need to be identified to assist with mine planning. Economic considerations preclude drilling at spacings similar to the lateral dimensions of potential mining hazards, but seismic reflection surveys constitute a viable alternative.

Two-dimensional seismic reflection surveys in the Illinois Basin, southern Illinois, USA (Gochioco, 1991, 1992), were conducted to define the structure of the Springfield 5 (S5, also called Illinois 5) and Illinois 6 (I6, also called Herrin 6) seams, which occur within the Carboniferous Carbondale Formation. The average thickness of the I6 seam is 3 m, and its depth in the survey area is 200–250 m. The S5 seam is 1–2 m thick and is usually 6–8 m below the I6 seam. In places the I6 seam divides into two seams, the lower called the Rider seam. Additional examples of seismic responses from the same coal basin are provided by Henson and Sexton (1991).

The density and velocity of coal is significantly less than those of the clastic and carbonate lithotypes with which it is associated (Fig. 6.35). This means that large acoustic impedance contrasts and reflection coefficients can be expected at both the top and base of a coal seam. The reflection coefficient at the top will be negative and that at the base will be positive. Coal seams as thin as 1/50 of the dominant wavelength have been reported to produce recognisable reflections as a result of the very large acoustic impedance contrast (up to ~0.5).

Figure 6.45 shows a density log through part of the Carbondale Formation incorporating the S5, I6 and Rider seams. As expected, the coal seams have very low density relative to the rest of the sequence. A reflectivity series has been computed based on this log, the lack of a sonic log being overcome by assuming a simple linear relationship between density and velocity (which is not ideal, but should allow the major velocity contrasts to be identified). The large reflection coefficients of opposite polarity at the top and base of the coal seams are evident. A synthetic seismogram has been computed using a 150 Hz Ricker wavelet. The three seams (I6, Rider, S5) give rise to high-amplitude negative deflections in the trace. Note how the maxima (the shaded peaks) coincide with the upper boundary of their respective seam, where the first major acoustic impedance contrast for the seam occurs. Note also the width of the seismic wavelet compared with the separation of features in the reflectivity series. The spacings between non-zero reflection coefficients are smaller than the duration of the wavelet, so all reflections are necessarily composite responses to multiple acoustic-impedance contrasts.

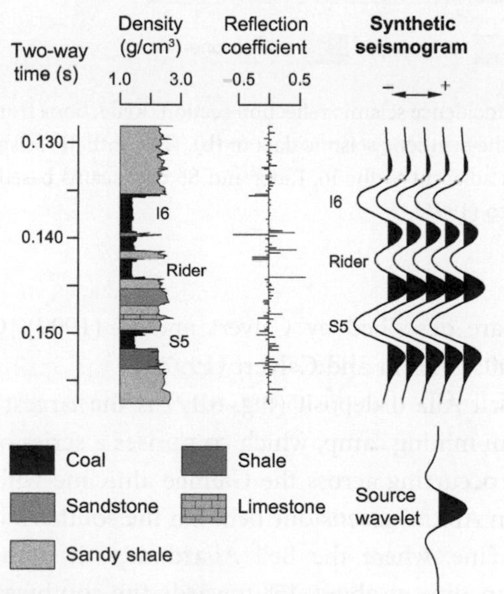

Figure 6.45 Lithology and density log from the southern Illinois coal basin and the computed reflectivity series and synthetic seismograms. Based on diagrams in Gochioco (1992).

Figure 6.46a shows a time-migrated normal-incidence seismic section from the study area. The section extends to a depth of about 400 m, which is comparatively shallow, but the requirement to image the seams in detail necessitated using a source with a central frequency of about 150 Hz. The source frequency spectrum extended from about 60 to 250 Hz. The general character of the seismic section is typical of a sedimentary terrain having little deformation. Distinct sub-horizontal reflections with good lateral continuity are prominent across the section. The reflection at 0.1 s TWT is from a major shale–limestone interface.

Figure 6.46c shows a simplified geological model of the area interpolated from drillhole intersections. This has been used, along with density data, to compute the synthetic seismic data shown in Fig. 6.46b. A 150 Hz Ricker wavelet was used as the source. The change from a simple layered form at the western end of the section to a more complex situation further east is of particular importance from a mining perspective. The change is due to a 'roll' between 120 and 150 m related to differential compaction associated with the sandstone lenses between 125 and 140 m. Across the roll, depth to the I6 seam decreases by about 9 m; it is thinner in the centre of the roll and to the east splits to form the Rider seam. Also, depth to the underlying S5 seam increases significantly to about 18 m, but it retains a near-constant thickness of 1.2 m. The 'roll' represents a significant impediment to mining, mostly owing to the changes it causes in the roof and floor conditions.

The synthetic and observed data show good agreement, with each coal seam associated with a distinct negative feature on the traces. The thickness of the seams, and the surrounding units, varies between 1/5 and 1/15 of the dominant seismic wavelength and is insufficient to produce separate reflections from their top and bottom surfaces. Instead, changes in the geology cause changes to the interference between reflections from the top and base of the seams and produce subtle variations in the wavelength and amplitude of 'reflections' from the seams. The laterally changing response from the S5 seam is due to changes in the geology above and below it. The response from the I6 seam gets broader towards the east and, eventually, forms a doublet (a pair of peaks or troughs in the trace) created by the response of the Rider seam interacting with the responses of the various thin beds in the vicinity. Although

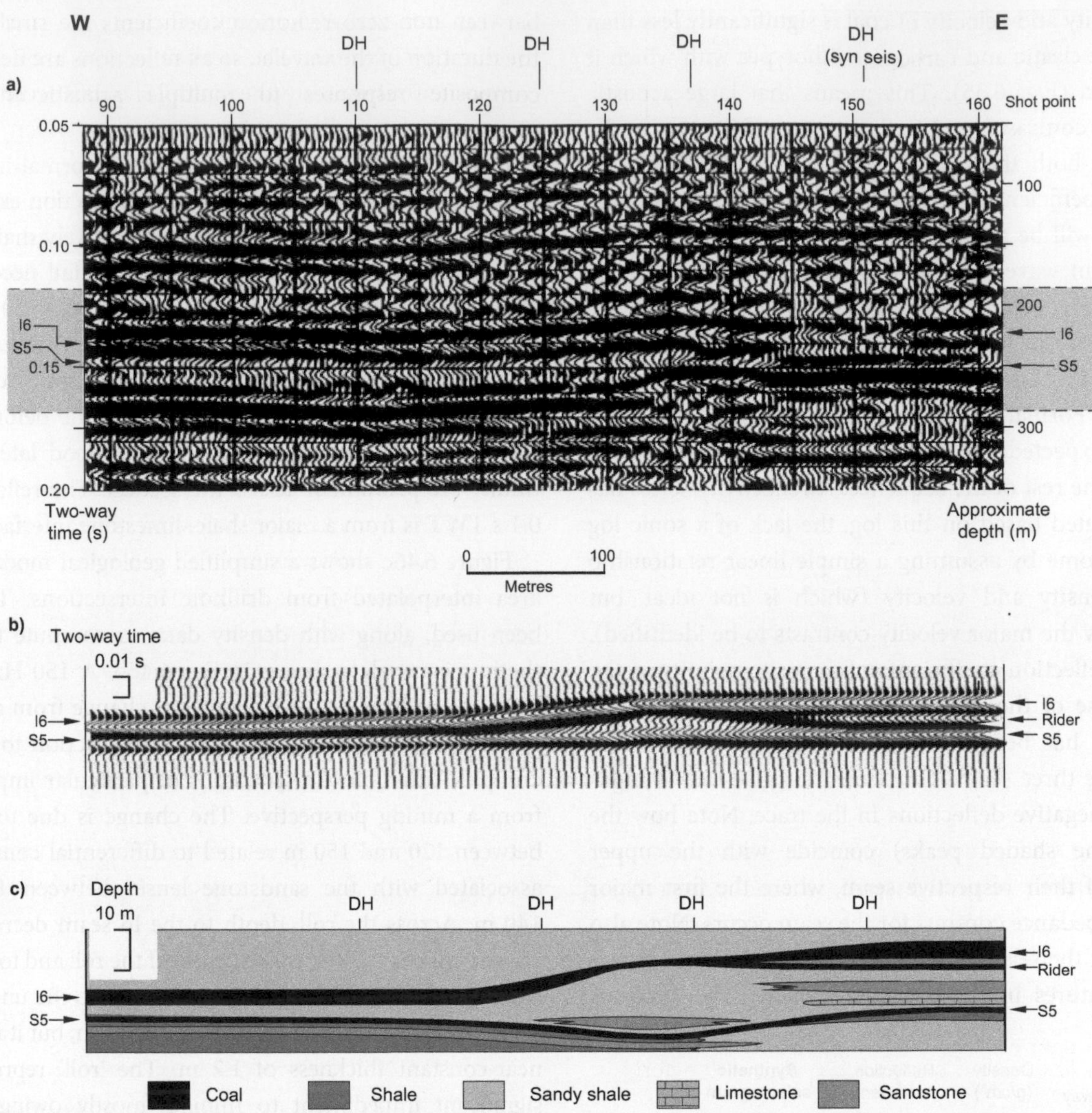

Figure 6.46 Seismic reflection data from the southern Illinois coal basin. (a) Normal-incidence seismic reflection section. Reflections from the I6 and S5 coal seams are labelled. The grey shading shows the approximate extent of the synthetic seismic data in (b). (b) Synthetic seismic data computed from the geological model in (c). (c) Geological model of the succession adjacent to the I6, Rider and S5 coal seams based on interpolations between drillhole (DH) intersections. Based on diagrams in Gochioco (1991).

the 'roll' is quite a large feature, its seismic response is subtle, primarily because of interference affects. The hazard itself is detected, rather than resolved.

6.7.4.2 Imaging structure and massive sulphide mineralisation in a Greenstone Belt

A seismic reflection survey in the vicinity of the Bell Allard Cu–Zn–Ag–Au deposit in Québec, Canada, illustrates the seismic reflection method's application to volcanogenic massive sulphide (VMS) deposits and the complex hard-rock volcanic formations that host these deposits. The surveys are described by Calvert and Li (1999), Calvert *et al.* (2003) and Li and Calvert (1997).

The Bell Allard deposit (Fig. 6.47) is the largest in the Matagami mining camp, which comprises a series of VMS deposits occurring across the Galinée anticline within the Archaean Abitibi greenstone belt. On the southern limb of the anticline, where the Bell Allard deposit occurs, the succession dips at about 45° towards the southwest. The local succession comprises the Bell River Complex which consists of layered mafic and ultramafic rocks, overlain by felsic and mafic volcanic units. The lower of the units, the

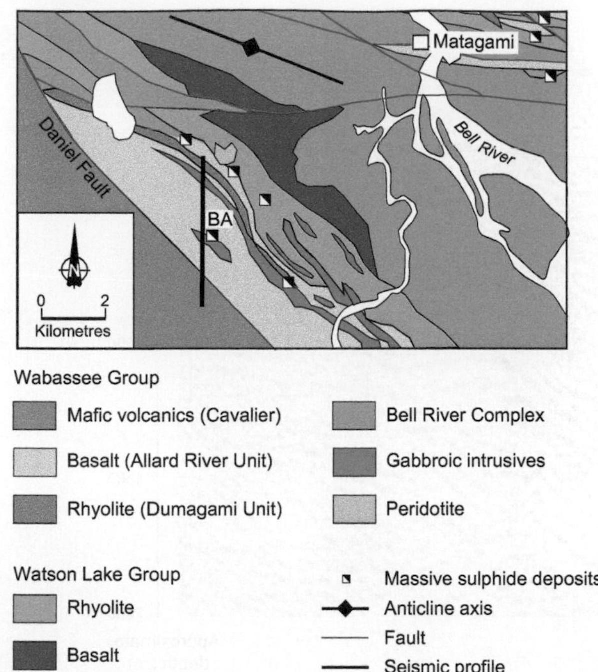

Figure 6.47 Simplified geological map of the Matagami mining camp. The location of the seismic section in Fig. 6.48 is also shown. BA – Bell Allard VMS deposit. Redrawn, with permission, from Calvert and Li (1999).

Watson Lake Group, consists mostly of felsic rocks. Overlying these are the mainly intermediate and mafic rocks of the Wabassee Group. The VMS deposits in the area occur at the contact between the two groups, where there is a 1–3 m thick cherty tuffaceous horizon called the Key Tuffite. The youngest rocks in the area are tonalite and gabbro sills. The seismic survey was designed to determine:

- Whether the economically important Key Tuffite produces a recognisable response that enables it to be mapped across the survey area;
- Whether the lithological contrasts within the felsic and mafic units produce recognisable reflections, especially in the Watson Lake Group (since these would allow identification of faults that potentially acted as conduits for mineralising fluids); and
- Whether massive sulphide mineralisation produces identifiable seismic responses. A seismic response was considered more likely from the sharp upper contacts of the sulphide bodies, with the most pyritic considered the more likely to be detected because of the increased acoustic impedance caused by the high velocity of pyrite (see Fig. 6.35).

A single 2D seismic profile was recorded along a highway which passes near (100–280 m) to the Bell Allard deposit. In this area the Watson Lake Group comprises a dacite overlain by rhyolite. The deposit is 900–1150 m below surface and dips 50° to the south. The data were acquired using a Vibroseis source with a 30–140 Hz sweep. The final section has a fold of 120%.

A time-migrated section is shown in Fig. 6.48. The response exhibits many short and discontinuous arrivals, and is typical of a deformed and metamorphosed hard-rock terrain; compare these data with those shown in Fig. 6.46. The strongest reflections were assigned to late intrusions. Synthetic seismograms suggested that these would give the strongest responses in the area with the comparatively large cross-cutting (less deformed) surfaces of sills favourable for producing good responses. The Key Tuffite, with a thickness of 1–3 m (corresponding to about 1/30 of the seismic wavelength), can be detected, but its thickness not resolved. Its location can also be inferred from changes in the character of the seismic responses associated with the contrasting lithotypes in the Watson Lake and Wabassee Groups. The synthetic seismogram confirmed that lithological contrasts in the felsic and mafic volcanic units would produce reflections, albeit weaker than those from the later intrusions. The most distinctive responses of this type are interpreted as Dumagami rhyolites, which occur in the lower part of Wabassee Group. The responses suggest they are conformable with the Key Tuffite. Many faults were interpreted from the displacement of distinctive reflectors. In particular, there appears to be an older set of faults largely restricted to the Watson Lake Group that are potential conduits for mineralising fluids. In addition, there is a post-mineralisation fault set mostly affecting the Wabassee Group.

Figure 6.49 shows an enlargement of the seismic section in the vicinity of the Bell Allard deposit with the local geology derived from drillholes. The weak reflections from lithological contrasts in the volcanic successions are confirmed. There is evidence of a response from the mineralisation itself, correlating with a lower and more pyritic lobe of the orebody. Although this is encouraging, it is obvious that the response is hard to recognise; and confusion with responses from faulted sills is likely.

6.8 In-seam and downhole seismic surveys

Small-scale seismic surveys at mine and prospect scale may use tomographic techniques or in-seam methods, usually to assist in delineating a resource prior to mining (see

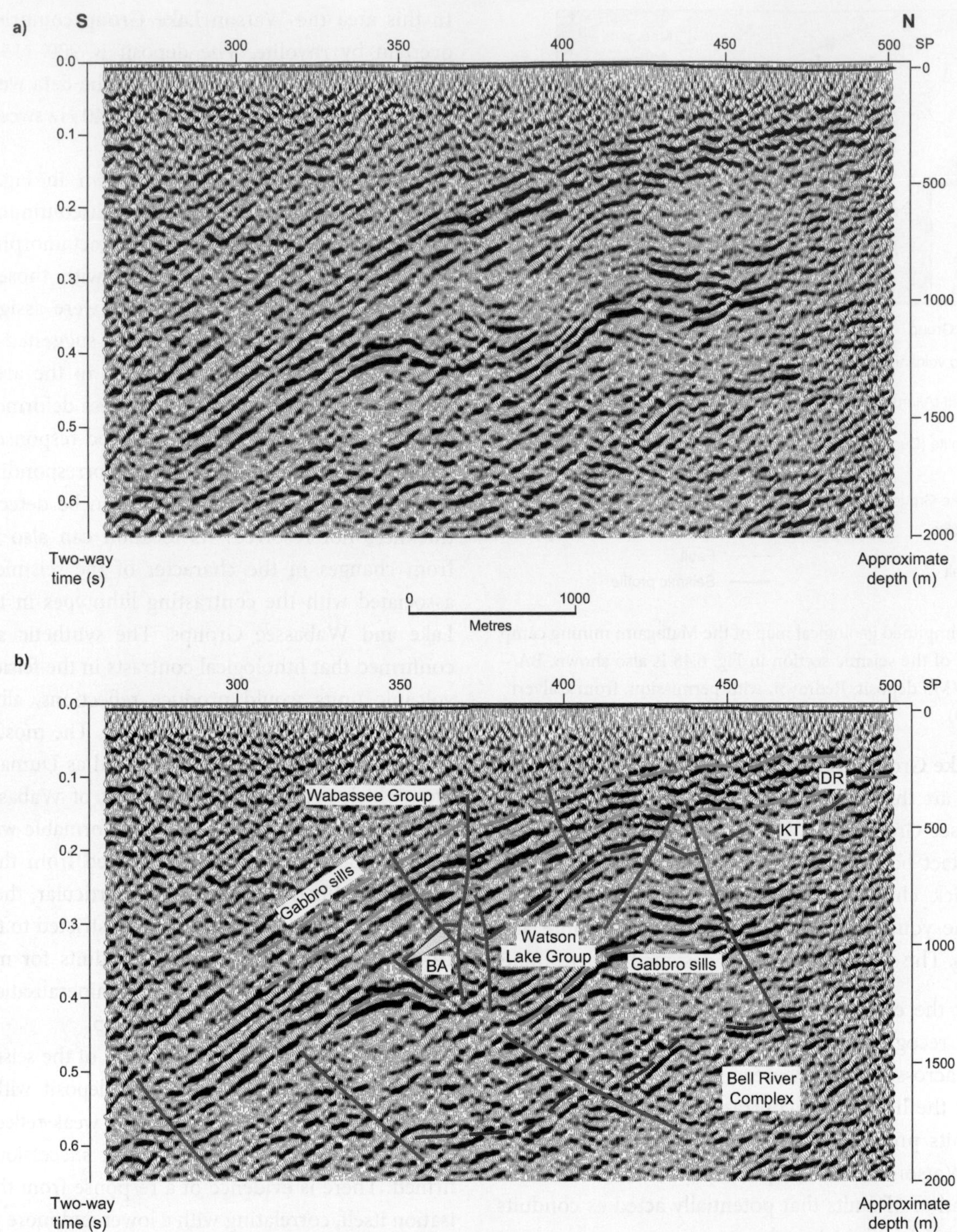

Figure 6.48 Migrated zero-offset seismic reflection section across part of the southwestern limb of the Galinée Anticline. (a) Uninterpreted data, and (b) major stratigraphic markers and faults (red) interpreted from the data. BA – Bell Allard VMS deposit, DR – Dumagami Rhyolite, KT – Key Tuffite. Redrawn, with permission, from Calvert and Li (1999).

Fig. 6.1). In-seam surveys, as the name suggests, are the domain of the coal industry. Drillhole-to-drillhole (downhole) tomographic surveys have been successfully applied to bodies of massive sulphide mineralisation.

6.8.1 In-seam surveys

Coal seams have seismic properties that are usually very different from the host strata (Fig. 6.35). Coal has lower density and lower seismic velocity than most rocks, which

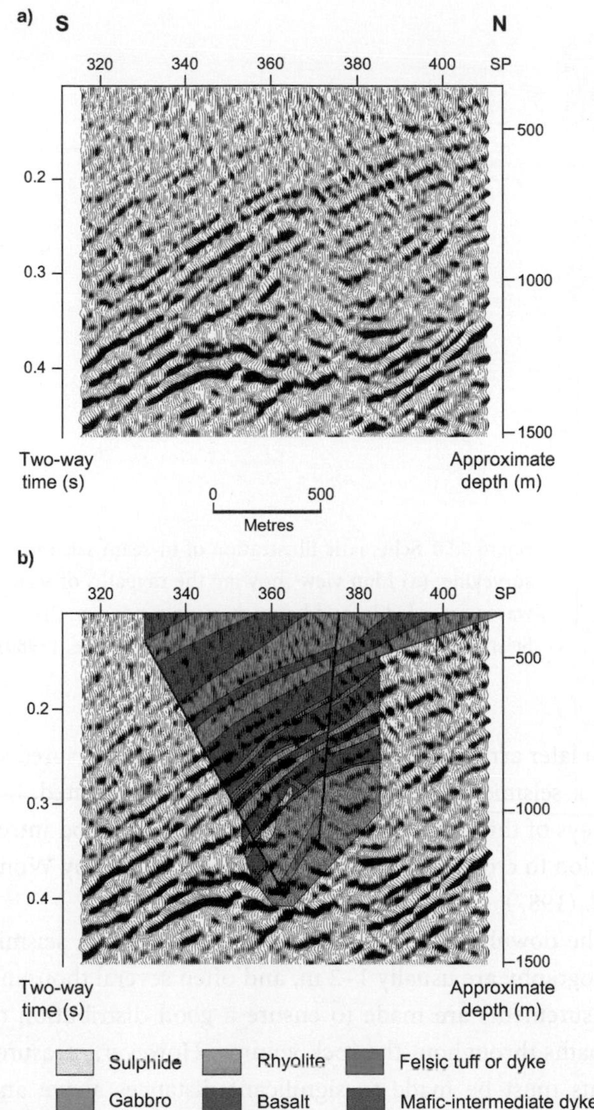

Figure 6.49 Detail of Fig. 6.48 in the vicinity of the Bell Allard VMS deposit. (a) Uninterpreted data, and (b) with simplified geology superimposed. Redrawn, with permission, from Calvert and Li (1999).

means that very large acoustic impedance contrasts, and therefore very large reflection coefficients, occur at the margins of the seams. Consequently, seismic waves generated by a source located in a coal seam tend to travel within the seam, repeatedly reflecting into the seam from its upper and lower surfaces, i.e. the seam acts as a waveguide (see Section 6.3.4.2). Seismic waves propagating within the seam are known as *channel waves* or *seam waves*. They are a form of surface waves and so they are dispersive, i.e. different frequency components travel at different velocities. Dispersion causes the arrivals to have longer and more complex waveforms than the more compact wavelets of a conventional seismic recording.

Seam waves are a useful source of information about the coal seam and its environment, in particular the nature of the roof and floor, and the seam thickness and continuity – all factors of great importance for longwall mining. In-seam seismic surveys have greater resolving power than conventional surface surveys, which usually cannot detect features of the size that seriously affect mining, for example faults with throws of less than the seam thickness. Most in-seam seismic surveys access the seam where it is exposed in underground drives etc., although surveys from drillholes intersecting the seam are also possible. There are two basic types of in-seam survey: *transmission surveys* and *reflection surveys* (Fig. 6.50a), analogous to the two types of in-mine radar survey described in online Appendix 5.

Transmission surveys record seam waves that have passed through an area of interest. Various characteristics of the seam wave can be analysed in terms of geological structure: for example, discontinuities such as faults and dykes can be identified because they are highly attenuating. Reflection surveys allow particular geological discontinuities to be mapped in more detail. They are fundamentally the same as surface reflection surveys, except that the sources, detectors, waves and reflectors all lie in the plane of the seam, with the discontinuity detected from the reflection and diffraction of the seam waves. Processing techniques can be applied to resolve reflections in the dispersed waves, after which interpretation is essentially the same as that for conventional surface surveys. In Fig. 6.50b, note the complex waves caused by dispersion of the seam waves and, for the transmission survey, note that the slower seam-wave arrives after the faster direct P-wave.

A concise summary of seam waves and their application to coal mining is provided by Regueiros (1990a, 1990b).

6.8.2 Tomographic surveys

Seismic tomography (see *Tomography* in Section 2.11.2.1) applied in the mining industry usually involves cross-hole surveys, where a series of sources in one drillhole is recorded by detectors located in another (see Fig. 6.1). The intention is to map the geology in the region between the drillholes. The target must have significantly different seismic properties to its surrounds. The property measured is usually seismic travel time, resulting in a *velocity* (or *slowness*) *tomogram*. Normally, only first arrivals are used in the analysis, but more sophisticated approaches are currently being developed to obtain more information

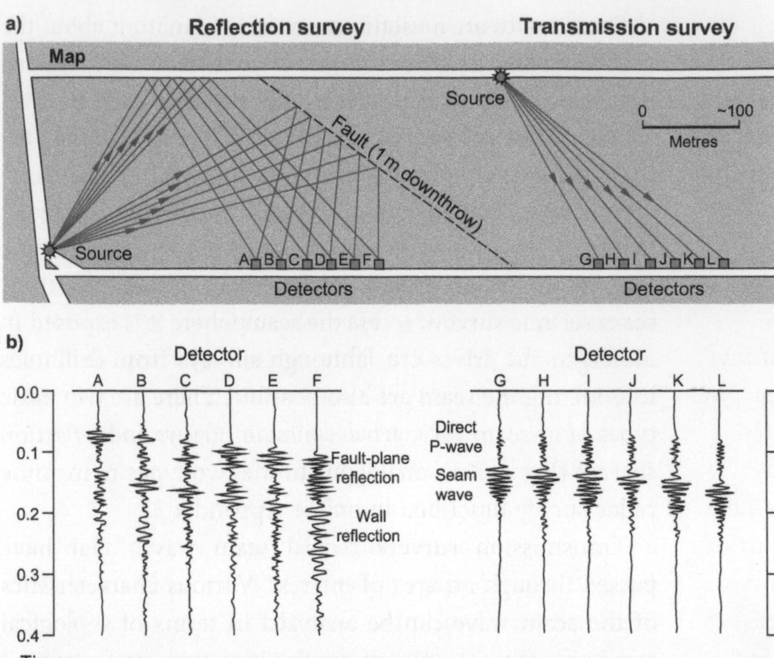

Figure 6.50 Schematic illustration of in-seam seismic surveying. (a) Map view showing the raypaths of seam waves recorded in and between mine roadways. (b) Seismic data. Based on diagrams in Mason *et al.* (1980).

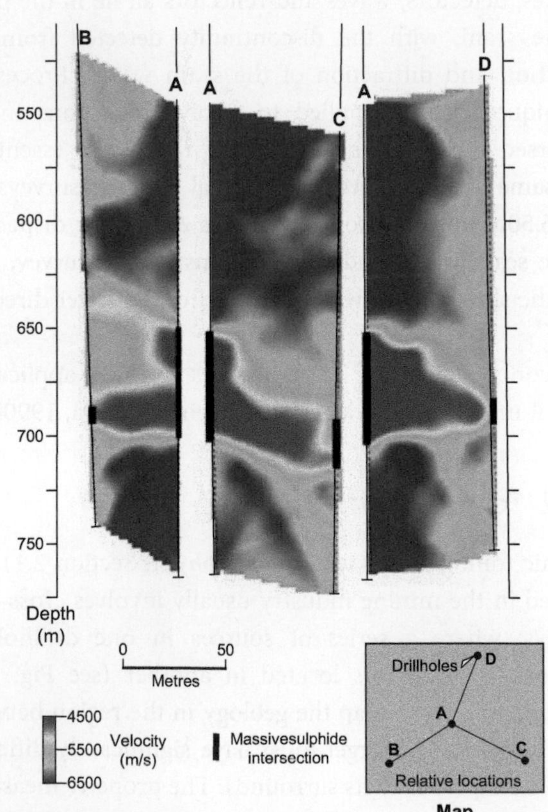

Figure 6.51 Velocity tomograms from a group of four drillholes in the Voisey's Bay nickel sulphide deposit. Redrawn, with permission, from Enescu *et al.* (2002).

from later arrivals. Amplitudes are sometimes measured so that a seismic attenuation tomogram can be created, but surveys of this type are much less common. A good introduction to cross-hole seismic surveys is provided by Wong *et al.* (1987).

The downhole detector and source spacings in seismic tomography are usually 1–2 m, and often several thousand measurements are made to ensure a good distribution of raypaths throughout the rock volume. However, measurements must be made to significant distances above and below the target to ensure adequate variation in ray orientation (see *Tomography* in Section 2.11.2.1). For drillhole-to-drillhole surveys this can be a major problem, since making measurements in directions parallel to the drillhole orientations is impossible (see Fig. 6.1). To resolve features on the scale of interest, which is usually a few metres, high-frequency waves of up to a few thousand hertz are used (see Fig. 6.7). This, and the comparatively low power of downhole seismic sources, limits the distance over which measurements can be made to a few tens of metres. It is essential that travel times are accurately measured and that the drillholes are accurately surveyed, otherwise the resulting errors in time and distance will strongly influence the computed velocity distribution.

The creation of the velocity tomogram itself also presents particular problems, because each raypath is refracted at points where the velocity changes. Sometimes straight raypaths are assumed, but this is unrealistic unless velocity variations are small, which is unlikely in the typically complex geology that hosts mineralisation. The usual approach is an iterative one where an initial velocity distribution is determined and the travel times along the various raypaths are computed. Differences between the computed model and the observations are used to modify the velocity model, and since this affects the raypaths these must also be recomputed. The process is repeated until a satisfactory solution is obtained.

Another problem associated with the refraction of raypaths is uneven survey coverage of the area between the drillholes. Raypaths of the first arrivals are concentrated in high-velocity regions, so lower-velocity areas may be poorly sampled. Also, the basic geometry of a cross-hole survey means that more raypaths cross the central part of the intervening area than the marginal zones of the survey, so the results are most reliable in the middle, decreasing outwards. Unless the drillholes are co-planar, the velocity distribution is determined on a best-fitting plane, which requires projecting the various locations onto the plane. This may introduce significant errors when the projection distance is large or the geology complex.

Despite these limitations, velocity tomograms can accurately reproduce the local geology in favourable environments. An excellent result was obtained, albeit in a simple geological setting, by Wong (2000) for the McConnell nickel sulphide deposit in Ontario, Canada. Figure 6.51 shows velocity tomograms recorded between three pairs of drillholes in the Eastern Deeps area of the Voisey's Bay Ni deposit, located in Labrador, Canada (Enescu *et al.*, 2002). Massive sulphide mineralisation occurs at the base of a large troctolite-gabbro unit that has intruded quartz-feldspar-biotite gneiss. Data were recorded between a central drillhole (A) and three adjacent drillholes (B, C, D). Downhole station spacing varied between 0.5 and 1 m, and each tomogram was obtained from 16,200 measurements. The seismic velocity of the massive sulphide mineralisation is 4–4.5 km/s and that of the host rocks 5–6 km/s. Note that it is not unusual for mineralisation to have a lower velocity than its host (Fig. 6.35). This pronounced velocity contrast allows the mineralisation to be mapped between the drillholes. Agreement between the tomograms and the known mineralisation is excellent.

Summary

- The seismic method uses the travel times and amplitudes of elastic waves to determine the structure of the subsurface.
- Seismic surveys are active surveys with the seismic waves being created by a source and detected by geophones. The data are displayed as traces, which are time series comprising graphs of the deformation of the ground following activation of the source. The arrival of seismic waves at a detector causes deflections of the trace.
- Seismic energy can travel as surface waves and body waves. The type of body waves most utilised in exploration are primary or P-waves. All other types of seismic wave are normally treated as noise.
- Changes in acoustic impedance cause seismic waves to be reflected and diffracted. Changes in velocity cause refraction. Continuous discontinuities (layers) can cause reflection and critical refraction of the waves. Local discontinuities (scatterers) cause diffraction. It is the location of seismic discontinuities in the subsurface that is obtained from the seismic method.
- The main controls on velocity and acoustic impedance In sedimentary rocks are porosity and pore fluids. Mineralogy is the main control in crystalline rocks.
- Massive metal oxide/sulphide mineralisation, evaporites and coal seams are all forms of mineralisation which may have significant acoustic impedance and velocity contrasts with their host rocks and therefore may be detectable using seismic methods.
- There are four main classes of seismic survey: reflection, refraction, in-seam and tomographic surveys. Seismic reflection surveys use reflected and diffracted waves to produce high-resolution pseudosection and pseudovolume displays of the subsurface. Refraction surveys use the arrival times of mainly critically refracted waves to map prominent changes in

seismic velocity in the subsurface, allowing construction of a velocity cross-section. In-seam and tomographic surveys are used at a local scale to map mineralisation.

- Complex data acquisition and processing are required in order to detect and resolve the weak signals in seismic reflection data. Important data processing techniques include stacking and migration. Stacking involves the use of repeat measurements to enhance the signal and migration corrects for distortions caused by reflections from points located away from the midpoint beneath source and detector.
- Seismic reflection surveys produce the most detailed images of the subsurface of any geophysical method, allowing structure and stratigraphy to be mapped at the scale of metres to tens of metres at depths of less than a few hundred metres.
- The resolution of subsurface features depends on the size of the feature relative to the wavelengths of the seismic waves. Features with dimensions less than ¼ of the dominant wavelength cannot be resolved.

Review questions

1. Describe what is meant by the terms wavefront, ray and Fresnel zone, and explain their relationship to each other.
2. Explain the principle of geometric spreading. By how much do the energy and amplitude of body waves and surface waves change when the wavefront expands to double, and four times, its radius?
3. Describe the main controls on variations of velocity and acoustic impedance across the three main rock classes.
4. Describe the effects of metamorphism, serpentinisation and weathering on seismic velocity and acoustic impedance.
5. Describe the process of stacking seismic reflection data. What assumptions are made about the seismic properties of the subsurface, and how might they affect the results in an area of complex geology?
6. How is a synthetic seismogram produced and how can it be used in the interpretation of seismic data?
7. Describe the fundamental phenomenon that controls the resolution of seismic reflection data.
8. Compare and contrast the seismic refraction and reflection methods describing the strengths and weaknesses of both methods.
9. What type of seismic survey would you use (a) to map the thickness of the sediments in a palaeochannel, (b) to detect massive sulphide mineralisation between drillholes and (c) to map a coal seam over a distance of several kilometres? Explain your choices.

FURTHER READING

Ashcroft, W., 2011. *A Petroleum Geologist's Guide to Seismic Reflection*. Wiley-Blackwell.

This book is an excellent summary of the seismic reflection method and, although written for the petroleum geologist, is highly recommended for anyone wishing to work with reflection data.

Eaton, D.W., Milkereit, B. and Salisbury, M.H., 2003. *Hardrock Seismic Exploration*. Society of Exploration Geophysicists, Geophysical Development Series 10.

A comprehensive set of papers describing the application of the seismic reflection method in hard-rock terrains. Included are datasets from various deposits and mineralised terrains.

Gochioco, L.M., 1991. Tuning effect and interference reflections from thin beds and coal seams. *Geophysics*, 56, 1288–1295.

Gochioco, L.M., 1992. Modelling studies of interference reflections in thin-layered media bounded by coal seams. *Geophysics*, 57, 1209–1216.

These two papers describe seismic responses from coal seams and are excellent illustrations of how synthetic seismograms can be used to understand the relationship between seismic wavelets, variations in the local geology, and the seismic reflection response.

Malehmir, A., Durrheim, R., Bellefleur, G. *et al.*, 2012. Seismic methods in mineral exploration and mine planning: A general overview of past and present case histories and a look into the future. *Geophysics*, 77, WC173–WC190.

Malehmir, A., Koivisto, E., Manzi, M. *et al.*, 2014. A review of reflection seismic investigations in three major metallogenic regions: the Kevista Ni-Cu-PGE district (Finland), Witwatersrand goldfields (South Africa) and the Bathurst mining camp (Canada). *Ore Geology Reviews*, 56, 423–441.

The first of these two papers is part of a thematic volume with numerous examples of minerals-oriented seismic surveys. The two papers together represent a review of the current state of hard-rock seismic surveys for exploration and mining purposes and contain numerous examples.

Yilmaz, O., 2001. *Seismic Data Analysis: Processing Inversion and Interpretation of Seismic Data*, Volumes 1 and 2. Society of Exploration Geophysicists, Investigations in Geophysics 10.

This is the definitive text on seismic processing and is orientated towards petroleum applications. It is dauntingly large and quite mathematical in places. However, the clarity of its written descriptions and numerous case studies make the material accessible to all readers. The 'Basic Data Processing Sequence' in Volume 1 is a good elementary introduction to seismic processing, with examples showing the effects of each processing action on a dataset.

REFERENCES

Adam, E., Arnold, G., Beaudry, C. *et al.*, 1997. Seismic exploration for VMS deposits, Matagami, Québec. In Gubins, A.G. (Ed.), *Proceedings of Exploration '97, Fourth Dicennial International Conference on Mineral Exploration.* Prospectors and Developers Association of Canada, 433–438.

Adams, D.C., Miller, K.C. and Kargi, H., 1997. Reconciling physical properties with surface seismic data from a layered mafic intrusion. *Tectonophysics*, 271, 59–74.

Agarwal, B.N.P. and Sivaji, Ch., 1992. Separation of regional and residual anomalies by least-squares orthogonal polynomial and relaxation techniques: A performance evaluation. *Geophysical Prospecting*, 40, 143–156.

Agocs, W.B., 1955. Line spacing effect and determination of optimum spacing illustrated by Marmora, Ontario, magnetic anomaly. *Geophysics*, 20, 871–885.

Airo, M.L. and Loukola-Ruskeeniemi, K., 2004. Characterization of sulfide deposits by airborne magnetic and gamma-ray responses in eastern Finland. *Ore Geology Reviews*, 24, 67–84.

Al-Chalabi, M., 1974. An analysis of stacking, rms, average and interval velocities over a horizontally layered ground. *Geophysical Prospecting*, 22, 458–475.

Alva-Valdivia, L.M. and Urrutia-Fucugauchi, J., 1998. Rock magnetic properties and ore microscopy of the iron deposit of Las Truchas, Michoacan, Mexico. *Journal of Applied Geophysics*, 38, 277–299.

Am, K., 1972. The arbitrarily magnetized dyke: Interpretation by characteristics. *Geoexploration*, 10, 63–90.

Anand, R.R. and Paine, M., 2002. Regolith geology of the Yilgarn Craton, Western Australia: implications for exploration. *Australian Journal of Earth Sciences*, 49, 3–162.

Anstie, J., Aravanis, T., Johnston, P. *et al.*, 2010. Preparation for flight testing the VK1 gravity gradiometer. In Lane, R.J.L. (Ed.), *Airborne Gravity 2010 –ASEG-PESA Airborne Gravity 2010 Workshop* (Abstracts). Published jointly by Geoscience Australia and the Geological Survey of New South Wales, Geoscience Australia Record 2010/23 and GSNSW File GS2010/0457, 5–12.

Archie, G.E., 1942. The electrical resistivity log as an aid in determining some reservoir characteristics. Transactions of the American Institute of Mining, *Metallurgical and Petroleum Engineers*, 146, 54–62.

Ashcroft, W., 2011. *A Petroleum Geologist's Guide to Seismic Reflection.* Wiley-Blackwell.

Asten, M.W., 2000. Interpretation of thick conductive zones using time-domain EM parasections. *Exploration Geophysics*, 31, 595–602.

Atchuta Rao, D. and Ram Babu, H.V., 1984. On the half-slope and straight slope methods of basement depth determination. *Geophysics*, 49, 1365–1368.

Baas Becking, L.G.M., Kaplan, I.R. and Moore, D., 1960. Limits of the natural environment in terms of pH and oxidation and reduction potentials. *Journal of Geology*, 68, 243–284.

Balch, S.J., Boyko, W.P. and Paterson, N.R., 2003. The AeroTEM airborne electromagnetic system. *The Leading Edge*, 22(6), 562–566.

Barrett, D. and Dentith, M., 2003. Geophysical exploration for graphite at Uley, South Australia. In Dentith, M.C. (Ed.), *Geophysical Signatures of South Australian Mineral Deposits.* Centre for Global Metallogeny, University of Western Australia, Publication 31; Australian Society of Exploration Geophysicists, Special Publication 12; and Primary Industries and Resources South Australia, 47–57.

Barsukov, P.O. and Fainberg, E.B., 2001. Superparamagnetic effect over gold and nickel deposits. *European Journal of Environmental and Engineering Geophysics*, 6, 61–72.

Bate, S.J., Thorsen, K.R. and Jones, D., 1989. The Casa Bernadi area: An exploration case history. In Garland, G.D. (Ed.), *Proceedings of Exploration '87. Third Dicennial International Conference on Geophysical and Geochemical Exploration for Minerals and Groundwater.* Ontario Geological Survey Special Volume 3, 855–870.

Becker, A. and Telford, W.M., 1965. Spontaneous polarization studies. *Geophysical Prospecting*, 13, 173–188.

Beltrão, J.F., Silva, J.B.C. and Costa, J.C., 1991. Robust polynomial fitting method for regional gravity estimation. *Geophysics*, 56, 80–89.

Berning, J., 1986. The Rossing uranium deposit, South West Africa, Namibia. In Anhaeusser, C.R. and Maske, S. (Eds.), *Mineral Deposits of South Africa*, Volume II, Geological Society of South Africa, 1819–1832.

Bishop, J.R., 1996. DHEM surveys of the 2K area, North Mine, Broken Hill. *Exploration Geophysics*, 27, 51–65.

Bishop, J.R. and Lewis, R.J.G., 1992. Geophysical signatures of Australian volcanic-hosted massive sulphide deposits. *Economic Geology*, 87, 913–930.

Blakely, R.J., 1995. *Potential Theory in Gravity and Magnetic Applications*. Cambridge University Press.

Bleil, U., Hall, J.M., Johnson, H.P., Levi, S. and Schonharting, G., 1982. The natural magnetization of a 3-kilometer section of Icelandic crust. *Journal of Geophysical Research*, 87, 6569–6589.

Bleil, U. and Petersen, N., 1982. Magnetic properties of natural minerals. In Angenheister, G. (Ed.), *Physical Properties of Rocks, Landolt-Börnstein: Numerical Data and Functional Relationships in Science and Technology, Group V: Geophysics and Space Research*, Volume 1b. Springer, 308–365.

Bloss, F.D., 1952. Relationship between density and composition and mol per cent for some solid solution series. *American Mineralogist*, 37, 966–981.

Bolt, B.A., 1976. *Nuclear Explosions and Earthquakes: The Parted Veil*. W.H. Freeman.

Booden, M.A., Mauk, J.L. and Simpson, M.P., 2011. Quantifying metasomatism in epithermal Au-Ag deposits: A case study from the Waitekauri area, New Zealand. *Economic Geology*, 106, 999–1030.

Borradaile, G.J. and Kukkee, K.K., 1996. Rock-magnetic study of gold mineralization near a weakly deformed Archean syenite, Thunder Bay, Canada. *Exploration Geophysics*, 27, 25–31.

Bosch, M. and McGaughey, J., 2001. Joint inversion of gravity and magnetic data under lithologic constraints. *The Leading Edge*, 20(8), 877–881.

Bourne, B.T., Trench, A., Dentith, M.C. and Ridley, J., 1993. Physical property variations within Archaean granite-greenstone terrane of the Yilgarn Craton, Western Australia: The influence of metamorphic grade. *Exploration Geophysics*, 24, 367–374.

Braile, L.W., 1978. Comparison of four random to grid methods. *Computers and Geosciences*, 4, 341–349.

Breiner, S., 1973. *Applications Manual for Portable Magnetometers*. Geometrics Inc.

Brescianini, R.F., Asten, M.W. and McLean, N., 1992. Geophysical characteristics of the Eloise Cu-Au deposit. *Exploration Geophysics*, 23, 33–42.

Brummer, J.J., MacFadyen, D.A. and Pegg, C.C., 1992. Discovery of kimberlites in the Kirkland Lake area, northern Ontario, Canada: Part II, Kimberlite discoveries, sampling, diamond content, ages and emplacement. *Exploration and Mining Geology*, 1, 351–370.

Buckingham, A., Dentith, M. and List, R., 2003. Towards a system for content-based magnetic image retrieval. *Exploration Geophysics*, 34, 195–206.

Buselli, G., 1982. The effect of near-surface superparamagnetic material on electromagnetic measurements. *Geophysics*, 47, 1315–1324.

Buselli, G. and Cameron, M., 1996. Robust statistical methods for reducing sferics noise contaminating transient electromagnetic measurement. *Geophysics*, 61, 1633–1646. doi: 10.1190/1.1444082

Calvert, A.J. and Li, Y., 1999. Seismic reflection imaging over a massive sulfide deposit at the Matagami mining camp, Québec. *Geophysics*, 64, 24–32. doi: 10.1190/1.1444521

Calvert, A.J., Perron, G. and Li, Y., 2003. A comparison of 2D seismic lines shot over the Ansil and Bell Allard Mines in the Abitibi Greenstone Belt. In Eaton, D.W., Milkereit, B. and Salisbury, M.H. (Eds.), *Hardrock Seismic Exploration*. Society of Exploration Geophysicists, Geophysical Development Series 10, 164–177.

Carter, D.C., 1989. Downhole geophysics as an aid to the interpretation of an evaporite sequence: examples from Nova Scotia. *CIM Bulletin*, 82 (925), 58–64.

Chapin, D., 1998. Gravity instruments: past, present, future. *The Leading Edge*, 17(1), 100–112.

Chapman, R.H., Clark, W.B. and Chase, G.W., 1980. A geophysical approach to locating Tertiary gold channels, Port Wine, Sierra County, California. *California Geology*, 33, 173–180.

Chappell, B.W. and White, A.J.R., 1974. Two contrasting granite types. *Pacific Geology*, 8, 173–174.

Charbonneau, B.W., Killeen, P.G., Carson, J.M., Cameron, G.W. and Richardson, K.A., 1976. Significance of radioelement concentration measurements made by airborne gamma-ray spectrometry over the Canadian Shield. In *Exploration for Uranium Ore Deposits*. International Atomic Energy Agency, 35–54.

Cheesman, S., MacLeod, I. and Hollyer, G., 1998. A new, rapid, automated grid stitching algorithm. *Exploration Geophysics*, 29, 301–305.

Clark, D.A., 1997. Magnetic petrophysics and magnetic petrology: aids to geological interpretation of magnetic surveys. *AGSO Journal of Australian Geology & Geophysics*, 17(2), 83–103.

Clark, D.A., 1999. Magnetic petrology of igneous intrusions: implications for exploration and magnetic interpretation. *Exploration Geophysics*, 30, 5–26.

Clark, D.A. and Emerson, D.W., 1991. Notes on rock magnetization characteristics in applied geophysical studies. *Exploration Geophysics*, 22, 547–555.

Clark, D.A. and Schmidt, P.W., 1994. Magnetic properties and magnetic signatures of BIFs of the Hamersley Basin and Yilgarn Block, Western Australia. In Dentith, M.C., Frankcombe, K.F., Ho, S.E. *et al.* (Eds.), *Geophysical Signatures of Western Australian Mineral Deposits*. University of Western Australia, Geology and Geophysics Department (Key Centre), Publication 26, and Australian Society of Exploration Geophysicists, Special Publication 7, 343–354.

Clark, D.A. and Tonkin, C., 1994. Magnetic anomalies due to pyrrhotite: examples from the Cobar area, N.S.W., Australia. *Journal of Applied Geophysics*, 32, 11–32.

Cobb, J.C. and Kulp, J.L., 1961. Isotopic geochemistry of uranium and lead in the Swedish kolm and its associated shale. *Geochimica et Cosmochimica Acta*, 24, 226–249.

Coggon, J., 2003. Magnetism – the key to the Wallaby gold deposit. *Exploration Geophysics*, 34, 125–130.

Collinson, D.W., 1983. *Methods in Rock Magnetism and Palaeomagnetism*. Chapman and Hall.

Cooper, G. and Cowan, D., 2003. The application of fractional calculus to potential field data. *Exploration Geophysics*, 34, 51–56.

Cordell, L. and McCafferty, A.E., 1989. A terracing operator for physical property mapping with potential field data. *Geophysics*, 54, 621–634.

Corry, C.E., 1985. Spontaneous polarization with porphyry sulfide mineralisation. *Geophysics*, 50, 1020–1034.

Cossey, S.P.J. and Frank, H.J., 1983. Uranium mineralization and use of resistance log character in deltaic point bars: Franklin Mine, Karnes County, Texas. *American Association of Petroleum Geologists Bulletin*, 67, 131–151.

Cowan, D. and Cooper, G., 2003a. Drape-related problems in aeromagnetic surveys: the need for tight-drape surveys. *Exploration Geophysics*, 34, 87–92.

Cowan, D. and Cooper, G., 2003b. Wavelet analysis of detailed drillhole magnetic susceptibility data, Brockman Iron Formation, Hamersley Basin, Western Australia. *Exploration Geophysics*, 34, 63–68.

Cowan, D. and Cooper, G., 2005. The Shuttle Radar Topography Mission – A new source of near global digital-elevation data. *Exploration Geophysics*, 36, 334–340.

Cowan, D.R. and Cowan, S., 1993. Separation filtering applied to aeromagnetic data. *Exploration Geophysics*, 24, 429–436.

Cowan, D.R., Kane, M.F., Merghelani, H. and Pitkin, J.A., 1985. Regional aeromagnetic/radiometric surveys: A perspective. *First Break*, 3(2), 17–21.

Criss, R.E., Champion, D.E. and McIntyre, D.H., 1985. Oxygen isotope, aeromagnetic and gravity anomalies associated with hydrothermally altered zones in the Yankee Fork Mining District, Custer County, Idaho. *Economic Geology*, 80, 1277–1296.

Cruden, A.R. and Launeau, P., 1994. Structure, magnetic fabric and emplacement of the Archean Lebel stock, SW Abitibi Greenstone Belt. *Journal of Structural Geology*, 16, 677–691.

Da Costa, A.J.M., 1989. Palmietfontein kimberlite pipe, South Africa – A case history. *Geophysics*, 54, 689–700.

Darnley, A.G., 1972. Airborne gamma-ray survey techniques. In Bowie, S.H.U., Davis, M. and Ostle, D. (Eds.), *Uranium Prospecting Handbook*. The Institution of Mining and Metallurgy, 174–211.

Davis, J.C., 1986. *Statistics and Data Analysis in Geology*. Wiley.

Davies, A.L. and McManus, D.A., 1990. Geotechnical applications of downhole sonic and neutron logging for surface coal mining. *Exploration Geophysics*, 21, 73–81.

De Boer, C., 2001. *A Practical Guide to Splines, Revised Edition. Applied Mathematical Sciences*, Volume 27. Springer.

De Hua, H., Nur, A. and Morgan, D., 1986. Effects of porosity and clay content on wave velocities in sandstones. *Geophysics*, 51, 2093–2107.

Degeling, P.R., Gilligan, L.B., Scheibner, E. and Suppel, D.W., 1986. Metallogeny and tectonic development of the Tasman Fold Belt system in New South Wales. *Ore Geology Reviews*, 1, 259–313.

Delius, H., Brewer, T.S. and Harvey, P.K., 2003. Evidence for textural and alteration changes in basaltic lava flows using variations in rock magnetic properties (ODP Leg 183). *Tectonophysics*, 371, 111–140.

Dentith, M.C., Frankcombe, K.F. and Trench, A., 1994. Geophysical signatures of Western Australia mineral deposits: An overview. In Dentith, M.C., Frankcombe, K.F., Ho, S.E. *et al.* (Eds.), *Geophysical Signatures of Western Australian Mineral Deposits*. University of Western Australia, Geology and Geophysics Department (Key Centre), Publication 26, and Australian Society of Exploration Geophysicists, Special Publication 7, 29–84.

Dickson, B. and Taylor, G., 2000. Maximum noise fraction method reveals detail in aerial gamma-ray surveys. *Exploration Geophysics*, 31, 73–77.

Dickson, B.L., Fraser, S.J. and Kinsey-Henderson, A., 1996. Interpreting aerial gamma-ray surveys utilising geomorphological and weathering models. *Journal of Geochemical Exploration*, 57, 75–88.

Dickson, B.L. and Scott, K.M., 1997. Interpretation of gamma-ray surveys – adding the geochemical factors. *AGSO Journal of Australian Geology & Geophysics*, 17(2), 187–200.

Diot, H., Bolle, O., Lambert, J-M., Launeau, P. and Duchesne, J-C., 2003. The Tellnes ilmenite deposit (Rogaland, South Norway): magnetic and petrofabric evidence for emplacement of a Ti-enriched noritic crystal mush in a fracture zone. *Journal of Structural Geology*, 25, 481–501.

Dix, C.H., 1955. Seismic velocities from surface measurements. *Geophysics*, 20, 68–86.

Dobrin, M.B. and Savit, C.H., 1988. *Introduction to Geophysical Prospecting*. McGraw Hill.

Doe, A.R.D., Carswell, J.T., Smith, C.K. and Erickson, M.E., 1990. Underground downhole geophysics at CSA Mine, Cobar. In Mutton, B.K. (Chairman), *Mine Geologists' Conference, Mt Isa, Queensland, 1990*. Australian Institute of Mining and Metallurgy, 9–13.

Dowsett, J.S., 1967. Exploration geophysical methods for nickel. In Morley, L.W. (Ed.), *Mining and Groundwater Geophysics – 1967*. Geological Survey of Canada, Economic Geology Report 26, 310–321.

Doyle, H.A. and Lindeman, F.W., 1985. The effect of deep weathering on geophysical exploration in Australia – a review. *Australian Journal of Earth Sciences*, 32, 125–135.

Doyle, M., Morrissey, C. and Sharp, G., 2004. The Las Cruces orebody, Seville Province, Andalucia, Spain. In Kelly, J.G., Andrew, C.J., Ashton, J.H., *et al.* (Eds.), *Europe's Major Base Metal Deposits*. Irish Association for Economic Geology, 381–389.

Dransfield, M., 2007. Airborne gravity gradiometry in the search for mineral deposits. In Milkereit, B. (Ed.), *Proceedings of Exploration '07: Fifth Decennial International Conference on Mineral Exploration*. Decennial Mineral Exploration Conferences, 341–354.

Drury, S.A., 1987. *Image Interpretation in Geology*. Allen and Unwin.

Dubuc, F., 1966. Geology of the Adams mine. *Canadian Mining and Metallurgical Bulletin*, 59, 176–181.

Duff, D., Hurich, C. and Deemer, S., 2012. Seismic properties of the Voisey's Bay massive sulfide deposit: Insights into approaches to seismic imaging. *Geophysics*, 77, WC59–WC68.

Dunlop, D.J. and Ozdemir, O., 1997. *Rock Magnetism: Fundamentals and Frontiers*. Cambridge University Press.

Dyck, A.V., 1991. Drill-hole electromagnetic methods. In Nabighian, M.N. (Ed.), *Electromagnetic Methods in Applied Geophysics, Volume 2, Applications: Parts A and Part B*. Society of Exploration Geophysicists, Investigations in Geophysics 3, 881–930.

Dyer, B.C. and Fawcett, A., 1994. The use of tomographic imaging in mineral exploration. *Exploration and Mining Geology*, 3, 383–387.

Eaton, D.W., Milkereit, B. and Adam, E., 1997. 3-D seismic exploration. In Gubins, A.G. (Ed.), *Proceedings of Exploration '97, Fourth Dicennial International Conference on Mineral Exploration*. Prospectors and Developers Association of Canada, 65–78.

Ehsan, S.A., Malehmir, A. and Dehghannejad, M., 2012. Re-processing and interpretation of 2D seismic data from the Kristineberg mining area, northern Sweden. *Journal of Applied Geophysics*, 80, 43–55.

El-Araby, H.M., 2004. A new method for complete quantitative interpretation of self-potential anomalies. *Journal of Applied Geophysics*, 55, 211–224.

Ellis, R.G., 1998. Inversion of airborne electromagnetic data. *Exploration Geophysics*, 29, 121–127.

Emerson, D., Macnae, J. and Sattel, D., 2000. Physical properties of the regolith in the Lawlers area, Western Australia. *Exploration Geophysics*, 31, 229–235.

Emerson, D.W., 1990. Notes on mass properties of rocks – density, porosity, permeability. *Exploration Geophysics*, 21, 209–216.

Emerson, D.W., Embleton, B.J.J. and Clark, D., 1979. The Flemington intrusion, Fifield NSW – Petrophysical and petrological notes. *Bulletin of the Australian Society of Exploration Geophysicists*, 10, 99–100.

Emerson, D.W. and Macnae, J., 2001. Further physical property data from the Archaean regolith, Western Australia. *Preview*, 92, 33–37.

Emery, D. and Myers, K., 1996. *Sequence Stratigraphy*. Blackwell Science.

Enescu, N., McDowell, G., Cosma, C. and Bell, C., 2002. Crosshole seismic investigation at Voisey's Bay, Canada. *Society of Exploration Geophysicists, 72nd Annual Conference* (Extended Abstracts) 21, 1472–1475. doi: 10.1190/1.1816942

Ernstson, K. and Scherer, H.U., 1986. Self-potential variations with time and their relation to hydrogeologic and meteorological parameters. *Geophysics*, 51, 1967–1977.

Esdale, D., Pridmore, D.F., Coggon, J. *et al.*, 2003. The Olympic Dam copper–uranium–gold–silver–REE deposit, South Australia: a geophysical case history. In Dentith, M.C. (Ed.), *Geophysical Signatures of South Australian Mineral Deposits*. Centre for Global Metallogeny, University of Western Australia, Publication 31; Australian Society of Exploration Geophysicists, Special Publication 12; and Primary Industries and Resources South Australia, 147–168.

Evans, B.J., 1997. A Handbook for Seismic Data Acquisition in Exploration. Society of Exploration Geophysicists, Geophysical Monograph Series 7, 164–177.

Evans, J.R., 1982. Running median filters and a general despiker. *Bulletin of the Seismological Society of America*, 72, 331–338.

Fallon, G.N., Fullagar, P.K. and Sheard, S.N., 1997. Application of geophysics in metalliferous mines. *Australian Journal of Earth Sciences*, 44, 391–409.

Farquharson, C.G., Ash, M.R. and Miller, H.G., 2008. Geologically constrained gravity inversion for the Voisey's Bay ovoid deposit. *The Leading Edge*, 27(1), 64–69.

Farr, T.G. and Kobrick, M., 2000. Shuttle radar topographic mission produces a wealth of data. *EOS, Transactions of the American Geophysical Union*, 81, 583–585.

Featherstone, W.E. and Dentith, M.C., 1997. A geodetic approach to gravity data reduction for geophysics. *Computers and Geosciences*, 23, 1063–1070.

Feebrey, C.A., Hishida, H., Yoshioka, K. and Nakayama, K., 1998. Geophysical expression of low sulphidation epithermal Au–Ag deposits and exploration implications – Examples from the Hokusatsu region of SW Kyushu, Japan. *Resource Geology*, 48, 75–86.

Fernandes, J.F., Iyer, S.S., Imakuma, K. and Choudhuri, A., 1987. Geochemical studies in the Proterozoic metamorphic terrane of the Guaxupé Massif, Minas Gerais, Brazil. A discussion on large ion lithophile element fractionation during high-grade metamorphism. *Precambrian Research*, 36, 65–79.

Flis, M.F. and Cowan, D.R., 2000. Aeromagnetic drape corrections applied to the Turner Syncline, Hamersley Basin. *Exploration Geophysics*, 31, 84–88.

Flis, M.F., Butt, A.L. and Hawke, P.J., 1998. Mapping the range front with gravity – are the corrections up to it? *Exploration Geophysics*, 29, 378–383.

Flis, M.F., Newman, G.A. and Hohmann, G.W., 1989. Induced-polarization effects in time-domain electromagnetic measurements. *Geophysics*, 54, 514–523. doi: 10.1190/1.1442678

Floyd, J.D. and Trench, A., 1989. Magnetic susceptibility contrasts in Ordovician greywackes of the Southern Uplands of Scotland. *Journal of the Geological Society of London*, 145, 77–83.

Ford, K.L, Dilabio, R.N.W. and Rencz, A.N., 1988. Geological, geophysical and geochemical studies around the Allan Lake carbonatite, Algonquin Park, Ontario. *Journal of Geochemical Exploration*, 30, 99–121.

Forman, J.M.A. and Angeiras, A.G., 1981. Poços de Calderas and Itataia: Two case histories of uranium exploration in Brazil. In *Uranium Exploration Case Histories*. International Atomic Energy Agency, 99–139.

Fox, P.J. and Opdyke, N.D., 1973. Geology of the oceanic crust: Magnetic properties of oceanic rocks. *Journal of Geophysical Research*, 78, 5139–5154.

Fox, P.J., Schreiber, E. and Peterson, J.J., 1973. The geology of the oceanic crust: Compressional wave velocities of oceanic rocks. *Journal of Geophysical Research*, 78, 5155–5172.

Fox, R.C., Hohmann, G.W., Killpack, T.J. and Rijo, L., 1980. Topographic effects in resistivity and induced-polarization surveys. *Geophysics*, 45, 75–93.

Frasheri, A., Lubonja, L. and Alikaj, P., 1995. On the application of geophysics in the exploration for copper and chrome ores in Albania. *Geophysical Prospecting*, 43, 743–757.

Fritz, F.P., 2000. The economics of geophysical applications. In Ellis, R.B., Irvine, R. and Fritz, F. (Eds.), *Practical Geophysics III for the Exploration Geologist*. Northwest Mining Association, Practical Geophysics Short Course 1998: Selected Papers.

Frost, B.R., 1991. Introduction to oxygen fugacity and its petrologic importance. In Lindsley, D.H. (Ed.), *Oxide Minerals: Petrologic and Magnetic Significance*. Mineralogical Society of America, Reviews in Mineralogy, 25, 1–9.

Fullagar, P.K. and Fallon, G.N., 1997. Geophysics in metalliferous mines for ore body delineation and rock mass characterisation. In Gubins, A.G. (Ed.), *Proceedings of Exploration '97, Fourth Dicennial International Conference on Mineral Exploration*. Prospectors and Developers Association of Canada, 573–584.

Fullagar, P.K. and Pears, G.A., 2007. Towards geologically realistic inversion. In Milkereit, B. (Ed.), *Proceedings of Exploration 07: Fifth Decennial International Conference on Mineral Exploration*. Decennial Mineral Exploration Conferences, 444–460.

Fullagar, P.K., Fallon, G.N., Hatherly, P.J. and Emerson, D.W., 1996. *Implementation of Geophysics at Metalliferous Mines. Application of Geophysics to Mine Planning and Operations.* Australian Mineral Industries Research Association, Project P436, Final Report.

Fullagar, P.K., Zhou, B. and Bourne, B., 2000. EM-coupling removal from time-domain IP data. *Exploration Geophysics*, 31, 134–139.

Galbraith, J.H. and Saunders, D.F., 1983. Rock classification by characteristics of aerial gamma-ray measurements. *Journal of Geochemical Exploration*, 18, 49–73.

Gallardo, L.A., 2007. Multiple cross-gradient joint inversion for geospectral imaging: *Geophysical Research Letters*, 34, L19301.

Galloway, W.E. and Hobday, D.K., 1983. *Terrigenous Clastic Depositional Systems. Applications to Petroleum, Coal and Uranium Exploration.* Springer-Verlag.

Galybin, K., Boschetti, F. and Dentith, M.C., 2007. A new approach to potential field data collection. *Journal of Geodynamics*, 43, 248–261.

Garrels, R.M. and Christ, C.L., 1965. *Solutions, Minerals and Equilibria*. Harper and Row.

Gascoyne, M., 1992. Geochemistry of the actinides and their daughters. In Invanovich, M. and Harmon, R.S. (Eds.), *Uranium-series Disequilibrium: Applications to Earth, Marine and Environmental Sciences*. Clarendon Press, 34–61.

Gibson, M.A.S., 2011. Application of seismic to mineral deposit exploration and evaluation. *The Leading Edge*, 30(6), 616–620.

Gibson, M.A.S., Jolley, S.J. and Barnicoat, A.C., 2000. Interpretation of the Western Ultra Deep Levels 3-D seismic survey. *The Leading Edge*, 19(7), 730–735.

Gillot, E., Gibson, M., Verneau, D. and Laroche, S., 2005. Application of high-resolution 3D seismic to mine planning in shallow platinum mines. *First Break*, 23(7), 59–66.

Giroux, B., Chouteau, M., Seigel, H.O. and Nind, C.J.M., 2007. A program to model and interpret borehole gravity data. In Milkereit, B. (Ed.), *Proceedings of Exploration '07: Fifth Decennial International Conference on Mineral Exploration*. Decennial Mineral Exploration Conferences, 1111–1114.

Glen, R.A, 1992. Thrust, extensional and strike-slip tectonics in an evolving Palaeozoic orogen – a structural synthesis of the Lachlan Orogen of southeastern Australia. *Tectonophysics*, 214, 341–380.

Gochioco, L.M., 1991. Tuning effect and interference reflections from thin beds and coal seams. *Geophysics*, 56, 1288–1295.

Gochioco, L.M., 1992. Modelling studies of interference reflections in thin-layered media bounded by coal seams. *Geophysics*, 57, 1209–1216.

Gochioco, L.M., 2000. High-resolution 3-D seismic survey over a coal mine reserve area in the U.S. – A case study. *Geophysics*, 65, 712–718.

Graham, E.K. and Barsch, G.R., 1969. Elastic constants of single-crystal forsterite as a function of temperature and pressure. *Journal of Geophysical Research*, 74, 5949–5960.

Grant, F.S., 1985. Aeromagnetics, geology and ore environments, I. Magnetite in igneous, sedimentary and metamorphic rocks: An overview. *Geoexploration*, 23, 303–333.

Grant, F.S and West, G.F., 1965. *Interpretation Theory in Applied Geophysics*. McGraw-Hill.

Grasty, R.L., 1979. Gamma ray spectrometric methods in uranium exploration – theory and operational procedures. In Hood, P.J. (Ed.), *Geophysics and Geochemistry in the Search for Metallic Ores*. Geological Survey of Canada, Economic Geology Report 31, 147–161.

Grauch, V.J.S. and Cordell, L., 1987. Limitations of determining density or magnetic boundaries from the horizontal gradient of gravity or pseudogravity data. *Geophysics*, 52, 118–121.

Gray, D.R. and Foster, D.A., 2004. Tectonic evolution of the Lachlan Orogen, southeast Australia: historical review, data synthesis and modern perspectives. *Australian Journal of Earth Sciences*, 51, 773–817.

Greenhalgh. S.A. and Emerson. D.W., 1986. Elastic properties of coal measure rocks from the Sydney Basin, New South Wales. *Exploration Geophysics*, 17, 157–163.

Greenhalgh, S. and Cao, S., 1998. Applied potential modelling of simple orebody structures. *Exploration Geophysics*, 29, 391–395.

Greenhalgh, S. and Mason, I., 1997. Seismic imaging with application to mine layout and development. In Gubins, A.G. (Ed.), *Proceedings of Exploration '97: Fourth Decennial International Conference on Mineral Exploration*. Prospectors and Developers Association of Canada, 585–598.

Griffiths, D.H. and King, R.F., 1981. *Applied Geophysics for Geologists and Engineers*. Pergamon Press.

Groves, D.I., Barley, M.E., Barnicoat, A.C. *et al.*, 1992. Sub-greenschist- to granulite-hosted Archaean lode-gold deposits of the Yilgarn Craton: A depositional continuum model from deep-seated hydrothermal fluids in crustal-scale plumbing systems. In Glover, J.R. and Ho, S.E. (Eds.), *The Archaean: Terrains, Processes and Metallogeny. Proceedings of the Third International Archaean Symposium*. Geology Department (Key Centre) & University Extension, The University of Western Australia, 325–338.

Gunn, P.J. and Chisholm, J., 1984. Non-conductive volcanogenic massive sulphide mineralization in the Pilbara area of Western Australia. *Exploration Geophysics*, 15, 143–153.

Gunn, P.J. and Dentith, M.C., 1997. Magnetic responses associated with mineral deposits. *AGSO Journal of Australian Geology & Geophysics*, 17(2), 145–158.

Haack, U., 1983. On the content and vertical distribution of K, Th and U in the continental crust. *Earth and Planetary Science Letters*, 62, 360–366.

Hageskov, B., 1984. Magnetic susceptibility used in mapping of amphibolite facies recrystallisation in basic dykes. *Tectonophysics*, 108, 339–351.

Haggerty, S.E., 1979. The aeromagnetic mineralogy of igneous rocks. *Canadian Journal of Earth Sciences*, 16, 1281–1293.

Hall, M., 2007. Smooth operator: Smoothing seismic interpretations and attributes. *The Leading Edge*, 26(1), 16–20.

Hansen, D.A., 1970. Iron ore exploration in North and South America. In Morley, L.W. (Ed.), *Mining and Groundwater Geophysics – 1967*. Geological Survey of Canada, Economic Geology Report 26, 371–380.

Haren, R., Liu, S., Gibson, G.M. *et al.*, 1997. Geological interpretation of high-resolution airborne geophysical data in the Broken Hill region. *Exploration Geophysics*, 28, 235–241.

Harman, P.G., 2004. Geophysical signatures of orebodies under cover. In Muhling, J., Goldfarb, R.J., Vielreicher, N. *et al.* (Eds.), *SEG 2004: Predictive Mineral Discovery Under Cover; Society of Economic Geologists Conference and Exhibition* (Extended Abstracts). Centre for Global Metallogeny, University of Western Australia, Publication 33, 85–89.

Harrison, R.J. and Freiburg, J.M., 2009. Mineral magnetism: Providing new insights into geoscience processes. *Elements*, 5, 209–215.

Hart, C.J.R., 2007. Reduced intrusion-related gold systems. In Goodfellow, W.D. (Ed.), *Mineral Deposits of Canada: A Synthesis of Major Deposit Types, District Metallogeny, the Evolution of Geological Provinces and Exploration Methods.* Geological Association of Canada, Mineral Deposits Division, Special Publication 5, 95–112.

Hatch, D., 2004. Evaluation of a full tensor gravity gradiometer for kimberlite exploration. In Lane, R.J.L. (Ed.), *Airborne Gravity 2004 – ASEG-PESA Airborne Gravity 2004 Workshop* (Abstracts). Geoscience Australia and the Geological Survey of New South Wales, Geoscience Australia Record 2004/18, 73–80.

Hatch, D. and Pitts, B., 2010. The De Beers airship gravity project. In Lane, R.J.L. (Ed.), *Airborne Gravity 2010 – ASEG-PESA Airborne Gravity 2010 Workshop* (Abstracts). Geoscience Australia and the Geological Survey of New South Wales, Geoscience Australia Record 2010/23 and GSNSW File GS2010/0457, 97–106.

Hatherly, P.J., Poole, G., Mason, I., Zhou, B. and Bassingthwaighte, H., 1998. 3D seismic surveying for coal mine applications at Appin Colliery, NSW. *Exploration Geophysics*, 29, 407–409.

Hattula, A., 1986. Magnetic 3-component borehole measurements in Finland. In Killeen, P.G. (Ed.), *Borehole Geophysics for Mining and Geotechnical Applications*. Geological Survey of Canada Paper 85-27, 237–250.

Hattula, A. and Rekola, T., 1997. The power and role of geophysics applied to regional and site-specific mineral exploration and mine grade control in Outokumpo Base Metals Oy. In Gubins, A.G. (Ed.), *Proceedings of Exploration '97, Fourth Dicennial International Conference on Mineral Exploration*, Prospectors and Developers Associations of Canada, 617–630.

Hearst, R.H. and Morris, W.A., 2001. Regional gravity setting of the Sudbury Structure. *Geophysics*, 66, 1680–1690.

Heinonen, S., Imaña, M., Snyder, D.B., Kukkonen, I.T. and Heikkinen, P.J., 2012. Seismic reflection profiling of the Pyhäsalmi VHMS-deposit: A complementary approach to the deep base metal exploration in Finland. *Geophysics*, 77, WC15-WC23.

Heithersay, P.S. and Walshe, J.L., 1995. Endeavour 26 North: a porphyry copper-gold deposit in the Late Ordovician, shoshonitic Goonumbla Volcanic Complex, New South Wales, Australia. *Economic Geology*, 90, 1506–1532.

Heithersay, P.S., O'Neill, W.J., van der Helder, P., Moore, C.R. and Harbon, P.G., 1990. Goonumbla porphyry copper district – Endeavour 26 North, Endeavour 22 and Endeavour 27 copper-gold deposits. In Hughes, F.E. (Ed.), *Geology of the Mineral Deposits of Australia and Papua New Guinea.* Australasian Institute of Mining and Metallurgy, Monograph Series 14, 1385–1398.

Henkel, H., 1991. Petrophysical properties (density and magnetization) of rocks from the northern part of the Baltic Shield. *Tectonophysics*, 192, 1–19.

Henkel, H., 1994. Standard diagrams of magnetic properties and density – a tool for understanding magnetic petrology. *Journal of Applied Geophysics*, 32, 43–53.

Henkel, H. and Guzmán, M., 1977. Magnetic features of fracture zones. *Geoexploration*, 15, 173–181.

Henson, H. and Sexton, J.L., 1991. Premine study of shallow coal seams using high-resolution seismic reflection methods. *Geophysics*, 56, 1494–1503.

Herron, D.A., 2000. Pitfalls in seismic interpretation: depth migration artefacts. *The Leading Edge*, 19(9), 1016–1017.

Hinze, W.J., 2003. Bouguer reduction density, why 2.67? *Geophysics*, 68, 1559–1560.

Hogg, S., 2004. Practicalities, pitfalls and new developments in airborne magnetic gradiometry. *First Break*, 22(7), 59–65.

Hoschke, T., 1985. A new drill hole magnetometer: preliminary results from the Tennant Creek area. *Exploration Geophysics*, 16, 365–374.

Hoschke, T., 1991. Geophysical discovery and evaluation of the West Peko copper–gold deposit, Tennant Creek. *Exploration Geophysics*, 22, 485–495.

Hoschke, T., 2011. *Geophysical Signatures of Copper–Gold Porphyry and Epithermal Gold Deposits, and Implications for Exploration.* ARC Centre of Excellence in Ore Deposits, University of Tasmania.

Hoschke, T. and Sexton, M., 2005. Geophysical exploration for epithermal gold deposits at Pajingo, North Queensland, Australia. *Exploration Geophysics*, 36, 401–406.

Hrouda, F., 1982. Magnetic anisotropy of rocks and its application in geology and geophysics. *Geophysical Surveys*, 5, 37–82.

Hsu, S.K., Sibuet, J.C. and Shyu, C.T., 1996. High-resolution detection of geologic boundaries from potential field anomalies: An enhanced analytic signal technique. *Geophysics*, 61, 373–386.

Huang, H. and Rudd, J., 2008. Conductivity-depth imaging of helicopter-borne TEM data based on a pseudolayer half-space model. *Geophysics*, 73, F115–F120.

Hughes, N.A. and Ravenhurst, W.R., 1996. Three component DHEM surveying at Balcooma. *Exploration Geophysics*, 27, 77–89.

Hunter, J.A., Pullan, S.E., Burns, R.A. *et al.*, 1998. Downhole seismic logging for high-resolution reflection surveying in unconsolidated overburden. *Geophysics*, 63, 1371–1384.

Hurich, C.A., Deemer, S.J., Indares, A. and Salisbury, M., 2001. Compositional and metamorphic controls on velocity and reflectivity in the continental crust: An example from the Grenville Province of eastern Quebec. *Journal of Geophysical Research*, B106, 665–682.

Huston, D.L. and Taylor, T.W., 1990. Dry River copper and lead-zinc-copper deposits. In Hughes, F.E. (Ed.), *Geology of the Mineral Deposits of Australia and Papua New Guinea*. Australasian Institute of Mining and Metallurgy, Monograph Series 14, 1519–1526.

IAEA, 1982. Borehole logging for uranium exploration—a manual. International Atomic Energy Agency, Technical Reports Series 212.

IAEA, 1991. Airborne gamma ray spectrometer surveying. International Atomic Energy Agency, Technical Reports Series 323.

Ildefonse, B. and Pezard, P., 2001. Electrical properties of slow-spreading ridge gabbros from ODP site 735, Southwest Indian Ridge. *Tectonophysics*, 330, 69–92.

Irvine, R.J. and Smith, M.J., 1990. Geophysical exploration for epithermal gold deposits. *Journal of Geochemical Exploration*, 36, 375–412.

Irvine, R.J. and Staltari, G., 1984. Case history illustrating interpretation problems in transient electromagnetic surveys. *Exploration Geophysics*, 15, 156–167.

Irvine, R.J., Hartley, J.S. and Mourot, A., 1985. The geophysical characteristics of the Thalanga volcanogenic sulphide deposit, Queensland. *Exploration Geophysics*, 16, 231–234.

Irving, E., Molyneux, L. and Runcorn, S.K., 1966. The analysis of remanent intensities and susceptibilities of rocks. *Geophysical Journal of the Royal Astronomical Society*, 10, 451–464.

Ishihara, S., 1981. The granitoid series and mineralization. *Economic Geology, 75th Anniversary Volume*, 458–484.

Ishikawa, K., Takahashi, K., Miyajima, K., Ochi, H. and Seko, T., 1981. Geological investigation utilizing borehole measurement system and judgement of engineering properties for weathered granite. In Akai, K., Hayashi, M. and Nishimatsu, Y. (Eds.), *Proceedings of the International Symposium on Weak Rock, Tokyo, 1981*, 387–392.

Ispolatov, V., Lafrance, B., Dube, B., Creaser, R. and Hamilton, M., 2008. Geologic and structural setting of gold mineralization in the Kirkland Lake–Larder Lake gold belt, Ontario. *Economic Geology*, 103, 1309–1340.

Jackson, J.C., Fallon, G.N. and Bishop, J.R., 1996. DHEM at Isa Mine. *Exploration Geophysics*, 27, 91–104.

Jackson, S.L. and Fyon, J.A., 1991. The Western Abitibi Subprovince in Ontario. In Thurston, P.C., Williams, H.R., Sutcliffe, R.H. and Stott, G.M. (Eds.), *Geology of Ontario*, Ontario Geological Survey, Special Volume 4, Part 1, 405–484.

Jayawardhana, P.M. and Sheard, S.N., 1997. The use of airborne gamma ray spectrometry by M.I.M. Exploration – A case study from the Mount Isa Inlier, northwest Queensland, Australia. In Gubins, A.G. (Ed.), *Proceedings of Exploration '97, Fourth Dicennial International Conference on Mineral Exploration*. Prospectors and Developers Association of Canada, 765–774.

Jenke, G. and Cowan, D.R., 1994. Geophysical signature of the Ellendale lamproite pipes, Western Australia. In Dentith, M.C., Frankcombe, K.F., Ho, S.E. *et al.* (Eds.), *Geophysical Signatures of Western Australian Mineral Deposits*. University of Western Australia, Geology and Geophysics Department (Key Centre), Publication 26, and Australian Society of Exploration Geophysicists, Special Publication 7, 403–414.

Johnston, D.H., Toksöz, M.N. and Timur, A., 1979. Attenuation of seismic waves in dry and saturated rocks: II. Mechanisms. *Geophysics*, 31, 691–711. doi: 10.1190/1.1440970

Juhlin, C. and Palm, H., 2003. Experiences with shallow reflection seismics over granitic rocks in Sweden. In Eaton, D.W., Milkereit, B. and Salisbury, M.H. (Eds.), *Hardrock Seismic Exploration*. Society of Exploration Geophysicists, Geophysical Development Series 10, 93–109. doi: 10.1190/1.9781560802396.ch6

Jurza, P., Campbell, I., Robinson, P. *et al.*, 2005. Use of ^{214}Pb photopeaks for radon removal: Utilising current airborne gamma-ray spectrometer technology and data processing. *Exploration Geophysics*, 36, 322–328.

Kannan, S. and Mallick, K., 2003. Accurate regional-residual separation by finite element approach, Bouguer gravity of a Precambrian mineral prospect in northwestern Ontario. *First Break*, 21(4), 39–42.

Karlsen, T.A. and Olesen, O., 1996. Airborne geophysical prospecting for ultramafite associated talc, Altermark, northern Norway. *Journal of Applied Geophysics*, 35, 215–236.

Katsube, T.J. and Hume, J.P., 1987. Permeability determination in crystalline rocks by standard geophysical logs. *Geophysics*, 52, 342–352.

Kay, A. and Duckworth, K., 1983. The effect of permafrost on the IP response of lead zinc ores. *Journal of the Canadian Society of Exploration Geophysicists*, 19, 75–83.

Kayal, J.R., 1979. Electrical, gamma-ray and temperature logging for coal in Gondwana Basin of Eastern India. In Laskar, B.

and Raja Rao, C.S. (Eds.), *Fourth International Gondwana Symposium: Papers*, Volume 2. Geological Survey of India, 875–885.

Kayal, J.R., Sukumar Datta and Madhusudan, I.C., 1982. Resistance and self-potential logging for copper deposits. *Geophysical Research Bulletin*, 20, 157–161.

Keele, R.A., 1994. Magnetic susceptibilities of rocks associated with some Archaean gold deposits in Western Australia. In Dentith, M.C., Frankcombe, K.F., Ho, S.E. *et al.* (Eds.), *Geophysical Signatures of Western Australian Mineral Deposits*. University of Western Australia, Geology and Geophysics Department (Key Centre), Publication 26, and Australian Society of Exploration Geophysicists, Special Publication 7, 315–329.

Keller, G.V., 1988. Rock and mineral properties. In Nabighian, M.N. (Ed.), *Electromagnetic Methods in Applied Geophysics* Volume 1, *Theory*. Society of Exploration Geophysicists, Investigations in Geophysics 3, 13–51. doi: 10.1190/1.9781560802631.ch2

Kenter, J.A.M., Podladchikov, F.F., Reiners, M., *et al.*, 1997. Parameters controlling sonic velocities in a mixed carbonate-siliciclastic Permian shelf-margin (upper San Andres Formation, Last Chance Canyon, New Mexico). *Geophysics*, 62, 505–520.

Kern, H., 1982. Elastic wave velocities and constants of elasticity of rocks at elevated pressures and temperatures. In Angenheister, G. (Ed.), *Physical Properties of Rocks, Landolt-Börnstein. Numerical Data and Functional Relationships in Science and Technology, Group V : Geophysics and Space Research*, Volume 1b. Springer, 99–140.

Kerr, T.L., O'Sullivan, A.P., Podmore, D.C., Turner, R. and Waters, P., 1994. Geophysics and iron ore exploration: Examples from the Jimblebar and Shay Gap–Yarrie regions, Western Australia. In Dentith, M.C., Frankcombe, K.F., Ho, S.E. *et al.* (Eds.), *Geophysical Signatures of Western Australian Mineral Deposits*. University of Western Australia, Geology and Geophysics Department (Key Centre), Publication 26, and Australian Society of Exploration Geophysicists, Special Publication 7, 355–367.

Ketola, M., 1972. Some points of view concerning mise-a-la-masse measurements. *Geoexploration*, 10, 1–21.

Ketola, M., Liimatainen, M. and Ahokas, T., 1976. Application of petrophysics to sulfide ore prospecting in Finland. *Pure and Applied Geophysics*, 114, 215–234.

Killeen, P.G., 1997a. Borehole Geophysics: Exploring the third dimension. In Gubins, A.G. (Ed.), *Proceedings of Exploration '97, Fourth Dicennial International Conference on Mineral Exploration*. Prospectors and Developers Association of Canada, 31–42.

Killeen, P.G., 1997b. Nuclear techniques for ore grade estimation. In Gubins, A.G. (Ed.), *Proceedings of Exploration '97, Fourth Dicennial International Conference on Mineral Exploration*. Prospectors and Developers Association of Canada, 677–684.

Killeen, P.G., Mwenifumbo, C.J., Elliot, B.E., and Chung, C.J., 1997. Improving exploration efficiency by predicting geological drill core logs with geophysical logs. In Gubins, A.G. (Ed.), *Proceedings of Exploration '97, Fourth Dicennial International Conference on Mineral Exploration*. Prospectors and Developers Association of Canada, 713–716.

King, A., 1996. Deep drillhole electromagnetic surveys for nickel/copper sulphides at Sudbury Canada. *Exploration Geophysics*, 27, 105–118.

King, A., 2007. Review of geophysical technology for Ni-Cu-PGE deposits. In Milkereit, B. (Ed.), *Proceedings of Exploration '07: Fifth Decennial International Conference on Mineral Exploration*, Decennial Mineral Exploration Conferences, 647–665.

King, M.S., 1966. Wave velocities in rocks as a function of change in overburden pressure and pore fluid saturants. *Geophysics*, 31, 50–73.

Kissin, S.A. and Scott, S.D., 1982. Phase relations involving pyrrhotite below 350 °C. *Economic Geology*, 77, 1739–1754.

Klein, J. and Lajoie, J.J., 1992. Electromagnetics: Electromagnetic prospecting for minerals. In Van Blaricom, R. (Compiler), *Practical Geophysics for the Exploration Geologist II*. Northwest Mining Association, 383–438.

Kleinkopf, M.D., Peterson, D.L. and Gott, G., 1970. Geophysical studies of the Cripple Creek mining district, Colorado. *Geophysics*, 35, 490–500.

Klingspor, A.M., 1966. Cyclic deposits of potash in Saskatchewan. *Bulletin of Canadian Petroleum Geology*, 14, 193–207.

Komor, S.C., Elthon, D. and Casey, J.F., 1985. Serpentinization of cumulate ultramafic rocks from the North Arm Mountain massif of the Bay of Islands ophiolite. *Geochimica et Cosmochimica Acta*, 49, 2331–2338.

Kontny, A. and Dietl, C., 2002. Relationships between contact metamorphism and magnetite formation and destruction in a pluton's aureole, White-Inyo Range, eastern California. *Geological Society of America Bulletin*, 114, 1438–1451.

Krauskopf, K.B. and Bird, D.K., 1995. *Introduction to Geochemistry*. McGraw-Hill.

Kumazawa, M. and Anderson, O.L., 1969. Elastic moduli, pressure derivatives, and temperature derivatives of single-crystal olivine and single-crystal forsterite. *Journal of Geophysical Research*, 74, 5961–5972.

Kvasnicka, M. and Cerveny, V., 1996. Analytical expression for Fresnel volumes and interface Fresnel zones of seismic body waves. Part 1: Direct and unconverted reflected waves. *Studia Geophysica et Geodaetica*, 40, 136–155.

LaFehr, T.R., 1991. An exact solution for the gravity curvature (Bullard B) correction. *Geophysics*, 56, 1179–1184.

Lajoie, J.J. and Klein, J., 1979. Geophysical exploration at the Pine Point Mines Ltd zinc–lead property, Northwest Territories, Canada. In Hood, P.J. (Ed.), *Geophysics and Geochemistry in the Search for Metallic Ores*. Geological Survey of Canada, Economic Geology Report 31, 653–664.

Lanne, E., 1986. Statistical multivariant analysis of airborne geophysical data on the SE border of the central Lapland greenstone complex. *Geophysical Prospecting*, 34, 1111–1128.

Lanning, E.N. and Johnson, D.M., 1983. Automated identification of rock boundaries: An application of the Walsh transform to geophysical well-log analysis. *Geophysics*, 48, 197–205.

Lapointe, P., Morris, W.A. and Harding, K.L., 1986. Interpretation of magnetic susceptibility: a new approach to geophysical evaluation of the degree of rock alteration. *Canadian Journal of Earth Sciences*, 23, 393–401.

Larroque, M., Postel, J-J., Slabbert, M. and Duweke, W., 2002. How 3D seismic can help enhance mining. *First Break*, 20(7), 472–475.

Larsson, L.O., 1977. *Statistical Treatment of In Situ Measurements of Magnetic Susceptibility*. Sveriges Geologiska Undersokning, Serie C 727.

Lawton, D.C. and Hochstein, M.P., 1993. Geophysical study of the Taharoa iron sand deposit, west coast, North Island, New Zealand. *New Zealand Journal of Geology and Geophysics*, 36, 141–160.

Leaman, D.E., 1994. Criteria for evaluation of potential field interpretations. *First Break*, 12(4), 181–191.

Leaman, D.E., 1998. The gravity terrain correction – practical considerations. *Exploration Geophysics*, 29, 467–471.

Leggatt, P.B., Klinkert, P.S. and Hage, T.B., 2000. The Spectrem airborne electromagnetic system – Further developments. *Geophysics*, 65, 1976–1982.

Lehmann, K., 2010. Environmental corrections to gamma-ray log data: Strategies for geophysical logging with geological and technical drilling. *Journal of Applied Geophysics*, 70, 17–26.

Lemoine, F.G., Kenyon, S.C., Factor, J.K. *et al.*, 1998. The Development of the Joint NASA GSFC and the National Imagery and Mapping Agency (NIMA) Geopotential Model EGM96, National Aeronautics and Space Administration, *Technical Paper* 1998–206861.

Levinson, A.A. and Coetzee, G.L., 1978. Implications of disequilibrium in exploration for uranium ores in the surficial environment using radiometric techniques – a review. *Minerals Science Engineering*, 10, 19–27.

Li, X., 2008. Magnetic reduction-to-the-pole at low latitudes: Observations and considerations. *The Leading Edge*, 27(8), 990–1002.

Li, X. and Götze, H-J., 1999. Comparison of some gridding methods. *The Leading Edge*, 18(8), 898–900.

Li, X. and Götze, H-J., 2001. Ellipsoid, geoid, gravity, geodesy and geophysics. *Geophysics*, 66, 1660–1668.

Li, Y. and Calvert, A.J., 1997. Seismic reflection imaging of a shallow, fault-controlled VMS deposit in the Matagami mining camp, Québec. In Gubins, A.G. (Ed.), *Proceedings of Exploration '97, Fourth Dicennial International Conference on Mineral Exploration*. Prospectors and Developers Association of Canada, 467–472.

Lilley, F.E.M., 1982. Geomagnetic field fluctuations over Australia in relation to magnetic surveys. *Exploration Geophysics*, 13, 68–76.

Lillie, R.J., 1999. *Whole Earth Geophysics: An Introductory Textbook for Geologists and Geophysicists*. Prentice Hall.

Lindsey, J.P., 1989. The Fresnel zone and its interpretive significance. *The Leading Edge*, 8(10), 33–39.

Lipton, G., 1997. Spectral and microwave remote sensing: An evolution from small scale regional studies to mineral mapping and ore deposit targeting. In Gubins, A.G. (Ed.), *Proceedings of Exploration '97, Fourth Dicennial International Conference on Mineral Exploration*. Prospectors and Developers Association of Canada, 43–58.

Lipton, I.T., 2001. Measurement of bulk density for resource estimation. In Edwards, A.C. (Ed.), *Mineral Resource and Ore Reserve Estimation – The AusIMM Guide to Good Practice*. The Australasian Institute of Mining and Metallurgy, 57–66.

Lockhart, G., Grutter, H. and Carlson, J., 2004. Temporal, geomagnetic and related attributes of kimberlite magmatism at Ekati, Northwest Territories, Canada. *Lithos*, 77, 665–682.

Logn, Ø. and Bølviken, B., 1974. Self-potentials at the Joma pyrite deposit, Norway. *Geoexploration*, 12, 11–28.

Longman, I.M., 1959. Formulas for computing the tidal accelerations due to the moon and the sun. *Journal of Geophysical Research*, 64, 2351–2355.

Løvborg, I., Nyegaard, P., Christiansen, E.M. and Nielsen, B.L., 1980. Borehole logging for uranium by gamma-ray spectrometry. *Geophysics*, 45, 1077–1090.

Lowrie, W., 2007. *Fundamentals of Geophysics*. Cambridge University Press.

Luyendyk, A.P.J., 1997. Processing of airborne magnetic data. *AGSO Journal of Australian Geology & Geophysics*, 17(2), 31–38.

Mach, K., 1997. A logging correlation scheme for the Main coal seam of the North Bohemian brown coal, and the implications for the palaeogeographical development of the basin. In Gayer, R. and Pesek, J. (Eds.), *European Coal Geology and Technology*. Geological Society of London, Special Publication 125, 309–320.

MacLeod, I.N., Jones, K. and Dai, T.F., 1993. 3-D analytic signal in the interpretation of total magnetic field data in low magnetic latitudes. *Exploration Geophysics*, 24, 679–688.

Macnae, J., 1995. Applications of geophysics for the detection and exploration of kimberlites and lamproites. *Journal of Geochemical Exploration*, 53, 213–243.

Macnae, J.C., 1979. Kimberlites and exploration geophysics. *Geophysics*, 44, 1395–1416.

Macnae, J.C., Lamontagne, Y. and West, G.F., 1984. Noise processing for time-domain EM systems. *Geophysics*, 49, 934–948. doi: 10.1190/1.1441739

Madhusudan, I.C., Pal, T. and Das, B., 1990. Geophysical surveys for the Sargipalli graphite, Sambalpur, Orissa. *Indian Minerals*, 44, 325–334.

Maidment, D.W., Gibson, G.M. and Giddings, J.W., 2000. Regional structure and distribution of magnetite: implications for the interpretation of aeromagnetic data in the Broken Hill region, New South Wales. *Exploration Geophysics*, 31, 8–16.

Malehmir, A., Koivisto, E., Manzi, M. *et al.*, 2014. A review of reflection seismic investigations in three major metallogenic regions: the Kevista Ni-Cu-PGE district (Finland), Witwatersrand goldfields (South Africa) and the Bathurst mining camp (Canada). *Ore Geology Reviews*, 56, 423–441.

Manzi, M.S.D., Durrheim, R.J., Hein, K.A.A. and King, N., 2012a. 3D edge detection seismic attributes used to map potential conduits for water and methane in deep gold mines in the Witwatersrand Basin, South Africa. *Geophysics*, 77, WC133–WC147.

Manzi, M.S.D., Gibson, M.A.S., Hein, K.A.A., King, N. and Durrheim, R.J., 2012b. Application of 3D seismic techniques to evaluate ore resources in the West Wits Line goldfield and portions of the West Rand goldfield, South Africa. *Geophysics*, 77, WC163–WC171. doi: 10.1190/geo2012-0135.1

Mason, I.M., Buchanan, D.J. and Booer, A.K., 1980. Channel wave mapping of coal seams in the United Kingdom. *Geophysics*, 45, 1131–1143.

Mauring, E., Beard, L.P., Kihle, O. and Smethurst, M.A., 2002. A comparison of aeromagnetic levelling techniques with an introduction to median levelling. *Geophysical Prospecting*, 50, 43–54.

McBarnet, A., 2005. A little goes a long way with Fugro's latest unmanned aeromagnetic survey flying machine. *First Break*, 23(20), 9–11.

McDowell, G.M., Stewart, R. and Monteiro, R.N., 2007. In-mine exploration and delineation using an integrated approach. In Milkereit, B. (Ed.), *Proceedings of Exploration '07: Fifth Decennial International Conference on Mineral Exploration*. Decennial Mineral Exploration Conferences 571–589.

McEnroe, S.A., Brown, L.L. and Robinson, P., 2009a. Remanent and induced magnetic anomalies over a layered intrusion: Effects from crystal fractionation and magma recharge. *Tectonophysics*, 478, 119–134.

McEnroe, S.A., Fabian, K., Brown, L.L. *et al.*, 2009b. Crustal magnetism, lamellar magnetism and rocks that remember. *Elements*, 5, 241–246.

McGaughey, J., 2007. Geological models, rock properties, and the 3D inversion of geophysical data. In Milkereit, B. (Ed.), *Proceedings of Exploration '07: Fifth Decennial International Conference on Mineral Exploration*. Decennial Mineral Exploration Conferences, 473–483.

McIntosh, S.M., Gill, J.P. and Mountford, A.J., 1999. The geophysical response of the Las Cruces massive sulphide deposit. *Exploration Geophysics*, 30, 123–134.

McKay, W.J. and Davies, R.H., 1990. Woodlawn: A synthesis of exploration and structural geology. In Glasson, K.R. and Rattigan, J.H. (Eds.), *Geological Aspects of the Discovery of Some Important Mineral Deposits in Australia*. Australasian Institute of Mining and Metallurgy, Monograph Series 17, 211–218.

McNeil, J.D., 1980. Applications of transient electromagnetic techniques. Technical Note TN-7, Geonics Ltd.

Mehegan, J.M., Robinson, P.T. and Delaney, J.R., 1982. Secondary mineralisation and hydrothermal alteration in the Reydarfjordur drill core, eastern Iceland. *Journal of Geophysical Research*, 87, 6511–6524.

Meju, M.A., 2002. Geoelectromagnetic exploration for natural resources; models, case studies and challenges. *Surveys in Geophysics*, 23, 133–205.

Miller, D.J. and Christensen, N.I., 1997. Seismic velocities of lower crustal and upper mantle rocks from the slow-spreading Mid-Atlantic Ridge, south of the Kane transform zone (MARK). In Karson, J.A., Cannat, M., Miller, D.J. and Elthon, D. (Eds.), *Proceedings of the Ocean Drilling Programme, Scientific Results*, 153, 437–454.

Miller, H.G. and Singh, V., 1994. Potential field tilt – a new concept for the location of potential field sources. *Journal of Applied Geophysics*, 32, 213–217.

Milligan, P.R. and Gunn, P.J., 1997. Enhancement and presentation of airborne geophysical data. *AGSO Journal of Australian Geology & Geophysics*, 17(2), 63–75.

Minty, B.R.S., 1991. Simple micro-levelling for aeromagnetic data. *Exploration Geophysics*, 22, 591–592.

Minty, B.R.S., 1992. Airborne gamma-ray spectrometric background estimation using full spectrum analysis. *Geophysics*, 57, 279–287.

Minty, B.R.S., 1997. Fundamentals of airborne gamma-ray spectrometry. *AGSO Journal of Australian Geology & Geophysics*, 17(2), 39–50.

Minty, B.R.S., 1998. Recent developments in the processing of airborne gamma-ray spectrometric data. *Preview*, 75, 12–24.

Minty, B.R.S., Luyendyk, A.P.J. and Brodie, R.C., 1997. Calibration and data processing for airborne gamma-ray spectrometry. *AGSO Journal of Australian Geology & Geophysics*, 17(2), 51–62.

Moore, D.W., Young, L.E., Modene, J.S. and Plahuta, J.T., 1986. Geological setting and genesis of the Red Dog zinc-lead-silver deposit, Western Brooks Range, Alaska. *Economic Geology*, 81, 1696–1727.

Moritz, H., 1980. Geodetic Reference System 1980. *Bulletin Géodésique*, 54, 395–405.

Morrell, A.E., Locke, C.A., Cassidy, J. and Mauk, J., 2011. Geophysical characteristics of adularia-sericite epithermal gold–silver deposits in the Waihi-Waitekauri region, New Zealand. *Economic Geology*, 106, 1031–1041.

Morris, B., Pozza, M., Boyce, J. and Leblanc, G., 2001. Enhancement of magnetic data by logarithmic transform. *The Leading Edge*, 20(8), 882–885.

Moxham, R.M., Foote, R.S. and Bunker, C.M., 1965. Gamma-ray spectrometer studies of hydrothermally altered rocks. *Economic Geology*, 60, 653–671.

Mudge, S., 1996. Helicopter magnetic surveys in rugged terrains: From planning to processing. *The Leading Edge*, 15(4), 305–308.

Mudge, S. and Teakle, M., 2003. Geophysical exploration for heavy-mineral sands near Mindarie, South Australia. In Dentith, M.C. (Ed.), *Geophysical Signatures of South Australian Mineral Deposits*. Centre for Global Metallogeny, University of Western Australia, Publication 31; Australian Society of Exploration Geophysicists, Special Publication 12; and Primary Industries and Resources South Australia, 249–255.

Mudge, S.T., 1988. The slope correction method of analysing magnetic data from a sloping survey plane. *Exploration Geophysics*, 19, 489–497.

Mudge, S.T., 1991. New developments in resolving detail in aeromagnetic data. *Exploration Geophysics*, 22, 277–284.

Mudge, S.T., 1996. The location of TEM transmitter loops underground and in rugged terrain. *Exploration Geophysics*, 27, 175–177.

Mudge, S.T., 1998. Crossing the borders: from plains to rugged terrains. *Exploration Geophysics*, 29, 531–534.

Mudge, S.T., 2004. Radial resistivity/IP surveying using a downhole current electrode. *Exploration Geophysics*, 35, 188–193.

Mueller, R.F., 1978. Iron. In Wedepohl, K.H. (Ed.), *Handbook of Geochemistry*, Vol. II/3. Springer, 26-A-1–M-5.

Murphy, C.A., 2004. The Air-FTG^{TM} airborne gravity gradiometer system. In Lane, R.J.L. (Ed.), *Airborne Gravity 2004 – ASEG-PESA Airborne Gravity 2004 Workshop* (Abstracts). Geoscience Australia and the Geological Survey of New South Wales, Geoscience Australia Record 2004/18, 7–14.

Mushayandebvu, M.F., van Driel, P., Reid, A.B. and Fairhead, J.D., 2001. Magnetic source parameters of two-dimensional structures using extended Euler deconvolution. *Geophysics*, 66, 814–823.

Mutton, A.J., 1994. Application of downhole geophysical logging to lithological correlation and resource assessment in base metal deposits. *Symposium on the Application of Borehole Logging to Mineral Exploration and Mining, Perth 1994.* Australian Mineral Industries Research Association.

Mwenifumbo, C.J., 1997. Electrical methods for ore body delineation. In Gubins, A.G. (Ed.), *Proceedings of Exploration '97, Fourth Dicennial International Conference on Mineral Exploration*. Prospectors and Developers Association of Canada, 667–767.

Mwenifumbo, C.J., Elliot, B.E., Jefferson, C.W., Bernius, G.R. and Pflug, K.A., 2004. Physical rock properties from the Athabasca Group: Designing geophysical exploration models for unconformity uranium deposits. *Journal of Applied Geophysics*, 55, 117–135.

Mwenifumbo, C.J., Killeen, P.G. and Elliot, B.E., 1998. Borehole geophysical signatures of kimberlites in Canada. *The Log Analyst*, 39, 38–52.

Nabighian, M.N., 1991. *Electromagnetic Methods in Applied Geophysics,* Volume 2, *Application: Parts A and B*. Society of Exploration Geophysicists, Investigations in Geophysics 3.

Nabighian, M.N. and Asten, M.W., 2002. Metalliferous mining geophysics – state of the art in the last decade of the 20th century and the beginning of the new millennium. *Geophysics*, 67, 964–978. doi: 10.1190/1.1484538

Nelson, P.H. and Van Voorhis, G.D., 1983. Estimation of sulfide content from induced polarization data. *Geophysics*, 48, 62–75. doi: 10.1190/1.1441408

Nemeth, B., Danyluk, T., Prugger, A. and Halabura, S., 2002. Benefits of 3D poststack depth migration: case study from the Potash Belt of Saskatchewan. *Canadian Society of Exploration Geophysicists National Convention 2002, Taking Exploration to the Edge* (Extended Abstracts).

Nowell, D.A.G., 1999. Gravity terrain corrections – an overview. *Journal of Applied Geophysics*, 42, 117–134.

O'Connell, M.D., Smith, R.S. and Vallee, M.A., 2005. Gridding aeromagnetic data using longitudinal and transverse horizontal gradients with the minimum curvature operator. *The Leading Edge*, 24(2), 142–145.

Oldenburg, D.W. and Li, Y., 1994. Inversion of induced polarization data. *Geophysics*, 59, 1327–1341.

Oldenburg, D.W. and Pratt, D.A., 2007. Geophysical inversion for mineral exploration: a decade of progress in theory and practice. In Milkereit, B. (Ed.), *Proceedings of Exploration '07: Fifth Decennial International Conference on Mineral Exploration*. Decennial Mineral Exploration Conferences, 61–95.

Oldenburg, D.W., Li, Y., Farquharson, C.G. *et al.*, 1998. Applications of geophysical inversion in mineral exploration. *The Leading Edge*, 17(4), 461–465.

Olesen, O., Henkel, H., Kaada, K. and Tveten, E., 1991. Petrophysical properties of a prograde amphibolite-granulite facies transition zone at Sigerfjord, Vesterålen, northern Norway. *Tectonophysics*, 192, 33–39.

Ontario Geological Survey, 2001. Physical rock property data from the Operation Treasure Hunt physical rock property study in the Matheson and Kirkland Lake areas. *Ontario Geological Survey Miscellaneous Release Data* – 91.

Ontario Geological Survey, 2003. *Ontario Airborne Geophysical Surveys, Magnetic and Electromagnetic Data, Kirkland Lake Area.* Ontario Geological Survey, Geophysical Dataset 1102 – Revised.

Oppliger, G.K., 1984. Three-dimensional terrain corrections for mise-a-la-masse and magnetometric resistivity surveys. *Geophysics*, 49, 1718–1729.

Osmond, R.T., Watts, A.H., Ravenhurst, W.R., Foley, C.P. and Leslie, K., 2002. Finding nickel from the B-field at Raglan – 'To B or not db'. *CSEG Recorder*, 27, 44–47.

Palacky, G.J., 1988. Resistivity characteristics of geologic targets. In Nabighian, M.N. (Ed.), *Electromagnetic Methods in Applied Geophysics*, Volume 1, Theory. Society of Exploration Geophysicists, Investigations in Geophysics 3, 53–129.

Palacky, G.J. and Kadekaru, K., 1979. Effect of tropical weathering on electrical and electromagnetic measurements. *Geophysics*, 44, 69–88.

Parasnis, D.S., 1970. Some recent geoelectrical measurements in the Swedish sulphide ore fields illustrating scope and limitations of the methods concerned. In Morley, L.W. (Ed.), *Mining and Groundwater Geophysics – 1967*. Geological Survey of Canada, Economic Geology Report 26, 290–301.

Parker Gay, S., 2004. Glacial till: A troublesome source of near-surface magnetic anomalies. *The Leading Edge*, 23(6), 542–547.

Parkhomenko, E.I., 1967. *Electrical Properties of Rocks*. Plenum Press.

Pelton, W.H., Ward, S.H., Hallof, P.G., Sill, W.R. and Nelson, P.H., 1978. Mineral discrimination and removal of inductive coupling with multifrequency IP. *Geophysics*, 43, 588–609.

Peric, M., 1981. Exploration of Burundi nickeliferous laterites by electrical methods. *Geophysical Prospecting*, 29, 274–287.

Pezard, P.A., Howard, J.J. and Goldberg, G., 1991. Electrical conduction in oceanic gabbros, Hole 735B, Southwest Indian Ridge. In Von Herzem, R., Robinson, P.T. *et al.*, *Fracture Zone Drilling on the Southwest Indian Ridge. Proceedings of Ocean Drilling Program Scientific Results*, 118, 323–331.

Pilkington, M. and Keating, P., 2004. Contact mapping from gridded magnetic data – a comparison of techniques. *Exploration Geophysics*, 35, 306–311.

Pires, A.C.B. and Harthill, N., 1989. Statistical analysis of airborne gamma-ray data for geological mapping purposes: Crixas-Itapaci area, Goias, Brazil. *Geophysics*, 54, 1326–1332.

Pirkle, F.L., Campbell, K. and Wecksung, G.W., 1980. Principal components analysis as a tool for interpreting NURE aerial radiometric survey data. *Journal of Geology*, 88, 57–67.

Pitkin, J.A. and Duval, J.S., 1980. Design parameters for aerial gamma-ray surveys. *Geophysics*, 45, 1427–1439. doi: 10.1190/1.1441131

Polzer, B., 2007. A new software system for data visualization and process integration. *First Break*, 25(7), 85–90.

Pous, J., Queralt, P. and Chavez, R., 1996. Lateral and topographic effects in geoelectric soundings. *Journal of Applied Geophysics*, 35, 237–248.

Powell, B., Wood, G. and Bzdel, L., 2007. Advances in geophysical exploration for uranium deposits in the Athabasca Basin. In Milkereit, B. (Ed.), *Proceedings of Exploration '07: Fifth Decennial International Conference on Mineral Exploration*. Decennial Mineral Exploration Conferences, 771–790.

Pratt, D. and Witherly, K., 2003. Integration of polyhedral and voxel-based inversion strategies. *Joint Annual Meeting of the Geological Association of Canada, Mineralogical Association of Canada and Society of Exploration Geophysicists, Vancouver* (Extended Abstracts), 28.

Pretorius, C.C., 2004. Use of geophysics for targeting: some useful lessons from the Witwatersrand paradigm. In Muhling, J., Goldfarb, R., Vielreicher, N. *et al.* (Eds.), *SEG*

2004: Predictive Mineral Discovery Under Cover. Society of Economic Geologists Conference and Exhibition (Extended Abstracts). Centre for Global Metallogeny, University of Western Australia, Publication 33, 134–138.

Pretorius, C.C., Muller, M.R., Larroque, M. and Wilkins, C., 2003. A review of 16 years of hardrock seismics on the Kappvaal Craton. In Eaton, D.W., Milkereit, B. and Salisbury, M.H. (Eds.), *Hardrock Seismic Exploration*, Society of Exploration Geophysicists, Geophysical Development Series 10, 247–268. doi: 10.1190/1.9781560802396.ch16

Prugger, A., Nemeth, B. and Danyluk, T., 2004. Detailed 3D seismic imaging of Paleozoic karst/collapse disturbances in Saskatchewan: Cases study from the Potash Belt. *Great Explorations – Canada and Beyond*, Canadian Society of Economic Geologists National Convention 2004 (Extended Abstracts).

Puranen, M., Marmo, V. and Hämäläinen, U., 1968. On the geology, aeromagnetic anomalies and susceptibilities of Precambrian rocks in the Virrat region (central Finland). *Geoexploration*, 6, 163–184.

Qinfan, Y., 1988. Application of magnetic anisotropy to magnetic anomaly interpretation in iron ore districts. In Jingxiang, Z. (Ed.), *An Overview of Exploration Geophysics in China 1988*. Society of Exploration Geophysicists, Geophysical References 3, 181–202. doi: 10.1190/1.9781560802662.ch3

Rajagopalan, S. and Milligan, P., 1995. Image enhancement of aeromagnetic data using automatic gain control. *Exploration Geophysics*, 25, 173–178.

Ramachandran, V.S., 1988. Perceiving shape from shading. *Scientific American*, 259(2), 76–83.

Ranjbar, H., Hassanzadeh, H., Torabi, M. and Ilaghi, O., 2001. Integration and analysis of airborne geophysical data of the Darrehzar area, Kerman Province, Iran, using principal component analysis. *Journal of Applied Geophysics*, 48, 33–41.

Read, H.H., 1970. *Rutley's Elements of Mineralogy*, 26th edition. Allen and Unwin.

Reed, L.E. and Witherly, K.E., 2007. 50 years of kimberlite geophysics, a review. In Milkereit, B. (Ed.), *Proceedings of Exploration '07: Fifth Decennial International Conference on Mineral Exploration*. Decennial Mineral Exploration Conferences, 679–689.

Reeve, J.S., Cross, K.C., Smith, R.N. and Oreskes, N., 1990. Olympic Dam copper–uranium–gold–silver deposit. In Hughes, F.E., (Ed.), *Geology of the Mineral Deposits of Australia and Papua New Guinea*. Australasian Institute of Mining and Metallurgy, Monograph Series 14, 1009–1035.

Regueiros, S.J., 1990a. Seam waves: What are they? *The Leading Edge*, 9(4), 19–23.

Regueiros, S.J., 1990b. Seam waves: What are they used for? *The Leading Edge*, 9(8), 32–34.

Reid, A.B., Allsop, J.M., Granser, H., Millett, A.J. and Somerton, I.W., 1990. Magnetic interpretation in three dimensions using Euler deconvolution. *Geophysics*, 55, 80–91.

Reid, J.E. and Macnae, J.C., 1998. Comments of the electromagnetic "smoke ring" concept. *Geophysics*, 63, 1908–1913.

Renwick, R.I., 1981. The uses and benefits of down-hole geophysical logging in coal exploration programs. *Australian Coal Geology*, 3, 37–50.

Reynolds, G.A., 1980. Influence of glacial overburden on geoelectrical prospecting in Ireland. *Transactions of the Institute of Mining and Metallurgy (Section B: Applied Earth Science)*, 89, B44–B49.

Rhodes, H., Lantos, E.A., Lantos, J.A., Webb, R.J. and Owens, D.C., 1984. Pine Point orebodies and their relationship to the stratigraphy, structure, dolomitization and karstification of the Middle Devonian Barrier Complex. *Economic Geology*, 79, 991–1055.

Rider, M., 1996. *The Geological Interpretation of Well Logs*. Whittles Publishing.

Robert, F. and Poulsen, K.H., 1997. World-class Archaean gold deposits in Canada: An overview. *Australian Journal of Earth Sciences*, 44, 329–351.

Roberts, B., Zaleski, E., Perron, G. *et al.*, 2003. Seismic exploration of the Manitouwadge Greenstone Belt, Ontario: A case history. In Eaton, D.W., Milkereit, B. and Salisbury, M.H. (Eds.), *Hardrock Seismic Exploration*. Society of Exploration Geophysicists, Geophysical Development Series 10, 110–126. doi: 10.1190/1.9781560802396.ch7

Robinson, E., Poland, P., Glover, L. and Speer, J., 1985. Some effects of regional metamorphism and geologic structure on magnetic anomalies over the Carolina Slate Belt near Roxboro, North Carolina. In Hinze, W.J. (Ed.), *The Utility of Regional Gravity and Magnetic Anomaly Maps*. Society of Exploration Geophysicists, 320–324.

Robinson, E.S. and Coruh, C., 1988. *Basic Exploration Geophysics*. Wiley.

Robinson, P.T., Hall, J.M., Christensen, N.I. *et al.*, 1982. The Iceland Research Drilling Project: Synthesis of results and implications for the nature of Icelandic and oceanic crust. *Journal of Geophysical Research*, 87, 6657–6667.

Robinson, R.F. and Cook, A., 1966. The Safford copper deposit, Lone Star mining district, Graham County, Arizona. In Titley, S.R. and Hicks, C.L. (Eds.), *Geology of the Porphyry Copper Deposits: Southwestern North America*, University of Arizona Press, 251–266.

Rochette, P., 1987. Metamorphic control of the magnetic mineralogy of black shales in the Swiss Alps: toward the use of "magnetic isogrades". *Earth and Planetary Science Letters*, 84, 446–456.

Rochette, P. and Lamarche, G., 1986. Evolution des propriétés magnétiques lors des transformations minérales dans les roches: example du Jurassique Dauphinois (Alpes Françaises). *Bulletin de Mineralogie*, 109, 687–696.

Roser, B.P. and Nathan, S., 1997. An evaluation of elemental mobility during metamorphism of a turbidite sequence (Greenland Group, New Zealand). *Geological Magazine*, 134, 219–234.

Russell, B., 1998. A simple seismic imaging exercise. *The Leading Edge*, 17(7), 885–889.

Saad, A.H., 1969. Magnetic properties of ultramafic rocks from Red Mountain, California. *Geophysics*, 34, 974–987.

Salem, A., Williams, S., Fairhead, J.D., Ravat, D. and Smith, R., 2007. Tilt-depth method: A simple depth estimation method using first-order magnetic derivatives. *The Leading Edge*, 26(12), 1502–1505.

Salier, B.P., Groves, D.I., McNaughton, N.J. and Fletcher, I.R., 2004. The world-class Wallaby gold deposit, Laverton, Western Australia: an orogenic-style overprint on a magmatic-hydrothermal magnetite–calcite alteration pipe? *Mineralium Deposita*, 39, 473–494.

Salisbury, M.H., Milkereit, B. and Bleeker, W., 1996. Seismic imaging of massive sulfide deposits: Part 1. Rock Properties. *Economic Geology*, 91, 821–828.

Sampson, L. and Bourne, B., 2001. The geophysical characteristics of the Trilogy massive sulphide deposit, Ravensthorpe, Western Australia. *Exploration Geophysics*, 32, 181–184.

Sander, L., 2003. How airborne gravity surveys can make sense for cost effective exploration. *First Break*, 21(11), 77–80.

Sato, M. and Mooney, H.M., 1960. The electrochemical mechanism of sulfide self-potentials. *Geophysics*, 25, 226–249.

Saul, S.J. and Pearson, M.J., 1998. Levelling of aeromagnetic data. *Canadian Journal of Exploration Geophysics*, 34, 9–15.

Scheibner, E. and Basden, H., 1998. *Geology of New South Wales – Synthesis*, Volume 2, *Geological Evolution*. Geological Survey of New South Wales, Memoir Geology 13.

Schlinger, C.M., 1985. Magnetization of lower crust and interpretation of regional magnetic anomalies: example from Lofoten and Vesterålen, Norway. *Journal of Geophysical Research*, 90, 11484–11504.

Schmidt, P.W. and Clark, D.A., 2000. Advantages of measuring the magnetic gradient tensor. *Preview*, 85, 26–30.

Schmitt, D.R., Mwenifumbo, C.J., Pflug, K.A. and Meglis, I.L., 2003. Geophysical logging for elastic properties in hard rock: A tutorial. In Eaton, D.W., Milkereit, B. and Salisbury, M.H. (Eds.), *Hardrock Seismic Exploration*. Society of Exploration Geophysicists, Geophysical Developments 10, 20–41. doi: 10.1190/1.9781560802396.ch2

Schön, J.H., 1996. *Physical Properties of Rocks,* Volume 18, *Fundamentals and Principles of Petrophysics. (Handbook of Geophysical Exploration, Seismic Exploration)*. Pergamon.

Schwarz, E.J., 1991. Magnetic expressions of intrusions including magnetic aureoles. *Tectonophysics*, 192, 191–200.

Schwarz, E.J. and Broome, J., 1994. Magnetic anomalies due to pyrrhotite in Palaeozoic metasediments in Nova Scotia, eastern Canada. *Journal of Applied Geophysics*, 32, 1–10.

Schwarz, E.J., Hood, P.J. and Teskey, D.J., 1987. Magnetic expression of Canadian diabase dykes and downward modelling. In Halls, H.C. and Fahrig, W.F. (Eds.), *Mafic Dyke Swarms*. Geological Association of Canada, Special Paper 34, 153–162.

Scott, R.G. and Spray, J.G., 1999. Magnetic fabric constraints on friction melt flow regimes and ore emplacement direction within the South Range Breccia Belt, Sudbury Impact Structure. *Tectonophysics*, 307, 163–189.

Scott, R.L., Whiting, T.H. and Turner, R., 1994. Role of geophysics in exploration for MVT lead-zinc deposits on the Lennard Shelf, Western Australia. In Dentith, M.C., Frankcombe, K.F., Ho, S.E. *et al.* (Eds.), *Geophysical Signatures of Western Australian Mineral Deposits*. University of Western Australia, Geology and Geophysics Department (Key Centre), Publication 26, and Australian Society of Exploration Geophysicists, Special Publication 7, 107–117.

Scott, W.J., Sellmann, P.V. and Hunter, J.A., 1990. Geophysics in the study of permafrost. In Ward, S.H. (Ed.), *Geotechnical and Environmental Geophysics,* Volume 1: *Review and Tutorial*. Society of Exploration Geophysicists, Investigations in Geophysics 5, 355–384. doi: 10.1190/1.9781560802785.ch13

Seguin, M.K., Allard, M. and Gahé, E., 1989. Surface and downhole geophysics for permafrost mapping in Ungava, Quebec. *Physical Geography*, 10, 201–232.

Seigel, H.O., 1965. Three recent Irish discovery case histories using pulse-type induced polarization. *Canadian Mining and Metallurgical Bulletin*, 58, 1179–1184.

Shermer, M., 2011. *The Believing Brain*. Times Books.

Shives, P.N., Frost, B.R. and Peretti, A., 1988. The magnetic properties of metaperidotitic rocks as a function of metamorphic grade: Implications for crustal magnetic anomalies. *Journal of Geophysical Research*, 93, 12,187–12,195.

Shives, R.B.K., Charbonneau, B.W. and Ford, K.L., 1997. The detection of potassic alteration by gamma-ray spectrometry – recognition of alteration related to mineralization. In Gubins, A.G. (Ed.), *Proceedings of Exploration '97, Fourth Dicennial International Conference on Mineral Exploration*. Prospectors and Developers Association of Canada, 741–752.

Shives, R.B.K., Ford, K.L. and Peter, J.M., 2003. Mapping and exploration applications of gamma ray spectrometry in the Bathurst Mining Camp, northeastern New Brunswick. In Goodfellow, W.D., McCutcheon, S.R. and Peter, J.M. (Eds.), *Massive Sulfide Deposits of the Bathurst Mining Camp, New Brunswick, and Northern Maine*. Society of Economic Geologists, Economic Geology Monograph 11, 819–840.

Sivarajah, Y., Holden, E.J., Togneri, R. and Dentith, M., 2013. Identifying effective interpretation methods for magnetic data by profiling and analyzing human data interactions. *Interpretation*, 1, T45–T55.

Skianis, G.A. and Papadopoulis, T.D., 1993. A contribution to the study of the production mechanism of sulphide mineralization self-potential. *First Break*, 11(4), 119–125.

Smee, B.W. and Sinha, A.K., 1979. Geological, geophysical and geochemical considerations for exploration in clay-covered areas; a review. *CIM Bulletin*, 72 (804), 67–82.

Smith, M.L., Scales, J.A. and Fischer, T.L., 1992. Global search and genetic algorithms. *The Leading Edge*, 11(1), 22–26.

Smith, R.J., 1985. Geophysics in Australian mineral exploration. *Geophysics*, 50, 2637–2665.

Smith, R.J. and Pridmore, D.F., 1989. Exploration in weathered terrains - 1989 perspective. *Exploration Geophysics*, 20, 411–434.

Smith, R.S. and Annan, A.P., 1997. Advances in airborne time-domain EM technology. In Gubins, A.G. (Ed.), *Proceedings of Exploration '97, Fourth Dicennial International Conference on Mineral Exploration*, Prospectors and Developers Association of Canada, 497–504.

Smith, W.H.F. and Wessel, P., 1990. Gridding with continuous curvature splines in tension. *Geophysics*, 55, 293–305.

Sobczak, L.W. and Long, D.G.F., 1980. Preliminary analysis of a gravity profile across the Bonnet Plume Basin, Yukon Territory, Canada: an aid to coal basin evaluation. *Canadian Journal of Earth Sciences*, 17, 43–51.

Sorenson, K.I., Christiansen, A.V. and Auken, E., 2006. The transient electromagnetic methods (TEM). In *Groundwater Resources in Buried Valleys. A Challenge for the Geosciences*. Burval Working Group, Leibniz Institute for Applied Geosciences, 65–76.

Speer, J.A., 1981. The nature and magnetic expression of isograds in the contact aureole of the Liberty Hill pluton, South Carolina: Summary. *Geological Society of America Bulletin*, 92, 603–609.

Srivastava, V.N. and Virenda Mohan, 1979. Geophysical investigations for graphite in Kalimat-Sirar-Pirsal areas of Almora District in U.P. Himalaya. *Indian Minerals*, 33(2), 31–35.

Subrahmanyam, C. and Verma, R.K., 1981. Densities and magnetic susceptibilities of Precambrian rocks of different metamorphic grade (southern Indian Shield). *Journal of Geophysics*, 49, 101–107.

Sykes, M.P. and Das, U.C., 2000. Directional filtering for linear feature enhancement in geophysical maps. *Geophysics*, 65, 1758–1768.

Symons, D.T.A., Kawasaki, K., Walther, S. and Borg, G., 2011. Palaeomagnetism of the Cu–Zn–Pb-bearing Kupferschiefer black shale (Upper Permian) at Sangerhausen, Germany. *Mineralium Deposita*, 46, 137–152.

Taleb, N.M., 2001. *Fooled by Randomness: The Hidden Role of Chance in Life and in the Markets*. Random House.

Tarling, D.H., 1966. The magnetic intensity and susceptibility distribution in some Cenozoic and Jurassic basalts. *Geophysical Journal of the Royal Astronomical Society*, 11, 423–432.

Telford, W.M. and Becker, A., 1979. Exploration case histories of the Iso and New Insco orebodies. In Hood, P.J. (Ed.), *Geophysics and Geochemistry in the Search for Metallic Ores*. Geological Survey of Canada, Economic Geology Report 31, 605–629.

Telford, W.M., Geldart, L.P. and Sheriff, R.E., 1990. *Applied Geophysics*. Cambridge University Press.

Templeton, R.J., Tyne, E.D. and Quick, K.P., 1981. Surface and downhole applied potential (mise-à-la-masse) surveys of the Woodlawn orebody. In Whiteley, R.J. (Ed.), *Geophysical Case Study of the Woodlawn Orebody, New South Wales, Australia*. Pergamon, 509–518.

Thomas, M.D., 1999. Application of the gravity method in mineral exploration: Fundamentals and recent developments. In Lowe, C., Thomas, M.D. and Morris, W.A. (Eds), *Geophysics in Mineral Exploration: Fundamentals and Case Histories*. Geological Association of Canada, Short Course Notes 14, 73–100.

Thompson, T.B., Trippel, A.D. and Dwelley, P.C., 1985. Mineralized veins and breccias of the Cripple Creek district, Colorado. *Economic Geology*, 80, 1669–1688.

Timur, A., 1968. Velocity of compressional waves in porous media at permafrost temperatures. *Geophysics*, 33, 584–595.

Timur, A., 1977. Temperature dependence of compressional and shear wave velocities in rocks. *Geophysics*, 42, 950–956.

Todd, T. and Simmons, G., 1972. Effect of pore pressure on the velocity of compressional waves in low-porosity rocks. *Journal of Geophysical Research*, 77, 3731–3743.

Toft, P.B., Arkani-Hamed, J. and Haggerty, S.E., 1990. The effects of serpentinization on density and magnetic susceptibility:

a petrophysical model. *Physics of Earth and Planetary Interiors*, 65, 137–157.

Trench, A., Dentith, M.C. and Li, Z.X., 1992. Palaeomagnetic dating of mineral deposits: A brief review and a Western Australian perspective. *Exploration Geophysics*, 23, 373–379.

Ugalde, H. and Morris, B., 2008. An assessment of topographic effects on airborne and ground magnetic data. *The Leading Edge*, 27(1), 76–79.

Urbancic, T.I. and Mwenifumbo, C.J., 1986. Multiparameter logging techniques applied to gold exploration. In Killen, P.G. (Ed.), *Borehole Geophysics for Mining and Geotechnical Applications*, Geological Survey of Canada, Paper 85-27, 13–28.

Vajk, R., 1956. Bouguer corrections with varying surface density. *Geophysics*, 21, 1004–1020.

Van Blaricom, R. and O'Connor, L.J., 1989. The geophysical response of the Red Dog deposit. In Garland, G.D. (Ed.), *Proceedings of Exploration '87. Third Dicennial International Conference on Geophysical and Geochemical Exploration for Minerals and Groundwater*, Ontario Geological Survey, Special Volume 3, 486–503.

van Zijl, J.S.V. and Köstlin, E.O., 1986. *The Electromagnetic Method, Field Manual for Technicians No. 3*. South African Geophysical Association.

Verma, R.K., Bhuin, N.C. and Mukhopadhyay, M., 1979. Geology, structure and tectonics of the Jharia coalfield, India – a three-dimensional model. *Geoexploration*, 17, 305–324.

Wahl, W.G. and Lake, S., 1957. Airborne magnetometer survey discovers Marmora magnetite deposit. In De Wet, J.P. (Ed.), *Methods and Case Histories in Mining Geophysics, Proceedings of the Sixth Commonwealth Mining and Metallurgical Congress*, 155–162.

Webb, M. and Rowston, P., 1995. The geophysics of the Ernest Henry Cu–Au deposit (NW) Queensland. *Exploration Geophysics*, 26, 51–59.

Welland, M., Donnelly, N. and Menneer, T., 2006. Are we properly using our brains in seismic interpretation? *The Leading Edge*, 25(2), 142–144.

Wessel, P., 1998. An empirical method for optimal robust regional-residual separation of geophysical data. *Mathematical Geology*, 30, 391–408.

West, G.F. and Macnae, J.C., 1991. The physics of the electromagnetic induction methods. In Nabighian, M.N. (Ed.), *Electromagnetic Methods in Applied Geophysics*, Volume 2, *Applications: Parts A and B*. Society of Exploration Geophysicists, Investigations in Geophysics 3, 5–46. doi: 10.1190/1.9781560802686.ch1

Wheildon, J., Evans, T.R. and Girdler, R.W., 1974. Thermal conductivity, density and sonic velocity measurements of samples of anhydrite and halite from sites 225 and 227. In Whitmarsh, R.B., Wesser, O.E., Ross, D.A. *et al.* (Eds.), *Initial Reports of the Deep Sea Drilling Project* Volume 23, *Columbo, Ceylon to Djibouti, FTAI*, US Government Printing Office, 909–911.

White, R.M.S., Collins, S., Denne, R., Hee, R. and Brown, P., 2001. A new survey design for 3D IP inversion modelling at Copper Hill. *Exploration Geophysics*, 32, 152–155.

Wilford, J.R., Bierwirth, P.N. and Craig, M.A., 1997. Application of airborne gamma-ray spectrometry in soil/regolith mapping and applied geomorphology. *AGSO Journal of Australian Geology & Geophysics*, 17(2), 201–216.

Wilkinson, L., Cruden, A.R. and Krogh, T.E., 1999. Timing and kinematics of post-Timiskaming deformation within the Larder Lake–Cadillac deformation zone, southwest Abitibi greenstone belt, Ontario, Canada. *Canadian Journal of Earth Sciences*, 36, 627–647.

Witherly, K., 2000. The quest for the Holy Grail in mining geophysics: A review of the development and application of airborne EM systems over the last 50 years. *The Leading Edge*, 19(3), 270–274.

Wohlenberg, J., 1982. Density. In Angenheister, G. (Ed.), *Physical Properties of Rocks, Landolt-Börnstein: Numerical Data and Functional Relationships in Science and Technology, Group V: Geophysics and Space Research*, Volume 1a, Springer, 66–183.

Wolfgram, P. and Golden, H., 2001. Airborne EM applied to sulphide nickel: examples and analysis. *Exploration Geophysics*, 32, 136–140.

Wolfgram, P., Sattel, D. and Christensen, N.B., 2003. Approximate 2D inversion of AEM data. *Exploration Geophysics*, 34, 29–33.

Wong, J., 2000. Crosshole seismic imaging for sulfide orebody delination near Sudbury, Ontario, Canada. *Geophysics*, 65, 1900–1907.

Wong, J., Bregman, N., West, G. and Hurley, P., 1987. Cross-hole seismic scanning and tomography. *The Leading Edge*, 6(1), 36–41.

Wood, D.A. and Gibson, I.L., 1976. The relationship between depth of burial and mean intensity of magnetization for basalts from eastern Iceland. *Geophysical Journal of the Royal Astronomical Society*, 46, 497–498.

Wyllie, M.R.J., Gardner, G.H.F. and Gregory, A.R., 1962. Studies of elastic wave attenuation in porous media. *Geophysics*, 27, 563–589.

Wynn, J.C. and Zonge, K.L., 1975. EM coupling, its intrinsic value, its removal and the cultural coupling problem. *Geophysics*, 40, 831–850.

Yeates, A.N., Wyatt, B.W. and Tucker, D.H., 1982. Application of gamma-ray spectrometry to prospecting for tin and tungsten granites, particularly within the Lachlan Fold Belt, New South Wales. *Economic Geology*, 77, 1725–1738.

Yilmaz, O., 2001. *Seismic Data Analysis: Processing Inversion and Interpretation of Seismic Data*, Volumes 1 and 2. Society of Exploration Geophysicists, Investigations in Geophysics, 10.

Yüngül, S., 1956. Prospecting for chromite with gravimeter and magnetometer over rugged topography in east Turkey. *Geophysics*, 21, 433–454. doi: 10.1190/1.1438245

Zhang, C., Mushayandebvu, M.F., Reid, A.B., Fairhead, J.D. and Odegard, M.E., 2000. Euler deconvolution of gravity tensor gradient data. *Geophysics*, 65, 512–520.

Zhdanov, M.S. and Keller, G.V., 1994. The geoelectrical methods in geophysical exploration. *Methods in Geochemistry and Geophysics*, 31, Elsevier.

Zhou, B. and Hatherly, P., 2004. Coal seismic depth conversion for mine data integration: A case study from the Sandy Creek 3D seismic survey. *Exploration Geophysics*, 35, 324–330.

INDEX